A FIRST COURSE IN SYSTEMS BIOLOGY

SECOND EDITION

To Ann,
Still the Hub of my Support System

A FIRST COURSE IN SYSTEMS BIOLOGY

SECOND EDITION

Eberhard O. Voit

NEW YORK AND LONDON

Garland Science
Vice President: Denise Schanck
Senior Editorial Assistant: Katie Laurentiev
Assistant Editor: David Borrowdale
Production Editor: Georgina Lucas
Illustrations: Nigel Orme
Copyeditor: H.M. (Mac) Clarke
Typesetting: NovaTechset Pvt Ltd
Proofreader: Sally Huish
Indexer: Nancy Newman
Cover Design: Andrew Magee

ISBN 978-0-8153-4568-8

Library of Congress Cataloging-in-Publication Data
Names: Voit, Eberhard O., author.
Title: A first course in systems biology / Eberhard O. Voit.
Description: Second edition. | New York : Garland Science, 2017.
Identifiers: LCCN 2017017580 | ISBN 9780815345688 (alk. paper)
Subjects: LCSH: Systems biology. | Computational biology.
Classification: LCC QH324.2 .V65 2017 | DDC 570.1/13–dc23
LC record available at https://lccn.loc.gov/2017017580

Published by Garland Science, Taylor & Francis Group, LLC, an informa business,
711 Third Avenue, 8th Floor, New York NY 10017, USA,
and 2 Park Square, Milton Park, Abingdon, OX14 4RN, UK.

Printed by Ashford Colour Press Ltd.

15 14 13 12 11 10 9 8 7 6 5 4 3 2 1

Visit our website at http://www.garlandscience.com

Front cover image. The beautiful geometric shape of the fractal is called self-similar because it has the same appearance at smaller and smaller scales. It reminds us of fundamental design features like feedback loops that we encounter at many organizational levels of biological systems. Fractals are generated with nonlinear recursive models, and they are discussed with simpler examples in Chapter 4. (Courtesy of Wolfgang Beyer under Creative Commons Attribution-Share Alike 3.0 Unported license.)

Preface

Hard to believe, but it is already time for the second edition! I am happy to report that the first edition of *A First Course in Systems Biology* has met with great success. The book has been a required or recommended text for over 70 courses worldwide, and it has even been translated into Korean. So why should a new edition be necessary after only five short years? Well, much has happened. Systems biology has come out of the shadows with gusto. Research is flourishing worldwide, quite a few new journals have been launched, and many institutions now offer courses in the field.

While the landscape of systems biology has evolved rapidly, the fundamental topics covered by the first edition are as important as they were five years ago and probably will be several decades from now. Thus, I decided to retain the structure of the first edition but have rearranged some items and added a few topics, along with new examples. At Georgia Tech we have used the book to teach well over 1000 students, mostly at the undergraduate level, but also for an introductory graduate-level course. Most of the additions and amendments to this new edition respond to feedback from these students and their instructors, who have pointed out aspects of the material where more or better explanations and illustrations would be helpful. New topics in this edition include: default modules for model design, limit cycles and chaos, parameter estimation in Excel, model representations of gene regulation through transcription factors, derivation of the Michaelis-Menten rate law from the original conceptual model, different types of inhibition, hysteresis, a model of differentiation, system adaptation to persistent signals, nonlinear nullclines, PBPK models, and elementary modes.

I would like to thank three undergraduates from my classes who helped me with the development of some of the new examples, namely Carla Kumbale, Kavya Muddukumar, and Gautam Rangavajla. Quite a few other students have helped me with the creation of new practice exercises, many of which are available on the book's support website. I also want to express my gratitude to David Borrowdale, Katie Laurentiev, Georgina Lucas, Denise Schanck, and Summers Scholl at Garland Science for shepherding this second edition through the review and production process.

It is my hope that this new edition retains the appeal of the original and has become even better through the alterations and twists it has taken, large and small.

Eberhard Voit
Georgia Tech
2017

Instructor Resources Website

The images from *A First Course in Systems Biology, Second Edition* are available on the Instructor Site in two convenient formats: PowerPoint® and JPEG. They have been optimized for display on a computer. Solutions to end-of-chapter exercises are also available. The resources may be browsed by individual chapters and there is a search engine. Figures are searchable by figure number, figure name, or by keywords used in the figure legend from the book.

Accessible from www.garlandscience.com, the Instructor's Resource Site requires registration and access is available only to qualified instructors. To access the Instructor Resource site, please email science@garland.com.

Acknowledgments

The author and publisher of *A First Course in Systems Biology, Second Edition* gratefully acknowledge the contributions of the following reviewers in the development of this book:

Guy Grant, University of Bedfordshire

Princess Imoukhuede, University of Illinois at Urbana-Champaign

Dimitrios Morikis, University of California at Riverside

Oliver Schildgen, University of Witten

Manuel Simões, University of Porto

Mark Speck, Chaminade University

Marios Stavridis, Ninewells Hospital & Medical School

Geraint Thomas, University College London

Floyd Wittink, Leiden University

Contents

Biological Systems

1

When you have read this chapter, you should be able to:

- Describe the generic features of biological systems
- Explain the goals of systems biology
- Identify the complementary roles of reductionism and systems biology
- List those challenges of systems biology that cannot be solved with intuition alone
- Assemble a "to-do" list for the field of systems biology

When we think of biological systems, our minds may immediately wander to the Amazon rainforest, brimming with thousands of plants and animals that live with each other, compete with each other, and depend on each other. We might think of the incredible expanse of the world's oceans, of colorful fish swimming through coral reefs, nibbling on algae. Two-meter-high African termite mounds may come to mind, with their huge colonies of individuals that have their specific roles and whose lives are controlled by an intricate social structure (**Figure 1.1**). We may think of an algae-covered pond with tadpoles and minnows that are about to restart yet another life cycle.

These examples are indeed beautiful manifestations of some of the fascinating systems nature has evolved. However, we don't have to look that far to find biological systems. Much, much smaller systems are in our own bodies and even within our cells. Kidneys are waste-disposal systems. Mitochondria are energy-production systems. Ribosomes are intracellular machines that make proteins from amino acids. Bacteria are amazingly complicated biological systems. Viruses interact with cells in a well-controlled, systemic way. Even seemingly modest tasks often involve an amazingly large number of processes that form complicated control systems (**Figure 1.2**). The more we learn about the most basic processes of life, such as cell division or the production of a metabolite, the more we have to marvel the incredible complexity of the systems that facilitate these processes. In our daily lives, we usually take these systems for granted and assume that they function adequately, and it is only when, for example, disease strikes or algal blooms kill fish that we realize how complex biology really is and how damaging the failure of just a single component can be.

We and our ancestors have been aware of biological systems since the beginning of human existence. Human birth, development, health, disease, and death have long been recognized as interwoven with those of plants and animals, and with the environment. For our forebears, securing food required an understanding of seasonal changes in the ecological systems of their surroundings. Even the earliest forays into agriculture depended on detailed concepts and ideas of when and what to

Figure 1.1 Biological systems abound at all size scales. Here, a termite mound in Namibia is visible evidence of a complex social system. This system is part of a larger ecological system, and it is at once the host to many systems at smaller scales. (Courtesy of Lothar Herzog under the Creative Commons Attribution 2.0 Generic license.)

plant, how and where to plant it, how many seeds to eat or to save for sowing, and when to expect returns on the investment. Several thousand years ago, the Egyptians managed to ferment sugars to alcohol and used the mash to bake bread. Early pharmaceutical treatments of diseases certainly contained a good dose of superstition, and we are no longer convinced that rubbing on the spit of a toad during full moon will cure warts, but the beginnings of pharmaceutical science in antiquity and the Middle Ages also demonstrate a growing recognition that particular plant products can have significant and specific effects on the well-being or malfunctioning of the systems within the human body.

In spite of our long history of dealing with biological systems, our mastery of engineered systems far outstrips our capability to manipulate biological systems. We send spaceships successfully to faraway places and predict correctly when they will arrive and where they will land. We build skyscrapers exceeding by hundreds of

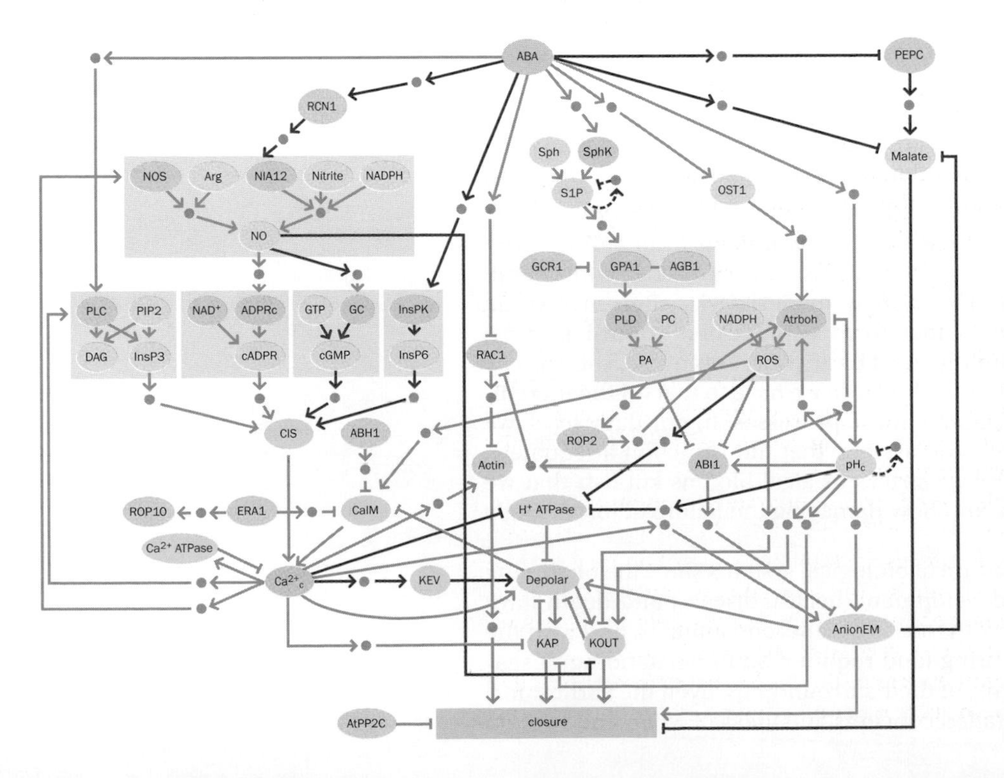

Figure 1.2 Diagram of a complicated system of molecules that coordinate the response of plants to drought. While the details are not important here, we can see that a key hormone, called abscisic acid (ABA), triggers a cascade of reactions that ultimately promote the closure of stomata and thereby reduce water evaporation [1]. Even a narrowly defined response like this closure process involves a complicated control system that contains a multitude of molecules and their interactions. In turn, this system is just one component within a much larger, physiological stress response system (cf. Figure 1.7). (From Saadatpour A, Albert I & Albert A. *J. Theor. Biol.* 266 [2010] 641–656. With permission from Elsevier.)

times the sizes of the biggest animals and plants. Our airplanes are faster, bigger, and more robust against turbulence than the most skillful birds. Yet, we cannot create new human cells or tissues from basic building blocks and we are seldom able to cure diseases except with rather primitive methods like cutting into the body or killing a lot of healthy tissue in the process, hoping that the body will heal itself afterwards. We can anticipate that our grandchildren will only shake their heads at such medieval-sounding, draconian measures. We have learned to create improved microorganisms, for instance for the bulk production of industrial alcohol or the generation of pure amino acids, but the methods for doing so rely on bacterial machinery that we do not fully understand and on artificially induced random mutations rather than targeted design strategies.

Before we discuss the roots of the many challenges associated with understanding and manipulating biological systems in a targeted fashion, and our problems predicting what biological systems will do under yet-untested conditions, we should ask whether the goal of a deeper understanding of biological systems is even worth the effort. The answer is a resounding "Yes!" In fact, it is impossible even to imagine the potential and scope of advances that might develop from biological systems analyses. Just as nobody during the eighteenth century could foresee the ramifications of the Industrial Revolution or of electricity, the Biological Revolution will usher in an entirely new world with incredible possibilities. Applications that are already emerging on the horizon are personalized medical treatments with minimal side effects, pills that will let the body regain control over a tumor that has run amok, prevention and treatment of neurodegenerative diseases, and the creation of spare organs from reprogrammed stem cells. A better understanding of ecological systems will yield pest- and drought-resistant food sources, as well as means for restoring polluted soil and water. It will help us understand why certain species are threatened and what could be done effectively to counteract their decline. Deeper insights into aquatic systems will lead to cleaner water and sustainable fisheries. Reprogrammed microbes or nonliving systems composed of biological components will dominate the production of chemical compounds from prescription drugs to large-scale industrial organics, and might create energy sources without equal. Modified viruses will become standard means for supplying cells with healthy proteins or replacement genes. The rewards of discovering and characterizing the general principles and the specifics of biological systems will truly be unlimited.

If it is possible to engineer very sophisticated machines and to predict exactly what they will do, why are biological systems so different and difficult? One crucial difference is that we have full control over engineered systems, but not over biological systems. As a society, we collectively know all details of all parts of engineered machines, because we made them. We know their properties and functions, and we can explain how and why some engineer put a machine together in a particular fashion. Furthermore, most engineered systems are modular, with each module being designed for a unique, specific task. While these modules interact with each other, they seldom have multiple roles in different parts of the system, in contrast to biology and medicine, where, for instance, the same lipids can be components of membranes and have complicated signaling functions, and where diseases are often not restricted to a single organ or tissue, but may affect the immune system and lead to changes in blood pressure and blood chemistry that secondarily cause kidney and heart problems. A chemical refinery looks overwhelmingly complicated to a layperson, but for an industrial engineer, every piece has a specific, well-defined role within the refinery, and every piece or module has properties that were optimized for this role. Moreover, should something go wrong, the machines and factories will have been equipped with sensors and warning signals pinpointing problems as soon as they arise and allowing corrective action.

In contrast to dealing with sophisticated, well-characterized engineered systems, the analysis of biological systems requires investigations in the opposite direction. This type of investigation resembles the task of looking at an unknown machine and predicting what it does (**Figure 1.3**). Adding to this challenge, all scientists collectively know only a fraction of the components of biological systems, and the specific roles and interactions between these components are often obscure and change over time. Even more than engineered systems, biological systems are full of sensors and signals that indicate smooth running or ensuing problems, but in most

Figure 1.3 Analyzing a biological system resembles the task of determining the function of a complicated machine that we have never seen before. Shown here as an example is the cesium fountain laser table of the United States Naval Observatory, which is used to measure time with extreme accuracy. This atomic clock is based on transitions in cesium, which have a frequency of 9,192,631,770 Hz and are used to define the second. See also [2].

cases our experiments cannot directly perceive and measure these signals and we can only indirectly deduce their existence and function. We observe organisms, cells, or intracellular structures as if from a large distance and must deduce from rather coarse observations how they might function or why they fail.

What exactly is it that makes biological systems so difficult to grasp? It is certainly not just size. **Figure 1.4** shows two networks. One shows the vast highway system of the continental United States, which covers several million miles of major

(A)

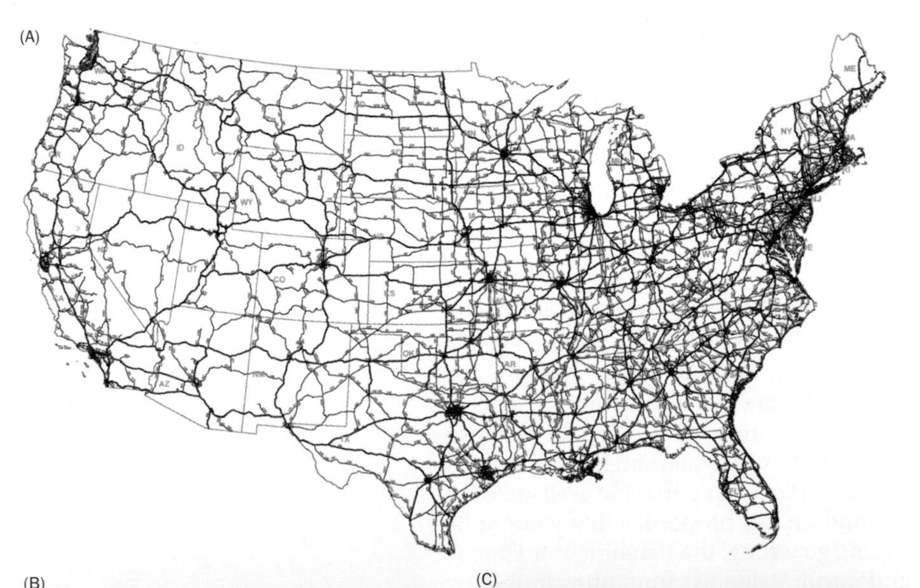

(B)

(C)

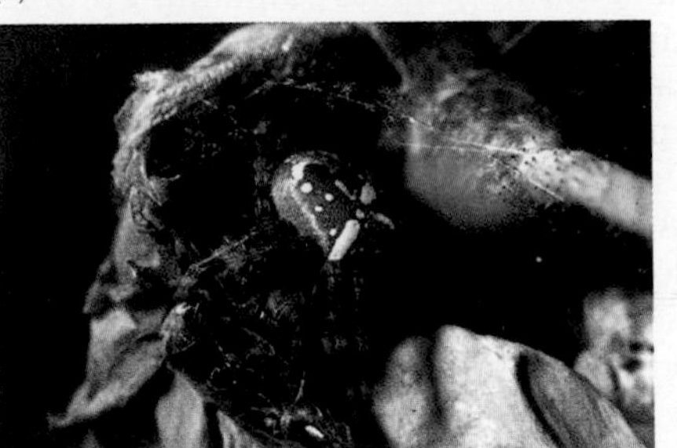

Figure 1.4 The size of a network or system is not necessarily correlated with its complexity. (A) The network of major highways in the continental United States covers over 3 million square miles. Nonetheless, its functionality is easy to grasp, and problems with a particular road are readily ameliorated with detours. (B) The web of the European diadem spider (*Araneus diadematus*) (C) is comparatively small, but the functional details of this little network are complex. Some lines are made of silk proteins that have the tensile strength of steel but can also be eaten and recycled by the spider; other lines are adhesive due to a multipurpose glue that may be sticky or rubbery depending on the situation; yet others are guide and signal lines that allow the spider to move about and sense prey. The creation of the web depends on different types of spinneret glands, whose development and function require the complex molecular machinery of the spider, and it is not yet clear how the instructions for the complicated construction, repair, and use of the web are encoded and inherited from one generation to the next. ((A) From the United States Department of Transportation.)

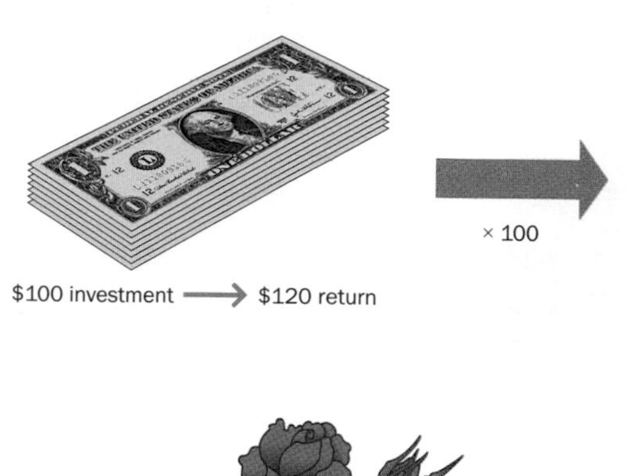

Figure 1.5 Biological phenomena are often difficult to understand, because our minds are trained to think linearly. (A) The return on an investment grows (or decreases) linearly with the amount invested. (B) In biology, more is not necessarily better. Biological responses often scale within a modest range, but lead to an entirely different response if the input is increased a lot.

highways. It is a very large system, but it is not difficult to understand its function or malfunction: if a highway is blocked, it does not take much ingenuity to figure out how to circumvent the obstacle. The other network is a comparably tiny system: the web of a diadem spider. While we can observe the process and pattern with which Ms. Spider spins her web, we do not know which neurons in her brain are responsible for different phases of the complicated web production process and how she is able to produce the right chemicals for the spider silk, which in itself is a marvel of material science, let alone how she manages to survive, multiply, and maybe even devour her husband.

Biological systems often consist of large numbers of components, but they pose an additional, formidable challenge to any analysis, because the processes that govern them are not linear. This is a problem, because we are trained to think in linear ways: if an investment of $100 leads to a return of $120, then an investment of $10,000 leads to a return of $12,000. Biology is different. If we fertilize our roses with 1 tablespoon of fertilizer and the rose bushes produce 50 blossoms, a little bit more fertilizer may increase the number of blossoms, but 100 tablespoons of fertilizer will not produce 5000 blossoms but almost certainly kill the plants (**Figure 1.5**). Just a small amount of additional sun exposure turns a tan into sunburn. Now imagine that thousands of components, many of which we do not know, respond in such a fashion, where a small input does not evoke any response, more input evokes a physiological response, and a little bit more input causes the component to fail or exhibit a totally different "stress" response. We will return to this issue later in this and other chapters with specific examples.

REDUCTIONISM AND SYSTEMS BIOLOGY

So the situation is complicated. But because we humans are a curious species, our forebears did not give up on biological analysis and instead did what was doable, namely collecting information on whatever could be measured with the best current methods (**Figure 1.6**). By now, this long-term effort has resulted in an amazing list of biological parts and their roles. Initially, this list contained new plant and animal

Figure 1.6 Collecting information is the first step in most systems analyses. The eighteenth-century British explorer Captain James Cook sailed the Pacific Ocean and catalogued many plants and animal species that had never been seen before in Europe.

species, along with descriptions of their leaves, berries, and roots, or their body shapes, legs, and color patterns. These external descriptions were valuable, but did not provide specific clues on how plants and animals function, why they live, and why they die. Thus, the next logical step was to look inside—even if this required stealing bodies from the cemetery under a full moon! Cutting bodies open revealed an entirely new research frontier. What were all those distinct body parts and what did they do? What were organs, muscles, and tendons composed of? Not surprisingly, this line of investigation eventually led to the grand-challenge quest of discovering and measuring *all* parts of a body, the parts of the parts (. . . of the parts), as well as their roles in the normal physiology or pathology of cells, organs, and organisms. The implicit assumption of this reductionist approach was that knowing the building blocks of life would lead us to a comprehensive understanding of how life works.

If we fast-forward to the twenty-first century, have we succeeded and assembled a complete parts catalog? Do we know the building blocks of life? The answer is a combination of yes's and no's. The catalog is most certainly not complete, even for relatively simple organisms. Yet, we have discovered and characterized genes, proteins, and metabolites as the major building blocks. Scientists were jubilant when the sequencing of the human genome in the early years of this millennium was declared complete: we had identified the ultimate building blocks, our entire blueprint. It turned out to consist of roughly three billion nucleotide pairs of DNA.

The sequencing of the human genome was without any doubt an incredible achievement. Alas, there is much more to a human body than genes. So, the race for building blocks extended to proteins and metabolites, toward individual gene variations and an assortment of molecules and processes affecting gene expression, which changes in response to external and internal stimuli, during the day, and throughout our lifetimes. As a direct consequence of these ongoing efforts, our parts list continues to grow at a rapid pace: A parts catalog that started with a few organs now contains over 20,000 human genes, many more genes from other organisms, and hundreds of thousands of proteins and metabolites along with their variants. In addition to merely looking at parts in isolation, we have begun to realize that most biological components are affected and regulated by a variety of other components. The expression of a gene may depend on several transcription factors, metabolites, and a variety of small RNAs, as well as molecular, epigenetic attachments to its DNA sequence. It is reasonable to expect that the list of processes within the body is much larger than the number of components on our parts list. Biologists will not have to worry about job security any time soon!

The large number of components and processes alone, however, is not the only obstacle to understanding how cells and organisms function. After all, modern computers can execute gazillions of operations within a second. Our billions of telephones worldwide are functionally connected. We can make very accurate predictions regarding a gas in a container, even if trillions of molecules are involved. If we increase the pressure on the gas without changing the volume of the container, we know that the temperature will rise, and we can predict by how much. Not so with a cell or organism. What will happen to it if the environmental temperature goes up? Nothing much may happen, the rise in temperature may trigger a host of physiological response processes that compensate for the new conditions, or the organism may die. The outcome depends on a variety of factors that collectively constitute a complex stress response system (**Figure 1.7**). Of course, the comparison to a gas is not

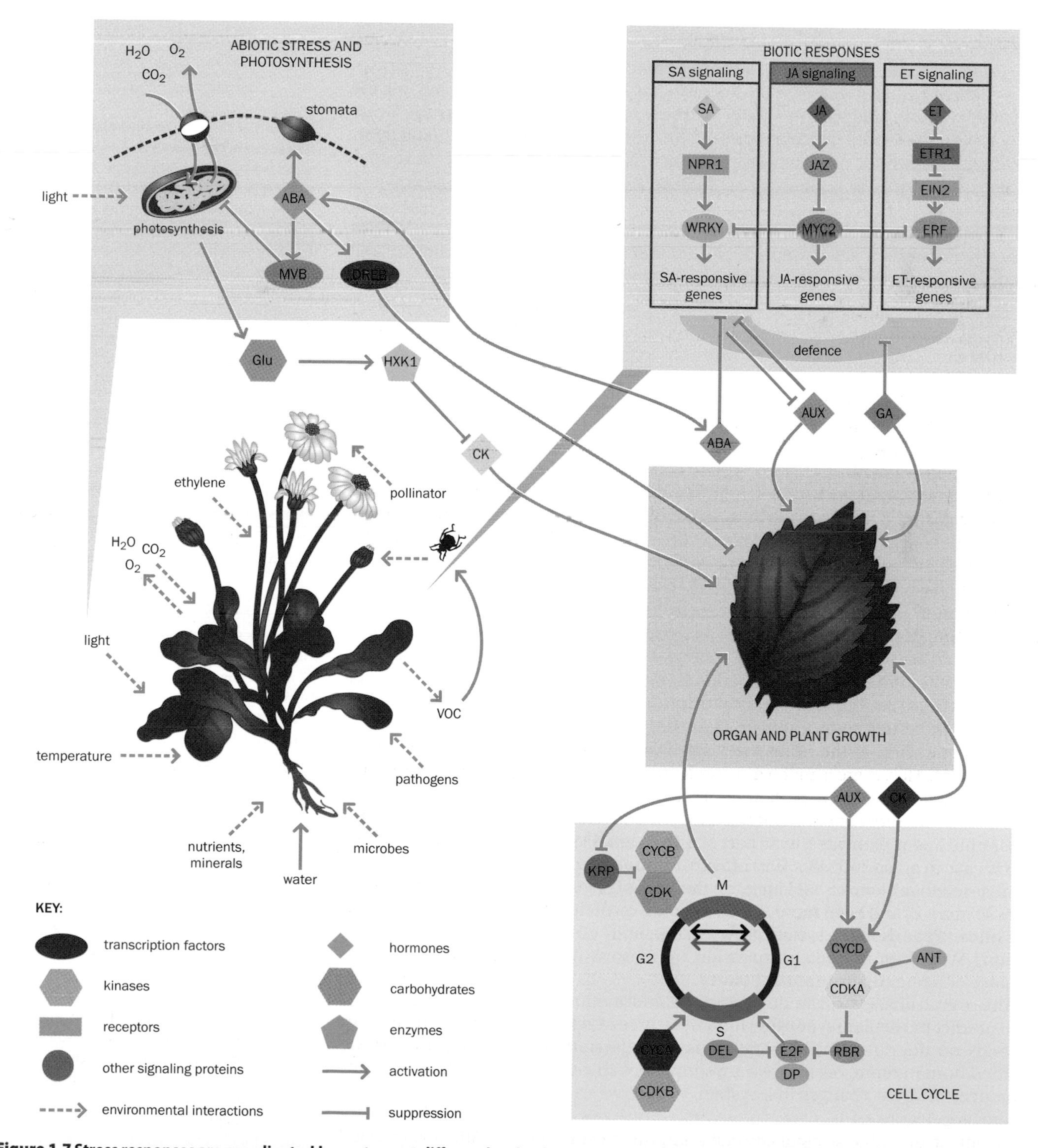

Figure 1.7 Stress responses are coordinated by systems at different levels of organization (cf. Figure 1.2). At the physiological level, the stress response system in plants includes changes at the cellular, organ, and whole-plant levels and also affects interactions of the plant with other species. (From Keurentjes JJB, Angenent GC, Dicke M, et al. *Trends Plant Sci.* 16 [2011] 183–190. With permission from Elsevier.)

quite fair, because, in addition to their large number, the components of a cell are not all the same, which drastically complicates matters. Furthermore, as mentioned earlier, the processes with which the components interact are nonlinear, and this permits an enormous repertoire of distinctly different behaviors with which an organism can respond to a perturbation.

EVEN SIMPLE SYSTEMS CAN CONFUSE US

It is easy to demonstrate how quickly our intuition can be overwhelmed by a few nonlinearities within a system. As an example, let's look at a simple chain of processes and compare it with a slightly more complicated chain that includes regulation [3]. The simple case merely consists of a chain of reactions, which is fed by an external input (**Figure 1.8**). It does not really matter what *X*, *Y*, and *Z* represent, but, for the sake of discussion, imagine a metabolic pathway such as glycolysis, where the input, glucose, is converted into glucose 6-phosphate, fructose 1,6-bisphosphate, and pyruvate, which is used for other purposes that are not of interest here. For illustrative purposes, let's explicitly account for an enzyme *E* that catalyzes the conversion of *X* into *Y*.

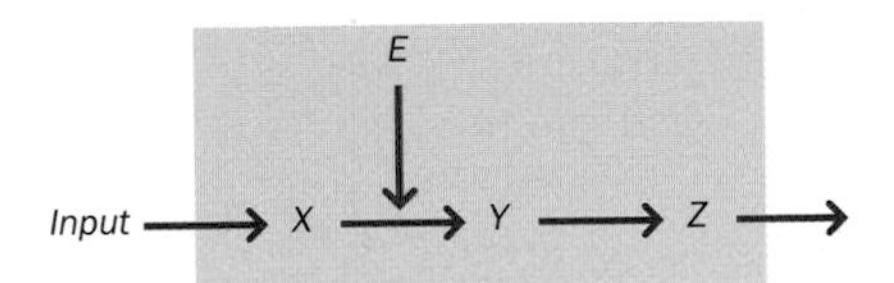

Figure 1.8 The human brain handles linear chains of causes and events very well. In this simple pathway, an external input is converted sequentially into *X*, *Y*, and *Z*, which leaves the system. The conversion of *X* into *Y* is catalyzed by an enzyme *E*. It is easy to imagine that any increase in *Input* will cause the levels of *X*, *Y*, and *Z* to rise.

We will learn in the following chapters how one can formulate a model of such a pathway system as a set of differential equations. And while the details are not important here, it does not hurt to show such a model, which might read

$$\begin{aligned} \dot{X} &= Input - aEX^{0.5}, \\ \dot{Y} &= aEX^{0.5} - bY^{0.5}, \\ \dot{Z} &= bY^{0.5} - cZ^{0.5}. \end{aligned} \tag{1.1}$$

Here, *X*, *Y*, and *Z* are concentrations, *E* is the enzyme activity, and *a*, *b*, and *c* are rate constants that respectively represent how fast *X* is converted into *Y*, how fast *Y* is converted into *Z*, and how quickly material from the metabolite pool *Z* leaves the system. The dotted quantities on the left of the equal signs are differentials that describe the change in each variable over time, but we need not worry about them at this point. In fact, we hardly have to analyze these equations mathematically to get an idea of what will happen if we change the input, because intuition tells us that any increase in *Input* should lead to a corresponding rise in the concentrations of the intermediates *X*, *Y*, and *Z*, whereas a decrease in *Input* should result in smaller values of *X*, *Y*, and *Z*. The increases or decreases in *X*, *Y*, and *Z* will not necessarily be exactly of the same extent as the change in *Input*, but the direction of the change should be the same. The mathematical solution of the system in (1.1) confirms this intuition. For instance, if we reduce *Input* from 1 to 0.75, the levels of *X*, *Y*, and *Z* decrease, one after another, from their initial value of 1 to 0.5625 (**Figure 1.9**).

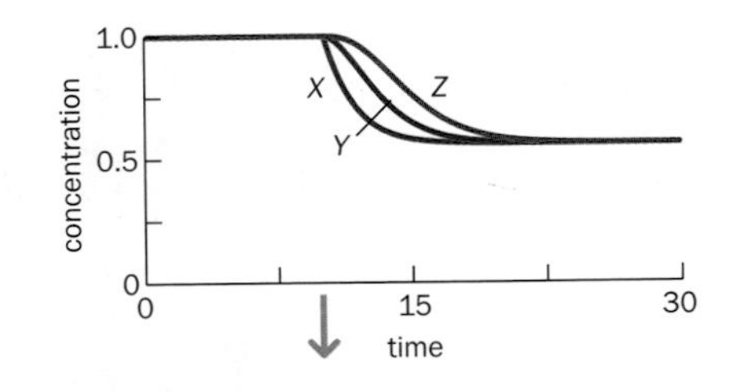

Figure 1.9 Simulations with the system in (1.1) confirm our intuition: *X*, *Y*, and *Z* reflect changes in *Input*. For instance, reducing *Input* in (1.1) to 75% at time 10 (arrow) leads to permanent decreases in *X*, *Y*, and *Z*.

Now suppose that *Z* is a signaling molecule, such as a hormone or a phospholipid, that activates a transcription factor *TF* that facilitates the up-regulation of a gene *G* that codes for the enzyme catalyzing the conversion of *X* into *Y* (**Figure 1.10**). The simple linear pathway is now part of a functional loop. The organization of this loop is easy to grasp, but what is its effect? Intuition might lead us to believe that the positive-feedback loop should increase the level of enzyme *E*, which would result in more *Y*, more *Z*, and even more *E*, which would result in even more *Y* and *Z*. Would the concentrations in the system grow without end? Can we be sure about this prediction? Would an unending expansion be reasonable? What will happen if we increase or decrease the input as before?

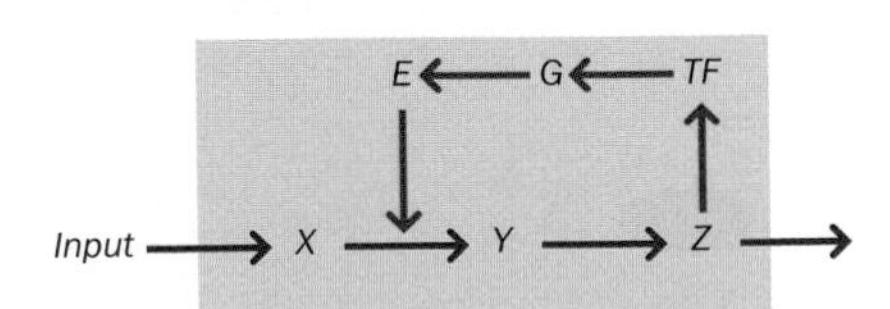

Figure 1.10 Even simple systems may not allow us to make reliable predictions regarding their responses to stimuli. Here, the linear pathway from Figure 1.8 is embedded into a functional loop consisting of a transcription factor *TF* and a gene *G* that codes for enzyme *E*. As described in the text, the responses to changes in *Input* are no longer obvious.

The overall answer will be surprising: the information given so far does not allow us to predict particular responses with any degree of reliability. Instead, the answer depends on the numerical specifications of the system. This is bad news for the unaided human mind, because we are simply not able to assess the numerical consequences of slight changes in a system, even if we can easily grasp the logic of a system as in Figure 1.10.

To get a feel for the system, one can compute a few examples with an expanded model that accounts for the new variables (for details, see [3]). Here, the results are more important than the technical details. If the effect of *Z* on *TF* is weak, the response to a decrease in *Input* is essentially the same as in Figure 1.9. This is not too surprising, because the systems in this case are very similar. However, if the effect of *Z* on *TF* is stronger, the concentrations in the system start to oscillate, and after a while these oscillations dampen away (**Figure 1.11A**). This behavior was not easy to predict. Interestingly, if the effect is further increased, the system enters a stable oscillation pattern that does not cease unless the system input is changed again (Figure 1.11B).

The hand-waving explanation of these results is that the increased enzyme activity leads to a depletion of *X*. A reduced level of *X* leads to lower levels of *Y* and *Z*, which in turn lead to a reduced effect on *TF*, *G*, and ultimately *E*. Depending on the numerical characteristics, the ups and downs in *X* may not be noticeable, they may be damped and disappear, or they may persist until another change is introduced. Intriguingly, even if we know that these alternative responses are possible, the unaided human mind is not equipped to integrate the numerical features of the model in such a way that we can predict which system response will ensue for a specific setting of parameters. A computational model, in contrast, reveals the answer in a fraction of a second.

The specific details of the example are not as important as the take-home message: If a system contains regulatory signals that form functional loops, we can no longer rely on our intuition for making reliable predictions. Alas, essentially all realistic systems in biology are regulated—and not just with one, but with many control loops. This leads to the direct and sobering deduction that intuition is not sufficient and that we instead need to utilize computational models to figure out how even small systems work and why they might show distinctly different responses or even fail, depending on the conditions under which they operate.

The previous sections have taught us that biological systems contain large numbers of different types of components that interact in potentially complicated ways and are controlled by regulatory signals. What else is special about biological systems? Many answers could be given, some of which are discussed throughout this book. For instance, two biological components are seldom 100% the same. They vary from one organism to the next and change over time. Sometimes these variations are inconsequential, at other times they lead to early aging and disease. In fact, most

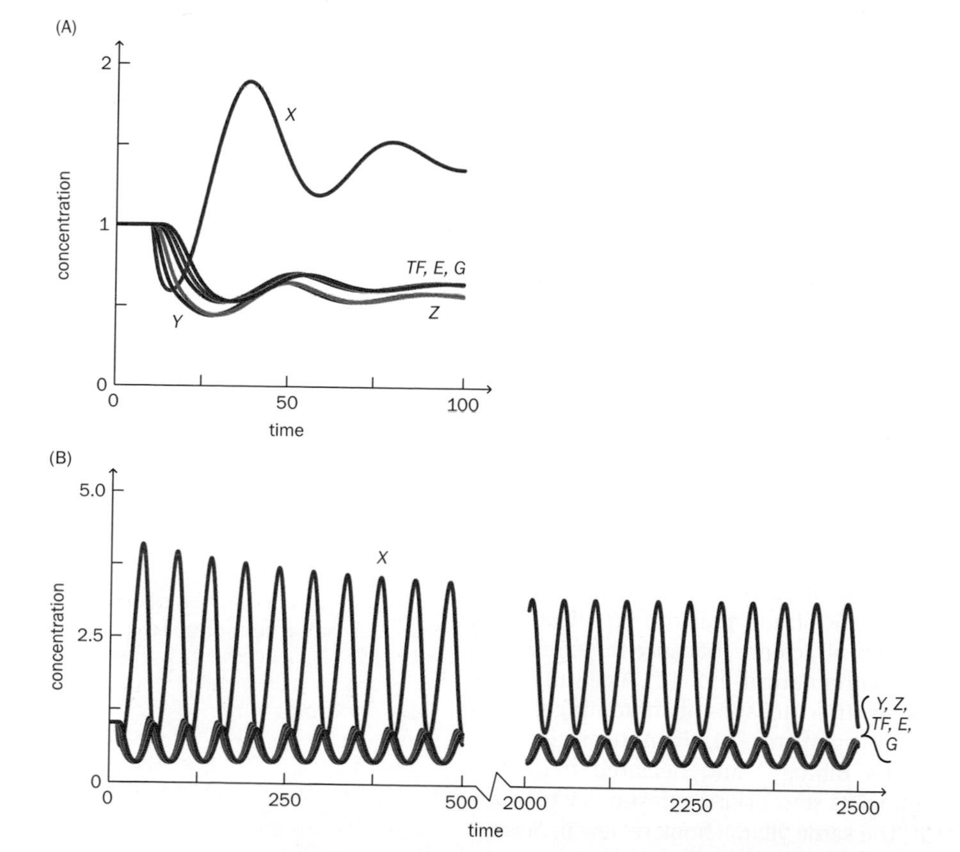

Figure 1.11 Simulation results demonstrate that the looped system in Figure 1.10 may exhibit drastically different responses. If the effect of *Z* on *TF* is very small, the response is essentially like that in Figure 1.9 (results not shown). (A) If the effect of *Z* on *TF* is relatively small, the functional feedback loop causes the system to go through damped oscillations before assuming a new stable state. (B) For stronger effects of *Z* on *TF*, the system response is a persistent oscillation.

diseases do not have a single cause, but are the consequence of an unfortunate combination of slight alterations in many components. Another feature that complicates intuition is the delay in many responses to stimuli. Such delays may be of the order of seconds, hours, or years, but they require the analyst to study not merely the present state of a biological system but also its history. For instance, recovery from a severe infection depends greatly on the preconditioning of the organism, which is the collective result of earlier infections and the body's responses [4].

Finally, it should be mentioned that different parts of biological systems may simultaneously operate at different scales, with respect to both time and space. These scales make some aspects of their analysis easier and some harder. Let's begin with the temporal scale. We know that biology at the most basic level is governed by physical and chemical processes. These occur on timescales of the order of milliseconds, if not faster. Biochemical processes usually run on a scale of seconds to minutes. Under favorable conditions, bacteria divide every 20–30 minutes. Our human lifespan extends to maybe 120 years, evolution can happen at the genetic level with lightning speed, for instance, when radiation causes a mutation, while the emergence of an entirely new species may take thousands or even millions of years. On one hand, the drastically different timescales make analyses complicated, because we simply cannot account for rapid changes in all molecules of an organism over an extended period of time. As an example, it is impossible to study aging by monitoring an organism's molecular state every second or minute. On the other hand, the differences in timescales justify a very valuable modeling "trick" [5, Chapter 5]. If we are interested in understanding some biochemical process, such as the generation of energy in the form of adenosine triphosphate (ATP) by means of the conversion of glucose into pyruvate, we can assume that developmental and evolutionary changes are so slow in comparison that they do not change during ATP production. Similarly, if we study the phylogenetic family tree of species, the biochemical processes in an individual organism are comparatively so fast that their details become irrelevant. Thus, by focusing on just the most relevant timescale and ignoring much faster and much slower processes, any modeling effort is dramatically simplified.

Biology also happens on many spatial scales. All processes have a molecular component, and their size scale is therefore of the order of ångströms and nanometers. If we consider a cell as the basic unit of life, we are dealing with a spatial scale of micrometers to millimeters, with some exceptions such as cotton "fiber" cells reaching the length of a few centimeters [6] and the afferent axons of nerve cells in giraffes, reaching from toe to neck, extending to 5 meters [7, p. 14]. The sizes of typical cells are dwarfed by higher plants and animals and by ecosystems such as our oceans, which may cover thousands of square kilometers. As with the different temporal scales, and using analogous arguments, models of biological systems often focus on one or two spatial scales at a time [5]. Nonetheless, such simplifications are not always applicable, and some processes, such as aging and algal blooms, may require the simultaneous consideration of several temporal and spatial scales. Such multiscale assessments are often very complicated and constitute a challenging frontier of current research (see Chapter 15).

WHY NOW?

Many of the features of biological systems have been known for quite a while, and, similarly, many concepts and methods of systems biology have their roots in its well-established parent disciplines, including physiology, molecular biology, biochemistry, mathematics, engineering, and computer science [8–11]. In fact, it has been suggested that the nineteenth-century scientist Claude Bernard might be considered the first systems biologist, since he proclaimed that the "application of mathematics to natural phenomena is the aim of all science, because the expression of the laws of phenomena should always be mathematical" [12, 13]. A century later, Ludwig von Bertalanffy reviewed in a book his three decades of attempting to convince biologists of the systemic nature of living organisms [14, 15]. At the same time, Mihajlo Mesarović used the term "Systems Biology" and declared that "real advance ... will come about only when biologists start asking questions which are based on systems-theoretic concepts" [16]. The same year, a book review in *Science* envisioned "... a field of systems biology with its own identity and in its own right" [17]. A few years later, Michael Savageau proposed an agenda for studying biological systems with mathematical and computational means [5].

In spite of these efforts, systems biology did not enter the mainstream for several more decades. Biology kept its distance from mathematics, computer science, and engineering, primarily because biological phenomena were seen as too complicated for rigorous mathematical analysis and mathematics was considered applicable only to very small systems of little biological relevance. The engineering of biological systems from scratch was impossible, and the budding field of computer science contributed to biology not much more than rudimentary data management.

So, why has systems biology all of the sudden moved to the fore? Any good detective will know the answer: motive and opportunity. The motive lies in the realization that reductionist thinking and experimentation alone are not sufficient if complex systems are involved. Reductionist experiments are very good in generating detailed information regarding specific components or processes of a system, but they often lack the ability to characterize, explain, or predict **emergent properties** that cannot be found in the parts of the system but only in their web of interactions. For instance, the emergence of oscillations in the example system represented by the equations in (1.1) cannot be credited to a single component of the system but is a function of its overall organization. Although we had complete knowledge of all details of the model pathway, it was very difficult to foresee its capacity either to saturate or oscillate in a damped or stable fashion. Biology is full of such examples.

A few years ago, Hirotada Mori's laboratory completed the assembly of a complete catalogue of single mutants in the bacterium *Escherichia coli* [18]. Yet, the scientific community is still not able to foresee which genes the bacterium will up- or down-regulate in response to new environmental conditions. Another very challenging example of emergent system properties is the central nervous system. Even though we understand quite well how action potentials are generated and propagated in individual neurons, we do not know how information flows, how memory works, and how diseases affect the normal functioning of the brain. It is not even clear how information in the brain is represented (see also Chapter 15). Thus, while reductionist biology has been extremely successful and will without any doubt continue to be the major driving force for future discovery, many biologists have come to recognize that the detailed pieces of information resulting from this approach need to be complemented with new methods of system integration and reconstruction [19].

The opportunity for systems biology is the result of the recent confluence and **synergism** of three scientific frontiers. The first is of course the rapid and vast accumulation of detailed biological information at the physiological, cellular, molecular, and submolecular levels. These targeted investigations of specific phenomena are accompanied by large-scale, high-throughput studies that were entirely infeasible just a couple of decades ago. They include quantification of genome-wide expression patterns, simultaneous identification of large arrays of expressed proteins, comprehensive profiling of cellular metabolites, characterization of networks of molecular interactions, global assessments of immune systems, and functional scans of nervous systems and the human brain. These exciting techniques are generating unprecedented amounts of high-quality data that are awaiting systemic interpretation and integration (**Figure 1.12**).

The second frontier is the result of ingenuity and innovation in engineering, chemistry, and material sciences, which have begun to provide us with a growing array of technologies for probing, sensing, imaging, and measuring biological systems that are at once very detailed, extremely specific, and usable *in vivo*. Many tools supporting these methods are in the process of being miniaturized, in some cases down to the nanoscale of molecules, which allows diagnoses with minute amounts of biological materials and one day maybe biopsies of individual, living cells. Devices at this scale will allow the insertion of sensing and disease treatment devices into the human body in an essentially noninvasive and harmless fashion [20–22]. Bioengineering and robotics are beginning to render it possible to measure hundreds or thousands of biomarkers from a single drop of blood. It is even becoming feasible to use molecular structures, prefabricated by nature, for new purposes in medicine, drug delivery, and biotechnology (**Figure 1.13**).

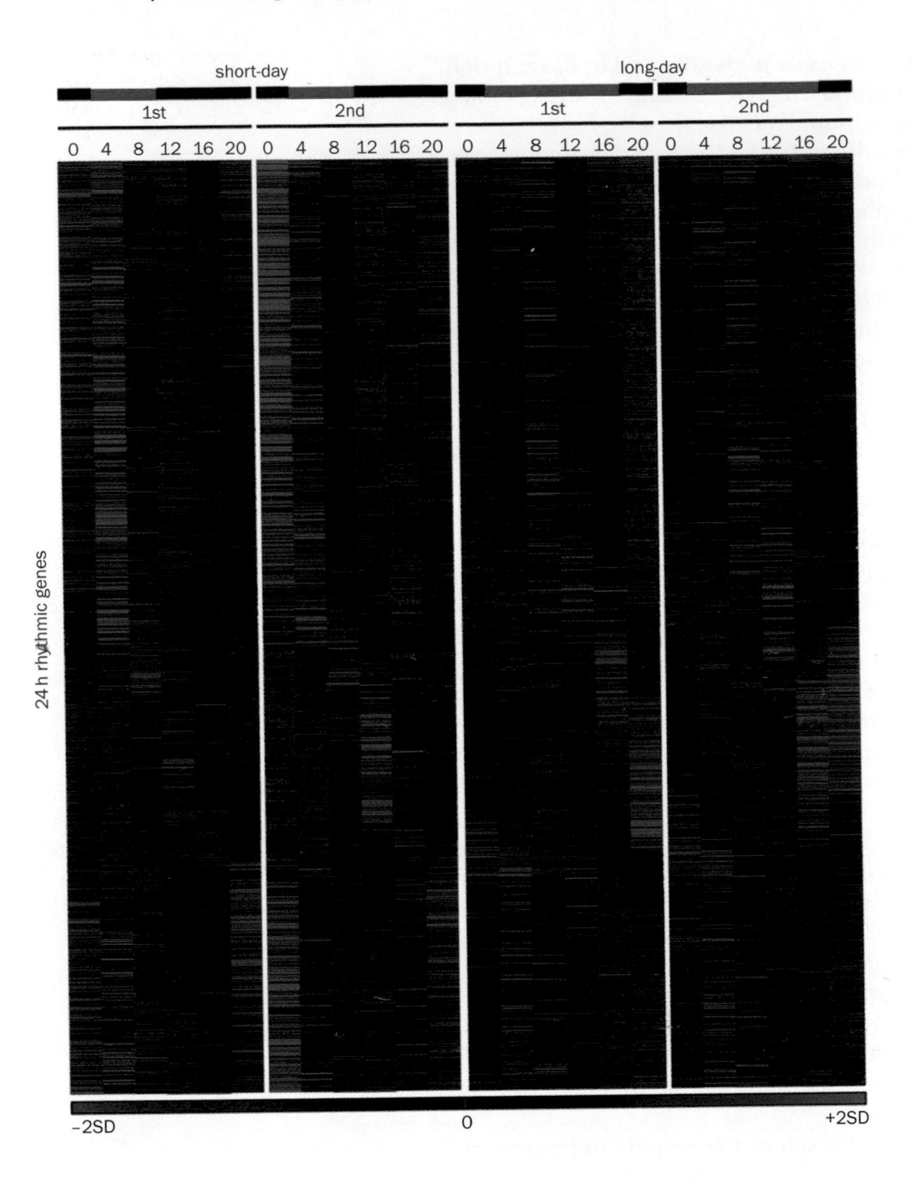

Figure 1.12 Modern high-throughput methods of molecular biology offer data in unprecedented quantity and quality. As an example, the heat map shown here represents a genome-wide expression profile of 24-hour-rhythmic genes in the mouse under chronic short-day (left two panels) and long-day (right two panels) conditions. (From Masumoto KM, Ukai-Tadenuma M, Kasukawa T, et al. *Curr. Biol.* 20 [2010] 2199–2206. With permission from Elsevier.)

The third frontier is the co-evolution of mathematical, physical, and computational techniques that are more powerful and accessible to a much wider audience than ever before. Imagine that only a few decades ago computer scientists used punch cards that were read by optical card readers (**Figure 1.14**)! Now, there are even specific computing environments, including Mathematica® and MATLAB®, as well as different types of customized mark-up languages (**XML**), such as the systems biology mark-up language **SBML** [23] and the mark-up language AGML, which was developed specifically for analyzing two-dimensional gels in proteomics [24].

Before today's much more effective computer science techniques were available, it was not even possible to keep track of the many components of biological systems, let alone analyze them. But over the past few decades, a solid theoretical and numerical foundation has been established for computational methods specifically tailored for the investigation of dynamic and adaptive systems in biology and medicine. These techniques are now at the verge of making it possible to represent and analyze large, organizationally complex systems and to study their emergent properties in a rigorous fashion. Methods of machine learning, numerical mathematics, and bioinformatics permit the efficient mining and analysis of the most useful data from within an overwhelming amount of data that are not pertinent for a given task. Algorithmic advances permit the simulation and optimization of very large biological flux distribution networks. Computer-aided approximation approaches yield ever-finer insights into the dynamics of complex nonlinear systems, such as the control of blood flow in healthy and diseased hearts. New mathematical, physical, and computational methods are beginning to make it possible to

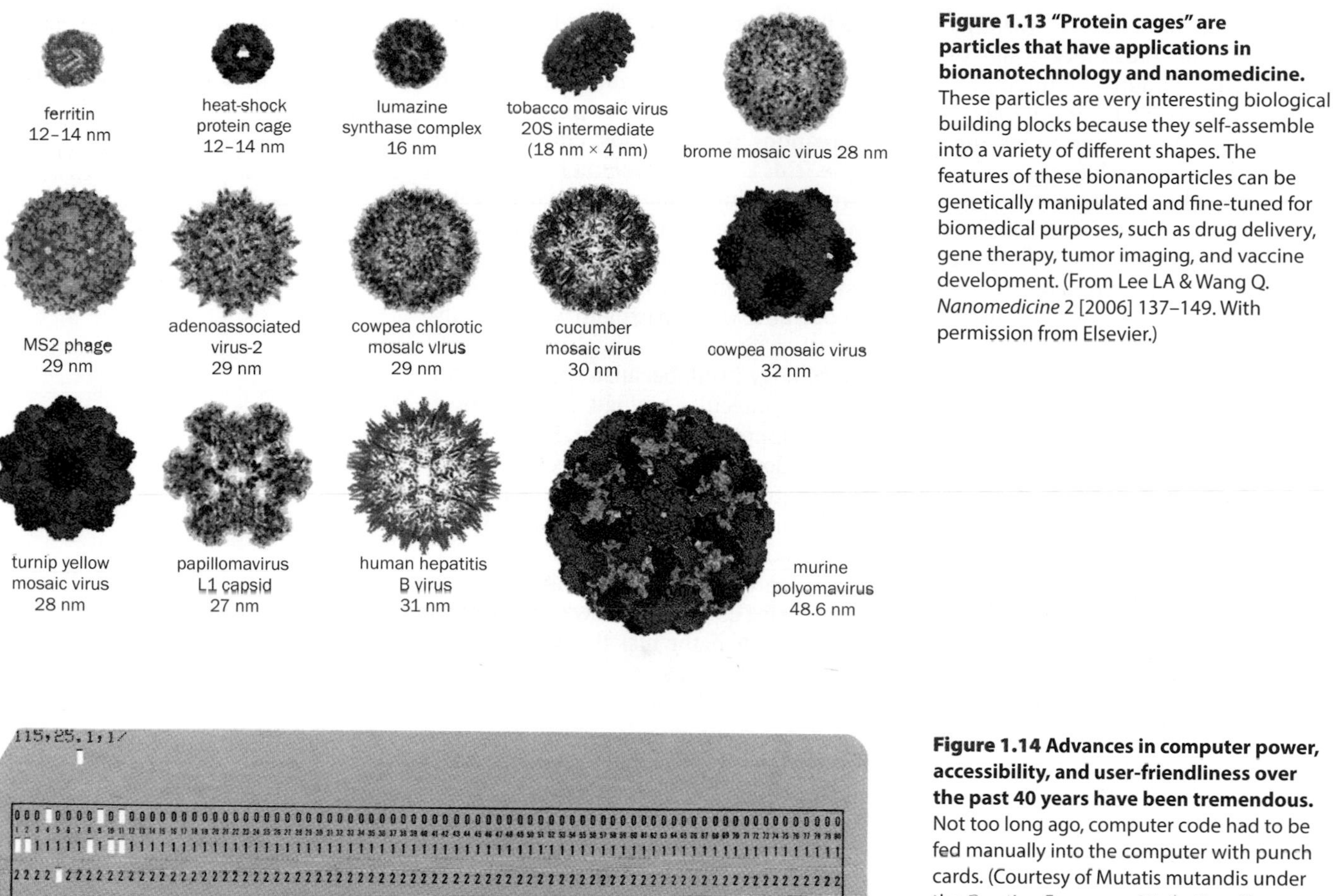

Figure 1.13 "Protein cages" are particles that have applications in bionanotechnology and nanomedicine. These particles are very interesting biological building blocks because they self-assemble into a variety of different shapes. The features of these bionanoparticles can be genetically manipulated and fine-tuned for biomedical purposes, such as drug delivery, gene therapy, tumor imaging, and vaccine development. (From Lee LA & Wang Q. *Nanomedicine* 2 [2006] 137–149. With permission from Elsevier.)

Figure 1.14 Advances in computer power, accessibility, and user-friendliness over the past 40 years have been tremendous. Not too long ago, computer code had to be fed manually into the computer with punch cards. (Courtesy of Mutatis mutandis under the Creative Commons Attribution-Share Alike 3.0 Unported license.)

predict the folding of proteins and the binding between target sites and ligands. These predictions, in turn, suggest insights into specific molecular interactions and promise the potential of targeted drug interventions that minimize toxic side effects.

Motive and opportunity have met to make systems biology attractive and feasible. It has become evident that the relevant disciplines complement each other in unique ways and that the synergism among them will revolutionize biology, medicine, and a host of other fields, including biotechnology, environmental science, food production, and drug development.

COMMUNICATING SYSTEMS BIOLOGY

It is not a trivial task to talk succinctly about 25,000 genes and their expression state or about the many processes occurring simultaneously in response to a signal that a cell receives at its outer surface. Our minds are ill equipped to characterize numerical relationships, let alone discuss complicated mathematical functions, especially if these depend on many variables. If we did not have numbers, we would even have problems describing everyday features such as temperature. Of course, we can say that it is cold or hot, and we have a dozen or so adjectives in between. But if we need more accuracy, common language ceases to be sufficient. Discerning 37.0°C from 38.4°C is not easy without a numerical scale, but it is necessary to have the tools to describe the difference, for instance, because the former reflects normal body temperature, while the latter is a sign of fever.

We may willingly or grudgingly accept the fact that we need mathematics, which comes with its own terminology, but communication is a two-way process. If we start talking about eigenvalues and Hopf bifurcations, we are almost guaranteed to lose mainstream biologists, let alone laypeople. This is a real problem, because our results must be conveyed to biologists, who are providing us data, and to the public that pays for our research and has a right to reap the fruit of the enormous investment of resources going into science [25]. The only true solution to this challenge is the bilingual education and nurturing of systems biologists who can translate biological phenomena into math and computer code and who can explain what it really means for the biological system if the real part of an eigenvalue is positive [19].

Communication is not trivial even within biology itself, because specialization has progressed so far that different fields such as molecular biology, immunology, and nanomedicine have developed their own terminology and jargon. Let's look at this issue in the form of a parable from Indian folklore that describes six blind men exploring an elephant (**Figure 1.15**). This story is quite old and usually ends in utter confusion, but it is useful to analyze it further than is usually done. The story has it that each of the blind men touched a different part of an elephant and came to a different conclusion concerning the object of his research. The man touching the side thought he was touching a wall, the one feeling the leg concluded he was touching a tree. The elephant's trunk gave the impression of a snake, the tusk that of a pointed scimitar, the tail felt like a rope, and the ear appeared to be like a large leaf or fan. It is not difficult to see the analogy to a complex biological system like the onset of Alzheimer's disease. The first scientist found "the Alzheimer gene," the second discovered "a strong association between the disease and former head injuries," another scientist detected "problems with fatty acid metabolism in the brain," and yet another suggested that "aluminum in cookware might be the culprit." As in the case of the elephant, the scientists were right, to some degree.

Let's analyze the elephant story a little further. The first problem among the six blind men might have been the homogeneous pool of researchers. Including a female or a child might have provided additional clues. Also, we have to feel sorry for the Indian men for being blind. However, they were apparently not mute or deaf, so that a little discussion among them might have gone a long way. While all six were blind, it is furthermore fair to assume that they had friends with working vision, who could have set them straight. They could have used not just their hands but also their other senses,

Figure 1.15 Information about isolated parts of a system alone does not always reveal the true nature of the system. An old story of six blind Indian men trying to determine what they touch is a parable for the dangers of scientific silos and the lack of good communication.

such as smell. Do tree trunks really smell like elephant feet? Finally, they apparently stayed in their one spot, thereby greatly limiting their experience base.

It is again easy to translate these issues into biology, especially when we think of purely reductionist strategies. Instead of a homogeneous pool of biologists analyzing biological systems, it is without doubt more effective to have a multidisciplinary team including different varieties of biologists, but also physicists, engineers, mathematicians, chemists, and smart people trained in the liberal arts or economics. Instead of only focusing on the one aspect right in front of our nose, communication with others provides context for singular findings. We don't know whether the Indian men spoke the same language, but we know that even if biologists, computer scientists, and physicists all use English to communicate, their technical languages and their views of the scientific world are often very different, so that communication may initially be superficial and ineffective. That is where multidisciplinary groups must engage in learning new terminologies and languages and include interdisciplinary translators. Just as the Indian men should have called upon their seeing friends, investigators need to call in experts who master techniques that have not been applied to the biological problem at hand. Finally, it behooves the trunk analyzer to take a few steps and touch the tusk and the side. Established scientific disciplines have in the past often become silos. Sometimes without even knowing it, researchers have kept themselves inside these silos, unable or unwilling to break out and to see the many other silos around, as well as a whole lot of space between them.

Systems biology does not ask the six blind men to abandon their methods and instead to run circles around the elephant. By focusing on one aspect, the reductionist "elephantologists" are poised to become true experts on their one chosen body part and to know everything there is to know about it. Without these experts, systems biology would have no data to work on. Instead, what systems biology suggests is a widening of the mindset and at least rudimentary knowledge of a second language, such as math. It also suggests the addition of other researchers, assisting the "trunkologist" and the "tuskologist" by developing new tools of analysis, by telling them in their language what others have found, by closing the gap between trunks and tusks and tails.

One strategy for accomplishing this synergism is to collect the diverse pieces of data and contextual information obtained by the six blind men and to merge them into a conceptual model. What kind of "thing" could consist of parts that feel like a snake, tree trunks, large walls, two scimitars, two fans, and a rope? How could an aberrant gene, former head injuries, and altered brain metabolism functionally interact to result in Alzheimer's disease? Well-trained systems biologists should be able to develop strategies for merging heterogeneous information into formal models that permit the generation of testable hypotheses, such as "tree-trunk-like things are connected by wall-like things." These hypotheses may be wrong, but they can nevertheless be very valuable, because they focus the scientific process on new, specific experiments that either confirm or refute the hypothesis. An experiment could be to walk along the "wall" as far as possible. Is there a "tree trunk" on the end? Are there "tree trunks" to the left and to the right? Is there a "pointed scimitar" at one or both ends? Is the "snake" connected to a "wall" or to a "tree trunk"? Does the "wall" reach the ground? Each answer to one of these questions constrains further and further what the unknown "thing" could possibly look like, and this is the reason that refuted hypotheses are often as valuable or even more valuable than confirmed hypotheses. "The wall does indeed not reach the ground!" Then, how is it supported? By "tree trunks"?

The story tells us that effective communication can solve a lot of complex questions. In systems biology, such communication is not always easy, and it requires not only mastering the terminology of several parent disciplines but also internalizing the mindset of biologists and clinicians on the one hand and of mathematicians, computer scientists, and engineers on the other. So, let's learn about biology. Let's study laboratory data and information and explore the mindset of biologists. Let's study graphs and networks with methods from computer science. Let's see how mathematicians approach a biological system, struggle with assumptions, make simplifications, and obtain solutions that are at first incomprehensible to the non-mathematician but do have real meaning once they are translated into the language of biology.

THE TASK BEFORE US

We have discussed the need to understand biological systems. But what does that really mean? Generically, it means that we should be able to explain how biological systems work and why they are constructed in the fashion as we observe them and not in a different one. Second, we should be able to make reliable predictions of responses of biological systems under yet-untested conditions. And third, we should be able to introduce targeted manipulations into biological systems that change their responses to our specifications.

This level of understanding is a tall order, and we will need many years to achieve it even for a narrowly focused domain within the huge realm of biological systems. An important component of the task is the conversion of actual biological systems into computational models, because this conversion, if it is valid, allows us almost unlimited and comparatively cheap analyses. The resulting models of biological systems come in two types. The first focuses on specific systems and includes all pertinent functional and numerical details—one might think of the analogy to a flight simulator. The second type of model is intended to help us understand the fundamental, generic features of the organization of biological systems; here one might think of elementary geometry, which offers us valuable insights into spatial features of the world by dealing with ideal triangles and circles that do not really exist in nature. The two model types point to two opposite ends of a large spectrum. The former models will be large and complex, while the latter will be as reduced and simple as feasible. In practice, many models will fall between these two extremes.

To pave the way toward these goals, this book is organized in three sections. The first of these introduces in four chapters a set of modeling tools for converting biological phenomena into mathematical and computational analog and for diagnosing, refining, and analyzing them. The second section describes in five chapters the molecular inventories that populate biological systems, and the five chapters in the third section are devoted to representative case studies and a look into the future.

The modeling approaches parallel two fundamental properties of biological systems, namely their static structure and their dynamics, that is, their changes over time. For static analyses, we will characterize and rationalize how nature put particular systems together and which parts are directly or loosely connected with each other. We will see that there are distinct types of connections and interactions. One important difference is that some connections allow the flow of material from a source to a target, while others serve the sole purpose of signaling the state of the system. In the latter case, no material changes location. Like a billboard that is not changed whether hundreds of people look at it or nobody at all, a signaling component may not be changed when it sends a signal. It is also to be expected that some connections are crucial, while others are of secondary importance. Finally, there is the very challenging question of how we can even determine the structure of a system. What types of data do we need to infer the structure of a system, and how reliable is such an inference?

The dynamics of a system is of the utmost importance, because all biological systems change over time. Organisms traverse a life cycle during which they undergo tremendous changes. Even a lowly yeast cell fills up with scars where it has given birth to daughter cells, and once its surface is full, the cell slides into the sunset of life. We can easily see these changes under a microscope, but an incomparably larger number of changes remain hidden from our sight. Gene expression patterns, amounts of proteins, profiles of metabolites, all of these change dramatically between birth and death. In addition to normal changes throughout its lifetime, every organism responds to fast and slow changes in the environment and adapts rather quickly to new situations. Today we may observe and characterize the gene expression network in a bacterium, but tomorrow it may already have changed in response to some environmental pressures. Indeed, bacteria evolve so quickly that the commonly used term "wild type" no longer has much meaning. Even more than static aspects, dynamic aspects of biological systems mandate the use of computational models. These models help us reveal how fascinating living systems are, with respect both to their overall efficiency and to the ingenuity with which dynamic responses and adaptations are coordinated.

Whether static or dynamic, some model designs and analyses will be performed in the bottom-up and others in the top-down direction. However, since we seldom start at the very bottom, that is, with individual atoms, or at the very top with models of complete organisms as they interact with their environment, most modeling strategies in systems biology are in truth "middle-out," to use Nobel Laureate Sydney Brenner's expression (cited in [26]). They begin somewhere in between the extremes, maybe with pathways or with cells. Over time, they may be incorporated into larger models, or they may become more and more refined in detail.

The second section of the book addresses the molecular inventories of biological systems. Paralleling the biological organization of organisms, one chapter is devoted to gene systems, one to proteins, and one to metabolites. A further chapter discusses signal transduction systems, and the final chapter of this section describes features of populations. Clearly, all these chapters are woefully incomplete and should not be thought of as substitutes for real biology books. Their purpose is merely to provide brief overviews of the main classes of biological components that are the foundation of all modeling strategies.

The third section contains case studies that in one way or another highlight aspects of biological systems that are in some sense representative. One chapter describes the coordinated stress response system in yeast, which operates simultaneously at the gene, protein, and metabolite levels. Another chapter provides a presentation of how very different models can be useful to focus attention on selected aspects of a multiscale system, the heart. A third chapter indicates how systems biology can contribute to medicine and drug development. The fourth chapter illuminates aspects of the natural design of biological systems and of the artificial design of synthetic biological systems. Finally, the last chapter discusses emerging and future trends in systems biology.

EXERCISES

1.1. Search the Internet, as well as different dictionaries, for definitions of a "system." Extract commonalities among these definitions and formulate your own definition.

1.2. Search the Internet for definitions of "systems biology." Extract commonalities among these definitions and formulate your own definition.

1.3. List 10 systems within the human body.

1.4. Exactly what features make the system in Figure 1.10 so much more complicated than the system in Figure 1.8?

1.5. Imagine that Figure 1.10 represents a system that has become faulty owing to disease. Describe the consequences of its complexity for any medical treatment strategy.

1.6. In Figure 1.2, are there control paths along which ABA either activates or inhibits closure of stomata? If so, list at least one path each. If both paths exist in parallel, discuss what you expect the effect of an increase in ABA to be on stomata closure?

1.7. Imagine a control system like that in Figure 1.2, but much simpler. Specifically, suppose there is only one activating and one inhibiting path in parallel. Suppose further that the activating path reacts much faster than the inhibiting path. What would be the consequence with respect to the effect of an increase in input (ABA) on output (stomata closure)? How could a difference in speed be implemented in a natural cell? Does it matter how strong or weak the activation and inhibition are? Discuss!

1.8. We have discussed that it is often difficult to infer the structure of a biological system from data. Is it possible that two different systems produce exactly the same input–output data? If you think it is impossible, discuss and defend your conclusion. If you think the answer is affirmative, construct a conceptual example.

1.9. List and discuss features supporting the claim that reductionism alone is not sufficient for understanding biological systems.

1.10. List those challenges of systems biology that cannot be solved with intuition alone.

1.11. Discuss why it is important to create terminology and tools for communicating systems biology.

1.12. Assemble a "to-do" list for the future of systems biology.

REFERENCES

[1] Li S, Assmann SM & Albert R. Predicting essential components of signal transduction networks: a dynamic model of guard cell abscisic acid signaling. *PLoS Biol.* 4 (2006) e312.

[2] USNO. United States Naval Observatory Cesium Fountain. http://tycho.usno.navy.mil/clockdev/cesium.html (2010).

[3] Voit EO, Alvarez-Vasquez F & Hannun YA. Computational analysis of sphingolipid pathway systems. *Adv. Exp. Med. Biol.* 688 (2010) 264–275.
[4] Vodovotz Y, Constantine G, Rubin J, et al. Mechanistic simulations of inflammation: current state and future prospects. *Math. Biosci.* 217 (2009) 1–10.
[5] Savageau MA. Biochemical Systems Analysis: A Study of Function and Design in Molecular Biology. Addison-Wesley, 1976.
[6] Kim HJ & Triplett BA. Cotton fiber growth *in planta* and *in vitro*. Models for plant cell elongation and cell wall biogenesis. *Plant Physiol.* 127 (2001) 1361–1366.
[7] Kumar A. Understanding Physiology. Discovery Publishing House, 2009.
[8] Wolkenhauer O, Kitano H & Cho K-H. An introduction to systems biology. *IEEE Control Syst. Mag.* 23 (2003) 38–48.
[9] Westerhoff HV & Palsson BO. The evolution of molecular biology into systems biology. *Nat. Biotechnol.* 22 (2004) 1249–1252.
[10] Strange K. The end of "naive reductionism": rise of systems biology or renaissance of physiology? *Am. J. Physiol. Cell Physiol.* 288 (2005) C968–C974.
[11] Voit EO & Schwacke JH. Understanding through modeling. In Systems Biology: Principles, Methods, and Concepts (AK Konopka ed.), pp 27–82. Taylor & Francis, 2007.
[12] Bernard C. Introduction a l'étude de la médecine expérimentale. JB Baillière, 1865 (reprinted by Éditions Garnier-Flammarion, 1966).
[13] Noble D. Claude Bernard, the first systems biologist, and the future of physiology. *Exp. Physiol.* 93 (2008) 16–26.
[14] Bertalanffy L von. Der Organismus als physikalisches System betrachtet. *Naturwissenschaften* 33 (1940) 521–531.
[15] Bertalanffy L von. General System Theory: Foundations, Development, Applications. G Braziller, 1969.
[16] Mesarović MD. Systems theory and biology—view of a theoretician. In Systems Theory and Biology (MD Mesarović, ed.), pp 59–87. Springer, 1968.
[17] Rosen R. A means toward a new holism: "Systems Theory and Biology. Proceedings of the 3rd Systems Symposium, Cleveland, Ohio, Oct. 1966. M. D. Mesarović, Ed. Springer-Verlag, New York, 1968. xii + 403 pp" [Book review]. *Science* 161 (1968) 34–35.
[18] Yamamoto N, Nakahigashi K, Nakamichi T, et al. Update on the Keio collection of *Escherichia coli* single-gene deletion mutants. *Mol. Syst. Biol.* 5 (2009) 335.
[19] Savageau MA. The challenge of reconstruction. *New Biol.* 3 (1991) 101–102.
[20] Freitas RAJ. Nanomedicine, Volume I: Basic Capabilities. Landis Bioscience, 1999.
[21] Lee LA & Wang Q. Adaptations of nanoscale viruses and other protein cages for medical applications. *Nanomedicine* 2 (2006) 137–149.
[22] Martel S, Mathieu J-B, Felfoul O, et al. Automatic navigation of an untethered device in the artery of a living animal using a conventional clinical magnetic resonance imaging system. *Appl. Phys. Lett.* 90 (2007) 114105.
[23] SBML. The systems biology markup language. http://sbml.org/Main_Page (2011).
[24] Stanislaus R, Jiang LH, Swartz M, et al. An XML standard for the dissemination of annotated 2D gel electrophoresis data complemented with mass spectrometry results. *BMC Bioinformatics* 5 (2004) 9.
[25] Voit EO. The Inner Workings of Life. Cambridge University Press, 2016.
[26] Noble D. The Music of Life: Biology Beyond Genes. Oxford University Press, 2006.

FURTHER READING

Alon U. An Introduction to Systems Biology: Design Principles of Biological Circuits. Chapman & Hall/CRC, 2006.
Kitano H (ed.). Foundations of Systems Biology. MIT Press, 2001.
Klipp E, Herwig R, Kowald A, et al. Systems Biology in Practice: Concepts, Implementation and Application. Wiley-VCH, 2005.
Noble D. The Music of Life: Biology Beyond Genes. Oxford University Press, 2006.
Szallasi Z, Periwal V & Stelling J (eds). System Modeling in Cellular Biology: From Concepts to Nuts and Bolts. MIT Press, 2006.
Voit EO. The Inner Workings of Life. Cambridge University Press, 2016.
Voit EO. Computational Analysis of Biochemical Systems: A Practical Guide for Biochemists and Molecular Biologists. Cambridge University Press, 2000.

Introduction to Mathematical Modeling

2

When you have read this chapter, you should be able to:

- Understand the challenges of mathematical modeling
- Describe the modeling process in generic terms
- Know some of the important types of mathematical models in systems biology
- Identify the ingredients needed to design a model
- Set up simple models of different types
- Perform basic diagnostics and exploratory simulations
- Implement changes in parameters and in the model structure
- Use a model for the exploration and manipulation of scenarios

At the core of any computational analysis in systems biology is a mathematical **model**. A model is an artificial construct in the language of mathematics that represents a process or phenomenon in biology. **Modeling** is the process of creating such a construct and squeezing new insights out of it. Mathematical models in systems biology can take many forms. Some are small and simple, others large and complicated. A model can be intuitive or very abstract. It may be mathematically elegant in its streamlined simplicity or account for very many details with a large number of equations and specifications that require sophisticated computer **simulations** for their analysis. The key commonality among all good models in biology is their ability to offer us insights into processes or systems that we would not be able to gain otherwise. Good models can make sense of a large number of isolated facts and observations. They explain why natural systems have evolved into the particular organizational and regulatory structures that we observe. They allow us to make **predictions** and **extrapolations** about experimentally untested situations and future trends. They can lead to the formulation of new hypotheses. Good models have the potential to guide new experiments and suggest specific recipes for manipulating biological systems to our advantage. Models can tell us how cell cycles are controlled and why plaques appear in certain sections of our arteries and not in others. Future models may help us understand how the brain works and prescribe specific interventions in tumor formation that will be infinitely more subtle and less crude than our current methods of radio- or chemotherapy.

Modeling in systems biology is really a special way of looking at the world. Whether we just conceptualize a situation in the form of components and interactions, or whether we actually set up equations and analyze them, we begin to see how complicated phenomena can be conceptually simplified and dissected into manageable submodels or **modules**, and we develop a feel for how such modules interact with each other and how they might respond to changes. Be aware! Modeling may become a life-changing adventure.

An analogy for a mathematical model is a geographical map. This map is obviously very different from reality, and many details are missing. What do the houses look like in some remote village? Are the lakes clean enough for swimming? Are the people friendly? Then again, it is exactly this simplification of the complicated real world that makes the map useful. Without being distracted by secondary details, the map allows us to look at the network of major roads throughout an entire country while sitting at home on our living room floor. The map provides us with distances and highway types that let us estimate travel times. Altitudes on the map indicate whether the area we want to drive through is hilly and could be cold even during the summer months. Once we learn how to interpret the map, we can infer what the countryside might look like. Empty spaces, like the Badlands in South Dakota, suggest that living in the area might be difficult or undesirable for humans. The density of cities and villages provides us with some idea of how easy it might be to find lodging, food, or gas. Clusters of cities along the ocean probably imply that living there is desirable. Mathematical models have the same generic properties: they reflect some aspects quite well, but barely touch on others or ignore them altogether. They allow us to infer certain properties of reality but not others. The inclusion or exclusion of certain features in models has a direct, important implication for the modeling process. Namely, the key task in designing a suitable model is a crisp focus on the aspects of primary interest and a representation of these aspects that retains their most important properties. At the same time, extraneous or distracting information must be minimized, and the model must be limited in complexity to permit analysis with reasonable efficiency.

Modeling is a mathematical and computational endeavor that in some sense resembles the fine arts (**Figure 2.1**). Basic math problems in calculus or linear algebra have the correct solutions in the back of the book, or at least somebody,

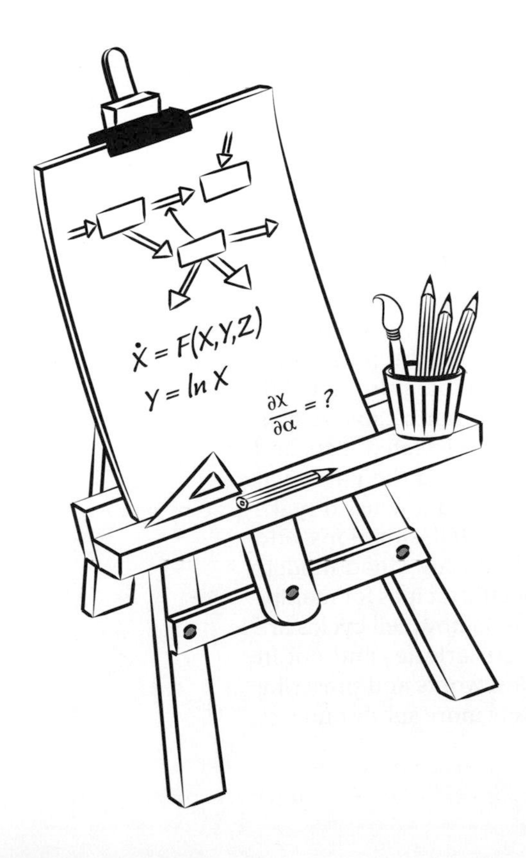

Figure 2.1 Modeling resembles a fine art. Modeling requires both mastery of techniques and creativity.

somewhere, knows these unique, correct solutions. In contrast to this rigor and crispness, many aspects of modeling allow a lot more flexibility. They permit different approaches, and often there are multiple, quite distinct solutions that are all valuable and correct in their own right. Who is to say that an oil painting is better than a watercolor? Or that an opera has higher artistic value than a symphony? Similarly, a mathematical model has aspects and properties in different dimensions and its value is not always easy to judge. For instance, is a very complicated, detailed model better than a simple, yet elegant model?

The answers to these questions depend critically on the goals and purposes of the model, and it is not just beneficial but indeed mandatory to specify the reasons for setting up a model and to ask a lot of questions before we embark on actually designing it. These questions must clarify the potential utilization of the model, its desired accuracy, the degree of complexity we are willing to accept, and many other details that we will touch on in this chapter. Indeed, it is prudent to invest quite a bit of time and exploration in pondering these aspects before one begins selecting a specific model structure and defining variables, parameters, and other model features, because technical challenges often distract from the big picture.

As in the fine arts, the creation of a good model requires two ingredients: technical expertise and creativity. The technical aspects can and must be learned through the acquisition of a solid repertoire of math and computer science skills. To make theoretical and numerical analyses meaningful, one also has to dig sufficiently deeply into biology in order to understand the background and terminology of the phenomenon of interest and to interpret the model results in a correct and sensible manner. As with the fine arts, the creative aspect of the process is more difficult to learn and teach. For instance, it requires more experience than one might expect to ask the right questions and reformulate them in such a way that they become amenable to mathematical and computational analysis. For a novice, all data may look equally valuable, but they really are not. In fact, it takes time and experience to develop a feel for what is doable with a particular set of data and what is probably out of reach. It is also important to learn that some data are simply not well suited for mathematical modeling. This insight is sometimes disappointing, but it can save a lot of effort and frustration.

The shaping and refining of specific questions mandates that we learn how to match the biological problem at hand with appropriate mathematical and computational techniques. This matching requires knowledge of what techniques are available, and it also means that the modeler must make an effort to understand how biologists, mathematicians, and computer scientists are trained to think. The differences in approach are quite pronounced in these three disciplines. Mathematicians are encouraged to simplify and abstract, and there is hardly anything more appealing than a tough problem that can be scribbled on the back of a napkin. Computer scientists look at problems algorithmically, that is, by dissecting them into very many, very tiny steps. Contrast that with biologists, who have learned to marvel at the complexity and intricacies of even the smallest biological item, and you can imagine the mental and conceptual tensions that may develop during the modeling process.

The formation of good questions requires decisions about what is really important in the model, which features can be ignored, and what inaccuracies we are willing to tolerate when we simplify or omit details. It is this complex of decisions on the inclusion and exclusion of components, facts, and processes that constitutes the artistic aspect of modeling. As in the fine arts, a good model may give an outsider the impression "Oh sure, I could have done that," because it is well designed and does not reveal the hard work and failed attempts that went into it. A good degree of proficiency with respect to these decision processes can be learned, but, like everything else, it requires practice with simple and then increasingly complex modeling tasks. This learning by doing can be exhilarating in success and depressing in failure, and while we have every right to enjoy our successes, it is often the failures that in the end are more valuable, because they point to aspects we do not truly understand.

The computational side of modeling draws from two classes of methods: mathematical analysis and computer simulation. Most practical applications use a combination of the two, because both have their own strengths. A purely mathematical analysis leads to general solutions that are always true, if the given assumptions and prerequisites are satisfied. This is different from even a large set of computer simulations, which can never confer the same degree of certainty and generality. For instance, we can mathematically prove that every real number has a real inverse (4 has 0.25, –0.1 has –10), with the notable exception of 0. But if we were to execute a mega-simulation on a computer, randomly picking a real number and checking whether its inverse is real, we would come to the conclusion that indeed all real numbers have a real inverse. We would miss zero, because the probability of randomly picking zero from an infinite range of numbers is nil. Thus, in principle, mathematics should always be preferred. However, in practical terms, the mathematics needed to analyze biological systems becomes very complicated very quickly, and even relatively simple-looking tasks may not have an explicit analytical solution and therefore require computational analysis.

Independent of the specific biological subject area and the ultimate choice of a particular mathematical and computational framework, the generic modeling process consists of several phases. Each phase is quite distinct both in input requirements and in techniques. Broadly categorized, the phases are shown in **Figure 2.2**, together with a coarse flow chart of the modeling process. Further details are presented in **Table 2.1**. The process starts with conceptual questions about the biological problem, becomes more and more technical, and gradually returns to the biological phenomenon under study. It is very important to devote sufficient effort to each of these phases, especially the early ones, because they ultimately determine whether the modeling effort has a good chance of succeeding.

One of the early choices addresses the explanatory character of the model. One may choose a comparatively simple regression model, which correlates two or more quantities to each other, without, however, suggesting a rationale for the correlation. As an example, one may find that cardiovascular disease is often associated with high blood pressure, but a regression model does not offer an explanation. As an alternative, one could develop a much more complex mechanistic model of cardiovascular disease, in which blood pressure is a component. If formulated validly, the model would explain the mechanistic connection, that is, the network of causes and effects leading from high blood pressure to the disease. Even if we disregard the difficulties in formulating a comprehensive mechanistic model, the choice of model type is not black and white. Regression models do not provide an explanation, but they often make correct, reliable predictions. An explanatory model may also be quite simple, but if it accounts for a large number of processes, it often becomes so

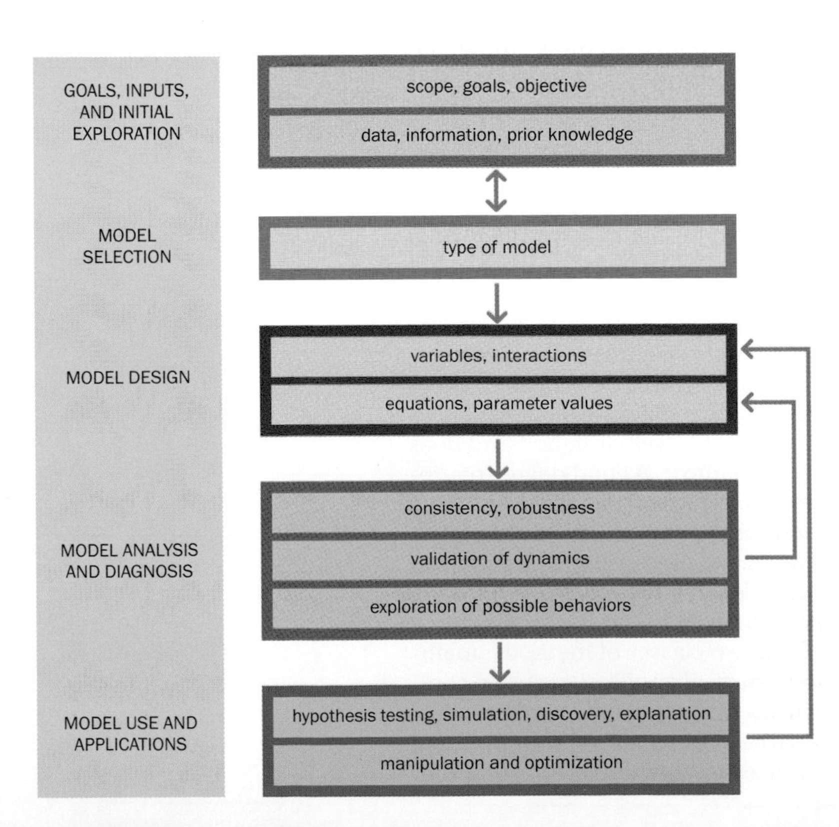

Figure 2.2 Flow chart of the modeling process. The modeling process consists of distinct phases, but is often iterative and involves refinements in model design and repeated diagnosis. See also Table 2.1. (Adapted from Voit EO, Qi Z & Miller GW. *Pharmacopsychiatry* 41(Suppl 1) [2008] S78–S84. With permission from Thieme Medical Publishers.)

TABLE 2.1: ISSUES TO PONDER DURING THE MODELING PROCESS

Goals, Inputs, and Initial Exploration

- Scope of the model, including goals, objectives, and possible applications
- Data needs and availability (types, quantity, quality)
- Other available information (nonquantitative heuristics, qualitative input from experts)
- Expected feasibility of the model
- Relevance and degree of interest within the scientific community

Model Selection

- Model structure
  - Explanatory (mechanistic) versus correlative (black box)
  - Static versus dynamic
  - Continuous versus discrete
  - Deterministic versus stochastic
  - Spatially distributed versus spatially homogeneous
  - Open versus closed
  - Most appropriate, feasible approximations

Model Design

- Model diagram and lists of components
  - Dependent variables
  - Independent variables
  - Processes and interactions
  - Signals and process modulations
  - Parameters
- Design of symbolic equations
- Parameter estimation
  - Bottom-up
  - Top-down

Model Analysis and Diagnosis

- Internal consistency (for example, conservation of mass)
- External consistency (for example, closeness to data and expert opinion)
- Reasonableness of steady state(s)
- Stability analysis
- Sensitivity and robustness analysis
- Structural and numerical boundaries and limitations
- Range of behaviors that can or cannot be modeled for (example, oscillations, multiple steady states)

Model Use and Applications

- Confirmation, validation of hypotheses
- Explanation of causes and effects; causes of failure, counterintuitive responses
- Best case, worst case, most likely scenarios
- Manipulation and optimization (for example, treatment of disease; avoidance of undesired states or dynamics; yield optimization in biotechnology)
- Discovery of design principles

complicated and requires so many assumptions that it may no longer describe or predict new scenarios with high reliability. In reality, most models are intermediates—explanatory in some aspects and correlative in others.

In the following sections, we discuss each phase of the modeling process and illustrate it with a variety of small, didactic "sandbox models" as well as a famous, very old model describing the spread of infectious diseases. The process begins with an exploration of the available information and the selection of a model that can utilize this information and ultimately yield new insights (see Figure 2.2 and

Table 2.1). Often, one will go back and forth between these two aspects before a suitable model design is chosen. Once an appropriate model type is determined, the actual model design phase begins. Components of the model are identified and characterized. The resulting model is diagnosed, and, should troublesome aspects emerge, one returns to the model design. A model that fares well in the diagnostic phase is used for explanations, simulations, and manipulations. It often turns out that model features are not optimal after all, requiring a return to the model design phase.

GOALS, INPUTS, AND INITIAL EXPLORATION

The initial phase of model development consists of defining the purpose of the model and of surveying and screening input data and contextual information. This setting of goals, combined with an exploration of available information and the context for the model, is arguably the most crucial of all phases. This assessment may sound counterintuitive, because one might think that the most effort would go into some complicated math. However, if insufficient consideration is given to this initial exploration, the modeling process may derail right away, and one may not even notice that something is not quite the way it was intended until the final phases of the modeling process.

The main task of the input and exploration phase is to establish the specific purposes and goals of the model, and to feel out whether the model may have a chance to achieve its goals, given the available data and information. "Feeling out" does not sound like rigorous mathematical terminology, but actually describes the task quite well. It is often here where experience is needed the most: right at the beginning of the effort!

2.1 Questions of Scale

The first important aspect of the exploration phase is to ask what are the most appropriate scales for the model, with respect to both time and space, and also with respect to the organizational level (**Figure 2.3**). These scales are typically tied to each other, because processes at the organizational level of a cell occur on a small spatial scale and usually run faster than processes at the scales of organisms or ecological systems. Suppose we are interested in trees and want to develop a mathematical model describing their **dynamics**. While that might sound like a fine idea, it is far too unspecific, because trees may be studied in many different ways. To model the growth of a forest or a tree plantation, the scale of organization is typically the individual tree, the spatial scale is in hectares or square kilometers, and the corresponding timescale is most likely in years. Contrast this with the photosynthesis in a leaf or needle, which truly drives tree growth. That organizational

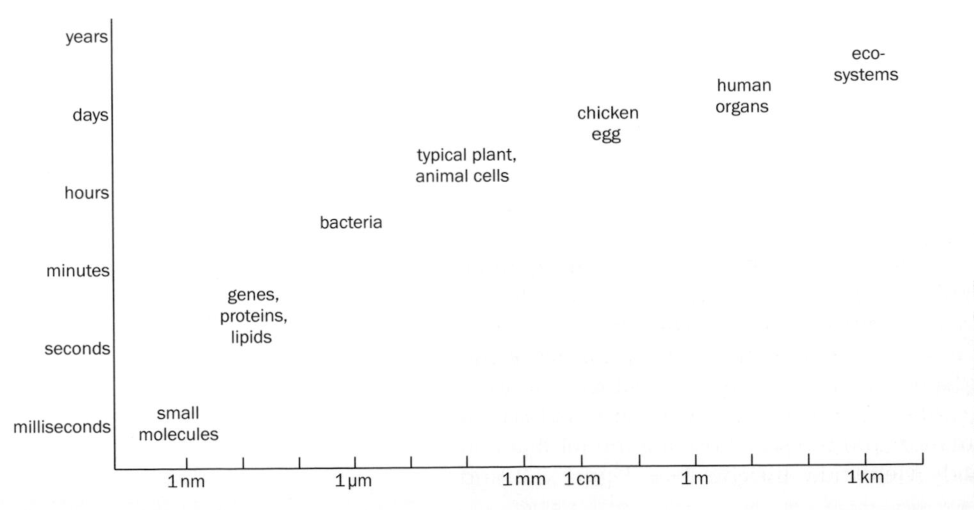

Figure 2.3 Scales in biology span several orders of magnitude in size and time. This plot shows typical sizes and time ranges. Of course, human organs and ecosystems contain small molecules and genes, so that they really span multiple size- and timescales, which makes multiscale modeling challenging and fascinating.

level is biochemical or physiological, and the corresponding timescale is that of seconds or minutes. It is easy to imagine other time and organizational scales between or outside these two scenarios.

While multiscale modeling is one of the challenging frontiers of systems biology (see Chapter 15), the attempt to accommodate too many different scales at once is very difficult and, at least at the beginning of model design, dangerous. Although there are exceptions (see for instance Chapter 12), it is usually not really feasible to account in the same model for the entire spectrum of processes, from the details of photosynthesis or wood formation to the long-term growth of forests. Nonetheless, the real strength of mathematical models lies in their ability to explain phenomena at one level through the analysis and integration of processes at a lower level. Thus, a good model, even of small size, may span two or maybe three levels of biological organization. For instance, it may describe the formation of a leaf in terms of genetic and biochemical processes. A model may also skip some levels. For instance, we could try to associate long-term tree growth and final size with alterations in gene expression. In this case, one might be able to translate the effects of gene modulations into the tree's growth rate, but a model of this type would not capture all the biochemical and physiological causes and effects that occur between gene expression and the total amount of wood in the forest after 50 years.

The determination of suitable scales depends very significantly on the available data that are needed to construct, support, and validate the model. If only yearly biomass measurements of the tree stand are available, we should not even attempt to model the biochemistry that deep down controls tree growth. Conversely, if the model uses light conductance as main input data, good as they may be, this input information would hardly support a model of the growth of trees in the forest over many years.

As a different example, suppose it is our task to improve the yield of a fermentation process in which yeast cells produce alcohol. Typical experimental techniques for this task are selections of mutants and changes in the growth medium. Thus, if we intend to use a model to guide and optimize this process, the model must clearly account for gene activities and their biochemical consequences, as well as for concentrations of nutrients in the medium and their uptake by cells.

2.2 Data Availability

The quantity of available data provides important clues for exploring options of model selection and design. If a complicated model with many **variables** and parameters is supported only by very scarce data, the model may capture these data very nicely, but as soon as we try to apply it to a slightly different scenario, such as wider spacing of the trees or a richer fertilization regimen, the model predictions may become unreliable or inaccurate. This over-fitting problem (that is, using a model that is too complicated for the given data) always looms when only one or a few datasets are available. A similar problem arises if the data show a lot of **variability**, or noise, as it is often called in the field of data analysis. In this case, many models may yield a good representation of the particular data, but predictions regarding new data may be unreliable. Counterintuitive as it may sound, it is often easier to model bad data than good data, because bad data grant us a lot more latitude and potentially let us get away with an inferior model. Of course, this inferior model may turn round and bite us later when the model is expected to fit new data and simply doesn't!

While crisp, numerical data are the gold standard for mathematical modeling, we should not scoff at qualitative data or information that contains a fair amount of uncertainty. For instance, if we are told that experiments show some output increasing over time (even though the exact degree is not given), this piece of information can provide a valuable constraint that the model must satisfy. Especially in a clinical setting, this type of qualitative and semiquantitative information is prevalent, and the more of it that is available, the more any model of the phenomenon is constrained. Expressed differently, this information can be extremely helpful for weeding out models that could otherwise appear plausible. In a study of Parkinson's disease, we asked clinicians for concentrations of dopamine and other brain metabolites associated with the disease. While they could not give us exact values, they provided us with rough, relative estimates of concentrations and metabolic fluxes through the system, and this information, combined with literature data, allowed us to set up a model that made surprisingly accurate predictions [1].

Finally, the exploration phase should ascertain whether experimental biologists or clinicians are really interested in the project. If so, their interest will help drive the project forward, and their direct insights and contacts with other experts can be invaluable, because much biological knowledge is not published, especially if it is semiquantitative or not yet one hundred percent supported by data and controls. It is, of course, possible to base models solely on information found in the literature, but if no experts are at hand, many pitfalls loom.

MODEL SELECTION AND DESIGN

To illustrate the process of model design, let's consider one specific example that we will carry through this chapter and revisit again in Chapter 4. This example concerns an infectious disease, such as tuberculosis or influenza, that easily spreads throughout a human population (**Figure 2.4**). Down the road, we would like to determine whether it is beneficial to vaccinate all or some individuals, whether infected persons should be quarantined, or whether the disease will eventually go away by itself because people recover and acquire immunity. Immediately, many questions come to mind. Do people die from the disease? If the death rate is small, can we ignore it in the model? How does the disease get started in the first place? For how long are people sick? How infectious are they? Are there individuals who show no symptoms yet are infectious? Do all people in our target population have potential contact with all other people? Do we need to account for air travel? Should we attempt to characterize waves of infection that start at the point of the first contact with an infected person and move through certain geographical areas or the entire population? Does our disease model have to be accurate enough to predict whether a specific person will get sick, or is our target the average infection level within a community or the nation? Clearly, questions are unlimited in number and type, and some of them may be important to ponder now while others are probably less relevant, at least at the beginning.

For our purposes of studying a very complex phenomenon with as simple a model as possible, let's suppose the following:

- The infectious disease is spread by human-to-human contact.
- Although bacteria or viruses are the true vectors of the disease, we ignore them.
- Individuals may acquire and lose immunity.
- The primary questions are how fast the disease progresses, whether it disappears without intervention due to immunity, and whether vaccination or quarantine would be effective.

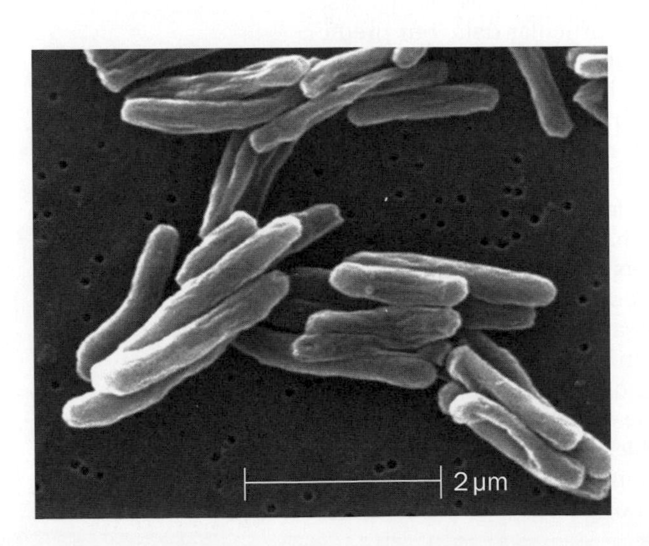

Figure 2.4 Mathematical modeling can be a powerful tool for studying the spread of bacterial diseases. Shown here is the bacterium *Mycobacterium tuberculosis*, which causes tuberculosis, one of the ongoing scourges of humankind.

- We do not intend to account for the spatial distribution of cases of infection throughout some geographical area.

2.3 Model Structure

Once we have decided what the model is supposed to accomplish, we need to determine its structure. Some decision points are given in Table 2.1. None of these are really associated with different degrees of sophistication, usefulness, or quality of the model. We are simply required to make decisions, again based on the data and the purposes of the model.

The first decision targets the question of how much detail the model is expected to explain. As mentioned before, the two extremes in this dimension are **correlative** and **explanatory models**; most models have aspects of both. A correlative model simply and directly relates one quantity to another. For instance, data may show that the infectious disease is particularly serious among the elderly; specifically, the number of infected individuals per 100,000 may increase with age (**Figure 2.5**). A **linear** or **nonlinear** regression could even quantify this relationship and allow us to make predictions about the average disease risk for a person of a certain age. However, even if the prediction were quite reliable, the model would not give us a hint as to why older people are more likely to become infected. The model does not account for health status, genetic predisposition, or lifestyle. Sometimes a prediction would be wrong, but the model would not be able to explain why.

An explanatory model, by contrast, relates an observation or outcome of an experiment to biological processes and mechanisms that drive the phenomenon under investigation. To illustrate, imagine an explanatory model of cancer. Cancer is associated with cells that are not dying when they should. Cell death normally occurs through a process called apoptosis, whose control involves the tumor suppressor protein p53, which in turn acts as a transcription factor for a number of pertinent genes. The p53 protein becomes activated under many stress conditions, such as ultraviolet radiation or oxidative stress. Pulling the pieces of information together allows us to compose a conceptual or mathematical model in the form of a chain of causes and effects: Suzie gets baked on the beach. She is exposed to enough ultraviolet radiation to activate the p53 proteins in her skin. The proteins change the normal activation of some genes. The genes cause some cells to escape apoptosis and proliferate out of control. Suzie is diagnosed with melanoma. The model explains to some degree (although not completely, of course) the development of cancer. This explanatory potential does not come for free: explanatory, mechanistic models are usually much more complicated than correlative models and require vastly more input in the form of data and assumptions. While explanatory models offer more insight, we should not discard correlative models offhand. They are much easier to construct and often make more reliable predictions than explanatory models, especially when assumptions are uncertain and data are scarce.

The second decision refers to the question of whether something associated with the phenomenon of interest changes over time. If the data show trends changing with time, we will probably prefer a dynamical model, which allows us to quantify these changes. By contrast, in a static model, such as a regression model, only the dependencies of one variable on others are explored. **Box 2.1** contrasts static

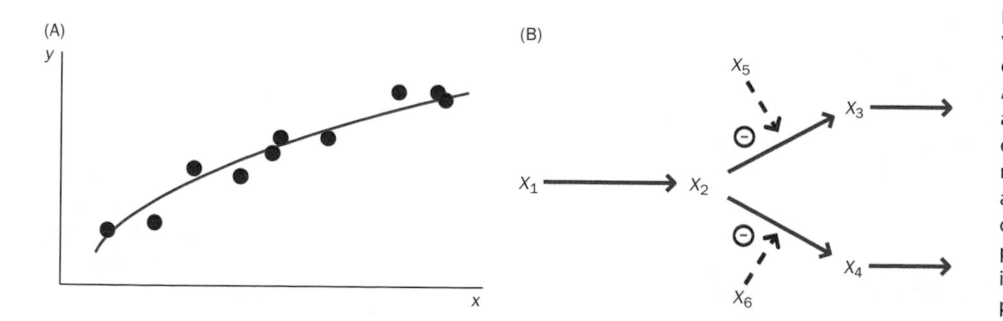

Figure 2.5 Correlative and explanatory models serve different purposes. The correlation between x and y in (A), whether linear or nonlinear, permits robust predictions of a value of y, given a new value of x, although the model does not explain the connection between x and y. An example could be disease prevalence (y) and age (x) among the elderly. The simple explanatory model in (B) could describe a metabolic pathway with two inhibitors (X_5 and X_6). It not only captures the (positive) correlation between X_3 and X_6, but also provides an explanation for why X_3 increases if the inhibitor X_6 of the other branch of the pathway is increased.

BOX 2.1: STATIC AND DYNAMIC MODELS

Suppose it is our task to compute the volume of an *Escherichia coli* cell. Studying a photograph of an *E. coli* culture (**Figure 1**), we see that each bacterium looks more or less like a hot dog with variable length and flattened ends, and since the volume doesn't change if the bacterium curves up, we can just study the shape when it is stretched out. Thus, a simple static model could be a cylinder with two "caps" at the end. The caps look a little flatter than hemispheres and we decide to represent them mathematically by halved ellipsoids that have the same axis radius in two directions (a circular cross section) and a shorter axis radius in the third direction (**Figure 2**). The formula for the volume of a whole ellipsoid is $\frac{4}{3}\pi r_1 r_2 r_3$, where r_1, r_2, r_3 are the axis radii. Thus, if the "straight" part of the cell is x µm long and has a diameter of y µm, and if each cap has a height of z µm, we can formulate a static model for the model of an *E. coli* cell (**Figure 3**) as

$$V(x,y,z) = \underbrace{x\pi\left(\frac{y}{2}\right)^2}_{\text{cylinder}} + \underbrace{2\left(\frac{4}{3}\pi\frac{y}{2}\frac{y}{2}\frac{z}{2}\right)}_{\text{two half-ellipsoildal caps}} \tag{1}$$

$$= \left(\frac{x}{4} + \frac{z}{3}\right)\pi y^2. \tag{2}$$

Our static model allows us to pick any length, width, and cap height and to compute the corresponding volume of the bacterium.

Let's now suppose the bacterium has just divided and is growing. The size parameters x, y, and z become variables that depend on time t, and we may write them as $x(t)$, $y(t)$, and $z(t)$. It may even happen that these variables do not develop independently from each other, but that the bacterium maintains some degree of proportionality in its shape, which would mathematically connect $x(t)$, $y(t)$, and $z(t)$ through constraint relationships. The main consequence is that the volume is no longer static, but a function of time. The model has become a dynamic or **dynamical** model.

In the typical understanding of dynamical models, one expects that the current state of the object or system will influence its further development. In the case of a bacterium, the growth in volume very often depends on how big the bacterium already is. If it is small, it grows faster, and as it reaches its final size before it divides, growth slows down. This "self-dependence" makes the mathematics more complicated. Specifically, one formulates the change in volume over time as a function of the current volume. The change in volume over time is given in mathematical terminology as its time derivative dV/dt. Thus, we come to the conclusion that volume-dependent growth should be formulated as

$$\frac{dV}{dt} = \text{some function of volume } V \text{ at time } t \tag{3}$$

$$= f\big(V(t)\big) = f(V). \tag{4}$$

This type of formulation, in which the derivative of a quantity (dV/dt) is expressed as a function of the quantity (V) itself, constitutes an ordinary differential equation, which is also lovingly called a diff. eq., o.d.e., or ODE. Very many models in systems biology are based on differential equations. The function f is not limited to a dependence on V; in fact, it most often contains other variables as well. For example, f could depend on the temperature and the substrate concentration in the medium, which will certainly affect the speed of growth, and on a lot of "internal" variables that describe the uptake mechanisms with which the bacterium internalizes the substrate. Looking more closely, many biochemical and physiological processes are involved in the conversion of substrate into bacterial volume and growth, so that a very detailed dynamical model can quickly become quite complicated. Setting up, analyzing, and interpreting these types of dynamical models is at the core of computational systems biology.

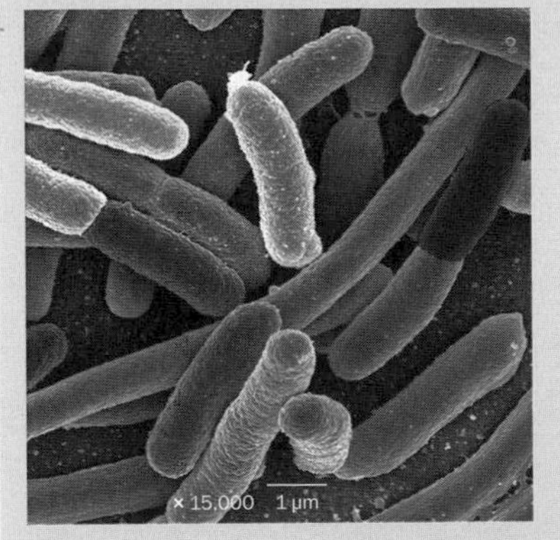

Figure 1 The shape of *E. coli* bacteria resembles cylinders with caps. The shape may be modeled approximately with the geometrical shapes shown in Figures 2 and 3.

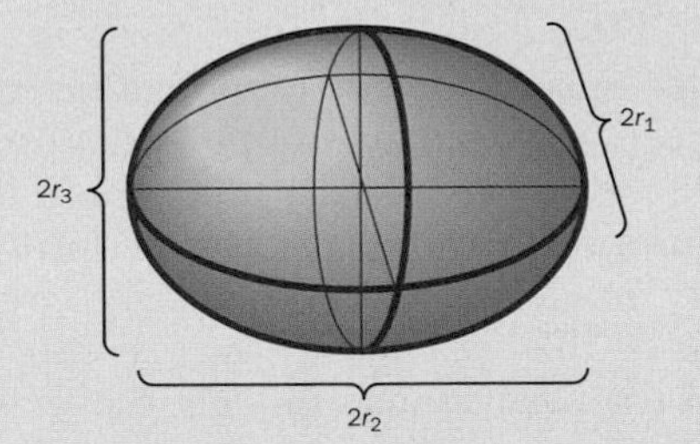

Figure 2 To model the shape of a bacterium (see Figure 1), we need a cylinder and "caps" at both ends. The caps can be modeled with flat ellipsoids. Here, such an ellipsoid is shown with two longer radii of the same length $r_1 = r_2$ and one shorter radius r_3.

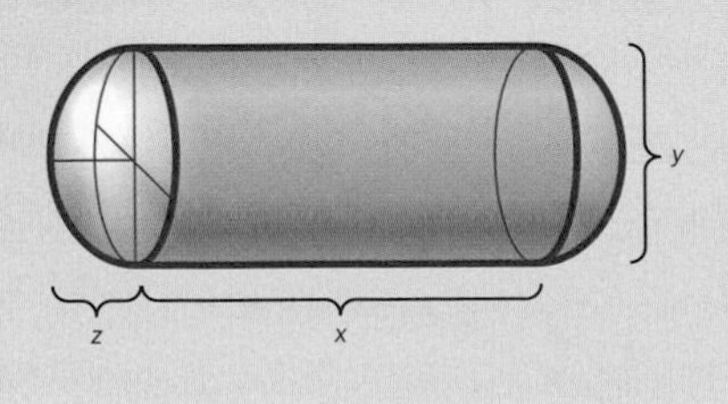

Figure 3 Static model of the volume of *E. coli*. The model is inspired by the actual cell shape (Figure 1) and composed of a cylinder with caps at both ends (Figure 2).

and dynamic models using a specific example. Static models may appear to be insufficient, because they ignore change, but they play important roles in systems biology, where they are, for instance, used to characterize the features of large networks (see Chapter 3).

Within the realm of dynamic models, one needs to decide whether a discrete or continuous timescale is better suited. An example of the discrete case is a digital clock, which displays time only with a certain resolution. If the resolution is in minutes, the clock does not give any indication whether the minute just started or is almost over. A continuous, analog clock would, in principle, show the time exactly. On one hand, the decision on a continuous or discrete timescale is very important, because the former usually leads to differential equations and the latter to iterative or recursive maps, which require different mathematical methods of analysis and sometimes lead to different types of insights (see Chapter 4). On the other hand, if we select very small time intervals, the **behaviors** of discrete models approach those of continuous models, and differential equations, which are often easier to analyze, may replace the difference equations.

The convergence of discrete and continuous models for increasingly smaller time steps is not the only instance where one formulation is replaced with another. More generally, we need to embrace the idea of **approximations** in other dimensions as well (**Figure 2.6**). We are certainly interested in "true" descriptions of all processes, but these do not really exist. Even the famous laws of physics are not strictly "true." For instance, Newton's law of gravity is an approximation in the context of Einstein's general relativity theory. Not that there would be anything wrong with using Newton's law: apples still fall down and not up. But we should realize very clearly that we are in truth dealing with approximations and that these are both necessary and indeed very useful. The need for approximations is much more pronounced in biology, where laws corresponding to those in physics are almost nonexistent. Chapter 4 discusses typical approximations used in systems biological modeling and Chapter 15 discusses more philosophical issues of laws in biology.

Somewhat of an approximation is the intentional omission of spatial considerations. In the real world, almost everything occurs in three spatial dimensions. Nonetheless, if the spatial aspects are not of particular importance, ignoring them makes life much easier. With diabetes, for instance, it may be important to study the distribution of glucose and insulin throughout specific organs and the cardiovascular system of the body. But for modeling the administration of insulin in a given situation, it may also be sufficient to consider merely the overall balance between glucose and insulin. A model that does not consider spatial aspects is often called (spatially) homogeneous. Models accounting for spatial aspects typically require partial differential equations or entirely different modeling approaches (see Chapter 15), which we will not discuss further at this point.

Finally, a systems model may be **open** or **closed**. In the former case, the system receives input from the **environment** and/or releases material as output, while in the latter case materials neither enter nor leave the system.

The model types we have discussed so far are called **deterministic**, which simply means that all information is completely known at the beginning of the computational experiment. Once the model and its numerical settings are defined, all properties and future responses are fully determined. In particular, a deterministic model does not allow any types of **random** features, such as environmental fluctuations or noise affecting its progress. The opposite of a deterministic model is a **probabilistic** or **stochastic** model. The latter term comes from the Greek word *stochos* meaning aim, target, or guess, and implies that a stochastic process or stochastic model always involves an aspect of chance. Somewhere, somebody in the system is rolling dice and the result of each roll is fed into the model. As a consequence, the variables become **random variables** and it is not possible to predict with any degree of certainty the outcome of the next experiment, or **trial** as it is often called. For instance, if we flip a coin, we cannot reliably predict whether the next flip will results in heads or tails. Nevertheless, even if we do not know exactly what will happen in future trials, we can compute the probability of each possible event (0.5 for each coin toss), and the more often we repeat the same random trial, the closer the collective outcome approaches the computed probability, even though each new trial again and again exhibits the same unpredictability.

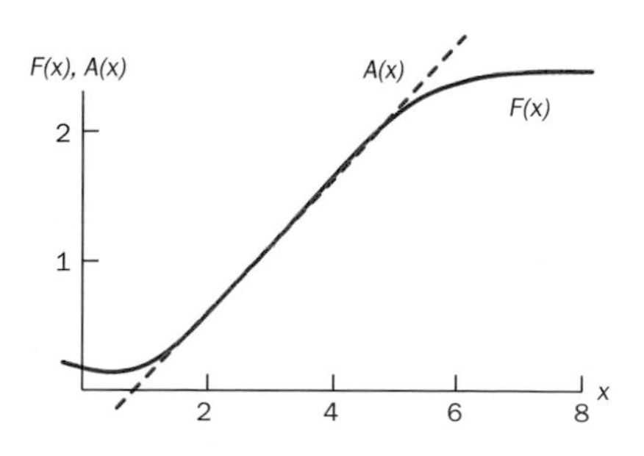

Figure 2.6 Approximations are necessary and very useful for simplified representations of reality in a mathematical model. In this illustration, $A(x)$ (dashed blue line) is a linear approximation of a more complicated nonlinear function $F(x)$ (red line). If only the range between 2 and 5 for the variable x is biologically relevant, the approximation is sufficiently accurate and easier to analyze.

It is no secret that biological data always contain variability, uncertainties, measurement inaccuracies, and other types of noise. Thus, as a final point to ponder when choosing a model: should we—must we—always use a model that somehow captures this randomness? Not necessarily. If we are primarily interested in average model responses, a deterministic model will generally do just fine. However, if we need to explore all possible outcomes, the worst and best cases, or the likelihood of specific outcomes from a system with a lot of uncertainty, we may be best advised to choose a stochastic model.

Because this is an important and subtle issue, let's study similarities and differences between deterministic and stochastic models in more detail. As we discussed, each trial in a stochastic system is unpredictable, but we can make very accurate predictions about average or long-term behaviors. By comparison, deterministic models present us with situations where everything is exactly and unambiguously defined. Yet, there are strange cases where the behavior of the system over time is so complicated that we cannot forecast it. A quite simple, but very instructive example is the blue sky catastrophe, which is discussed in **Box 2.2**. This system is completely determined and does not contain any noise or stochastic features. All information about the system is fixed in its definitions and settings, and yet the system is able to generate chaotic time courses that are impossible to anticipate. Chapter 4 discusses more cases of this nature.

As an example of a stochastic process, suppose we irradiate a culture of microorganisms with ultraviolet light (see, for example, [4]) in order to introduce random mutations. This method of random mutagenesis is often used to create strains that out-compete the wild type in a specific situation of interest. For instance, we may want to find a strain with improved tolerance to acid in its environment. We know from genetics that most mutations are neutral or decrease rather than increase fitness. Nonetheless, one mutation in a million or a billion may improve the microorganism's survival or growth rate, and this particular mutation will rise to the top and multiply in a properly designed random mutagenesis experiment. For simplicity, let's suppose that the mutations occur randomly throughout the genome with a rate of between three and four mutations per kilobase (1000 base pairs of DNA) [5].

A typical question in this scenario is: What is the probability of actually finding three or four mutations in a given DNA segment that is one kilobase long? Our first inclination might be to claim (wrongly) that the probability is 1, because isn't that what we just assumed? Not quite. We assumed that the *average* number of mutations is between three and four, and need to realize that it could happen that there are only one or two mutations, or five or six, owing to the randomness of the mutation process. Models from probability theory allow us to assess situations of this type. For instance, we can forecast the probability of encountering one, two, or three mutations in a given piece of DNA. Or we can make statements like "with p percent probability, we will find at least four mutations." Would it be possible to find 200 random mutations within 1000 base pairs? In a true stochastic process, this outcome is indeed theoretically possible, although the probability is extremely low.

Even in this clear-cut scenario with precise conditions and well-defined questions, different models are available. For instance, we could use the binomial model or the Poisson model. Both are applicable in principle. The binomial model would be better suited for small numbers (here, short segments of DNA), while it would be very cumbersome for large numbers (long segments of DNA). In fact, a pocket calculator would not be able to manage the numbers that would have to be computed for 1000 kilobases. The Poisson model uses a clever approximation, which is not very accurate for small numbers (short segments of DNA), but is incomparably easier to use and misses the correct results for large numbers (long segments of DNA) by only very small errors. **Boxes 2.3** and **2.4** discuss these two models as typical representatives of simple stochastic models.

2.4 System Components

Biological systems models contain different classes of components, which we discuss in this section. The most prominent components are variables, which represent biological entities of interest, such as genes, cells, or individuals. Because biological

BOX 2.2: AN UNPREDICTABLE DETERMINISTIC MODEL

Thompson and Stewart [2] describe a very interesting system that can be formulated as a pair of ordinary differential equations. The system, called the blue sky catastrophe, is completely deterministic, yet can produce responses that are utterly unpredictable, unless one actually solves the differential equations. The system has the form

$$\frac{dx}{dt} = y,$$
$$\frac{dy}{dt} = x - x^3 - 0.25y + A\sin t. \quad (1)$$

If we set $A = 0.2645$ and initiate the system with $x_0 = 0.9$, $y_0 = 0.4$, both variables quickly enter an oscillation that is quite regular; the time course for x for the first 100 time units is shown in **Figure 1A**.

However, if we think we understand what the system is doing, we are wrong. Continuing the same example until $t = 500$, without any changes to the model or its parameters, presents big surprises: out of the blue, the system "crashes" and undergoes regular or irregular oscillations at a much lower level, before it eventually returns to something like the original oscillation (Figure 1B). Continuing further in time shows irregular switches between oscillations around 1 and oscillations around −1. Intriguingly, even minute changes in the initial values of x and y or the parameter A totally change the appearance of the oscillation (Figure 1C, D). In fact, this chaotic system is so fickle that even settings in the numerical solver affect the solution. The plots in Figure 1 were computed in the software PLAS [3] with the standard local error tolerance of 10^{-6}. If one reduces the tolerance to 10^{-9} or 10^{-12}, the appearance of the figures is strikingly different.

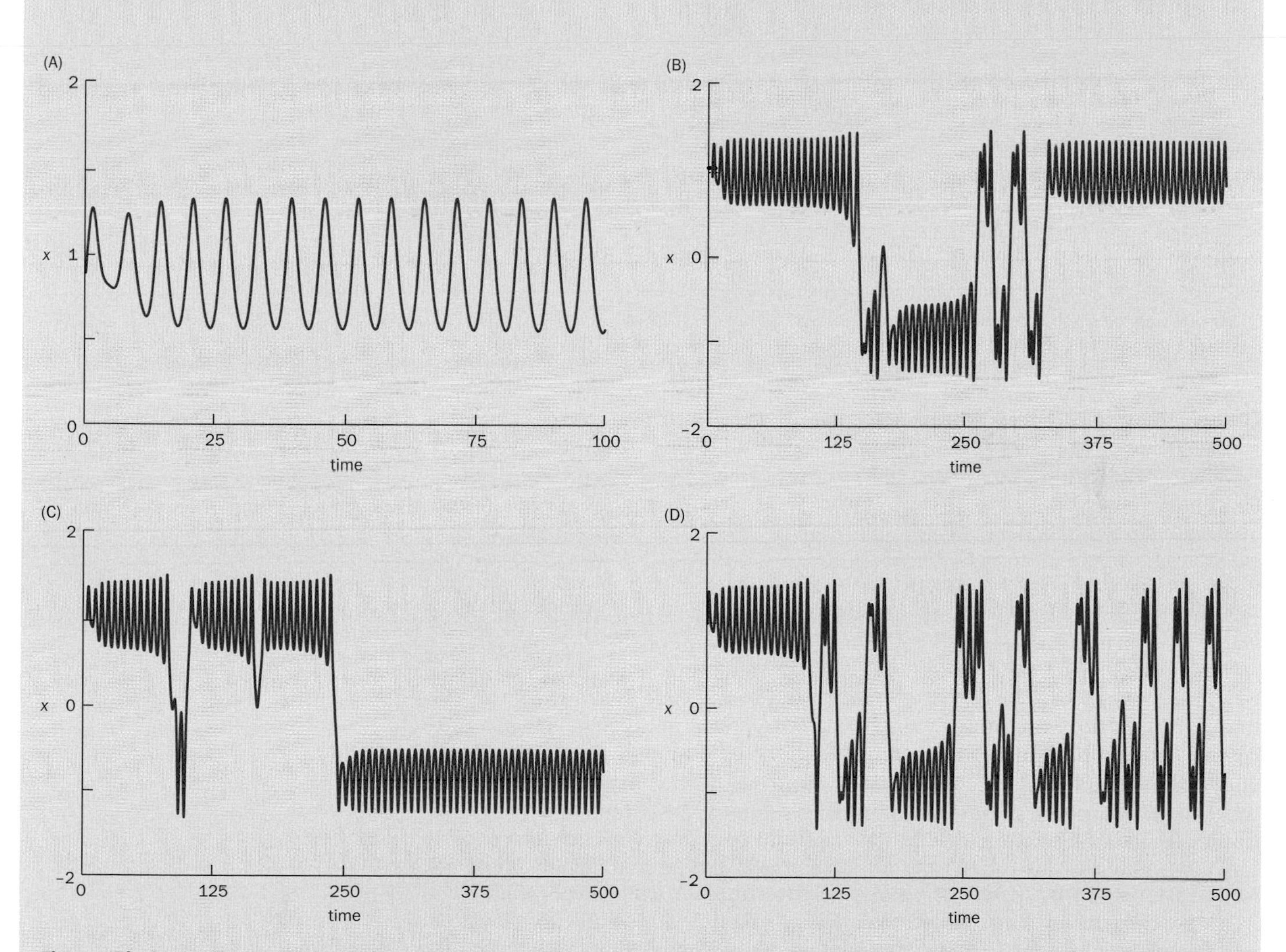

Figure 1 The intriguing model of a blue sky catastrophe. The system of two differential equations (with $x_0 = 0.9$, $y_0 = 0.4$, and $A = 0.2645$) appears to oscillate quite regularly about a baseline value of 1 (A). However, without any changes, the oscillations later become irregular (B: note the different scales). Even very slight changes in numerical features change the appearance of the oscillations dramatically. (C) $A = 0.265$, $x_0 = 0.9$, $y_0 = 0.4$. (D) $A = 0.2645$, $x_0 = 0.91$, $y_0 = 0.4$.

systems usually contain many variables, they are called **multivariate**. Variables may represent single entities or collections or pools of entities, such as all insulin molecules within a body, no matter where they are located. The second class of components consists of processes and interactions, which in some fashion involve and connect the variables. The third class contains **parameters**. These are numerical characteristics of the system, such as pH or temperature, or the turnover rate of a

BOX 2.3: THE BINOMIAL PROCESS: A TYPICAL STOCHASTIC MODEL

In the text, we asked: What is the probability of finding a certain number of mutations in a DNA segment, if one typically observes three or four mutations per 1000 base pairs? Let's do some math to find out. We formulate the problem as a binomial process, for which we need two ingredients, namely the average mutation rate, which we assume to be 3.5 (instead of "3 or 4") mutations per kilobase, and the number of nucleotides within the DNA segment of interest. From this information, we can compute how likely it is to find 3, 4, 1, 20, or however many mutations within some stretch of DNA.

In the language of probability theory, a mutation in our experiment is called a success and its (average) probability is usually termed p; in our case, p is 3.5 in 1000, that is, 0.0035. A failure means that the base is not mutated; its probability is called q. Because nothing else can happen in our experiment (either a base is mutated or it is not; let's ignore deletions and insertions), we can immediately infer $p + q = 1$. Put into words: "for a given nucleotide, the probability of a mutation plus the probability of no mutation together are a sure bet, because nothing else is allowed to happen."

In order to deal with more manageable numbers, let's first assume that the DNA piece contains only $n = 10$ bases. Then the probability P to find k mutations is given as

$$P(k;n,p) = \frac{n!}{k!(n-k)!} p^k (1-p)^{n-k}, \qquad (1)$$

according to the binomial formula of probability theory, which we will not discuss here further. In this formulation, the term $n!$ (pronounced "n factorial") is shorthand for the product $1 \cdot 2 \cdot 3 \cdot \cdots \cdot n$. The semicolon within the function indicates that k is the independent variable, while n and p are constant parameters, which may, however, be changed from one example to the next. So, what is the probability of finding exactly one mutation in the 10-nucleotide segment? The answer results from substituting values for the parameters and for k, namely,

$$P(1;10,0.0035) = \frac{10!}{1!9!} 0.0035^1 (1-0.0035)^9 = 0.0339. \qquad (2)$$

That is not a very big probability, so the result is rather unlikely: roughly 3%. What about two mutations? The answer is

$$P(2;10,0.0035) = \frac{10!}{2!8!} 0.0035^2 (1-0.0035)^8 = 0.000537, \qquad (3)$$

which is even smaller. It is not difficult to see that the probabilities become smaller and smaller for higher numbers of mutations. So, what's the most likely event? It is that there is no mutation at all. This probability is given by the binomial **probability distribution** (1) just like the others, namely,

$$P(0;10,0.0035) = \frac{10!}{0!10!} 0.0035^0 (1-0.0035)^{10} = 0.9655. \qquad (4)$$

In over 96% of all 10-nucleotide DNA segments, we expect to find no mutation at all!

From these three results, we can estimate the chance of finding more than two mutations. Because all probabilities taken together must sum to 1, the probability of more than two mutations is the same as the probability of *not* having zero, one, or two mutations. Thus, we obtain

$$P(k > 2) = 1 - 0.9655 - 0.0339 - 0.000537 \approx 0.00002, \qquad (5)$$

which is 2 in 100,000!

If we increase the length of the DNA piece, we should intuitively expect the probabilities of finding mutations to increase. Using the same formula, the probability of exactly one mutation in a 1000-nucleotide segment can be written as

$$P(1;1000,0.0035) = \frac{1000!}{1!999!} 0.0035^1 (1-0.0035)^{999} \qquad (6)$$

This is a perfectly fine formulation, but our pocket calculator goes on strike. 1000! is a huge number. To ameliorate the situation, we can play some tricks. For instance, we see from the definition of factorials ($n! = 1 \cdot 2 \cdot 3 \cdot \cdots \cdot n$) that the first 999 terms in 1000! and 999! are exactly the same and cancel out from the fraction, leaving us simply with 1000 for the first term. With that, we can compute the probability of 1 mutation as 0.1054 and the probability of no mutations at all as 0.0300 (0! is defined as 1). Indeed, in contrast to the scenario with 10 bases above, we are now about three times more likely to encounter one mutation than no mutation at all. Is it imaginable that we would find 10, 20, or 30 mutations? The probability of finding 10 mutations is about 0.00226. Formulate the corresponding probabilities for 20 and 30. It is not hard to do, but even with the tricks above we run into problems computing the numerical results from the formulae. Box 2.4 offers a solution!

chemical reaction, which is the number of molecules the reaction can process within a given amount of time. The particular values of the parameters depend on the system and its environment. They are constant during a given computer experiment, but may assume different values for the next experiment. Finally, there are (universal) constants, such as π, e, and Avogadro's number, which never change.

In order to design a specific formulation of a model, we begin with a diagram and several lists, which are often developed in parallel and help us with our bookkeeping. The diagram contains all entities of the biological system that are of interest and indicates their relationships. These entities are represented by variables and drawn as nodes. Connections between nodes, called edges, represent the flow of material from one node to another. For instance, a protein in a signaling cascade may be unphosphorylated or phosphorylated in one or two positions. Within a short timespan, the total amount of the protein is constant, but the distribution among the three forms changes in response to some signal. Thus, a diagram of this small protein phosphorylation system might look like **Figure 2.7**.

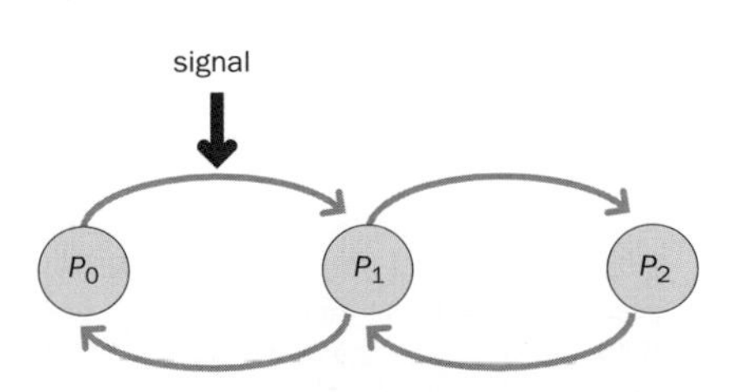

Figure 2.7 Different states of a protein in a signaling cascade. The protein may be unphosphorylated (P_0) or singly (P_1) or doubly (P_2) phosphorylated. The total amount of the protein remains the same, but, during signal transduction, material flows among the pools, driven by kinases and phosphatases, which are not shown here (see Chapter 9).

In contrast to the flow of material, a diagram may also contain the flow of information. For instance, the end product of a pathway may signal—through **feedback** inhibition—that no further substrate should be used. This signal is distinctly

BOX 2.4: THE POISSON PROCESS: AN ALTERNATIVE STOCHASTIC MODEL

The binomial model runs into problems for large numbers k and n. As a demonstration that modeling problems often permit different solutions, we provide here an alternative to the binomial model. The mathematician Siméon-Denis Poisson (1781–1840) developed a formula, nowadays called the Poisson distribution, that lets us compute binomial-type probabilities for large numbers k and n. The two ingredients we need are the mutation rate λ and the number k of mutations whose probability we want to compute. How can Monsieur Poisson get away with only two pieces of information instead of three? The answer is that he no longer looks at the success or failure at each nucleotide, but substitutes the step-by-step checking with a fixed rate that is really only valid for a reasonably long stretches or DNA. The mutation rate with respect to 1000 nucleotides is 3.5, whereas the rate for 100 nucleotides is one-tenth of that, namely, 0.35. Using a fixed rate makes life simple for large numbers of nucleotides, but does not make much sense for very short DNA pieces.

Poisson's formula, which is again based on probability theory [6], is

$$P(k;\lambda)=\frac{e^{-\lambda}\lambda^k}{k!}. \quad (1)$$

Thus, the probability of 10 mutations (in a 1000-nucleotide segment) is

$$P(10;3.5)=\frac{e^{-3.5}3.5^{10}}{10!}=0.00230. \quad (2)$$

The probabilities of 3 or 4 mutations are 0.216 and 0.189, respectively. The two together make up about 40% of all cases, which appears reasonable, because the overall mutation rate is 3.5 per 1000 nucleotides. The probability of no mutation is 0.0302, which is very similar to the computation with the binomial model. With the Poisson formula, we can compute all kinds of related features. For instance, we may ask for the probability of more mutations (anywhere between 5 and 1000 mutations). Similar to the binomial case, this number is the same as the probability of *not* finding 0, 1, 2, 3, or 4 mutations. The probabilities of 1 or 2 mutations are 0.106 and 0.185, respectively; the others we have already computed. Thus, the probability of finding 5 or more mutations is approximately $1 - 0.0302 - 0.106 - 0.185 - 0.216 - 0.189 = 0.274$, which corresponds to almost a third of all 1000-nucleotide strings.

different from the flow of material, because sending out the signal does not affect the sender. It does not matter to a billboard how many people look at it. Because of this difference, we use a different type of arrow. This arrow does not connect pools, but points from a pool to an edge (flow arrow). As an illustration, the end product X_4 in **Figure 2.8** inhibits the process that leads to its own generation and activates an alternative branch toward X_5. These modulation processes are represented with differently colored or other types of arrows to make their distinction from edges clear.

Of course, there are many situations where we do not even know all the pertinent components or interactions in the system. Different strategies are possible to deal with such situations. In opportune cases, it is possible to infer components, interactions, or signals from experimental data. We will discuss such methods in later chapters. If such methods are not feasible, we design the model with the best information we have and explore to what degree it answers our questions and where it fails. While we understandably do not like failures, we can often learn much from them. Indeed, we typically learn more from failures than from successes, especially if the model diagnosis points toward specific parts of the model that are most likely the cause of the failure.

Parallel to drawing out the diagram, it is useful to establish four lists that help us organize the components of the model. The first list contains the key players of the system, such as metabolites in a biochemical system, genes and transcription factors in a gene regulatory network, or population sizes of different species in an ecological study. We expect that these players will change over time in their concentrations, numbers, or amounts, and therefore call them **dependent variables**, because their fate depends on other components of the system. For instance, it is easy to imagine that variable X_2 in Figure 2.8 depends on X_1 and other variables. Typically, dependent variables can only be manipulated indirectly by the experimenter.

The system may also include **independent variables**, which we collect in the second list. An independent variable has an effect on the system, but its value or concentration is not affected by the system. For example, independent variables are often used in metabolic systems to model enzyme activities, constant substrate inputs, and co-factors. In many cases, independent variables do not change in quantity or amount over time, at least not within the time period of the mathematical experiments we anticipate to execute. However, it may also happen that forces outside the system cause an independent variable to change over time. For instance, a metabolic engineer might change the amount of nutrients flowing into a bioreactor. The variable is still considered independent, if its changes are not caused by the system.

The distinctions between dependent and independent variables on the one hand and between independent variables and parameters on the other are not always clear-cut. First, a dependent variable may indeed not change during an

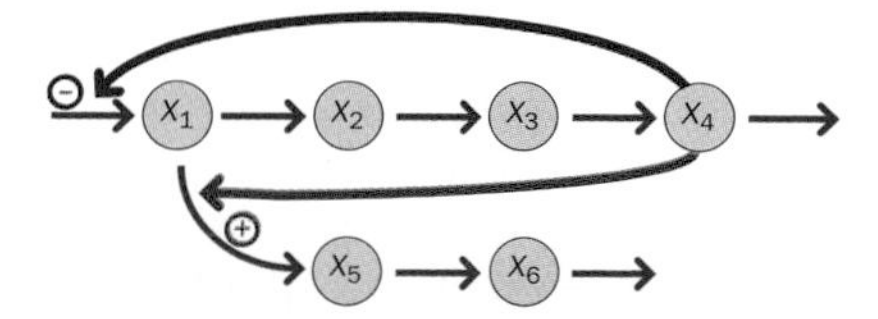

Figure 2.8 Generic pathway system, in which X_4 inhibits its own production process and activates the generation of an alternative pathway toward X_5 and X_6. The flow of material is represented with blue arrows, while modulation processes are shown with red (inhibition) and green (activation) arrows that distinguish them from flow arrows.

experiment and could therefore be considered independent. This replacement would make the model simpler, but might preclude analyses of other model settings where the variable does change. It actually happens more frequently that independent variables are replaced by dependent variables in later model extensions. A reason could be that the extended system affects the dynamics of a formerly independent variable. Second, if an independent variable is constant throughout an experiment, we could replace it with a parameter. However, it is often beneficial to keep parameters and independent variables separate. First, variables and parameters have a different biological meaning, which may help us with the estimation of appropriate values for them. And second, the computational effort of including an independent variable or a parameter is the same.

The third list is actually closer to a table or spreadsheet. It shows which of the (dependent or independent) variables have a direct effect on any of the processes in the system. As in the example of Figure 2.8, an effect may be positive (enlarging or activating) or negative (diminishing or inhibiting). The list accounts for all edges (flow of material between pools), as well as all signals that affect any of the edges.

The final list contains parameters that contribute to the external or internal conditions of the system, such as pH, temperature, and other physical and chemical determinants, which will ultimately require numerical specification.

Throughout the model design process, we will add to these lists quantitative information about pool sizes, magnitudes of fluxes, strengths of signals, and normal values and ranges of parameters.

The lists are the basis for establishing a diagram of the system. For the design of this diagram, it is beneficial to begin by representing each variable with one of the core modules in **Figure 2.9**. In the simpler case (Figure. 2.9A), each variable is displayed as a box with one process entering and one exiting. In the end, there may be more or fewer arrows, but using one influx and one efflux is a good default, since it reminds us later to include these processes when we set up equations. Indeed, a dependent variable tends to deplete without influx and to keep accumulating without efflux. Of course there are situations where depletion or accumulation is the desired behavior, and, if so, we can secondarily remove the corresponding process. However, these situations are quite rare. We may also use the default in Figure 2.9(B), which reminds us that many processes are regulated by other variables and that the variable in question may send out signals that could affect the influxes or effluxes of other variables. Independent variables typically do not have an influx, but they have an efflux if they feed material into the system. Once all variables are defined, they need to be functionally connected: the efflux of one variable may become the influx of another, and the signal sent by one variable may affect the influxes or effluxes of other variables.

Let's illustrate this construction of a diagram with the example of a population that is experiencing the outbreak of an infection. The model is intended to answer some of the questions mentioned earlier. Following some old ideas of Kermack and McKendrick [7], we use a similar terminology and keep things as simple as possible; one should, however, note that thousands of variations on this model have been proposed since Kermack and McKendrick's days. We begin by defining just three dependent variables, namely, the number of individuals susceptible to the disease (S), the number of infected individuals (I), and the number of individuals that are "removed" (R) from the two pools, because they have acquired immunity. We suppose that all individuals could be in contact with each other, at least in principle, and assume that a certain percentage of the immune individuals (R) lose their immunity and become susceptible again. We also allow for the possibility that individuals are born or immigrate and that individuals may die while infected. A diagram summarizing this population dynamics is shown in **Figure 2.10**. Processes

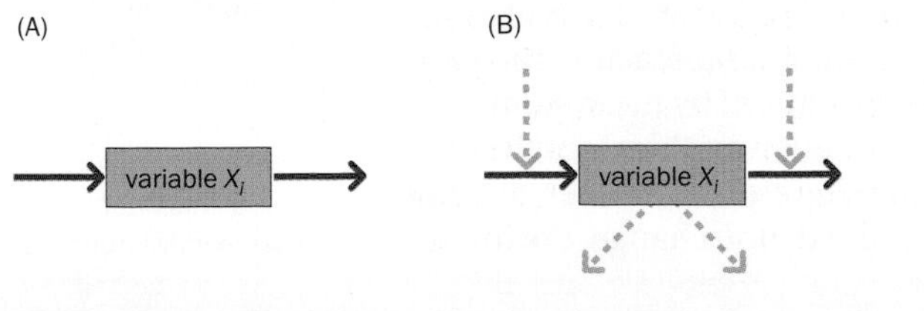

Figure 2.9 Default core modules. When designing a diagram of a system, it is beneficial to augment the display of each variable with an influx and efflux (A) and possibly with modification signals that affect the variable or are sent out by the variable (B).

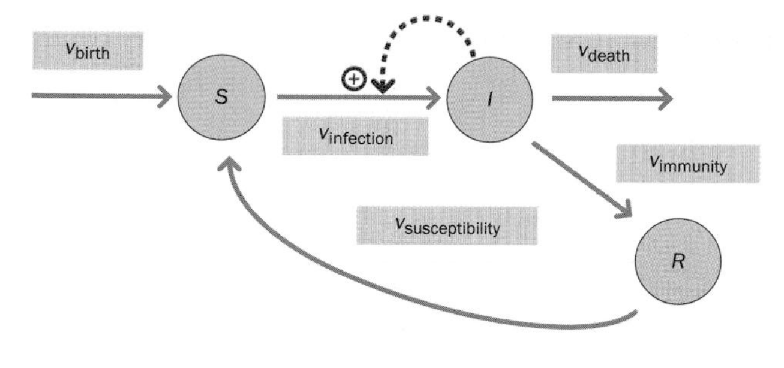

Figure 2.10 Diagram of a model for the spread of an infectious disease within a population. The model contains three dependent variables: S, the number of individuals susceptible to the disease; I, the number of infected individuals; and R, the number of immune ("removed" in the original terminology) individuals. Each process, in which people move from one pool to another, is shown as an edge and coded as v with a specific index. The process v_{birth} represents birth or immigration, $v_{infection}$ is the infection process, which requires contact between a susceptible and an infected individual. The process $v_{immunity}$ models the fact that infected individuals may acquire immunity. Infected people may also die from the disease (v_{death}). Finally, $v_{susceptibility}$ represents the rate with which individuals lose their immunity and become susceptible again. An immediately visible simplification is that S and R do not die, unless they are first infected. The model does not include independent variables.

such as infection, recovery, and death are denoted with indexed variables v. For instance, the infection process is coded as $v_{infection}$ and involves the contact between individuals of types S and I. The label v is often used in systems models, because it characterizes the velocity of a process. Some arrows have no source (v_{birth}), which means that we do not model explicitly where the individuals (or material) come from. Similarly, arrows may not have a target (v_{death}) because we do not account for their fate; individuals or molecules simply leave the system.

2.5 Model Equations

Once we have completed our model diagram and established the lists of components and processes, we need to translate this information into equations that reflect the chosen model structure. At first, these equations are symbolic, which means that no numerical values are yet assigned to the processes that characterize the system.

For the infectious disease problem, we choose a simple differential equation model, called an **SIR** model, which by its nature is moderately explanatory, dynamic, continuous, deterministic, and independent of spatial aspects. More complicated choices could include random processes associated with the infection of susceptible individuals or age distributions within the population with different degrees of susceptibility and recovery.

Like Kermack and McKendrick [7], we use as a model structure the simple and intuitive process description of mass action kinetics. This structure entails that the terms in each equation are set up either as constants (birth and immigration) or as a rate constant multiplied with the dependent variables that are directly involved in the process. This strategy reflects that the number of individuals moving from one pool to another depends on the size of the source pool. The more infected individuals there are, for instance, the more people will acquire immunity.

The specific construction is implemented one differential equation at a time and requires us to introduce new quantities, namely, the rates with which the various processes occur. The change in S is governed by two augmenting processes, namely birth and immigration (v_{birth}) with a constant rate r_B, and replenishment of the susceptible pool with individuals losing immunity ($v_{susceptibility}$) with rate r_S. At the beginning, no immune individuals may be around ($R = 0$), but the model accounts for the possibility, which may become reality later. The pool S is diminished by the infection process ($v_{infection}$), which has a rate r_I and depends on both S and I, because an infection requires that a susceptible and an infected individual come into contact. Thus, the first equation reads

$$\frac{dS}{dt} = \dot{S} = r_B + r_S R - r_I S I, \quad S(t_0) = S_0. \tag{2.1}$$

Here we have snuck in a new symbol, namely a variable with a dot, $\dot{S}$. This notation is shorthand for the derivative of this variable with respect to time. We have also added a second equation, or rather an assignment, namely the value of S at the beginning of the computational experiment, that is, at time t_0. Because we don't want to commit to a numerical value yet, we assign it the symbol S_0. This initial value must eventually be specified, but we do not have to do it right now. The system does not contain independent variables.

The equations for I and R are constructed in a similar fashion, and the entire model therefore reads

$$\dot{S} = r_B + r_S R - r_I SI, \quad S(t_0) = S_0, \tag{2.2}$$

$$\dot{I} = r_I SI - r_R I - r_D I, \quad I(t_0) = I_0, \tag{2.3}$$

$$\dot{R} = r_R I - r_S R, \quad R(t_0) = R_0. \tag{2.4}$$

Here, r_R is the rate of acquiring immunity and r_D is the rate of death. The set-up implies that only infected individuals may die, which may be a matter of debate, or of a later model extension.

We can see that several terms appear twice in the set of equations, once with a positive and once with a negative sign. This is a very common occurrence, because they describe in one case the number of individuals leaving a particular pool and in the other case the same number of individuals entering another pool.

The equations thus constructed are symbolic, because we have not yet committed to specific values for the various rates. Therefore, this set of symbolic equations really describes infinitely many models. Many of these will have similar characteristics, but it may also happen that one set of parameter values leads to very different responses than a different set.

2.6 Parameter Estimation

No matter which particular type of model we select, the model structure alone is usually not sufficient for a comprehensive assessment. For most specific analyses, we also need numerical values for the model parameters. In a statistical model, such a parameter may be the arithmetic mean. In a stochastic system, it might be the average rate of an event in a binomial trial. In a deterministic model, we usually need sizes or quantities of variables, rates of processes, and characteristic features of the model components. As an example, let's consider a mathematical function that is often used in biochemistry to describe the conversion of one type of molecule, called the substrate, into a different type, called the product. This conversion is usually facilitated by an enzyme. We will discuss biochemical conversion processes, along with their context and rationale, in Chapters 4, 7, and 8. Here it is sufficient to state that such a conversion is frequently modeled either with a so-called Michaelis–Menten function or with a Hill function. Both have the mathematical form

$$v(S) = \frac{V_{max} S^n}{K_M^n + S^n}. \tag{2.5}$$

The parameter V_{max} describes the maximal speed (velocity) of the conversion, which is attained as S becomes very large; in mathematical jargon, V_{max} is the asymptote of the function $v(S)$ as S goes to infinity. The second parameter is the Michaelis constant K_M. This describes a property of the enzyme facilitating the conversion and represents the value of S where the substrate-to-product conversion $v(S)$ runs at half-maximal speed: $v(K_M) = V_{max}/2$. The third parameter is the Hill coefficient n. In the case of a Michaelis–Menten function, we have $n = 1$, while true Hill functions are typically characterized by $n = 2$ or $n = 4$; however, the Hill coefficient may also take other positive values, including non-integer values such as 1.25. Depending on the numerical values of these three parameters, the function has a different appearance (**Figure 2.11**; Exercises 2.1–2.4).

Parameter values may be obtained in various ways, but their determination is almost always a complicated process. Indeed, in very many cases, parameter estimation for a systems model from actual data is the most challenging bottleneck in the entire modeling process. Two extreme classes for obtaining parameter values are from the bottom up and from the top down; in reality, the estimation is often a mixture of the two.

In a bottom-up estimation, parameter values are collected for each system component and process. For instance, experimental methods of enzyme kinetics could

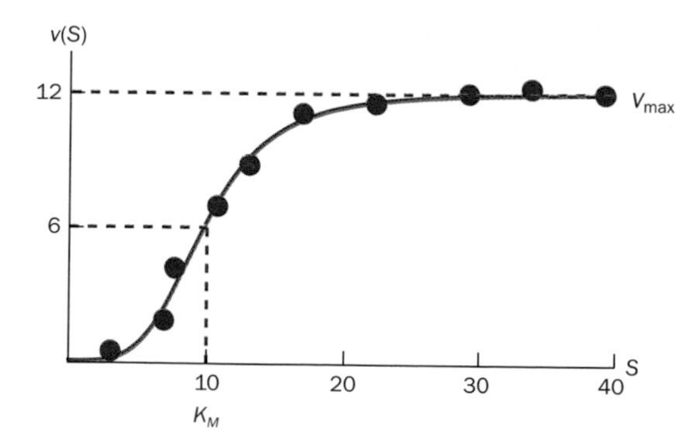

Figure 2.11 The parameters of a function determine its shape. In this case of a Hill function with $n = 4$ ($v(S) = V_{max}S^4/(K_M^4 + S^4)$; green line), the unknown parameters V_{max} and K_M can be determined from experimental data (red symbols) of $v(S)$ plotted against the substrate concentration (S). They are given respectively as the asymptote for large S and the value of S where $v(S)$ equals $V_{max}/2$. In the example, $V_{max} = 12$ and $K_M = 10$ (see Chapters 4, 5, and 8 for details).

be used to determine the K_M of an enzyme in the example above (see Chapter 8). Similarly, in the SIR example, one could tally how many people die from the disease within a given period of time and translate this measurement into the parameter r_D in (2.3). Thus, parameter values are measured or collected one by one, until the model is fully specified.

Top-down estimation methods are very different. In this case, one needs experimental measurements of all dependent variables under different conditions or at successive time points. In the case of a Hill function, the top-down estimation requires data such as the red dots in Figure 2.11, which measure $v(S)$ for different values of S. In the SIR example, one would need measurements of S, I, and R at many time points, following the outbreak of the disease. In either top-down estimation, the data and the symbolic equations are entered into an optimization algorithm that simultaneously determines those values of all parameters for which the model matches the data best.

Chapter 5 discusses methods of parameter estimation in detail. In our example of an infectious disease outbreak, parameter values could be derived directly or indirectly from the disease statistics established by a community health center in a bottom-up or top-down fashion.

Essentially all parameters in our case are rates (which generically describe the number of events per time), so that it is necessary to decide on a time unit. For our SIR example, we use days and set the rates correspondingly. Babies are born (or people immigrate) at a rate of 3 per day, 2% of the infected individuals actually die per day, and individuals lose immunity at a rate of 1% of R. The other rates are self-explanatory. The initial values say that, at the start of the model period, 99% of the individuals in a population of 1000 are healthy, yet susceptible to the disease ($S = 990$), that 1% are infected ($I = 10$), for reasons we do not know, and that nobody is initially immune ($R = 0$). These settings are entered into the symbolic equations (2.1)–(2.4), and the resulting parameterized equations are thus

$$\begin{aligned} \dot{S} &= 3 + 0.01R - 0.0005SI, & S_0 &= 990, \\ \dot{I} &= 0.0005SI - 0.05I - 0.02I, & I_0 &= 10, \\ \dot{R} &= 0.05I - 0.01R, & R_0 &= 0. \end{aligned} \tag{2.6}$$

In general, parameter estimation is complicated, and Chapter 5 is dedicated to the description of standard and ad hoc estimation methods. For the remainder of this chapter, we simply assume that all parameter values are available.

MODEL ANALYSIS AND DIAGNOSIS

A typical model analysis contains two phases. The first consists of model diagnostics, which attempts to ensure that nothing is obviously wrong with the model and assesses whether the model has a chance of being useful. For example, a model in which a variable disappears altogether when some physiological feature is changed by just a few percent is not very robust and the actual natural system would not survive in the rough-and-tumble outside world for very long. After we have received a green light from the diagnostics, we enter the second phase of exploring what the model is able or unable to do. Can it oscillate? Can some variable of interest reach a level of 100 units? What would it take to reduce a variable to 10% of its normal value? How long does it take until a system recovers from a **perturbation**?

Because we are dealing with mathematics, we might expect that we could directly compute all properties of a model with calculus or algebra. However, this is not necessarily so, even in apparently simple situations. As an illustration, suppose our task is to determine one or more values of x that satisfy the equation $e^x - 4x = 0$. Is there a solution? We can draw e^x and $4x$ on the same plot (**Figure 2.12**) and see immediately that the two intersect twice. Hence, the equation is satisfied at these two intersection points (see Exercise 2.5). Interestingly, and probably surprisingly, while there are two clear solutions, there is no algebraic method that would let us compute these solutions exactly. Our only systematic alternative is a numerical algorithm that (quickly) finds approximate solutions. More generally, situations in modeling where mathematical analysis must be supplemented with computer methods are the rule rather than the exception.

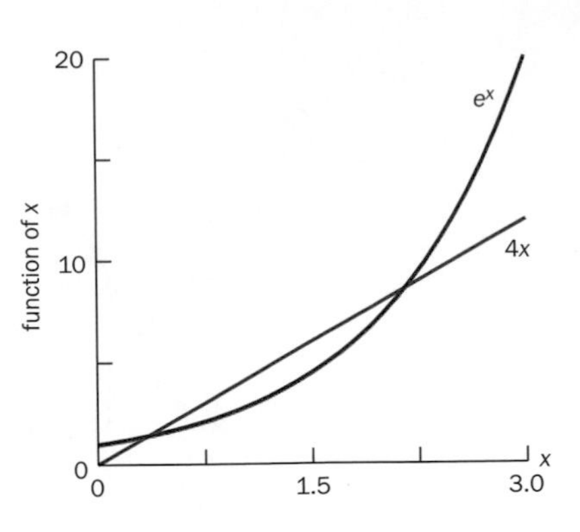

Figure 2.12 Computer simulation can be a very useful tool for model evaluation. The equation $e^x - 4x = 0$ obviously has two solutions, where the red and blue lines intersect, but we cannot compute them directly with algebraic methods. A suitable computer algorithm easily finds both solutions, although only approximately.

2.7 Consistency and Robustness

Before we can rely on a model, we need to do some diagnostics. This model checking can be a lengthy process, especially if we find flaws or inconsistencies that must be ameliorated by changing or refining the model structure. The targets of the diagnostics are partly biological and partly mathematical. As a biological example, assume that we are analyzing a metabolic network of pools and channels through which material flows. As a minimum requirement for consistency, we need to ensure that all material is accounted for at every pool and every time point, that effluxes from each pool actually reach one of the other pools, or correctly leave the system, and that the biochemical moieties of interest are preserved.

Many of the mathematical aspects of diagnostics follow rather strict guidelines, but the techniques are not always entirely trivial. For instance, we need to assess the robustness of the model, which essentially means that the model should be able to tolerate small to modest changes in its structure and environment. Techniques for these purposes are discussed in Chapter 4. In lieu of—or in addition to—assessing the model with purely mathematical techniques, we can learn a lot about the robustness and some of the features of the model through computational exploration. This is accomplished by using software to solve (simulate) the model under normal and slightly changed conditions. The order of items to diagnose (see Table 2.1) is not important.

For our illustration, let's start by solving the equations using a computer simulation with the baseline parameters that we set above. We may use for this purpose software like Mathematica®, MATLAB®, or the freeware PLAS [3]. **Figure 2.13** shows the result, which consists of one time course for each dependent variable. In our example, the changes within the population are quite dramatic. The subpopulation of susceptible people almost disappears, because almost everyone gets sick, but the number of immune individuals eventually exceeds half the population size. Yet, the infection does not seem to disappear within the first 100 days since the outbreak. Even after a whole year, some infectives are still present (Figure 2.13B); the situation is reminiscent of the persistence of diseases like tuberculosis. In general, we should always check whether the simulation results appear to make sense or are obviously wrong, before we launch any comprehensive mathematical diagnostics. In our example, we cannot judge whether the model yields the correct dynamics, but at least there is nothing obviously wrong or worrisome.

Let's do some computational exploration. The first question is whether the model in (2.6) reaches a **state** where none of the variables changes anymore. The answer is *yes*: we can show such a **steady state** by solving the equations for a sufficient time range (note that the number of immune individuals still climbs after one year). After a long time, the variables eventually reach the values $S = 140$, $I = 150$, and $R = 750$, and this steady state is stable. How we assess stability is discussed in Chapter 4. If we compute the total population size after several years, we note that there are now more people (1040) than at the beginning (1000). Is that reason to worry? No, it just means that the initial values were below the "normal" steady-state

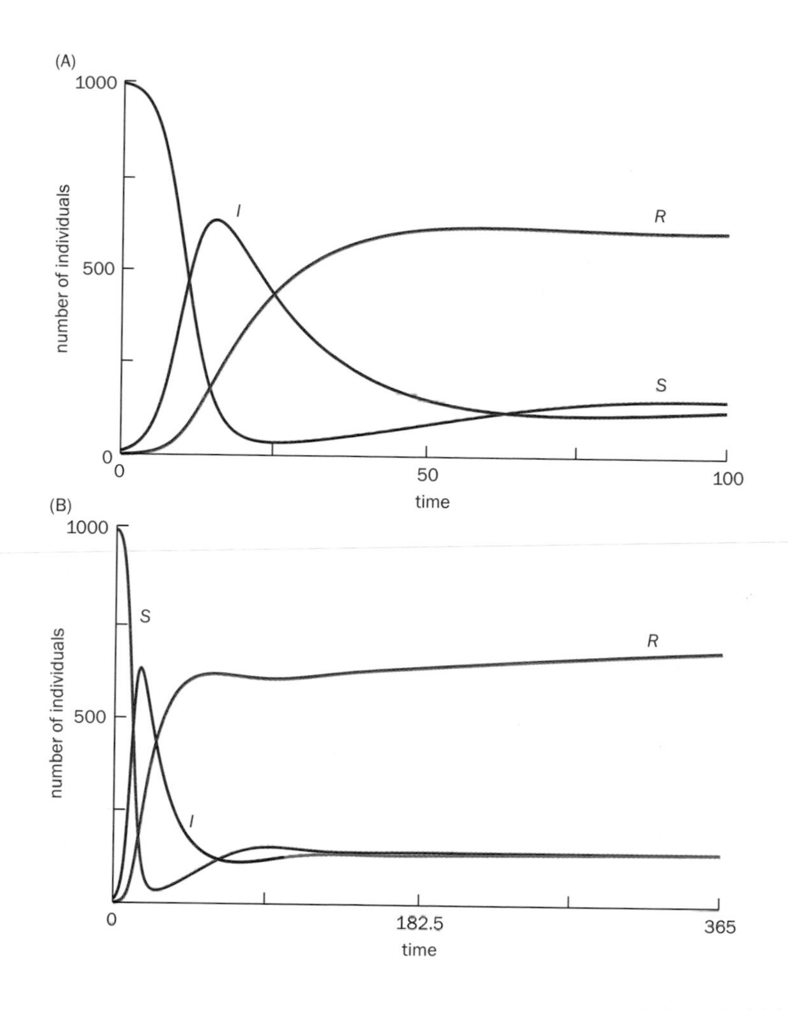

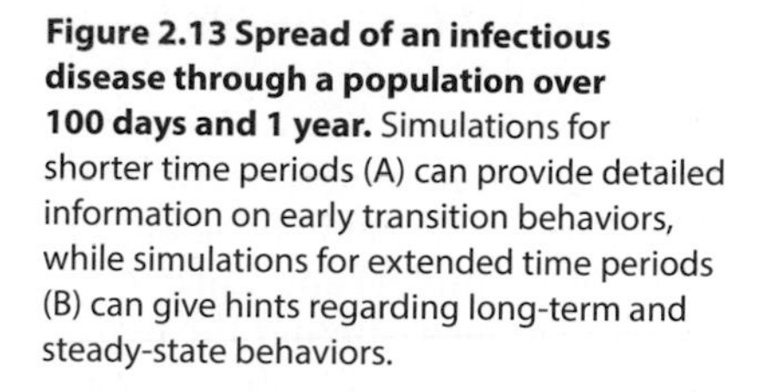
Figure 2.13 Spread of an infectious disease through a population over 100 days and 1 year. Simulations for shorter time periods (A) can provide detailed information on early transition behaviors, while simulations for extended time periods (B) can give hints regarding long-term and steady-state behaviors.

size of the population, which is driven by the birth and death rates. What happens when we start with $S = 1500$? Because the steady state is stable, the population will shrink and in the end approach 1040 again (confirm this!).

Much of the remaining diagnostics characterizes the sensitivity of the model. The generic question here is: If one of the parameters is changed a little bit, what is the effect? Sensitivity analysis can be done mathematically in one big swoop (see Chapter 4). Specifically, all changes in steady-state values, caused by a small (for example, 1%) change in each of the parameters, are computed with methods of differentiation and linear algebra. However, we can also explore important sensitivity features by changing a parameter value manually and checking what happens. For instance, we could lower the infection rate to 20% of the present value: $r_I = 0.0001$. The consequences are dramatic, as **Figure 2.14** demonstrates. Two pieces of good

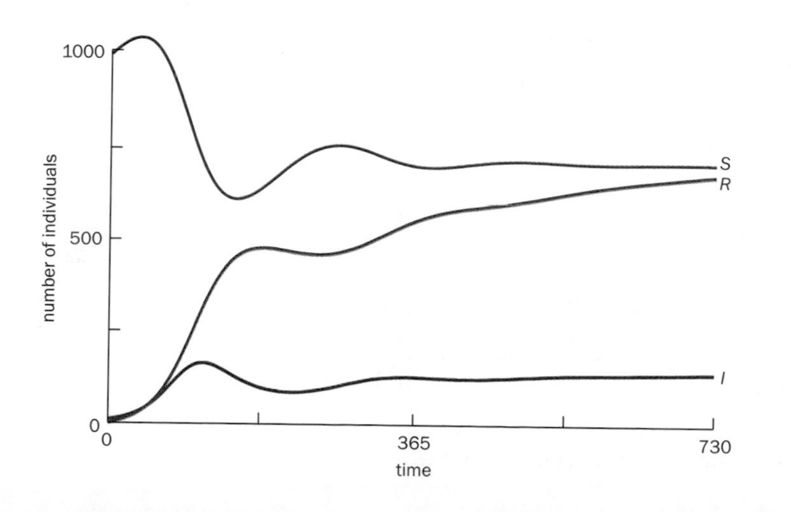

Figure 2.14 A model can be sensitive to changes in some parameters and robust against changes in others. As an example, the spread of the disease is rather different if the infection rate is smaller (here, $r_I = 0.0001$; compare with Figure 2.12).

news are that fewer people become infected and that the total population size increases to 1650. The bad news is that more people remain susceptible.

If one has the luxury of many data, it is beneficial not to use them all for model design and parameterization and instead to keep some data for external **validation**. This validation may be accomplished intuitively or statistically. In the former (obviously simpler) approach, one compares the results of simulations with actual data and judges whether the two are sufficiently close. Often, but not always, such a simulation begins with the system at steady state and a **stimulus** that perturbs this state. Sometimes, one might be satisfied with an agreement in **qualitative behavior**, such as "If the input to the system is increased, these variables should go up, while those should decrease in value." In other cases, one may expect a model to be semiquantitative: "If the input to the system is doubled, variable 3 should increase to between 120% and 140%." A much more stringent criterion is true numerical agreement. In addition to this type of intuitive assessment, one may use statistical methods to evaluate the quality of fit between a model prediction and observed data.

One should note that a good fit is not the only criterion of quality. A particularly prominent counterexample is overfitting, where the model contains more parameters than can be estimated reliably from the data. While the model fit for the training data may be excellent, the model tends to fail if new data are analyzed. Methods have been developed to diagnose different types of residual errors and problems of overfitting [8, 9].

2.8 Exploration and Validation of Dynamical Features

Assuming that the model has survived the diagnostics of the previous phase, it is now time to explore its range of dynamical features, keeping in mind that even a comprehensive analysis might not detect all possible flaws and failures. In fact, since every model is at best a good approximation of reality, it is only a matter of time before this approximation leads to discrepancies with actual observations.

The exploration of the model consists of testing hypotheses regarding scenarios that had not been used for the model design. This process is often called a simulation. Some of the scenarios may correspond to situations that had already been tested biologically. In this case, a successful simulation may be used for supporting the validity of the model or for explanations of the internal operations that connect a particular input to the observed (and simulated) output. Such an explanation may lead to new hypotheses of the type "since process A is apparently more influential than process B, inhibition of A should have a stronger effect than inhibition of B." The new hypothesis may be tested mathematically, biologically, or both. It is also possible that the model makes predictions that are counterintuitive. In this case we need to check both our intuition and the model. If the model is correct, it could lead to new explanations that were not evident before.

Continuing beyond the diagnostic explorations, the scenarios of this phase are typically a little bit more complicated, and many will require changes in the model structure, that is, in the equations. These types of explorations can be used to evaluate the importance of assumptions, simplifications, and omissions. For instance, we assumed in the infectious disease model that only infected individuals die. What happens if we add mortality terms to the equations for S and R? What happens if immunity is permanent, as it is for most people in the case of chickenpox? Is it of consequence if some individuals are infected, but do not show symptoms?

As a specific example, let's investigate what happens if we vaccinate before the outbreak. The first task is to figure out where to enter vaccination into the equations. Any effective vaccination before the outbreak moves susceptible individuals into the R pool. In other words, the only change occurs in the initial values, where S_0 is lowered and R_0 correspondingly elevated. The exact numbers depend on the efficacy of the vaccination campaign. It is not difficult to imagine that complete and perfect vaccination would render $S_0 = 0$, $R_0 = 990$. One could also model that the vaccination would not necessarily prevent the disease altogether but only reduce the infection rate, allow faster immunity, and reduce the number of individuals in the R pool who become susceptible again.

Another scenario involves quarantine for infected individuals. Ideally, the entire pool I would immediately be subjected to quarantine until they recover, thus changing the initial value to $I_0 = 0$. It is easy to see in this extreme case that no infection can take place, because the infection term $r_I SI$ equals zero. Therefore, no susceptibles move from S to I, and $r_I SI$ remains zero. A more interesting question is: Is it really necessary to quarantine every infected person? Or is it sufficient to quarantine only a certain percentage? Let's explore this question with simulations. Suppose we continuously quarantine some of the infected individuals. Of course, actually sending infected individuals into quarantine is a **discrete** process, but we can model the process approximately by subtracting infectives with a constant rate from (2.3), which becomes

$$\dot{I} = r_I SI - r_R I - r_D I - r_Q I, \tag{2.7}$$

where r_Q is the rate of quarantining (**Box 2.5**).

Simulating the system reveals that quarantining even at a rate of $r_Q = 0.1$ has noticeable, positive effects: the number of infected individuals is roughly halved throughout the first 100 days of disease (**Figure 2.15A**). If we quarantine with a higher rate of $r_Q = 0.4$, a few individuals still become infected, but the infection level is very low (Figure 2.15B). If we quarantine an even higher percentage, the number of infectives quickly goes to zero and, perhaps surprisingly at first, the susceptible pool increases. Why is this? It is due to births and immigration, and to the fact that nobody in the model dies, unless she or he is first infected.

Case closed? Not quite. Something interesting happens if we let the disease process continue for a longer period of time. Because the quarantine is not perfect, some infectives remain in the population, and this can lead to repeated, smaller outbreaks. For instance, for $r_Q = 0.1$, a secondary outbreak occurs after about 165 days. Smaller outbreaks follow before the system finally settles in a state that is still not disease-free, although the infected subpopulation is smaller than in the case without quarantine (**Figure 2.16**).

BOX 2.5: THE RATE OF QUARANTINING IN THE SIR MODEL

The formulation in (2.7) is only an approximation of real-world, repeated quarantining, because the change in I is continuous, whereas the actual quarantining process is discrete. For instance, one might imagine an idealized situation where, every morning exactly at 8:00, one would send 10% of all infected individuals into quarantine. The model does not represent this daily routine but, instead, subtracts infectives at a constant rate throughout the day. Nonetheless, the approximation is sufficient for our exploration, as we can see from the following argument with a simplified situation. Suppose we start with 100 infectives, nobody becomes infected, nobody dies or becomes immune, and the only change in I is quarantining. In this simplified scenario, (2.7) becomes

$$\dot{I} = -r_Q I. \tag{1}$$

Starting with $I_0 = 100$ and $r_Q = 0.1$, the number of infectives over time is reduced roughly by 10% per time unit (see **Table 1**). For larger values of r_Q, the correspondence is less accurate. For instance, for $r_Q = 0.5$, the number of infectives decreases from 100 only to 60.65 and 36.79 between times 0, 1, and 2, respectively, which roughly corresponds to 40% decreases per time step, rather than 50% decreases. The basis for the discrepancy is similar to continuously compounding interest.

TABLE 1: DECREASE IN THE NUMBER OF INFECTIVES IN THE SYSTEM WHEN THE CONTINUOUS QUARANTINING RATE IS $r_Q = 0.1$; THE DECREASE ROUGHLY CORRESPONDS TO 10% PER TIME UNIT

t	0	1	2	3	4	5	6	7	8	9	10
I	100	90.48	81.87	74.08	67.03	60.65	54.88	49.66	44.93	40.66	36.79

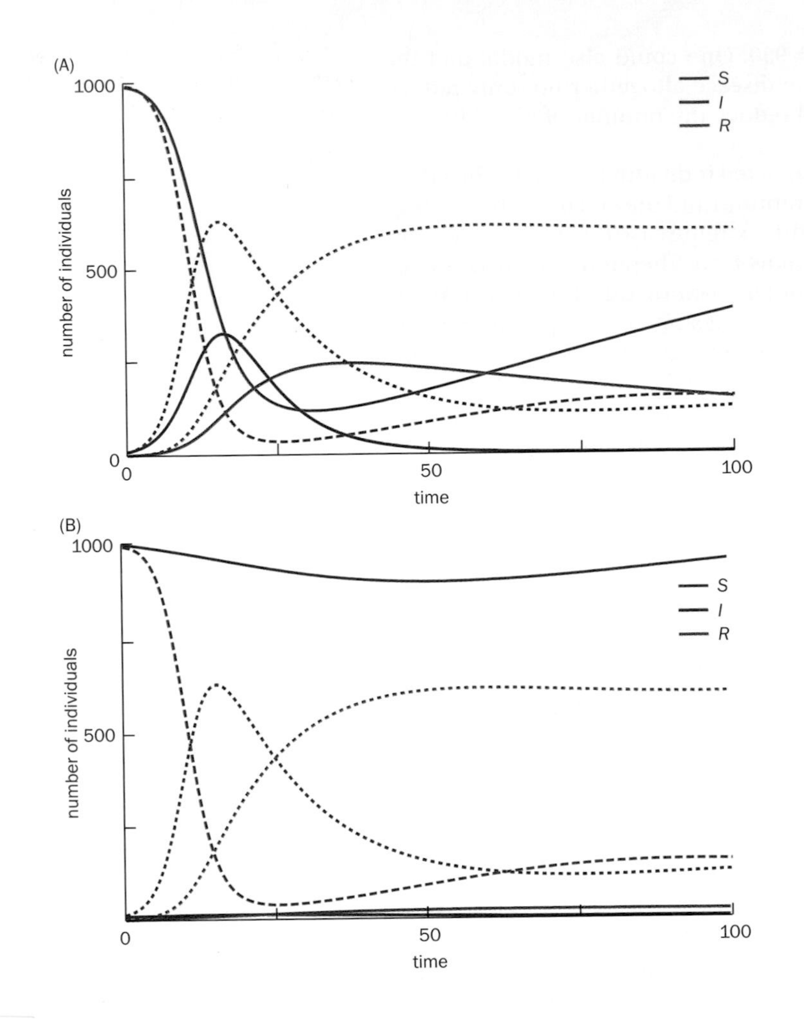

Figure 2.15 Quarantining infectives has a positive effect on the population. (A) Results of continuous quarantining at a rate of $r_Q = 0.1$ (solid lines) in comparison with no quarantine (dotted lines). (B) The corresponding result for $r_Q = 0.4$.

Now that we have seen the positive effects of quarantine, we can explore questions such as: Is it better to quarantine or to reduce the infection rate? Reduction of the infection rate by a certain percentage is easy to implement: we simple reduce the value of the rate r_I. **Figure 2.17** shows two results: the outcomes of quarantining and reducing the infection rate are quite different.

It is clear that the possibilities for new scenarios are endless. Thankfully, once the model is set up, many slightly altered scenarios are very easy to implement, and the process can even become addictive, if we try to find just the right parameter combinations for a desired outcome.

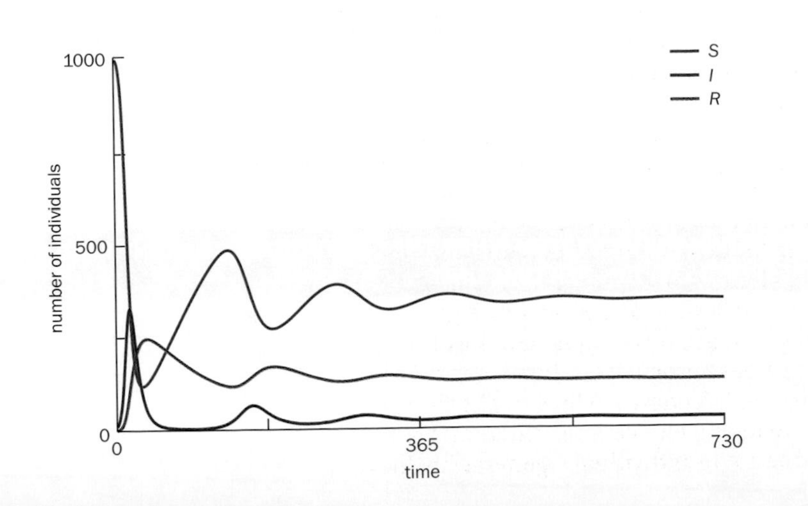

Figure 2.16 Quarantining may lead to further, milder outbreaks. For $r_Q = 0.1$, the disease seems to be under control initially, but a second outbreak occurs after about 165 days, followed by later, smaller outbreaks.

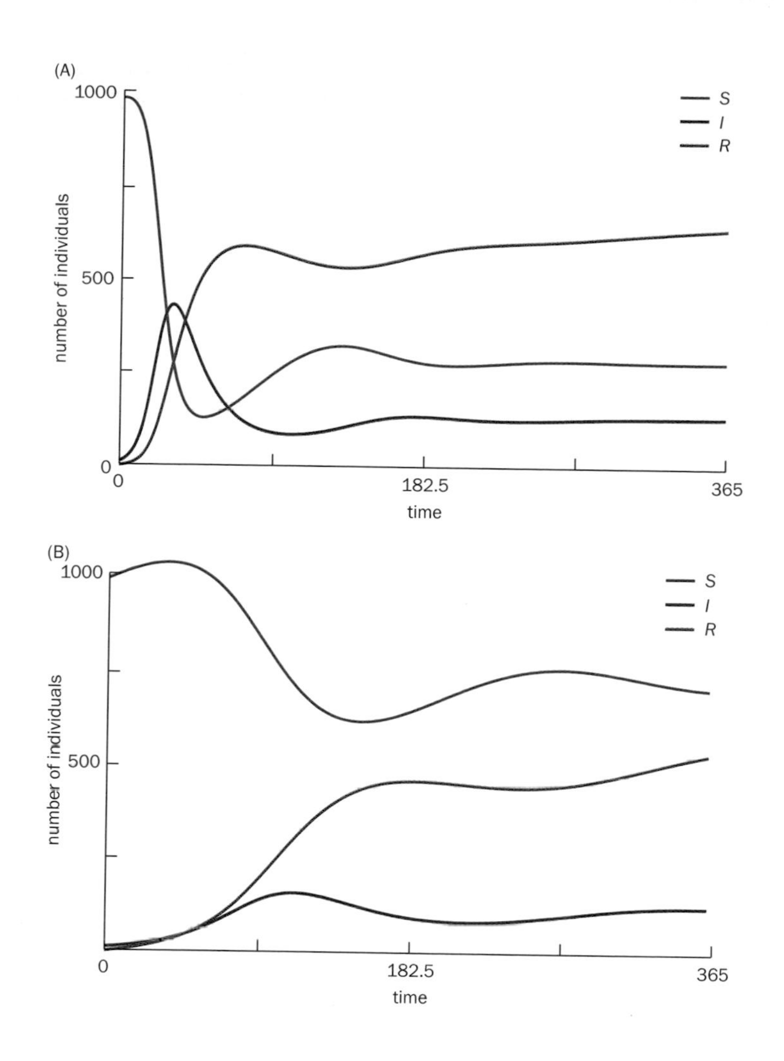

Figure 2.17 Reduction in the infection rate has a drastically different effect than quarantining. (A) The infection rate r_I has been lowered to half the original value. (B) r_I has been lowered to 20% of the original value.

MODEL USE AND APPLICATIONS

The most common uses of a model (see Table 2.1) may be classified in two categories:

- Applications of the model as it is, including slight extension.
- Targeted manipulation and optimization of the model structure toward a specific goal.

2.9 Model Extensions and Refinements

As an example of a model extension, suppose that it becomes clear that many infected individuals do not show any serious symptoms and often do not even know that they carry the disease. In the jargon of SIR models, such individuals are called asymptomatics. How can the model account for them? The extension is really a mini-version of the model design and analysis phases. First, we define a new dependent variable, A, and identify how it interacts with the other variables in the system (2.2)–(2.4). As for the initial model design, it is a good idea to sketch these interactions in a diagram. At first, we might treat A just like I, except that we should use different names for the processes, because they are indeed different (**Figure 2.18A**). However, should we also consider that the contact between S and I could lead to A, and that the contact between S and A might lead to I (Figure 2.18B)? The answer cannot be given with mathematical arguments and has to be based on biological or medical features of the disease process. For instance, if all A result from a weaker strain of the disease vector, an infection of S by A might always result in A and an infection by I might always lead to I; let's call this situation Case 1. But if the difference between A and I is due to an individual's general health

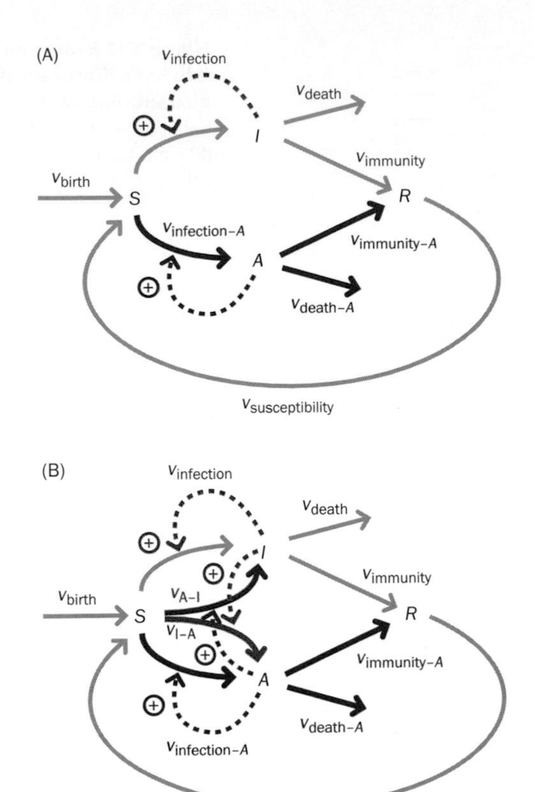

Figure 2.18 Diagrams accounting for asymptomatics. (A) The inclusion of *A* requires new processes (red) with their own names. (B) Further interactions (blue and green) are needed if an infection of *S* by *I* can result in *A* or vice versa.

status and his or her age and genetic predisposition, we should probably consider cases where an infection of S by A can indeed result in I, and vice versa; let's call this situation Case 2.

For Case 1, the equation for the asymptomatics is similar to that for the infectives, (2.3), namely,

$$\dot{A} = r_A SA - r_{R\text{-}A} A - r_{D\text{-}A} A, \qquad A(t_0) = A_0. \tag{2.8}$$

While the structure of the equation is the same as in (2.3) for I, the parameters have different names and potentially different values. For instance, the death rate $r_{D\text{-}A}$ might be different, because the disease in A is apparently not as bad as for I. Similar arguments might hold for r_A and $r_{R\text{-}A}$. Does that cover Case 1? No, for sure we also need to adjust the equations for S and R. It could even be that R is different for asymptomatics and for infectives (why?); but let's assume that that is not the case. The adjusted equations for S and R are

$$\dot{S} = r_B + r_S R - r_I SI - r_A SA, \qquad S(t_0) = S_0, \tag{2.9}$$

$$\dot{R} = r_R I + r_{R\text{-}A} A - r_S R, \qquad R(t_0) = R_0. \tag{2.10}$$

For Case 2, we need to add more terms: one each to the equations for A and I and two terms to the equation for S; the equation of R is not affected:

$$\dot{A} = r_A SA + r_{I\text{-}A} SI - r_{R\text{-}A} A - r_{D\text{-}A} A, \qquad A(t_0) = A_0, \tag{2.11}$$

$$\dot{I} = r_I SI + r_{A\text{-}I} SA - r_R I - r_D I, \qquad I(t_0) = I_0, \tag{2.12}$$

$$\dot{S} = r_B + r_S R - r_I SI - r_A SA - r_{I\text{-}A} SI - r_{A\text{-}I} SA, \qquad S(t_0) = S_0. \tag{2.13}$$

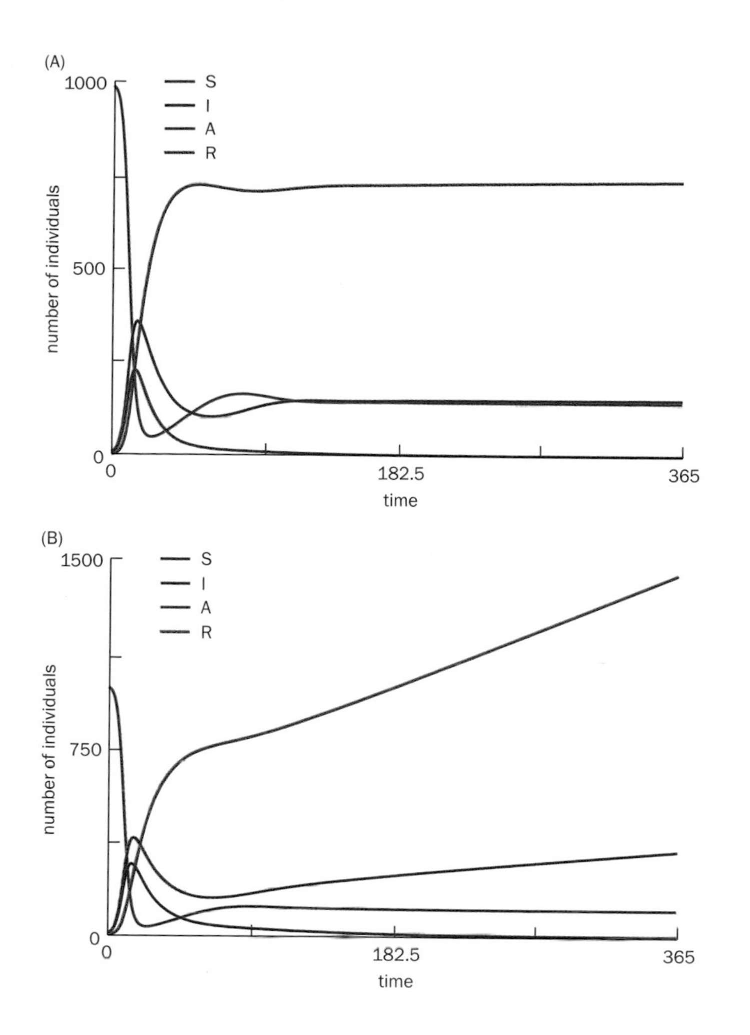

Figure 2.19 Effect on asymptomatics *A* on the disease progression. (A) If asymptomatics recover twice as fast as *I*, they quickly disappear from the population, whether their death rate is zero (as shown) or not. (B) If asymptomatics recover with the same rate as *I*, and if they do not die, the infectives essentially disappear, and the population grows (note different scale). However, the disease does not disappear; it is present in a growing number of asymptomatics.

As an illustration, let's simulate Case 1. For the numerical settings, we will retain most values from our earlier example (2.6). Regarding *A*, we assume that initially half of the infected individuals are asymptomatic, so that $A(t_0) = I(t_0) = 5$. Furthermore, we assume that the infection rate for *A* and *I* is the same ($r_A = r_I = 0.0005$), that no *A* die from the disease ($r_{D\text{-}A} = 0$), and that they recover more quickly than *I* ($r_{R\text{-}A} = 2r_R = 0.1$). The result of these settings is that the asymptomatics disappear quickly and the disease ultimately assumes the same profile as we saw in Figure 2.13 (**Figure 2.19A**). Interestingly, the fact that none of the asymptomatics die does not have much effect (results not shown). By contrast, if no *A* die, but the recovery rate for *A* and *I* is the same, the outcome is very different: *I* essentially disappears, while the *R* and *A* subpopulations grow to much higher levels (Figure 2.19B) The explanation is that essentially nobody in the model dies anymore. To make the model more realistic, one would need to introduce death rates for *S*, *R*, and *A*.

2.10 Large-Scale Model Assessments

In reality, most parameters of biological systems are not entirely fixed but can vary within certain ranges, for instance, due to inter-individual variability. If the biologically relevant ranges of all parameter values are known or can be estimated, the model may be used to evaluate best- and worst-case scenarios, along with the most likely situations. This analysis can be accomplished with targeted **stochastic simulations** in which parameters are varied, as above, or with large-scale, automated simulation studies, in which thousands of parameter combinations are screened. The latter method is often referred to as **Monte Carlo simulation**. The principle is the following. Suppose there are 10 parameters of interest and the range for each is more or less known. For instance, there could be information that the infection rate in our example is between 0 and 0.002 and that the most likely value is 0.0005, as we set it in our baseline model. Suppose that this type of information is available for the other nine parameters as well. A Monte Carlo simulation consists of the following **iteration**:

1. Automatically draw a random value for each parameter, according to the available information about its range.
2. Solve the equations for the given set of values.
3. Repeat this procedure thousands of times.
4. Collect all outputs.

The overall result of such a Monte Carlo simulation is a visualization or listing of a very large number of possible outcomes and their likelihoods. Furthermore, statistical analysis of the output conveys insights into which parameters or parameter combinations are most or least influential.

As an example, let's use our infection model (2.6) again and test the effects of different infection and death rates. We use a very simple Monte Carlo simulation, consisting of just 20 random choices for the parameter r_I, which is selected randomly from the interval [0, 0.001] and, simultaneously, 20 random choices for r_D, which is selected randomly from the interval [0, 0.04]. The lower bound of these intervals (0 in both cases) means no infection (death) at all, while the upper bounds (0.001 and 0.04) correspond to twice the corresponding rates in (2.6). The simulation proceeds as follows:

1. Select a random number for r_I from the interval [0, 0.001].
2. Select a random number for r_D from the interval [0, 0.04].
3. Plug these two values into (2.6).
4. Simulate the model for a time period of 365 days.
5. Repeat Steps 1–4 many (here just 20) times.
6. Superimpose all plots and analyze results.

The 20 time courses of S are shown in **Figure 2.20** for two Monte Carlo simulations. Owing to the random nature of the 20 picks of r_I and r_D, these two simulation runs of exactly the same model structure lead to different results. We can see that most results lead to final values of S between 100 and 500, but that some combinations of r_I and r_D lead to much higher values. A statistical analysis of this type of simulation would reveal that very low death rates lead to high (and sometimes even increasing) values of S; note one such case in Figure 2.20B. Other steady-state values, peak values, and time courses may be analyzed in the same fashion. In a more comprehensive Monte Carlo simulation, one would simultaneously select random values for several parameters and execute so many simulations that the results would no longer change much from one run to the next. Uncounted variations on this theme are possible. As a notable example, a parameter value does not have to be chosen from its interval with equal probability, and one could instead specify that values close to the original (0.0005 and 0.02 in our example) are chosen more frequently than extreme values like 0.0000001.

The model may also be used to screen for particularly desirable outcomes. For instance, health officials will be interested in shortening the period during which many individuals are sick, and in minimizing the chance of infection. To develop ideas for combined interventions, a Monte Carlo simulation could be a first step. It is very efficient to try out alternative strategies toward these goals with a model and, possibly, to optimize them with numerical methods from the field of operations research. Of course, predicted results always have to be considered with caution, because the model is a vast simplification of reality. Nonetheless, a good model is able to guide and correct our intuition, and the more reliable the model becomes through testing and refinements, the more competent and important its predictions will be.

2.11 Questions of Design

An academically important use of a model is to explain design and operating principles, which offer the closest resemblance to "laws" that biology can offer. The

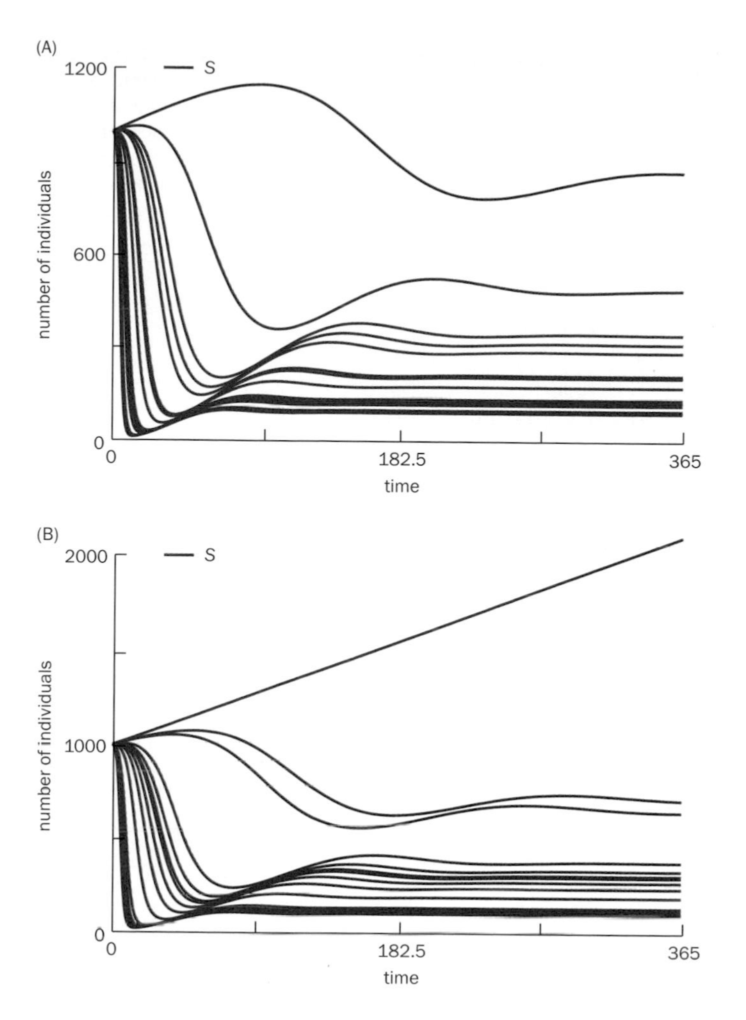

Figure 2.20 Two results of a Monte Carlo simulation. The set-up in the simulations in (A) and (B) is exactly the same; namely, (2.6) is simulated with randomly chosen values for the infection and death rates. Owing to the probabilistic nature of this process, the results are quite different in some respects, but similar in others. Notably, the simulation in (B) includes the random choice of a very low death rate, which allows *S* to grow. However, most other patterns in time courses are rather similar, although they differ numerically. Note the different scales on the vertical axes.

generic question to be asked is: Why is this system organized in this particular fashion and not in an alternative fashion? For instance, to increase the concentration of a metabolite, one could increase the amount of substrate that is used for its production, increase the production rate, or decrease the degradation of the metabolite. By analyzing these options comparatively with two models (for example, one showing increased production and the other showing decreased degradation), it is possible to identify advantages and disadvantages of one design over the other. For instance, we might find that a model with one design responds faster to a sudden demand for some product than the alternative. This interesting topic will be discussed in much more detail in Chapters 14 and 15.

2.12 Simplicity versus Complexity

This chapter has given you a flavor of what modeling means and what we can do with mathematical models in systems biology. Chapters 3 and 4 will describe standard techniques for setting up, diagnosing, analyzing, and using models of different types. Before we dive into the nitty-gritty details of the mathematics of modeling, we should realize that the complexity of the mathematical representations and techniques is not necessarily correlated with the usefulness of a model. Certainly, some models have to be complicated, but sometimes very simple models are "better" than complicated counterparts. For instance, many growth models are rather simple, even if they describe very complicated biological systems, such as the human body. There is no chance that we will come up with a complete physiological model of the human body in the foreseeable future, yet the simple growth curves of children show amazingly small variation from one child to the next (**Figure 2.21**). Predictions based on these curves are much more reliable than predictions from any large physiological model that currently exists.

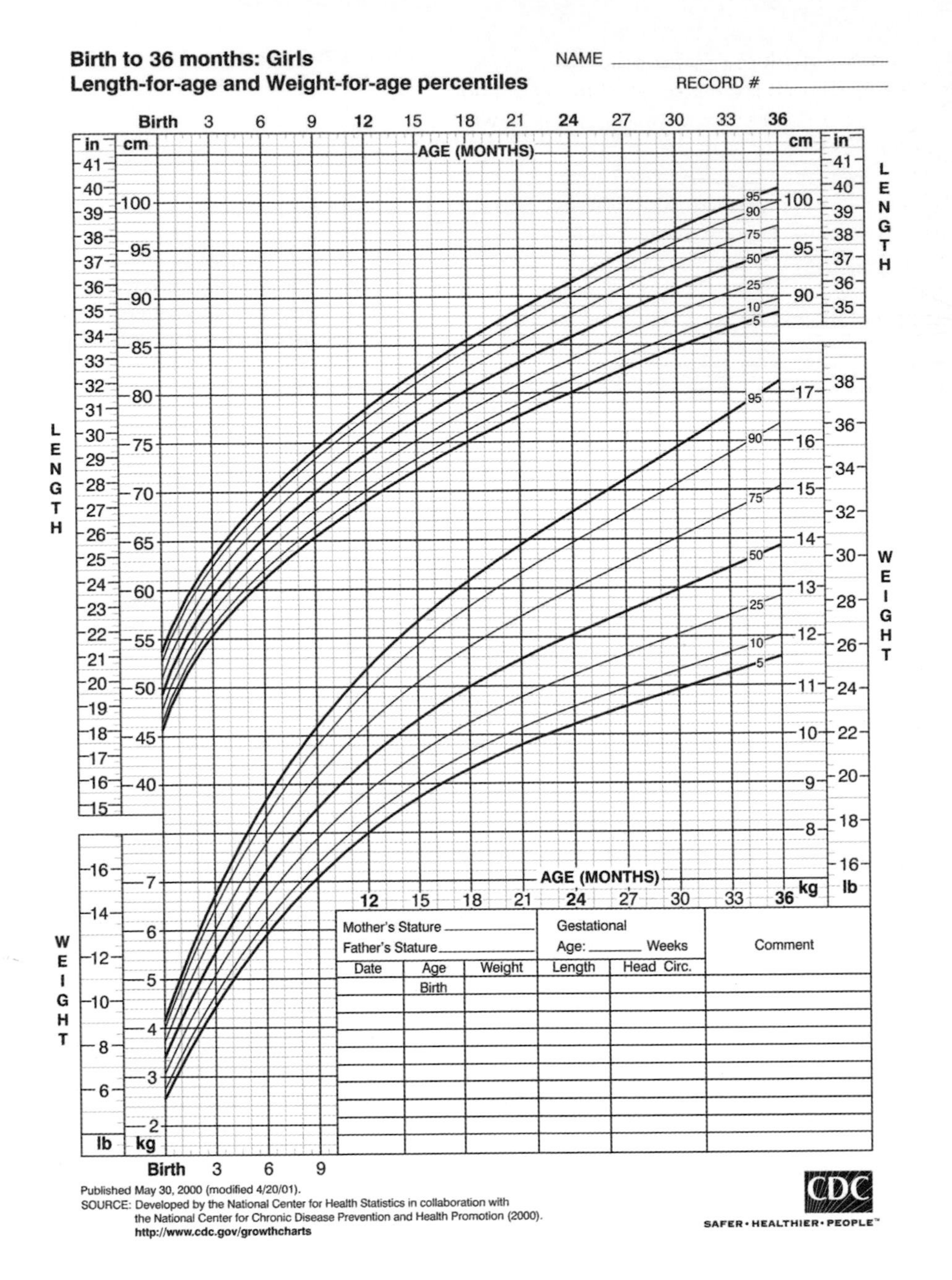

Figure 2.21 Growth curves of girls from birth to 3 years. Considering how complicated the human body is, the normal growth trends of 50% of the girls are within surprisingly narrow bounds. (From the National Center for Health Statistics.)

An example at the other end of the spectrum is the mathematical model in **Figure 2.22**. It adorns a T-shirt and the artist has even provided the correct answer, upside down below the formula, as *age* = *50*. The display is certainly intimidating to mere mortals, but does that mean that we should judge this "model for age" as good? The answer is a resounding *No*, for the following (and many more) reasons. First, the model does not explain or predict anything. Most importantly, does the T-shirt show the age of the wearer? Only if he or she happens to be 50 at the time; wearing the shirt next year, the implied age will be wrong. Age is obviously a function of time, but the T-shirt does not account for time. In fact, there is no true variable, except for the integration variable x, which is merely a placeholder and disappears as soon as the integration has been executed. There are no true parameters either, because the only candidates, α and n, disappear in the summation, and the answer is always 50. In other words, the formula does not permit any adaptation to the wearer and his or her true age. Instead, the display is a mathematically unnecessary complication of the simple statement *age* = 50 and not much more. If we are interested in a better T-shirt model of age, which correctly captures the wearer's age, this model must certainly contain time as a variable. Barring magic or serendipity, the formula must also contain a place for explicit or implicit information about the wearer's year of birth. This

Figure 2.22 Awe-inspiring, useless model on a T-shirt. While the formula looks intimidating, it is simply a complicated way of representing the number 50. The "model" contains no real parameters and only shows the correct age if the wearer happens to be 50 years old.

information may be given directly or in some convoluted fashion, but it must be there as some (potentially explicit or hidden) parameter. We also need to decide on the accuracy of the improved T-shirt model. If age in years is sufficient, the model will certainly be simpler than if age has to be computed to the minute.

The message from these two examples is that the simplicity or complexity of a model is not indicative of its quality. Of course, there are situations where complicated mathematics is needed, but there are also many cases where simplicity will do just fine, if not better. The key criterion of quality of a model is, instead, its usefulness. Mathematical models are tools, and tools are only good and helpful if they allow us to solve problems. Sometimes, simple tools are our best choice, and at other times the complexity of the model must match the complexity of the system we want to analyze. Because usefulness is the most important criterion, every model analysis should begin with a clear understanding of why we want to design a model and what this model is supposed to accomplish. Time and effort spent on these two questions are the most prudent investments in the realm of modeling.

EXERCISES

2.1. Explore the range of shapes that a Hill function (2.5) can attain. First, keep the settings $V_{max} = 12$ and $K_M = 10$ and use different values for n ($n = 4, 6, 8; 3, 2, 1, 0.5, 0.2$). Then, set $n = 4$ and vary V_{max} or K_M individually or both simultaneously. Discuss differences and similarities in the shapes of the resulting graphs.

2.2. The function $L(S) = 1/(1 + e^{10-S})$ is a shifted logistic function, whose graph is an S-shaped curve. Through trial and error, determine parameter values for the Hill function (2.5) that approximate $L(S)$ as closely as possible.

2.3. Many processes in biology can be approximated quite accurately with power-law functions of the type $P(X) = \gamma X^f$ (one variable) or $P(X_1, X_2, \ldots, X_n) = \gamma X_1^{f_1} X_2^{f_2} \cdots X_n^{f_n}$ (n variables; $n = 2, 3, 4, \ldots$). The positive parameter γ is a rate constant and the real-valued parameters f are kinetic orders. Explore the shapes of these power-law functions for different values of γ and f. For γ use (at minimum) the values 0.1, 1, 10, and 100. For the parameter f use (as a minimum) the values −2, −1, −0.5, −0.1, 0, +0.1, +0.5, +1, +2. Display your results as tables and/or graphs. Summarize the results of your analysis in a short report.

2.4. Compare the Michaelis–Menten function $V(S) = V_{max}S/(K_M + S)$, with $V_{max} = 1$ and $K_M = 2$, with the power-law function $P(S) = \gamma S^f$, where $\gamma = 0.3536$ and $f = 0.5$. Discuss your findings. Choose different values for V_{max} and K_M and determine values of γ and f such that the power-law approximation matches the Michaelis–Menten function well for some range of values of S. Write a brief report about your results.

2.5. Compute the two solutions for the equation $e^x - 4x = 0$ in Figure 2.12 by trying out different values. Write down your thought processes when selecting the next value to be tried. Remember these thought processes when you study Chapter 5.

2.6. Write equations for the small protein system in the signaling cascade of Figure 2.7. Model each process as a rate constant times the concentration of the variable involved. For instance, the phosphorylation of P_0 toward P_1 should be formulated as $r_{01}P_0$. Begin with $P_0 = P_1 = P_2 = 1$ and set all rate constants to 0.5. Implement the system in a software program and simulate it. Explore the dynamics of the system by changing the initial values of P_0, P_1, and P_2. Discuss the results. Now change the rate constants one by one or in combination, until you develop a feel for the system. Discuss the results and summarize your insights into the system.

For Exercises 2.7–2.19, use the infectious disease model in (2.2)–(2.4) and (2.6) and be aware that several of these exercises are open-ended. Before you execute a particular simulation, make a prediction of what the model will do. After the simulation, check how good your predictions were. Consider a problem "done" when your predictions are consistently correct.

2.7. Change each initial value in (2.6) (up and down) and study the consequences.

2.8. What would it mean biologically if $r_I = 0$ or if $r_S = 0$? Test these scenarios with simulations.

2.9. What happens if no babies are born and nobody immigrates? Explore what happens if nobody in the model dies.

2.10. Double the death rate in the model and compare the results with the original simulation.

2.11. Alter the model in (2.2)–(2.4) so that susceptible (S) and immune (R) individuals may die from causes other than the infection. Discuss whether it is reasonable to use the same rate constant for the death of S and R individuals. Implement your proposed alterations in (2.6) and execute simulations demonstrating their effects.

2.12. Change the infection rate, starting from the value of 0.0005 in (2.6), by raising it to 0.00075, 0.001, 0.01, and 0.1, or by lowering it stepwise toward 0. Discuss trends in consequences on the dynamics of the system.

2.13. Change the infection rate or the rate of acquiring immunity to different values and solve the system until it reaches its steady state, where no variable changes anymore. For each scenario, check the number of deaths per unit time, which is given by $r_D I$. Record and discuss your findings.

2.14. Predict the consequences of partial vaccination. Implement a vaccination strategy in the model and test your predictions with simulations.

2.15. Compare the effectiveness of quarantine and vaccination.

2.16. Implement scenarios where certain percentages of asymptomatics become infectives. Test the consequences with simulations.

2.17. How could the model in (2.2)–(2.4) be expanded to allow for different levels of disease susceptibility that depend on age and gender? Without actually implementing these changes, list what data would be needed to construct such a model.

2.18. Describe in words what data would be required to develop an infectious disease model like that in (2.2)–(2.4) that accounts for the gradual spread of the disease throughout a geographical area.

2.19. Without actually implementing them in equations, propose extensions that would make the SIR model in (2.2)–(2.4) more realistic. For each extension, sketch how the equations would have to be changed.

2.20. Set up symbolic equations for a two-stage cancer model that consists of the following processes. Normal cells divide; some of them die. In a rare case, a normal cell becomes "initiated," which means that it is a possible progenitor for tumor formation. The initiated cells divide; some die. In a rare case, an initiated cell becomes a tumor cell. Tumor cells divide; some die.

2.21. Implement the system in Box 2.2 in computer software such as Mathematica®, MATLAB®, or PLAS [3], and compute solutions with different values for A, and with slightly changed initial values of x or y. Explore what happens if A is close to or equal to zero.

2.22. Use the binomial model in Box 2.3 to compute the probability that exactly 5 out of 10 flips of a coin are heads and 5 are tails. Compute the probability that exactly all 10 flips of a coin are heads.

2.23. Use the Poisson model in Box 2.4 to compute the probability that exactly 5 out of 10 flips of a coin are heads and 5 are tails. Compute the probability that exactly 50 out of 100 flips of a coin are heads and 50 are tails.

2.24. Use the Poisson model in Box 2.4 to compute the probability that at most 3 out of 20 flips of a coin are heads.

2.25. Implement the equation from Box 2.5 and compute the numbers of infectives over time for different rates r_Q. In parallel, design a model where a certain, fixed percentage of infectives is quarantined at $t = 0$, 1, 2, . . ., but not in between. Compare the results of the two modeling strategies.

REFERENCES

[1] Qi Z, Miller GW & Voit EO. Computational systems analysis of dopamine metabolism. *PLoS One* 3 (2008) e2444.

[2] Thompson JMT & Stewart HB. Nonlinear Dynamics and Chaos, 2nd ed. Wiley, 2002.

[3] PLAS: Power Law Analysis and Simulation. http://enzymology.fc.ul.pt/software.htm.

[4] Meng C, Shia X, Lina H, et al. UV induced mutations in *Acidianus brierleyi* growing in a continuous stirred tank reactor generated a strain with improved bioleaching capabilities. *Enzyme Microb. Technol.* 40 (2007) 1136–1140.

[5] Fujii R, Kitaoka M & Hayashi K. One-step random mutagenesis by error-prone rolling circle amplification. *Nucleic Acids Res.* 32 (2004) e145.

[6] Mood AM, Graybill FA & Boes DC. Introduction to the Theory of Statistics, 3rd ed. McGraw-Hill, 1974.

[7] Kermack WO & McKendrick AG. Contributions to the mathematical theory of epidemics. *Proc. R. Soc. Lond. A* 115 (1927) 700–721.

[8] Raue A, Kreutz C, Maiwald T, et al. Structural and practical identifiability analysis of partially observed dynamical models by exploiting the profile likelihood. *Bioinformatics* 25 (2009) 1923–1929.

[9] Voit EO. What if the fit is unfit? Criteria for biological systems estimation beyond residual errors. In Applied Statistics for Biological Networks (M Dehmer, F Emmert-Streib & A Salvador, eds), pp 183–200. Wiley, 2011.

FURTHER READING

Adler FR. Modeling the Dynamics of Life: Calculus and Probability for Life Scientists, 3rd ed. Brooks/Cole, 2013.

Batschelet E. Introduction to Mathematics for Life Scientists, 3rd ed. Springer, 1979.

Britton NF. Essential Mathematical Biology. Springer-Verlag, 2004.

Edelstein-Keshet L. Mathematical Models in Biology. McGraw-Hill, 1988 (reprinted by SIAM, 2005).

Haefner JW. Modeling Biological Systems: Principles and Applications, 2nd ed. Springer, 2005.

Murray JD. Mathematical Biology I: Introduction, 3rd ed. Springer, 2002.

Murray JD. Mathematical Biology II: Spatial Models and Biomedical Applications, Springer, 2003.

Yeagers EK, Shonkwiler RW & Herod JV. An Introduction to the Mathematics of Biology. Birkhäuser, 1996.

Static Network Models

3

When you have read this chapter, you should be able to:

- Identify typical components of static networks
- Understand the basic concepts of graphs and how they are applied to biological networks
- Describe different structures of networks and their features
- Discuss the concepts and challenges of network inference
- Describe typical examples of static networks in different fields of biology
- Characterize the relationships between static networks and dynamic systems

All molecules in biological systems are components of **networks**. They are linked to other molecules through connections that may take a variety of forms, depending on what the network represents. Molecules may associate loosely with each other, bind to each other, interact with each other, be converted into each other, or directly or indirectly affect each other's state, dynamics, or function. Rather than focusing on one particular molecular process at a time, the computational analysis of biological networks has the goal of characterizing and interpreting comprehensive collections of connections. The methods for these analyses can be automated and applied, using computers, to very large networks. Typical results are characteristic features of the connectivity patterns within the investigated networks, which in turn provide clues regarding the functions of particular components or subnetworks. For instance, an analysis may reveal which components of a network are particularly strongly connected, and this insight may suggest important roles of these components in different situations. The analysis may also show which variables act in concert with each other, and whether the connectivity pattern within the network changes, for example, during development and aging or in response to perturbations, such as disease.

Many concepts and methods of network analysis are actually quite simple and intuitive when studied in small networks. This simplicity is crucial for the real power of network analyses, namely, their scalability to very large assemblages of molecules and their ability to reveal patterns that the unaided human mind would simply not be able to recognize. Thus, computers can be trained to use the basic methods over and over again, establish network models from experimental data, identify their global and local features, and ultimately lead to hypotheses regarding their functionality. In this chapter, we will study these types of analyses with small networks, because they show most clearly what features to look for and how to characterize them. Several web tools are available for computational analyses of large biological networks. Prominent examples include the free open-source platform Cytoscape [1, 2], the commercial package IPA from Ingenuity Systems [3], and other software packages, such as BiologicalNetworks [4] and Pajek [5], all of which offer a wide variety of tools for data integration, visualization, and analysis.

STRATEGIES OF ANALYSIS

The ancient Greeks had a saying that all is in flux. Heraclitus of Ephesus (*c.* 535–475 BC), to whom this wisdom is attributed, considered change as a central force in the universe and once allegedly asserted that one can never step into the same river twice. Indeed, everything in the living world is moving, rather than **static**—if not macroscopically, then at least at the molecular level. So we must ask ourselves whether there is even a point in studying static networks, which by definition do *not* move. The answer is a resounding *Yes*, for a number of reasons. First, even if biological networks change over time, changes in interesting components are often relatively slow. For example, even if an organism is growing, its size is almost constant during a biochemical experiment that takes a few minutes. Second, much can be learned from assuming that a network is at least temporarily static. A typical example may be the activity profile of enzymes in the liver, which certainly changes throughout the life of an organism and even within a 24-hour period, but which may be more or less constant within a period of a few seconds or minutes. And third, even if systems are dynamically changing, certain important features of these same systems may be static. The most prominent examples are steady states, which we have already discussed in Chapter 2 (see also Chapter 4), where material is moving through a system and control signals are active but, for instance, the concentrations of metabolite pools are constant. Steady states are usually easier to characterize than dynamic changes and, once established, can provide a solid starting point for looking into dynamic features of a system [6–8].

A very practical reason for focusing on static networks or static aspects of dynamic systems is that the mathematics needed for static analyses is incomparably simpler than that required for dealing with temporally changing networks or fully regulated dynamic systems. As a consequence, even large static systems with thousands of components can be analyzed very efficiently, which is not really the case for dynamic systems of the same size. This relative simplicity is due to two aspects of static networks. First, because there is no change with time, the time derivatives in the differential equations describing the network are all equal to 0. Consequently, these equations become algebraic equations, which are easier to handle. Second, the structures of very many static networks are (at least approximately) linear, and linear algebra and computer science have blessed us with many elegant and powerful methods for analyzing linear phenomena. In some cases, these methods may even be applied to aspects of dynamic systems, if their subsystems are connected in a static fashion. A surprising example is a group of pacemaker cells in the sino-atrial node of the heart that can oscillate independently but are synchronized by their static interaction network, thereby enabling them to generate coordinated impulses that are required exactly once during each cardiac cycle (see [9] and Chapter 12).

The analysis of static networks can be subdivided in various ways. First, networks fall into distinct classes, which permit different methods of analysis. In some cases, for instance, actual material flows through the network, and accounting for this material permits accurate bookkeeping of all amounts or masses in different parts of the networks at different points in time. Metabolic pathway systems are prime examples in this category and therefore play a special role in this chapter. To illustrate, suppose there is a single reaction between a substrate and a product within a metabolic network. We immediately know that the amount of product generated in this reaction is directly related to the amount of substrate consumed; in fact, the two must be the same. This equivalence is very beneficial for mathematical analyses, because it eliminates a lot of uncertainty. By contrast, a signaling network transduces information, and whether or not this information is actually received and used is typically not relevant to the source of the information. Furthermore, the main purpose of such a signaling network is often an all-or-nothing response to an external signal, such as switching on the expression of a gene in response to some stress. There is no accounting for masses or specific amounts, and even if the signal were much stronger, the effect would still be the same. Similarly, simply the presence of some component in a regulatory network may affect the function of the network somewhere else. An example is the inhibition of an enzyme, which may alter the steady state and dynamics of a metabolic pathway, but which does not change the amount of the inhibitor itself.

A second distinguishing criterion in the treatment of static networks is the type of research activity, which can broadly be divided into analysis and construction (or reconstruction, as it is usually referred to). A typical analysis assumes that the structure of a network is known and investigates features such as the connectivity among its components and their contributions to the functioning of the network. By contrast, (re-)construction studies attempt to infer the connectivity of an unknown or ill-characterized network from experimental observations. At first glance, this distinction may appear to be splitting hairs, but the two aspects require distinctly different approaches, and the latter is incomparably more difficult than the former. It might seem logical to study the construction of a network first, before launching into its analysis, but it is easier to discuss and learn the two aspects in the opposite order. To do so, we will initially assume that we know the structure of the network and later worry about where our knowledge of that structure may have come from.

Finally, rigorous network analyses have led to the discovery of network **motifs**, which are small, yet prominent connectivity patterns that are found time and again in diverse biological systems and therefore are expected to be particularly effective for solving certain tasks. Large sections of Chapter 14 are devoted to this fascinating topic (see also [10]).

INTERACTION GRAPHS

When we are interested in the structure of a network, the specific interpretation of its components becomes secondary, and it does not truly matter for the mathematical or computational analysis whether the interacting components are molecules, species, or subpopulations: they simply become **nodes**. Nodes are sometimes also called vertices (singular: **vertex**), components, pools, or species, or they may be termed or described in a variety of other specific or generic ways. The second set of ingredients of a network consists of the connections between nodes, which are typically called **edges**, but may also be termed links, interactions, processes, reactions, arrows, or line segments. Among the edges, we actually have a bit more flexibility than with the nodes, because edges may be directed, undirected, or bidirectional. All three cases are relevant in one biological situation or another, and they may even appear within the same network (**Figure 3.1**). No features apart from nodes and edges are usually admissible.

Taken together, the nodes and edges form **graphs**, which are the principal representations of networks. Quite intuitively, the nodes of a network can only be connected by edges, and two edges can only be connected with a node between them. Two nodes connected by a single edge are called neighbors. Typical biological examples of directed graphs are transcriptional regulatory networks, where nodes represent genes and edges denote interactions between them. The choice of a directed graph seems reasonable in this case, because gene A might regulate the expression of gene B, but the opposite might not be true. Protein–protein interaction (PPI) networks, by contrast, are usually formulated as undirected graphs.

One might be tempted to surmise that all networks and systems could be represented in this fashion. However, we must be careful with such a sweeping inference. For instance, metabolic pathway systems do contain nodes (metabolite pools) and edges (chemical or enzymatic reactions and transport processes). However, they also

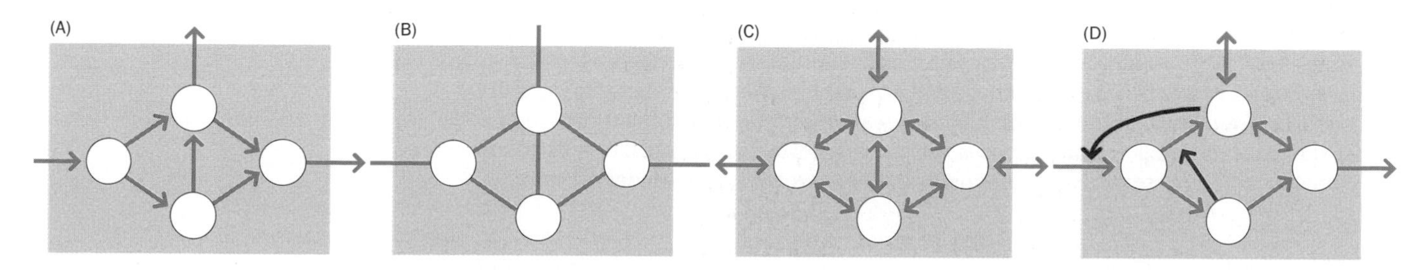

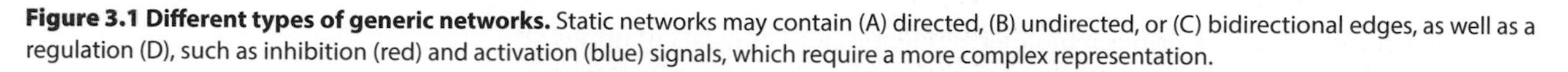
Figure 3.1 Different types of generic networks. Static networks may contain (A) directed, (B) undirected, or (C) bidirectional edges, as well as a regulation (D), such as inhibition (red) and activation (blue) signals, which require a more complex representation.

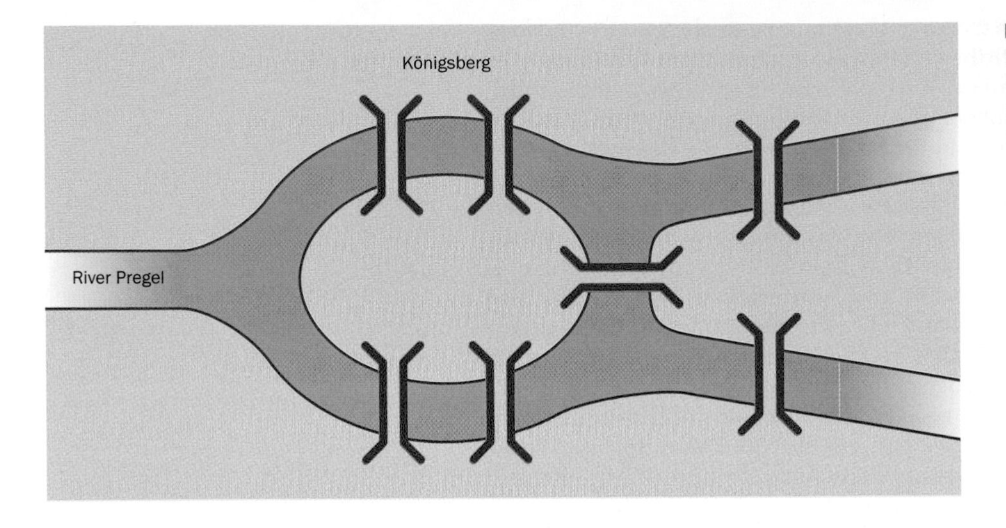

Figure 3.2 Euler's problem of the Seven Bridges of Königsberg. Euler answered the question whether it is possible to find a path crossing each bridge exactly once, and in so doing initiated the field of graph theory.

exhibit regulatory features, with which metabolites may affect the strength, capacity, or amount of material flow through the edges of a pathway. These regulatory effects are genuinely different from metabolic conversions and therefore require other types of representations, for instance, in the form of arrows pointing from a node to an edge, rather than to another node (see Figure 3.1D) [11]. As an alternative representation, it is possible to formulate such regulated networks with bipartite graphs, which contain two sets of different types of nodes. In the case of metabolic networks, one set represents metabolites, while the other set represents the physical or virtual locations where the enzymatic conversions between metabolites take place. A modeling approach following this paradigm of bipartite graphs is **Petri net** analysis [12, 13].

The subdiscipline of mathematics and computer science addressing static networks is **graph theory**, which has a long and esteemed history that reaches back to Leonard Euler in the eighteenth century. Euler's first interest in graph theory was piqued by a puzzle he posited for himself in his adopted home town of Königsberg in Prussia. The River Pregel dissected this city in a peculiar fashion and at the time was crossed by seven bridges (**Figure 3.2**). Euler wondered whether it was possible to find a route that would cross each bridge exactly once. Of course, being a mathematician, Euler not only solved this particular puzzle but also developed an entire theoretical framework, the foundation of graph theory, to decide in which cases a route can be found, and when this is impossible.

Generically speaking, graph theory is concerned with the properties of graphs, such as their connectivity, cycles of nodes and edges, shortest paths between two nodes, paths connecting all nodes, and the most parsimonious manner of cutting edges in such a way that the graph is split into two unconnected subgraphs. Many famous mathematical problems are associated with graphs. One example, arguably the first to be proven using a computer, is the four-color theorem, which postulates that it is always possible to color a (geographical) map with four colors in such a fashion that neighboring states or countries never have the same color. Another famous problem, which combines graphs with **optimization**, is the traveling salesman problem, which asks for the determination of the shortest path through a country such that every city is visited exactly once.

Most graph analyses in biology address questions of how a network is connected and what impact this connectivity has on the functionality of the network. Thus, these analyses characterize structural features, such as the numbers and densities of connections in different parts of the graph. These types of features can be subjected to rigorous mathematical analyses, some of which lead to watertight proofs, while others are of a more statistical nature. For our purposes, we can minimize the technical details, focus on concepts and representative scenarios, and provide some key references.

3.1 Properties of Graphs

Let us start with a graph G that describes a biological network and consists of nodes and edges, and denote the edge connecting nodes N_i and N_j by $e(N_i, N_j)$. A first important and rather intuitive characteristic of a node is its degree, denoted by $\deg(N_i)$, which represents the number of associated edges. For an undirected graph, $\deg(N_i)$ is simply the total number of edges at N_i, and this number is the same as the number of N_i's neighbors. In a directed graph, it is useful to define the in-degree $\deg_{in}(N_i)$ and out-degree $\deg_{out}(N_i)$, which refer to the numbers of edges terminating or starting at N_i, respectively (**Figure 3.3**).

For studies of the entire connectivity structure of a graph G with m nodes $N_1, \ldots, N_m$, one defines the adjacency matrix $\mathbf{A}$, which indicates with 1's and 0's whether two nodes are connected or not. This matrix has the form

$$\mathbf{A} = (a_{ij}) = \begin{cases} 1 & \text{if } e(N_i, N_j) \text{ is an edge in } G, \\ 0 & \text{otherwise.} \end{cases} \tag{3.1}$$

Here, undirected edges are counted as two directed edges that represent the forward and reverse directions. Examples of $\mathbf{A}$ are given in **Figure 3.4**. The adjacency matrix contains all pertinent information regarding a graph in a very compact representation that permits numerous types of analyses with methods of linear algebra. **Box 3.1** discusses an interesting use of $\mathbf{A}$.

In addition to investigating degrees of nodes, it is of interest to study how edges are distributed within a graph G. The principal measure of this feature is the **clustering coefficient** C_N, which characterizes the density of edges associated with the neighborhood of node N. Specifically, C_N quantifies how close the neighbors of N are to forming a **clique**, which is a fully connected graph. Let's begin with an undirected graph and suppose that some node N has k neighbors. Then, there can at most be $k(k-1)/2$ edges among them. Now, if e is the actual number of edges among N's neighbors, then C_N is defined as the ratio

$$C_N = \frac{e}{k(k-1)/2} = \frac{2e}{k(k-1)} \quad \text{(if } G \text{ is undirected).} \tag{3.2}$$

For directed graphs, the maximal number of edges among neighbors is twice as high, because each pair of neighbors could be connected with directed edges in both directions. Thus, the maximum number of edges, among k neighbors is $k(k-1)$, and the clustering coefficient is defined as

$$C_N = \frac{e}{k(k-1)} \quad \text{(if } G \text{ is directed).} \tag{3.3}$$

Note that the minimum of C_N in both cases is 0 and the maximum is 1.

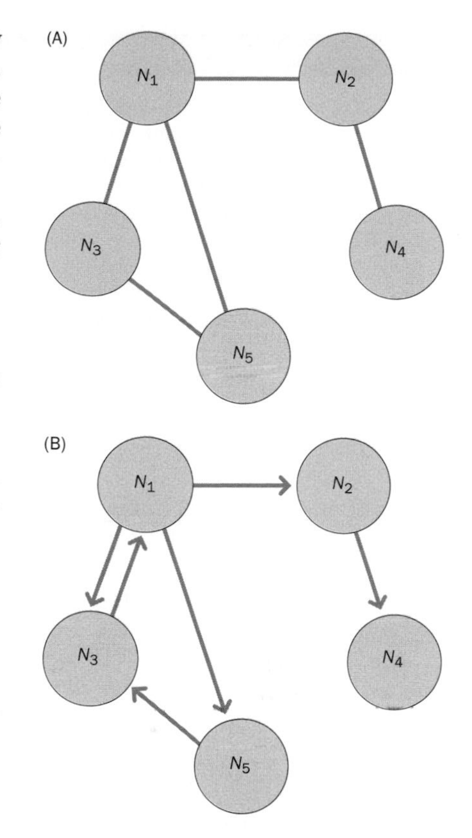

Figure 3.3 Visualization of the concept of degrees associated with nodes. (A) In this undirected graph, node N_1 has neighbors N_2, N_3, and N_5, and therefore its degree is $\deg(N_1) = 3$. Node N_4 has only one neighbor, N_2, and therefore $\deg(N_4) = 1$. (B) In this directed graph, node N_1 has an in-degree $\deg_{in}(N_1) = 1$ and an out-degree of $\deg_{out}(N_1) = 3$, while node N_4 has $\deg_{in}(N_4) = 1$ and $\deg_{out}(N_4) = 0$.

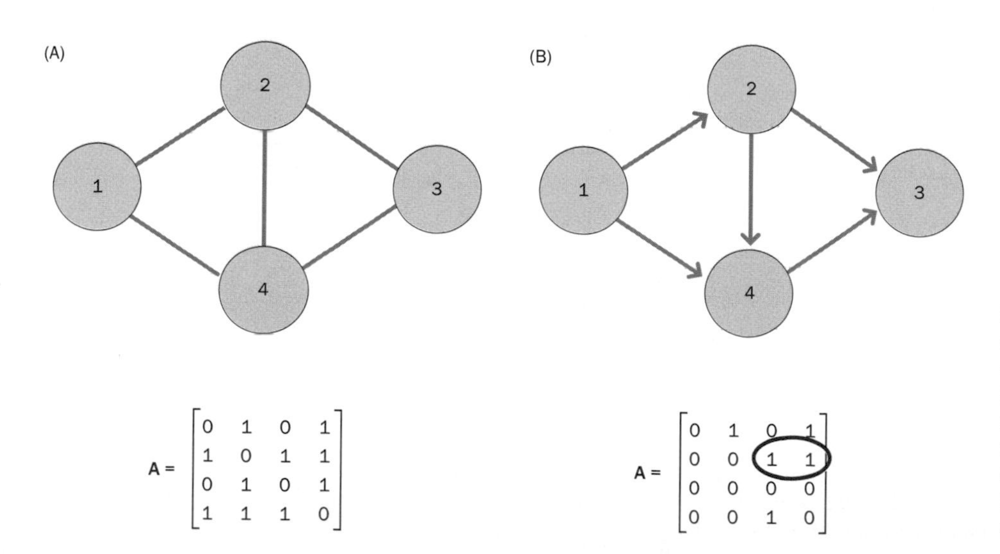

Figure 3.4 Two graphs and their adjacency matrices. In (A), each edge is considered bidirectional, which leads to twice as many 1's as edges. (B) shows a directed graph and its adjacency matrix. As an example, the red ellipse indicates that directed edges point from node 2 to nodes 3 and 4. The third row of the matrix contains only 0's because no arrows leave node 3.

BOX 3.1: COMPUTATION OF THE NUMBER OF PATHS BETWEEN NODES

Interestingly, one can use adjacency matrices, as shown in (3.1), to compute the number of paths between two nodes that consist of a fixed number of steps. These numbers are given by powers of the matrices. For instance, to compute all paths of length 2 in the matrix on the right-hand side of Figure 3.4, we multiply the matrix by itself, which results in a new matrix **B**. According to the rules for this operation, each element B_{ij} in matrix **B** is given as the sum of the products of the elements in row i of matrix **A** and the elements in column j of matrix **A**. For instance, element B_{13} is computed as

$$B_{13} = A_{11}A_{31} + A_{12}A_{32} + A_{13}A_{33} + A_{14}A_{34} = 0\cdot 0 + 1\cdot 1 + 0\cdot 0 + 1\cdot 1 = 2. \quad (1)$$

Executing the analogous operations for all B_{ij} yields

$$\mathbf{A}^2 = \begin{bmatrix} 0 & 1 & 0 & 1 \\ 0 & 0 & 1 & 1 \\ 0 & 0 & 0 & 0 \\ 0 & 0 & 1 & 0 \end{bmatrix} \begin{bmatrix} 0 & 1 & 0 & 1 \\ 0 & 0 & 1 & 1 \\ 0 & 0 & 0 & 0 \\ 0 & 0 & 1 & 0 \end{bmatrix} = \begin{bmatrix} 0 & 0 & 2 & 1 \\ 0 & 0 & 1 & 0 \\ 0 & 0 & 0 & 0 \\ 0 & 0 & 0 & 0 \end{bmatrix} = \mathbf{B}. \quad (2)$$

According to this result, there are two alternative two-step paths from node 1 to node 3, while there is one two-step path from 1 to 4 and one from 2 to 3. These three solutions exhaust all possible two-step paths. In this simple case, the results are easily checked from the graph. The reader should confirm these results.

An example is shown in **Figure 3.5**. The graph in (A) is undirected, and because N_1 has 8 neighbors and 12 edges among them, $C_{N_1} = 12/(8\times 7/2) = 0.4286$. The directed graph in Panel B also has 8 neighbors, but 15 edges; in this case, $C_{N_1} = 15/(8\times 7) = 0.2679$. It has a lower clustering coefficient, because there could theoretically be 56 edges among the neighbors of N_1 instead of the actual 15.

It is useful to define an overall measure of the network's clustering. This is accomplished with the graph's average clustering coefficient C_G, which is defined simply as the arithmetic mean of the C_N for all nodes [14]:

$$C_G = \frac{1}{m}\sum_{N=1}^{m} C_N. \quad (3.4)$$

Intriguingly, biological systems typically have clustering coefficients that are higher than those of randomly connected graphs. Furthermore, with an increasing size of the network, the clustering coefficient tends to decrease. This implies that biological networks contain many nodes with a low degree, but only relatively small numbers of **hubs**, which are highly connected nodes that are surrounded by dense clusters of nodes with a relatively low degree [15]. For example, in both the protein interaction network and the transcription regulatory network in yeast, the hubs are significantly more often connected to nodes of low degree, rather than to other hubs (**Figure 3.6**) [16, 17]. We will return to this feature later in this chapter.

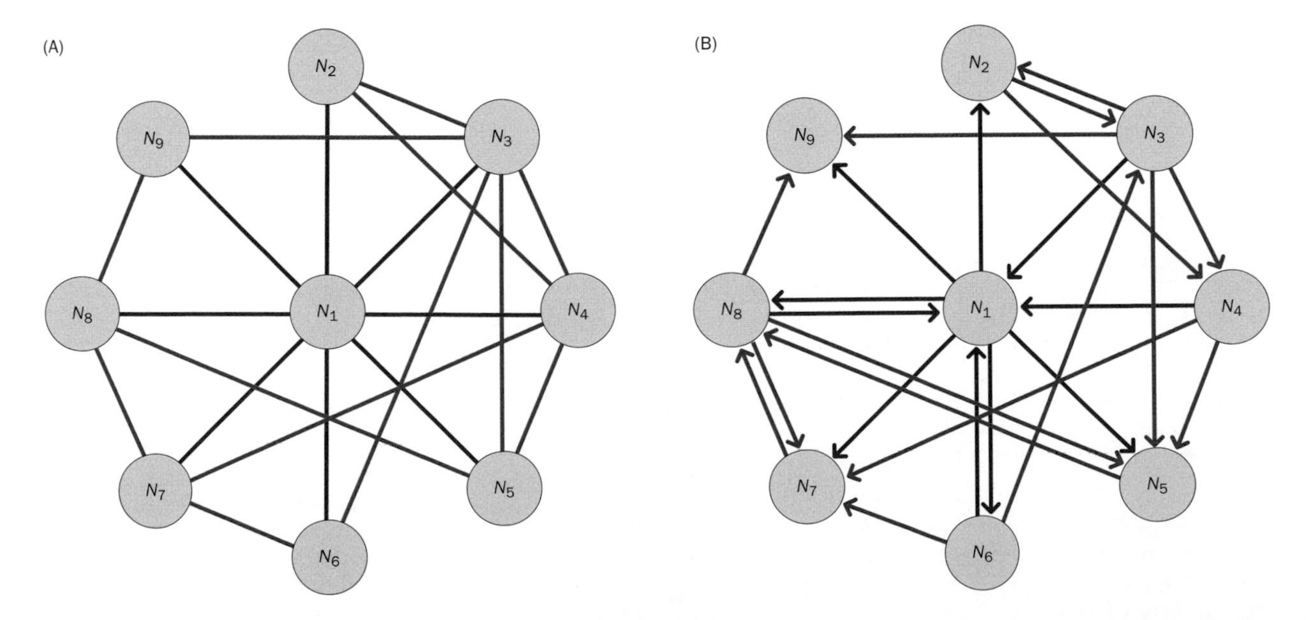

Figure 3.5 Visualization of the clustering coefficient of a node. Node N_1 has 8 neighbors. In the undirected graph in (A), the nodes are connected among themselves through 12 (blue) edges. Thus, $C_{N_1} = (2\times 12)/(8\times 7) = 0.4286$. In the directed graph in panel (B) the nodes are connected among themselves through 15 directed (blue) edges. Therefore, $C_{N_1} = 15/(8\times 7) = 0.2679$.

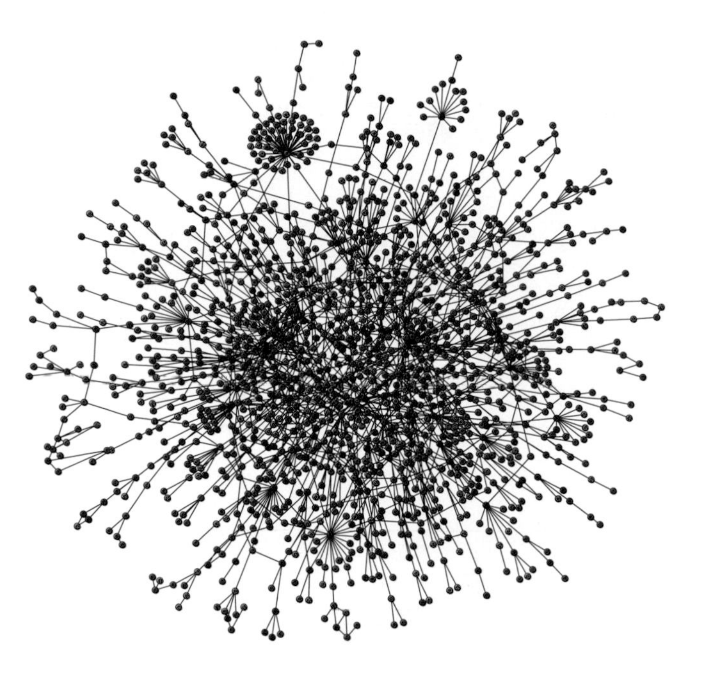

Figure 3.6 One large cluster within the protein–protein interaction network of the yeast *Saccharomyces cerevisiae*. The entire network consists of 1870 proteins as nodes, which are connected by 2240 direct physical interactions. The cluster presented here contains about 80% of this network. The graph clearly shows a number of hubs, which are highly connected, as well as the vast majority of nodes with a low degree. Each node is colored to signify the phenotypic result if the corresponding protein is removed (red, lethal; green, nonlethal; orange, slow growth; yellow, unknown). (From Jeong JP, Mason SP, Barabási A-L & Oltvai ZN. *Nature* 411 [2001] 41–42. With permission from Macmillan Publishers Ltd.)

A very important characteristic of a graph as a whole is its **degree distribution** $P(k)$, which measures the proportion of nodes within G that have degree k. If m_k is the number of nodes of degree k, and m is the total number of all nodes, then

$$P(k) = m_k/m. \tag{3.5}$$

Note that $P(k)$ is not a single number, but a set of numbers, with one degree value for each k ($k = 0, 1, 2, \ldots$). The division by m assures that the sum of $P(k)$ over all values of k is equal to 1. For the computation of $P(k)$, the directions of edges are typically ignored.

Many characterizations of graphs, including degree distributions, are performed by using comparisons with random graphs, which were first studied by Erdős and Rényi in the late 1950s [18]. In such an artificially created graph, the edges are randomly associated with nodes. One can show with mathematical rigor that the degree distribution in this case is a binomial distribution, which resembles a skinny bell curve with a small variance. In other words, most nodes have a degree that is close to average.

The binomial degree distribution within random graphs is very different from what is frequently observed in biological and other real-world systems, where a few hubs are connected to disproportionally many other nodes, while most other nodes are associated with much fewer edges than in a random graph (see Figure 3.6). Specifically, $P(k)$ often follows a **power-law distribution** of the form

$$P(k) \propto k^{-\gamma}, \tag{3.6}$$

where the exponent γ has a value between 2 and 3 [19]. Plotting such a power-law degree distribution in logarithmic coordinates results in a straight line with slope γ (**Figure 3.7**). A network whose degree distribution more or less follows a power law is called a **scale-free** network. This terminology was chosen because the power-law property is independent of the size (or scale) of the network.

While a power-law distribution is often observed for most nodes in a large biological network, the plot is seldom truly linear for very highly and very sparsely connected nodes within such a network. Nonetheless, except for these extremes, the power-law function often holds quite nicely, which demonstrates, unsurprisingly, that most biological systems are clearly not organized randomly (see, for example, [20]).

An interesting feature of scale-free networks is that the shortest paths between two randomly chosen nodes are distinctly shorter than those in random graphs. Here, a path is an alternating sequence of nodes and edges in a graph, leading from a start node to an end node, such as

$$N_6;\ e(N_6,N_1);\ N_1;\ e(N_1,N_4);\ N_4;\ e(N_4,N_5);\ N_5;\ e(N_5,N_2);\ N_2 \tag{3.7}$$

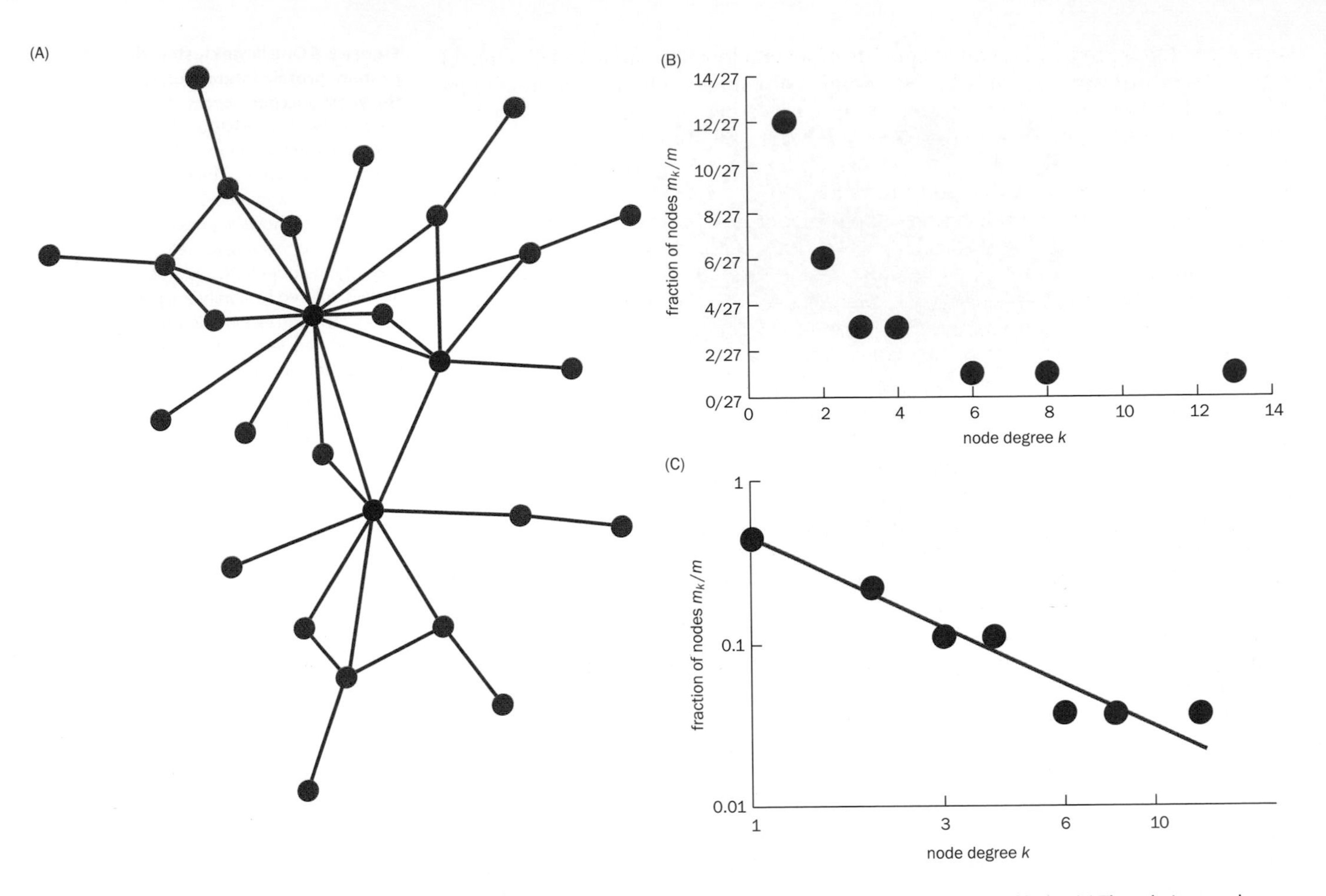

Figure 3.7 Example of a small scale-free network. (A) Among the 27 nodes, three (red) are the most highly connected hubs. (B) The relative number of nodes with k edges, plotted against the number of nodes with a given number of edges, k, shows a decreasing power-law relationship. (C) If the relationship in (B) is plotted in logarithmic coordinates, the result is approximately linear, with negative slope γ. (Adapted from Barabási A-L & Oltvai ZN. *Nat. Rev. Genet.* 5 [2004] 101–113. With permission from Macmillan Publishers Limited, part of Springer Nature.)

in **Figure 3.8**. Of particular interest is the shortest path (or distance) between two nodes, because it contains important clues about the structure of the network. In Erdős and Rényi's random graphs [18], the average shortest distance is proportional to the logarithm of the number of nodes, log m. By contrast, the average shortest distance in a scale-free network with a power-law degree distribution is shorter and increases more slowly for larger networks [21]. Indeed, analyzing over 40 metabolic networks with between 200 and 500 nodes, Jeong and collaborators found that a few hubs dominated the connectivity patterns and that, as a consequence, the average shortest distance was essentially independent of the size of the network and always had an amazingly low value of about three reaction steps [22]. In a different interpretation, one may also consider the average distance as an indicator of how quickly information can be sent through the network. According to this criterion, biological networks are indeed very efficient.

Most authors define the diameter of a network as the minimum number of steps that must be traversed to connect the two most distant nodes in the network, and we will follow this convention. In other words, the diameter is the shortest path between the farthest two nodes. However, one must be cautious, because other authors use the same term for the average shortest distance between any two nodes.

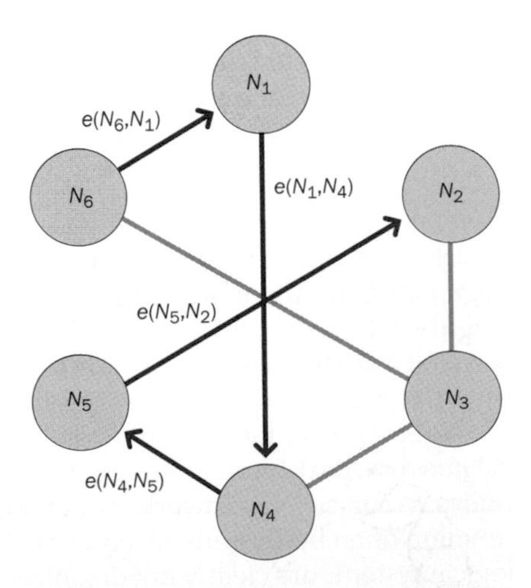

Figure 3.8 A possible path from start node N_6 to end node N_2. The path uses the edges $e(N_6, N_1)$, $e(N_1, N_4)$, $e(N_4, N_5)$, and $e(N_5, N_2)$. The path is not the shortest possible path, which would use $e(N_6, N_3)$ and $e(N_3, N_2)$. It is customary not to allow the same node or edge to appear twice in a path.

3.2 Small-World Networks

If a network is scale-free (that is, if it has a power-law degree distribution and therefore a small average shortest distance) and if it furthermore has a high clustering coefficient, it is called a **small-world** network. This terminology reflects the special structure of these networks, which consists of several loosely connected clusters of nodes, each surrounding a hub. In this organization, most nodes are not neighbors of each other, but moving from any one node to any other node in the network does not take many steps, because of the existence of the well-connected hubs. We have already mentioned metabolic networks, which have average **path lengths** in the range of 2–5 steps. Ecological studies have found the same length ranges in many food webs [23]. For other biological examples and further discussions, see [24, 25]. Outside biology, the situation is similar in commercial flight routes, which use major hubs combined with connecting flights to smaller airports. The small-world property is furthermore reminiscent of John Guare's famous play *Six Degrees of Separation*, whose premise is that essentially every human is connected to every other human by a chain of six or fewer acquaintances.

Interestingly, the small-world property emerges automatically when a network grows by the addition of new links preferentially to nodes that are already highly connected. Specifically, Barabási and collaborators (see, for example, [26]), who have devoted a lot of effort to these types of graph analyses, showed that a distribution with $\gamma = 3$ is the result of a growth process where, for each addition of new nodes, the edges are connected to an existing node N with a probability that is directly proportional to $\deg(N)$ [27]. Variations on the mechanism of preferential attachment lead to other γ values that can range anywhere between 2 and ∞. One should note, however, that entirely different mechanisms for the evolution of scale-free and small-world networks have been proposed. For instance, according to the duplication–divergence (DD) model, a gene or protein network evolves owing to the occasional copying of individual genes or proteins, which is followed by mutations and pruning [28]. The DD model can lead to $\gamma < 2$. See also **Box 3.2**.

BOX 3.2: CONSTRUCTION OF REGULAR, LOW-DIAMETER, AND RANDOM NETWORKS

Interesting connectivity patterns in networks can be constructed in various ways. Watts and collaborators [14, 29] started with a regularly patterned ring of nodes, in which each node was associated with four edges. Namely, the node was connected to its two immediate neighbors and also to their immediate neighbors (**Figure 1**). They then selected a constant probability p anywhere between 0 and 1, with which each edge could be disrupted and reconnected to a new, randomly chosen, node. Of course, for the extreme case of $p = 0$, the structure of the network remained unchanged. The other extreme case, $p = 1$, was shown to result in a random network. For values of p strictly between 0 and 1, other networks with much reduced diameters were obtained. Watts and collaborators called them small-world networks, but the terminology was different at the time. Today, we might call them short-distance or low-diameter networks, but they are no longer considered small-world because they do not exhibit high clustering coefficients.

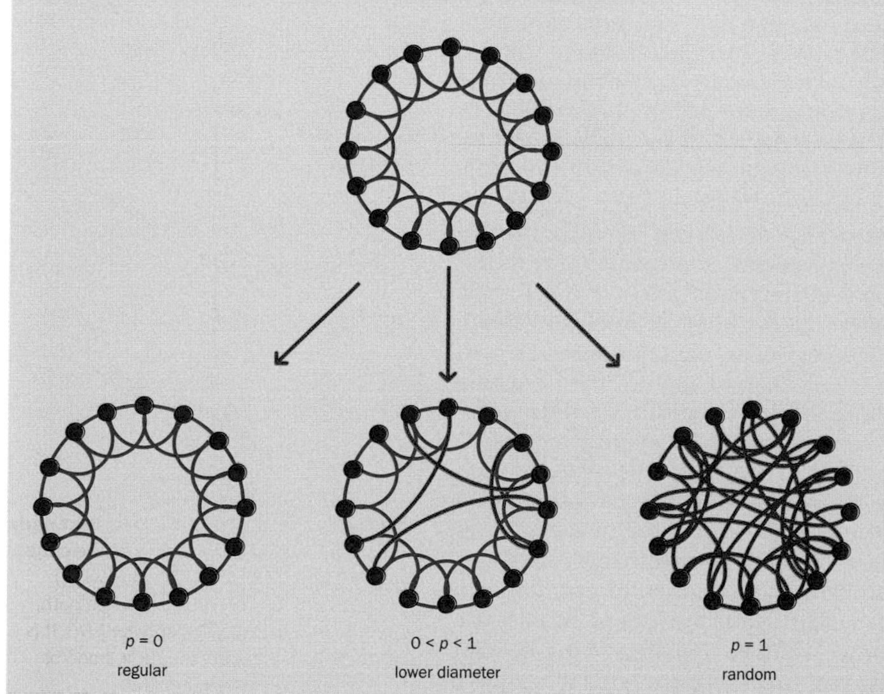

Figure 1: Construction of different types of networks, according to Newman and Watts [29]. The construction begins with a regular network, in which each node is connected to its immediate neighbors and their immediate neighbors (top of figure). Considering one node after another, the edge to one of its nearest neighbors is removed with a probability p ($0 < p < 1$) and replaced with a new edge to a randomly selected node; double edges are not allowed. In a second round, the process is repeated for second-nearest neighbors (bottom of figure). For $p = 0$, no edges are changed and the regular network is retained; for p between 0 and 1, a few edges (green) are replaced and the result is typically a network with shorter diameter; for $p = 1$, all edges are replaced and the network exhibits random connectivity.

It is not surprising that the hubs in a gene or protein network are often among the most important components and that their destruction or mutation is frequently fatal. In fact, it seems that the degree of a protein within a PPI network may be directly correlated with the fitness of the gene that codes for it [30]. This feature correlates in an interesting fashion with the robustness of a small-world network, or specifically its tolerance to attacks: attacks on randomly selected nodes are easily tolerated in most cases, because the likelihood of hitting a hub is low; by contrast, a single, targeted hit on a hub can break up the entire network [16, 22]. For these reasons, the identification of hubs may be important, for instance, for drug target identification. The interesting co-occurrence of robustness and fragility has been observed in many complex networks and systems, including both biological and engineered examples, and may be a characteristic design principle of such systems [31–33]. Other design principles and network motifs will be discussed in Chapter 14.

While molecular networks have recently received the most attention, it should be mentioned that static networks have played a role in systems ecology for a long time. Beginning with the pioneering work of Joel Cohen [34], graph representations have been used to study food webs, the flow of material through them, and the robustness and reliability of their connectivity structures. It was shown, for instance, that typical food webs have maximal path lengths between 3 and 5, that marine paths tend to be longer than those in terrestrial systems, and that the reliability of a food web decreases with increasing length [23]. Food webs often exhibit loops, where *A* feeds on *B*, *B* on *C*, and *C* on *A* [35].

While biological systems are often scale-free, their subnetworks sometimes have a different structure. For instance, sets of nodes within a small-world network are highly connected around hubs and may even form cliques, within which each node is connected to every other node. Clearly, the connectivity patterns of a clique and the larger network are different, and caution is in order when only parts of larger networks are investigated.

One should also keep in mind that most networks are in reality dynamic. For instance, PPI networks change during development and in response to disease. They sometimes even change periodically on a shorter timescale, as is the case during the **cell cycle** in yeast [36]. Han and collaborators [37] analyzed the temporal properties of PPI hubs in yeast during the cell cycle and in response to stress, and postulated two types of hubs: party hubs, which interact simultaneously with most of their neighbors; and date hubs, which connect to different neighbors at different times and possibly in different cellular locations. The dynamic changes associated with hubs may indicate a control structure of organized modularity, where party hubs reside inside defined, semi-autonomous modules, which perform some biological functionality, while date hubs connect these modules to each other and organize their concerted function. However, the distinction between party and date hubs is not always clear-cut [38].

As a final example of graph-based network analysis, let us look at the galactose utilization gene network in yeast [39], which serves as a documentation example in the Cytoscape network analysis platform [2], where the following analyses can be retraced. The network consists of 331 nodes, which have an average of 2.18 neighbors. Once loaded into Cytoscape, it can be visualized in a number of ways, for instance, in the so-called force-directed layout, whose details can be customized to the user's taste (**Figure 3.9**). Each node can be clicked on screen, which reveals the gene identifier. A separate spreadsheet contains further information. For instance, the hub (YDR395W) has a role in the transport of mRNA from the nucleus into the cytosol, while its neighbors code for structural components of the ribosome; a possible exception could be YER056CA, whose role is unknown. Thus, the closeness within the network is reflected by functional relationships. By contrast, the gene at hub YNL216W, which is a few steps removed, codes for a repressor activator protein. With a few clicks, Cytoscape reveals typical numerical features of the network, such as the clustering coefficient (0.072) and the average (9.025) and the longest (27) lengths of the shortest paths. It is furthermore easy to display the distribution of shortest paths (**Figure 3.10A**) and the degree distribution (Figure 3.10B). The latter distribution more or less follows a power law, with the exception of very low and very high degrees, as we discussed before. Many other static features of the network can be displayed. Beyond these typical network features, Cytoscape plug-ins can even be used for transitions to dynamic analyses of small subnetworks [40].

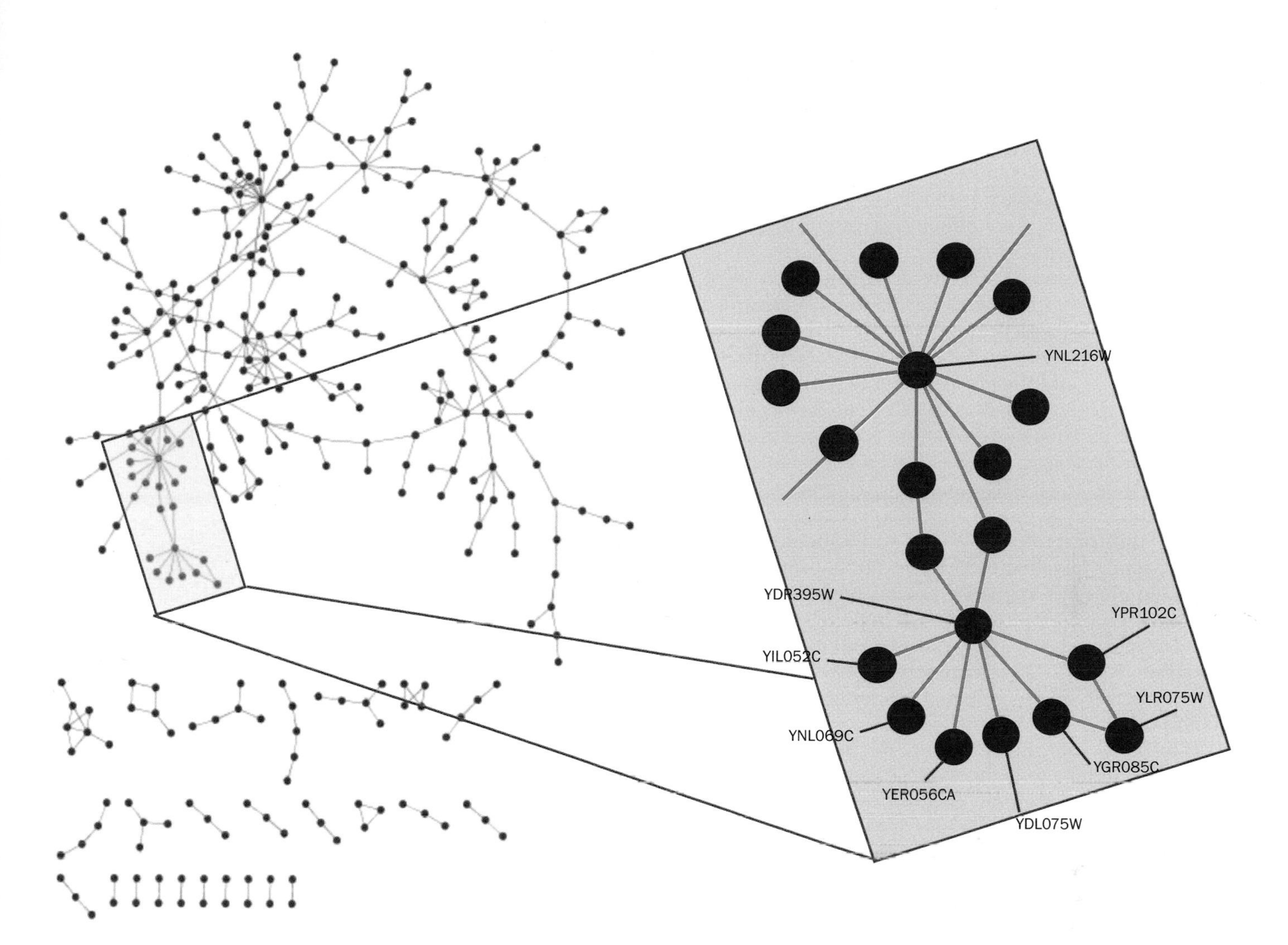

Figure 3.9 Visualization of a network. Every node in Cytoscape's network visualization of the galactose utilization gene network in yeast can be clicked on screen to reveal its identity and features. Neighboring nodes are often functionally related. See the text for further details.

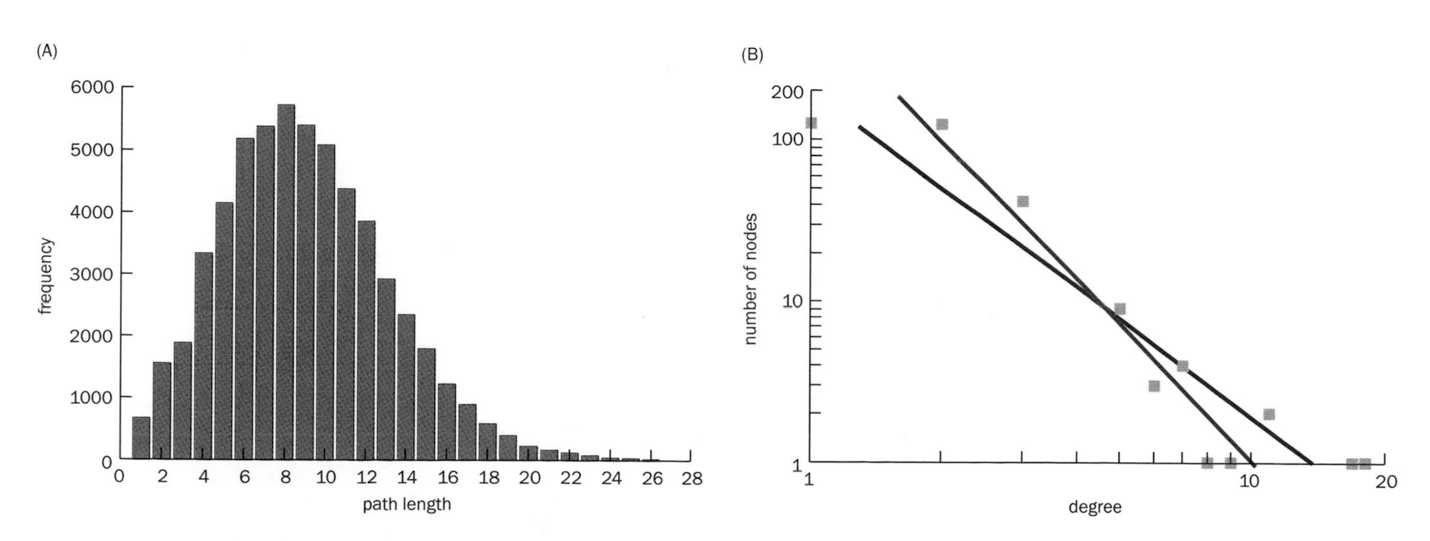

Figure 3.10 Typical output from Cytoscape. (A) Distribution of shortest paths. (B) Degree distribution with regression line computed in Cytoscape (red) and regression line ignoring very low and very high degrees (blue); the latter leads to a steeper slope (–2.45) than the former (–2.08).

DEPENDENCIES AMONG NETWORK COMPONENTS

3.3 Causality Analysis

So far, we have discussed the connectivity among nodes in a network, as well as paths leading from a source to a target. Especially in undirected networks, the connections represent associations. However, they do not necessarily imply causation, which is much more difficult to establish, but yields greater insight into the functionality of the network.

The search for causes and effects is probably as old as humankind. We know that Aristotle stated that the cause has to precede an effect, while Galileo in his 1638 book *Two New Sciences* postulated that a true cause (*vera causa*) must not only precede the effect but also be both necessary and sufficient. Two hundred years later, John Stuart Mill additionally required of a true cause that other explanations for a cause–effect relationship must be eliminated, and, in the 1920s, the geneticist Sewall Wright formulated systems of linear equations and analytical tools for addressing indirect effects. Wright's studies can be considered the foundation of modern **causality analysis** (cf. [41]). Nowadays, causes and effects are often studied with statistical methods. A prominent approach is Granger causality, which requires that if A causes B, then past values of both A and B should be more informative for predicting B than past information on B alone. Such a situation can be evaluated with linear regression methods applied to stochastic processes. The techniques have recently been extended to nonlinear cases, but the corresponding computations become quite complicated. While a warning is often voiced that correlation does not imply causation, techniques of path analysis, structural equation modeling, and Bayesian statistics permit to some degree the reliable inference of true causes of observed effects [42].

Typical questions are whether effects are direct or indirect and how to address the role of hidden variables, which cannot be measured. In many cases, the methods involve a cause–effect diagram, which consists of a directed graph without cycles, which means that one can never return to the same node. The problem with causality analysis in biology is that realistic systems almost always contain a considerable degree of circularity, which makes most statistical approaches to analyzing or inferring network structures and their causality troublesome. In systems with many cycles, causes may be distributed throughout the whole system, and multiple related or unrelated factors may contribute to the observed effect with different weights. As a consequence of these typical complications, causality analysis has found only limited applications in biology (for a successful exception in a relatively simple case, see [43, 44]), with the exception of Bayesian methods, which are often used for the reconstruction of genetic and signaling networks, as we will see in a later section.

3.4 Mutual Information

A simpler task than identifying true causality concerns the mutual dependence of two variables. As an example, suppose we would like to quantify how much more likely it is that some gene of interest is expressed if we know that certain other genes are expressed. It is not difficult to see that such an analysis of the likely co-expression of sets of genes could give important clues regarding the structure of pathways and might shed light on diseases such as cancer that involve many changes in gene expression [45]. Questions of this nature may be addressed with statistical methods characterizing the mutual information between two or more quantities. These methods are derived from Claude Shannon's and Warren Weaver's ideas concerning the reliability of communication through noisy channels [46], which led to what is now known as information theory. In this field, one studies how likely it is that one variable X is *on* (present), while another variable Y is either *on* (present) or *off* (absent). If X and Y are perfectly coupled to each other, the two will always be both *on* or both *off*, or it could be that one is always *on* if the other is *off*. However, in most cases, the coupling is not that clear-cut and one often wonders whether two variables influence each other at all. The mutual information between X and Y characterizes their dependence in cases where the coupling is perfect, strong, loose, or absent. Expressed differently, the mutual information is a measure of the extent to which knowledge regarding one of the two variables reduces uncertainties in the other. If X and Y are totally independent of each other, the mutual information is 0, because, even if we knew everything about X, we would be unable to make reliable predictions regarding Y. By contrast, a strong dependence between X and Y corresponds to a high value of the mutual information criterion. Specific examples of the use of mutual information in molecular systems biology include inferences regarding the functional relevance of gene–gene interactions [47] and the coordination between gene expression and the organization of transcription factors into regulatory motifs [48].

A key concept for quantifying mutual information is the entropy $H(X)$, which measures the uncertainty associated with a random variable X. Here, the term random variable means that X may take different values with certain probabilities; for example, the roll of a die may result in any number between 1 and 6. Depending on the situation, the uncertainty surrounding X may depend on how much we know about another variable, Y. For instance, if two genes G_1 and G_2 are often co-expressed, and if we observe that G_2 is indeed expressed in a certain situation, we are more confident (although still not certain) that G_1 will be expressed as well. The uncertainty of X, given that we know Y, is expressed as the conditional entropy $H(X \mid Y)$. Furthermore, the uncertainty of a pair of random variables is quantified as the joint entropy $H(X, Y)$, which is defined as the sum of the entropy $H(X)$ and the conditional entropy $H(Y \mid X)$. Since $H(X, Y)$ is the same as $H(Y, X)$, we obtain the relationship

$$H(X, Y) = H(X) + H(Y \mid X) = H(Y) + H(X \mid Y).$$

The mutual information $I(X; Y)$ is the sum of the individual entropies minus the joint entropy:

$$I(X; Y) = \left[H(X) + H(Y)\right] - H(X, Y).$$

These relationships can be visualized in a Venn diagram of entropies (**Figure 3.11**). Shannon, Weaver, and others worked out the mathematical formulations of the various entropies, and these can directly be extended to uncertainties and the mutual information among more than two variables. A good introduction to the topic is [45].

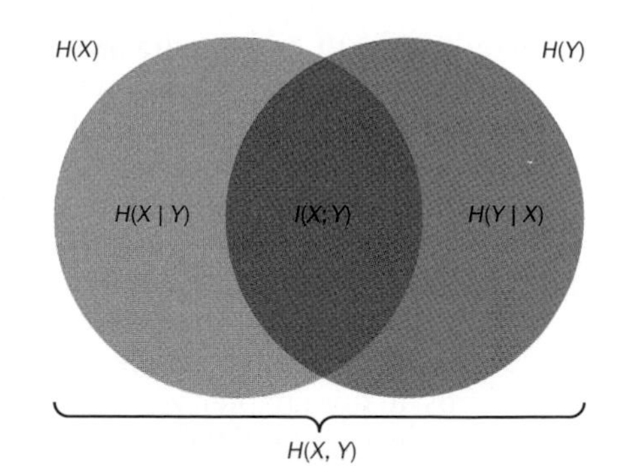

Figure 3.11 Mutual information and entropy. The Venn diagram visualizes the relationships between mutual information $I(X; Y)$ and different types of entropy, namely the entropies $H(X)$ and $H(Y)$ associated with the random variables X and Y, the joint entropy $H(X, Y)$, and the conditional entropies $H(X \mid Y)$ and $H(Y \mid X)$.

BAYESIAN RECONSTRUCTION OF INTERACTION NETWORKS

While it is relatively easy to characterize graphs using measures such as their degree distribution or clustering coefficients, it is much more difficult to find the connectivity within an unknown graph. This section introduces the basic concepts needed to execute **reverse engineering** techniques in ill-characterized graphs. The important message of this section is the confirmation that it is indeed possible to infer the structure of unknown static networks with some reliability, at least under favorable conditions. The technical details, which are not always simple and are unfortunately difficult to avoid, should not cloud this message.

It is easy to imagine that one can infer the structure of a network only if one has suitable data—lots of them. It is also clear that such data in reality are essentially always corrupted by noise, which may come from genuine randomness or from inaccuracies during the experimental measurements. For instance, the components of a signaling system that are active at a particular point in time may depend on all kinds of factors, including environmental perturbations, inbuilt redundancies, and the recent history of the system. Overlaid on these perturbations are the experimental difficulties of obtaining accurate characterizations of active and inactive states. Given such enormous challenges, is it possible—and indeed valid—to infer an active, causative role of two components A and B in the stimulation of C, if C is usually active when A and B are both turned on? It is not difficult to imagine that the reliability of such an inference grows the more often we observe the same pattern of

C being on if both A and B are on, and C being off if A or B or both are off. In other words, it is easy to imagine that network reconstruction tasks require many data that permit the application of statistical tools.

Useful data can come in two forms. In the first case, one has many independent observations of co-occurrences of component activities, as in the case of A, B, and C. In the second case, one may have measurements of many or all components of the network, which were obtained as a sequence of many time points, following some stimulus. For example, one could expose a cell culture to an environmental stress and study the expression profile of a set of genes over many minutes, hours, or days afterwards. In this chapter, we discuss only the former case of many co-occurrences. We will return to the latter case in Chapter 5, which addresses inverse modeling tasks for dynamic systems.

The strategy for attempting to infer the structure of a network from co-occurrence data is of a statistical nature and goes back almost 250 years to Thomas Bayes [49]; it is still called **Bayesian network inference**. If the data are plentiful and reasonably good, the strategy is rather robust and addresses direct and indirect causal effects. However, it also has its limitations, as we will discuss later.

Before we can outline the principles of Bayesian network inference, it is necessary to review the very basics of probabilities. We do this in a somewhat nonchalant, conceptual fashion, but there are many books that treat these ideas in a rigorous mathematical fashion (for example, [50]).

By convention, probabilities are real numbers between 0 and 1 (or 100%), where 0 signifies the impossibility of an event and 1 denotes the certainty of an event. Suppose we have nine new $1 bills and one new $20 bill in our wallet and they are not sorted in any particular way. Because the bills are new and of the same size, weight, and material, we cannot feel differences between them, and, pulling a bill blindly out of the wallet, the probability of a $1 bill is 90% or 0.9, whereas the probability of the $20 bill is 10% or 0.1. The probability of pulling a $5, $100, or $7.50 bill is 0, and the probability of pulling either a $1 or a $20 bill is 1. All this makes intuitive sense and can be proven with mathematical rigor.

Imagine now that we are repeating the experiment many times, always returning the bill to the wallet, and recording every time what dollar amount we pulled. In the previous example, if we pull a bill often enough, we expect to pick the $20 bill roughly every tenth time. Now suppose we do not know how many $1 or $20 bills are among the 10 bills are in our wallet. Pulling only one bill does not really answer the question, but if we repeat the experiment many times, the numbers of $1 or $20 pulls begin to give us an increasingly clearer picture. The argument is as follows: If we blindly pull a bill 50 times, and if there is only one $20 bill, we would expect a total of five $20 pulls. It might be four, it might be six, but five would be our best estimate. By contrast, if there were six $20 bills among the 10 bills, we would expect a much higher return, namely something like 30 pulls of $20 bills, corresponding to the 60% chance for each pull. Thus, just by observing the outcome of many experiments, we can infer something about the unknown internal structure of the "system." **Bayesian inference** is a sophisticated formalization of such a strategy. We are not 100% sure about our results, because they are probabilistic, but our confidence grows with the number of observations.

Continuing with the example, suppose we are in a game show where we have to choose one of three wallets and then blindly pull one bill out of it. The wallets have different colors, and, for whatever reason, we have a slight preference for gray. To be specific, let's say the probability to select the gray wallet is 40% and the other two probabilities are 30% each. Of course, we don't know what each wallet contains. Suppose the brown wallet contains two $100 bills and eight $1 bills, the gray wallet has six $20 bills and four $1 bills, and the green wallet contains ten $10 bills (**Figure 3.12**).

Clearly, our personal outcome of the game will ultimately be $1, $10, $20, or $100, but what is the likelihood of each? It is easy to see that the probability of pulling a particular dollar amount out of a wallet depends on which wallet we select in the first place. In statistical terms, the probability is conditional, because it depends (is conditioned) on the selection of a particular wallet. Figure 3.12 illustrates the situation.

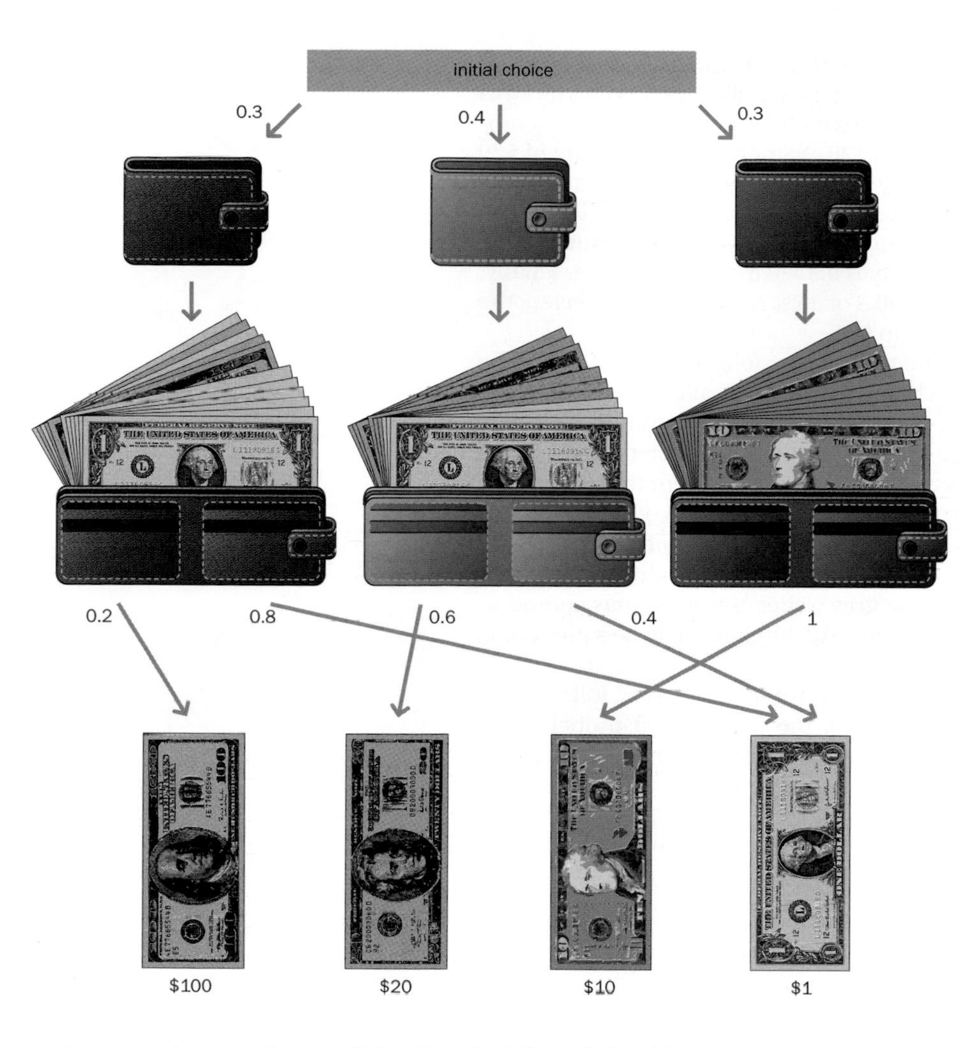

Figure 3.12 Conditional probabilities can be explored intuitively with playing cards, dice, and games. Shown here is the schematic of a fictitious game show, where the contestant selects a wallet and pulls a bill out of it.

In general terms, the conditional probability $P(B|A)$ (that is, the probability of B given A) of an event B, given that event A is true (is occurring or has happened before) is formally given as

$$P(B|A) = \frac{P(B \text{ and } A)}{P(A)}. \tag{3.8}$$

Here $P(B \text{ and } A)$ denotes the (joint) probability that both A and B are simultaneously true, and it is tacitly assumed that the probability of A, $P(A)$, is not zero. Because $P(B$ and $A)$ is the same as $P(A$ and $B)$, we can formally rearrange the terms and write a sequence of equations as

$$P(B|A)P(A) = P(B \text{ and } A) = P(A \text{ and } B) = P(A|B)P(B). \tag{3.9}$$

The equivalence of the outer expressions here is known as Bayes' theorem, Bayes' formula, or the Bayes rule, and is typically written as

$$P(A|B) = \frac{P(A \text{ and } B)}{P(B)} = \frac{P(B|A)P(A)}{P(B)}. \tag{3.10}$$

The formula says that we can make a statement of the (unknown) occurrence of event A given an (observed) outcome B, assuming that $P(B)$ is not zero.

It is not difficult to make various forward computations with the formulae for conditional probabilities, for instance, for our game show example. Because the choice of the wallet and the pulling of a bill from it are independent events, their probabilities simply multiply. Thus, the probability of choosing the brown wallet and securing the grand prize of $100 is $0.3 \times 0.2 = 0.06$, which corresponds to a measly 6% probability. The probability of choosing the gray wallet and ending up with $100 is nil. In this fashion, we can exhaust all possibilities and, for instance, compute what the probability is of ending up with $20. Namely, the probability of $20 from the brown or red wallets is zero. The gray wallet is chosen with probability 0.4 and choosing a $20 bill among the bills is 0.6, so that the overall probability is 0.24. To compute the probability of the meager outcome of $1, we need to consider the brown and the gray wallets, and the probability is $0.3 \times 0.8 + 0.4 \times 0.4 = 0.4$. Finally, the probability of an overall outcome of $10 is 0.3 or 30%. As should be expected, the probabilities for $1, $10, $20, and $100 sum up to 1.

Another question we can answer quite easily is the following: If we knew the contents of the three wallets, would the choice of the gray wallet be optimal? The answer lies in the expected value, which we may imagine as the average over very many runs of the game. The easiest case for computing the expected values is the red wallet, which, when chosen, will always reward us with $10. For the brown wallet, the expected value is $20.80, because in 100 runs of the game we would expect to pull about twenty $100s and eighty $1s. The total dollar amount of these is $2080, which corresponds to an average of $20.80 per run (even though we can never actually get $20.80 in any given run!). For the gray wallet, the analogous computation yields (60 × $20 + 40 × $1)/100, which is only $12.40. Gray might be a nice color, but it is a bad choice in this game!

All these types of computations are straightforward, because they follow well-established probability theory and, in particular, the law of total probabilities. This law says that the probability of an event B that is conditioned on distinct events $A_1, A_2, \ldots, A_n$ can be computed as

$$P(B) = \sum_{i=1}^{n} P(B \mid A_i)P(A_i). \tag{3.11}$$

We can see that this result is directly related to Bayes' formula, if we consider all possible and distinct events A_i and compute all conditional probabilities $P(B|A_i)$. The probability of event B is then the sum of all these $P(B|A_i)$, given that A_i actually happens, which it does with probability $P(A_i)$. A direct example is the preceding computation of the least desirable outcome of $1 in the game show example. Each $P(A_i)$ is called a prior probability, and $P(B|A_i)$ is again the conditional probability of B, given A_i.

3.5 Application to Signaling Networks

Many of the concepts of Bayesian analysis are applicable to genomic, proteomic, signaling, and other networks. Here we use signaling as the prime example for illustration, but it should be clear that gene and protein networks can be analyzed in a similar fashion.

Signal transduction is a key element for the proper function of cells and organisms. Simplistically speaking, a cell receives a chemical, electrical, mechanical, or other signal at the outside of its membrane. This signal is internalized in some fashion, for instance with the help of G-protein-coupled receptors, and typically triggers an amplification mechanism that is often referred to as a signaling cascade. Ultimately, the received and amplified signal leads to some cellular response, such as the expression of appropriate genes. Fundamental aspects of these fascinating systems are discussed in Chapter 9.

For an illustration here, let us apply our insights on conditional probabilities to a signaling network like the one in **Figure 3.13**. Suppose two receptors, R and S, are anchored in the cell membrane and respond independently to appropriate ligands. The response to the binding of L consists of sending a signal to a signaling cascade C, whose details are unimportant here, but which ultimately triggers some genomic response G. For the cascade to be activated, both R and S should fire. Furthermore, if a ligand binds only to S, the system triggers a different response Z. In reality, the interactions are corrupted by noise, such as the binding of nonspecific ligands, so that the cascade may sometimes be triggered even if only R or S is active, or it may not be triggered, even if both are occupied by ligands. The terminology of noise is commonly used, even though signaling cascades don't make audible sounds.

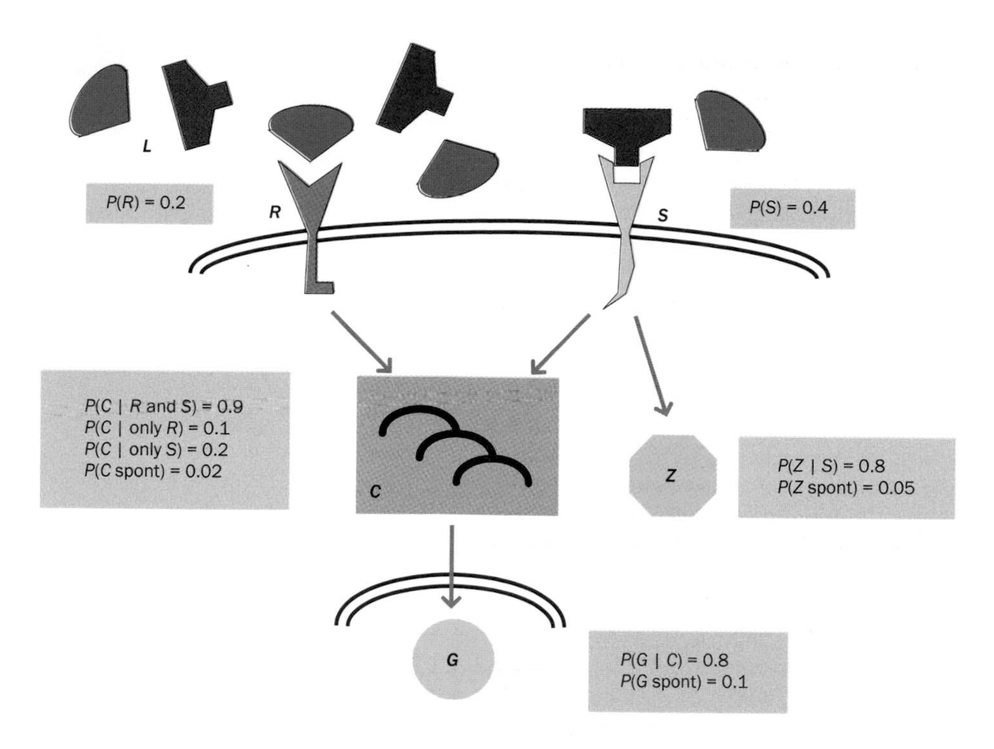

Figure 3.13 A generic, simplified signaling system responds to the binding of two different ligands. The expected numbers of correct or faulty responses can be computed with conditional probabilities. See the text for details.

The connectivity of the system may be sketched out in a diagram, as shown in Figure 3.13, which also shows probability tables of various events. For instance, the probability that receptor R is active (occupied by a specific ligand L) is 0.2. The probability that cascade C is triggered if both R and S are occupied by ligands, is $P(C|R \text{ and } S) = 0.9$, whereas spontaneous activation of the cascade has a very low probability of $P(C \text{ spont}) = 0.02$. The probability tables allow us to compute all kinds of probabilities, such as the probability that a ligand binds to S and triggers Z. Or we might compute the probability that (1) ligands bind to R and S; (2) the cascade is activated; and (3) the result is a genomic response G. This latter scenario is modeled as

$$\begin{aligned} P(G, C, R, S \text{ are "on"}) &= P(G|C)P(C|R \text{ and } S)\,P(R)P(S) \\ &= 0.8 \times 0.9 \times 0.2 \times 0.4 = 0.0576 \end{aligned} \tag{3.12}$$

The roughly 6% probability of the response may seem low, but it is mainly dictated by the relatively low binding probabilities of ligands of R and S, which both must happen for a normal response.

A typical situation for Bayesian inference is the following. If we observe that gene G is expressed, can we make inferences on whether the cascade is activated and/or whether R and/or S are present? Knowing the probability tables, we can infer with high probability that C is active, because the probability that G is on, given that C is active, is high (0.8), and spontaneous gene expression is not overly likely (0.1). It is not difficult to imagine that the inferences become better and better as more observations on the on–off states of the nodes in the graph become available.

Bayesian inference really becomes interesting if we do not know the probability tables, but must deduce them from observations on the states of the system. Not surprisingly, the execution of such an inference is rather complicated and requires sophisticated computer algorithms. Nonetheless, it is useful to understand the underlying conceptual ideas. In general, a Bayesian network is formulated as a **directed acyclic graph (DAG)**, which means that its nodes X_i are connected by arrows heading in only one direction and that there is no possibility of returning to any of the nodes, once one leaves this node through any of its outgoing arrows; Figure 3.13 is an example. The parents of a node X_i, denoted by $\text{Par}_j(X_i)$, are defined as those nodes that directly send a signal to X_i through a single connecting arrow. The state of a parent may again be *on* or *off*, 1 or 0, or take other values, just as X_i. If the graph consists of n nodes, then the probability P that the state of the system has particular values $(x_1, x_2, \ldots, x_n)$ can be computed, with a dependence on the state of the parents, as

$$P(X_1 = x_1 \wedge X_2 = x_2 \wedge \ldots \wedge X_n = x_n) = \prod_{i=1}^{n} P\left(X_i = x_i \mid \left\langle \mathrm{Par}_j(X_i) = x_j \right\rangle\right). \quad (3.13)$$

Here, the notation $\wedge$ is the logical "and." Thus, the expression on the left-hand side formalizes the probability P that all the nodes X_i simultaneously have the particular values x_i. The notation $\langle \mathrm{Par}_j(X_i) = x_j \rangle$ on the right-hand side refers to the set of all parents P_j of X_i and their particular states x_j. The probabilities in the product are therefore probabilities that X_i has the value x_i, conditioned on the states of all its parents. The states of the parents can in turn be expressed in terms of their own parents, and so on, as we exemplified above. This stepwise dependence is possible because the graph is acyclic; if it had loops, we would be in trouble. With this formalization, we can compute any conditional probability in the system. These computations are often very tedious, but they are straightforward and can be executed with customized software.

The computation of conditional probabilities requires that we know (a) the connectivity of the network and (b) the probability tables at each node, as it was the case in Figure 3.13. For many actual signal transduction networks in biology, we may have a rough idea about their connectivity, but we usually do not know the probabilities. Sometimes, we have information on whether the probabilities are high or rather low, but we seldom know precise numerical values. That's where Bayesian inference comes in. Instead of being able to compute numerically forward, as we did so far, one needs a large dataset of observations consisting of *on* or *off* states of all or at least many of the nodes. The focus of the analysis is now on correlations of *on* or *off* values among the nodes. For instance, two observations regarding the signal transduction system in Figure 3.13 could be

$$(R = \text{off};\ S = \text{on};\ C = \text{off};\ G = \text{off};\ Z = \text{on})$$

and

$$(R = \text{on};\ S = \text{on};\ C = \text{on};\ G = \text{on};\ Z = \text{off}).$$

If C and G are strongly correlated, as they are in the above example, many observations will have them either both *on* or both *off*. By contrast, such a correlation would not be expected between R and Z.

The inference strategy begins with composing a network graph that contains all known, alleged, or hypothesized connections, the direction of causality between any two nodes, and, if possible, likely ranges of probabilities. All probabilities are represented by symbolic parameters $p_1, p_2, \ldots, p_m$ rather than numerical values, and the goal is to determine the unknown values for these parameters in such a manner that the parameterized model reflects all (or at least most) observed combinations of states as accurately as possible. Needless to say, figuring out such a well-matching configuration of many parameters constitutes an optimization task that is anything but trivial, especially if there are many nodes. Several methods have been developed for this purpose in recent times, including Markov chain Monte Carlo (MCMC) algorithms such as the Gibbs sampler and the Metropolis–Hastings method, and they are all quite involved [8]. The basic concept of these optimization methods consists of randomly choosing probabilities (first blindly or within known or assumed ranges, and later within shrinking ranges that are determined by the algorithm) and computing the corresponding joint distributions of all nodes (recall the Monte Carlo simulations in Chapter 2). This forward computation with putative probabilities is relatively simple and is executed thousands of times. A statistical comparison between the computed joint distributions and the observed distributions is then used to determine probability values that slowly improve during the algorithmic iterations and, if successful, ultimately yield a constellation of probability values with which the model matches the observations best. For further details, the interested reader may consult [8] and [51]. The same type of analysis is sometimes used to decide which set of network connections among several candidates is most likely, given observations on input and output nodes.

One theoretical and practical disadvantage of the Bayesian inference strategy is that it does not permit graphs that contain any types of loops. For instance, chains of arrows starting and ending at the same node, as well as feedback signals, are not permitted. Furthermore, graphs or probability tables that change over time make inferences much more complicated. Current research efforts are focusing on these challenges.

It may seem that Bayesian inferences of actual signal transduction networks might run into insurmountable problems. However, several success stories have been documented. A premier example is the identification of a signaling network in $CD4^+$ T cells of the human immune system [52]. The data consisted of thousands of single-cell flow-cytometric measurements of multiply phosphorylated protein and phospholipid components in individual cells that followed external molecular perturbations. Specifically, the data consisted of knowledge of the molecular stimulus (inhibition or stimulation of a specific component) and the corresponding state or condition of each of 11 phosphorylated molecules within a given cell, 15 minutes after the stimulus. The Bayesian inference successfully confirmed known relationships between some of the signaling components and revealed additional, previously unknown, causal relationships. Not surprisingly, the analysis missed three causal relationships that corresponded to known loops in the signal transduction network.

3.6 Applications to Other Biological Networks

Outside of signaling systems, graph methods and Bayesian approaches have been used to reconstruct protein–protein interaction networks, where the main purpose was to identify which proteins are interacting with each other, and which of these interactions are functionally meaningful [53, 54]. As just one example, Rhodes and collaborators [55] proposed a model of the human protein–protein interaction network that integrated information from protein domain data with gene expression and functional annotation data. Using Bayesian methods, they constructed and partially validated a network consisting of nearly 40,000 protein–protein interactions. Beyond the presence or absence of particular interactions, one is also often interested in the structure of these networks and the features of the underlying graphs, such as types of hubs, cliques, and small-world or random structure, as we discussed earlier in this chapter. Since large-scale protein–protein interaction networks often contain hundreds of nodes, their visualization is a great challenge.

Bayesian methods have also been applied to gene regulatory networks and the control of transcription [56, 57]. In genomic networks, one asks which genes are functionally co-regulated and whether the expression of one set of genes might depend on the expression of another set of genes. Answers can be found in microarray data from perturbation experiments, from the interactions between transcription factors and promoters, or from time series data, in which gene expression is measured multiple times following a stimulus (see, for example, [58, 59]). A sophisticated example is the analysis of genetic variations among individuals and their influence on the occurrence and severity of complex diseases such as obesity, diabetes, and cancer [60].

Bayesian inference methods are seldom applied to metabolic networks. The main reason is that the interactions in these networks are strictly dictated by biochemistry, so that it is usually more or less clear how these networks are connected. Because of this strong biochemical basis, metabolic networks allow a whole class of other analyses, which we will discuss next.

STATIC METABOLIC NETWORKS AND THEIR ANALYSIS

Metabolic networks will be described in greater detail in Chapter 8, but they are intuitive enough to serve here as an illustration. Metabolic networks are uniquely suited for analyses beyond clustering coefficients and degree distributions, because

The system of linear differential equations describing the reaction network can be written in matrix form as

$$\dot{\mathbf{S}} = \mathbf{NR}, \tag{3.17}$$

where the vector **R** contains the fluxes between pools. Thus, the dynamics of metabolites in the vector **S** is dissected into a matrix of stoichiometric coefficients, to which a vector of fluxes is multiplied.

In most stoichiometric investigations, the dynamical solutions of these differential equations, which would reveal all metabolites S_i as functions of time, are actually not of primary interest, and one focuses instead on the distribution of fluxes at steady state. In this situation, all the time derivatives in the vector on the left-hand side of (3.17) are zero, so that the ith differential equation becomes an algebraic linear equation of the type

$$0 = N_{i1}R_1 + N_{i2}R_2 + \ldots + N_{im}R_m \tag{3.18}$$

which has the same structure as (3.16) in the simple example above. Now the fluxes R_i are considered variables, which have to satisfy (3.18). More precisely, they must collectively satisfy one equation of this type for each metabolite pool.

In the ideal case, we would be able to measure many fluxes within a network, which would allow us to compute the remaining fluxes. In most realistic cases, however, only a few influxes and effluxes can be measured, and they are typically not sufficient to identify the entire network. But with the exclusive focus on the steady state, we have entered the realm of linear algebra, which offers methods for a great variety of tasks. For instance, it allows us to characterize the solution of systems of equations like (3.18), whether or not we have sufficiently many measured fluxes for a full identification.

As an example, Kayser and co-workers [69] analyzed the metabolic flux distribution in *Escherichia coli* under different glucose-limited growth conditions (**Figure 3.16**). It is easy to check that the incoming and outgoing fluxes balance exactly for each metabolite pool and for both steady-state growth conditions (left and right numbers in beige boxes). For instance, 100 units flow from glucose into glucose 6-phosphate (G6P), and these are forwarded in three directions: depending on the experimental conditions, 58 or 31 units flow toward fructose 6-phosphate (F6P), 41 or 67 units are used for the production of ribulose

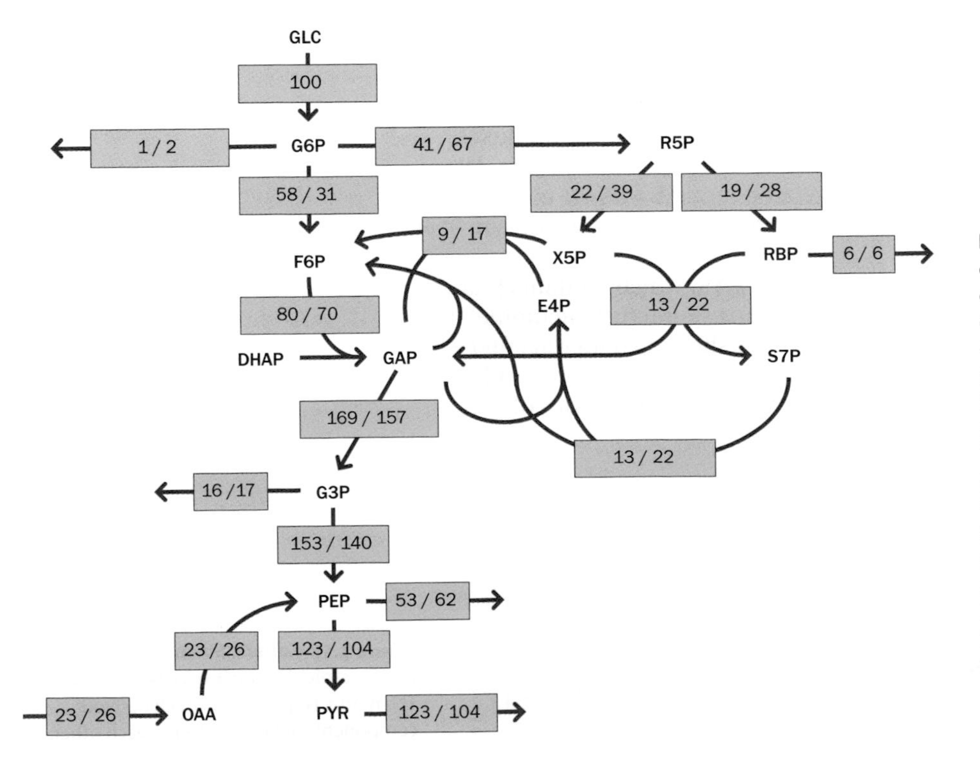

Figure 3.16 Flux distributions in cultures of *E. coli* growing under different conditions. The blue arrows represent enzyme-catalyzed reactions, while the boxes contain flux measurements for two different growth conditions. All magnitudes of fluxes are represented in relationship to the glucose consumption step, which is set as 100. Abbreviations: DHAP, dihydroxyacetone phosphate; E4P, erythrose 4-phosphate; F6P, fructose 6-phosphate; GAP, glyceraldehyde 3-phosphate; GLC, glucose; G6P, glucose 6-phosphate; G3P, 3-phosphoglycerate; OAA, oxaloacetate; PEP, phosphoenolpyruvate; PYR, pyruvate; R5P, ribulose 5-phosphate; RBP, ribose 5-phosphate; S7P, sedoheptulose 7-phosphate; X5P, xylulose 5-phosphate. (Data from Kayser A, Weber J, Hecht V & Rinas U. *Microbiology* 151 [2005] 693–706.)

5-phosphate (R5P), and 1 or 2 units move toward unspecified uses, respectively. Note the apparent doubling between F6P and glyceraldehyde 3-phosphate (GAP), which is due to the fact that F6P contains 6 carbons and GAP only 3. The system also produces carbon dioxide and uses other compounds, such as ATP, which are not explicitly listed.

The balances at each metabolite pool can be formulated as a set of simple (stoichiometric) equations. If not all fluxes can be measured, these equations can be used to infer them. For instance, the efflux of 16 or 17 units of 3-phosphoglycerate (G3P) in Figure 3.16 simply balances the known influx from GAP and the efflux toward phosphoenolpyruvate (PEP).

3.8 Variants of Stoichiometric Analysis

If only a few fluxes can be measured, the solution of the stoichiometric equations is unique only in rare cases. The reason is that essentially all metabolic networks contain more reactions than metabolite pools. As a mathematical consequence, the system of equations is underdetermined and permits many different solutions, which can all be characterized with methods of linear algebra. Indeed, all these solutions form a solution space of sometimes high dimension. To determine which of the many possible solutions is most likely to exist in nature, it has become customary to posit a suitable criterion of optimality. The most typical criterion is that a microorganism strives to optimize its growth rate. With such an objective, one uses optimization methods to identify the one solution that satisfies all stoichiometric steady-state equations (that is, of type (3.18)) and at the same time optimizes the growth rate. Different variants have been proposed to address underdetermined stoichiometric systems [70]. The most widely used is **flux balance analysis (FBA)** [64, 71]. Besides the balance between incoming and outgoing fluxes at each node and the use of an objective to be optimized, FBA accounts for biologically motivated **constraints** on some or all fluxes, which the system must satisfy. These constraints may be thermodynamic or of some other physicochemical nature, and the core method of analysis is **constrained optimization.** Other variants include elementary mode analysis [72], which will be discussed in Chapter 14, extreme pathway analysis [64, 73], and the minimization of metabolic adjustment [74, 75].

It is worth noting that the flux distribution within a static metabolic network depends strongly not only on the optimization objective, but also on the experimental conditions. For instance, the flux distributions in Figure 3.16 differ quite a bit for different substrate availabilities. As an example, algae of strain *Synechocystis* sp. PCC 6803 can either draw energy from substrates such as glucose (heterotrophic growth) or produce energy through photosynthesis (autotrophic growth). While the metabolic network is the same in both cases, the two modes of growth correspond to distinctly different flux distributions [76].

No matter which variation of stoichiometric analysis is applied, flux analysis does not tell us what the metabolite concentrations in the system are. The focus is entirely on fluxes, and if information on metabolites is of interest, it must come from additional data. In the most straightforward case, all concentrations can be measured at the steady state. In other cases, it is sometimes but not always possible to deduce the steady-state concentration if features of the catalyzing enzymes are known.

FBA can to some degree be extended to situations where the system is not in a steady state [77]. However, this type of extended analysis is more complicated.

3.9 Metabolic Network Reconstruction

In the cases above, we assumed that we knew the structure of the metabolic network. What needs to be done if we don't know it? The answer is that the scope of the problem depends on the availability and type of data. Traditionally, biochemists have been using wet-lab experiments, trying to identify reactions, regulators, and kinetic properties associated with a particular metabolite. These studies are very labor-intensive and slow, but have collectively produced an enormous body of knowledge, which was originally depicted in the Boehringer Map that adorned many hallways of

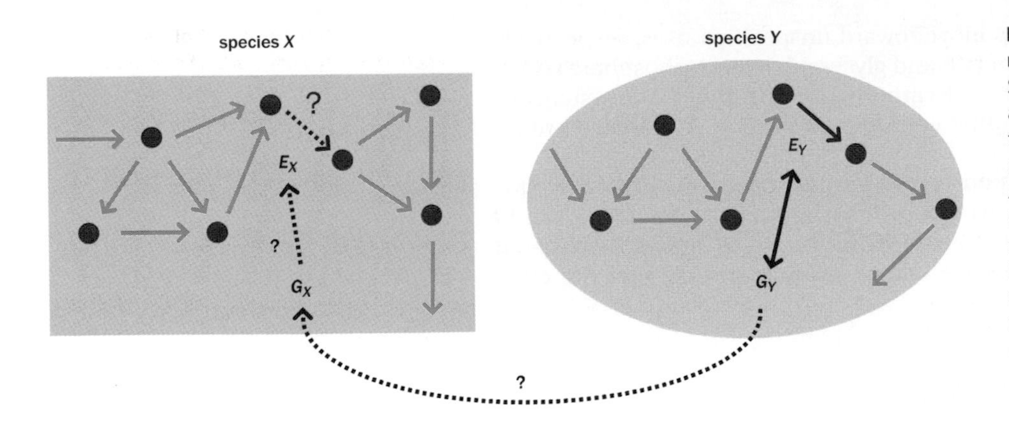

Figure 3.17 Filling gaps in a metabolic map through genome comparison. Suppose that metabolic and proteomic analysis has so far not revealed an enzyme E_X that would be able to connect two separate parts of the metabolic network of species *X*. Suppose that an enzyme E_Y catalyzing the missing reaction is known in the related species *Y*, and that its gene G_Y has been identified. Sequence comparisons may reveal a homologous gene G_X in species *X* and identify it as a candidate for coding for the unknown enzyme E_X.

biochemistry departments. This map is now available electronically [78], along with other databases such as KEGG [79] and MetaCyc [80].

In more recent times, high-throughput methods of molecular biology have become available for metabolic network characterization. Many of these focus on the actual existence of specific metabolites in the target network and use methods such as mass spectrometry and nuclear magnetic resonance (see, for example, Chapter 8 and [81]). Since genome data are usually easier to obtain than information on the connectivity of metabolic pathway systems, methods have also been developed to infer pathways from the presence of genes, homologies with other organisms, and the genomic organization into operons and other regulatory structures (see, for example, Chapter 6). For instance, consider the situation in which there is a gap in an incomplete metabolic network of species *X*, where some metabolite apparently cannot be produced with any of the enzymes known in *X* (**Figure 3.17**). Suppose that a related species *Y* possesses a gene G_Y that codes for an enzyme E_Y catalyzing the needed reaction and that could potentially bridge the gap in *X*. If so, one uses homology methods of **bioinformatics** in an attempt to determine whether *X*'s genome contains a gene G_X that has sufficient similarity with G_Y. If such a gene can be found, one infers that G_X might code for an enzyme E_X that is similar to E_Y and fills the gap [61]. Free software associated with BioCyc and MetaCyc, such as BioCyc's *PathoLogic* and KEGG's KAAS [82], serve this specific purpose [83–85]. An intriguing case study using this approach is [86].

A different situation occurs when the connectivity and regulation of a metabolic map is known, but specific kinetic parameters are not. Many methods for estimating such parameters have been developed in recent years (see, for example, [87]). None of these methods are foolproof, however, and all require quite a bit of computing experience. We will return to this question in Chapter 5.

3.10 Metabolic Control Analysis

An interesting transition from static to dynamic models (Chapter 4) is **metabolic control analysis (MCA)**, which characterizes the effects of small perturbations in a metabolic pathway that operates at a steady state [88–91]. MCA was originally conceived to replace the formerly widespread notion that every pathway has one rate-limiting step, which is a slow reaction that by itself is credited with determining the magnitude of flux through the pathway. The rate-limiting step could be compared to the neck of a funnel, which controls how much liquid can run through the funnel per second, independent of how much liquid has accumulated in the funnel. In linear pathway sections without branches, the rate-limiting step was traditionally thought to be positioned at the first reaction, where it was possibly inhibited through feedback exerted by the end product. In MCA, this concept of a rate-limiting step was supplanted with the concept of shared control, which posits that every step in a pathway contributes, to some degree, to the control of the steady-state flux. MCA formalizes this concept primarily with quantities called **control coefficients** and **elasticities**, and with mathematical relationships that permit certain insights into

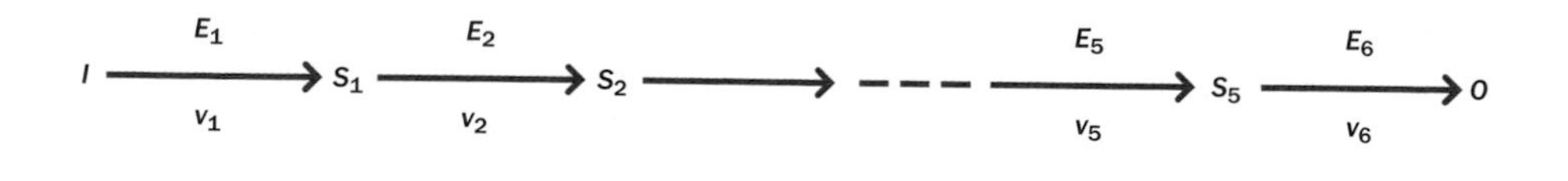

Figure 3.18 Generic metabolic pathway used to explain concepts of metabolic control analysis (MCA). The linear pathway consists of input I, output O, intermediate substrates $S_1, \ldots, S_5$, and reactions $v_1, \ldots, v_6$ that are catalyzed by enzymes $E_1, \ldots, E_6$.

the control structure of the pathway. While many variations of the original quantities in MCA have been defined over the years [88, 91], one elasticity coefficient and two primary control coefficients stand out: one with respect to the flux through the pathway and one with respect to concentrations.

It might be most intuitive to illustrate the basic concepts of MCA with the linear pathway in **Figure 3.18**. This generic pathway consists of an input I, output O, intermediate substrate concentrations $S_1, \ldots, S_n$, and reactions $v_1, \ldots, v_{n+1}$ that are catalyzed by enzymes $E_1, \ldots, E_{n+1}$; in the example, $n = 5$. The pathway is assumed to operate at a steady state, and all perturbations are required to be small; strictly speaking, they must be infinitesimally small. In the generic terminology of systems analysis, the control coefficients are sensitivities, as we will encounter in Chapter 4. They quantify the effects of small changes in parameters on features of the system as a whole. Outside of the minute perturbations in enzymes or other quantities, dynamical aspects are not addressed in the original MCA, although later studies have moved into this direction (see, for example, [92, 93]). As we discussed in Chapter 2, the steady state corresponds to a situation where $\dot{S}_i = 0$ for all substrates, as well as $\dot{I} = 0$ and $\dot{O} = 0$. It is easy to see that in this situation all reaction rates must have the same magnitude as the overall flux J, namely $v_1 = v_2 = \cdots = v_6 = J$; the reader should confirm this conclusion.

The control coefficients measure the relative change in a flux or substrate concentration at a steady state that results from a relative or percent change in a key parameter, such as an enzyme activity. In other words, one asks a question such as: "If one alters the activity of one of the enzymes, E_i, by a small amount such as 2%, how much will the flux J through the pathway change?" It is assumed that each v_i is directly proportional to the corresponding E_i, so that the control coefficients may be equivalently expressed in terms of v_i or E_i. This assumption holds in most cases, although one should be aware that there are exceptions [94]. Furthermore, it is mathematically convenient to replace the relative change with a change in the logarithm of the quantity in question, which is valid if the changes are small (the reader should confirm this). The flux control coefficient is thus defined as the derivative

$$C_{v_i}^{J} = \frac{\partial J}{J} \bigg/ \frac{\partial v_i}{v_i} = \frac{v_i}{J}\frac{\partial J}{\partial v_i} = \frac{\partial \ln J}{\partial \ln v_i}. \tag{3.19}$$

Similarly, the concentration control coefficient quantifies the effect of a change in enzyme E_i on the steady-state concentration of metabolite S_k and is defined as

$$C_{v_i}^{S_k} = \frac{v_i}{S_k}\frac{\partial S_k}{\partial v_i} = \frac{\partial \ln S_k}{\partial \ln v_i}. \tag{3.20}$$

In a slight variation to the usual sensitivities in systems analysis, the control coefficients are thus defined as relative sensitivities, which are expressed in terms of logarithms. This use of relative (or logarithmic) quantities is generally accepted as more suitable in metabolic studies, because it removes the dependence of the results on volumes and concentrations. For instance, according to (3.20), the change in S_k, written as ∂S_k, is divided by the actual amount of S_k, and it is assessed as a consequence of a change in v_i, written as ∂v_i, that is divided by the actual magnitude of v_i. Thus, relative changes are being compared.

The control exerted by a given enzyme either on a flux or on a metabolite concentration can be distinctly different, and it is indeed possible that a concentration is strongly affected but the pathway flux is not. Therefore, the distinction between the two types of control coefficients is important. The distinction is also pertinent for practical considerations, for instance in biotechnology, where gene and enzyme manipulations are often targeted either toward an increased flux or toward an increased concentration of a desirable compound [91] (see Chapter 14). Finally, one should note that several variations on this concept of control coefficients have been explored. For instance, response coefficients may be defined for the dependence of fluxes, concentrations, or other pathway features on other, external or internal, parameters [88, 91].

Each elasticity coefficient measures how a reaction rate v_i changes in response to a perturbation in a metabolite S_k or some other parameter. With respect to S_k, it is defined as

$$\varepsilon_{S_k}^{v_i} = \frac{S_k}{v_i}\frac{\partial v_i}{\partial S_k} = \frac{\partial \ln v_i}{\partial \ln S_k}. \tag{3.21}$$

Because only one metabolite (or parameter) and one reaction are involved in this definition, but not the entire pathway, each elasticity is a local property, which can in principle be measured *in vitro*. Closely related to the elasticity with respect to a variable is the elasticity with respect to the Michaelis constant K_k of an enzyme for the substrate S_k (see Chapters 2, 4, and 8). It is simply the complement of the metabolite elasticity [95]:

$$\varepsilon_{S_k}^{v_i} = -\varepsilon_{K_k}^{v_i}. \tag{3.22}$$

The main insights provided by MCA are gained from relationships among the control and elasticity coefficients. Returning to the example in Figure 3.18, we may be interested in the overall change in the flux J, which is mathematically determined by the sum of responses to possible changes in all six enzymes. It has been shown that all effects, in the form of flux control coefficients, sum to 1, and that all concentration control coefficients with respect to a given substrate S sum to 0:

$$\sum_{i=1}^{n+1} C_{v_i}^{J} = 1, \tag{3.23}$$

$$\sum_{i=1}^{n+1} C_{v_i}^{S} = 0 \tag{3.24}$$

[89, 90]. These summation relationships hold for arbitrarily large pathways under the most pertinent conditions (for exceptions, see [94, 96]).

The summation relationship with respect to flux control, (3.23), has two interesting implications. First, one sees directly that the control of metabolic flux is shared by all reactions in the system; this global aspect identifies the control coefficients as systemic properties. And second, if a single reaction is altered and its contribution to the control of the flux changes, the effect is compensated by changes in flux control by the remaining reactions. The summation relationship with respect to flux control has often been used experimentally in the following fashion: one measures in the laboratory the effects of changes in various reactions of a pathway, and if the sum of flux control coefficients is below 1, then one knows that one or more contributions to the control structure are missing.

A second type of insight comes from connectivity relationships, which establish constraints between control coefficients and elasticities. These relationships have been used to characterize the close connection between the kinetic features of individual reactions and the overall responses of a pathway to perturbations. The most important of these relationships is

$$\sum_{i=1}^{n} C_{v_i}^{J} \varepsilon_{S_k}^{v_i} = 0. \tag{3.25}$$

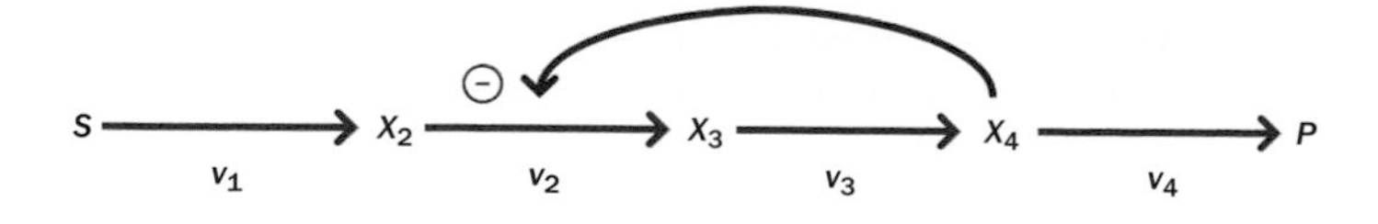

Figure 3.19 Simple linear pathway with feedback. If the goal is to increase the flux through the pathway, MCA provides a tool for identifying the best manipulation strategies.

A comprehensive example of the use of these relationships is the analysis of mitochondrial oxidative phosphorylation, which clearly demonstrated that the flux control is distributed among a number of enzymes, thereby refuting the formerly expected existence of a rate-limiting step (for a review, see [91]). A much simpler yet instructive example is the linear pathway with feedback in **Figure 3.19**, which has been adapted from [95]. The following elasticities are assumed to be known, maybe from *in vitro* experiments:

$$\varepsilon_{X_2}^{v_1} = -0.9,\ \varepsilon_{X_2}^{v_2} = 0.5,\ \varepsilon_{X_3}^{v_2} = -0.2,\ \varepsilon_{X_3}^{v_3} = 0.7, \varepsilon_{X_4}^{v_2} = -1, \text{and } \varepsilon_{X_4}^{v_4} = 0.9;$$

all other elasticities are equal to 0. For this unbranched pathway, the information on elasticities, together with the connectivity theorem in (3.25), is actually sufficient to compute the flux control coefficients. For instance, the connectivity relationship for X_2 is

$$C_{v_1}^J \varepsilon_{X_2}^{v_1} + C_{v_2}^J \varepsilon_{X_2}^{v_2} = 0.19 \times (-0.9) + 0.34 \times 0.5 = 0.$$

Solving the set of these equations yields

$$C_{v_1}^J = 0.19,\ C_{v_2}^J = 0.34,\ C_{v_3}^J = 0.10, \text{and } C_{v_4}^J = 0.38.$$

Suppose there is reason to believe that the reaction step v_2 is a bottleneck and that a goal of the analysis is to propose strategies for increasing the flux through the pathway. A reasonable strategy might be to increase the amount of the enzyme in v_2, for instance, using modern tools of genetic engineering. Or would it be better to modify the enzyme in such a manner that it is less affected by the inhibition exerted by X_4? Suppose we could experimentally alter the effect of X_4 by changing the binding constant K_4^2 of the enzyme in v_2 by p%. For relatively small values of p, this change corresponds to a change in ln K_4^2, as we discussed before. Owing to the inhibition signal, K_4^2 has an effect on v_2, which is quantified by the elasticity in (3.21) and (3.22). Furthermore, the effect on changes in v_2 on the pathway flux J is given by the flux control coefficient in (3.19). Multiplication yields

$$C_{v_2}^J \varepsilon_{K_4^2}^{v_i} = -\frac{\partial \ln J}{\partial \ln v_2} \frac{\partial \ln v_2}{\partial \ln K_4^2} = -\frac{\partial \ln J}{\partial \ln K_4^2}, \tag{3.26}$$

rearrangement of which leads to

$$\partial \ln J = -C_{v_2}^J \varepsilon_{K_4^2}^{v_i} \partial \ln K_4^2 \approx -C_{v_2}^J \varepsilon_{K_4^2}^{v_i} p\%. \tag{3.27}$$

Substituting numerical values yields a relative change in flux J of $0.34 \times (-1)$ per percent change in the binding constant K_4^2 of enzyme v_2. The predicted change has a negative value; thus, in order to increase the pathway flux, K_4^2 must be decreased.

We could also explore a strategy of changing the activity of the enzyme in v_4, arguing that a higher activity would convert more X_4, thereby reduce the inhibition of v_2, and possibly lead to a higher overall flux. Equation (3.19) quantifies this effect:

$$\partial \ln J = C_{v_4}^J \partial \ln v_4 \approx C_{v_4}^J q\% = 0.38. \tag{3.28}$$

Indeed, the effect is about 10% stronger than in the previous strategy of decreasing the binding constant K_4^2. In this fashion, changes in kinetic constants or enzymes can be computed in a comprehensive manner [95].

For this small pathway, the computations are quite simple. For larger systems, matrix methods have been developed that permit a more structured assessment of the control within pathways [97, 98]. It has also been shown that the fundamental features of MCA are directly reflected and generalized in the steady-state properties of fully dynamic models in biochemical systems theory (BST) [99, 100] and in the lin–log formalism [92, 93]. Thus, MCA can be seen as a direct transition from static networks to dynamic systems analyses.

EXERCISES

3.1. Try to figure out Euler's problem of the Seven Bridges of Königsberg (see Figure 3.2): Is there a path that uses every bridge exactly once? Does the solution change if any one of the bridges is removed? If so, does it matter which bridge is removed?

3.2. Are there connected paths in the graphs in **Figure 3.20** that use every edge exactly once? Does it matter where one starts?

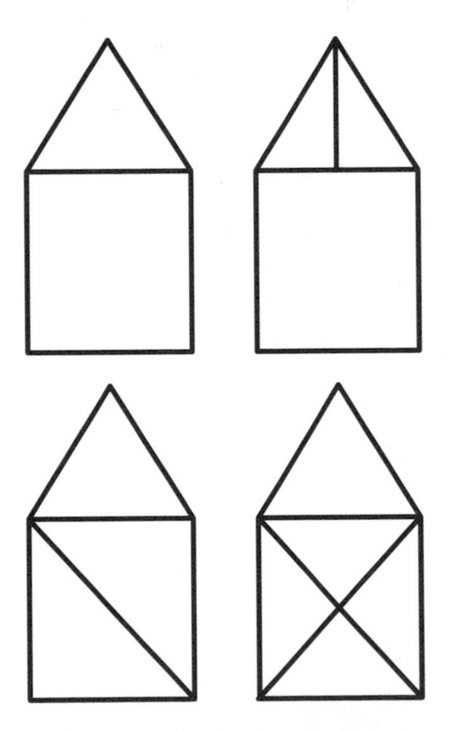

Figure 3.20 Simple graphs with different connectivities. Is it possible to find paths that use every edge exactly once?

3.3. Construct 10 undirected random graphs with m nodes and a total of mk edges, using different combinations of m and k. As a simple start, study a graph with six nodes and determine random edges with two dice, which identify the nodes to be connected. Record the degree distributions. As a more efficient alternative, write a computer algorithm and compute the degree distribution for each graph.

3.4. Construct 10 directed graphs with five nodes that in each case form a single loop, such as $N_1 \to N_5 \to N_3 \to N_4 \to N_2 \to N_1$, and contain no other edges. How many different graphs of this type are there? Study similarities in their adjacency matrices and summarize them in a brief report.

3.5. How are hubs reflected in the adjacency matrix of an undirected network?

3.6. How are cliques reflected in the adjacency matrix of an undirected network?

3.7. What happens to the adjacency matrix of a graph if this graph consists of two subgraphs that are not connected to each other? Can you make a general statement? Does this statement cover both directed and undirected graphs?

3.8. Using the method in Box 3.1, compute all three-step paths of the matrix on the left of Figure 3.4.

3.9. Confirm that in an undirected graph the number of edges between all neighbors of a node is at most $k(k-1)/2$.

3.10. Compute the clustering coefficients of the red nodes in the undirected graph of **Figure 3.21**.

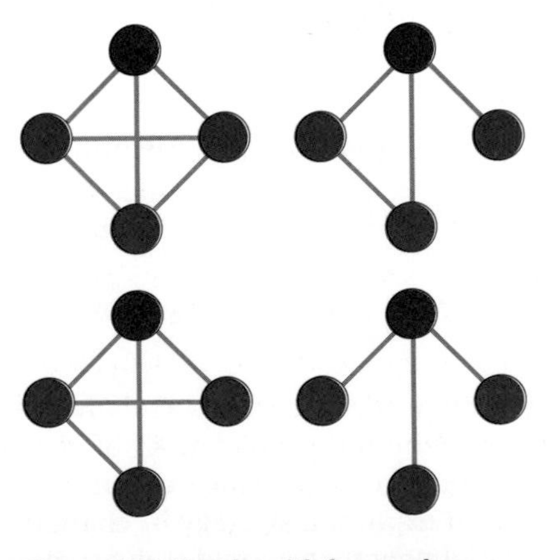

Figure 3.21 Graphs with four nodes. The connectivity patterns determine their clustering coefficients.

3.11. What are the minimum and maximum values of C_N in a network without multiple edges between two nodes?

3.12. What is the average clustering coefficient of a clique?

3.13. Provide rationale for the two different definitions of C_N for directed and undirected graphs.

3.14. Construct five random graphs with 6 nodes and 12 directed edges. Write a computer program or use two dice to determine source and end nodes; do not allow double edges. Compute for each case all shortest path lengths and the average shortest distance.

3.15. Search the Internet for actual biological networks with power-law degree distributions. Study γ and discuss what happens to the power-law relationship for very high and very low k.

3.16. Search the Internet for two studies that discuss party hubs and date hubs. Write a short report about the goals, methods, and results of each study.

3.17. Box 3.2 discusses the construction of different connectivity pattern from an initial ring-like structure. Execute this construction yourself with different probabilities p $(0 \le p \le 1)$. Specifically, study a ring with 12 nodes, which is connected as shown in Box 3.2, and perform the analysis with $p = 1/6, 2/6, \ldots,$ $5/6$. Start with the first node and go around the ring in one direction. For each node, roll a die to determine whether one of its edges should be replaced or not. For instance, for $p = 2/6$, replace an edge only if you roll a 1 or 2. If an edge is to be replaced, use a coin to determine whether to remove the node's connection to its immediate right or left neighbor. Use a coin and a die to determine the node to which a new edge is to be connected (head: nodes between 1 and 6; tail: nodes between 7 and 12). In cases where you identify the source node itself or where the edge already exists, roll again. After one lap, repeat the procedure, this time using second-nearest neighbors. Once you have finished the second lap, report the degree distribution, the network's clustering coefficient, and the network's diameter. As a more efficient alternative, write an algorithm that implements the above procedure with m nodes and a probability p $(0 \le p \le 1)$.

3.18. Discuss the tolerance of random and small-world networks to random attacks and to targeted attacks, where one node is eliminated in each case.

3.19. Load the galactose utilization network in Cytoscape. Explore different modes of visualization.

3.20. Load a different sample network in Cytoscape and determine its numerical features, such as the number of nodes, and clustering information.

3.21. Load the galactose utilization network in Cytoscape. Select two hubs and find out whether they and their neighbors are functionally related.

3.22. Compute the probability that the cascade in Figure 3.13 is active even though R and S are not active. Interpret $P(C \text{ spont})$ as the probability that C is on, although neither R nor S is active. Compute the probabilities that R or S is not active as $1 - P(R)$ and $1 - P(S)$, respectively.

3.23. Compute the probability that there is a genomic response in the system in Figure 3.13 even though R and S are not active. Interpret $P(G \text{ spont})$ as the probability that G is on, although C is not active.

3.24. Search the Internet for an investigation of a protein–protein interaction network. Write a short report about the goals, methods, and results of the study. Pay particular attention to the visualization of the network.

3.25. Search the Internet for an investigation of a gene interaction network. Write a short report about the goals, methods, and results of the study.

3.26. Search the Internet for an example of a stoichiometric or flux balance analysis. Write a short report about the goals, methods, and results of the study.

3.27. If you are familiar with linear algebra and the features of matrices, discuss under what conditions (3.17) can be solved uniquely or not at all.

3.28. Set up a stoichiometric model for the pentose pathway shown in **Figure 3.22**. This pathway exchanges carbohydrates with different numbers of carbon atoms, as shown in **Table 3.1**. Thus, two units of X_1 are needed to produce one unit of X_2 and X_3, and each reaction $V_{5,3}$ uses one unit of X_5 to generate two units of X_3. Note that X_3 appears in two locations, but represents the same pool.

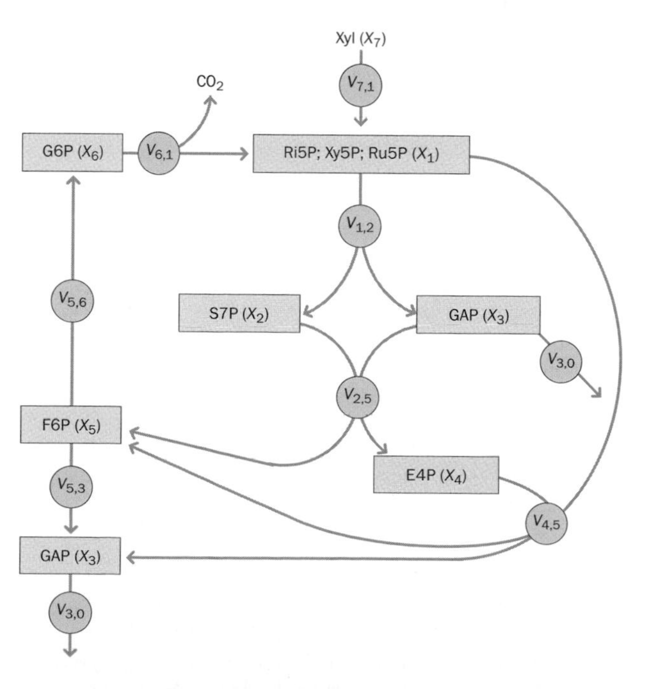

Figure 3.22 Diagram of the pentose phosphate pathway. Given influxes and effluxes, internal fluxes can be computed. Table 3.1 shows the number of carbon atoms in each metabolite. (From Wiechert W & de Graaf AA. *Biotechnol. Bioeng.* 55 [1997] 101–117. With permission from John Wiley & Sons.)

TABLE 3.1: NUMBER OF CARBON ATOMS IN EACH METABOLITE IN THE PATHWAY IN FIGURE 3.22

Compound Pool	Number of Carbon Atoms
X_1	5
X_2	7
X_3	3
X_4	4
X_5	6
X_6	6
X_7	5

3.29. For the same pathway as shown in Figure 3.22, assume that the value for the input flux $V_{7,1}$ is 20 units and try different values for $V_{3,0}$. Compute the corresponding flux distributions. How much CO_2 is produced?

3.30. If the influx in the pathway systems in Figures 3.16 or 3.22 were doubled, would doubling all other fluxes in the system lead to a valid solution? Argue from a mathematical and a biological point of view.

3.31. If the rate of a reaction is given as $v_2(S_1) = kS_1^p$, where k and p are constant parameters, what are the concentration control coefficient $C_{v_2}^{S_1}$ and the elasticity $\varepsilon_{S_1}^{v_2}$?

3.32. If the rate v of a reaction is given as a Michaelis–Menten function of the type

$$v(S) = \frac{VS}{K+S},$$

where V and K are parameters, what is the corresponding elasticity? Does the elasticity depend directly on the substrate concentration?

REFERENCES

[1] Cytoscape. http://www.cytoscape.org/.
[2] Shannon P, Markiel A, Ozier O, et al. Cytoscape: a software environment for integrated models of biomolecular interaction networks. *Genome Res.* 13 (2003) 2498–2504.
[3] Ingenuity. http://www.ingenuity.com.
[4] BiologicalNetworks. http://biologicalnetworks.net.
[5] Pajek. http://pajek.imfm.si/doku.php?id=pajek.
[6] Purvis JE, Radhakrishnan R & Diamond SL. Steady-state kinetic modeling constrains cellular resting states and dynamic behavior. *PLoS Comput. Biol.* 5 (2009) e1000298.
[7] Lee Y & Voit EO. Mathematical modeling of monolignol biosynthesis in *Populus* xylem. *Math. Biosci.* 228 (2010) 78–89.
[8] Wilkinson DJ. Stochastic Modelling for Systems Biology, 2nd ed. Chapman & Hall/CRC Press, 2011.
[9] Keener J & Sneyd J. Mathematical Physiology. II: Systems Physiology, 2nd ed. Springer, 2009.
[10] Alon U. An Introduction to Systems Biology: Design Principles of Biological Circuits. Chapman & Hall/CRC, 2006.
[11] Voit EO. Computational Analysis of Biochemical Systems: A Practical Guide for Biochemists and Molecular Biologists. Cambridge University Press, 2000.
[12] Chaouiya C. Petri net modelling of biological networks. *Brief Bioinform.* 8 (2007) 210–219.
[13] Hardy S & Robillard PN. Modeling and simulation of molecular biology systems using Petri nets: modeling goals of various approaches. *J. Bioinform. Comput. Biol.* 2 (2004) 619–637.
[14] Watts DJ & Strogatz SH. Collective dynamics of "small-world" networks. *Nature* 393 (1998) 440–442.
[15] Ravasz E, Somera AL, Mongru DA, et al. Hierarchical organization of modularity in metabolic networks. *Science* 297 (2002) 1551–1555.
[16] Jeong JP, Mason SP, Barabási A-L & Oltvai ZN. Lethality and centrality in protein networks. *Nature* 411 (2001) 41–42.
[17] Maslov S & Sneppen K. Specificity and stability in topology of protein networks. *Science* 296 (2002) 910–913.
[18] Erdős P & Rényi A. On random graphs. *Publ. Math.* 6 (1959) 290–297.
[19] Barabási A-L & Oltvai ZN. Network biology: understanding the cell's functional organization. *Nat. Rev. Genet.* 5 (2004) 101–113.
[20] Wagner A & Fell DA. The small world inside large metabolic networks. *Proc. R. Soc. Lond. B* 268 (2001) 1803–1810.
[21] Chung F & Lu L. The average distance in random graphs with given expected degrees. *Proc. Natl Acad. Sci. USA* 99 (2002) 15879–15882.
[22] Jeong H, Tombor B, Albert R, et al. The large-scale organization of metabolic networks. *Nature* 407 (2000) 651–654.
[23] Jordán F & Molnár I. Reliable flows and preferred patterns in food webs. *Evol. Ecol. Res.* 1 (1999) 591–609.
[24] Lima-Mendez G & van Helden J. The powerful law of the power law and other myths in network biology. *Mol. Biosyst.* 5 (2009) 1482–1493.
[25] Nacher JC & Akutsu T. Recent progress on the analysis of power-law features in complex cellular networks. *Cell Biochem. Biophys.* 49 (2007) 37–47.
[26] Barabási A-L. Linked: The New Science of Networks. Perseus, 2002.
[27] Albert R & Barabási AL. The statistical mechanics of complex networks. *Rev. Mod. Phys.* 74 (2002) 47–97.
[28] Vázquez A, Flammini A, Maritan A & Vespignani A. Modeling of protein interaction networks. *ComPlexUs* 1 (2003) 38–44.
[29] Newman ME & Watts DJ. Scaling and percolation in the small-world network model. *Phys. Rev. E* 60 (1999) 7332–7342.
[30] Fraser HB, Hirsh AE, Steinmetz LM, et al. Evolutionary rate in the protein interaction network. *Science* 296 (2002) 750–752.
[31] Carlson JM & Doyle J. Complexity and robustness. *Proc. Natl Acad. Sci. USA* 99 (2002) 2538–2545.
[32] Csete ME & Doyle JC. Reverse engineering of biological complexity. *Science* 295 (2002) 1664–1669.
[33] Doyle JC, Alderson DL, Li L, et al. The "robust yet fragile" nature of the Internet. *Proc. Natl Acad. Sci. USA* 102 (2005) 14497–14502.
[34] Cohen JE. Graph theoretic models of food webs. *Rocky Mountain J. Math.* 9 (1979) 29–30.
[35] Pimm SL. Food Webs. University of Chicago Press, 2002.
[36] de Lichtenberg U, Jensen LJ, Brunak S & Bork P. Dynamic complex formation during the yeast cell cycle. *Science* 307 (2005) 724–727.

[37] Han JD, Bertin N, Hao T, et al. Evidence for dynamically organized modularity in the yeast protein–protein interaction network. *Nature* 430 (2004) 88–93.
[38] Agarwal S, Deane CM, Porter MA & Jones NS. Revisiting date and party hubs: novel approaches to role assignment in protein interaction networks. *PLoS Comput. Biol.* 6 (2010) e1000817.
[39] Ideker T, Thorsson V, Ranish JA, et al. Integrated genomic and proteomic analyses of a systematically perturbed metabolic network. *Science* 292 (2001) 929–934.
[40] Xia T, Van Hemert J & Dickerson JA. CytoModeler: a tool for bridging large-scale network analysis and dynamic quantitative modeling. *Bioinformatics* 27 (2011) 1578–1580.
[41] Olobatuyi ME. A User's Guide to Path Analysis. University Press of America, 2006.
[42] Shipley B. Cause and Correlation in Biology: A User's Guide to Path Analysis, Structural Equations and Causal Inference. Cambridge University Press, 2000.
[43] Torralba AS, Yu K, Shen P, et al. Experimental test of a method for determining causal connectivities of species in reactions. *Proc. Natl Acad. Sci. USA* 100 (2003) 1494–1498.
[44] Vance W, Arkin AP & Ross J. Determination of causal connectivities of species in reaction networks. *Proc. Natl Acad. Sci. USA* 99 (2002) 5816–5821.
[45] Anastassiou D. Computational analysis of the synergy among multiple interacting genes. *Mol. Syst. Biol.* 3 (2007) 83.
[46] Shannon CE & Weaver W. The Mathematical Theory of Communication. University of Illinois Press, 1949.
[47] Butte AJ & Kohane IS. Mutual information relevance networks: functional genomic clustering using pairwise entropy measurements. *Pac. Symp. Biocomput.* 5 (2000) 415–426.
[48] Komili S & Silver PA. Coupling and coordination in gene expression processes: a systems biology view. *Nat. Rev. Genet.* 9 (2008) 38–48.
[49] Bayes T. An essay towards solving a problem in the doctrine of chances. *Philos. Trans. R. Soc. Lond.* 53 (1763) 370–418.
[50] Pearl J. Causality, Models, Reasoning, and Inference. Cambridge University Press, 2000.
[51] Mitra S, Datta S, Perkins T & Michailidis G. Introduction to Machine Learning and Bioinformatics. Chapman & Hall/CRC Press, 2008.
[52] Sachs K, Perez O, Pe'er S, et al. Causal protein-signaling networks derived from multiparameter single-cell data. *Science* 308 (2005) 523–529.
[53] Bork P, Jensen LJ, Mering C von, et al. Protein interaction networks from yeast to human. *Curr. Opin. Struct. Biol.* 14 (2004) 292–299.
[54] Zhang A. Protein Interaction Networks: Computational Analysis. Cambridge University Press, 2009.
[55] Rhodes DR, Tomlins SA, Varambally S, et al. Probabilistic model of the human protein–protein interaction network. *Nat. Biotechnol.* 23 (2005) 951–959.
[56] Libby E, Perkins TJ & Swain PS. Noisy information processing through transcriptional regulation. *Proc. Natl Acad. Sci. USA* 104 (2007) 7151–7156.
[57] Zhu J, Wiener MC, Zhang C, et al. Increasing the power to detect causal associations by combining genotypic and expression data in segregating populations. *PLoS Comput. Biol.* 3 (2007) e69.
[58] Hecker M, Lambeck S, Toepfer S, et al. Gene regulatory network inference: data integration in dynamic models—a review. *Biosystems* 96 (2009) 86–103.
[59] Gasch AP, Spellman PT, Kao CM, et al. Genomic expression programs in the response of yeast cells to environmental changes. *Mol. Biol. Cell* 11 (2000) 4241–4257.
[60] Sieberts SK & Schadt EE. Moving toward a system genetics view of disease. *Mamm. Genome* 18 (2007) 389–401.
[61] Reed JL, Vo TD, Schilling CH & Palsson BØ. An expanded genome-scale model of *Escherichia coli* K-12 (iJR904 GSM/GPR). *Genome Biol.* 4 (2003) R54.
[62] Wagner A. Evolutionary constraints permeate large metabolic networks. *BMC Evol. Biol.* 9 (2009) 231.
[63] Schryer DW, Vendelin M & Peterson P. Symbolic flux analysis for genome-scale metabolic networks. *BMC Syst. Biol.* 5 (2011) 81.
[64] Palsson BØ. Systems Biology: Properties of Reconstructed Networks. Cambridge University Press, 2006.
[65] Stephanopoulos GN, Aristidou AA & Nielsen J. Metabolic Engineering. Principles and Methodologies. Academic Press, 1998.
[66] Torres NV & Voit EO. Pathway Analysis and Optimization in Metabolic Engineering. Cambridge University Press, 2002.
[67] Clarke BL. Complete set of steady states for the general stoichiometric dynamical system. *J. Chem. Phys.* 75 (1981) 4970–4979.
[68] Gavalas GR. Nonlinear Differential Equations of Chemically Reacting Systems. Springer, 1968.
[69] Kayser A, Weber J, Hecht V & Rinas U. Metabolic flux analysis of *Escherichia coli* in glucose-limited continuous culture. I. Growth-rate-dependent metabolic efficiency at steady state. *Microbiology* 151 (2005) 693–706.
[70] Trinh CT, Wlaschin A & Srienc F. Elementary mode analysis: a useful metabolic pathway analysis tool for characterizing cellular metabolism. *Appl. Microbiol. Biotechnol.* 81 (2009) 813–826.
[71] Varma A, Boesch BW & Palsson BØ. Metabolic flux balancing: basic concepts, scientific and practical use. *Nat. Biotechnol.* 12 (1994) 994–998.
[72] Schuster S & Hilgetag S. On elementary flux modes in biochemical reaction systems at steady state. *J. Biol. Syst.* 2 (1994) 165–182.
[73] Schilling CH, Letscher D & Palsson BØ. Theory for the systemic definition of metabolic pathways and their use in interpreting metabolic function from a pathway-oriented perspective. *J. Theor. Biol.* 203 (2000) 229–248.
[74] Lee Y, Chen F, Gallego-Giraldo L, et al. Integrative analysis of transgenic alfalfa (*Medicago sativa* L.) suggests new metabolic control mechanisms for monolignol biosynthesis. *PLoS Comput. Biol.* 7 (2011) e1002047.
[75] Segrè D, Vitkup D & Church GM. Analysis of optimality in natural and perturbed metabolic networks. *Proc. Natl Acad. Sci. USA* 99 (2002) 15112–15117.
[76] Navarro E, Montagud A, Fernández de Córdoba P & Urchueguía JF. Metabolic flux analysis of the hydrogen production potential in *Synechocystis* sp. PCC6803. *Int. J. Hydrogen Energy* 34 (2009) 8828–8838.
[77] Goel G, Chou I-C & Voit EO. System estimation from metabolic time series data. *Bioinformatics* 24 (2008) 2505–2511.
[78] Expasy. http://web.expasy.org/pathways/.
[79] KEGG: The Kyoto Encyclopedia of Genes and Genomes. http://www.genome.jp/kegg/pathway.html.
[80] MetaCyc. http://biocyc.org/metacyc/index.shtml.
[81] Harrigan GG & Goodacre R (eds). Metabolic Profiling: Its Role in Biomarker Discovery and Gene Function Analysis. Kluwer, 2003.
[82] KAAS. http://www.genome.jp/tools/kaas/.
[83] Pathologic. http://biocyc.org/intro.shtml#pathologic.
[84] Karp PD, Paley SM, Krummenacker M, et al. Pathway Tools version 13.0: integrated software for pathway/genome informatics and systems biology. *Brief Bioinform.* 11 (2010) 40–79.
[85] Paley SM & Karp PD. Evaluation of computational metabolic-pathway predictions for *Helicobacter pylori*. *Bioinformatics* 18 (2002) 715–724.

[86] Yus E, Maier T, Michalodimitrakis K, et al. Impact of genome reduction on bacterial metabolism and its regulation. *Science* 326 (2009) 1263–1268.
[87] Chou I-C & Voit EO. Recent developments in parameter estimation and structure identification of biochemical and genomic systems. *Math. Biosci.* 219 (2009) 57–83.
[88] Fell DA. Understanding the Control of Metabolism. Portland Press, 1997.
[89] Heinrich R & Rapoport TA. A linear steady-state treatment of enzymatic chains. Critique of the crossover theorem and a general procedure to identify interaction sites with an effector. *Eur. J. Biochem.* 42 (1974) 97–105.
[90] Kacser H & Burns JA. The control of flux. *Symp. Soc. Exp. Biol.* 27 (1973) 65–104.
[91] Moreno-Sánchez R, Saavedra E, Rodríguez-Enríquez S & Olín-Sandoval V. Metabolic control analysis: a tool for designing strategies to manipulate metabolic pathways. *J. Biomed. Biotechnol.* 2008 (2008) 597913.
[92] Hatzimanikatis V & Bailey JE. MCA has more to say. *J. Theor. Biol.* 182 (1996) 233–242.
[93] Heijnen JJ. New experimental and theoretical tools for metabolic engineerlng of microorganisms. *Meded. Rijksuniv. Gent Fak. Landbouwkd. Toegep. Biol. Wet.* 66 (2001) 11–30.
[94] Savageau MA. Dominance according to metabolic control analysis: major achievement or house of cards? *J. Theor. Biol.* 154 (1992) 131–136.
[95] Kell DB & Westerhoff HV. Metabolic control theory: its role in microbiology and biotechnology. *FEBS Microbiol. Rev.* 39 (1986) 305–320.
[96] Savageau MA. Biochemical systems theory: operational differences among variant representations and their significance. *J. Theor. Biol.* 151 (1991) 509–530.
[97] Reder C. Metabolic control theory: a structural approach. *J. Theor. Biol.* 135 (1988) 175–201.
[98] Visser D & Heijnen JJ. The mathematics of metabolic control analysis revisited. *Metab. Eng.* 4 (2002) 114–123.
[99] Savageau MA, Voit EO & Irvine DH. Biochemical systems theory and metabolic control theory. I. Fundamental similarities and differences. *Math. Biosci.* 86 (1987) 127–145.
[100] Savageau MA, Voit EO & Irvine DH. Biochemical systems theory and metabolic control theory. II. The role of summation and connectivity relationships. *Math. Biosci.* 86 (1987) 147–169.

FURTHER READING

Alon U. An Introduction to Systems Biology: Design Principles of Biological Circuits. Chapman & Hall/CRC, 2006.
Barabási A-L. Linked: The New Science of Networks. Perseus, 2002.
Davidson EH. The Regulatory Genome: Gene Regulatory Networks In Development and Evolution. Academic Press, 2006.
Fell DA. Understanding the Control of Metabolism. Portland Press, 1997.
Palsson BØ. Systems Biology: Properties of Reconstructed Networks. Cambridge University Press, 2006.
Stephanopoulos GN, Aristidou AA & Nielsen J. Metabolic Engineering. Principles and Methodologies. Academic Press, 1998.
Zhang A. Protein Interaction Networks: Computational Analysis. Cambridge University Press, 2009.

The Mathematics of Biological Systems

When you have read this chapter, you should be able to:

- Understand basic modeling approaches
- Discuss the need for approximations
- Explain concepts and features of approximations
- Linearize nonlinear systems
- Approximate nonlinear systems with power-law functions
- Describe the differences and relationships between linear and nonlinear models
- Describe different types of attractors and repellers
- Design and analyze basic models with one or more variables in the following formats:
  - Discrete linear
  - Continuous linear
  - Discrete nonlinear
  - Continuous nonlinear
- Execute the steps of a typical biological systems analysis

Chapter 2 introduced some general concepts for modeling biological systems in an intuitive fashion, while skipping a lot of technical detail. Chapter 3 continued the discussion by addressing static networks, which do not change over time. This chapter presents mathematical and computational methods for analyzing dynamic systems. The range of such methods is almost unlimited, and not even experts master them all in detail. Nonetheless, there are various standard methods that are sufficient for setting up and analyzing basic models in a variety of formats, even if they may not have all the "bells and whistles" of the most modern mathematical techniques and software packages. If you are satisfied with understanding, albeit from a bird's eye view, what mathematical modeling is all about, and do not necessarily intend to engage in its minutiae, you may just browse through some sections or skip the technical details in this chapter, as long as you have a good grasp of the material in Chapters 2 and 3. However, if you love to "look under the hood," design your own models, and enjoy the thrills of exploring biology on your computer, this chapter is for you. It will put you on the right track, and, even if it is not comprehensive, it will provide you with a solid platform from which to dive into the rich literature of modeling, which spans a vast range from relatively easy introductions to very specialized texts and a growing number of journal articles. All analyses shown in this chapter can be executed in Mathematica® and MATLAB®, and many can be done with the easy-to-learn software PLAS [1] or even in Excel®.

As discussed in the introduction to modeling in Chapter 2, there are numerous types and implementations of mathematical and computational models. Because of their prominence in systems biology, this chapter is heavily biased toward dynamic, deterministic systems models and discusses only a few stochastic models that depend on random processes. The chapter requires, here and there, basic working knowledge of ordinary and partial **differentiation**, **vectors**, and **matrices**, but most material can be understood without deep knowledge of these topics. The main goal of the chapter is to formalize concepts of model design, diagnostics, and analysis in more depth than in Chapter 2 and to provide the skills to set up and analyze models from scratch. In contrast to a typical math book, which uses the famous lemma–theorem–proof–corollary–example style, this chapter introduces concepts and definitions on the fly, uses examples as motivation, and stops before things get too complicated. It is subdivided into discussions of **discrete** and **continuous** models, as well as of **linear** and **nonlinear** models.

Linear systems models are of utmost importance for two reasons. First, they permit more mathematical analyses than any other format, and, second, it is possible to analyze many features of **nonlinear systems** with methods of linear analysis.

A very interesting feature germane to linear systems is the principle of **superposition**. This principle says the following. If we analyze a system by giving it an input I_1, to which it responds with output O_1, and if we do a second experiment with input I_2 and obtain output O_2, then we know that the output response to a combined input $I_1 + I_2$ will be $O_1 + O_2$. Maybe this superposition sounds obvious, trivial, or unassuming, but it has enormous consequences. Only if this principle holds is it possible to analyze the behavior of a system by studying each component separately and then validly summing the individual results. As a banal example, we could establish in one experiment a relationship between the pressure in the tires of a car and its performance, and in a separate experiment a relationship between performance and different grades of gas. Under the assumption that tire pressure and gas quality independently affect performance, the results are additive. The reason why we have not chosen a biological example here is that most biological systems are indeed not linear, thereby violating the principle of superposition. For instance, the growth of a plant depends on genetics as well as environmental conditions, including nutrients. However, the two are not independent of each other, and separate experiments with different strains and different experimental conditions may not allow us to make valid predictions of how a particular strain will grow under untested conditions.

If we are able to forge a system representation into a linear format, an incredible repertoire of mathematical and computational tools becomes available. In particular, the entire field of linear algebra addresses these systems and has developed concepts such as vectors and matrices that have become absolutely fundamental throughout mathematics and computer science. For instance, if the right-hand sides of a system of differential equations are linear, the computation of the steady state of this system becomes a matter of solving a system of linear algebraic equations, for which there are many methods.

Linear representations can come in discrete and continuous varieties. Linear (matrix) models in biology can effectively describe discrete, deterministic aging processes, as well as some stochastic processes (such as the Markov chain models discussed later in this chapter) that capture the probabilities with which a system develops over time. Discrete linear models are also at the heart of much of traditional statistics, such as regression, analysis of variance, and other techniques, which are efficiently formulated using matrix algebra. These types of statistical models are very important, but are outside our scope here.

Continuous linear representations are crucially important for biology as well, even though most systems in biology are nonlinear. The reason is that many analyses depend on the fact that nonlinear systems can be linearized, which essentially means that one may flatten out the nonlinear system close to a point of interest. The justification for this strategy comes from an old theorem attributed to Hartman and Grobman, which says that, under most pertinent conditions, important clues for the behavior of a nonlinear system can be deduced from the linearized system [2]. We will make use of this insight when we discuss stability and sensitivities.

DISCRETE LINEAR SYSTEMS MODELS

4.1 Recursive Deterministic Models

We begin our illustration of linear systems with discrete models that are constructed recursively. This terminology means that time is only considered in discrete steps and that time periods in between are ignored (as on the display of a digital clock). Furthermore, the state of the system at time t is represented as a linear function of the state of the system at time $t-1$. In the simplest, single-variable, case, the state is characterized by a single feature, whereas a multivariate state is characterized by several, simultaneous features. For instance, the blood pressure "state" of a person consists of two variables, namely, a systolic and a diastolic pressure reading. The state of an ecosystem might be describable in terms of the numbers of organisms within each species at a given time, but it could also include temperature, pH, and other factors.

A typical single-variable example is the size P_t of a bacterial population at time points $t = 0, 1, 2, \ldots$. The time points may refer to minutes, hours, days, years, or any other fixed time periods, such as 20-minute intervals. The index notation P_t is commonly used for discrete models, whereas the notation $P(t)$ usually refers to a continuous model, where any values of t are permitted.

For simplicity let us assume that the bacteria are synchronized and all divide simultaneously every τ minutes. Without these assumptions, the situation quickly becomes complicated [3, 4]. Then, we can formulate a simple recursive model of the type

$$P_{t+\tau} = 2P_t, \tag{4.1}$$

which simply says that at time point $t+\tau$ there are twice as many bacteria as at time t, and this is true for any time point t. Note that, in this formulation, t may progress from 1 to 2 to 3, or maybe from 60 to 120 to 180, while τ remains fixed at 1 or 60, respectively, throughout the history of the population. Also note that the process really describes exponential growth, although the equation is called a linear equation, because its right-hand side is linear.

Suppose the population starts with 10 cells at time $t = 0$ and the division time is $\tau = 30$ minutes. Then we can easily project the growth of this population, according to **Table 4.1**: every 30 minutes, the population size doubles. Using (4.1) for the second doubling step, we immediately see

$$P_{(t+\tau)+\tau} = 2P_{t+\tau} = 2 \cdot 2 \cdot P_t = 2^2 \cdot P_t, \tag{4.2}$$

and this relationship is true no matter how far t has progressed in its discrete steps. By extension, if we move forward n doubling steps of 30 minutes each, we find

$$P_{t+n\tau} = 2^n P_t. \tag{4.3}$$

This formulation contains two components: the initial population size at some time point t, which is P_t, and the actual doubling process, which is represented by the term 2^n. Because the **power function** 2^n is defined as $\exp(n \ln 2)$, this process is called exponential growth. Much has been written about this process, and further exploration is left for the Exercises. A feature to ponder is what happens (biologically and mathematically) at time points between t and $t+\tau$ or between $t+3\tau$ and

TABLE 4.1: GROWTH OF A BACTERIAL POPULATION ACCORDING TO (4.1)

Time (min)	0	30	60	90	120	150	180
Population size	10	20	40	80	160	320	640

$t + 4\tau$. Of course the bacterial population does not cease to exist, but the model does not address these time points. The reader should think about how the model could be altered to accommodate these missing time points.

A slightly more complicated recursive growth model, which has found wide application in population studies, was proposed by Leslie [5]. The key difference from the bacterial growth process in (4.1) is that the population in this model consists of individuals with different ages and different proliferation rates. Thus, two processes overlap: individuals age and also bear children, with a fecundity rate that depends on their age. Furthermore, a certain portion of each age group dies. We will discuss this model in Chapter 10.

Leah Edelstein-Keshet introduced a small, yet informative example of a recursive model [6]. It describes, in a very simplified form, the dynamics of red blood cells. These cells are constantly formed by the bone marrow and destroyed by the spleen, but their number should remain more or less constant from day to day. In the simple model, the loss is modeled as proportional to the current number of cells in circulation, and the bone marrow produces a number of new cells that is proportional to the number of cells lost on the previous day. Thus, there are two critical numbers, namely,

R_n = number of red blood cells in circulation on day n

and

M_n = number of red blood cells produced by the bone marrow on day n.

These numbers depend recursively on cell numbers during the previous day. Specifically, R_n is given as the number of red blood cells in circulation on the previous day, and this number is diminished by losses through degradation in the spleen and increased by cells that were newly produced by the bone marrow on the previous day. If the fraction of cells removed by the spleen is f, which typically is a small number between 0 and 1, and if the production rate is denoted by g, which is defined as the number of cells produced per number of cells lost, we can formulate the following model equations:

$$R_n = (1-f)R_{n-1} + M_{n-1}, \tag{4.4a}$$

$$M_n = g\,f\,R_{n-1}. \tag{4.4b}$$

Similar to the time counter in the previous example, n is simply a counter representing the day under investigation. As a consequence, we can write these recursive equations also with shifted indices as

$$M_{n+1} = g\,f\,R_n, \tag{4.5a}$$

$$R_{n+1} = (1-f)R_n + M_n, \tag{4.5b}$$

and the two formulations (4.4) and (4.5) are exactly equivalent.

We can now ask interesting questions. For instance, given starting values R_0 and M_0, and f and g, how do the numbers R_n and M_n develop over time? The problem is simple enough to allow an analysis in Excel®. As a specific example, suppose $R_0 = 1000$, $M_0 = 10$, $f = 0.01$, and $g = 2$. Solving the equations from $n = 0$ to $n = 1$, $n = 2$, and so forth shows that both R_n and M_n continue to increase (**Figure 4.1**). Resetting $g = 1$, the plot is different: both data trends are flat. Setting $f = 0.05$, $g = 1$ causes an initial jump, followed by flat trends at $R_n \approx 962$ and $M_n \approx 48$. Thus, the initial 1010 cells become differently distributed (the reader should confirm these results).

The model allows us to compute under what conditions the system exhibits a steady state, which is defined by the situation where neither R_n nor M_n changes in value from one iteration to the next. Specifically, the condition of a steady state requires $R_{n+1} = R_n$ and $M_{n+1} = M_n$. The easiest way to assess these conditions is to call

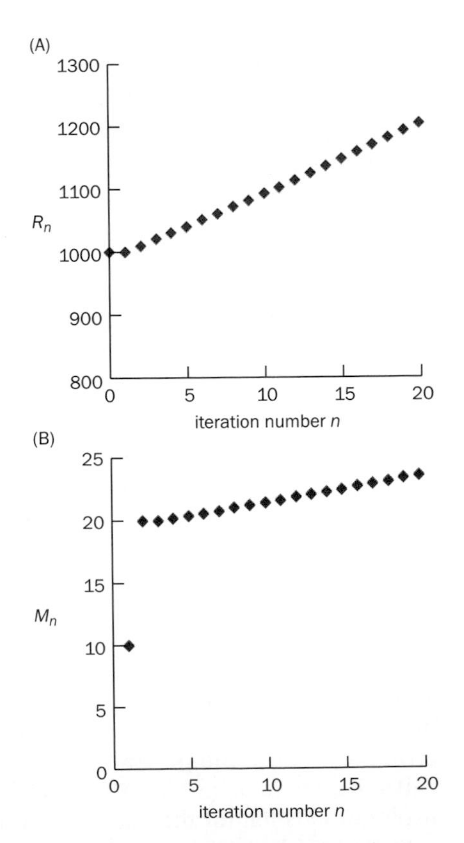

Figure 4.1 Trends in red blood cells over time. (A) The number of cells in circulation, R_n, and (B) the number of cells produced by the bone marrow, M_n, versus number of iterations, n. The model settings were R_0 = 1000, M_0 = 10, f = 0.01, and g = 2.

up the equation for M_n, (4.4b), and plug the result into the equation for R_{n+1}, (4.5a). The result is

$$R_{n+1} = (1-f)R_n + M_n = (1-f)R_n + g\,f\,R_{n-1}, \tag{4.6}$$

which contains R at three time points, but not M. For R and M to be in a steady state on days $n-1$, n, and $n+1$, we must demand that $R_{n+1} = R_n = R_{n-1}$, because if this condition is given, then R_{n+2}, R_{n+3} and all later values of R will remain the same. This pair of equations has two solutions, namely, $R_{n+1} = R_n = R_{n-1} = 0$, which is not very interesting, and

$$1 = (1-f) + g\,f, \tag{4.7}$$

which, for any positive f, is satisfied exclusively for $g = 1$. What does that mean biologically? The condition mandates that the bone marrow produces exactly as many cells as are lost, which in retrospect makes a lot of sense. Interestingly, this is true no matter the value of f.

So, what are the actual steady-state values for R and M? We know that g must be equal to 1; so let us assume that that is true. Furthermore, let us pick some (yet unknown) value for f and try to compute the steady-state values R_n and M_n from (4.5) by again requiring $R_{n+1} = R_n$ and $M_{n+1} = M_n$. The second equation yields $M_n = M_{n+1} = fR_n$; thus, the number of red blood cells produced by the bone marrow on day n must equal the cells lost on day n. Plugging this result into the first equation of the set just yields $R_n = R_{n+1} = (1-f)R_n + fR_n = R_n$, which means that the system is at a steady state for *any* values of f and R_n, as long as $g = 1$ and $M_n = fR_n$. The reader should confirm this result with numerical examples and ponder whether it makes sense biologically.

It is beneficial for many analyses to rewrite these types of linear recursive equations as a single matrix equation. In our case, this equation is

$$\begin{pmatrix} R \\ M \end{pmatrix}_{n+1} = \begin{pmatrix} 1-f & 1 \\ gf & 0 \end{pmatrix} \begin{pmatrix} R \\ M \end{pmatrix}_n. \tag{4.8}$$

According to the rules for multiplying a vector by a matrix, we obtain the value for R_{n+1} by multiplying the first row of the matrix with the vector $\begin{pmatrix} R \\ M \end{pmatrix}_n$, which yields $R_{n+1} = (1-f)R_n + M_n$, and we obtain M_{n+1} by multiplying the second row of the matrix with the same vector, which yields $M_{n+1} = gfR_n$. Thus, the matrix equation is a compact equivalent of the recursive equations (4.5a, b). This representation is particularly useful for large systems.

The matrix formulation makes it easy to compute future states of the system, such as R_{n+2} or M_{n+4}. First, we know that

$$\begin{pmatrix} R \\ M \end{pmatrix}_{n+2} = \begin{pmatrix} 1-f & 1 \\ gf & 0 \end{pmatrix} \begin{pmatrix} R \\ M \end{pmatrix}_{n+1}, \tag{4.9}$$

because n is only an index. Secondly, we can express the vector $\begin{pmatrix} R \\ M \end{pmatrix}_{n+1}$ according to the matrix equation (4.8). Combining these two aspects, the result is

$$\begin{aligned} \begin{pmatrix} R \\ M \end{pmatrix}_{n+2} &= \begin{pmatrix} 1-f & 1 \\ gf & 0 \end{pmatrix} \begin{pmatrix} R \\ M \end{pmatrix}_{n+1} = \begin{pmatrix} 1-f & 1 \\ gf & 0 \end{pmatrix} \begin{pmatrix} 1-f & 1 \\ gf & 0 \end{pmatrix} \begin{pmatrix} R \\ M \end{pmatrix}_n \\ &= \begin{pmatrix} 1-f & 1 \\ gf & 0 \end{pmatrix}^2 \begin{pmatrix} R \\ M \end{pmatrix}_n. \end{aligned} \tag{4.10}$$

According to the rules of matrix multiplication,

$$\begin{pmatrix} A & B \\ C & D \end{pmatrix}\begin{pmatrix} a & b \\ c & d \end{pmatrix}=\begin{pmatrix} Aa+Bc & Ab+Bd \\ Ca+Dc & Cb+Dd \end{pmatrix}, \quad (4.11)$$

we obtain

$$\begin{pmatrix} 1-f & 1 \\ gf & 0 \end{pmatrix}^2=\begin{pmatrix} (1-f)^2+gf & 1-f \\ gf(1-f) & gf \end{pmatrix}. \quad (4.12)$$

The matrix equation is not limited to two steps, and, indeed, any higher values of R_{n+k} and M_{n+k} can be computed directly from R_n and M_n, when we use the corresponding power of the matrix:

$$\begin{pmatrix} R \\ M \end{pmatrix}_{n+k}=\begin{pmatrix} 1-f & 1 \\ gf & 0 \end{pmatrix}^k\begin{pmatrix} R \\ M \end{pmatrix}_n. \quad (4.13)$$

Matrix methods for recursive systems can also be used to analyze the stability of steady states or to make statements regarding the long-term trends of the system [6]. We will discuss a popular matrix model of the growth of an age-structured population in Chapter 10.

The red blood cell model is of course extremely simplified. Nonetheless, the same structure of a recursive model can be used for realistic models of the dynamics of red blood cells in health and disease [7, 8].

4.2 Recursive Stochastic Models

The linear growth models discussed so far are entirely deterministic. Once they have been defined and started, their fate is sealed. This is not so in a certain class of probabilistic models, called **Markov models** or Markov chain models, which superficially have a format similar to a larger version of (4.8); a good introduction to these models is [9]. The crucial difference is that they have stochastic features, which means that they account for random events (see Chapter 2). Generically, a Markov model describes a system that can assume m different states. At each (discrete) time point, the system is in exactly one of these states, from where it transitions with some probability into a different state or stays where it is.

An intuitive example is a pinball machine, where the states are the typical flippers and bumpers. An adequate time step in this case is the time for the ball to move from one state to the next. For simplicity let us assume that the travel times between states are all the same and call the time points 0, 1, ..., t, $t+1$, ... At each time point, the one and only ball is in one state, from where it transitions, with some probability, to another state or stays in the same state. A Markov chain model is constructed by looking at very many games with the same pinball machine. Imagine we could record all movements of the ball from all games. If so, we would find balls and movements all over the place. However, some movements would certainly be observed more often than others, and, focusing on some given state i, each transition to some other state k would be observed with a specific frequency. The idea behind a Markov model is to use this frequency and interpret it as the probability of future transitions from state i to state k. Thus, the Markov chain process is defined by a fixed set of probabilities of transitions between all pairs of states: if a ball is in state i, the probability of moving to state j is p_{ij}, and this is true no matter where the ball was previously; the Markov model has no memory. For a process with m states, we can put all probabilities into an $m \times m$ matrix, which is called the Markov matrix:

$$\mathbf{M}=\begin{pmatrix} p_{11} & p_{12} & \cdots & p_{1,m-1} & p_{1m} \\ p_{21} & p_{22} & \cdots & p_{2,m-1} & p_{2m} \\ p_{31} & p_{32} & \cdots & p_{3,m-1} & p_{3m} \\ \vdots & \vdots & \ddots & \vdots & \vdots \\ p_{m1} & p_{m2} & \cdots & p_{m,m-1} & p_{mm} \end{pmatrix}. \quad (4.14)$$

We can use **M** to compute the dynamics of the Markov chain. Let us assume that at the beginning ($t = 0$) the ball is in state 1, which could be interpreted as the start position. Thus, $X_1 = 1$ and all other $X_i = 0$. At the next observation time point ($t = 1$) in game 1, the ball has moved to another state (or stayed in its start position). In the next game, the ball might have moved from the start position to a different state, and, in a third game, it may have moved to yet another position. This process is random for each game, but collectively governed by probabilities in **M**: p_{13} is the probability of moving from state 1 to state 3 in one time unit, and p_{16} is the probability of moving to state 6. Thus, the first row of the matrix describes what percentages of games are expected to move the ball from state 1 into any of the states i, namely p_{1i}. With the next move, the ball moves again, according to the probabilities in (4.14). Tracking all these expectations and transitions amounts to a big bookkeeping problem, which the matrix equation (4.14) solves beautifully.

What do we know about the matrix elements p_{ij}? They describe the probabilities of moving from state i to state j, where j must be one of the m states in the equation; there is no other option. Of course, every probability is between 0 and 1. It is actually possible that, for instance, $p_{13} = 0$, which would mean that it is impossible to move from state 1 to state 3 in one step. The ball could get to 3 in several steps, but not in one. If a probability p_{ij} is equal to 1, this means that the ball, if in state i, has no choice but to move to state j.

Suppose a ball is in state 4. Because it must move to one of the other states (or stay in state 4), all available probabilities, namely p_{4j} ($j = 1, \ldots, m$), summed together must equal 1. Generally, we find, for each i,

$$\sum_{j=1}^{m} p_{ij} = 1. \qquad (4.15)$$

It is possible that one of the probabilities for leaving state i is 1 and all others are 0. For instance, if $p_{i2} = 1$, then it is certain that every ball in state i in every pinball game of this type moves to state 2 next. In particular, if $p_{ii} = 1$, the ball can never again leave state i. Game over! Note that the summation over the first index (a column), $p_{1j} + p_{2j} + \cdots + p_{mj}$, does not necessarily yield 1 (the reader should explain why).

Markov chain models can be found in diverse corners of systems biology. For instance, they may describe the dynamics of networks, they can be used to compute the probability of extinction of a population, and they appear in the form of hidden Markov models in bioinformatics [10]. The adjective "hidden" in this case refers to the situation that one cannot tell whether the process is in one hidden state or another.

As an example, consider the transition diagram in **Figure 4.2**, which corresponds to the matrix

$$\mathbf{M} = \begin{pmatrix} 0.1 & 0.5 & 0.4 \\ 0.2 & 0.3 & 0.5 \\ 0.8 & 0.2 & 0 \end{pmatrix}. \qquad (4.16)$$

Suppose the process starts at time $t = 1$ in state X_2, which we denote as $X_1 = 0$, $X_2 = 1$, $X_3 = 0$. Linear algebra allows us to determine the probabilities of finding the process in each of the three states at time $t = 2$. This computation is accomplished by entering the initial values into a start vector and multiplying this vector by the transposed matrix $\mathbf{M}^{\mathrm{T}}$, whose rows are the columns of **M** and whose columns are the rows of **M**. Thus,

$$\begin{pmatrix} X_1 \\ X_2 \\ X_3 \end{pmatrix}_2 = \mathbf{M}^{\mathrm{T}} \begin{pmatrix} X_1 \\ X_2 \\ X_3 \end{pmatrix}_1 = \mathbf{M}^{\mathrm{T}} \begin{pmatrix} 0 \\ 1 \\ 0 \end{pmatrix} = \begin{pmatrix} 0.1 & 0.2 & 0.8 \\ 0.5 & 0.3 & 0.2 \\ 0.4 & 0.5 & 0 \end{pmatrix} \begin{pmatrix} 0 \\ 1 \\ 0 \end{pmatrix} = \begin{pmatrix} 0.2 \\ 0.3 \\ 0.5 \end{pmatrix}. \qquad (4.17)$$

Intuitively, the flipping of the matrix is necessary because M contains probabilities of moving forward in time, whereas (4.17) in a sense describes the "backwards" computation of the state vector at time 2 from the state vector at time 1.

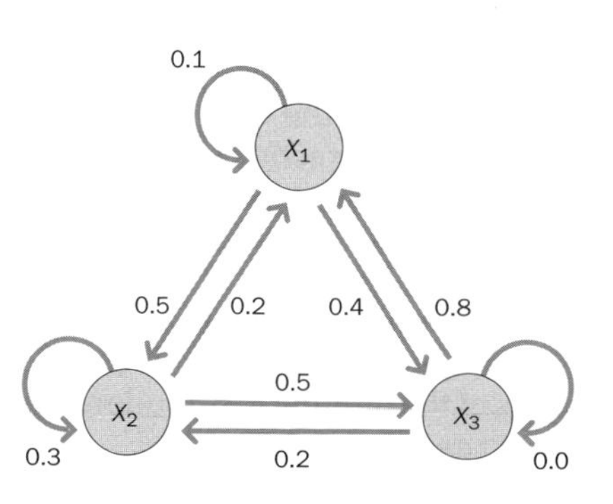

Figure 4.2 Diagram of a Markov process. The process is in one state, X_1, X_2, or X_3, at each time point and moves among these states with the probabilities shown on the figure.

4.3 Linear Differential Equations

The simplest example of this type is the linear differential equation

$$\frac{dX}{dt} = \dot{X} = aX, \tag{4.26}$$

which describes exponential growth or decay. Expressed in words from left to right, the equation says that change (growth), as expressed by the derivative on the left-hand side, is directly proportional to the current size or amount of the quantity X, with a proportionality constant a. If X and a are positive, the change is positive, and X continues to grow. If a is negative, an initially positive X keeps decreasing. Ponder what happens if both a and X are negative.

Note that we write X rather than $X(t)$, even though X depends on time, and that time t is no longer seen on the right-hand side. This omission is a convention of simplicity, and it is implicitly understood that X is a function of time.

In the simple case of (4.26), we can actually compute the solution of the differential equation, which explicitly describes the dependence of X on time. It consists of the exponential function $X(t) = X_0e^{at}$, because differentiation of X yields $dX/dt = aX_0e^{at}$, which equals $aX(t)$ and thus the right-hand side of (4.26). We see that X_0 is visible in the time-dependent solution, but not in the differential equation. Nevertheless, X_0 has a clearly defined meaning as the **initial value** of X at time $t = 0$. This is easy to confirm, when we substitute 0 for t in $X(t) = X_0e^{at}$, which directly returns the result $X(0) = X_0$. In fact, this last statement is often added to the definition of the differential equation in (4.26). As in the case of the recursive growth process, the differential equation with a positive initial value goes either to infinity (positive a) or to zero (negative a), unless a happens to be 0, which means that X remains at its initial value X_0 forever. If X_0 is negative, the solution can go to minus infinity or zero, depending on a. Because it takes infinitely long to reach the final value exactly, exponential decay processes are often characterized by their **half-life**, which is the time it takes for half of some amount of material to be lost. In addition to growth or decay processes, equations like (4.26) are used to describe transport processes in biochemical and physiological systems and for physical phenomena such as heating and cooling. We will return to this important equation in Chapter 8.

More interesting than a single linear differential equation is a system of linear differential equations. These appear to have the same general form as (4.26), namely,

$$\frac{d\mathbf{X}}{dt} = \dot{\mathbf{X}} = \mathbf{AX}, \tag{4.27}$$

but here $\mathbf{X}$ is now a vector and $\mathbf{A}$ is a matrix with **coefficients** A_{ij}. We have already encountered these models when we discussed stoichiometric models of metabolic pathways (Chapter 3). They are also used to study physiological **compartment models**, where nutrients, drugs, or toxic substances can move between the bloodstream and the organs and tissues in the body (for details see, for example, [12] and Chapter 13).

The main feature of systems like (4.27) is that they allow us to compute explicit solutions, instead of requiring numerical simulation that are almost always the only solution for nonlinear systems. If $\mathbf{X}$ has n elements, these solutions consist of n time-dependent functions $X_i(t)$ that are determined by the **eigenvalues** of $\mathbf{A}$, which are real or complex numbers that characterize a matrix. For small systems, they can be computed algebraically, and for larger systems, there are numerous computational methods and software programs. Assuming that the eigenvalues of $\mathbf{A}$ are all different, the solution of (4.27) consists of a simple sum of exponential functions whose **rate constants** equal the eigenvalues of the system matrix. However, because these exponential functions may have complex arguments, the solutions also include trigonometric functions, such as sine and cosine. An example was given in (4.25).

Many methods are available for studying linear differential equations, including the so-called Laplace transform, which permits analyses without the need to integrate the differential equations [13].

4.4 Linearized Models

Someone once said that the distinction between linear and nonlinear models is like a classification of the animal kingdom into elephants and non-elephants. It is true: there is essentially only one linear structure, but there are uncounted—and in fact infinitely many—nonlinear structures. Nevertheless, linear models are of utmost importance even for nonlinear systems, because some analyses of nonlinear systems can be reduced to a linearized analog. The strategy is to flatten the nonlinear system in a small region around a point of particular interest, called the **operating point, *OP***. This point can be chosen arbitrarily, as long as it is possible to compute derivatives of the nonlinear system there. In practical investigations, *OP* is often the steady state.

The graph in **Figure 4.4** shows the **linearization** of a curved surface at some point *P*. The linearization is visualized as the tangent plane, which at *P* has exactly the same value as the curved surface and also the same slopes. As a consequence, all paths C_i on the surface that go through *P* have the same local directions T_i in the tangent plane, and many analyses can validly be done in this tangent plane.

As an illustration, suppose a system has the form

$$\frac{dX}{dt} = f(X,Y), \qquad \frac{dY}{dt} = g(X,Y), \tag{4.28}$$

where *f* and *g* are nonlinear functions. The first order of business is the selection of an operating point, where we anchor the linearization. The main ingredient of the linearization process is differentiation, and, because the system has two independent variables, *X* and *Y*, the derivatives must be partial derivatives. These are computed like regular derivatives under the assumption that the other variable is kept constant. The linearization is based on a theorem of the English mathematician Brook Taylor (1685–1731), who discovered that a function may be approximated with a polynomial, which in the simplest case is a linear function (see Box 4.1).

For our demonstration, we approximate *f* and *g* in the vicinity of our operating point of choice, (X_{OP}, Y_{OP}), which, from now on, we always assume to be a steady-state point. This assumption is of course not always true, but it reflects by far the most important situation. We denote a solution in the vicinity of the steady-state operating point by $[X_{OP} + X'(t), Y_{OP} + Y'(t)]$, so that $X'(t)$ and $Y'(t)$ are deviations in the *X*- and *Y*-directions. Because we intend to stay fairly close to the operating point, this solution is called a **perturbation**. The strategy is now to formulate this perturbation as a system of differential equations. Simply plugging in the perturbed values into (4.28) yields

$$\frac{d}{dt}(X_{OP} + X') = f(X_{OP} + X', Y_{OP} + Y'), \qquad \frac{d}{dt}(Y_{OP} + Y') = g(X_{OP} + X', Y_{OP} + Y'), \tag{4.29}$$

where the argument (t) is again not shown.

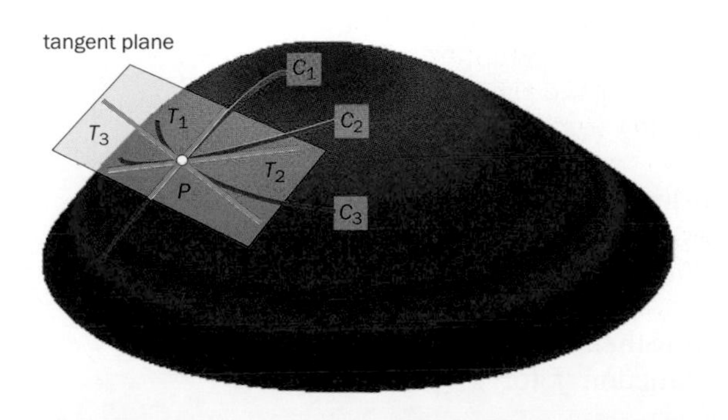

Figure 4.4 Illustration of the linearization of a curved surface. At *P*, the tangent plane and the curved surface touch and have the same slopes.

BOX 4.1 APPROXIMATION

Very many approximations in mathematics are based on an old theorem by the English mathematician Brook Taylor (1685–1731), who showed that any smooth function (one that has sufficiently many continuous derivatives) can be represented as a power series. If this series has infinitely many terms, it is an exact representation of the original function within some region of convergence. However, if the series is truncated, that is, if one ignores all terms beyond a certain number, the resulting Taylor polynomial is an approximation of the function. This approximation is still exact at a chosen operating point and very similar to the original function close to this operating point. However, for values distant from the operating point, the approximation error may be large. The coefficients of the Taylor polynomial are computed through differentiation of the original function.

Specifically, if a function $f(x)$ has at least n continuous derivatives, it is approximated at the operating point p by

$$\begin{aligned} f(x) \approx {} & f(p) + f'(p)(x-p) + \frac{1}{2!} f^{(2)}(p)(x-p)^2 \\ & + \frac{1}{3!} f^{(3)}(p)(x-p)^3 + \cdots + \frac{1}{n!} f^{(n)}(p)(x-p)^n. \end{aligned} \tag{1}$$

The expressions 2!, 3!, and $n!$ are factorials, which we have already encountered in Box 2.3, and $f^{(1)}$, $f^{(2)}$, $f^{(3)}$, and $f^{(n)}$ denote the first, second, third, and nth derivatives of $f(x)$, respectively. It is not difficult to see that the original function f and the Taylor approximation are exactly the same at the operating point p: just substitute p for x.

An important special case is linearization, where $n = 1$. That is, one keeps merely the constant term and the term with the first derivative and ignores everything else. This strategy sounds quite radical, but it is extremely useful, because the linearity of the result permits many simplified analyses close to the operating point. The result of the linearization is

$$L(x) = f(p) + f^{(1)}(p)(x-p), \tag{2}$$

which has the familiar form of the equation $y = mx + b$ for a straight line, with slope m and intercept b (**Figure 1**). Here, m corresponds to the derivative $f^{(1)}(p)$, and the vertical-axis intersect b is $f(p) - p f^{(1)}(p)$.

As an example, consider the shifted exponential function

$$f(x) = e^x + 1 \tag{3}$$

and choose $x_0 = 0$ as the operating point. The linearization includes the value of f at x_0, which is 2, and the first derivative, evaluated at the operating point, which is

$$f^{(1)}(0) = e^x = 1. \tag{4}$$

Thus, the linearization of f at (x_0) is

$$L(x) = 2 + 1 \cdot (x - 0) = 2 + x. \tag{5}$$

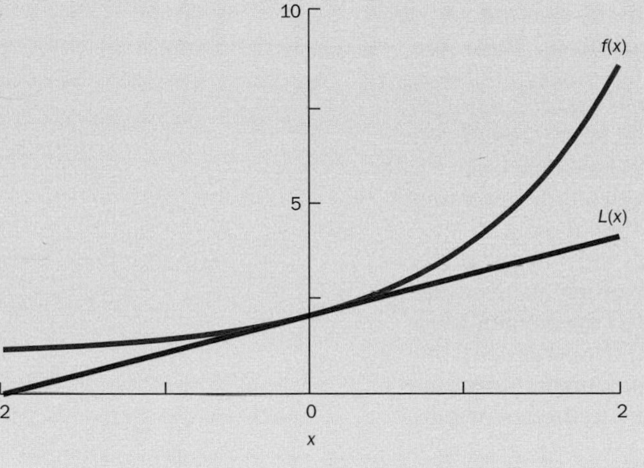

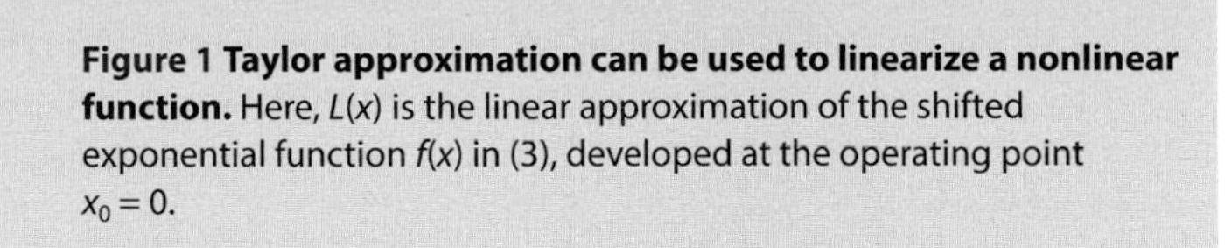

Figure 1 Taylor approximation can be used to linearize a nonlinear function. Here, $L(x)$ is the linear approximation of the shifted exponential function $f(x)$ in (3), developed at the operating point $x_0 = 0$.

To analyze the right-hand sides of this equation, we use Taylor's insight (**Boxes 4.1** and **4.2**) and obtain the linear approximation

$$f(X_{OP}+X', Y_{OP}+Y') \approx f(X_{OP}, Y_{OP}) + \left.\frac{\partial f}{\partial X}\right|_{(X_{OP},Y_{OP})} X' + \left.\frac{\partial f}{\partial Y}\right|_{(X_{OP},Y_{OP})} Y', \tag{4.30a}$$

$$g(X_{OP}+X', Y_{OP}+Y') \approx g(X_{OP}, Y_{OP}) + \left.\frac{\partial g}{\partial X}\right|_{(X_{OP},Y_{OP})} X' + \left.\frac{\partial g}{\partial Y}\right|_{(X_{OP},Y_{OP})} Y'. \tag{4.30b}$$

The curly ∂ signifies partial differentiation and the vertical lines with subscripts indicate that this differentiation is evaluated at the operating point. Because we chose the operating point as a steady state, where $dX/dt = dY/dt = 0$ in (4.28), the terms $f(X_{OP}, Y_{OP})$ and $g(X_{OP}, Y_{OP})$ are 0 and drop out.

We can gain some intuition about (4.30a, b) by dissecting them into recognizable components (**Figure 4.5**). Expressed in words, the function f (or g) can be

BOX 4.2 LINEARIZATION OF NONLINEAR FUNCTIONS IN TWO VARIABLES

Intriguingly, an approximation analogous to the Taylor representation in Box 4.1 can be performed in higher dimensions. For the case of two dependent variables, one obtains the approximation

$$\begin{aligned} A(x,y) &= f(x_0,y_0)+(x-x_0)f_x(x_0,y_0)+(y-y_0)f_y(x_0,y_0) \\ &+\frac{1}{2!}[(x-x_0)^2 f_{xx}(x_0,y_0)+2(x-x_0)(y-y_0)f_{xy}(x_0,y_0) \\ &+(y-y_0)^2 f_{yy}(x_0,y_0)]+\cdots. \end{aligned} \quad (1)$$

The first term on the right-hand side is the value of f at the chosen operating point (x_0, y_0), the expressions f_x and f_y are the first derivatives of f with respect to x and y, respectively, and f_{xx}, f_{xy}, and f_{yy} are the corresponding second derivatives of f, which are all evaluated at the operating point.

Ignoring the second and all higher derivatives, we obtain the linearization $L(x,y)$. Specifically, we can write

$$L(x,y)=f(x_0,y_0)+(x-x_0)f_x(x_0,y_0)+(y-y_0)f_y(x_0,y_0). \quad (2)$$

At the operating point, L has the same value and the same slopes as f.

As an example, consider the function $f(x, y) = ye^x$ and choose the operating point $(x_0, y_0) = (0, 0)$. The linearization includes the value of f at (x_0, y_0), which is 0, and the first derivatives, evaluated at the operating point. These are

$$f_x(0,0)=ye^x=0\cdot 1=0 \quad \text{and} \quad f_y(0,0)=e^x=1. \quad (3)$$

Thus, the linearization of f at (x_0, y_0) is

$$L(x,y)=0+(x-0)\cdot 0+(y-0)\cdot 1=y. \quad (4)$$

The linearization is flat with respect to the x-direction and has a slope of 1 in the y-direction. Some values of f and L are given in **Table 1**. The approximation for small x-values is good, even if y is large, because both f and L are linear with respect to y. By contrast, large x-values lead to much higher approximation errors, because L does not even depend on x, owing to the choice of $x_0 = 0$ for the operating point, while f increases exponentially with respect to x.

TABLE 1: SELECTED VALUES OF $f(x, y)$ AND $L(x, y)$ FOR DIFFERENT COMBINATIONS OF x AND y

(x, y)	(0, 0)	(0, 1)	(0.1, 0.1)	(0.1, 1)	(0.2, 0.2)	(0.1, 0.5)	(0.5, 0.1)	(0.5, 0.5)	(1, 0.1)
f(x, y)	0	1	0.115	1.105	0.244	0.553	0.165	0.824	0.272
L(x, y)	0	1	0.1	1	0.2	0.5	0.1	0.5	0.1

approximated at some point by using the value at this point and adding to it deviations in the directions of the dependent variables. Each deviation consists of the slope in that direction (expressed by the partial derivative) and the distance of the deviation along the corresponding axis (here X' and Y', respectively). This method works for any number of dependent variables.

The left-hand sides of (4.30a, b) can be simplified as well. Namely, we are allowed to split the differential of each sum into a sum of two differentials. Furthermore, $dX_{OP}/dt = dY_{OP}/dt = 0$, because X_{OP} and Y_{OP} are constant values. Thus, only dX'/dt and dY'/dt are left. Taking the simplifications of the two sides together, the statements in (4.29) and (4.30) become

$$\frac{d}{dt}X' \approx \left.\frac{\partial f}{\partial X}\right|_{(X_{OP},Y_{OP})} X' + \left.\frac{\partial f}{\partial Y}\right|_{(X_{OP},Y_{OP})} Y', \quad (4.31a)$$

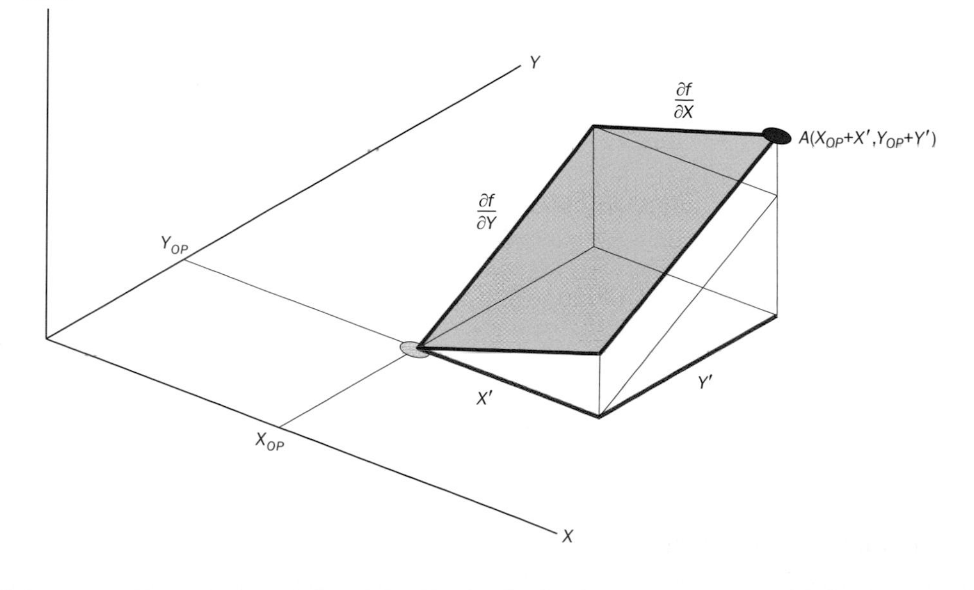

Figure 4.5 Visualization of part of a plane $A(X,Y)$ (pink) approximating some function $f(X,Y)$, which is not shown, at some operating point (X_{OP} and Y_{OP}) (green). The plane is constructed according to (4.30a). The blue lines have the slope of f at the operating point in the X-direction, while the red lines have the slope of f at the operating point in the Y-direction. The green lines indicate distances in the two directions from the operating point (X' and Y', respectively). The point $A(X_{OP} + X', Y_{OP} + Y')$ (red) lies on the plane but is only approximately equal to $f(X_{OP} + X', Y_{OP} + Y')$ (not shown). The two are equivalent for $X' = Y' = 0$ and similar for small values of X' and Y'.

$$\frac{d}{dt}Y' \approx \left.\frac{\partial g}{\partial X}\right|_{(X_{OP},Y_{OP})} X' + \left.\frac{\partial g}{\partial Y}\right|_{(X_{OP},Y_{OP})} Y'. \quad (4.31b)$$

Note that this linearization could be a rather crude approximation if f and g are strongly curved. But the approximation may also be very good for moderate deviations from the operating point. Without further information about f and g, we just don't know, even for rather large perturbations. We do know that the linearization is good if the perturbation is sufficiently small (whatever that means in a specific situation). A second important result is that the right-hand sides are linear.

We can rename the partial derivatives in (4.31), evaluated at the operating point, as

$$\begin{aligned} a_{11} &= \left.\frac{\partial f}{\partial X}\right|_{(X_{OP},Y_{OP})}, & a_{12} &= \left.\frac{\partial f}{\partial Y}\right|_{(X_{OP},Y_{OP})}, \\ a_{21} &= \left.\frac{\partial g}{\partial X}\right|_{(X_{OP},Y_{OP})}, & a_{22} &= \left.\frac{\partial g}{\partial Y}\right|_{(X_{OP},Y_{OP})}, \end{aligned} \quad (4.32)$$

which identifies the linearization of (4.28) as

$$\begin{aligned} \frac{dX'}{dt} &= a_{11}X' + a_{12}Y', \\ \frac{dY'}{dt} &= a_{21}X' + a_{22}Y'. \end{aligned} \quad (4.33)$$

In general, we may write the linearization of a system with m variables in matrix form as

$$d\begin{pmatrix} X_1' \\ X_2' \\ \vdots \\ X_m' \end{pmatrix} / dt = \begin{pmatrix} \frac{\partial f_1}{\partial X_1} & \frac{\partial f_1}{\partial X_2} & \cdots & \frac{\partial f_1}{\partial X_m} \\ \frac{\partial f_2}{\partial X_1} & \frac{\partial f_2}{\partial X_2} & \cdots & \frac{\partial f_2}{\partial X_m} \\ \vdots & \vdots & \ddots & \vdots \\ \frac{\partial f_m}{\partial X_1} & \frac{\partial f_m}{\partial X_2} & \cdots & \frac{\partial f_m}{\partial X_m} \end{pmatrix}_{OP} \begin{pmatrix} X_1' \\ X_2' \\ \vdots \\ X_m' \end{pmatrix}. \quad (4.34)$$

The matrix on the right-hand side is so important that it has its own name: the **Jacobian** of the system (at the operating point OP).

The linearization describes approximately the distance of the nonlinear system from its steady-state operating point. Specifically in our case,

$$\begin{aligned} X'(t) &\approx X(t) - X_{OP}, \\ Y'(t) &\approx Y(t) - Y_{OP}. \end{aligned} \quad (4.35)$$

Close to the operating point, the linear representation is very good, but on moving away from this point, the approximation error typically becomes larger. Important investigations into the stability and sensitivity of linear and nonlinear systems are executed in a close vicinity of the steady state, so that the accuracy of the approximation becomes secondary for these purposes (see later in this chapter).

As a numerical demonstration of the linearization process, consider the system

$$\begin{aligned} \dot{X} &= f(X,Y) = 2Y - 3X^4, \\ \dot{Y} &= g(X,Y) = 0.5e^X Y - 2Y^2. \end{aligned} \quad (4.36)$$

The system has the steady state ($X_{ss} \approx 0.776$, $Y_{ss} \approx 0.543$), which we use as our operating point (OP). The values of the partial derivatives, evaluated at OP, are

$$
\begin{aligned}
a_{11} &= \left.\frac{\partial f}{\partial X}\right|_{(X_{OP},Y_{OP})} = -12X^3\Big|_{(X_{OP},Y_{OP})} = -5.608, \\
a_{12} &= \left.\frac{\partial f}{\partial Y}\right|_{(X_{OP},Y_{OP})} = 2, \\
a_{21} &= \left.\frac{\partial g}{\partial X}\right|_{(X_{OP},Y_{OP})} = 0.5e^{X}Y\Big|_{(X_{OP},Y_{OP})} = 0.590, \\
a_{22} &= \left.\frac{\partial g}{\partial Y}\right|_{(X_{OP},Y_{OP})} = \left(0.5e^{X} - 4Y\right)\Big|_{(X_{OP},Y_{OP})} = -1.086.
\end{aligned}
\tag{4.37}
$$

The linearization is thus

$$
\begin{aligned}
\frac{dX'}{dt} &= -5.608X' + 2Y', \\
\frac{dY'}{dt} &= 0.590X' - 1.086Y'.
\end{aligned}
\tag{4.38}
$$

The variables X' and Y' are functions of t that describe approximately how far $X(t)$ and $Y(t)$ deviate from the steady state ($X_{SS} = 0.776$, $Y_{SS} = 0.543$).

Suppose the original nonlinear system uses as initial values $X_0 = X_{ss} + 0.75$ and $Y_0 = Y_{ss} - (-0.2)$. The linearized system (4.38) may start with any initial values, of course, but for a direct comparison with the original nonlinear system we choose 0.75 and −0.2. We solve this ODE system and then shift the solution to the operating point: $\tilde{X}(t) = X'(t) + X_{ss}$, $\tilde{Y}(t) = Y'(t) + Y_{ss}$. **Figure 4.6** shows that $\tilde{X}$ and $\tilde{Y}$ approximate X and Y very well close to the operating point, which equals the steady state. At the beginning of the simulation, the approximation of X is not quite as good.

As a second example, recall the SIR model (2.6) that we analyzed in Chapter 2. It has the form

$$
\begin{aligned}
\dot{S} &= 3 + 0.01R - 0.0005SI, & S_0 &= 170 = S_{ss} + 30, \\
\dot{I} &= 0.0005SI - 0.05I - 0.02I, & I_0 &= 120 = I_{ss} - 30, \\
\dot{R} &= 0.05I - 0.01R, & R_0 &= 700 = R_{ss} - 50.
\end{aligned}
\tag{4.39}
$$

Note that, for convenience, the initial values are expressed here as deviations from the steady state, which is $(S_{ss}, I_{ss}, R_{ss}) = (140, 150, 750)$, and which we use as our operating point (OP). Without going through the theoretical derivation, we can immediately compute the Jacobian, which is

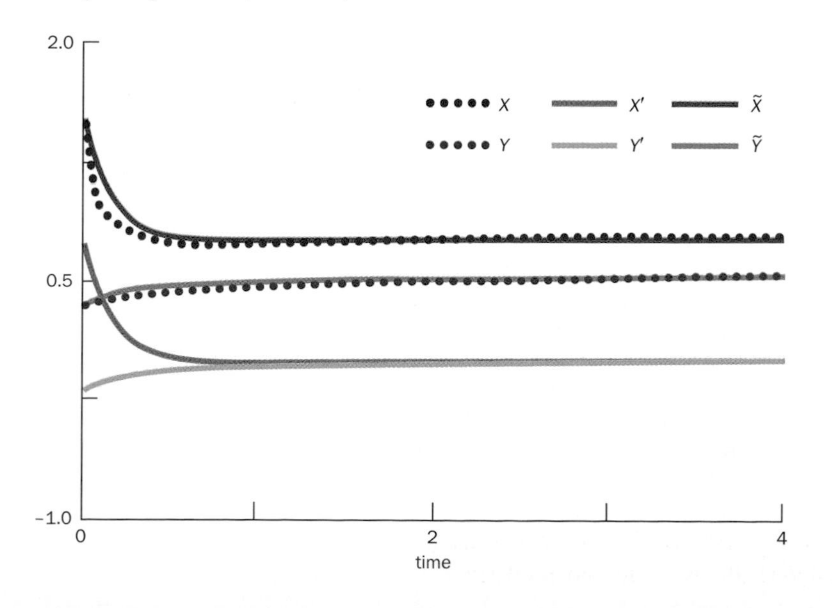

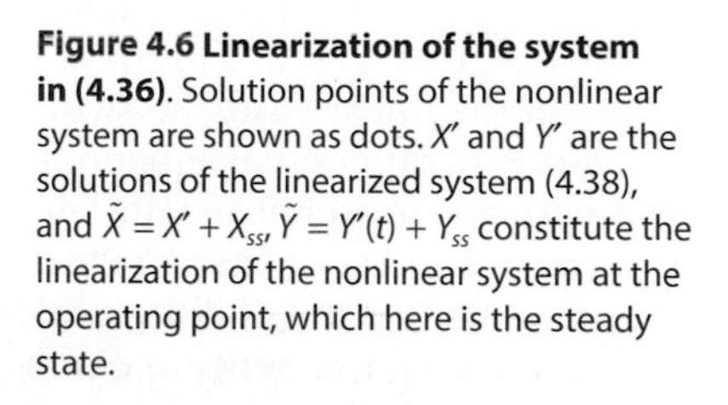

Figure 4.6 Linearization of the system in (4.36). Solution points of the nonlinear system are shown as dots. X' and Y' are the solutions of the linearized system (4.38), and $\tilde{X} = X' + X_{ss}$, $\tilde{Y} = Y'(t) + Y_{ss}$ constitute the linearization of the nonlinear system at the operating point, which here is the steady state.

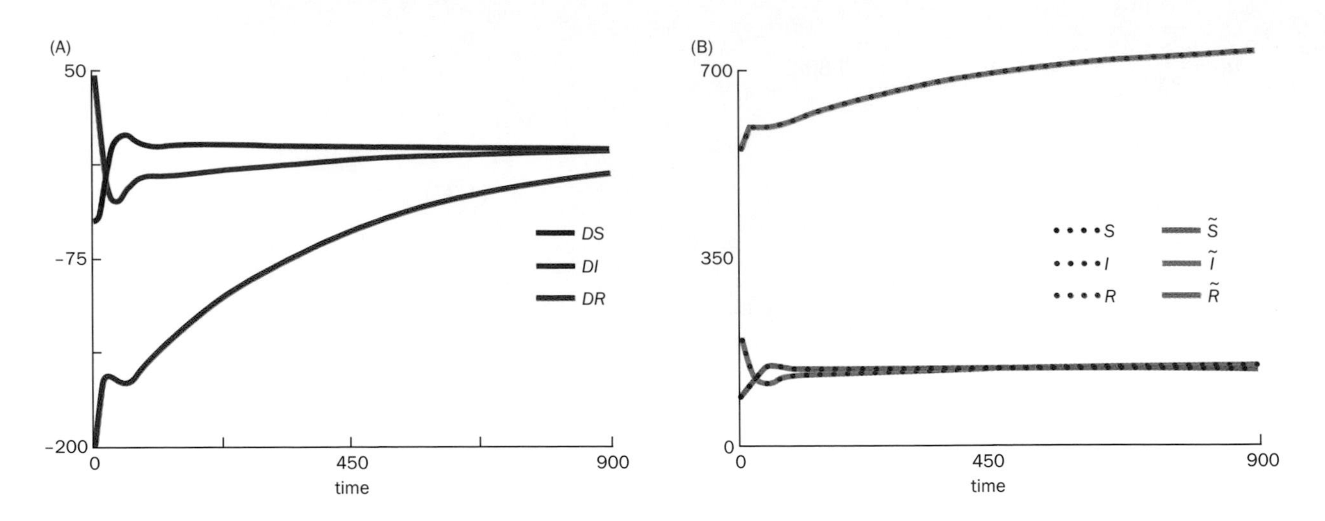

Figure 4.7 Linearization of the SIR system in (4.39). (A) The solution to the linearization in (4.41). (B) The same solution, shifted to the operating point *OP*. The solution points of the nonlinear system are shown as dots. $\tilde{S}$, $\tilde{I}$, and $\tilde{R}$ constitute the linearization of the nonlinear system at the steady-state operating point. The linearization is excellent, because the nonlinear system does not deviate very far from the steady state. See Exercise 4.21 for a less accurate approximation.

$$\mathbf{J}=\begin{pmatrix}\frac{\partial f_1}{\partial S} & \frac{\partial f_1}{\partial I} & \frac{\partial f_1}{\partial R}\\ \frac{\partial f_2}{\partial S} & \frac{\partial f_2}{\partial I} & \frac{\partial f_2}{\partial R}\\ \frac{\partial f_3}{\partial S} & \frac{\partial f_3}{\partial I} & \frac{\partial f_3}{\partial R}\end{pmatrix}_{OP}=\begin{pmatrix}-0.075 & -0.07 & 0.01\\ 0.075 & 0 & 0\\ 0 & 0.05 & -0.01\end{pmatrix}, \tag{4.40}$$

where f_1, f_2, and f_3 are the right-hand sides of the system in (4.39), and the derivatives are evaluated at the stead-state operating point *OP*. Therefore, the linearized equations are given as

$$\begin{pmatrix}\frac{dS'}{dt}\\ \frac{dI'}{dt}\\ \frac{dR'}{dt}\end{pmatrix}=\begin{pmatrix}-0.075 & -0.07 & 0.01\\ 0.075 & 0 & 0\\ 0 & 0.05 & -0.01\end{pmatrix}\begin{pmatrix}S'\\ I'\\ R'\end{pmatrix}. \tag{4.41}$$

We solve these equations with initial values that correspond to the deviations of the nonlinear system from the steady state, namely, (30, –30, –50); see (4.39) and **Figure 4.7**. Once computed, the solution is shifted to the operating point *OP*, which corresponds to the steady state: $\tilde{S}(t) = S'(t) + S_{SS}$, $\tilde{I}(t) = I'(t) + I_{SS}$, $\tilde{R}(t) = R'(t) + R_{SS}$. Figure 4.7 shows the result: the linearization reflects the nonlinear system very well, because the deviation from the steady state is relatively minor.

CONTINUOUS NONLINEAR SYSTEMS

As in the context of linear systems, continuous representations account for entire time intervals and not just for discrete time points. Furthermore, continuous systems become the more convenient choice if the number of players in the system becomes large. Who wants to count (or account for) every cell in a culture or organism? As in the linear case, the switch to continuous variables leads directly to the formulation of differential equations, but in the nonlinear case these afford us with incomparably more choices: How do we choose the right-hand side? Should we use polynomials? Or exponentials? Or combinations of all kinds of functions? Clearly, there are unlimited possibilities, and nature has not provided us with an instruction manual. As a consequence, nobody will be able to tell what the best or most appropriate functions are, except for a minute class of very special situations. This fact poses a significant challenge and is the subject of extensive ongoing research. In most cases, one of two strategies is used. Either one makes assumptions regarding functions that appear to be reasonable, for instance, because they somehow model the mechanisms that drive the phenomenon under investigation. These functions are sometimes called **ad hoc**, indicating that they were chosen for this particular situation and are not claimed to have wider application. Or one uses a generic nonlinear approximation that does not require specific assumptions and offers sufficient flexibility in the repertoire of responses it can represent. An approximation of this type is called **canonical**.

4.5 Ad hoc Models

Ad hoc models use functions that for some reason seem to suit the given task the best. Sometimes they are based on the principles of physics and sometimes they simply seem to have just the right shapes and other properties. As an example of the latter, the logistic function

$$F(x)=\frac{a}{b+e^{-cx}} \tag{4.42}$$

is frequently used, sometimes even with $a=b=c=1$, because it describes a sigmoidal function as it is often encountered in biology (**Figure 4.8**). Altering a and/or b changes the output range and horizontal shift of the function, and c affects its steepness. The reader should try out a few combinations of a, b, and c.

In the field of population dynamics, the same logistic function is more widely used in the equivalent format of a differential equation of the type

$$\dot{N}=\alpha N-\beta N^2, \tag{4.43}$$

where N is the size of a population and α and β are parameters. The first parameter, α, is the growth rate, and α/β is the final size or carrying capacity of the population (see Chapter 10).

At first glance, (4.42) and (4.43) don't seem to have much in common. However, if we rename F in (4.42) as N and x as t and differentiate N with respect to t, the result can be manipulated with simple algebra to attain the form (4.43), where the parameters α and β and the initial value N_0 are combinations of a, b, and c. We should also note that (4.43), slightly rewritten as $\dot{N}=\beta N(\gamma-N)$, with $\gamma=\alpha/\beta$, looks a lot like the logistic map, thereby explaining the similarity in name. However, the two are quite different, because the left-hand sides have a different meaning.

The logistic model in (4.43) is the launch pad for very many models of population dynamics, where different species compete for resources or where one species feeds on another. It is also the base case for canonical Lotka-Volterra systems (see later), which can describe an incredible variety of nonlinearities.

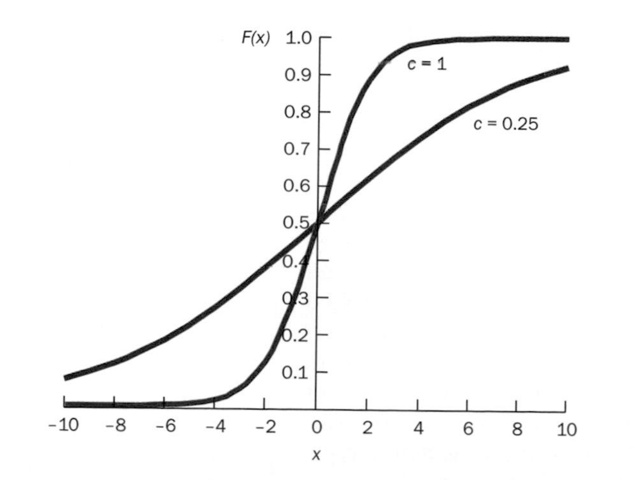

Figure 4.8 Graphs of the logistic function (4.42). Depending on its parameter values, the logistic function models different slopes, final values, and horizontal shifts.

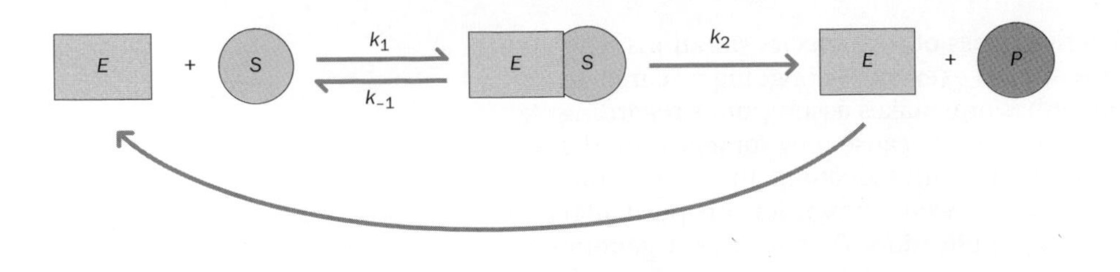

Figure 4.9 Schematic diagram of an enzyme-catalyzed reaction. Substrate S and enzyme E form a complex, which may revert to S and E or break down to yield product P while releasing the enzyme. See Chapter 8 for details.

There is hardly a limit to the possible choices for ad hoc functions, and the selection of a particular function depends heavily on the context and knowledge of the biological phenomenon. Possibilities include trigonometric functions, simple exponentials, or more unusual functions such as spikes or square waves. Indeed, there are catalogs of functions with various shapes [14].

A widely used example of an ad hoc function based on alleged mechanics is the Michaelis–Menten function of enzyme kinetics, which has the form

$$v_P = \frac{V_{max} S}{K_M + S}. \tag{4.44}$$

It describes the speed v_P with which a metabolic substrate S is converted into a product P with the help of an enzyme E (**Figure 4.9**). The function has two parameters: V_{max} is the maximum speed v_P can reach, while K_M is the substrate concentration for which v_P runs with half the maximum speed (see Chapters 2 and 8).

4.6 Canonical Models

A rather different strategy for model selection is the use of canonical models, of which there are several types. The designation *canonical* implies that the processes of setting up and analyzing a model follow strict rules. Forced adherence to rules could be seen as a real limitation, but it has significant advantages, as we will see later in this chapter and in some of the application chapters. One attraction of canonical models is that they allow us to get started very quickly even if we are faced with a lot of uncertainty and scarce information on the details of a system. This advantage becomes most evident if we ask what would happen if we did not have rules and guidelines. As an example scenario, imagine being tasked to come up with a mathematical model describing how the introduction of a new species affects the dynamics of an ecosystem. It is immediately clear that all kinds of players are involved in the case, but it is very much unclear how exactly they interact. All interactions must follow the laws of physics and chemistry, but accounting for every physical and chemical detail in the system would quite obviously be impossible—and is also often unnecessary. Canonical modeling provides a compromise between general applicability and simplicity that circumvents some of these problems by using mathematically rigorous and convenient approximations. As a rule of thumb, the less one knows about the details of the system, the more one benefits from canonical models.

The ultimate canonical model is again the linear model, and we would love to use it. Alas, linearity is often too limiting for biological phenomena that deviate a bit from their normal steady state and exhibit responses that are very clearly nonlinear. For instance, a linear model cannot represent stable **limit-cycle oscillations**, as we see them in circadian rhythms and our own heartbeat (see later in this chapter). These limitations mean that we have to look for nonlinear canonical models.

The oldest example is the **Lotka–Volterra (LV) model** [15–19]. Interestingly, this model was proposed around the same time as the infectious disease model that we introduced in Chapter 2, namely, in the 1920s. Its independent authors were Alfred Lotka, an American pioneer in mathematical biology, and Vito Volterra, an applied mathematician from Italy. The rules for setting up an LV model are simple. Enumerate all dependent variables of interest; independent variables are merged with parameters. Assume that each variable may potentially interact with every other variable in the system and formulate each interaction as the product of the two variables and a rate constant. In other word, this mechanism is of mass-action type, as we encountered it in Chapter 2. LV models also permit a linear term in each equation, which could represent birth, death, degradation, or transport out of the system. Thus, each differential equation of an LV model has exactly the same format:

$$\dot{X}_i = \sum_{j=1}^{n} a_{ij} X_i X_j + b_i X_i, \tag{4.45}$$

where some (or many) of the parameters may have values of 0. If the system has only one variable, the LV model with negative a_{11} and positive b_1 becomes the logistic function (4.43).

The LV model bears good news and bad news. The main good news is that it is easy and intuitive to set up the equations and to compute the steady state (see later). Second, the model has found very many applications in ecological systems analyses, where much of the dynamics, like competition for food or predation, is driven by one-to-one interactions (see [17] and Chapter 10). A further piece of very surprising and intriguing news is that LV equations are extremely flexible: they are capable of modeling any type of differentiable nonlinearities and **complexities**, including all kinds of **damped** or stable oscillations and chaos, if one includes enough auxiliary variables and does some mathematical trickery with them [18–20]. The bad news is that, without such trickery, many classes of biological systems are not really compatible with the LV format. For instance, suppose we want to model a simple pathway as shown in **Figure 4.10**. We define four variables $X_1, \ldots, X_4$ and want to formulate the equations. However, this is a problem for the LV format. For instance, the production of X_3 should depend only on X_2, but we are not allowed to include a term like kX_2 in the equation for X_3. Furthermore, the only choice to account for the feedback by X_4 in the equation for X_2 would be a term like $\kappa X_1 X_4$, but such a term is not permitted in the equation. As soon as we try to ameliorate this issue, we increase the applicability of the model but compromise the mathematical advantages of the LV format. For instance, if we allow all possible linear and bilinear terms, we arrive at a mass-action system, as we used for the infectious disease model in Chapter 2. We could also permit multilinear terms of the type $\alpha X_1 X_2 X_4$. In either case, the extension would compromise the analytical advantages of the LV model. In particular, the true LV format allows us to compute steady states with methods of linear algebra (see later in this chapter), whereas the extensions would require the use of nonlinear methods and search algorithms.

Greater flexibility than in LV and mass-action models is afforded by canonical **generalized mass action (GMA)** systems within the modeling framework of **biochemical systems theory (BST)** [21–25]. In this type of canonical model, each process v_i is formulated as a product of **power-law functions** of the type

$$v_i = \gamma_{ik} X_1^{f_{ik1}} X_2^{f_{ik2}} \cdots X_n^{f_{ikn}}. \tag{4.46}$$

As in mass-action and LV models, this format includes a non-negative rate constant γ_{ik}, which describes the turnover rate of the process, and a product of variables. In this case, the product may include all, some, or none of the systems variables. Each variable has its own exponent f_{ikj}, which is called a **kinetic order** and may take any real value. If a kinetic order is positive, it signifies a positive or augmenting effect, and if it is negative, it signifies an inhibitory or diminishing effect. If a kinetic order is zero, then the variable, raised to zero, becomes 1 and is therefore neutral in the product. Typical kinetic orders are between –1 and +2. The variables in the term include dependent and independent variables (see Chapter 2) in exactly the same format.

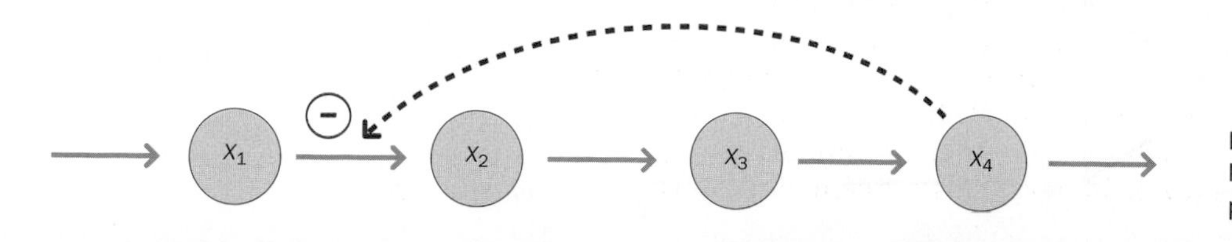

Figure 4.10 Linear pathway with feedback. It is not directly possible to represent this pathway as a Lotka–Volterra system.

With this type of formulation, each equation in a full GMA model with n dependent and m independent variables has the form

$$\dot{X}_i = \sum_{k=1}^{T_i} \pm \gamma_{ik} \prod_{j=1}^{n+m} X_j^{f_{ikj}}, \tag{4.47}$$

where T_i is the number of terms in the ith equation. Here, the large "pi" denotes a product; for instance,

$$\prod_{j=1}^{4} X_j = X_1 X_2 X_3 X_4. \tag{4.48}$$

The format of (4.47) looks complicated, but is actually quite intuitive once it becomes more familiar. One simply enumerates all processes (terms) affecting a variable, decides which variable has a direct effect on a given process, and includes this variable in the term. The variable receives an exponent, and the whole term is multiplied with a rate constant.

As a small example with three dependent variables, consider a branched pathway with one feedback signal, input from the independent variable X_0, and activation by an independent modulator X_4, as shown in **Figure 4.11**. Each equation contains power-law representations of all fluxes v_i that are directly involved. Thus, the GMA equations are

$$\begin{aligned}
\dot{X}_1 &= \gamma_{11} X_0^{f_{110}} X_2^{f_{112}} - \gamma_{12} X_1^{f_{121}} - \gamma_{13} X_1^{f_{131}}, \\
\dot{X}_2 &= \gamma_{21} X_1^{f_{211}} - \gamma_{22} X_2^{f_{222}} X_4^{f_{224}}, \\
\dot{X}_3 &= \gamma_{31} X_1^{f_{311}} - \gamma_{32} X_3^{f_{323}}.
\end{aligned} \tag{4.49}$$

The indices of the rate constants reflect, first, the equation and, second, the term. The kinetic orders have a further index, which refers to the variable with which it is associated. This convention is not always followed, however.

Some of the terms are actually the same, even though they seem different at first glance. For instance, the flux v_2 out of X_1 is the same as the flux into X_2. This equality translates into properties of power-law functions, which in the given case can only be equivalent if $\gamma_{12} = \gamma_{21}$ and $f_{121} = f_{211}$. Similar equalities hold for the two descriptions of the process between X_1 and X_3.

Very similar to the GMA format is the so-called **S-system** format [25, 26]. In this format, all processes entering a variable are merged into a single power-law term, and similarly for all processes leaving the variable. Biologically, this strategy corresponds to a focus on pools (dependent variables) in S-systems rather than on fluxes in GMA systems. Owing to the merging of influxes and effluxes in one net flux each, S-systems have at most one positive and one negative term in each equation, which is very significant for further analyses (see later). Therefore, their general form, with its own names for parameters, is always

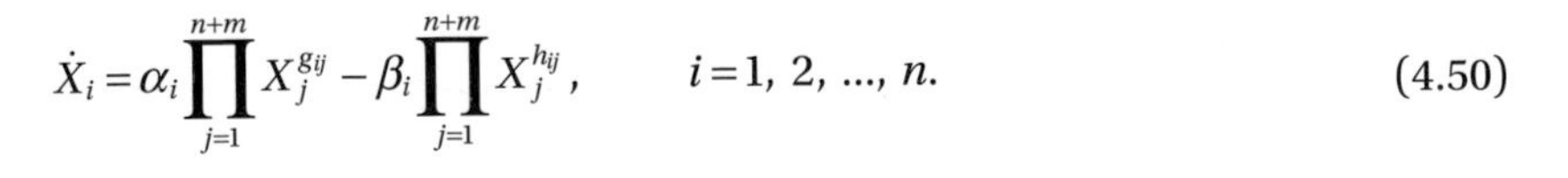

$$\dot{X}_i = \alpha_i \prod_{j=1}^{n+m} X_j^{g_{ij}} - \beta_i \prod_{j=1}^{n+m} X_j^{h_{ij}}, \qquad i = 1, 2, \ldots, n. \tag{4.50}$$

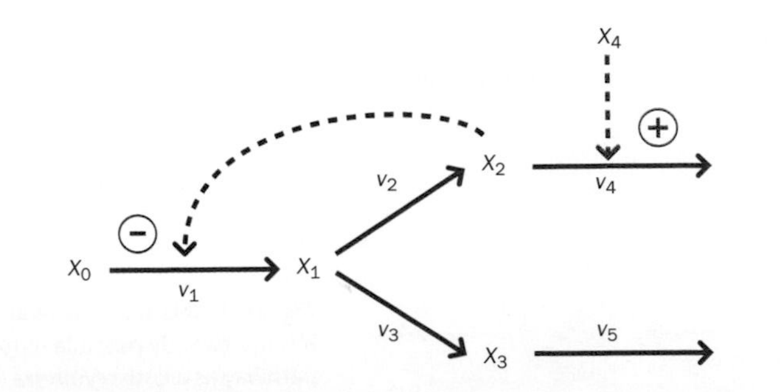

Figure 4.11 Generic pathway with one feedback signal and one external activator. The dependent variable X_1 is produced from a precursor X_0, which is modeled as an independent variable, and is used as substrate for the production of X_2 and X_3. X_2 exerts feedback inhibition on the production of X_1, and X_4 activates the degradation of X_2. Each process is modeled by a function v_i, which in a GMA model (4.49) has the form of a product of power-law functions (4.46).

For the example in Figure 4.11, the production of X_1 is exactly the same in GMA and S-systems. Also, the production and degradation processes of X_2 and X_3 remain unchanged. The only difference between the GMA and S-system models occurs for the efflux from X_1: in the GMA system, this efflux consists of two branches, while it consists of one net efflux in the S-system. Thus, the example in Figure 4.11 is described by the S-system model

$$\begin{aligned} \dot{X}_1 &= \alpha_1 X_0^{g_{10}} X_2^{g_{12}} - \beta_1 X_1^{h_{11}}, \\ \dot{X}_2 &= \alpha_2 X_1^{g_{21}} - \beta_2 X_2^{h_{22}} X_4^{h_{24}}, \\ \dot{X}_3 &= \alpha_3 X_1^{g_{31}} - \beta_3 X_3^{h_{33}}. \end{aligned} \tag{4.51}$$

Note the similarities and differences between this formulation and the GMA form in (4.49). If the S-system and the GMA system are identical at some operating point, such as a steady state, then the equation $\beta_1 X_1^{h_{11}} = \alpha_2 X_1^{g_{21}} + \alpha_3 X_1^{g_{31}}$ holds at this point. Generally, the two types of models differ only at converging or diverging branch points and are otherwise identical.

As a more interesting example, which has the same equations in both the GMA and S-system formats, consider a toggle switch model for the SOS pathway in *Escherichia coli*, with which the bacterium responds to stress situations where DNA is damaged by sending out SOS signals [27]; it is shown diagrammatically in **Figure 4.12A**. The main components are two genes, *lacI* and *λcI*, which are transcribed into the regulator proteins LacR and λCI, respectively. The most interesting feature of the system is cross inhibition: the expression of *lacI* is high when *λcI* has a low expression level, and vice versa. Specifically, the promoter P_L of the *lacI* gene is repressed by λCI and the promoter P_{trc} of the *λcI* gene is repressed by LacR,

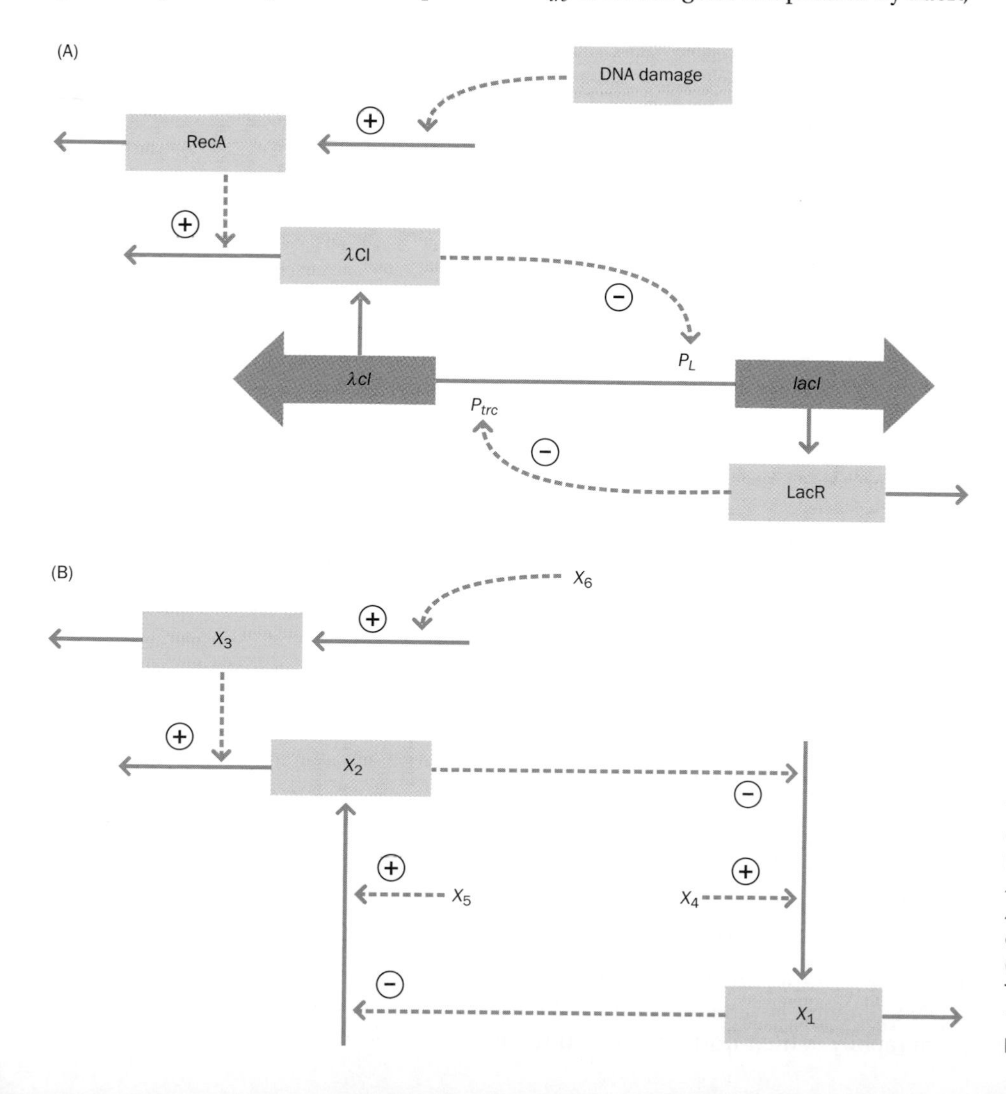

Figure 4.12 Cross inhibition in the SOS pathway of *E. coli*. (A) Diagrammatic representation of the SOS system. (B) Model representation, where the dependent variables are X_1 = LacR, X_2 = λCI, X_3 = RecA, and the independent variables are the promoters $X_4 = P_L$ and $X_5 = P_{trc}$, as well as X_6 = DNA damage. In the simplified model, X_1 and X_2 inhibit each other's production directly, rather than through the promoters (see the text for further discussion). ((A) from Tian T & Burrage K. *Proc. Natl Acad. Sci. USA* 103 [2006] 8372–8377. With permission from National Academy of Sciences, USA.)

subsequently leading to elevated expression of the gene *lacI*. DNA damage leads to activation of the protein RecA, which increasingly cleaves the λCI repressor protein. Reduced amounts of λCI relieve the inhibition of P_L and thus cause higher expression of *lacI*.

As discussed in Chapter 2, we identify the main players and categorize them as dependent and independent variables (see Figure 4.12B). Clear dependent variables, whose dynamics is driven by the system, are the two regulator proteins X_1 = LacR and X_2 = λCI, as well as X_3 = RecA. The genes, while obviously important biologically, are not modeled, because they serve in the model only as precursors of the proteins. A more difficult debate is whether the promoters should be dependent or independent variables. This is one of those decisions that need to be made during the model design phase that we discussed in Chapter 2. On one hand, the promoters are involved in the dynamics of the system. On the other hand, we have no information on how they become available or how they are removed from the system. Weighing our options in favor of simplicity, we make the decision to consider the promoters as independent variables $X_4 = P_L$ and $X_5 = P_{trc}$ with constant values. They are included in the production terms of X_1 and X_2. Because the promoters do not respond to changes in the system, we let X_2 directly inhibit the production of X_1 and we let X_1 directly inhibit the production of X_2. In addition to the promoters, X_6 = DNA is modeled as an independent variable.

To set up the equation for X_1, we identify all production and degradation processes, as well as the variables that directly affect each. Indirect effects are not included; they are realized through the dynamics of the system. The production is affected by its gene's promoter P_L and the inhibition by λCI. Its natural degradation, for instance, through proteases, only depends on its concentration. Thus, the first equation is

$$\dot{X}_1 = \alpha_1 X_2^{g_{12}} X_4^{g_{14}} - \beta_1 X_1^{h_{11}}. \tag{4.52}$$

We do not know the parameter values, but we know already that all are positive, except for g_{12}, which represents the inhibition by λCI.

The second equation is constructed in a similar fashion. A slight difference here is that its degradation is hastened by X_3:

$$\dot{X}_2 = \alpha_2 X_1^{g_{21}} X_5^{g_{25}} - \beta_2 X_2^{h_{22}} X_3^{h_{23}}. \tag{4.53}$$

The third equation describes the dynamics of RecA. Let us begin with its degradation, which depends on its concentration; thus, we formulate the power-law term $\beta_3 X_3^{h_{33}}$. For the production of RecA, we assume as a default that it is constant. However, DNA damage activates this production process, so that X_6 should be included with a kinetic order. Thus, the equation reads

$$\dot{X}_3 = \alpha_3 X_6^{g_{36}} - \beta_3 X_3^{h_{33}}. \tag{4.54}$$

The formulation of the symbolic model (without numerical values) is now complete. Imagine how difficult this step would have been without canonical modeling, that is, if we had been forced to select nonlinear functions for all processes.

For further analysis and for simulations, we need to assign numerical values to all parameters, based on experimental data. This parameterization step is generally rather difficult (see Chapter 5), and we shall skip it for this illustration. Instead, we choose values for all parameters that are typical in BST (see Chapter 5 of [23]) and explore the functionality of the model with these values. The chosen parameter values are $\alpha_1 = 10$, $\alpha_2 = 20$, $\alpha_3 = 3$, $\beta_1 = \beta_2 = \beta_3 = 1$, $g_{12} = -0.4$, $g_{14} = 0.2$, $h_{11} = 0.5$, $g_{21} = -0.4$, $g_{25} = 0.2$, $h_{22} = 0.5$, $h_{23} = 0.2$, $g_{36} = 0.4$, $h_{33} = 0.5$, $X_1(0) = 7$, $X_2(0) = 88$, $X_3(0) = 9$, $X_4 = 10$, $X_5 = 10$, and $X_6 = 1$ (no DNA damage) or $X_6 = 25$ (DNA damage).

We begin the simulation under normal conditions. One hour into the simulation experiment, we induce DNA damage by increasing X_6. The system responds to the damage immediately with a fast increase in X_1 and a rapid drop in X_2 (**Figure 4.13**). We furthermore take into account that DNA damage is repaired after 10 hours. In response, λCI and LacR expression return rapidly to their normal states. Thus, as

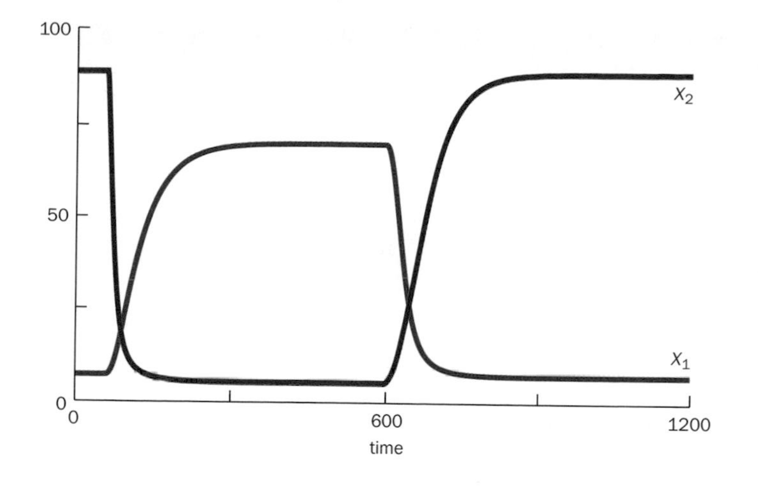

Figure 4.13 Simulation of switches between λCI and LacR expression states. The simulation begins with normal conditions, where the expression levels of λCI and LacR are 88 and 7, respectively. After 60 minutes, DNA is damaged. As a consequence, X_3 increases rapidly (not shown) and cleaves λCI, whose concentration drops in response. The inhibition of P_L is greatly reduced, and LacR rises. After 10 hours, DNA damage in the simulation is repaired, and λCI and LacR expression return to their normal states. Parameter values are given in the text.

these representative results demonstrate, the model generates high expression of λCI under normal conditions and high expression of LacR under conditions of DNA damage. Moreover, the model exhibits sharp switches from one steady state to another, as has been observed in experiments. While the term "switch" has no firm definition, it is used to distinguish a step-like change from a more gradual transition.

The format for processes in GMA and S-system models was not developed arbitrarily. It is the result of a general approximation strategy that is very convenient in biological systems analysis and at the same time is based on solid mathematical principles, namely Taylor's theorem (see Boxes 4.1 and 4.2). The approximation is very similar to the linearization we discussed earlier in the chapter (**Boxes 4.3** and **4.4**). However, the result is nonlinear and usually offers a wider range within which the approximation is good. Moreover, the power-law format of these models is so flexible that it admits models of essentially arbitrary complexity [20].

A more recent alternative for a canonical representation of metabolic processes is the lin–log model [28]. The name stems from the fact that the model is linear in logarithmically transformed variables. Specifically, all metabolites and modulators X_j and all enzymes e_i are expressed in relation to their normal reference values X_j^0 and e_i^0, respectively. Furthermore, the rate is expressed in relation to the reference steady-state flux j_i^0 through the pathway. Thus, the lin–log model has the form

$$\frac{v_i}{J_i^0}=\frac{e_i}{e_i^0}\left[1+\sum_{j=1}^{n+m}\varepsilon_{ij}^0\ln\left(\frac{X_j}{X_j^0}\right)\right], \tag{4.55}$$

where ε_{ij}^0 is the reference elasticity, which plays the same role as kinetic orders in BST.

The lin–log formulation has convenient features with respect to the steady state of the system and models situations with high substrate concentrations better than power-law models. Alas, nothing comes for free: power-law models are much more accurate than lin–log models for small substrate concentrations [29–31]. The lin–log model has been developed out of the framework of **metabolic control analysis (MCA)** (see [32–34] and Chapter 3). The collective experience with lin–log systems is at this point dwarfed by the experience gained for LV systems and models within BST.

Another recent idea is the use of S-shaped modules, rather than power-law functions, for the description of the role of variables within a system [35]. This formulation permits greater flexibility but also requires more parameter values.

An evident question is why we need so many different model structures. The answer is that each model structure has different advantages and disadvantages, for instance, with respect to the range within which the approximation is superb, good, or at least acceptable. The models also differ in the ease with which standard analyses are executable. For instance, LV, lin–log, and S-systems permit the explicit computation of the steady state, which is not the case for GMA systems. While the

BOX 4.3 POWER-LAW APPROXIMATION

The power-law approximation of a nonlinear function F results in a representation that consists of a product of power-law functions (4.46). This approximation is obtained in the following fashion:

1. Take the logarithm of all variables and also of the approximated function F; it is assumed that they are all positive.
2. In the second step, compute the linearization of the result (see Box 4.1).
3. Take the exponential of the linearization result.

The result is a function v_i of the form (4.46). This procedure may sound complicated, but there is a mathematically valid and correct shortcut, which says:

1. Each kinetic order (exponent) is equal to the partial derivative of the approximated function F with respect to the corresponding X_j, multiplied by X_j, and divided by F.
2. The rate constant is subsequently computed by equating the power-law term and the approximated function F at the operating point.

Let us illustrate the power-law approximation procedure with an example, where the approximated function F has one argument (see Box 4.4 for multivariate cases). As an example, we choose again the logistic function (4.42),

$$F(x) = \frac{a}{b + e^{-cx}}, \tag{1}$$

and assume that its parameters are $a = 0.2$, $b = 0.1$, and $c = 0.3$. We also suppose that x is some physical quantity and that its relevant range is therefore restricted to positive values. The task is to find parameters γ and f (we omit indices for simplicity) so that γX^f is a power-law approximation of $F(x)$.

Using the shortcut, we have to compute the derivative of $F(x)$, and, because there is only one dependent variable, this is an ordinary derivative:

$$\frac{dF}{dx} = \frac{ace^{-cx}}{(b + e^{-cx})^2}. \tag{2}$$

Now we multiply this derivative by x and divide by F, and evaluate this expression at some operating point of our choosing:

$$f = \left.\frac{dF}{dx}\frac{x}{F}\right|_{OP} = \left.\frac{ace^{-cx}}{(b + e^{-cx})^2} X \frac{b + e^{-cx}}{a}\right|_{OP}. \tag{3}$$

Algebraic simplification yields

$$f = \left.\frac{cxe^{-cx}}{b + e^{-cx}}\right|_{OP} = \left.\frac{cx}{1 + be^{cx}}\right|_{OP}. \tag{4}$$

Finally, we pick an operating point, such as $x = 5$, and substitute numerical values for the parameters. The result is the desired kinetic order f:

$$f = \frac{0.3 \cdot 5}{1 + 0.1e^{0.3 \cdot 5}} = 1.036. \tag{5}$$

The rate constant is computed in a second step. The function and the approximation must have the exact same value at the operating point ($x = 5$), and this value is $F(5) = 0.619$, according to (4.32) with the parameters chosen above. Thus, we equate F and γX^f for $x = 5$ and solve for γ:

$$F(x) = \gamma x^f\Big|_{OP}, \tag{6}$$

and so

$$\gamma = F(x)x^{-f}\Big|_{OP} = 0.619 \cdot 5^{-1.036} = 0.1168. \tag{7}$$

If we choose a different operating point, the approximation is different. For instance, for $x = 15$, we obtain

$$f = \frac{0.3 \cdot 15}{1 + 0.1e^{0.3 \cdot 15}} = 0.4499, \tag{8}$$

and so

$$\gamma = 1.8 \cdot 15^{-0.4499} = 0.5323. \tag{9}$$

Figure 1 shows $F(x)$ and the power-law approximations at the two chosen operating points. The ranges of the two approximations where the approximation is acceptable are quite different.

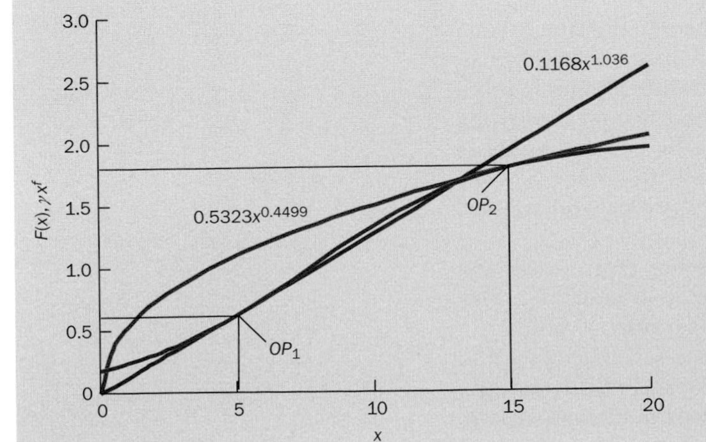

Figure 1 Function $F(x)$ (blue) in (1) with parameters $a = 0.2$, $b = 0.1$ and $c = 0.3$ and two power-law approximations at $x = 5$ (red) and $x = 15$ (green). At the operating points, $F(x)$ and the respective approximations have exactly the same value and slope. Depending on the purpose of the model and the required accuracy, the first approximation might be sufficiently accurate within the interval $3 \le x \le 14$ and the second within the interval $12 \le x \le 20$.

BOX 4.4 POWER-LAW APPROXIMATIONS OF MULTIVARIATE FUNCTIONS

Power-law approximations of functions with more than one dependent variable are constructed in an analogous fashion, except that we now use partial derivatives. We demonstrate the procedure for functions with two arguments; it is then easy to see how functions with more arguments are to be approximated.

As an illustration, suppose we are interested in a Michaelis–Menten process (4.44) that includes an enzyme whose activity changes as a consequence of altered gene expression and protein degradation. The first task is to make the enzyme activity E explicit in the Michaelis–Menten function. It is actually embedded in the parameter V_{max}, which equals the product of a turnover rate of the process and the total enzyme concentration, kE, so that the function with explicit E reads

$$v_p(S,E) = \frac{kES}{K_M + S}. \tag{1}$$

The corresponding power-law representation will have the form

$$V = \alpha S^{g_1} E^{g_2}, \tag{2}$$

and our task is to compute its parameter values α, g_1, and g_2 such that V and v_P have the same value at some operating point of our choice and such that the slopes in the directions of S and E are also the same. Only then is V a proper approximation of v_P. According to the shortcut for kinetic orders, which is the same as in Box 4.3, we:

1. Compute partial derivatives.
2. Multiply them by the corresponding variable.
3. Divide by v_P at the operating point.

These algebraic operations yield the following:

$$\begin{aligned} g_1 &= \left.\frac{\partial v_P}{\partial S}\frac{S}{v_P}\right|_{OP} \\ &= \left.\frac{kE(K_M+S) - kES}{(K_M+S)^2}\frac{S}{v_P}\right|_{OP} \\ &= \left.\frac{kEK_M}{(K_M+S)^2}\frac{S(K_M+S)}{kES}\right|_{OP} \\ &= \left.\frac{K_M}{(K_M+S)}\right|_{OP}, \end{aligned} \tag{3}$$

and

$$\begin{aligned} g_2 &= \left.\frac{\partial v_P}{\partial E}\frac{E}{v_P}\right|_{OP} \\ &= \left.\frac{kS(K_M+S)}{(K_M+S)^2}\frac{E}{v_P}\right|_{OP} \\ &= \left.\frac{kS(K_M+S)}{(K_M+S)^2}\frac{E(K_M+S)}{kES}\right|_{OP} = 1. \end{aligned} \tag{4}$$

The two results are different in several ways. Most importantly, g_1 depends on the operating point, and its value changes depending on where we want to anchor the approximation. For substrate concentrations close to 0, g_1 is close to 1; for $S = K_M$, $g_1 = 0.5$; and for very large substrate concentrations, g_1 approaches 0. Thus, the kinetic order g_1 in this application is always between 0 and 1. By contrast, g_2 does not depend on the operating point, and its value is always 1. The reason for this particular situation is that E contributes to the Michaelis–Menten process (1) as a linear function, which corresponds to a power-law function with kinetic order 1. Thus, with respect to E, the two functions V and v_P are the same for all operating values.

The rate constant is again computed following the rationale that if V Is to be a proper approximation of v_P, then v_P and V must have exactly the same value at the chosen operating point. Thus, equating the two at a chosen operating point, for instance, $S = K_M$, leads to

$$v_p(S,E) = V(S,E) = \alpha S^{g_1} E^{g_2} \quad \text{(for } S = K_M \text{ and some value of } E\text{)}, \tag{5}$$

and so

$$\begin{aligned} \alpha &= v_P S^{-g_1} E^{-g_2} \\ &= \frac{kEK_M}{2K_M} K_M^{-0.5} E^{-1} \\ &= \frac{k}{2} K_M^{-0.5}. \end{aligned} \tag{6}$$

The original rate law and its power-law approximation are shown in **Figure 1** for parameter values $k = 1.5$, $K_M = 10$, and $E = 2$. With respect to E, the power-law representation is exact. In other words, if the substrate concentration S is set at some operating point, the original and the power law are identical for all values of E.

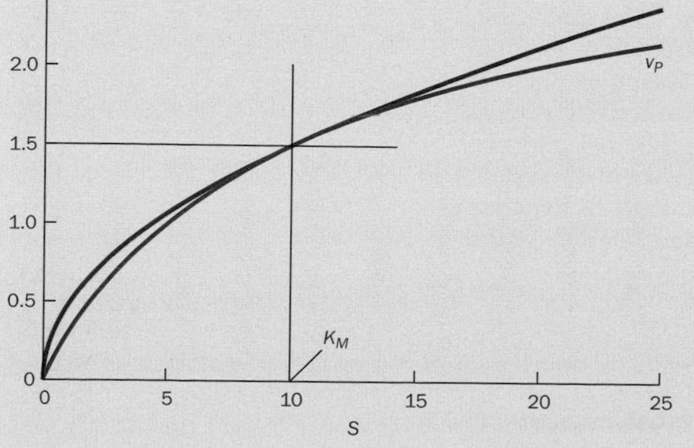

Figure 1 Power-law approximation of a Michaelis–Menten rate law (1) where E is also a dependent variable. At the operating point $S = K_M$, $E = 2$, the original function and the corresponding power-law approximation have the same value and slope.

mathematical structures differ, experience has shown that the choice of a particular canonical model is often (but not always) less influential than the connectivity and regulatory structure of the model. If this structure correctly captures the essence of the biological system, several of the alternative models may perform equally well. However, there is no guarantee, and the modeler is forced to decide on a model structure, possibly based on specific biological insights as well as on matters of mathematical and computational convenience, or to perform comparative analyses with several models (see, for example, [36]).

4.7 More Complicated Dynamical Systems Descriptions

Most biological systems are too complicated to be describable by explicit functions, and even **ordinary differential equation (ODE)** models are not always sufficient. One important limitation of ODEs is their failure to represent spatial components, in addition to changes over time. The mathematician's first answer is often the use of **partial differential equations (PDEs)**, which capture not just dynamic trends but also allow us to analyze variables with respect to changes in their location (see, for example, [6, 37]). For instance, it is rather evident that we need spatial features if we want to describe the formation of plaques in arteries, which critically depend on the details of blood flow and are especially often found at branches in arteries where backflow can form complex eddies. PDEs are very well suited for capturing the spatial–temporal dynamics of such a system. Unfortunately, their analysis requires significant investment in math, and we do not pursue them further in this text. An alternative is **agent-based modeling (ABM)** [38, 39], which we will describe in Chapter 15. A second issue with ODEs is that they do not directly allow for delays in the dynamics of a system. This situation is simpler, because mathematical tricks can be used to permit the modeling of delays [40].

Finally, ODEs are entirely deterministic and do not permit any account of stochasticity. Accounting for stochastic effects can be crucial, as beautiful experiments have shown that exact copies of the same cell may develop very differently over time and that the differences are caused by stochastic variability at the molecular level (see, for example, [41]). Alas, capturing molecular processes in such detail requires much more complicated approaches, such as stochastic differential equations or hybrid systems (see, for example, [42–44]), and it is again a question of judgment and of the modeling purposes as to whether the effort is worthwhile.

STANDARD ANALYSES OF BIOLOGICAL SYSTEMS MODELS

The preceding sections have demonstrated that the choice of a good model for a biological system is neither trivial nor unique. We have encountered alternative models even for simple growth processes, and the choices obviously increase for more complicated systems. Although the various representations differ quite a bit, there are general strategies of analysis. Indeed, much of this analysis is to some degree straightforward and follows guidelines that are more or less independent of the particular system under investigation. Thus, it is time to discuss the most pertinent standard methods for analyzing biological systems. They fall into two broad classes. One pertains to static features and is based on analyses of the system within the neighborhood of a steady state, while the other class contains dynamic features, such as the exploration of possible model responses to perturbations or the characterization of oscillations.

4.8 Steady-State Analysis

A steady state is a condition of a system in which none of the variables change in number, amount, or concentration. This lack of change does not imply that nothing is happening in the system. It simply means that all influxes and effluxes at all pools and variables are in balance. As an example, consider the concentrations of chemical components in the blood. There is ongoing metabolism, and blood cells take up and deliver oxygen. Nevertheless, many of the components of the serum remain within very close ranges about their normal states. These states are of special interest in the analysis of most biological systems.

Steady-state analyses related to dynamical models address three basic questions. First, does the system have one, many, or no steady states, and can we compute them? Second, is the system stable at a given steady state? And third, how sensitive to perturbations is the system at a given steady state? In addition to these more diagnostic topics, one might be interested in controlling, manipulating, or optimizing the steady state of the system. Interestingly, most steady-state analyses are executed with the use of linear mathematics, motivated and supported by the insight of Hartman and Grobman that many behaviors of a nonlinear system can be studied with the corresponding linear system [2]. In fact, now that we have gained some experience with approximations, we can say more exactly what that means. It refers to the (univariate or multivariate) linearization of the nonlinear system, usually with the steady state chosen as the operating point.

In a linear system, the steady state is relatively easy to assess, because the derivatives are zero, by definition of a steady state (that is, no overall change in any of the variables), so that the system of differential equations becomes a system of linear algebraic equations. Recalling (4.27), but allowing for an additional constant input **u**, and restricting the analysis to the steady state yields

$$\frac{d\mathbf{X}}{dt} = \dot{\mathbf{X}} = \mathbf{AX} + \mathbf{u} = 0. \tag{4.56}$$

Linear algebra tells us that there are three possibilities. The system may have no steady-state solution, or exactly one solution, or it may happen that whole lines, planes, or higher-dimensional analogs of planes satisfy (4.56). As a simple two-variable example, let us look at the following five systems S1–S5, which are quite similar:

S1:
$$\dot{X}_1 = 2X_2 - 2X_1,$$
$$\dot{X}_2 = X_1 - 2X_2.$$

In this example, there is no input, and both variables are decreasing toward the "trivial" steady-state solution $X_1 = X_2 = 0$. We can confirm this easily and directly, either by solving the system numerically or by setting the derivatives in both equations equal to zero and adding them together (which we are allowed to do according to linear algebra). The result of the latter is $X_1 - 2X_1 = 0$, which is only possible for $X_1 = 0$. Plugging this value into one of the equations yields $X_2 = 0$. Thus, the solution $(X_1 = X_2 = 0)$ is unique and not particularly interesting.

S2:
$$\dot{X}_1 = 2X_2 - 0.5X_1,$$
$$\dot{X}_2 = X_1 - 2X_2.$$

This system is very similar, except that the coefficient of X_1 in the first equation is –0.5 instead of –2. The solution is again exponential, but this time both variables grow without ceasing. The system again satisfies that trivial solution $X_1 = X_2 = 0$, but, in this case, the dynamical solution does not approach this steady-state solution. In fact, if we start a simulation anywhere but at (0, 0), the system leads to unbounded growth.

S3:
$$\dot{X}_1 = 2X_2 - X_1,$$
$$\dot{X}_2 = X_1 - 2X_2.$$

In this case, the coefficient of X_1 in the first equation is –1. Setting the derivatives in both equations equal to zero and multiplying one of them by –1 shows immediately that they are equal. Both require that X_1 be twice as big as X_2, but there are no other conditions. A bit of analysis (or computer simulation) confirms that any

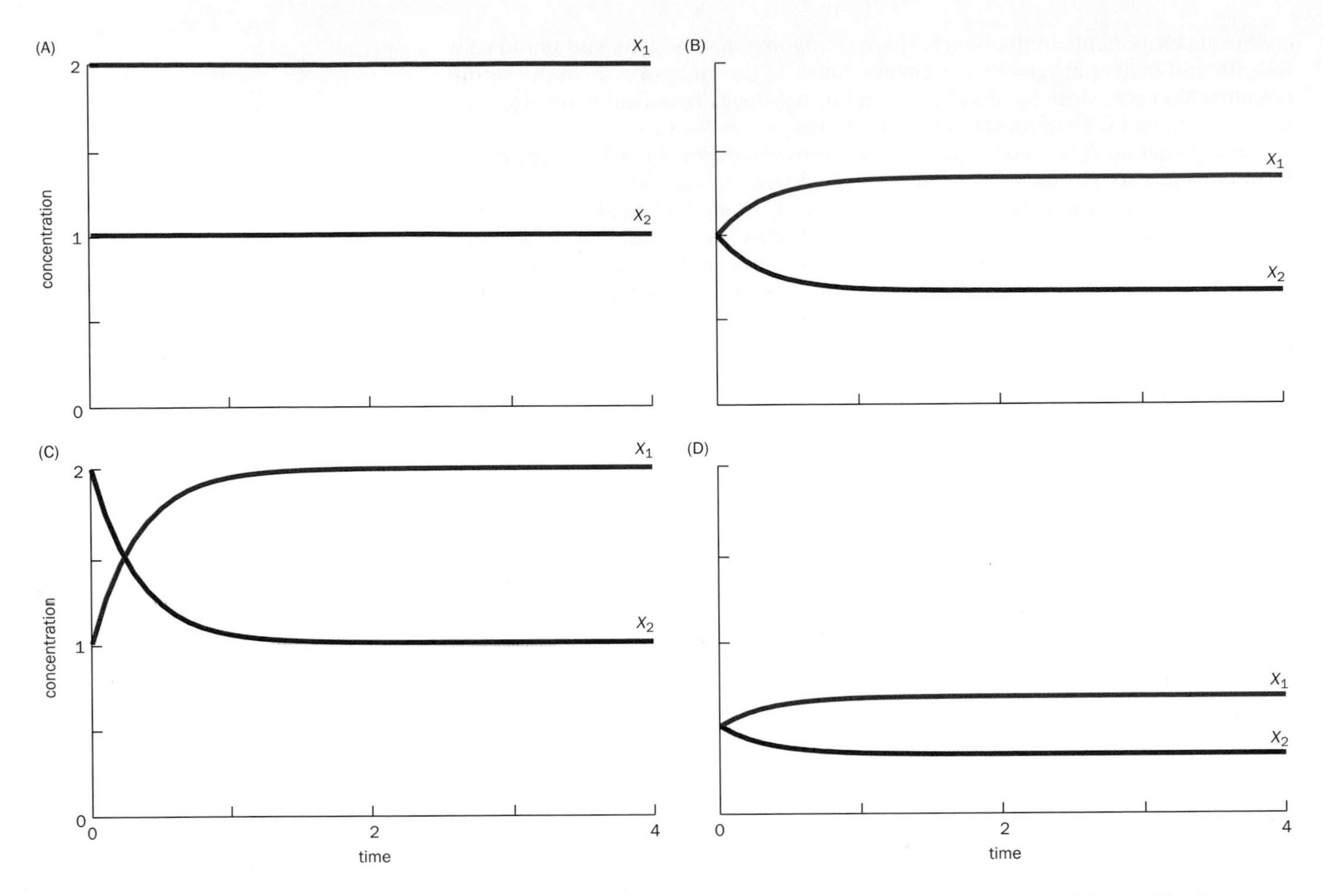

Figure 4.14 Four solutions of the system S3, which differ solely in the initial conditions. (A) $X_1 = 2, X_2 = 1$; (B) $X_1 = 1, X_2 = 1$; (C) $X_1 = 1, X_2 = 2$; (D) $X_1 = 0.5, X_2 = 0.5$.

combination consisting of an arbitrary value of X_1 and half that value for X_2 satisfies the steady state. The system has infinitely many solutions, including (0, 0) (**Figure 4.14**); all these solutions lie on a straight line, defined by $X_1 = 2X_2$.

S4:
$$\dot{X}_1 = 2X_2 - 2X_1 + 1,$$
$$\dot{X}_2 = X_1 - 2X_2.$$

The only difference between S4 and S1 is the constant input with value 1. This change drastically changes the dynamic and the steady-state solutions. The latter is no longer trivial ($X_1 = X_2 = 0$), but has the unique values $X_1 = 1$, $X_2 = 0.5$. In contrast to the decreasing processes toward 0 in S1, an input is feeding the system, and the system quickly finds a balance at (1, 0.5).

S5:
$$\dot{X}_1 = 2X_2 - X_1 + 1,$$
$$\dot{X}_2 = X_1 - 2X_2.$$

This system also has an input of 1, and the only difference with S4 is the coefficient of X_1 in the first equation. Setting the derivatives in both equations equal to zero and adding them yields the requirement 0 = 1, which we know cannot be satisfied. The system has no steady state—not even a trivial one.

The computation of steady states in nonlinear systems is in general much harder. In fact, even if we know that the system has a steady-state solution, we may not be able to compute it with simple analytical means and need to approximate the solution with some iterative algorithm. We already saw a harmless-looking example in Chapter 2, namely the equation $e^x - 4x = 0$. We could easily envision that this equation described the steady state of the differential equation $dx/dt = e^x - 4x$. If we cannot even solve one nonlinear equation, imagine how little is possible in a system with dozens of nonlinear equations! You might ask why we don't use linearization and compute the steady state of the linearized system. The reason this seemingly clever idea won't work is that we need an operating point for the linearization. But if we don't know what the steady state is, we cannot use it as operating point, and choosing a different point is not helpful.

We may obtain the steady state(s) of a nonlinear system with two numerical strategies. First, we can do a simulation and check where the solution stabilizes. This strategy is very simple and often useful. It has two drawbacks, however. First, if the steady state is unstable, then the simulation will avoid it and diverge away from it (see the system S2 above and later). Second, in contrast to linear systems, nonlinear systems may possess (many) different, isolated steady states. Isolated means that points in between are not steady states. As an example, consider the single nonlinear differential equation

$$\dot{x} = -0.1(x-1)(x-2)(x-4). \tag{4.57}$$

We can convince ourselves easily that 1, 2, and 4 are steady-state solutions, because for each of these values for x the right-hand side becomes zero. However, values in between, such as $x = 3$, are not steady-state solutions. The simulation strategy for computing the steady states would yield $x = 4$ for all initial values above 2 (see the red lines in **Figure 4.15**). For all initial values below 2, the simulation would yield $x = 1$ as the steady state, and, unless we started exactly at 2 (or used insight and imagination), the (unstable) steady state $x = 2$ would slip through the cracks. In the case presented here, the failure would be easy to spot, but in realistic models of larger size that is not always so, and numerical solution strategies can easily miss a steady state.

The second strategy for finding steady-state solutions uses **search algorithms**, of which there are many different types and software implementations. We will discuss some of them in the context of parameter estimation (see Chapter 5). The generic idea of many of these algorithms is that we start with a wild guess (the literature calls it an educated guess or a guesstimate). In the example of (4.57), shown in Figure 4.15, we might begin with the (wrong) guess that 3.4 could be a steady state. The algorithm computes how good this guess is by evaluating all steady-state equations of the system and computing how different they are from zero; in our case, there is only one equation, and the **residual error**, that is, the difference from the best possible solution, is $-0.1(3.4 - 1)(3.4 - 2)(3.4 - 4) = 0.2016$. Next, the

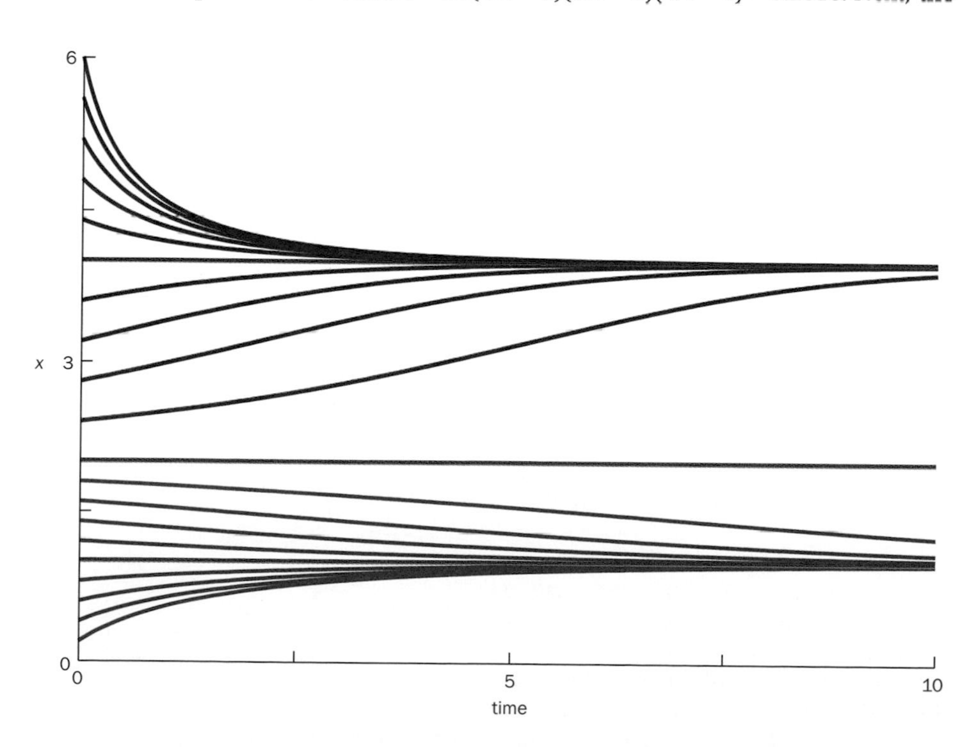

Figure 4.15 Some dynamical (red, blue) and steady-state (green) solutions of (4.57). The steady state at 2 is never reached unless the simulation starts exactly at 2, because it is unstable.

algorithm checks solutions close to the initial guess (for instance, 3.3 and 3.5) and determines the direction where the solution has the highest chance of improving. In our case, $x = 3.3$ yields an increased residual error of 0.2093, while $x = 3.5$ corresponds to a lower (= better) residual error of 0.1875. The algorithm now assumes that, because the error at $x = 3.5$ is lower, the best chances of improving the solution lie with even higher values of x. Thus, the algorithm updates the current guess with a higher number, such as 3.6. The technical details of this type of gradient search (such as the exact direction and distance for the next move) can be very complicated, especially for systems with many equations. More details are presented in Chapter 5.

While almost no nonlinear models permit an easy computation of their steady states, some canonical forms are shining exceptions. The first is the LV system that we discussed before (see (4.45)). Setting the derivatives equal to zero yields

$$\sum_{j=1}^{n} a_{ij} X_i X_j + b_i X_i = 0. \tag{4.58}$$

It is not hard to see that $X_i = 0$ is a solution. Because all n equations have to be satisfied for a steady state, $X_i = 0$ must be true for all variables. We are again encountering a trivial solution. This solution is not very interesting in most cases, but it is nevertheless useful: employing an old math trick, we record—but then intentionally exclude—this trivial solution, which allows us to divide each equation by X_i, an operation that was not permitted earlier owing to the possibility of $X_i = 0$. For simplicity, let us assume that all variables should be different from 0 at the steady state. Then we divide each equation by the corresponding X_i, and the nontrivial steady-state equations of the LV system become

$$\sum_{j=1}^{n} a_{ij} X_j = -b_i. \tag{4.59}$$

This is a system of linear algebraic equations, and we can apply all the tools of linear algebra. Of course, it is possible that some variables are 0 and others are not. Some people, but not all, call these solutions trivial as well.

The fact that we can use linear algebra for a nonlinear system is indeed a rare event, but LV systems are not the only cases. Something similar occurs with S-systems (4.50) [45]. Suppose for simplicity that the system does not contain independent variables and set the derivatives equal to zero, which yields

$$\alpha_i \prod_{j=1}^{n} X_j^{g_{ij}} - \beta_i \prod_{j=1}^{n} X_j^{h_{ij}} = 0, \qquad i = 1, 2, \ldots, n. \tag{4.60}$$

This result does not look linear at all. However, once we move the β-term to the right-hand side and insist that all variables are positive, we can take logarithms on both sides:

$$\ln \alpha_i + \ln\left(\prod_{j=1}^{n} X_j^{g_{ij}} \right) = \ln \beta_i + \ln\left(\prod_{j=1}^{n} X_j^{h_{ij}} \right). \tag{4.61}$$

Now remember that

$$\ln\left(\prod_{j=1}^{n} Z_j \right) = \sum_{j=1}^{n} \ln Z_j$$

is true for any positive quantities Z_j and that $\ln X^p = \ln e^{p \ln X} = p \ln X$ is true for positive X and any real parameter p. Thus, we obtain

$$\ln\left(\prod_{j=1}^{n} X_j^{g_{ij}}\right) = \sum_{j=1}^{n} g_{ij} \ln X_j \tag{4.62}$$

and the corresponding equations for terms with h_{ij}. When we rename the variables $y_i = \ln X_i$ and rearrange the equation, we obtain the result

$$\sum_{j=1}^{n} g_{ij} y_j - \sum_{j=1}^{n} h_{ij} y_j = \ln \beta_i - \ln \alpha_i. \tag{4.63}$$

Finally, we rename $a_{ij} = g_{ij} - h_{ij}$ for all i and j and define $b_i = \ln(\beta_i/\alpha_i)$, which reveals the intriguing truth that the steady-state equations of an S-system, which look very nonlinear, are indeed linear:

$$\begin{aligned}
a_{11}y_1 + a_{12}y_2 + a_{13}y_3 + \cdots + a_{1n}y_n &= b_1, \\
a_{21}y_1 + a_{22}y_2 + a_{23}y_3 + \cdots + a_{2n}y_n &= b_2, \\
a_{31}y_1 + a_{32}y_2 + a_{33}y_3 + \cdots + a_{3n}y_n &= b_3, \\
&\vdots \\
a_{n1}y_1 + a_{n2}y_2 + a_{n3}y_3 + \cdots + a_{nn}y_n &= b_n.
\end{aligned} \tag{4.64}$$

For later biological interpretations, we just must remember to translate back from y's to X's with an exponential transformation, once the analysis is completed. The same procedure works for systems with independent variables. The reader should confirm this or see [23].

As a final exceptional case of a nonlinear system with linear steady-state equations, consider the lin–log model (4.55) [28]. Setting the rates on the left-hand side of (4.55) equal to zero yields

$$\frac{e_i}{e_i^0}\left[1 + \sum_{j=1}^{n+m} \varepsilon_{ij}^0 \ln\left(\frac{X_j}{X_j^0}\right)\right] = 0, \tag{4.65}$$

and, justifiably assuming that the reference enzyme activities e_i^0 are not zero, we obtain

$$\sum_{j=1}^{n+m} \varepsilon_{ij}^0 \ln\left(\frac{X_j}{X_j^0}\right) = -1. \tag{4.66}$$

If we rename the variables as $y_j = \ln(X_j/X_j^0)$, we see that the steady-state equations in y-variables are linear.

With the exception of these systems, very few classes of nonlinear models have steady-state solutions that can be computed explicitly with methods of calculus and linear algebra. Even mass-action and GMA systems do not allow us to do this.

Once we have computed a steady state, either analytically or numerically, we can use it as an operating point for linearization and analyze important features such as stability and parameter sensitivities, as we will discuss next.

4.9 Stability Analysis

Stability analysis assesses the degree to which a system can tolerate external perturbations. In the simplest case of local stability analysis, we can ask whether the system will return to a steady state after a small perturbation. An analogy often used is a ball in a half-spherical bowl (**Figure 4.16A**). Left to its own devices, the ball will roll to the bottom of the bowl, maybe roll up and down a few times, and finally stay at the bottom. If we move the ball away from the bottom (that is, if we perturb it), it

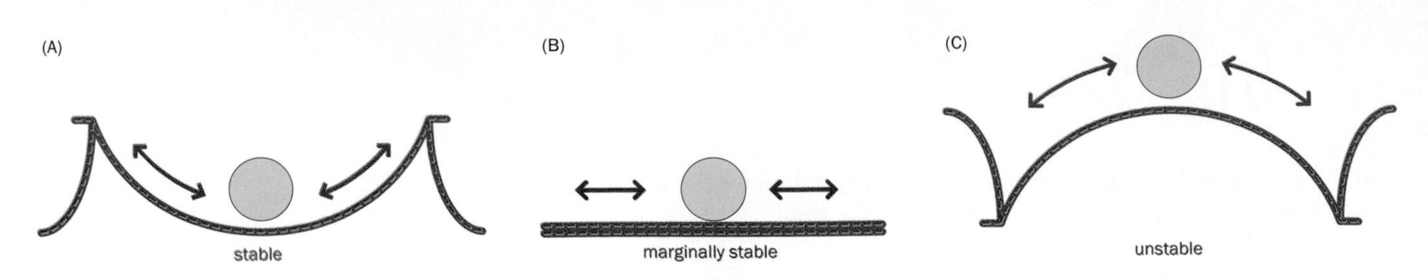

Figure 4.16 Illustration of different degrees of stability. (A) A ball placed in a bowl rolls to the bottom and returns to this position if it is slightly moved. (B) A ball on a perfectly flat surface rolls to a new position when it is tapped. (C) A ball placed perfectly on top of an inverted bowl might stay there for a while, but the slightest perturbation will cause it to roll to one of the sides.

will roll back to the stable steady state at the bottom of the bowl. Now suppose the bowl is turned upside down (Figure 4.16C). With a very steady hand, we might be able to place the ball on the top of the inverted bowl, but even the slightest perturbation will cause it to roll down. The top of the bowl is a steady state, but it is unstable. As an intermediate case, imagine a ball on a perfectly flat, horizontal surface (Figure 4.16B). Now any perturbation will cause the ball to roll and, owing to friction, come to rest at a new location. All points of the flat surface are steady-state points, but all are only marginally stable. Note that these considerations are true for relatively small perturbations. If we were to really whack the ball in the example in Figure 4.16A, it might fly out of the bowl and land on the floor. The same is true for local stability analysis: it only addresses small perturbations. Before we discuss technical details of stability, we should note that the terminology "the system is stable" is unfortunate, if not wrong. The reason is that a system often has several steady states. Thus, the correct terminology is "the system is stable at the steady state ..." or "the steady state ... of the system is stable."

Whether the steady state of a dynamical system is stable or not is not a trivial question. However, it can be answered computationally in two ways. First, we may actually start a simulation with the system at the steady state and perturb it slightly. If it returns to the same steady state, this state is locally stable. Second, we can study the system matrix. In the case of a linear system, the matrix is given directly, while nonlinear systems must first be linearized, so that the matrix of interest is the Jacobian (4.34), which contains partial derivatives of the system. In either case, the decision on stability rests with the eigenvalues of the matrix, and while these (real- or complex-valued) eigenvalues are complicated features of matrices, the rule for stability is simple. If any of the eigenvalues has a positive real part, even if it is just one out of three dozen or so eigenvalues, the system is locally unstable. For stability, all real parts have to be negative. Cases with real parts equal to zero are complicated and require additional analysis or simulation studies. Fortunately, software packages such as Mathematica®, MATLAB®, and PLAS [1] compute eigenvalues automatically.

In the case of a two-variable linear system without input (4.27), the stability analysis is particularly instructive, because one can visually distinguish all different behaviors of the system close to its steady state (0, 0). Specifically, the stability of (0, 0) is determined by three features of the matrix **A**, namely, its trace $\mathrm{tr}\,\mathbf{A} = A_{11} + A_{22}$, determinant $\det \mathbf{A} = A_{11}A_{22} - A_{12}A_{21}$, and discriminant $d(\mathbf{A}) = (\mathrm{tr}\,\mathbf{A})^2 - 4\det \mathbf{A}$, which are related to the two eigenvalues of **A**. Depending on the combination of these three features of **A**, the behaviors of the system close to (0, 0) can be quite varied. They may look like sinks, sources, attracting or repelling stars, inward or outward spirals, saddles, or collections of lines, as sketched in **Figure 4.17**. To reveal these images with a simulation, one solves the same system (4.27) with the same matrix **A** from many different initial values and superimposes all plots (**Figure 4.18**). Similar behaviors occur for higher-dimensional systems, where they are, however, more difficult to visualize.

The top half of the plot in Figure 4.17 contains on the left the stable systems, where both eigenvalues have negative real parts, and on the right the corresponding unstable systems, where the eigenvalues have positive real parts. Special situations occur on the axes and on the parabola given by the discriminant $d(\mathbf{A})$. For example, on crossing the vertical axis, the behaviors are transitions from stability

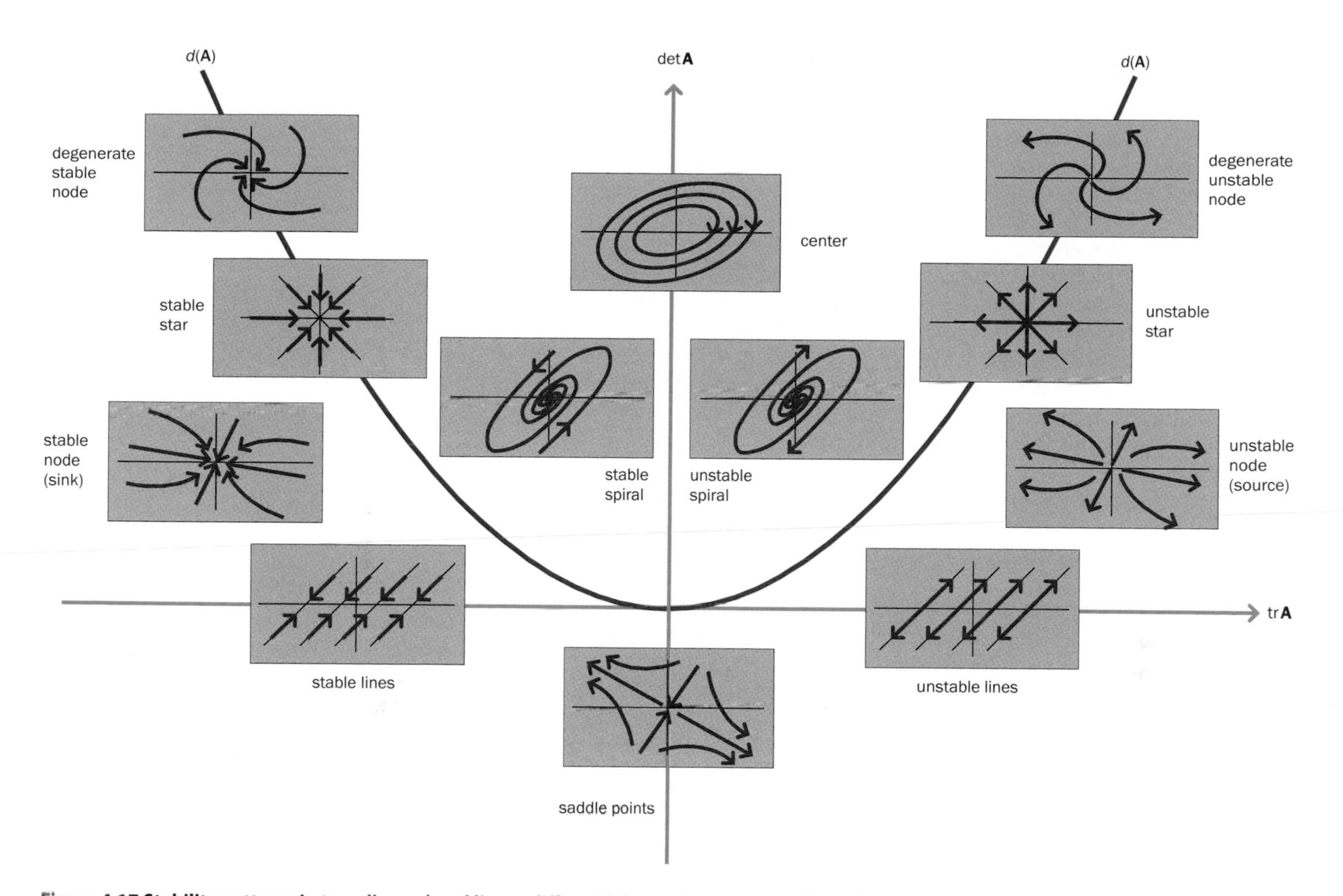

Figure 4.17 Stability patterns in two-dimensional linear differential equation systems without input—see (4.27). The elements of the system matrix **A** determine its trace, tr **A**, determinant, det **A**, and discriminant, $d(\mathbf{A})$. Different combinations of these features, in turn, lead to different system behaviors close to the steady state (0, 0).

to instability. In the particular case of a center, which is the transition case between stable and unstable spirals, all solutions collectively form an infinite set of concentric ellipses that cover the entire plane. The interesting case of a saddle point lies in the center of the bottom half of the plot: coming exactly from one of two opposite directions, the steady state appears to be stable; coming exactly from a second set of two opposite directions, the steady state appears to be unstable; coming from any other directions, the trajectories of the system seem to approach, but then curve away from, the origin. A unique situation occurs for tr $\mathbf{A}$ = det $\mathbf{A} = d(\mathbf{A}) = 0$. Here, the two eigenvalues are the same (so there is really only one eigenvalue), with a value of zero. It could be in this case that **A** contains four zero coefficients, so that both derivatives equal zero, and wherever the system starts, it stays. It could also be that **A** is not filled with zeros, and the system contains one line of steady states and all other trajectories are unstable. Further details can be found in [46].

The eigenvalue analysis just described is very important, but it is also limited. First, in the case of nonlinear systems, the deduction of stability is only guaranteed to be correct for infinitesimally small perturbations. In reality, stable systems tolerate larger perturbations, but there is no mathematical guarantee. Second, definite conclusions are difficult if one or more eigenvalues have real parts of 0. There are other types of stability that cannot be addressed with eigenvalue analysis alone. For instance, if one changes the parameter values in a system, it may happen that a stable steady state becomes unstable and that the system begins to oscillate around the unstable steady state. An obvious question is under what conditions something like that might happen. The answer to this question is complicated and constitutes an ongoing research topic in the field of dynamical systems analysis. We will show an example in the section on systems dynamics.

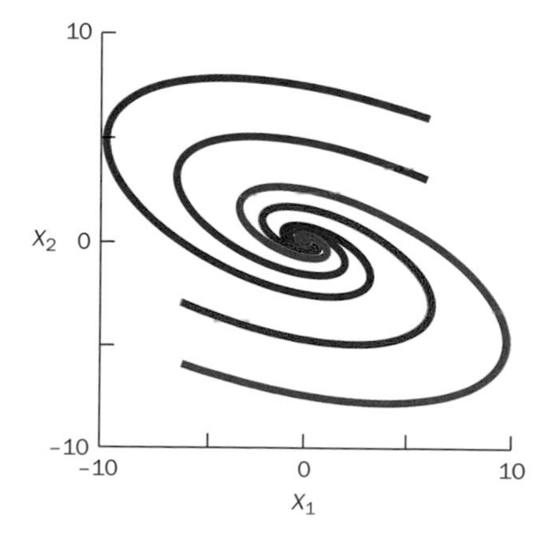

Figure 4.18 Visualization of a stable spiral. The coefficients of **A** are $(A_{11}, A_{12}, A_{21}, A_{22}) = (-3, -6, 2, 1)$ and initial values for the four superimposed trajectories are red, (6, 6); purple, (6, 3); green, (–6, –3); and blue, (–6, –6). Other initial values would lead to similar trajectories in between, collectively covering the entire plane.

4.10 Parameter Sensitivity

Sensitivity analysis is concerned with the generic question of how much a system is affected by small alterations in parameter values. In a biological context, such an alteration could represent a mutation or some permanent change in an enzyme activity that could be due to aging or disease. Questions of sensitivity are fundamentally different from local stability analysis. In stability analysis, one or more of the dependent variables are perturbed, the system responds to this perturbation, and one studies the response, which is usually a return to the steady state. In sensitivity and **gain** analysis, by contrast, parameters or independent variables are changed, respectively. These changes are permanent in the sense that the system cannot counteract them and they almost always cause the system to assume a new steady state. Thus, the question is how different the new steady state is from the old. It may happen that even large variations in parameters do not change the location of the steady state much, but it can also happen that a minute change in some crucial parameter throws the system off course. In a biomedical application, two individuals may be exposed to the same environmental toxicant (perturbation), but respond differently. One person may not be visibly affected, while the other may become very sick. The difference could be due to different expression levels (parameter values) of some key genes.

Sensitivity analysis is a crucial component of any systems analysis, because it can quickly show if a model wrong. Good, **robust** models usually have low sensitivities, which means that they are quite tolerant to small, persistent alterations, in which a parameter remains altered. Thus, if some of the sensitivities in the model are very high, we typically have cause for concern. However, there are exceptions, for instance in signal transduction systems, where even small changes in signal intensity are greatly amplified (see Chapter 9). Typical sensitivity analyses are again executed by computing the derivative of a system feature with respect to a parameter. While steady-state sensitivities are the most prevalent, it is also possible to compute sensitivities of other features, such as trajectories or amplitudes of stable oscillations.

The easiest way to see how sensitivity analysis works is to study an explicit function. As a specific illustration, let us look again at the logistic function $F(x) = a/(b + e^{-cx})$, (4.42). We have already seen that changing the parameter c affects the steepness of the sigmoidal function (see Figure 4.8). Let us now look at the sensitivity of the function with respect to the parameter b. As with c, we can just change b to some arbitrary different value, and maybe repeat this exploration a few times with different numbers. This strategy gives us some idea in this simple case, but we are interested in a more systematic way to compute the effect of b on F. So, we shall study the effect of a generic, small change in b. To make our intent of testing the effects of b explicit, we write the function as $F(x; b)$. The result of a small change Δb in $F(x; b)$ is given as $F(x; b + \Delta b)$. Using the **Taylor approximation** theorem (see Box 4.1), we can linearize this expression by using the operating point $F(x; b)$, the slope, and the distance of the deviation (see (4.29) and (4.30)). The result in this case of a single-variable function is

$$F(x;b+\Delta b) \approx F(x;b) + \frac{\partial F(x;b)}{\partial b}\Delta b, \tag{4.67}$$

where the expression with the ∂'s indicates the partial derivative of F with respect to b. The change in F caused by a small change in b is

$$\Delta F \approx F(x;b+\Delta b) - F(x;b). \tag{4.68}$$

Rearranging this equation and using (4.67) gives us the desired answer, namely, a quantification of how much F changes if b is altered:

$$\Delta F \approx \frac{\partial F(x;b)}{\partial b}\Delta b. \tag{4.69}$$

To put this result in words: a small change in b is translated into a change in F with magnitude $\partial F(x; b)/\partial b$. If this expression is equal to 1, the change in F is equal to the change in b; if it is larger than 1, the perturbation is amplified; and if it is smaller than 1, the perturbation is attenuated. According to Taylor's theorem, these answers are absolutely accurate only for infinitesimally small changes in b, but they are sufficiently good for modest—or sometimes even large—alterations in b. The take-home message is that the effect of b on F is driven by the partial derivative with respect to b, and this message is unchanged even for differential equations and systems of differential equations.

As a numerical example with the logistic function, let us assume that $a = 2$ and $c = 0.25$, and that the normal value of b is 1. The partial derivative of F with respect to b is computed while keeping all other parameters and the variable x constant. Thus,

$$\frac{\partial F(x;b)}{\partial b} = \frac{-a}{(b+e^{-cx})^2}. \tag{4.70}$$

Obviously, the right-hand side depends on x, which means that the sensitivity of F with respect to b is different for different values of x. A typical example is the sensitivity of the nontrivial steady state. In the case of F, the function approaches this state for x going to $+\infty$. Maybe we would prefer a finite number for our computation, but we can still compute the sensitivity, because for x going to $+\infty$, the exponential function e^{-cx} goes to zero. Therefore, the sensitivity is

$$\frac{\partial F(x;b)}{\partial b} = -\frac{a}{b^2} \quad \text{for} \quad x \to \infty. \tag{4.71}$$

Plugging this result into (4.69), along with the values $a = 2$ and $b = 1$, we obtain an estimate for how much the nontrivial steady state of F will change if b is altered. For instance, if $\Delta b = 0.1$, which would mean a 10% perturbation from its normal value of 1, the steady state of F is predicted to change by $-2 \times 0.1 = -0.2$. In fact, if we numerically compute $F(x; b = 1.1)$ for x going to $+\infty$, the result is $-2/1.1 = -1.81818$, which is close to the predicted 0.2 decrease.

It is not difficult to imagine that our function $F(x; b)$ could be the steady-state solution of a differential equation. Reinterpreting our procedure and results then tells us how to do steady-state sensitivity analysis for a single differential equation, and systems of equations follow the same principles. Software like PLAS [1] automates this computation for S-systems and GMA systems.

The profile of sensitivities is an important metric for assessing biological systems. If the sensitivities with respect to parameter p_1 are relatively high in magnitude, while they are very close to zero for parameter p_2, then it is important to obtain the numerical value of p_1 as accurately as possible, whereas a small error in p_2 has little effect on the system.

4.11 Analysis of Systems Dynamics

Because the differential equations describing the dynamical system are typically nonlinear, we seldom have the luxury of being able to compute explicit solutions, in which each dependent variable can be formulated as a function of time. The variables are indeed functions of time, and we can grind them out on a computer with so-called numerical integrators, but we cannot write them down, for instance, as $X(t) = X_0e^{at}$, as we did in the case of linear differential equations. Nonetheless, there is an entire branch of mathematics that studies dynamic phenomena, such as damped and stable oscillations, **bifurcations**, and chaos with analytical and numerical methods. Even introductions to these types of investigations tend to require quite a bit of effort and mastery of a variety of mathematical techniques (see, for example, [47]), but we will discuss some of these issues at a somewhat superficial level toward the end of this chapter.

While there are some nonlinear ODEs that do have explicit solutions, they are typically restricted to such special constellations of mathematical structures and parameters that they must be considered extremely rare exceptions in the real world. A consequence is that most dynamic analyses in systems biology are done by means of computer simulations. As we have already discussed, rigorous mathematical results are always to be preferred over a collection of numerical results, but in biological applications we can seldom get around computational simulation analyses.

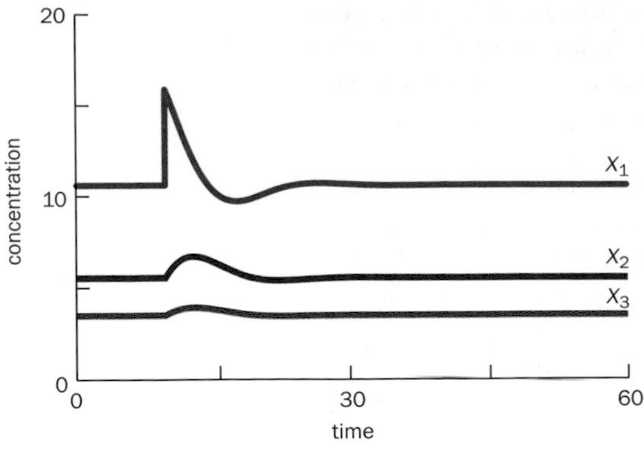

Figure 4.19 Simulation of a bolus experiment with a dependent variable. A 50% bolus of X_1 is supplied at time $t = 10$. All variables change in response, but the system quickly returns to its steady state.

Typical analyses fall into four classes:

1. Bolus experiments
2. Persistent changes in structure or input
3. Comprehensive screening experiments
4. Analyses of critical points, where the system behavior changes qualitatively

Bolus experiments usually begin with the system in a stable steady state. Some time period into the experiment, one of the variables is altered. If this variable is a dependent variable, the simulation shows whether and how the system returns to the steady state. The "how" here refers to the time it takes to return and the shape of the transient, which is the collective name for the time course between the alteration (stimulus) and the move back to the old steady state, to a new one, or to an oscillatory behavior. Note that the time it takes to return to the steady state is, strictly speaking, infinite in most dynamical models. For instance, if the solution is a decreasing exponential function, the final value of 0 is mathematically only reached when x reaches infinity. A solution to this predicament is to study how long it takes the system to return to within a small percentage of the steady state, such as 1%.

As an example, consider again the simple generic pathway in Figure 4.11, where time units are minutes. Using the GMA representation with parameter values $\gamma_{11} = 20$, $f_{110} = 1$, $f_{112} = -0.75$, $\gamma_{12} = 0.4$, $f_{121} = 0.75$, $\gamma_{13} = 2$, $f_{131} = 0.2$, $\gamma_{22} = 0.5$, $f_{222} = 0.5$, $f_{224} = 1$, $\gamma_{32} = 2.5$, $f_{323} = 0.2$, $X_0 = 1$, $X_4 = 2$, the steady state is $X_1 = 10.59$, $X_2 = 5.52$, $X_3 = 3.47$. As a bolus experiment with a dependent variable, we raise X_1 at time 10 minutes by 50%. Immediately following, the other variables respond, showing slight **overshoots**, and the system returns to within 1% of its steady state in under 20 minutes (**Figure 4.19**).

If the altered variable in the bolus experiment is an independent variable, such as an input, then any alteration in its value will likely affect the steady state. Therefore, the bolus is usually applied for only a few time units. Afterwards, the altered variable is reset to its value at steady state. The simulation shows how the system responds to this temporary change. As an example, let us increase X_4 in the branched pathway by 50% at time $t = 10$ minutes and reset X_4 after at $t = 45$ minutes. The responses are quite different (**Figure 4.20**). In particular, the system essentially reaches a new steady state after about 35 minutes.

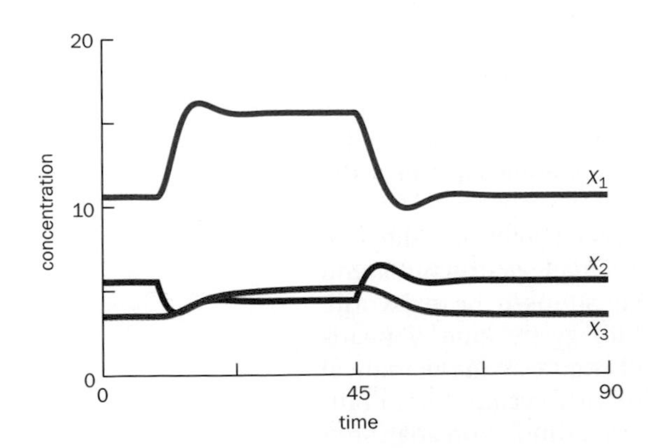

Figure 4.20 Simulation of a bolus experiment with an independent variable. A 50% bolus of X_4 is supplied at time $t = 10$ minutes and stopped at $t = 45$ minutes. During the altered setting, the system assumes a different steady state.

Simulations targeting persistent changes in structure or input are implemented with permanent alterations in parameter values or independent variables. A change in a parameter value could model a mutation or disease, which for instance could affect the rate in a growth model or an enzyme activity in a metabolic model. An alteration in an independent variable could be interpreted as a change in environmental conditions, which could affect the availability of a substrate or other essential factor. In these types of simulations, one may study transients between the change and the completion of the response, but the target is usually the effect of the alteration on the steady state. An illustration is the previous example of a change in X_4. However, here one would not reset X_4 at $t = 45$ minutes, and the system would stay close to the elevated state we observe at $t = 45$ minutes. A different example would be the change in the parameter γ_{32}, which represents the rate of degradation of X_3.

Comprehensive screening experiments are usually launched if one is not entirely sure what to look for. Many options are possible. In a grid search, every parameter is allowed to take on a certain number of possible values; let us say 6. A complete simulation now computes the system (transients and/or steady state) for every combination of parameter settings, thereby producing a coarse, but in a way comprehensive, picture of possible responses. The problem with this approach is that it easily gets out of hand, because a system with n parameters requires 6^n simulations, and, for instance, $6^{20} = 3.66 \times 10^{15}$. Even for a computer executing 10 simulations per second, it would take over 100 billion hours (more than 11½ million years) to complete the grid search! The rapid increase in numbers of simulations in these types of studies even has a name: combinatorial explosion. Various statistical techniques for overcoming this problem have been developed. They prescribe how to sample the vast space in a fashion that is manageable and at the same time statistically valid. Probably the best known methods are the Latin square and Latin hypercube techniques [48]. An alternative strategy is Monte Carlo simulation, which we discussed in Chapter 2.

Analyses of critical points address situations where the system behavior changes qualitatively and often abruptly. Most changes in parameter values have an effect on the steady state and on transients between stimuli and responses, but the effects are only gradual: the values go up or down a little bit, or the transient overshoots more and takes longer to come close to the steady state. However, even some extremely small alterations can throw a system very definitively off kilter. A qualitative change of this type may mean that the system becomes unstable, that one of the variables goes to zero or infinity, or that the system enters a stable oscillation that continues with the same amplitude infinitely. While these are typical examples, many weird things can happen in nonlinear systems [49], and some people have actually referred to the collection as a zoo of nonlinear phenomena.

Points within the space of all parameters (meaning specific combinations of parameter values) that have the characteristic of a threshold where the qualitative behavior changes are called bifurcation points, because there is a fork in the parameter space: above the bifurcation value the system does one thing and below the value the system does something else. Much of the field of nonlinear dynamics is concerned with mathematical ways of characterizing such bifurcation points. As a less elegant, but often easier, alternative, one can execute a large number of simulations, maybe with Monte Carlo methods, trying to establish domains where all parameter combinations lead to similar solutions, along with boundaries between domains where the system changes qualitatively.

As an example, consider a small system that could describe the mutual activation pattern of two genes (**Figure 4.21A**) [50]. Corresponding power-law equations with typical parameter values are shown in Figure 4.21B. Notice that one parameter (α_1) is left unspecified. It will be our critical parameter and could represent the strength of activation by a transcription factor. Even though the system has only two dependent variables, its responses can be quite complicated. For values of α_1 below 1, the system has a stable steady state, to which it returns after perturbations (see, for example, the exogenous increase in X_1 at time 10 in Figure 4.21C). As soon as α_1 is increased above 1, the steady state becomes unstable and the system enters an oscillation that will continue until some external force stops it (Figure 4.21D). The mathematical analysis of dynamical systems attempts to characterize these bifurcation points at which the system dynamics changes drastically. This analysis is

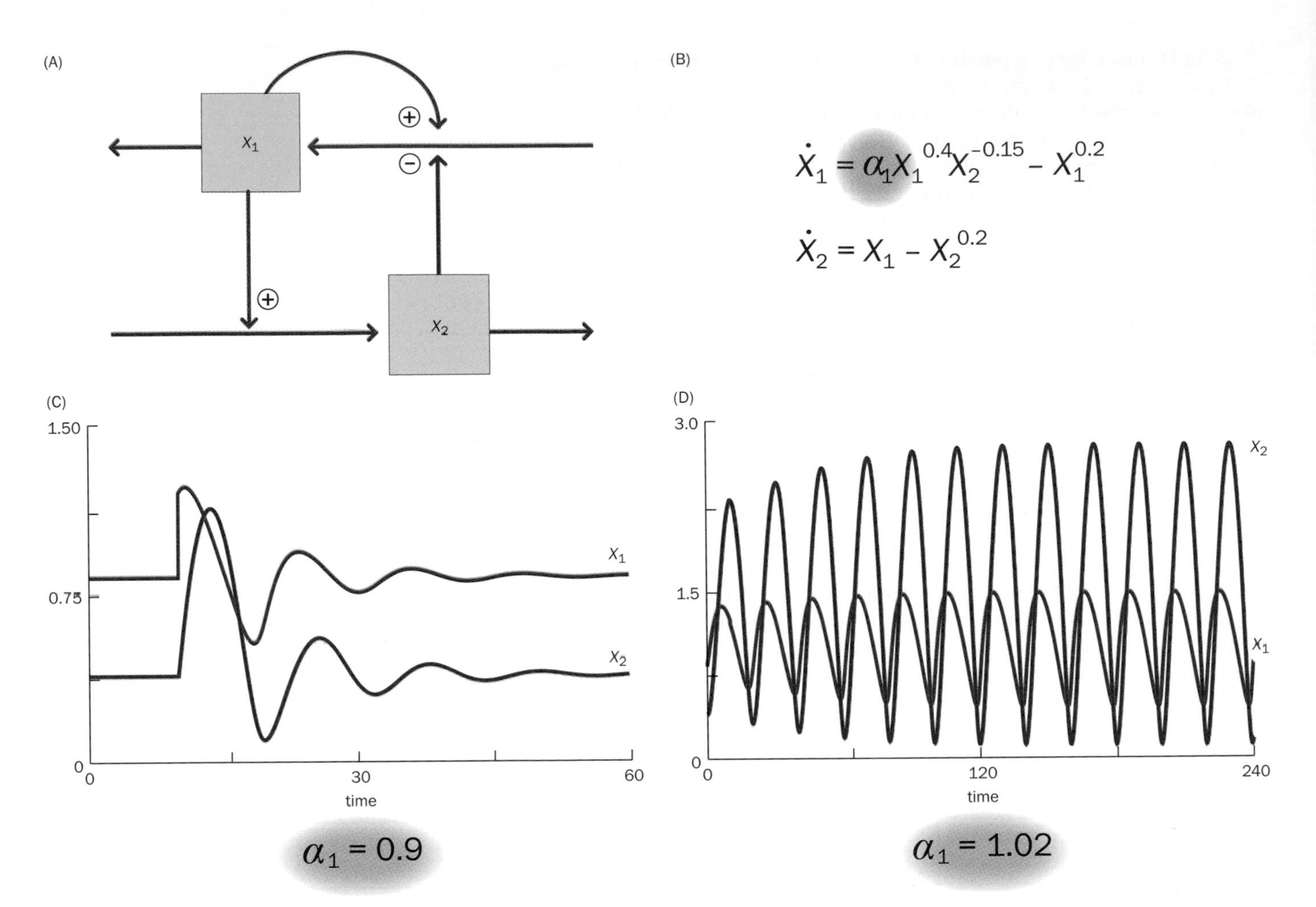

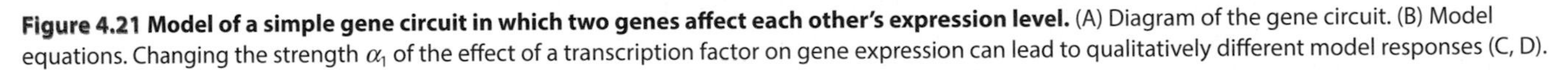

Figure 4.21 Model of a simple gene circuit in which two genes affect each other's expression level. (A) Diagram of the gene circuit. (B) Model equations. Changing the strength α_1 of the effect of a transcription factor on gene expression can lead to qualitatively different model responses (C, D).

typically quite involved, but happens to be simple for the specific form of two-variable S-systems [51, 52].

OTHER ATTRACTORS

As we discussed in Section 4.8, a stable steady state is a point to which trajectories of a system converge. In the language of nonlinear dynamics, a stable steady state is called an **attractor**, whereas an unstable steady state is called a **repeller**, because close-by trajectories are seemingly pushed away from it. Of interest now is that these more generic terms of attractors and repellers also include more complicated sets of points that draw trajectories in or push them away. These more complicated attractors or repellers consist of lines or volumes in two, three, or more dimensions. They may even be fractal, which means that their patterns are the same at different magnifications, like the cover illustration of this book. Indeed, some attractors are as difficult to comprehend as they are stunningly "attractive," as a quick Google search for images of "strange attractors" attests.

Two classes of attractors, beyond steady-state points, are particularly important for systems biology. The first consists of limit cycles, which are oscillations with a constant amplitude that attract nearby trajectories. The second contains various chaotic systems, which, however, are not "all over the place" but move unpredictably although they never leave some constrained, often oddly shaped, domain. The literature on both types is enormous, and the purely mathematical analysis of chaotic oscillators is quite complicated [53]. However, the basic concepts are easy to explore with simulations. A good introduction can be found in [11].

4.12 Limit Cycles

We have several times used the term "trajectory," referring to the solution of a system between two points in time and, specifically, between an initial value and a steady-state point. Trajectories may be linear or nonlinear, simply curved, oscillating, or very complicated. One particular type of trajectory is special in that it leaves and returns to the same point again and again. A simple example is the sine–cosine oscillation. As we discussed earlier—see (4.25)—it can be represented as a pair of linear differential equations:

$$\begin{aligned} \dot{x} &= y, \qquad x(0)=0, \\ \dot{y} &= -x, \qquad y(0)=1. \end{aligned} \tag{4.72}$$

Instead of displaying this oscillation as a function of time (**Figure 4.22A**), it is often more instructive to show it as a phase-plane plot, where y (cosine) is plotted against x (sine). In the case of a sine oscillator, this plot is a circle (Figure 4.22B). Note that this representation does not directly show time, so one cannot see how fast the system completes a cycle. Nor can we directly see in which direction the solution progresses or where it starts. However, if we look at the time-course plot (Figure 4.22A), we see that the solution starts at (0, 1) and that x increases, while y decreases. Identifying these trends in the phase-plane plot shows that the solution starts at 12 o'clock and moves clockwise. The phase-plane representation explains the terminology of a "periodic closed trajectory," because, after one period, the trajectory closes and continues on the same path time and again. Such a closed trajectory is also called an **orbit**.

A key feature of this oscillation is that it is not stable. We can easily test this by perturbing the solution. Specifically, we solve the differential equations, stop at some point in time, alter the numerical value of either x or y or both by a bit, and record what happens. A typical result is shown in Figure 4.22C, where the variable x is artificially changed from its current value of about –0.959 to a value of –0.25 at time $t = 5$. Owing to the coupling of x and y, y follows essentially instantaneously. Continuing the simulation, x and y still show sine and cosine oscillations, but the perturbation has done permanent damage: the amplitude has been altered and does not recover. The corresponding phase-plane plot is shown in Figure 4.22D. The solution starts again clockwise at 12 o'clock. At time $t = 5$, it has not yet completed one cycle, which it would do at $t = 2\pi$, but it is artificially perturbed when x is set to –0.25. This perturbation corresponds to the dashed horizontal line segment. From

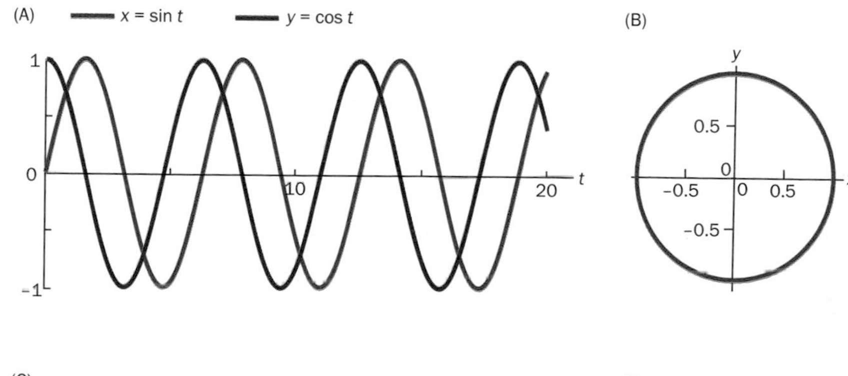

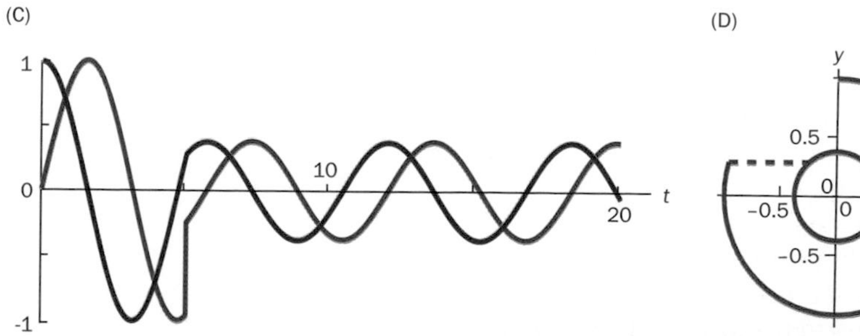

Figure 4.22 Sine–cosine oscillation. (A) The pair of linear ODEs in (4.72) corresponds to a sine–cosine oscillation that repeats every 2π time units. (B) The same oscillation is represented in the phase plane as a circle. (C) When the oscillation is perturbed, the oscillations retain the same frequency, possibly with a shift, but assume a new amplitude. (D) The corresponding phase-plane trajectory jumps (dashed line) to a new circle, which here has a smaller diameter.

then on, the oscillation continues clockwise along the inner circle and never returns to the original circle, unless it is forced to do so.

More interesting periodic closed trajectories are limit cycles, because they attract (or repel) close-by trajectories. If they attract, they are called stable oscillations or stable limit cycles, whereas repellers in this case refer to unstable limit cycles. Limit cycles are most easily discussed with examples. One famous case was proposed in the 1920s by the Dutch physicist Balthasar van der Pol (1889–1959) as a very simple model of a heartbeat [54, 55]. Since his early proposal, many variations of this van der Pol oscillator have been developed and applied to a wide range of oscillatory phenomena [11].

The prototype van der Pol oscillator is described by a second-order differential equation,

$$\ddot{v} - k(1-v^2)\dot{v} + v = 0, \tag{4.73}$$

where k is a positive parameter. The variable v with two dots represents the second derivative with respect to time, while v with one dot is the typical first derivative that we have used many times before. It is easy to convert this second-order equation into a pair of first-order differential equations, without losing any features of the original. This is done by defining a new variable

$$w = \dot{v}. \tag{4.74}$$

Because the derivative of the derivative of a variable is the second derivative, we immediately obtain

$$\dot{w} = \ddot{v}. \tag{4.75}$$

Rearranging (4.73) and substituting the results into (4.74) and (4.75) for $\dot{v}$ and $\ddot{v}$ results in a first-order system, provided the appropriate initial values are chosen, is exactly equivalent to (4.73):

$$\begin{aligned} \dot{w} &= k(1-v^2)w - v, \qquad & w(0) &= \dot{v}_0, \\ \dot{v} &= w, & v(0) &= v_0. \end{aligned} \tag{4.76}$$

To start the system, initial values must be specified at time 0 for v as well as $\dot{v}$.

Figure 4.23A confirms that the oscillation in v repeats with the same pattern. The secondary variable, w, oscillates with the same frequency, but quite a different shape (Figure 4.23B). Instead of displaying the oscillations as functions of time, we can again use the phase plane, where w is plotted against v. The result is shown in Figure 4.23C. Starting at the point on the far right, the solution initially decreases clockwise. Displaying the solution only for time points spaced by 0.05 shows that the points are dense in some sections, but not in others (Figure 4.23D). In the dense sections, the solution is comparatively slow, while it is fast for points that are farther apart.

Intriguingly, if the van der Pol oscillation is perturbed, it recovers and returns to the same oscillatory pattern. This feature is the key characteristic of a stable limit cycle. For instance, suppose the system is perturbed at $t = 18.6$. At this point, the system without perturbation has the values $v \approx 0.26$ and $w \approx 10$. Artificially resetting v to 0 yields the plots shown in **Figures 4.24** and **4.25.** It is evident that the original oscillation is regained, although with some shift in the direction of time.

Limit cycles may have very different shapes. A flexible way to create such shapes is to start with a limit-cycle model in the form of a two-variable S-system, for which Lewis [51] computed mathematical conditions. An example, adapted from [52], is

$$\begin{aligned} \dot{X}_1 &= 1.01(X_2^8 - X_1^{-4}), \qquad & X_1 &= 1.7, \\ \dot{X}_2 &= (X_1^{-3} - X_2^4), & X_2 &= 0.9. \end{aligned} \tag{4.77}$$

Figure 4.26 shows the oscillations of this system in both time-course and phase-plane format. If one selects initial values inside the limit cycle, but not exactly at the unstable steady state (1, 1), all trajectories are attracted to the limit cycle. Similarly,

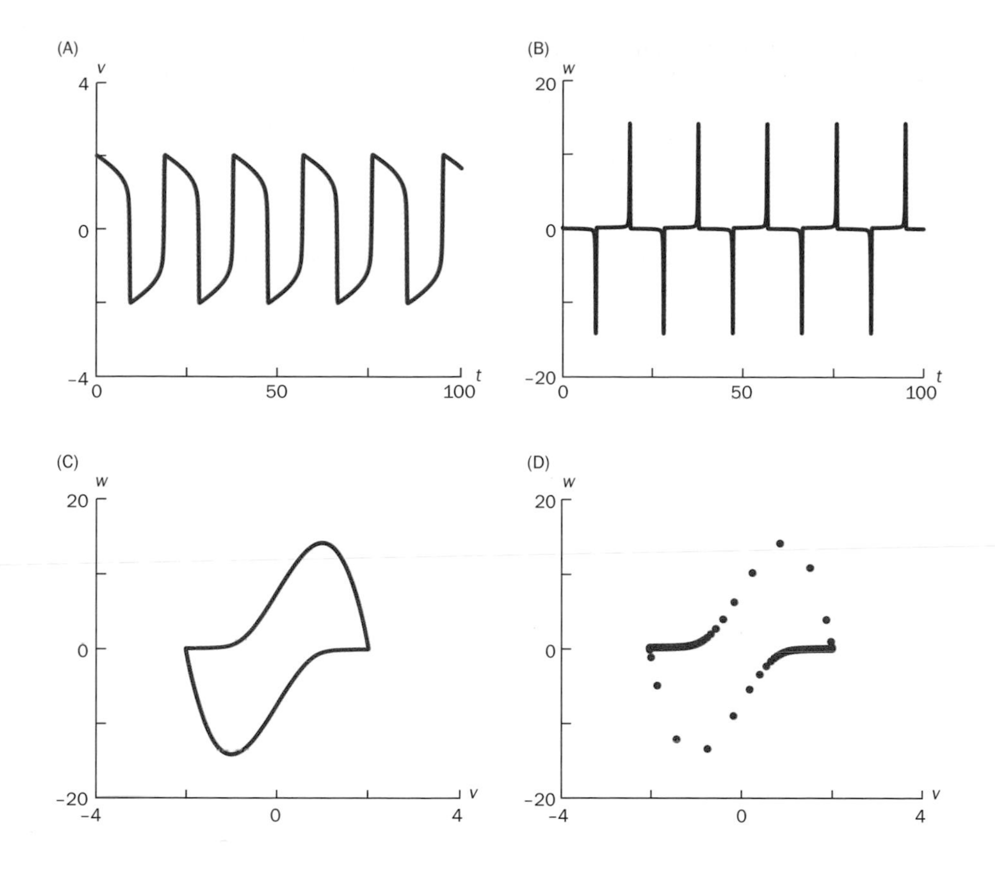

Figure 4.23 Visualization of the van der Pol oscillator (4.76). (A) The original variable *v* represents a limit-cycle oscillation that was designed to mimic a heartbeat. (B) The plot of the auxiliary variable *w* has the same frequency, but quite a different shape. (C) Phase-plane plot of *w* versus *v*. (D) Plotting the solution shown in (C) at spaced-out time points gives an idea of the speed of the oscillation in different phases of the oscillation. Dense points correspond to slow speed, while more widely separated points indicate a higher speed.

oscillations beginning close enough outside the limit cycle are attracted as well. If one starts further away, one of the variables may become zero, and the equations are no longer defined.

By multiplying both equations of the limit-cycle system by the same positive functions of X_1 and/or X_2, the appearance of the limit-cycle oscillation can be changed quite dramatically (**Figure 4.27**), although the phase-plane plot remains unchanged (Exercise 4.61). In all cases, the limit cycle remains stable and attracts close-by trajectories.

With the same methods as for systems without limit cycles, we can compute the steady states of the system in (4.77). Since a negative power is not really defined for $X = 0$, we will not discuss a possibly trivial steady state. The nontrivial steady state is (1, 1). Checking the stability of this state, we find a pair of complex-conjugate eigenvalues with a real part close to 0, which means that the steady state is unstable. This observation reflects the fact that inside a stable limit cycle there lies either at least one unstable steady state or an unstable limit cycle. We also note that the imaginary parts are nonzero, which indicates the potential for oscillations. If the rate constant 1.01 is changed to 1.00, the real parts become zero. At this bifurcation point, the stable limit cycle is "born" when the rate constant is slowly increased from below to above 1. For values smaller than 1, the system is stable.

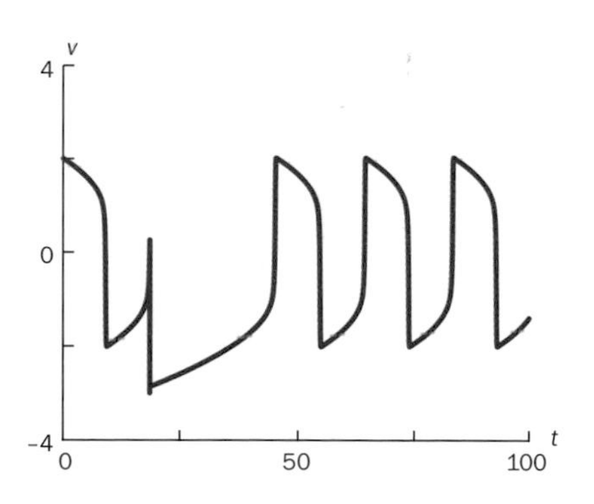

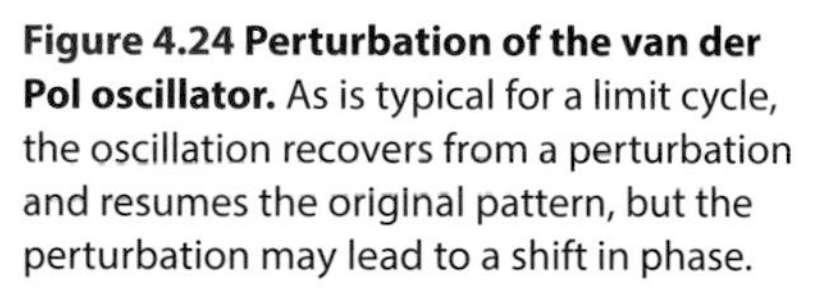

Figure 4.24 Perturbation of the van der Pol oscillator. As is typical for a limit cycle, the oscillation recovers from a perturbation and resumes the original pattern, but the perturbation may lead to a shift in phase.

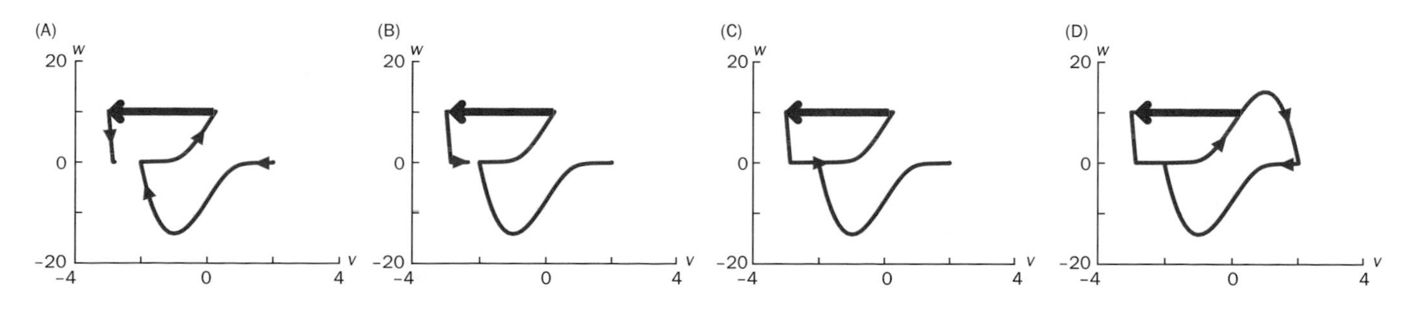

Figure 4.25 Phase-plane visualization of the perturbation in Figure 4.24. The van der Pol oscillator is perturbed at $t = 18.6$, where *v* is artificially changed from 0.26 to –3 (red arrow). (A), (B), (C), and (D) show, in the phase plane, the same trajectory, which is stopped at $t = 20$, 30, 40, and 50, respectively. The arrowheads have been added for illustration.

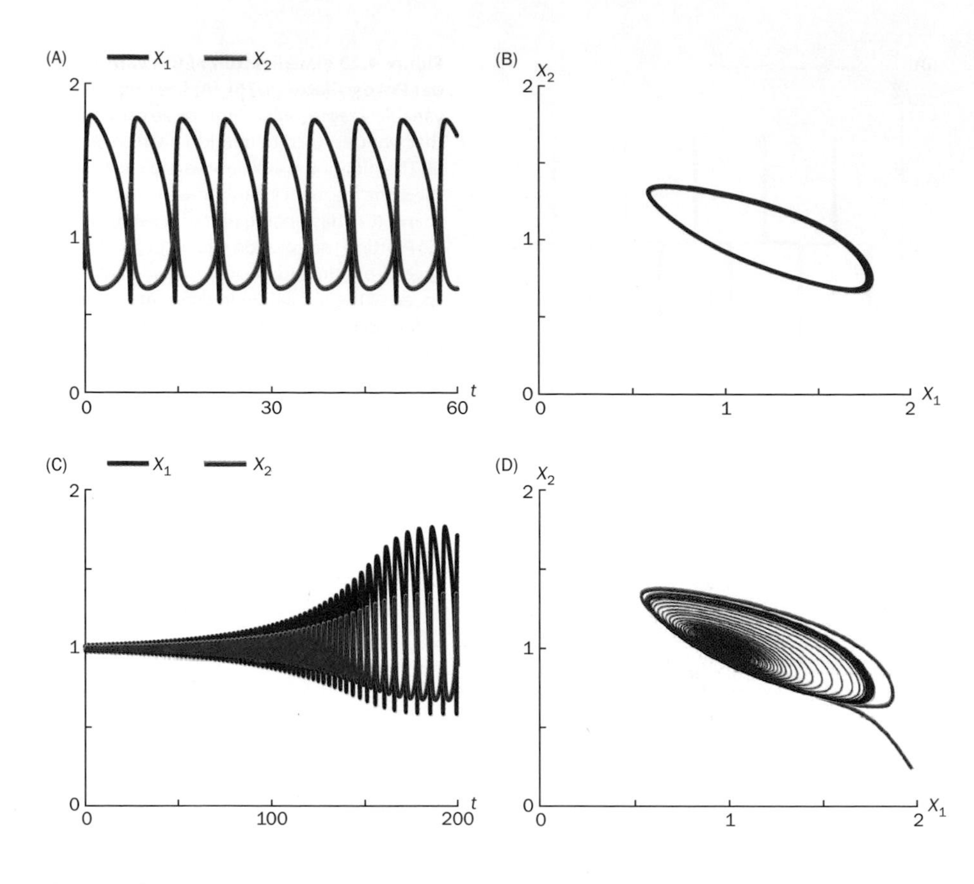

Figure 4.26 Limit cycles can have very different shapes. (A) Time-course representations of the limit cycle in (4.77). (B) The corresponding phase-plane plot. (C) Starting with initial values $(X_1, X_2) = (1.01, 0.99)$, which are close to the unstable steady state (1, 1), the oscillation spirals outward, eventually approaching the limit cycle. (D) Approach to the limit cycle from inside (green) and outside (blue) the limit cycle.

4.13 Chaotic Attractors

In 1963, the American mathematician and meteorologist Edward Lorenz [56] studied weather patterns and noticed that even minute changes in initial values could lead to totally different outcomes if one waited for a while. It was Lorenz who introduced the concept of the butterfly effect. Using a simplified model of air flow, Lorenz was able to reproduce this effect with three differential equations and thereby ushered in an intensive exploration of systems that had features similar to what Lorenz called a **strange attractor**. Trajectories in this and other, similar systems are confined to some volume in three-dimensional space, but their specific values cannot be predicted unless one solves the equations. The amazing aspect is that it is very easy to simulate these equations, which in Lorenz's notation read

$$
\begin{aligned}
\dot{x} &= P(y-x), \\
\dot{y} &= Rx - y - xz, \\
\dot{z} &= xy - Bz.
\end{aligned}
\tag{4.78}
$$

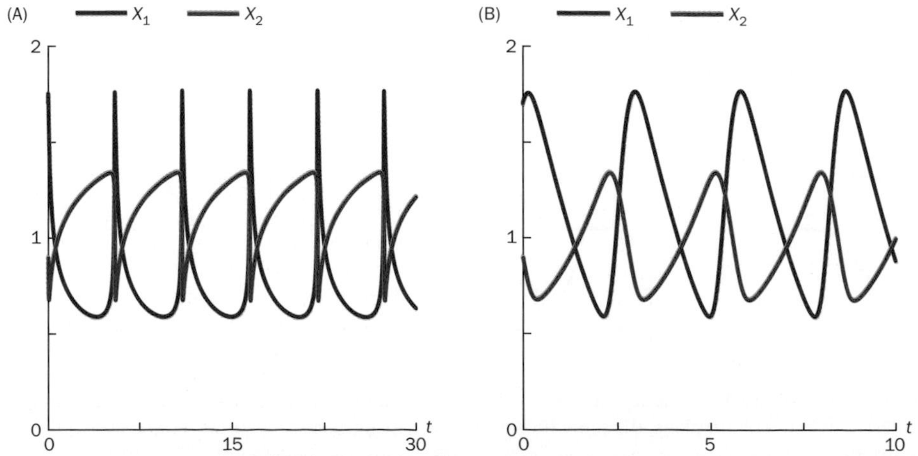

Figure 4.27 Variants of limit-cycle oscillations in (4.77). (A) Both equations were multiplied by $X_1^6 X_2^{-2}$ (B) Both equations were multiplied by $X_1 X_2^{-4}$. Note the different shapes and scales for the time axes and thus the different speeds of the oscillations.

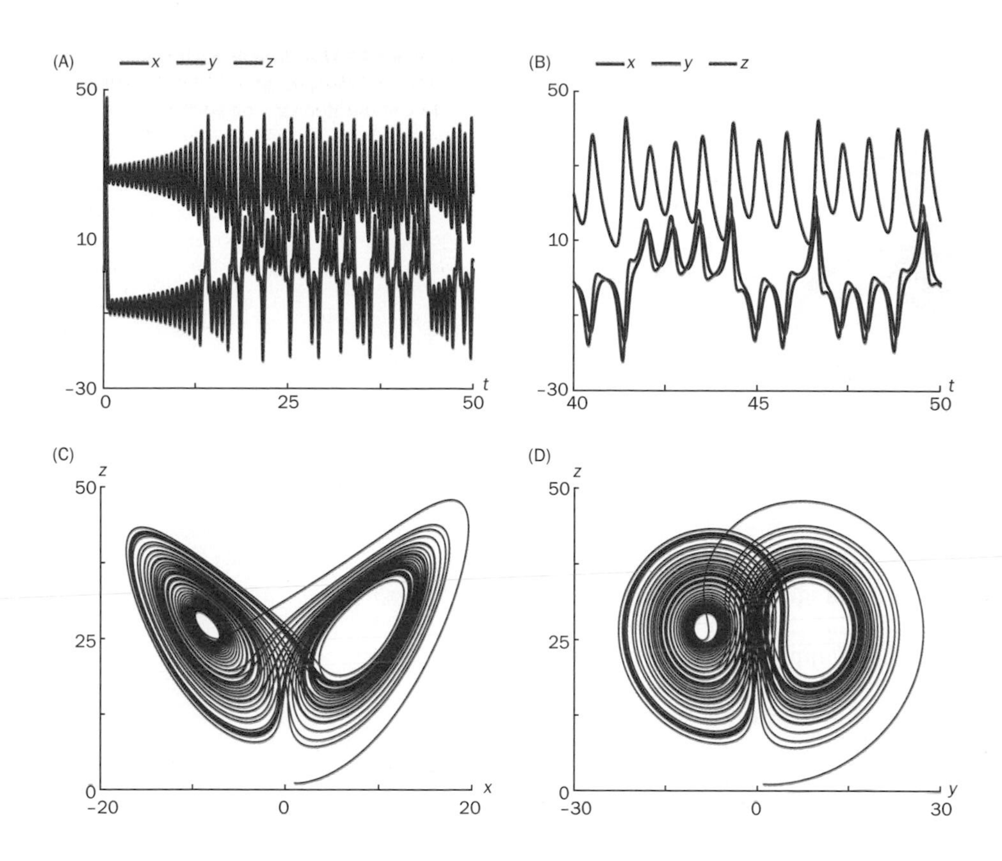

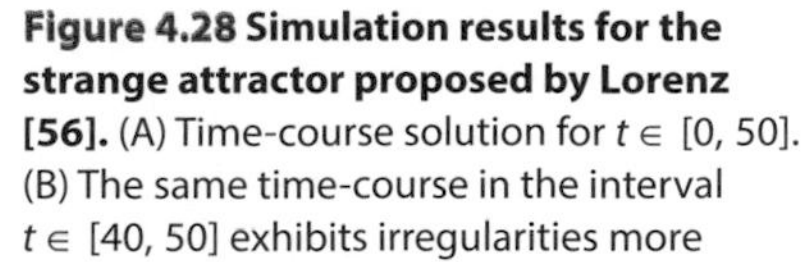

Figure 4.28 Simulation results for the strange attractor proposed by Lorenz [56]. (A) Time-course solution for $t \in [0, 50]$. (B) The same time-course in the interval $t \in [40, 50]$ exhibits irregularities more clearly. (C) and (D) Phase-plane plots corresponding to projections of the three-variable oscillator onto the x–z and y–z planes, respectively. See also Figure 4.29.

Lorenz used the parameter values $P = 10$, $R = 28$, and $B = 8/3$, along with the initial values (1, 1, 1). **Figure 4.28** shows some output that demonstrates the erratic, chaotic trajectories of the system, which never exactly repeat. A common characteristic of chaotic oscillators is the fact that they are extremely sensitive to changes in initial values, and even a change to (1.001, 1, 1) results in a different temporal pattern. With such high sensitivity, even the solver settings can lead to different results. Interestingly, long-term simulations lead to an attractor in three dimensions that is overall always very similar, but differs in details. Expressed differently, any trajectory lies somewhere within the attractor, but one cannot predict where exactly it will be without executing a numerical simulation (Exercise 4.66).

The Lorenz attractor actually has three steady states. They are easy to compute and, as it turns out, they are all unstable (Exercise 4.69). Starting a simulation close to one of the steady states fills one of the holes in the attractor with outward spirals that eventually merge into the attractor (**Figure 4.29**).

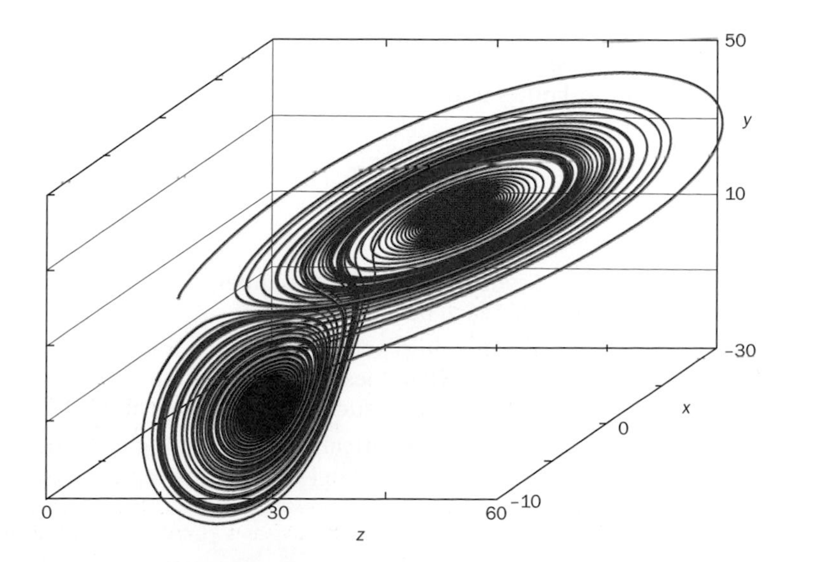

Figure 4.29 Filling the holes of the Lorenz attractor. Starting simulations close to the nontrivial unstable steady-state points yields outward spirals (red and green, respectively) that eventually merge into the attractor. The pseudo-three dimensional plot demonstrates that the solutions are confined to two "disks." Where exactly the future solution will be within this attractor cannot be predicted; it can only be determined through simulations.

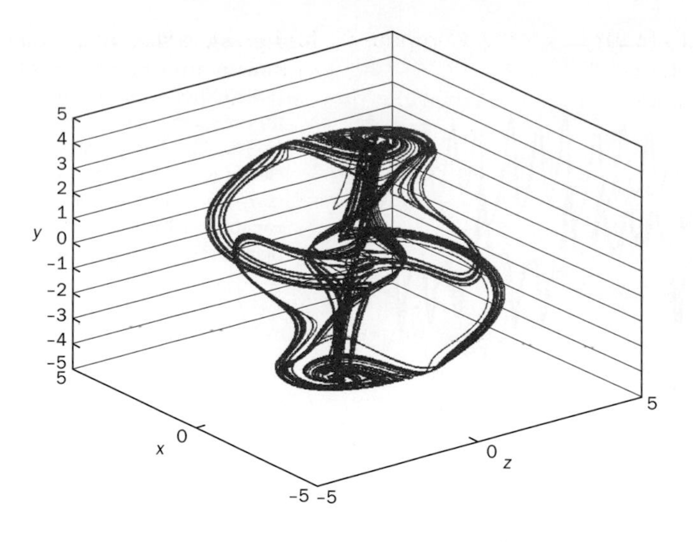

Figure 4.30 The Thomas oscillator. The deceivingly simple set of ODEs in (4.79) leads to a complicated strange attractor.

The Lorenz system is probably the best known strange attractor, but many others have been documented. One strategy for creating chaotic oscillators is to start with a limit cycle and add a small sine function to it. An example was shown with the blue sky catastrophe in Chapter 2 (see Box 2.2). A second example is the addition of a function like $0.00165 \sin(1.5t)$ to the first example of the limit cycle in (4.77). This situation is considered in Chapter 12. Such an addition sometimes leads to chaos, but not always. It is possible that the oscillation will become regular after some time or that one of the variables will goes to $\pm\infty$ or 0 (Exercise 4.62). Even systems of differential equations that have a simple structure, such as BST and Lotka–Volterra models, can exhibit chaos (see Chapter 10 and [57–59]).

A particularly stunning example, because of its apparent simplicity, is the Thomas oscillator

$$\begin{aligned} x' &= -ax + \sin y, \\ y' &= -ay + \sin z, \\ z' &= -az + \sin x, \end{aligned} \tag{4.79}$$

whose appearance changes dramatically depending on the choice of the parameter a [60]. The complicated phase plane plot for $a = 0.18$ and initial values $(x, y, z) = (0, 4, 0.1)$ is shown in **Figure 4.30** for $t \in [0, 2000]$. Several other examples of chaotic systems, including the continuous Rössler band [61] and variations of the Thomas oscillator, as well as the discrete Hénon map [11], are discussed in Exercises 4.63–4.72.

Seeing how complicated the trajectories of seemingly simple systems can become, one may ask how overwhelmingly complicated a large system could be. Indeed, the sky is the limit. At the same time, large systems often contain mechanisms that keep their subsystems from "misbehaving," and it can easily turn out that the behavior of a large system is quite dull. However, one never knows until one studies a system with rigorous mathematical techniques.

EXERCISES

4.1. Describe in words, as well as mathematically, what happens if the initial size P_t in the recursive model $P_{t+\tau} = 2P_t$ in (4.1) is different from 10.

4.2. What happens (biologically and mathematically) in (4.1) at time points between t and $t + \tau$ or between $t = 3\tau$ and $t = 4\tau$?

4.3. Is it important to distinguish whether or not the bacterial culture in the text (Table 4.1) is synchronized? What happens if τ is slightly different for each bacterium?

4.4. Is anything special about the number 2 in (4.1)–(4.3)? What does it mean if 2 is replaced by some other integer, such as 3 or 4? What about 1?

4.5. Is it mathematically and/or biologically meaningful if the number 2 in (4.1)–(4.3) is replaced by −2, −1, or 0?

4.6. What is the final size of the population in (4.1)–(4.3)? What is the final size if 2 is replaced with 0.5, 1, or −1? What effect does the initial population size P_1 have on the final size in each case?

4.7. Construct a recursive model that approaches a constant final value (steady state) that is not 0, $+\infty$ or $-\infty$. Is this possible with a linear model? If so, construct an example. If not, make the model nonlinear.

4.8. For the example in (4.4), compute R_{n+1} from R_n and then R_{n+2} from R_{n+1} for different values of f, g, R_n, and M_n.

4.9. For the example in (4.4), compute R_{n+2} directly from R_n, using the matrix formulation, and compare the results with those in Exercise 4.8. Compute M_{n+4}.

4.10. Compute the steady state(s) of the following two recursive systems:

A: $$X_{n+1} = (1-p)X_n + Y_n, \quad Y_{n+1} = qX_n + 1;$$

B: $$X_{n+1} = (1-p)X_n + Y_n, \quad Y_{n+1} = qX_n.$$

4.11. Discuss the steady state(s) of the so-called Fibonacci series, which is given by $a_{n+2} = a_{n+1} + a_n$ and is usually started with $a_1 = 1$ and $a_2 = 1$. Also discuss the steady states of the associated series of ratios $b_{n+1} = a_{n+1}/a_n$.

4.12. Discuss the implications of the assumption that the travel times between states in the pinball example are all the same.

4.13. What is the behavior of a Markov chain model in which all matrix elements are the same? What is the behavior if all matrix elements on the diagonal are 1?

4.14. Describe the dynamics of a Markov model with the matrix

$$\mathbf{M} = \begin{pmatrix} 0 & 1 & 0 & 0 & 0 \\ 1 & 0 & 0 & 0 & 0 \\ 0 & 0 & 0 & 0 & 1 \\ 0 & 0 & 0 & 1 & 0 \\ 0 & 0 & 1 & 0 & 0 \end{pmatrix}.$$

4.15. Do the matrix elements in each column of a Markov matrix necessarily sum to 1? Why or why not?

4.16. Compute $\begin{pmatrix} X_1 \\ X_2 \\ X_3 \end{pmatrix}_4$ for the example in (4.16)–(4.19) in two ways. First, multiply $\begin{pmatrix} X_1 \\ X_2 \\ X_3 \end{pmatrix}_3$ by $\mathbf{M}^{\mathrm{T}}$. Second, compute the square of $\mathbf{M}^{\mathrm{T}}$ and multiply $\begin{pmatrix} X_1 \\ X_2 \\ X_3 \end{pmatrix}_2$ by it. Explain why both results are identical.

4.17. Compute the limiting distribution for the example in (4.16)–(4.19) with methods of linear algebra. This distribution is the solution of the matrix equation

$$\begin{pmatrix} \pi_1 \\ \pi_2 \\ \pi_3 \end{pmatrix} = \mathbf{M}^{\mathrm{T}} \begin{pmatrix} \pi_1 \\ \pi_2 \\ \pi_3 \end{pmatrix}.$$

Does the limiting distribution contain the same amount of information as $\mathbf{M}$?

4.18. What happens with the logistic map (4.24) if r is greater than 4? What happens if r is less than 0?

4.19. Simulate the Gauss map model

$$P_{t+1} = e^{aP_t^2} + b.$$

Start with the values $a = -6.2$, $b = -1$, and $P_0 = 0.2$. Change P_0 and record what happens. Increase b in small steps (for example, 0.1) toward 1. Does the system have a steady state? Can it be computed algebraically?

4.20. Linearize a system of the form (4.28), with

$$f(X,Y) = c_1 - \frac{c_2 Y X^n}{c_3^n + X^n},$$
$$g(X,Y) = c_4 - c_5 Y,$$

where c_i are positive parameters and $n = 1, 2,$ or 4. Define initial values. Give a possible interpretation for these equations in terms of a metabolic or cellular system. Compute the functions and their approximations for different sets of parameter values. Consider different operating points. Write a brief report about your findings.

4.21. Linearize the system in (4.39) at the steady state. Select initial values of 50, 10, 100 for S, I, and R and assess the quality of the representation.

4.22. Compute the Jacobian of a generic two-variable linear system without input.

4.23. Show that $y = ce^x + e^{2x}$ is the solution of the differential equation $y' - y = e^{2x}$.

4.24. Show that the logistic function (4.42) is indeed the solution of the logistic differential equation (4.43). For this purpose, differentiate $N(t) = a/(b + e^{-ct})$ with respect to t and express this derivative as a function of $N(t)$ itself.

4.25. Explore the effects of changes in α, β, and the initial size N_0 on the shape of the logistic growth function (4.43).

4.26. Does it make sense biologically and/or mathematically if N_0 in (4.43) is greater than the carrying capacity? If feasible, simulate this case.

4.27. What happens mathematically in the logistic equation (4.43) if both α and β are negative? Is this situation biologically relevant?

4.28. Linearize the logistic function in (4.43) at its steady state(s) and determine stability.

4.29. Study the roles of K_M and V_{max} in (4.44). How do K_M and V_{max} relate to each other?

4.30. Revisit the SOS system (4.52)–(4.54) and consider P_L and P_{trc} as dependent variables. Change the model equations correspondingly. Redo the simulation in Figure 4.13 with the expanded system. Write a brief report on your findings.

4.31. Prove the shortcut in Boxes 4.3 and 4.4 for computing power-law approximations.

4.32. Compute power-law approximations for the Hill function

$$v_H = \frac{V_{max} S^n}{K_M^n + S^n}$$

for $n = 1$, 2, and 4 at different operating points. Select V_{max} and K_M as you like. Plot the approximations together with the original Hill functions.

4.33. Compute power-law approximations for the function

$$F(X_1, X_2, X_3) = X_1 e^{X_2 + 2X_3},$$

using different operating points. Assess the approximation accuracy for each operating point. Describe your findings in a report.

4.34. Compute the power-law approximation of the function

$$F(X,Y) = 3.2X^2 e^{2Y+1}$$

at the operating point $X = 1$, $Y = -0.5$.

4.35. Develop a power-law approximation of the function

$$f(x,y) = 4x^3 y^{-1}.$$

at the operating point $(x, y) = (1, 2)$.

4.36. Develop a power-law approximation of the function

$$f(x,y) = 4x^3 \sin y$$

at the operating point $(x, y) = (1, 2)$.

4.37. If a Hill function with a Hill coefficient of 2 is approximated by a power-law function, in which numerical ranges do the kinetic order and the rate constant fall?

4.38. Implement the branched pathway system in (4.49) with the parameter values $X_0 = 2$, $\gamma_{11} = 2$, $\gamma_{12} = 0.5$, $\gamma_{13} = 1.5$, $\gamma_{22} = 0.4$, $\gamma_{32} = 2.4$, $f_{110} = 1$, $f_{112} = -0.6$, $f_{121} = 0.5$, $f_{131} = 0.9$, $f_{222} = 0.5$, $f_{224} = 1$, and $f_{323} = 0.8$. Start with $X_1 = X_2 = X_3 = X_4 = 1$, but explore different values between 0.1 and 4 for X_4.

4.39. Determine the S-system that corresponds to the GMA system in (4.49) with the parameter values in Exercise 4.38. The two systems should be equivalent at the nontrivial steady state and as similar as possible otherwise. Perform comparative simulations.

4.40. Construct a diagram that could correspond to the following GMA system:

$$\dot{X}_1 = \alpha_1 X_2^{g_{12}} - \beta_1 X_1^{h_{111}} - \beta_2 X_1^{h_{112}},$$
$$\dot{X}_2 = \beta_1 X_1^{h_{111}} - \beta_3 X_2^{h_{22}}.$$

Is the diagram unique? Discuss your answer. Approximate the GMA system with an S-system. Discuss all conditions under which the S-system is equivalent to the GMA system for all values of X_1 and X_2.

4.41. Compute the steady state of the S-system in (4.51) for arbitrary kinetic orders and for rate constants that are all equal to 1; assume $X_0 = 1$. Show all steps of your computations. Discuss which features (steady state, stability, dynamics, ...) of the system change if all rate constants are set to 10.

4.42. Compute the steady state of the S-system

$$\dot{X}_1 = \alpha(X_2^8 - X_1^{-4}),$$
$$\dot{X}_2 = X_1^{-3} - X_2^4$$

for $\alpha = 0.8$, 1, and 1.01. Show all steps of your computations. In each case, determine the stability of the system. Simulate the system in PLAS.

4.43. Compute all steady states (trivial and nontrivial) of the SIR model

$$\dot{S} = 0.01R - 0.0005SI,$$
$$\dot{I} = 0.0005SI - 0.05I,$$
$$\dot{R} = 0.05I - 0.01R.$$

4.44. Consider the following ethanol (E)-producing fermentation system, in which the growth of bacteria B is driven by the availability of substrate (glucose, G):

$$\dot{B} = \mu B - aB^2,$$
$$\dot{E} = \frac{V_{max} B}{K_M + B} - cE.$$

In this formulation,

$$\mu = \mu_{max} \frac{G^2}{K_I^2 + G^2},$$

and μ_{max} K_I, V_{max}, and K_M are positive parameters. Explain the biological meaning of each term in the two ODEs. Assuming constant glucose, as well as G, E, $B \neq 0$, compute the power-law approximation of the fermentation system at its steady state. Execute comparative simulations with the original system and its power-law approximation.

4.45. Compute the trivial and nontrivial steady state(s) of the Lotka–Volterra system

$$\dot{N}_1 = r_1 N_1 K_1 (K_1 - N_1 - aN_2),$$
$$\dot{N}_2 = r_2 N_2 K_2 (K_2 - N_2 - bN_1),$$

where $r_1 = 0.15$, $K_1 = 50$, $a = 0.2$, $r_2 = 0.3$, $K_2 = 60$, $b = 0.6$, $N_1(0) = 1$, and $N_2(0) = 1$. Assess the stability of the steady state.

4.46. Enumerate the trivial and nontrivial steady state(s) of a Lotka–Volterra system with three dependent variables.

4.47. Is it possible to construct a model that is at once a Lotka–Volterra, GMA, and S-system? If your answer is no, say why. If your answer is yes, provide an example.

4.48. Is every Lotka–Volterra system also a GMA system? Is every GMA system also a Lotka–Volterra system? Is every S-system a GMA system? Is every lin–log system also a GMA system?

4.49. Determine matrices **A** for the different regions in Figure 4.17. Implement the systems and solve them, starting with different initial values, as illustrated in Figure 4.18.

4.50. Study the stability of the differential equation (4.27) with $(A_{11}, A_{12}, A_{21}, A_{22}) = (-1, 2.5, -1.5, 1)$. Compute trace, determinant, and discriminant and locate the expected behavior of the system close to (0, 0) on the plot of Figure 4.17. Use simulations to explore the consequences of slight changes in any of the A_{ij}. Is it possible to create different stability patterns by varying just one of the A_{ij} up and down?

4.51. Study the stability of the differential equation (4.27) with $(A_{11}, A_{12}, A_{21}, A_{22}) = (-1, 2.5, -0.4, 1)$. Explore the consequences of slight changes in any of the A_{ij}. Summarize your findings in a report.

4.52. Calculate the symbolic steady-state equations of the SOS systems (4.52)–(4.54). Enter numerical values and compare with simulation results.

4.53. Describe and explain the behavior of the following system at its nontrivial steady state:

$$\dot{x} = f(x,y) = 1 - x^2 y,$$
$$\dot{y} = g(x,y) = xy^2 + y.$$

4.54. Use the numerical system from Exercise 4.52 and multiply all the right-hand sides by −1. How would you interpret this operation? Do simulations and interpret the results.

4.55. Use the numerical system from Exercise 4.52 and vary each parameter individually by 5% up and down. Which parameter is most influential with respect to the system's steady state?

4.56. Analyze the system shown in **Figure 4.31**, using the S-system model

$$\dot{X}_1 = X_2^{-2} X_3 - X_1^{0.5} X_2,$$
$$\dot{X}_2 = X_1^{0.5} X_2 - X_2^{0.5}.$$

The system contains the independent variable X_3, which is constant. Begin the analysis by simulating the system with $X_1(0) = X_2(0) = X_3 = 1$. Compute all steady states and assess their stability. In the next step, vary $X_1(0)$ and $X_2(0)$ up and down and study the system responses. Finally, vary the constant value of X_3 up and down and study the system for different initial values $X_1(0)$ and $X_2(0)$.

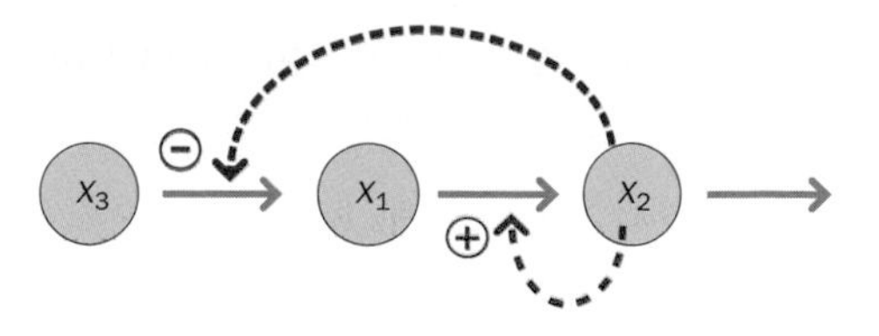

Figure 4.31 Generic pathway with two modulation signals. Even a seemingly simple pathway can exhibit quite complex responses.

4.57. Suppose all terms in a three-variable ODE system are approximated by power-law functions at a nontrivial steady state. Discuss the similarities and differences of the Jacobians of the two systems (the original ODE system and the approximating power-law system) at this steady state.

4.58. Why should one typically expect the model of a biological system to be flawed if some of its sensitivities are in a range between 10 and 1000?

4.59. What does it imply (1) mathematically or (2) for biological experiments associated with the system if a parameter value has very low sensitivities?

4.60. Modify the limit-cycle system (4.77) by multiplying only one of the two equations by a power function of X_1 and/or X_2. Record the effects on the oscillation and the phase-plane plot.

4.61. Explain why multiplication of the limit-cycle system (4.77) by a function of X_1 and/or X_2 changes the appearance of a limit-cycle oscillation in the time domain, but leaves the phase-plane plot unchanged.

4.62. Add to the first equation of the limit-cycle system (4.77) the function $c \sin(bt)$. Start with $c = 0.00165$, test different values for b, and record the shapes of the oscillations in time-course and phase-plane representations. Subsequently, alter c as well and record the effects.

4.63. Explore the discrete strange attractor proposed by Hénon, which is defined recursively by

$$x_{n+1} = y_n - 1.4x_n^2 + 1,$$
$$y_{n+1} = 0.3x_n.$$

Typical starting values are $(x_0, y_0) = (1, 1)$. To see the true shape of the chaotic attractor, execute several hundred iterations, for instance, in Excel®.

4.64. Explore the discrete attractor proposed by Hénon and defined in Exercise 4.63 after replacing the factor 1.4 with 1.25.

4.65. Compute the steady state(s) of the Hénon map in Exercises 4.63 and 4.64.

4.66. Simulate the so-called Rössler attractor, which is defined by the differential equations

$$x' = -y - z,$$

$$y' = x + ay,$$

$$z' = bx - cz + xz,$$

with $a = 0.36$, $b = 0.4$, and $c = 4.5$. Choose different initial values, beginning with $x = 0$, $y = -3$, and $z = 0$. Summarize your findings in a report.

4.67. Compute the steady state(s) of the Rössler system in Exercise 4.66 and assess its/their stability.

4.68. Explore the Lorenz oscillator for slightly changed initial values and parameter values.

4.69. Compute the steady states of the Lorenz attractor. Start simulations at or close to these steady states and study how the trajectories interact with the attractor itself.

4.70. Predict what the phase-plane plot of the blue sky catastrophe in Chapter 2 (see Box 2.2) looks like. Implement the system in PLAS and check your prediction.

4.71. Explore the Thomas attractor in (4.79) with different, small positive values of a. Solve the system with $x(0) = 0$, $y(0) = 4$, and $z(0) = 0.5$ and study the phase plane of x and y and the pseudo-three-dimensional representation of x, y, and z.

4.72. Explore the Thomas attractor in (4.79) with $a = 0$. Start the system with $x(0) = 0$, $y(0) = 4$, and $z(0) = 0.5$ and solve the system first for 500 time units and then for much longer time periods. Study the phase plane of x and y and the pseudo-three-dimensional representation of x, y, and z.

REFERENCES

[1] PLAS: Power Law Analysis and Simulation. http://enzymology.fc.ul.pt/software.htm.
[2] Guckenheimer J & Holmes P. Nonlinear Oscillations, Dynamical Systems, and Bifurcations of Vector Fields. Springer, 1983.
[3] Voit EO & Dick G. Growth of cell populations with arbitrarily distributed cycle durations. I. Basic model. *Mathem. Biosci.* 66 (1983) 229–246.
[4] Voit EO & Dick G. Growth of cell populations with arbitrarily distributed cycle durations. II. Extended model for correlated cycle durations of mother and daughter cells. *Mathem. Biosci.* 66 (1983) 247–262.
[5] Leslie PH. On the use of matrices in certain population mathematics. *Biometrika* 33 (1945) 183–212.
[6] Edelstein-Keshet L. Mathematical Models in Biology. SIAM, 2005.
[7] Fonseca L & Voit EO. Comparison of mathematical frameworks for modeling erythropoiesis in the context of malaria infection. *Math. Biosci.* 270 (2015) 224–236.
[8] Fonseca LL, Alezi HA, Moreno A, Barnwell JW, Galinski MR & Voit EO. Quantifying the removal of red blood cells in *Macaca mulatta* during a *Plasmodium coatneyi* infection. *Malaria J.* 15 (2016) 410.
[9] Ross SM. Introduction to Probability Models, 11th ed. Academic Press, 2014.
[10] Durbin R, Eddy S, Krogh A & Mitchison G. Biological Sequence Analysis: Probabilistic Models of Proteins and Nucleic Acids. Cambridge University Press, 1998.
[11] Thompson JMT & Stewart HB. Nonlinear Dynamics and Chaos, 2nd ed. Wiley, 2002.
[12] Jacquez JA. Compartmental Analysis in Biology and Medicine, 3rd ed. BioMedware, 1996.
[13] Chen C-T. Linear System Theory and Design, 4th ed. Oxford University Press, 2013.
[14] Von Seggern DH. CRC Handbook of Mathematical Curves and Surfaces. CRC Press, 1990.
[15] Lotka A. Elements of Physical Biology. Williams & Wilkins, 1924 (reprinted as Elements of Mathematical Biology. Dover, 1956).
[16] Volterra V. Variazioni e fluttuazioni del numero d'individui in specie animali conviventi. *Mem. R. Accad. dei Lincei* 2 (1926) 31–113.
[17] May RM. Stability and Complexity in Model Ecosystems. Princeton University Press, 1973 (reprinted, with a new introduction by the author, 2001).
[18] Peschel M & Mende W. The Predator–Prey Model: Do We Live in a Volterra World? Akademie-Verlag, 1986.
[19] Voit EO & Savageau MA. Equivalence between S-systems and Volterra-systems. *Math. Biosci.* 78 (1986) 47–55.
[20] Savageau MA & Voit EO. Recasting nonlinear differential equations as S-systems: a canonical nonlinear form. *Math. Biosci.* 87 (1987) 83–115.
[21] Savageau MA. Biochemical systems analysis. I. Some mathematical properties of the rate law for the component enzymatic reactions. *J. Theor. Biol.* 25 (1969) 365–369.
[22] Savageau MA. Biochemical Systems Analysis: A Study of Function and Design in Molecular Biology. Addison-Wesley, 1976.
[23] Voit EO. Computational Analysis of Biochemical Systems: A Practical Guide for Biochemists and Molecular Biologists. Cambridge University Press, 2000.
[24] Torres NV & Voit EO. Pathway Analysis and Optimization in Metabolic Engineering. Cambridge University Press, 2002.
[25] Voit EO. Biochemical systems theory: a review. *ISRN Biomath.* 2013 (2013) Article ID 897658.
[26] Shiraishi F & Savageau MA. The tricarboxylic-acid cycle in *Dictyostelium discoideum*. 1. Formulation of alternative kinetic representations. *J. Biol. Chem.* 267 (1992) 22912–22918.
[27] Tian T & Burrage K. Stochastic models for regulatory networks of the genetic toggle switch. *Proc. Natl Acad. Sci. USA* 103 (2006) 8372–8377.
[28] Visser D & Heijnen JJ. The mathematics of metabolic control analysis revisited. *Metab. Eng.* 4 (2002) 114–123.
[29] Heijnen JJ. Approximative kinetic formats used in metabolic network. *Biotechnol. Bioeng.* 91 (2005) 534–545.
[30] Wang F-S, Ko C-L & Voit EO. Kinetic modeling using S-systems and lin–log approaches. *Biochem. Eng.* 33 (2007) 238–247.
[31] del Rosario RC, Mendoza E & Voit EO. Challenges in lin–log modelling of glycolysis in *Lactococcus lactis*. *IET Syst. Biol.* 2 (2008) 136–149.
[32] Heinrich R & Rapoport TA. A linear steady-state treatment of enzymatic chains. General properties, control and effector strength. *Eur. J. Biochem.* 42 (1974) 89–95.

[33] Kacser H & Burns JA. The control of flux. *Symp. Soc. Exp. Biol.* 27 (1973) 65–104.
[34] Fell DA. Understanding the Control of Metabolism. Portland Press, 1997.
[35] Sorribas A, Hernandez-Bermejo B, Vilaprinyo E & Alves R. Cooperativity and saturation in biochemical networks: a saturable formalism using Taylor series approximations. *Biotechnol. Bioeng.* 97 (2007) 1259–1277.
[36] Curto R, Voit EO, Sorribas A & Cascante M. Mathematical models of purine metabolism in man. *Math. Biosci.* 151 (1998) 1–49.
[37] Kreyszig E. Advanced Engineering Mathematics, 10th ed. Wiley, 2011.
[38] Bonabeau E. Agent-based modeling: methods and techniques for simulating human systems. *Proc. Natl Acad. Sci. USA* 14 (2002) 7280–7287.
[39] Macal CM & North M. Tutorial on agent-based modeling and simulation. Part 2: How to model with agents. In Proceedings of the Winter Simulation Conference, December 2006, Monterey, CA (LF Perrone, FP Wieland, J Liu, et al., eds), pp 73–83. IEEE, 2006.
[40] Mocek WT, Rudnicki R & Voit EO. Approximation of delays in biochemical systems. *Math. Biosci.* 198 (2005) 190–216.
[41] Elowitz MB & Leibler S. A synthetic oscillatory network of transcriptional regulators. *Nature* 403 (2000) 335–338.
[42] Wilkinson DJ. Stochastic Modelling for Systems Biology, 2nd ed. CRC Press, 2011.
[43] Gillespie DT. Stochastic simulation of chemical kinetics. *Annu. Rev. Phys. Chem.* 58 (2007) 35–55.
[44] Wu J & Voit E. Hybrid modeling in biochemical systems theory by means of functional Petri nets. *J. Bioinform. Comput. Biol.* 7 (2009) 107–134.
[45] Savageau MA. Biochemical systems analysis. II. The steady-state solutions for an *n*-pool system using a power-law approximation. *J. Theor. Biol.* 25 (1969) 370–379.
[46] Wiens EG. Egwald Mathematics: Nonlinear Dynamics: Two Dimensional Flows and Phase Diagrams. http://www.egwald.ca/nonlineardynamics/twodimensionaldynamics.php.
[47] Strogatz SH. Nonlinear Dynamics and Chaos, 2nd ed. Westview Press, 2014.
[48] Hinkelmann K & Kempthorne O. Design and Analysis of Experiments. Volume I: Introduction to Experimental Design, 2nd ed. Wiley, 2008.
[49] Epstein IR & Pojman JA. An Introduction to Nonlinear Chemical Dynamics. Oscillations, Waves, Patterns, and Chaos. Oxford University Press, 1998.
[50] Voit EO. Modelling metabolic networks using power-laws and S-systems. *Essays Biochem.* 45 (2008) 29–40.
[51] Lewis DC. A qualitative analysis of S-systems: Hopf bifurcations. In Canonical Nonlinear Modeling. S-System Approach to Understanding Complexity (EO Voit, ed.), Chapter 16. Van Nostrand Reinhold, 1991.
[52] Yin W & Voit EO. Construction and customization of stable oscillation models in biology. *J. Biol. Syst.* 16 (2008) 463–478.
[53] Sprott JC & Linz SJ. Algebraically simple chaotic flows. *Int. J. Chaos Theory Appl.* 5(2) (2000) 1–20.
[54] van der Pol B & van der Mark J. Frequency demultiplication. *Nature* 120 (1927) 363–364.
[55] van der Pol B. & van der Mark J. The heart beat considered as a relaxation oscillation. *Philos. Mag.* 6(Suppl) (1928) 763–775.
[56] Lorenz EN. Deterministic nonperiodic flow. *J. Atmos. Sci.* 20 (1963) 130–141.
[57] Vano JA, Wildenberg J C, Anderson MB, Noel JK & Sprott JC. Chaos in low-dimensional Lotka–Volterra models of competition. *Nonlinearity* 19 (2006) 2391–2404.
[58] Sprott JC, Vano JA, Wildenberg JC, Anderson MB & Noel JK. Coexistence and chaos in complex ecologies. *Phys. Lett. A* 335 (2005) 207–212.
[59] Voit EO. S-system modeling of complex systems with chaotic input. *Environmetrics* 4 (1993) 153–186.
[60] Thomas R. Deterministic chaos seen in terms of feedback circuits: analysis, synthesis, "labyrinth chaos." *Int. J. Bifurcation Chaos* 9 (1999) 1889–1905.
[61] Rössler OE. An equation for continuous chaos. *Phys. Lett. A* 57 (1976) 397–398.

FURTHER READING

Easy Introductions

Batschelet E. Introduction to Mathematics for Life Scientists, 3rd ed. Springer, 1979.
Edelstein-Keshet L. Mathematical Models in Biology. SIAM, 2005.

General Texts on Mathematical Biology

Adler FR. Modeling the Dynamics of Life: Calculus and Probability for Life Scientists, 3rd ed. Brooks/Cole, 2012.
Beltrami E. Mathematics for Dynamic Modeling, 2nd ed. Academic Press, 1998.
Britton NF. Essential Mathematical Biology. Springer, 2004.
Ellner SP & Guckenheimer. J. Dynamic Models in Biology. Princeton University Press, 2006.
Haefner JW. Modeling Biological Systems: Principles and Applications, 2nd ed. Chapman & Hall, 2005.
Keener J & Sneyd J. Mathematical Physiology. I: Cellular Physiology, 2nd ed. Springer, 2009.
Keener J & Sneyd J. Mathematical Physiology. II: Systems Physiology, 2nd ed. Springer, 2009.
Klipp E, Herwig R, Kowald A, Wierling C & Lehrach H. Systems Biology in Practice: Concepts, Implementation and Application. Wiley-VCH, 2005.
Kremling, A. Systems Biology. Mathematical Modeling and Model Analysis. Chapman & Hall/CRC Press, 2013.
Kreyszig E. Advanced Engineering Mathematics, 10th ed. Wiley, 2011.
Murray JD. Mathematical Biology I: Introduction, 3rd ed. Springer, 2002.
Murray JD. Mathematical Biology II: Spatial Models and Biomedical Applications, Springer, 2003.
Strang G. Introduction to Applied Mathematics. Wellesley-Cambridge Press, 1986.
Strogatz SH. Nonlinear Dynamics and Chaos, 2nd ed. Westview Press, 1994.
Yeargers EK, Shonkwiler RW & Herod JV. An Introduction to the Mathematics of Biology. Birkhäuser, 1996.

Biochemical and Cellular Systems

Fell DA. Understanding the Control of Metabolism. Portland Press, 1997.
Heinrich R & Schuster S. The Regulation of Cellular Systems. Chapman & Hall, 1996.
Jacquez JA. Compartmental Analysis in Biology and Medicine, 3rd ed. BioMedware, 1996.
Palsson BØ. Systems Biology: Properties of Reconstructed Networks. Cambridge University Press, 2006.

Savageau MA. Biochemical Systems Analysis: A Study of Function and Design in Molecular Biology. Addison-Wesley, 1976.
Segel LA. Modeling Dynamic Phenomena in Molecular and Cellular Biology. Cambridge University Press, 1984.
Torres NV & Voit EO. Pathway Analysis and Optimization in Metabolic Engineering. Cambridge University Press, 2002.
Voit EO. Computational Analysis of Biochemical Systems: A Practical Guide for Biochemists and Molecular Biologists. Cambridge University Press, 2000.

Stochastic Processes and Techniques

Pinsky M & Karlin S. An Introduction to Stochastic Modeling, 4th ed. Academic Press, 2011.
Ross SM. Introduction to Probability Models, 11th ed. Academic Press, 2014.
Wilkinson DJ. Stochastic Modelling for Systems Biology, 2nd ed. CRC Press, 2011.

Parameter Estimation

5

When you have read this chapter, you should be able to:

- Understand and explain the concepts of linear and nonlinear regression
- Describe the concepts of various search algorithms
- Discuss the challenges arising in parameter estimation
- Execute ordinary and multiple linear regressions
- Estimate parameters in linearized models
- Estimate parameters for explicit nonlinear functions
- Estimate parameters for small systems of differential equations

Several earlier chapters have discussed the construction and analysis of mathematical and computational models. Almost absent from these discussions was a crucial component, namely, the determination of values for all model parameters. Clearly, these values are very important, because they connect an abstract structure, the symbolic model, with the reality of biological data. Even the best model will not reproduce, explain, or predict biological functionality if its parameter values are wrong. Furthermore, very few analyses can be performed with a model when its parameter values are unknown. The reason for dedicating an entire chapter to parameter estimation is that this key step of modeling is complicated. There are some situations that are easy to solve, but if the investigated systems become even moderately large, parameter estimation often emerges as the most severe bottleneck of the entire modeling effort. Moreover, in spite of enormous efforts, there is still no silver bullet solution for finding the best parameters for a model in a straightforward and efficacious manner. Thus, we will go through this chapter, classifying and dissecting the estimation problem into manageable tasks, presenting solution strategies, and discussing some of the reasons why computer **algorithms** tend to fail. Even if you have no intention to execute actual parameter estimation tasks in your future work, you should at least skim through the chapter, in order to gain an impression of the difficulties of this task, while skipping some of the technical details.

Generically speaking, parameter estimation is an **optimization** task. The goal is to determine the parameter vector (that is, a set of numerical values, one for each parameter of a model) that minimizes the difference between experimental data and the model. In order to avoid cancellation between positive and negative differences, the **objective function** to be minimized is usually the sum of the squared differences between data and model. This function is commonly called **SSE**, which stands for **sum of squared errors**, and methods searching for SSE are called **least-squares methods**.

As in other modeling contexts, the greatest divide in parameter estimation is between linearity and nonlinearity. If a model is linear, or can be transformed so that it is (approximately) linear, then life is good. Many methods are available, most of which are quite effective, and there are numerous statistical tools characterizing the quality of the solution. By contrast, as soon as a few nonlinearities are introduced, the vast majority of these effective methods become inapplicable. The two dominant solution strategies in these cases are the use of brute computational force and the development of methods that work with reasonable efficiency at least for some classes of problems.

The field has a second divide. First, it is sometimes possible to estimate parameters for each process in the system (or model) and then to merge all these local descriptions in the hope of achieving a well-functioning model that encompasses the interplay of all processes in an adequate fashion. This bottom-up strategy has dominated parameter estimation for biological systems in the past, and is still of great importance. However, it often fails, for a variety of suspected or unknown reasons. For instance, kinetic data may have come from *in vitro* experiments that were carried out under different conditions, or may even have been obtained from different organisms. The result of this unfortunately typical situation is that the resulting model, which was constructed with the best available local information, clearly does not generate reasonable results for an all-encompassing model. The discrepancy requires revisiting all assumptions, as well as the information used to parameterize the model. Sensitivity analysis (Chapter 4) is a useful diagnostic tool for this step, because it identifies parameters that affect the solution most and therefore need to be estimated most precisely. Experience shows that any successes in applying the bottom-up strategy to moderately large systems with maybe 40 or 50 parameters are almost always the product of an excruciatingly slow and cumbersome effort that sometimes takes months to complete.

A modern, very appealing alternative is becoming increasingly feasible. This alternative depends on experimentally determined **time series** of genomic, proteomic, or metabolomic data. A general setting for measuring these data is as follows. Typically, the biological system is in its normal state. At the beginning of the experiment, the system is perturbed, and one measures the responses of several components at many time points afterwards. Hidden in these time profiles is very valuable information about the structure, regulation, and dynamics of the system, and the trend toward using time series data is to be welcomed, because these data are obtained from the same organism under the same experimental conditions. The catch with time-series-based estimation is that the extraction of information is a difficult task that requires not only powerful computers but also ingenuity and insight into the biological system and into the intricacies of computational algorithms designed to execute the extraction in an effective fashion.

PARAMETER ESTIMATION FOR LINEAR SYSTEMS

5.1 Linear Regression Involving a Single Variable

In the simplest case of a single variable y depending on only one other variable x, the relationship between the two can be plotted on an x–y graph. This **scatter plot** gives an immediate impression of whether the relationship might be linear. If so, the method of linear **regression** quickly produces the straight line that optimally describes this relationship (**Figure 5.1**). The reason that this works so well, even on a moderately sophisticated pocket calculator, is that it is straightforward to express the difference Δ_i between each data point (x_i, y_i) and an imagined straight line traversing the data set and to compute the two characteristic features of the line, namely its slope and intercept, such that this line simultaneously minimizes all differences. Specifically, a linear regression algorithm solves the problem sketched in Figure 5.1, which shows data points (symbols) forming a stretched-out cloud, as well as the optimized straight line going through this cloud. This regression line is mathematically defined as the uniquely defined straight line that minimizes the sum of all squared residuals or errors Δ_i. The reason to square these quantities is simply that we do not want a positive Δ_i to compensate a negative Δ_j, and squaring is mathematically more

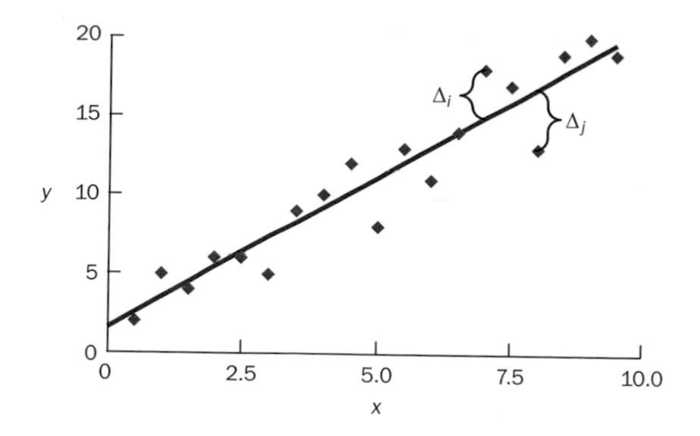

Figure 5.1 Simple linear regression. A more or less linear cloud is transected by the regression line, which is optimized in a sense that SSE, the sum of squared errors Δ_i, is minimized.

convenient than using absolute values. SSE is also called an error function, and parameter estimation tasks are therefore equivalent with tasks of minimizing error functions. The line **fitting** the data in Figure 5.1 has the representation

$$y = 1.899248x + 1.628571, \tag{5.1}$$

which is easily obtained from the data, for instance with Microsoft Excel®.

Linear regression is a straightforward exercise, and its result allows us to predict the expected value of y for some x that has never been analyzed. For instance, in Figure 5.1, we can rather reliably predict the approximate y-value that is expected for $x = 6.75$. Namely, plugging $x = 6.75$ into (5.1) yields $y = 14.4485$, which fits nicely into the picture.

Two issues require caution. First, extrapolations or predictions of y-values beyond the observed range of x are not reliable. While we can reliably compute the expected y-value for $x = 6.75$, computing y for $x = 250$ suggests $y = 476.4406$, which is a correct computation but may or may not be a good prediction; we simply don't know. If the example is biological, the solution may be utterly wrong, because biological phenomena usually deviate from linearity for large x values and saturate instead.

The second issue to keep in mind is that the algorithm blindly yields linear regression lines whether or not the relationship between x and y is really linear. As an example, the relationship in **Figure 5.2** clearly appears to be nonlinear, maybe following a trend as indicated by the green line. Nonetheless, it is easy to perform a linear regression, which quickly yields the relationship

$$y = 0.707068x + 8.151429. \tag{5.2}$$

It is evident that this model (the red line in Figure 5.2) does not make much sense. Thus, vigilance is in order. Often, simple inspection as in Figure 5.2 is sufficient. But one should also consider assessing a linear regression result with some mathematical diagnostics [1]. For instance, one might analyze the residual errors, which should form a normal distribution if the data really follow a linear trend. One may also study the lengths of "runs" of subsequent data falling below or above the regression line. For instance, in Figure 5.2, all low-valued data fall below the regression line, almost all data in the center lie above the line, and almost all high-valued data are

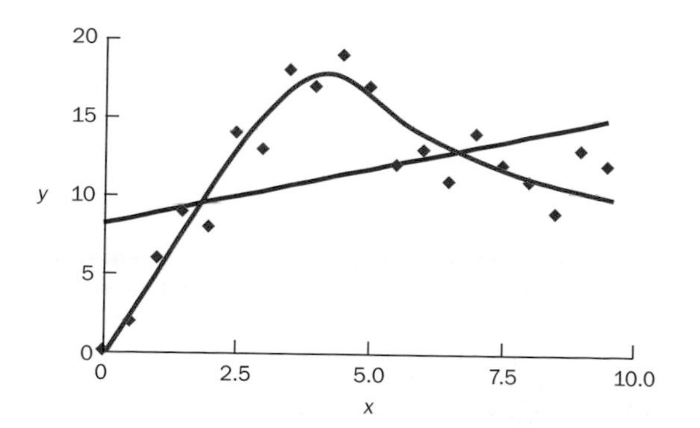

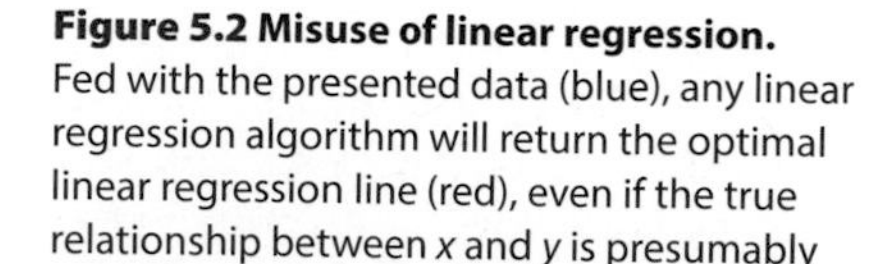
Figure 5.2 Misuse of linear regression. Fed with the presented data (blue), any linear regression algorithm will return the optimal linear regression line (red), even if the true relationship between x and y is presumably nonlinear, as indicated by the green line.

below the line, which would be statistically highly unlikely if the data more or less followed the assumed linear relationship.

5.2 Linear Regression Involving Several Variables

Linear regression can also be executed quite easily in cases of more variables (see, for example, [1]). As the most intuitive example, suppose that some quantity Z depends on the two variables X and Y. For instance, the risk of disease may increase with blood pressure and body mass index. Thus, for every pair (X, Y), there is a value of Z. If the dependence is linear, the triplets (X, Y, Z) are located close to a sloping plane in three-dimensional space (**Figure 5.3**), and one may, for instance, use the `regress` function in MATLAB® for a multiple linear regression, which means linear regression where Z is a linear function of several variables. For more than two variables, the result is difficult to visualize, but the `regress` function nevertheless computes the higher-dimensional analog of a plane, which is called a hyperplane.

An example for two independent variables is given in Figure 5.3, which shows the data from **Table 5.1** in different perspectives. Red and blue dots in (A) and (C) correspond to data above and below the regression plane, respectively. The regression plane, computed with `regress`, is characterized by the function

$$Z = -0.0423 + 0.4344X + 1.1300Y. \tag{5.3}$$

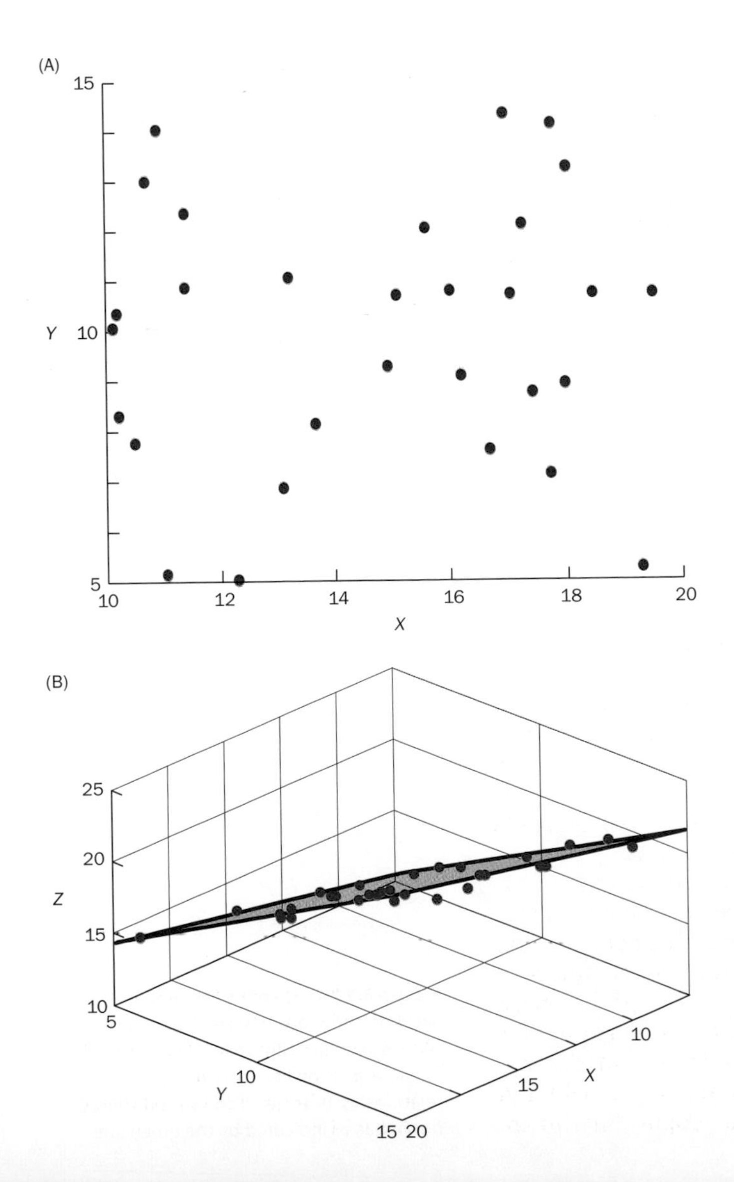

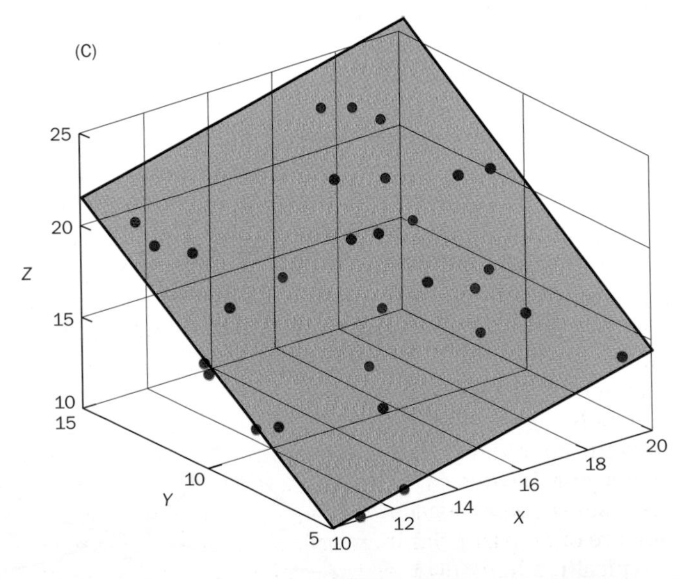

Figure 5.3 Result of a multiple linear regression analysis with the Matlab function `regress` and the data from Table 5.1. (A) The data are distributed throughout the ranges of X and Y. (B) A different perspective shows that all data points are close to a sloping plane, which in this perspective almost looks like a line. (C) In this third view, red dots lie above the regression plane (beige) and blue dots below.

TABLE 5.1: DATA FOR MULTIPLE LINEAR REGRESSION AND CORRESPONDING VALUES OF THE REGRESSION PLANE			
X	*Y*	*Z* (data)	*Z* (regression)
17.43	8.72	16.73	17.38
10.49	7.75	13.58	13.28
16.69	7.58	15.47	15.77
17.06	10.69	19.32	19.46
13.10	6.85	13.82	13.39
18.46	10.71	21.06	20.07
17.27	12.10	20.55	21.13
16.04	10.77	19.15	19.10
14.95	9.25	16.68	16.90
17.74	7.11	16.30	15.70
17.98	8.89	17.38	17.82
13.67	8.12	14.89	15.07
15.09	10.68	19.42	18.58
10.22	8.28	13.27	13.75
11.04	5.12	10.08	10.54
11.39	12.34	19.60	18.85
11.40	10.87	17.57	17.19
10.71	12.98	19.93	19.28
10.20	10.33	15.45	16.06
12.30	5.02	10.95	10.98
16.97	14.32	22.96	23.51
17.79	14.10	22.68	23.62
13.22	11.05	18.06	18.19
10.91	14.01	20.35	20.53
19.30	5.26	14.31	14.29
15.60	12.04	21.42	20.34
18.02	13.23	22.52	22.73
10.12	10.05	15.15	15.71
19.50	10.73	20.77	20.55
16.22	9.07	17.56	17.25

It may be surprising that linear regression is applicable to some nonlinear functions. This option is available if the functions permit a mathematical transformation that makes them linear. A well-known example is an exponential function, which becomes linear upon logarithmic transformation. An interesting case of linearization is the Michaelis–Menten rate law (MMRL) for enzyme-catalyzed reactions (see Chapters 2, 4, and 8). Suppose that the system of interest consists of a very short metabolic pathway with just two reactions, as shown in **Figure 5.4**. The pathway converts an initial substrate *S* into a metabolite *M* in a reaction that is catalyzed by an enzyme *E*, whose activity does not change during our experiment. The metabolite is subsequently used up or degraded, and the degradation products of the reaction are of no particular interest and therefore not explicitly included in the model. A typical mathematical representation for the first process is a Michaelis–Menten function, which is characterized by three quantities: the substrate concentration *S*, the maximally possible rate or velocity of the reaction, V_{max}, which depends on the (constant) enzyme concentration, and the Michaelis constant K_M, which quantifies the **affinity** between the enzyme and its substrate (**Figure 5.5**). The default description for the degradation of the metabolite is a so-called first-order process with rate constant *c*, which implies that the speed of degradation is proportional to the

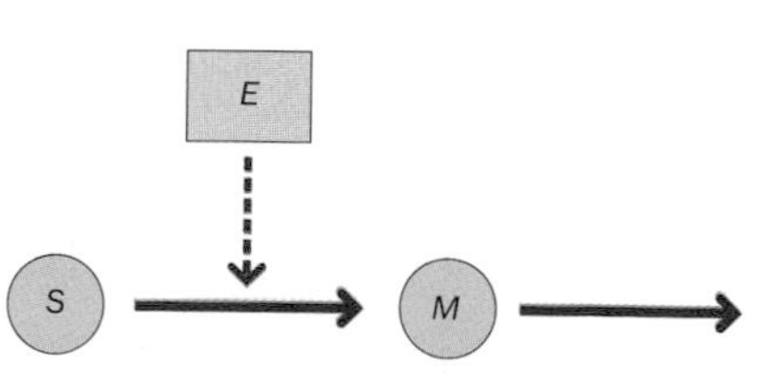

Figure 5.4 A simple model for the dynamics of a metabolite *M*. *M* is the product of a reaction, which is catalyzed by enzyme *E* and uses substrate *S*. *M* is degraded, but the product of this process is not specified.

current metabolite concentration M (**Figure 5.6**). Thus, the overall change in metabolite concentration, $\dot{M}$, is given by the difference between production and degradation, and, with the typical functions discussed above, it reads

$$\dot{M} = \frac{V_{max} S}{K_M + S} - cM. \tag{5.4}$$

The model contains three parameters, V_{max}, K_M, and c, as well as the concentration of S, which may or may not be constant. Suppose someone had executed experiments with varying levels of S and measured the production of metabolite (without degradation), thereby generating data consisting of pairs (S, v), where $v = V_{max}S/(K_M + S)$. For clarity of illustration, let us suppose these data are error-free (**Table 5.2** and Figure 5.5A). Our task is to estimate optimal values for the parameters V_{max} and K_M. Clearly, the process is nonlinear, and linear regression seems to be out of the question. However, an old trick for dealing with Michaelis–Menten functions is based on the observation that the function becomes linear if we plot $1/v$ against $1/S$. This inverse plot of the data is shown in Figure 5.5B. Since the data in this representation follow a linear function, we can use linear regression to determine the best-fitting straight line through the data. It turns out from the action of taking the reciprocal of the Michaelis–Menten function that the slope of the regression line corresponds to K_M/V_{max}, whose value here is 2.91, and that the intercept is $1/V_{max}$ with a value of 0.45. The two values allow us to compute V_{max} and K_M as 2.2 and 6.4, respectively.

Suppose the degradation process was estimated in a separate step. Artificial data are given in **Table 5.3**. These data exhibit exponential decay, which perfectly matches the formulation of the degradation term in (5.4). Thus, a logarithmic transformation of M into $\ln M$ yields a straight-line decay function (see Figure 5.6B) whose parameter values can be estimated by linear regression. The result of this procedure is the value of c that best fits the degradation data, namely, $c = 0.88$.

Taken together with the estimation of V_{max} and K_M, we have now estimated all parameters of the system by bottom-up analysis and can substitute them into (5.4) to perform numerical simulations or other analyses. Validation of the estimates and the model requires additional data, for instance in the form of responses of M to changes in S.

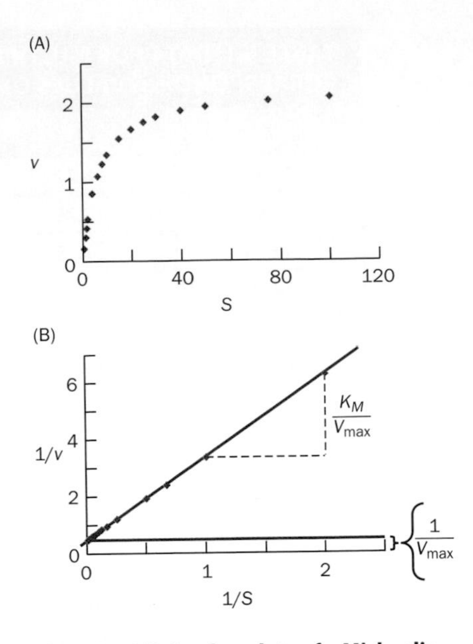

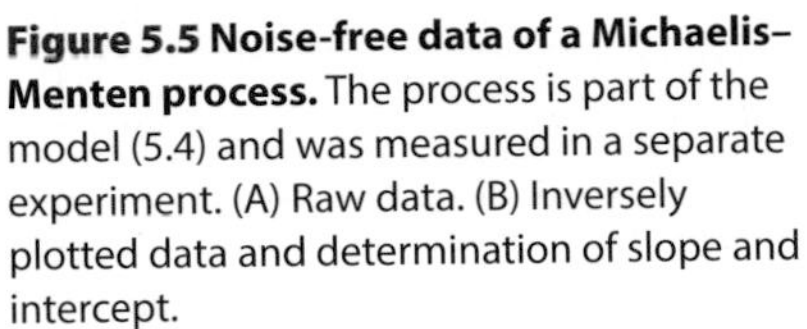
Figure 5.5 Noise-free data of a Michaelis–Menten process. The process is part of the model (5.4) and was measured in a separate experiment. (A) Raw data. (B) Inversely plotted data and determination of slope and intercept.

TABLE 5.2: MEASUREMENTS OF REACTION SPEED *V* VERSUS SUBSTRATE CONCENTRATION *S*

S	*V*
0.5	0.16
1	0.30
1.5	0.42
2	0.52
4	0.85
6	1.06
8	1.22
10	1.34
15	1.54
20	1.67
25	1.75
30	1.81
40	1.90
50	1.95
75	2.03
100	2.07

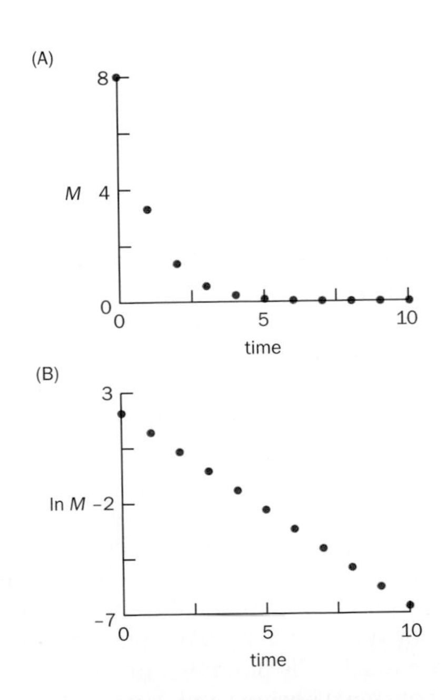

Figure 5.6 Degradation of *M* in (5.4). (A) In Cartesian coordinates, the function is exponential. (B) It becomes linear for the logarithm of *M*.

TABLE 5.3: TIME MEASUREMENTS OF M AND ln M FOR ESTIMATING PARAMETERS IN (5.4)

t	M	ln M
0	8.000	2.079
1	3.318	1.199
2	1.376	0.319
3	0.571	−0.561
4	0.237	−1.441
5	0.098	−2.321
6	0.041	−3.201
7	0.017	−4.081
8	0.007	−4.961
9	0.003	−5.841
10	0.001	−6.721

Now suppose that no kinetic data are available, but that we have instead time-course data for the system in the form of M as a function of time (**Figure 5.7**). In this case, we cannot use the bottom-up approach to estimate systems features from local information (kinetic data, linear degradation), but instead need to use a top-down estimation that works more or less in the opposite direction by using observations on the entire system (namely, time course data $M(t)$) to estimate all unknown kinetic features (V_{max}, K_M, and c) simultaneously. This top-down estimation requires a (nonlinear) regression procedure that uses the time course data, together with information on S, and all at once determines values for V_{max}, K_M, and c that capture the data the best. In this simple case, a decent nonlinear regression algorithm quickly returns the optimal values. More methods for this purpose will be discussed later in this chapter.

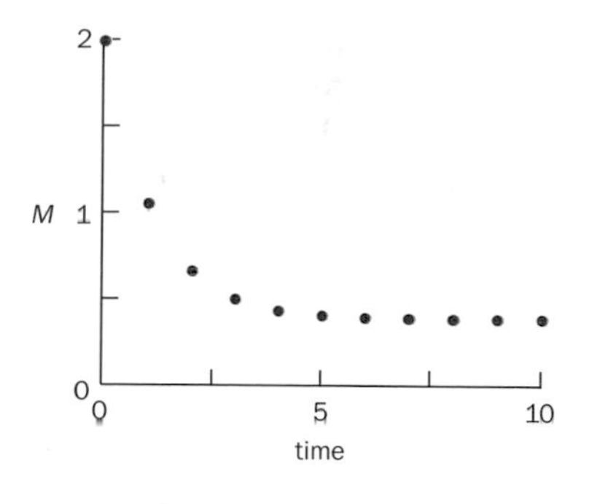

Figure 5.7 Time-course data for the model (5.4). The data cannot be linearized in any obvious fashion and require methods of nonlinear parameter estimation.

One of the great features of linear regression is that it also works for higher-dimensional problems, as we discussed before, and multiple linear regression also applies directly to linearized systems. For instance, consider a process described by the two-variable power-law function

$$V = \alpha X^g Y^h. \tag{5.5}$$

To linearize the function, we take logarithms of both sides, with the result

$$\ln V = \ln \alpha + g \ln X + h \ln Y, \tag{5.6}$$

which permits the application of multiple linear regression to the new variables ln X, ln Y, and ln V.

When using this strategy, one needs to keep in mind that transformations of whichever type also affect the noise in the data. As a consequence, residual errors that are normally distributed around the true nonlinear function are no longer normal in the linearized form. In many practical applications, though, this problem is considered secondary in comparison with the simplification afforded by linearization. If an accurate account of the error structure is important, one may first estimate parameters by linear regression and then use them as start values for a nonlinear regression that does not distort the errors.

PARAMETER ESTIMATION FOR NONLINEAR SYSTEMS

Parameter estimation for nonlinear systems is incomparably more complicated than linear regression. The main reason is that there is just one linear structure but there are infinitely many different nonlinear functions, and it is not known per se which nonlinear function(s) might model some data sufficiently well, let alone optimally, especially if the data are noisy. Even if a specific function has been selected, there are no simple methods for computing the optimal parameter values as they exist for linear systems. Moreover, the solution of a nonlinear estimation task may not be unique. It is possible that two different parameterizations yield exactly the same residual error or that many solutions are found, but none of them is truly good, let alone optimal.

Because the estimation of optimal parameter values for nonlinear models is challenging, many algorithms for optimization, and thus for parameter estimation, have been developed over many decades. All of them work well in some situations, but fail in others, especially for larger systems where many parameters have to be estimated. In contrast to linear regression, it is typically impossible to compute an explicit, optimal solution for a nonlinear estimation task. So one may ask: if there is no explicit mathematical solution, how is it possible that a computer algorithm can find a solution? The answer is that optimization algorithms iteratively search for ever better solutions, and, if they succeed, the solution is very close to the best possible solution, although it is usually not precisely the truly best solution. The trick of developing good search algorithms is therefore to guide the search in an efficient manner toward the optimal parameter set and to abandon or drastically alter search strategies early when they enter parameter ranges that do not show much promise.

A crucial question regarding the success of nonlinear estimation algorithms is whether the task calls for finding the parameters of a mathematical function or of a system of differential equations. In the former case, many methods work quite well, and we can even use software like the Solver option with the Data Analysis group of Excel® while the latter case is much more complicated, although the basic concepts for both tasks are the same. We will discuss different types of methods, exemplify some of them with the estimation of parameters in explicit functions, and toward the end discuss parameter estimation for dynamical systems.

The currently available search algorithms may be divided into several classes. The first consists of attempts to exhaust all possible parameter combinations and to select the best among them. A second class consists of gradient, steepest-descent, or hill-climbing methods. The term "hill-climbing" implies finding the maximum of a function, while parameter-estimation searches for the minimum use steepest-descent methods. Mathematically, these two tasks are equivalent, because finding the maximum of a function F is the same as finding the minimum of $-F$. The basic idea is simple. Imagine finding yourself in a hilly terrain and that it is very foggy so you can only see a few feet in each direction. You are thirsty, and you imagine that water is most likely to be found down in the valley. So, you look around (although you can't see very far) and check which direction is heading down. You walk in this direction until you come to a point where a different direction leads down even more steeply. With time, you have a good chance of getting to the bottom of the valley. Gradient methods imitate this procedure. As a warning, it is easy to realize that even in the hiking example this strategy is by no means failsafe. In fact, gradient search algorithms tend to have problems with rough terrains and, in particular, when they find themselves in a valley that is not as deep as some of the neighboring valleys. As a result, the algorithms often get trapped in a local minimum, which may be better than other points close-by, but is not as good as the desired solution of the true (global) minimum, which is located in a different valley (**Figure 5.8**).

A third class of parameter estimation methods consists of **evolutionary algorithms**, which operate in a fashion that is entirely different from the previous methods. The concepts here are gleaned from the natural processes of fitness-based selection, which we believe has time and again promoted the best-adapted individuals and species from among their competitors. The best-known evolutionary method, the **genetic algorithm** (**GA**), begins with a population of maybe 50–100 parameter vectors, which consist of default values or of values supplied by the user. Some of these values may be good and others not so. The GA solves the model with each parameter vector and computes the residual error associated with each case. The vectors are then ranked by their **fitness** (how well the model with the parameter values of the vector matches the data), and the fitter a vector, the higher is its probability to mate. In this mating process, one part of one vector (the mother) is merged with the complementary part of another vector (the father) to produce a new vector (the newborn baby vector). The vectors are furthermore subject to small mutations

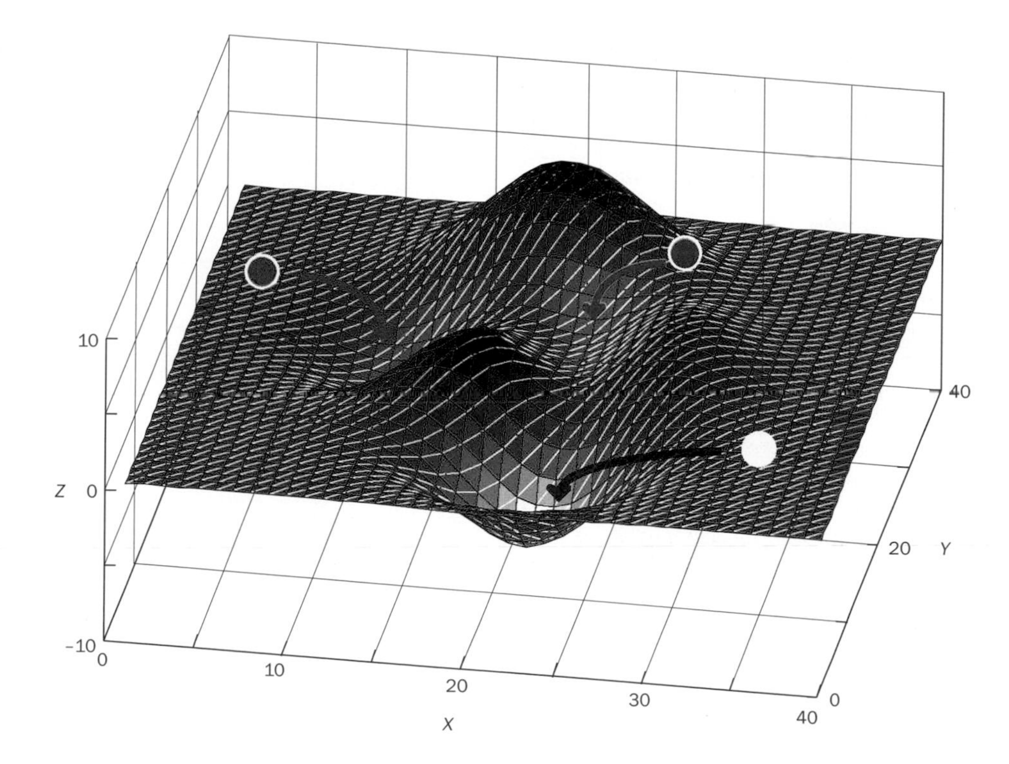

Figure 5.8 Gradient searches may fail in rough terrains. The function shown here (a Laplace function from MATLAB®) has three minima, of which the one in the center front ($X = 22, Y = 11, Z = -6.2$) is the lowest. A gradient search initiated at the white location will likely find this global minimum. However, gradient searches starting at the green or blue locations will easily get trapped in local minima with $Z = 0$ and $Z = -3$, respectively.

that permit the evolution of the population. Many mating processes within the parent population lead to a new generation of vectors, the model is again solved for each vector, residual errors are computed, the best vectors are selected, and the process continues to the next generation until a suitable solution is found or the algorithm runs out of steam. Gradient methods and genetic algorithms presently dominate the field of search methods for parameter estimation, but there are other alternatives, which we will briefly mention later in this chapter.

5.3 Comprehensive Grid Search

The idea of exhaustive searches is really simple. One determines the possible numerical range for each parameter (for instance, from biological knowledge of the process), evaluates the model with very many or even all combinations of parameter values within these ranges, and selects the parameter set that produces the best result. Because the parameter combinations are usually selected in regular intervals, this type of estimation is called a grid search.

As an illustration, let us return to the example of the Michaelis–Menten rate law (MMRL) $v = V_{max}S/(K_M + S)$ and pretend that our earlier linearization was not possible. Suppose we have measurements of the rate v for several substrate concentrations S. We know from our understanding of the MMRL that the two parameters V_{max} and K_M must be positive. Let us suppose that we also know that $V_{max} \leq 10$ and $K_M \leq 8$. Thus, we have admissible ranges, which we subdivide into even intervals of 1. For the 80 parameter combinations, it is now straightforward to compute the value of v for each substrate concentration S, subtract from this value the corresponding experimental measurement, square the differences, and divide the sum by the number of data points used. Thus, for $S = 0.5$, the experimental $v = 0.16$ from Table 5.2 and the pair $(V_{max}, K_M) = (1, 1)$, the first quantity of the sum is computed as

$$\left(\frac{1\times0.5}{1+0.5}-0.16\right)^2. \tag{5.7}$$

Figure 5.9A shows a plot of the sum of squared errors for the given ranges of V_{max} and K_M. The plot indicates two important results. First, many combinations of V_{max} and K_M are clearly not contenders for the optimum, because the associated errors are very high. Thus, a considerable amount of computation time has been wasted.

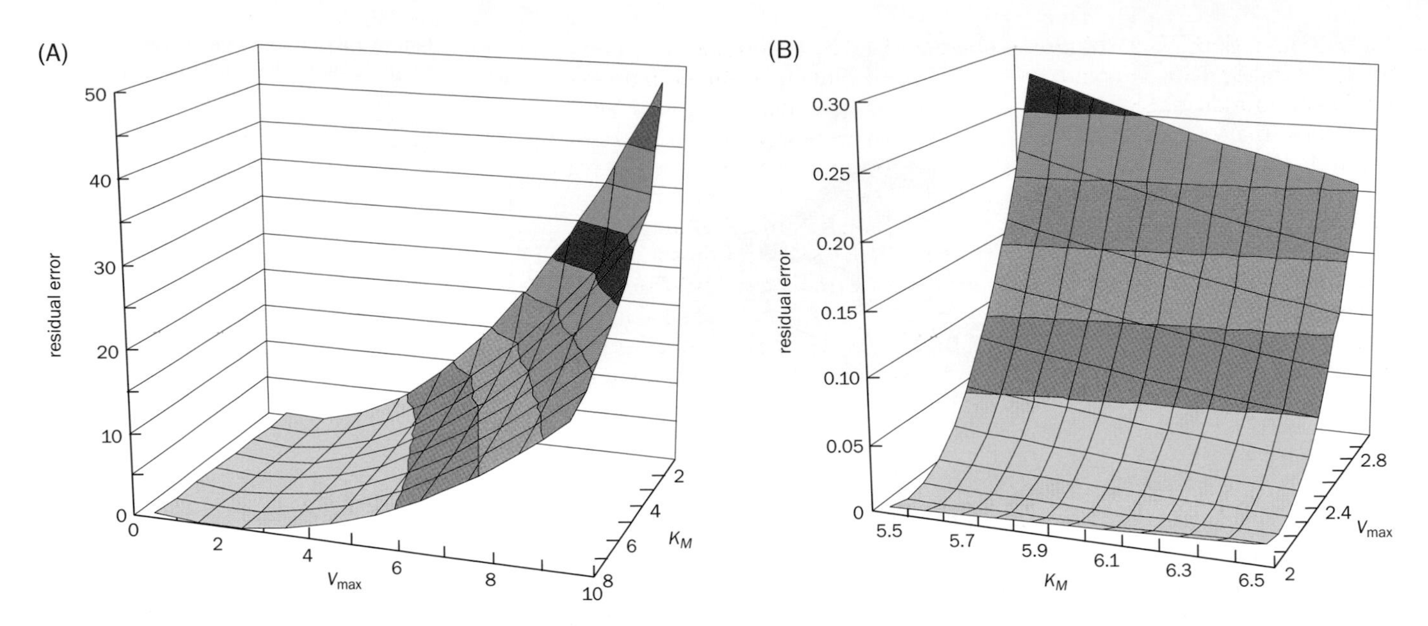

Figure 5.9 Residual errors (SSEs) in a grid search. The search targeted the MMRL parameter values K_M and V_{max} in (5.4), using the data in Table 5.2. (A) Coarse search. (B) Refined search within smaller parameter ranges.

Second, it is difficult to discern where exactly the parameter combination with the minimal error is located. It looks to be somewhere close to V_{max} between 2 and 3 and K_M around 6. We can use this information to refine the ranges for V_{max} and K_M and repeat the grid search with smaller intervals. Figure 5.9B indeed gives a more focused view and identifies the accurate solution $V_{max} = 2.2$ and $K_M = 6.4$. This solution is accurate, because the true values happened to be exactly on the grid we used. If the true values had been $V_{max} = 2.01234567$, $K_M = 6.54321098$, we would at best have found an approximate, relatively close solution.

This simple example already exhibits the main problems with grid searches. First, one has to know admissible ranges for all parameters; otherwise one might miss the optimal solution. Second, the method is not likely to produce very precise results, unless one iterates the search many times with smaller and smaller intervals. Third, the method wastes a lot of computation time for parameter combinations that are clearly irrelevant (in our case, high V_{max} and low K_M). Fourth, and maybe most important, the number of combinations grows very rapidly when many unknown parameters are involved. For instance, suppose the model has eight parameters and we begin by trying 10 values for each parameter. Then the number of combinations is already 10^8, which is 100 million. For more parameters and more values to be tried, these numbers become astronomical. The field of experimental design in statistics has developed possible means of taming this combinatorial explosion [1, 2]. Arguably most popular among them is **Latin hypercube sampling**. The adjective Latin here comes from the Latin square, which, for size 5 for instance, contains the numbers 1, 2, 3, 4, and 5 exactly once in each row and each column of a 5×5 grid. The term hypercube refers to the analog of a regular cube in more than three dimensions. The key idea of Latin square sampling is that one analyzes only enough points to cover each row and each column of a grid exactly once. In a higher-dimensional space, one searches a hypercube in an analogous fashion.

While grid searches are seldom effective, the attraction of obtaining a **global optimum**, at least in an approximate sense, has triggered research into improving grid searches without being overwhelmed by combinatorial explosion. As an example, Donahue and collaborators [3] developed methods that search a multidimensional grid more intensively in the most promising domains of the parameter space and only coarsely in domains that are unlikely to contain the optimal solution.

Branch-and-bound methods are significant improvements over grid searches, because they ideally discard large numbers of inferior solutions in each step [4, 5]. Branch-and-bound methods do this with two tools. The first is a splitting or branching procedure that divides the set of candidate solutions into two non-overlapping "partitions" A and B, so that all solutions are accounted for and each one is either in

A or *B* but not both. Each solution is characterized by a score for the quantity that is being optimized, which in our case is SSE. The second tool is the estimation of upper and lower bounds for SSE. A lower bound, say for partition *A*, is a number that is as small as or smaller than the SSE of any candidate solution (fitness of a parameter vector) in *A*, and the analogous definition holds for upper bounds. The key concept of the branch-and-bound method is that if the lower bound of partition *A* is greater than the upper bound of partition *B*, then the entire partition *A* may be discarded, because no solution in *A* can possibly have a lower SSE (and thus a higher fitness) than any of the solutions in *B*. The trick is therefore to split the candidate solution sets successively and effectively into partitions and to compute tight lower and upper bounds. In the end, branch-and-bound methods find globally optimal solutions, but their implementation is difficult and they require considerable computational effort.

5.4 Nonlinear Regression

Most nonlinear functions cannot be transformed into linear functions, and they are too complicated for exhaustive explorations of their parameter space. Moreover, a direct computation of optimal parameter values, which is the hallmark of linear regression, is no longer possible, and entirely different approaches must be used. The most prominent among these are iterative methods that search in a guided or random fashion for good parameter vectors (see, for example, [6]). Because the parameter space that encompasses all parameter vectors is usually of high dimension and is difficult to fathom, the algorithms initiate the search, assess the quality of fit with the current parameter values, search again, assess the quality of fit again, and thereby operate in an iterative mode that cycles between searching and assessing thousands of times. The algorithm starts with a default or user-provided parameter vector (sometimes collectively called a guesstimate), which may be quite good or rather inadequate, solves the model with this parameter vector, and computes the residual error. In the next step, the algorithm evaluates the model for a small number of other parameter vectors in a close neighborhood of the start vector—remember the search for water in a hilly, foggy terrain (see Figure 5.8)? The algorithm then uses this information to determine in which direction the error decreases the most. Pointing into this direction, the algorithm selects a new trial parameter vector at some distance from the original vector. This procedure is repeated thousands of times until, ideally, the algorithm converges to a parameter vector that yields a lower SSE than all vectors in its neighborhood as well as all earlier vectors. Understanding these concepts is more important than technical details, because it reveals why search algorithms sometimes do not succeed and because the technical aspects have been implemented in many software packages.

The iterative search strategy is most easily demonstrated in the case of a single parameter. Suppose the error function *R* has the shape shown in **Figure 5.10**. The plot shows on the horizontal axis the value of the parameter *p* and on the vertical axis the residual error *R*, which is a squared quantity and therefore not negative. Our task is to find the position along the horizontal axis (that is, the optimal parameter *p*) where *R* is closest to zero, because this position minimizes the error.

Suppose we are searching for the minimum inside the interval [0, +8] and our initial guesstimate is $p = 7$. The algorithm enters $p = 7$ into the model and evaluates the residual error *R* based on experimental data; in the figure, $R = 7.2$ (the circled 1). The algorithm now evaluates the model with two values of *p* that are a little smaller or a little greater than 7, respectively. From this information, the algorithm concludes

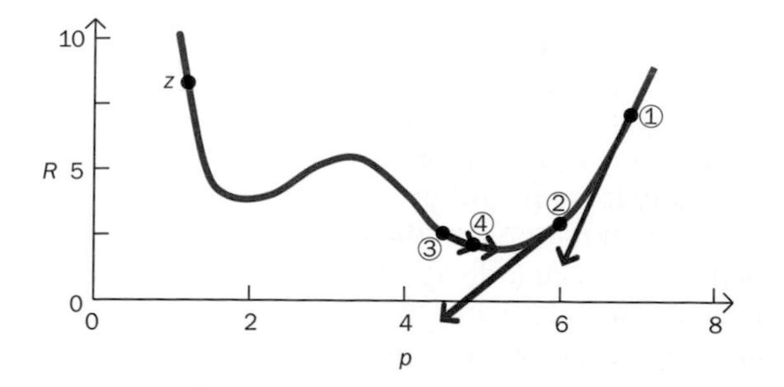

Figure 5.10 Fictitious error landscape for the search for the optimal value(s) of the parameter *p*. The residual error associated with each value of *p* is denoted by *R*. Numbers in circles indicate the starting points for different iterations, and arrows show each search direction, with the length of each arrow representing speed. If the search is started at position *z*, the algorithm may get stuck in a local minimum close to $p = 1.9$.

that lowering the guesstimate has a better chance of reducing R than increasing it. It also estimates the slope of the error function and uses it to determine by how much the original estimate should be lowered; how this determination is accomplished is not so important here. In our example, the algorithm determines the next candidate point as $p = 6$ and computes R at this point as about 2.6 (the circled 2). The slope at this point still suggests lower values for p, and because the same direction is taken again, the speed is increased and the next candidate is about $p = 4.5$ (the circled 3). Now the slope points into the opposite direction, which suggests raising the value of p. The algorithm does this and reaches the point indicated by the circled 4. The process continues in this fashion until the algorithm detects that the slope of the error function is essentially zero, which means that the error can no longer be lowered by moving p a little bit to the left or right. The algorithm terminates and the result is the approximate minimum. It should be kept in mind that the algorithm does not know where the true minimum is. Thus, it has to estimate how far to go at every step and may overshoot or undershoot the target for quite some while.

Figure 5.10 also indicates a potential problem with this type of algorithm. Imagine the initial guesstimate is $p = z$. Using the same rationale as above, it is easy to see that the algorithm might converge to a value close to $p = 1.9$ with a residual error of about $R = 3.75$. Clearly, this is not the optimal value, but it satisfies the criterion that the slope is zero and that other parameter choices close to $p = 1.9$ are even worse.

Exactly the same principles apply to estimation tasks involving two or more parameters. In higher dimensions, we humans have a hard time visualizing the situation, but iterative search algorithms function in exactly the same fashion as discussed above. For two parameters, we can still conceptualize the error function geographically: it is a surface that shows the residual error for each (x, y), where x and y are the two parameters with unknown values (see Figure 5.9 for a simple example). As in Figure 5.8, this error landscape potentially has many hills and valleys, as well as many mountain ridges and craters, and it is easy to imagine how a search algorithm might end up in a **local minimum** that is not, however, the true lowest point. This situation of being trapped in a local minimum is unfortunately quite common. In contrast to our two figures (Figures 5.8 and 5.10), where the mistake is easy to spot, a local minimum is difficult to diagnose in more complicated cases, where for instance six or eight parameters are to be optimized. One obvious, though not failsafe, solution is to start the algorithm many times with different guesstimates.

5.5 Genetic Algorithms

Using nonlinear regression for parameter estimation, one is often unsure whether the solution is truly the optimum or whether the algorithm might have converged to a local minimum (see Figures 5.8 and 5.10). Since this issue is directly tied to the core principles of iterative search algorithms, it is unlikely that small changes in implementation would solve the problem. Instead, an entirely different class of global search heuristics has been developed over the years. Global refers to their ability to avoid getting stuck in local minima, as nonlinear regression algorithms tend to do. Heuristics means that it is often impossible to prove with mathematical rigor that these algorithms will indeed succeed.

The best known global search heuristics are genetic algorithms (see, for example, [7, 8]). They fall into the larger category of evolutionary algorithms, where the attribute of evolution is due to their inspiration by natural selection of the fittest members in a population [9]. Genetic algorithms often find good approximate solutions, but they tend to have difficulties determining very precise solutions. Because this situation is opposite to that in nonlinear regression, it is sometime useful to start a parameter search with a genetic algorithm and to refine the solution with a subsequent nonlinear regression (we will discuss an example of this type later).

Just like in natural evolution, genetic algorithms are based on individuals with their specific fitness, and this fitness corresponds to their ability to mate and produce offspring. Furthermore, the evolution process is subject to mutations. In the specific application of genetic algorithms to parameter estimation, each individual is a parameter vector. In a simplified description, such a vector might look like the diagram in **Figure 5.11**: it lines up the values of all parameters that need to be

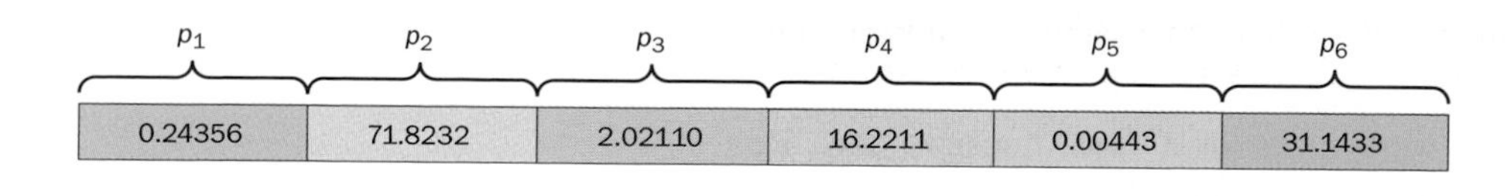

Figure 5.11 A typical individual in the application of a genetic algorithm to parameter estimation. The parameters are lined up either as floating point variables, as they are shown here, in binary code (see Figure 5.12), or in some other mathematical representation.

estimated. In most cases, the numbers are actually converted into binary code, but other encodings, such as floating point representations, have also been explored.

The starting population contains many such individuals (maybe 20, 50, or 100), which are defined through a random assignment of parameter values that collectively represent the admissible parameter ranges or are "seeded" based on prior knowledge of the problem. Each individual corresponds to a solution of the model, because entering the parameter values into the symbolic model allows us to compute any desired features of this model. In general, the feature of interest could be anything that can be quantitatively characterized with the model, such as the steady state, how fast the model reaches a point of interest, how much the transients of the model deviate from the normal state, or how fast the system oscillates. In the case of parameter estimation, the feature of prime interest is the fit to a set of experimental data, and thus SSE. For each individual in a given population, one simulates the model with its parameters and computes the residual error between the data and the corresponding values of the model. This error is used as the measure of fitness: the smaller the error, the fitter the individual. Now comes the time of reckoning. Out of the 20, 50, or 100 individuals in the population, a small number of individuals (maybe 20%) are permitted to mate. They are randomly selected, but the probability of being chosen is proportional to each individual's fitness and thus the SSE. Individuals that were not chosen are discarded. In this scheme, fit individuals have a higher chance of mating, but even unfit individuals may get lucky. Interestingly, the inclusion of a small number of less-fit individuals retains higher genetic diversity and to some degree prevents early termination at a local minimum, which corresponds to a suboptimal parameter vector.

The mating process occurs as follows. Two individuals (parameter vectors) are lined up next to each other, and offspring is produced by using the first few parameters (or parts of the parameter vector) from the first individual and the rest from the second individual, as sketched in **Figure 5.12**. Any positions within a parent vector may also be mutated; that is, from 1 to 0 or from 0 to 1 in the case of binary coding. Experience has shown that mutations are very important, but that their rate has to be rather low, lest the algorithm wander through the parameter space without ever converging. The result of the mating and mutation process is again an individual in the form of a new parameter vector. The parameter values from this vector are entered in the model and the fitness is computed. The mating process is usually organized such that the population size remains constant. Exceptionally fit individuals may mate several times, and most algorithms automatically move the fittest two or three individuals of the parent generation unchanged into the offspring generation. The process terminates for one of several reasons. The following are the most typical among these:

- A solution has been obtained that satisfies preset quality criteria. Success!
- Solutions with the highest fitness have reached a plateau where they do not change for many iterations from one generation to the next. This may be good or bad news.
- The number of generations has reached a preset limit. It could be started again with the last population of vectors.

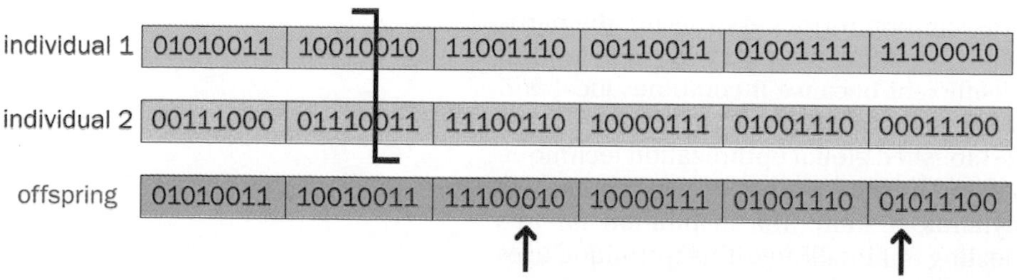

individual 1	01010011	10010010	11001110	00110011	01001111	11100010
individual 2	00111000	01110011	11100110	10000111	01001110	00011100
offspring	01010011	10010011	11100010	10000111	01001110	01011100

Figure 5.12 Generation of offspring in a generic genetic algorithm. Individuals (parameter vectors) are coded here as binary strings. The new offspring contains a part of individual 1 and the complementary part of individual 2. The newly combined offspring is furthermore subject to mutations before the next iteration (arrows).

- The algorithm encounters some numerical problem, for instance, while integrating differential equations.

In many cases, genetic algorithms work quite well, and, because they are extremely flexible, for instance in terms of fitness criteria, they have become very popular and are implemented in many software packages. Nonetheless, they are certainly no panacea. In some cases, they never converge to a stable population; in other cases, they may not find an acceptable solution; and in almost all cases, these algorithms are not particularly fast. If problems arise, the first attempt at resolution is to reset the algorithmic parameters, such as the population size, the mating probabilities of fit or less-fit individuals, the mutation rate, the number of fittest individuals that directly survive into the next generation, bounds on parameters, and criteria for termination. A specific example will be presented later in this chapter.

5.6 Other Stochastic Algorithms

In addition to genetic algorithms, several other classes of methods are available for global optimization and parameter estimation. They include evolutionary programming, for instance implemented as ant colony and particle swarm optimization, as well as simulated annealing and the branch-and-bound methods that we discussed before. In this section, we will discuss the concepts of these methods without giving a lot of detail.

As in genetic algorithms, evolutionary programming is a **machine learning technique** based on evolving populations of solutions. The main mechanisms for improving fitness are mutation and self-adaptation to adjust parameters, and variations may include operations such as combining information from more than two parents. One intriguing method within the class of evolutionary programming is ant colony optimization (ACO) (see, for example, [10]). ACO was inspired by the behavior of ants seeking a food source while wandering about their colony. The original exploration is more or less random, but once an ant finds food, it returns to the nest in a more or less direct fashion. Along the way, the ant leaves a pheromone trail that can be perceived by other ants. This type of chemical communication allows other ants preferably to follow the established pheromone trails rather than to follow random paths. If a path indeed leads to food, more and more ants follow it and release pheromone, thereby reinforcing the successful paths. However, if the path does not lead to success or if it is unduly long, it is less traveled, the pheromone evaporates, and its attraction dissipates. Over time, the short, efficient paths incur more traffic and thus enjoy higher concentrations of pheromone. ACO attempts to mimic this dynamic, adaptive mechanism of optimizing the path to a desired destination. Thus, parameters that show up time and again in good solutions are rewarded by increasing their probability of being selected in future generations, while other infrequent parameters become less and less prevalent in the population. Algorithmic issues related to ACO are discussed in [11].

Similar ideas also form the basis for particle swarm optimization (PSO), which was inspired by the flight patterns of birds (see, for example, [12]). Like genetic algorithms and ACO, PSO is a population-based stochastic search procedure, where each particle in the swarm represents a candidate solution to the optimization or estimation problem. The flight of each particle is influenced by the best positions of the particle and the swarm within a high-dimensional space, and the fitness is measured according to each particle's own experience, which is augmented by communication with its neighboring particles. Through this communication within the swarm, all particles share some of the information regarding high-fitness solutions. Specifically, as soon as a particle detects a promising solution, the swarm explores the vicinity of this solution further. As a result, the particles generally tend toward optimal positions, while searching a wide area around them. Expressed differently, PSO is quite efficient because it combines local and global search methods. PSO methods are reviewed in [13].

Simulated annealing (SA) is a more established global optimization technique that grew out of a Monte Carlo randomization method from the 1950s [14, 15] for generating sample states of a thermodynamic system. The inspiration for this method came from the technique of annealing in metallurgy. This technique uses heating and controlled cooling of a material to reduce defects. The heat provides energy that allows atoms to leave their current positions, which are states of locally minimal energy, and to reach states of higher energy. During cooling, the atoms have a better chance than before of finding lower-energy states. Therefore, the system eventually becomes more ordered and approaches a frozen state of minimal energy in a semi-guided random process. The analogy of this heating–cooling process in SA is that a population of solutions traverses the search space randomly, thereby allowing the current solutions to change a bit. The random mutations of each individual solution are tested (that is, the model is computed with the given parameter values), and mutants that increase fitness always replace the original solutions. Mutations with lower fitness are not immediately discarded but are accepted probabilistically, based on the difference in fitness and a decreasing temperature parameter. If this temperature is high, the current solutions may change almost randomly. This feature prevents the method from getting stuck in local minima. During cooling, the temperature gradually decreases and toward the end the solutions can only vary in a close neighborhood and finally assume local minima. Either the former solutions are regained in the process or new solutions are found that have a higher fitness. Because many solutions settle in their individual minima, it is hoped that the global minimum is among them. The key to a successful implementation is the choice of adequate algorithmic parameters. For instance, if the temperature of the heated system is too low at the beginning or if cooling occurs too rapidly, the system may not have enough time to explore sufficiently many solutions and may become stuck in a state of locally but not globally minimal energy. In other words, SA only finds a local minimum.

It is possible to combine several of these methods. Indeed, it is often useful to start a search with a genetic algorithm in order to obtain one or more coarse solutions, which are subsequently refined with a nonlinear regression. It is also possible to use SA within a standard genetic algorithm by starting with a relatively high mutation rate, which slowly decreases over time. Similarly, Li et al. [16] combined SA with ACO. The advantage of this combination was again that ACO is a global search method, while SA has features characteristic of probabilistic hill climbing.

5.7 Typical Challenges

No matter which search algorithm is used to determine optimal parameter values for a nonlinear model, some challenges come up time and time again [17]. One ubiquitous challenge in computational systems biology is **noise** in the data. Although data don't usually come in decibels, the expression is used to describe deviations from an expected (usually rather smooth) trend line, which may be due to variability among beakers, Petri dishes, cells, or organisms, and/or to inaccuracies in measurements. It does not require much imagination to infer from plots such as **Figure 5.13** that noise may become a challenge for parameter estimation, especially if it is so large that it obscures or hides the true trend line.

Another challenge, which occurs more often than one might think and which is actually quite hideous, is the fact that the estimation task may have many solutions with the same SSE. These solutions may all be connected in the parameter space or they may be distinctly separated. An intuitive example arises when two parameters always appear in the model in the same formation or combination, such as p_1p_2. Clearly, for every p_1 that some algorithm might determine incorrectly, there is a p_2 that perfectly makes up for the error (with the exception of $p_1 = 0$). It is useful to express the situation in a slightly different fashion. Suppose that, in the true solution, $p_1 = 3$ and $p_1 = 4$, so that $p_1p_2 = 12$. Then, any solution that satisfies $p_2 = 12/p_1$ has exactly the same SSE as the true solution.

In the case of p_1p_2, the culprit is easy to identify, but in other cases it might not be so obvious. For instance, consider the power-law function

$$f = pX^aY^b \tag{5.8}$$

with three parameters. Suppose data are available for X, Y, and f, as shown in **Table 5.4**. For clarity of illustration, these data are assumed to be noise-free. Initiating a

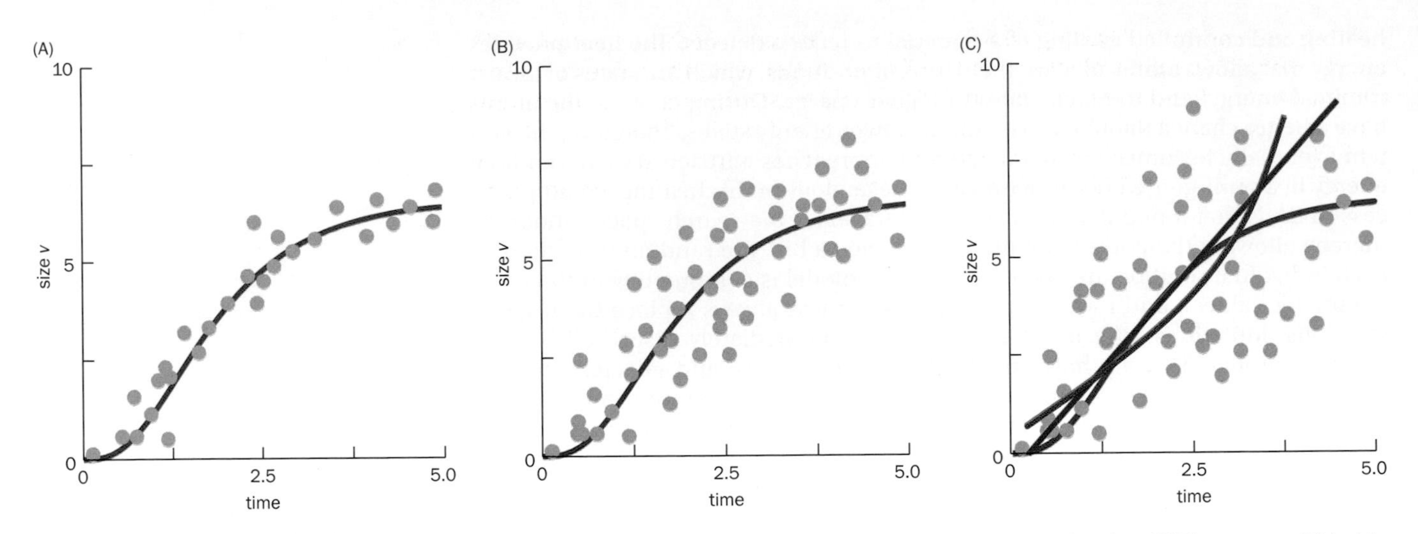

Figure 5.13 Too much noise may hide the true functional relationship and create problems with parameter estimation. (A) The true trend (blue) is clearly perceivable in spite of moderate noise. (B) In the presence of stronger noise, the true trend is more ambiguous; in fact, a straight line could possibly fit the data. (C) Even more noise makes it impossible to identify the true trend, and linear as well as drastically different nonlinear trends could be most appropriate.

search algorithm with different guesstimates for p, a, and b may lead to distinctly different solutions that are all optimal, which here means that the residual error is 0. The differences in solutions are not the consequence of noise, because the data in this constructed example are noise-free. So, what's the problem? In this case, the redundancy is actually due to the fact that the data for X and Y in Table 5.4 happen to be related by a power-law function, that is,

$$Y = \alpha X^{\gamma} \tag{5.9}$$

TABLE 5.4: DATA TO BE MODELED WITH THE POWER-LAW FUNCTION (5.8)

X	Y	f
1	1.75	2.07
2	3.05	4.03
3	4.21	5.95
4	5.31	7.84
5	6.34	9.71
6	7.34	11.57
7	8.30	13.41
8	9.24	15.25
9	10.15	17.07
10	11.04	18.89
11	11.92	20.70
12	12.78	22.50
13	13.62	24.30
14	14.45	26.09
15	15.27	27.88
16	16.08	29.66
17	16.88	31.44
18	17.67	33.21
19	18.45	34.98
20	19.22	36.75

with parameters $\alpha = 1.75$ and $\gamma = 0.8$ (**Figure 5.14**). Because of this relationship, Y^b may be substituted equivalently with $(\alpha X^\gamma)^b$, no matter what the value of b. As a consequence, if the parameters of f in (5.8) are to be estimated and the algorithm finds some incorrect value for b, the parameters p and a are able to compensate perfectly. As a numerical example, suppose the true parameters of f in (5.8) are $p = 2.45$, $a = 1.2$, $b = -0.3$. By applying the transformation $Y = 1.75X^{0.8}$, according to (5.9), we obtain

$$\begin{aligned} f &= 2.45X^{1.2}Y^{-0.3} \\ &= 2.45X^{1.2}YY^{-1.3} \\ &= 2.45X^{1.2}(1.75X^{0.8})Y^{-1.3} \\ &= 4.2875X^{2}Y^{-1.3}. \end{aligned} \tag{5.10}$$

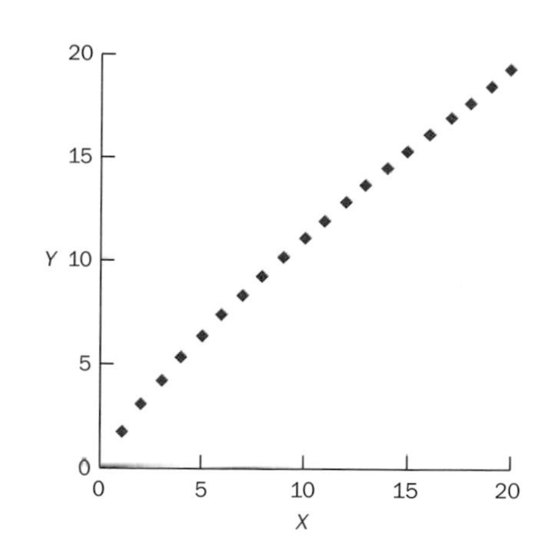

Figure 5.14 Dependences between variables may lead to redundant parameter estimates. The error-free (X, Y) data of Table 5.4 are related to each other by the power-law function (5.9) with $\alpha = 1.75$ and $\gamma = 0.8$.

Thus, we have two distinct solutions to the parameter estimation task, namely the parameter vectors (2.45, 1.2, −0.3) and (4.2875, 2, −1.3), which produce exactly the same values of f. Moreover, these are not the only solutions. Instead of replacing Y with $1.75X^{0.8}$, we could replace any power of Y with the corresponding representation of $1.75X^{0.8}$, and the resulting f would be characterized differently, but generate exactly the same fit to the data.

In the cases discussed so far, all equivalent solutions taken together formed straight or curved lines within the parameter space. A distinctly different situation is possible where two or more solutions are equivalent, and yet the solutions in between are not. The easiest demonstration of this situation is graphical: imagine that the error function R in **Figure 5.15** depends on only one parameter p. Clearly, there are two solutions where the residual error is minimal (with $R = 2$), namely for $p \approx \pm 3.16$, and all other values of the parameter p give solutions with higher errors. It is easy to imagine similar situations for error functions that depend on several parameters.

It is possible that estimating parameters for various subsets of the same data may lead to major differences in the optimized parameter values, an observation called **sloppiness**. Furthermore, and probably surprising, just computing averages of the values from different estimations does not necessarily provide a good overall fit. As an example, consider the task of fitting the function

$$w(t) = (p_1 - p_2e^{-p_3t})^{p_4} \tag{5.11}$$

to experimental data. This function, which has been used to describe the growth of animals [18], has four parameters that describe its sigmoid shape over time. Suppose two datasets were measured, and we fit them separately. The results (**Figure 5.16A, B**) are both good, and the optimized parameter values are given in **Table 5.5**. Next we average the parameter values obtained from the estimations for the two datasets. Surprisingly, the fit is rather bad (Figure 5.16C). Indeed, much better parameter values are obtained with the earlier fits for only one dataset or from simultaneously fitting the two datasets (Figure 5.16D). How is that possible? At a coarse level, the answer is again that the parameters in many models are not independent of each other. Thus, if one parameter value is artificially changed, it is to some degree possible to compensate the resulting error by changing a second parameter value, as we encountered with the extreme case of parameters that always form a product p_1p_2. In more general terms, the acceptable parameter vectors in these cases are actually part of a curved (possibly high-dimensional) cloud whose shape is governed by the dependences among the parameters, and all other points within this cloud are similarly acceptable parameter vectors [19, 20]. Averaging of results does not preserve the relationships within this cloud, and the short take-home message is that it is often not a good idea to average parameter values one by one from two or more fits (**Figure 5.17**).

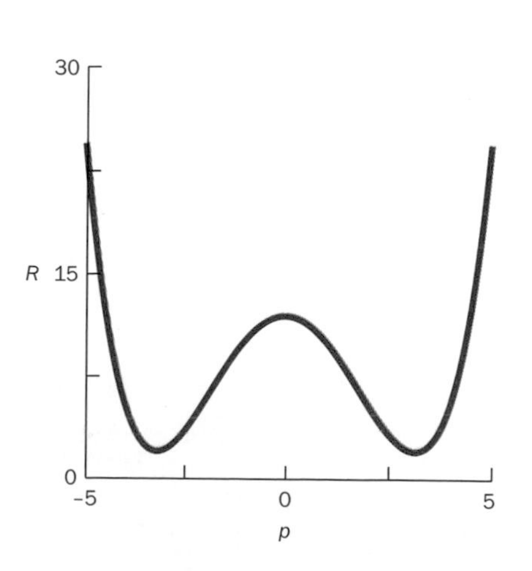

Figure 5.15 Two optimal solutions may be isolated. Here, the error function R assumes the same minimum value of 2 for $p \approx -3.16$ and $p \approx +3.16$. The error is higher for all other values of p.

Fitting two or more datasets separately does have some advantages. For instance, it indicates how stable an optimized parameter vector is, or how much it is affected by natural variability and experimental **uncertainty**. If the parameter values change drastically from one dataset to the next, none of the parameter vectors is likely to be

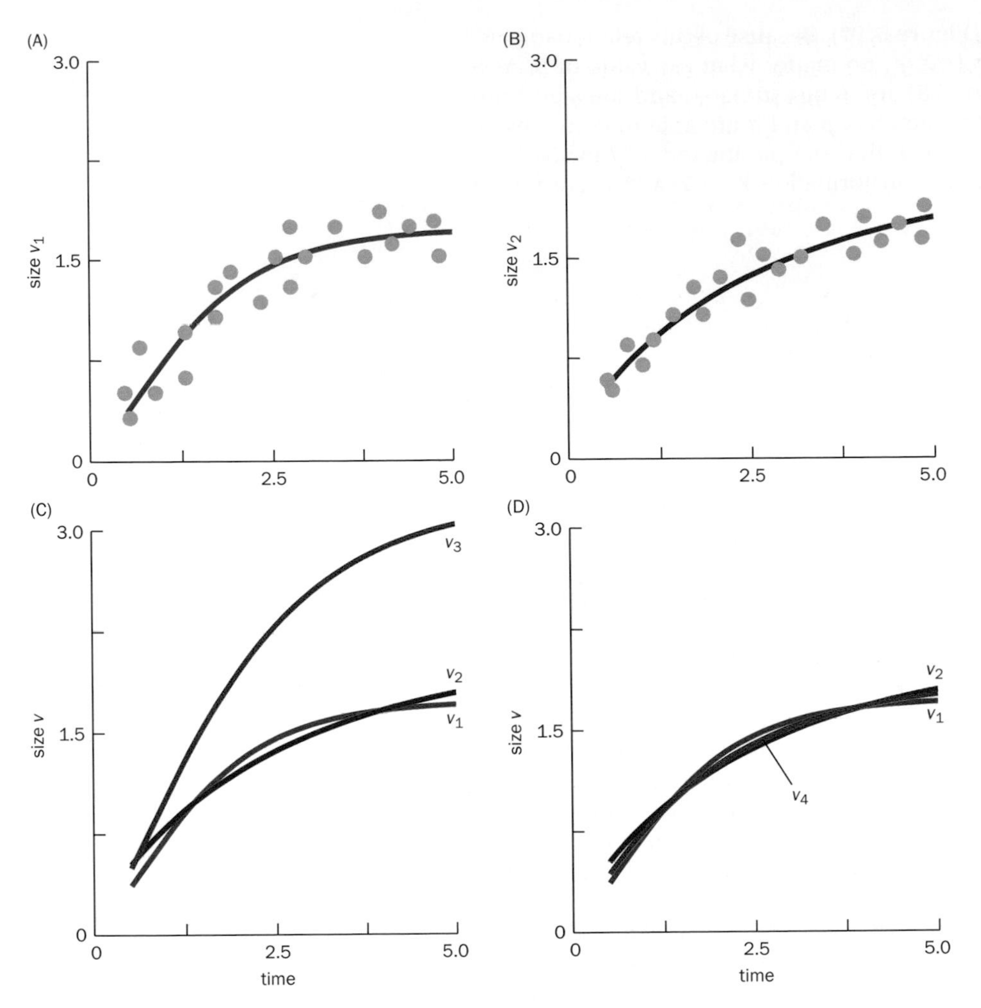

Figure 5.16 Data fits for the model (5.11) with parameter values listed in Table 5.5. (A) Fit v_1 for dataset 1. (B) Fit v_2 for dataset 2. (C) Fit v_3 with averaged parameter values. (D) Simultaneous fit v_4 to datasets 1 and 2.

reliable for predictions of untested scenarios. If many datasets are available, it may be beneficial to use some of the data as a **training set** for estimating parameters, while retaining the remaining, unused data for validation. In other words, one obtains parameters with some data and checks with the validation data how reliable the estimates from the training phase are.

If only one dataset is available, it is still to some degree possible to explore the reliability of the estimates. One option is sensitivity analysis (Chapter 4), while the other is the construction of clouds of candidate solutions. The basic concept is a resampling scheme such as the **bootstrap** or **jackknife** method [21–23]. In bootstrapping, one draws a random sample from the data hundreds of times, estimates the parameters for each sample, and collects all results, which again form clouds of acceptable parameter vectors. In the jackknifing technique, one data point at a time is randomly eliminated and the parameter estimation is executed with the remaining points. Repeating the estimation many times, while leaving out a different point every time, leads to different solutions, which provide an indication of how much

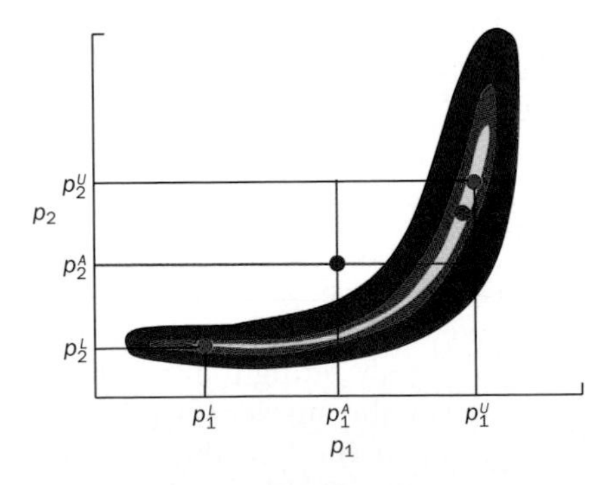

Figure 5.17 Averaging suitable parameter values from different estimations is not necessarily valid. Suppose suitable parameter combinations (p_1, p_2) (with low SSEs) are located within the pale orange region, while the orange and red regions are acceptable and barely satisfactory, respectively. All solutions outside the red region are unacceptable owing to high SSEs. The optimal solution is indicated by the green dot. Even if the solutions at the two blue dots, (p_1^L, p_2^L) and (p_1^U, p_2^U), are both suitable, averaging their parameter values (p_1^A, p_2^A) leads to an unacceptable solution (purple).

TABLE 5.5: PARAMETER VALUES FOR DATA FITS WITH THE MODEL (5.11) IN FIGURE 5.16

Dataset/ parameter set	Figure part	p_1	p_2	p_3	p_4
1	A	1.2	0.8	1	3
2	B	2.5	2.45	0.35	0.8
Average of 1 and 2	C	1.85	1.625	0.675	1.9
Simultaneous fit	D	1.68	1.6	0.6	1.2

the acceptable parameter vectors may vary. The obvious caveat with these techniques is that all information is literally based on a single dataset and therefore does not account for natural variability that comes from different experiments.

PARAMETER ESTIMATION FOR SYSTEMS OF DIFFERENTIAL EQUATIONS

Essentially all principles and techniques of parameter estimation discussed so far are applicable to explicit functions as well as to dynamical systems consisting of sets of differential equations. However, there are two main reasons why the estimation of dynamical systems is much harder. First, systems of differential equations usually contain many more parameters than individual functions, and we have already had a glimpse into the challenges that come with larger numbers of parameters and the potential for a combinatorial explosion of possible sets of parameters. We saw this issue most clearly for grid searches, but also for evolutionary methods like genetic algorithms that attempt to cover an exhaustive representation of the parameter space. The second issue making the estimation of dynamic systems difficult is that the necessary experimental data now consist of sets of data measured at a series of time points and these are to be compared with the solutions of the differential equations, which therefore require numerical integration of the equations during every parameter updating step. Typical data of this type consist of the expression of genes or the concentrations of metabolites, which are measured every so many seconds, minutes, or hours, following some stimulus. A specific example is a population of bacteria that have been starving for some while. At the beginning of the experiment ($t = 0$), a substrate such as glucose is added to the medium, and the measurements consist of concentrations of glycolytic metabolites every 30 seconds [24].

To estimate parameters from time-series data, one starts again with some parameter vector (or a population of vectors), but, before a comparison between model and data is possible, one must solve the system of differential equations. It does not sound like much effort to perform this solution step, but computation-time studies have shown that this step may use more than 95% of the entire time needed to optimize a parameter vector [25]. To make matters worse, unreasonable parameter vectors, upon which a search algorithm may easily stumble once in a while, sometimes cause the numerical integration of differential equations to become so slow that the algorithm eventually runs out of time before any decent parameter vector has been obtained. These computational issues, added to the combinatorial explosion of large numbers of parameters to be estimated and other issues mentioned previously, render the estimation of dynamical systems a very challenging topic that continues to be the object of much attention within the community of computational biologists [26–30]. Since no general method is presently capable of solving all estimation issues related to dynamical systems, several diverse shortcuts and means of simplification and amelioration have been proposed and are discussed in the following.

A most effective strategy to counteract the computational challenges of estimating parameters in dynamical systems is the avoidance of integrating the differential equations. The rationale is the following. The derivative on the left-hand side of a differential equation is really the slope of the time course of the variable at a given point. Thus, if we can estimate this slope, we can circumvent integrating the differential equation. It may be best to demonstrate the approach with a simple example.

Suppose we are studying the growth of a bacterial population, which begins with 2 units (maybe 2 million cells) and grows to 100 units. As a mathematical description, we assume the famous logistic growth function, which was proposed over 150 years ago [31] and has the mathematical formulation

$$\dot{N} = aN - bN^2, \tag{5.12}$$

which contains two parameters, a and b (see also Chapters 4 and 10). For ease of discussion, we pretend to have noise-free data N, which are shown in **Figure 5.18A**. Because the data are so clean, it is easy to estimate, at each time point, the slope S of

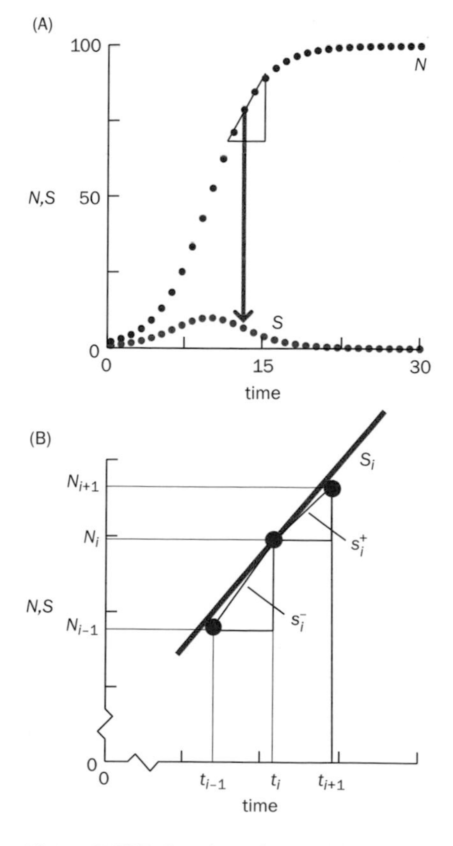

Figure 5.18 Estimation of parameters in (5.12). (A) Growth curve of a bacterial population with size N. At each time point, the slope of the growth curve is estimated and plotted as S. (B) The slope estimation may be performed by averaging the quantities s_i^- and s_i^+, as explained in the text. The result is the estimated slope S_i of the green line at the point (t_i, N_i) of the growth curve.

the growth curve; the slopes are also shown in the figure. The slope estimation in this case may be done in the following fashion, which is sometimes called the three-point method (Figure 5.18B). For the two subsequent data points N_{i-1} and N_i, which are measured at time points t_{i-1} and t_i, compute the difference $N_i - N_{i-1}$ and divide it by $t_i - t_{i-1}$; denote the result by s_i^-. Do the analogous computation for N_i and N_{i+1}, which are measured at t_i and t_{i+1}, and call the result s_i^+. The average of s_i^- and s_i^+ is an estimate of the slope S_i at (t_i, N_i). This method does not allow the computation of slopes at the first and last data points, but one can often do with two fewer points. Much more sophisticated and accurate methods are available for **smoothing** time courses and estimating slopes even in cases where the data are noisy [32, 33].

The slope estimation leads to an augmented dataset, which consists of three variables, namely t_i, N_i, and S_i (**Table 5.6**). Substituting the estimated slopes S_i for $\dot{N}(t_i)$ transforms the estimation task with one differential equation into as many decoupled algebraic equations as there are data points (possibly without the first and last points). The original task is now computationally much easier, because the algorithm no longer needs to solve the differential equation numerically. In the new system, the *i*th algebraic equation looks like

$$S_i = aN_i - bN_i^2 \tag{5.13}$$

in symbolic form. For the parameter estimation, the values for N and S are entered into this set of equations. As an illustration, the first two equations of this set, using S from the three-point method, are

$$\begin{aligned} 1.80 &= a4.34 - b4.34^2, \\ 3.51 &= a9.18 - b9.18^2. \end{aligned} \tag{5.14}$$

Using a gradient method, such as `nlinfit` in MATLAB®, we obtain parameter values very quickly as $a = 0.3899$ and $b = 0.0039$, which are quite close to the true values $a = 0.4$ and $b = 0.004$, which we know from constructing the example.

Importantly, the method of estimating slopes and converting systems of differential equations into algebraic equations also works for systems of more than one

TABLE 5.6: DATASET USED FOR ESTIMATING THE PARAMETERS IN (5.12)*

t	*N*	*S* (true)	*S* (three-point)
0	2	0.78	†
2	4.34	1.66	1.80
4	9.18	3.33	3.51
6	18.36	6.00	6.05
8	33.36	8.89	8.58
10	52.70	9.97	9.47
12	71.26	8.19	7.99
14	84.66	5.19	5.30
16	92.47	2.78	2.95
18	96.47	1.36	1.48
20	98.38	0.64	0.70
22	99.27	0.29	0.32
24	99.67	0.13	0.15
26	99.85	0.06	0.07
28	99.93	0.03	0.03
30	99.97	0.01	†

* The data consist of the numbers of bacteria (in units of millions), the true slopes, which are usually not known, and the slopes estimated with the three-point method.

† Slopes at times $t = 0$ and $t = 30$ cannot be obtained with this method.

variable. Let us illustrate the method for a dynamic system with two variables. For a change, this time the example comes from ecology. Suppose two species N_1 and N_2 are living in the same habitat. N_1 is an herbivore that once in a while is eaten by a predatory carnivore from population N_2. One standard description is a Lotka–Volterra model (see Chapters 4 and 10) of the form

$$\begin{aligned} \dot{N}_1 &= N_1(a_1 - b_1 N_1 - b_2 N_2), \\ \dot{N}_2 &= N_2(a_2 N_1 + a_3 - b_3 N_2). \end{aligned} \tag{5.15}$$

Following standard techniques for setting up the model, N_1 grows exponentially with rate a_1 and has a natural death rate of b_1. It is typical to represent the death term by the square of N_1, which is sometimes interpreted as crowding within the population. The population is also subject to decimation by population N_2 with rate b_2. This process depends on the number of encounters between the two species and is therefore represented by the product of N_1 and N_2. N_2 feeds on N_1 with a rate a_2, and this process augments its growth from other food sources, which is assumed to occur with rate a_3. The predators die with the rate b_3. Note that a_2 and b_2 typically have different values, because the growth benefit to the predator is seldom exactly equivalent to the effect on the prey.

Suppose we have reason to believe that the model formulation in (5.15) is adequate, but that we do not know appropriate parameter values. We do have data, which stem from a series of measurements around some devastating event for the prey. Specifically, the populations are both at steady state, with about four times as many predators as prey. At time $t = 1$, about 80% of the prey population succumb to some disease. Subsequent measurements of N_1 and N_2 show how both populations respond to the catastrophic event: initially, N_2 lacks food and decreases a little bit, while N_1 begins to recover. Subsequently, both N_1 and N_2 increase and decrease with a few damped oscillations and finally return to their original steady-state values. Since this is just a demonstration, let us begin by assuming that we have comprehensive, error-free measurements, which are given in **Table 5.7**. In addition to the population sizes, the table also lists the slopes of N_1 and N_2, which are denoted by S_1 and S_2, respectively. At time $t = 1$, the abrupt disease event happens, and the slopes at this point are not defined.

The estimation now proceeds by setting up two sets of algebraic equations: one for N_1 and one for N_2. The first two equations of the first set are

$$\begin{aligned} S_1(1.1) &= N_1(1.1)\left[a_1 - b_1 N_1(1.1) - b_2 N_2(1.1)\right], \\ S_1(1.2) &= N_1(1.2)\left[a_1 - b_1 N_1(1.2) - b_2 N_2(1.2)\right]. \end{aligned} \tag{5.16}$$

Substituting numerical values for S_1, N_1, and N_2 in the first equation yields

$$798.41 = 63.35(a_1 - b_1 63.35 - b_2 26.89). \tag{5.17}$$

Similar equations with numerical values listed in Table 5.7 are composed for time points 1.2, ..., 2.9. Again using `nlinfit` in MATLAB®, we obtain the exact parameters for the first equation in less than a second. Indeed, although the residual error SSE for the initial guesstimate (2, 2, 2) is huge at about 5×10^{10}, the algorithm converges to the correct solution ($a_1 = 40$, $b_1 = 0.008$, $b_2 = 1.0$) within only seven iterations. In exactly the same fashion, the parameter values for the equation for N_2 are computed very quickly as ($a_2 = 0.05$, $a_3 = 0.01$, $b_3 = 0.18$). The optimized solution perfectly matches the noise-free data (**Figure 5.19**).

It should be mentioned that this specific example would have allowed a simpler method of estimation. Namely, we could have divided all equations in (5.16) by N_1, which would have made the estimation linear [34]. This strategy is left as Exercise 5.17. It is very powerful, because linear regression provides an analytical solution and therefore does not require search algorithms, as we discussed before. In fact, this strategy can be used even for very large systems, such as bacterial metapopulations [35] (see Chapter 15). However, the Lotka–Volterra system is a special case, and the slope estimation method usually does not convert nonlinear into linear models.

TABLE 5.7: DATA FOR THE ESTIMATION OF PARAMETERS IN (5.15)*

t	N_1	N_2	S_1	S_2
0	139.77	38.88	0	0
0.25	139.77	38.88	0	0
0.5	139.77	38.88	0	0
0.75	139.77	38.88	0	0
1	139.77	38.88	—	—
†				
1	27.95	38.88	—	—
1.1	63.35	26.89	798.41	−44.71
1.2	204.04	30.86	1531.64	143.72
1.3	182.09	45.96	−1350.61	38.67
1.4	107.57	40.61	−158.29	−78.03
1.5	123.83	35.97	376.76	−9.81
1.6	156.03	37.95	125.05	37.21
1.7	145.80	40.31	−214.54	1.83
1.8	132.38	39.13	−25.07	−16.21
1.9	137.55	38.25	89.86	0.12
2	143.28	38.80	7.54	7.37
2.1	140.51	39.17	−41.59	−0.61
2.2	138.19	38.90	−0.11	−3.18
2.3	139.55	38.75	18.30	0.47
2.4	140.50	38.88	−1.14	1.39
2.5	139.82	38.94	−8.07	−0.31
2.6	139.45	38.88	1.06	−0.59
2.7	139.77	38.86	3.51	0.18
2.8	139.92	38.89	−0.71	0.25
2.9	139.76	38.89	−1.51	−0.10

* The data consist of error-free measurements of the sizes of the two populations N_1 and N_2, as well as slopes at each time point.

† At time $t = 1$, population N_1 abruptly falls to about 20% of its steady-state value, for reasons not represented in the model. At this point, slopes are not defined.

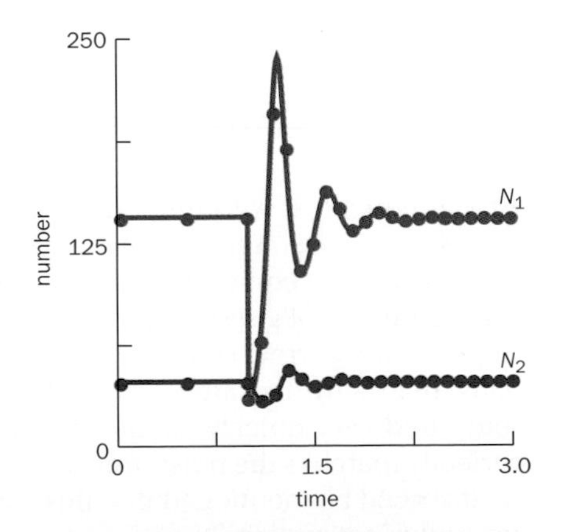

Figure 5.19 Error-free data (symbols) corresponding to the model (5.15) and solution with optimized parameters (lines). At time $t = 1$, 80% of population N_1 succumb to an unexplained disease. By time $t = 3$, both populations have returned to their steady-state values.

Of course, actual biological data contain noise, and this noise is usually amplified when slopes are computed. Nonetheless, the slope method is often a good start, especially if the noise is moderate. As an example, suppose noisy data are given, with values starting after the crash of population N_1 (**Table 5.8**). Using the same techniques of replacing the derivatives with slopes (which are also noisy now), we use `nlinfit` to compute parameters. Again, it does not take long to estimate the parameters, and the results are ($a_1 = 24.0126$, $b_1 = 0.0031$, $b_2 = 0.6159$) and ($a_2 = 0.0376$, $a_3 = -1.6032$, $b_3 = 0.0933$). The first observation is that these numbers are different from the true values, which is to be expected because of the noise in the data and the slopes. Some values are off by quite a bit, and it can be seen in particular that a_3 is actually negative rather than positive, which would change its interpretation from a birth to a death process. However, it is difficult to judge how good these parameter values are just by looking at the numbers. The first test is therefore a plot of the slopes as functions for N_1 and N_2. An example is shown in **Figure 5.20**, where S_1 is plotted against N_1. At first, the spiral in this plot may be confusing. It is due to the fact that N_1 increases and decreases, thereby moving back and forth along the horizontal axis. At the same time, S_1 moves up and down, thereby yielding the spiral pattern. The red spiral shows the true dynamics (with correct parameter values), the blue line with symbols represents the noisy data in Table 5.8, and the green line shows the

TABLE 5.8: DATA FOR ASSESSING THE ROLE OF NOISE IN THE ESTIMATION OF PARAMETERS IN (5.15)*

t	N_1 with noise	N_2 with noise	S_1 with noise	S_2 with noise
1	20	44	—	—
1.1	50	25	690	−60
1.2	210	30	1310	132
1.3	195	48	−1520	30
1.4	97	37	−120	−52
1.5	120	33	510	−18
1.6	144	41	112	50
1.7	148	41	−255	0
1.8	130	44	−10	−22
1.9	136	37	70	0
2	148	35	10	10
2.1	130	36	−50	0
2.2	142	40	0	−2
2.3	140	41	27	4
2.4	136	34	0	2
2.5	137	36	−12	−1
2.6	135	38	4	0
2.7	144	35	4	0
2.8	142	41	0	0
2.9	138	38	0	0

* Noisy measurements of the sizes of the two populations N_1 and N_2, as well as the slopes measured at each time point, starting at time $t = 1$, after population N_1 had abruptly fallen to about 20% of its steady-state value.

dynamics of the model with optimized parameter values. Visual inspection suggests that the fit is not so bad. The real test is the use of the optimized parameter values in the model equations (5.15). Three results are shown in **Figure 5.21**: in (A), noisy data and slopes are used only for N_1, while the parameter values for N_2 are taken as the true values; (B) shows the analogous situation for noisy N_2 and correct N_1 data; and (C) shows the results of adding noise to both N_1 and N_2. The main observations are that the overall dynamic patterns are retained, but that the timing of the curves is not quite right. This deviation in time is a typical outcome of using the slope method. The reason is that when derivatives are substituted with slopes, the regression does not explicitly use time. If there is no noise, the fit is usually very good. However, the consequences of noise often appear in the form of some time warp. This is the cost of the ease and speed of the slope substitution method.

If biological considerations mandate that a parameter value should be positive, as we would suspect in the case of a_3, the parameter estimation procedure can be set up to enforce solutions to stay within admissible ranges.

As a control, we estimate parameter values from the noise-free and noisy data, using the differential equations (5.15) directly. Again, the result is typical. Assuming the algorithm converges to the correct solution, this solution is usually better than the solution obtained with the slope substitution method. However, it often takes much, much longer to obtain this solution, especially for larger systems, and sometimes the algorithms do not converge at all to an acceptable solution.

The website supporting the book (http://www.garlandscience.com/product/isbn/9780815345688) contains MATLAB® code for a parameter estimation program that first uses a genetic algorithm to obtain a coarse global solution and then hands the results over to a gradient method to refine this solution. The important settings for the genetic algorithm are the population size (PopSize), the number of generations (GenSize), the number of best genes retained for the next generation

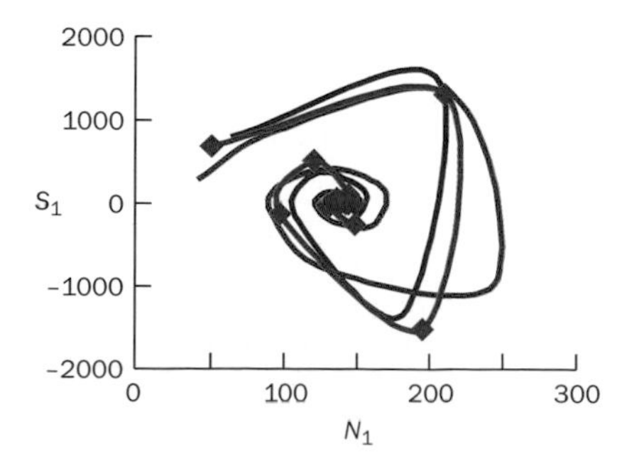

Figure 5.20 Assessing the oscillatory behavior of the estimated system (5.15). Shown here is the plot of the slope S_1 versus the size of population N_1. The red curve shows the true, noise-free relationship, the blue-connected symbols represent noisy data, and the green curve shows the relationship optimized for noisy data.

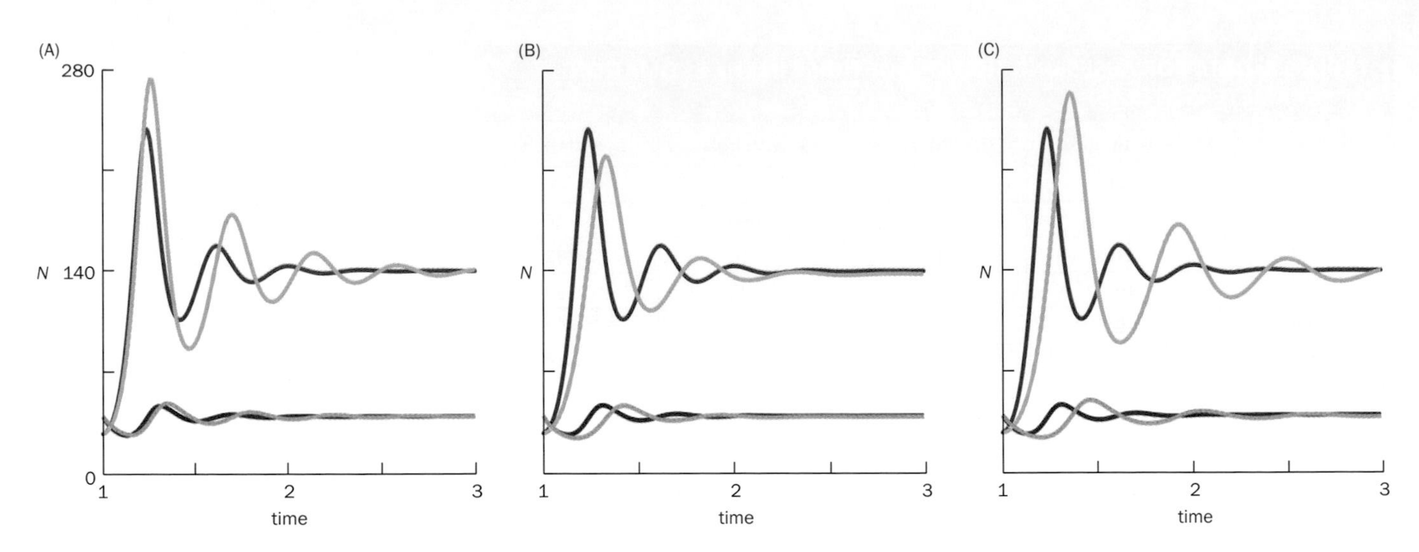

Figure 5.21 Simulations of the population model with parameter values estimated from noisy data. (A) Only N_1 data are noisy. (B) Only N_2 data are noisy. (C) Both N_1 and N_2 data are noisy. The plots start at the same perturbed value at time 1 as in Figure 5.19.

(EliteCount), a limit to the number of consecutive generations if there is no significant improvement in fit (GenLimit), and sets of admissible lower and upper bounds for the parameters. These settings allow a lot of flexibility, and, even in our case, they can make quite a bit of difference. With the settings on the website, the algorithm often actually converges to a good solution. However, because genetic algorithms are stochastic, running the same program with the same settings and data several times typically leads to different solutions. Five examples are given in **Table 5.9**.

The initial SSE of the best solution automatically created by the GA is of the order of 10^5. It decreases to somewhere between 10^3 and 10^4 at completion of the GA, depending on the random numbers the algorithm uses during the stochastic evolution. Under opportune conditions, the subsequent gradient method reduces this error to between 0 and 10. Interestingly, even if the residual error at the end of the GA phase is very similar in two runs, the gradient method sometimes converges to the correct solution and sometimes to an unsatisfactory local minimum, which corresponds to the simple shoulder curve for N_1 in **Figure 5.22**. The combined process may take several minutes to complete, which is much longer than the time needed for the decoupled system with estimated slopes. The residual error is almost never 0, even in cases of convergence, but a solution with SSE ≈ 7 is essentially perfect in this case (see Figure 5.22).

Depending on the upper bounds for the unknown parameters, randomly chosen start values may be so far off that the integration of the differential equations stops immediately with an error message. If the GA does get started, it often terminates with a higher SSE than in the cases before. In many cases, the gradient method terminates with a bad solution, but there are also cases where the iterative search is

TABLE 5.9: TRUE AND OPTIMIZED SOLUTIONS FOR NOISE-FREE POPULATION DATA (SEE (5.15))

Run	a_1	b_1	b_2	a_2	a_3	b_3	SSE
1	89.6965	0.6422	0.0000	0.4296	2.4108	1.6351	13,664.2
2	99.8746	0.0000	2.5988	0.4181	0.0000	1.5318	11,939.6
3	38.4554	0.0064	0.9601	0.0525	0.0048	0.1879	7.44405
4	38.5346	0.0065	0.9623	0.0524	0.0012	0.1874	6.99142
5	20.7977	0.0000	0.5359	1.2265	4.8767	4.6515	9162.04
True	40	0.008	1	0.05	0.01	0.18	0

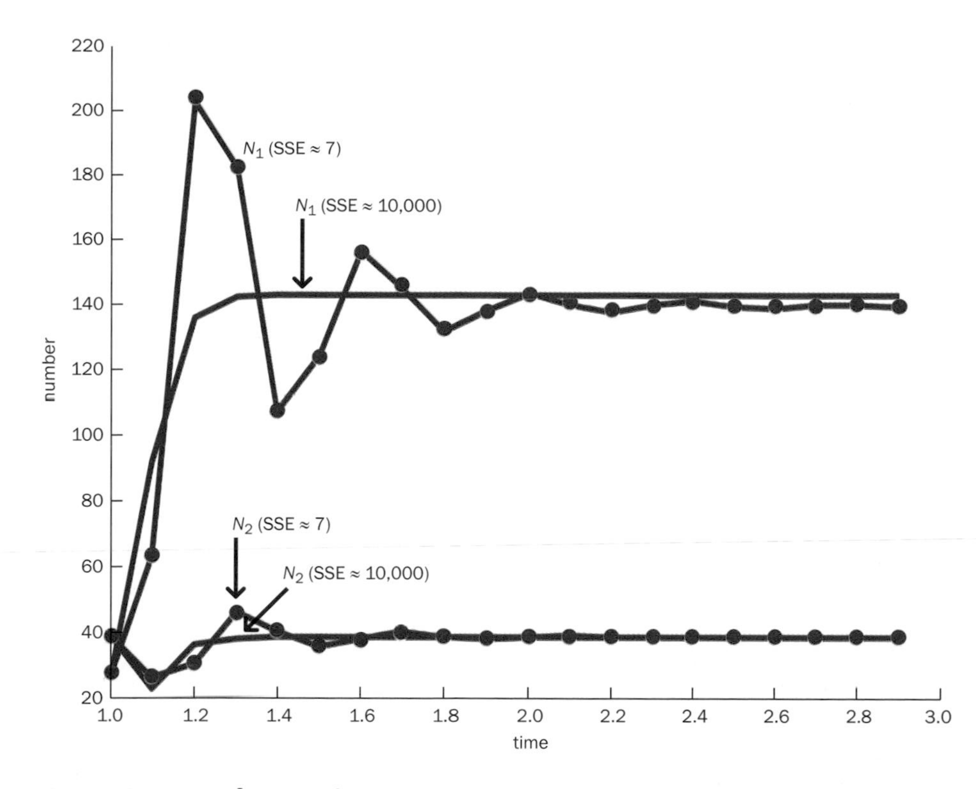

Figure 5.22 MATLAB® output of two optimization runs. Noise-free data are represented as symbols. A "correct" solution with SSE ≈ 7, connected with straight-line segments, essentially hits all points perfectly. However, the same optimization settings may run into a local minimum with SSE ≈ 10,000, which corresponds to an unsatisfactory solution, at least for N_1.

successful. Finally, if the data are noisy, we can of course not expect the SSE to decrease to single digits. If the algorithm converges to the right neighborhood, the SSE in the example is between 900 and 1000, which looks like a big number but in fact is not a bad solution (**Figure 5.23**).

In the previous examples, we have implicitly assumed that we know the true initial values for integrating the differential equations. This is a common assumption, which may or may not be acceptable. Instead of making the assumption, the initial values may be considered as parameters that are to be optimized along with the parameters we discussed above.

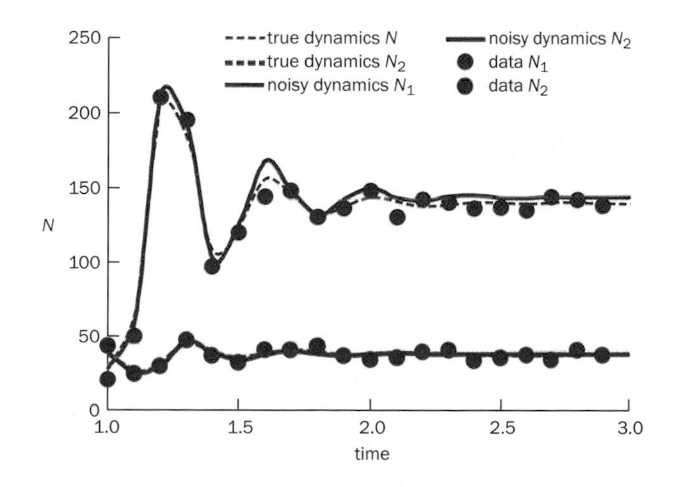

Figure 5.23 Effect of noise in the data on parameter estimates. The figure shows the true dynamics of the population system (dashed lines), noisy data (dots), and the dynamics estimated from these noisy data (solid lines). Even with an SSE of about 1000, the optimized solutions are not bad. In fact, the fit for N_2 (red) is essentially indistinguishable from the true solution.

STRUCTURE IDENTIFICATION

Related to the task of estimating parameter values is that of structure identification. In this case, it is not even known what the structure of the model looks like. For instance, one could have the same data as above, but the system would be represented as

$$\begin{aligned} \dot{N}_1 &= F_1(N_1, N_2), \\ \dot{N}_2 &= F_2(N_1, N_2), \end{aligned} \tag{5.18}$$

with unknown functions F_1 and F_2. Two approaches may be pursued to attack this complicated problem. In the most obvious, although cumbersome, approach, one could explore a number of candidate models that appear to have a chance of succeeding, based on experience or some biological rationale. For instance, in the context of enzyme kinetics, one could use Michaelis–Menten rate laws, sigmoidal Hill functions, or more general constructions [36]. It does not take much imagination to see that the numbers of possible candidates and their combinations are limitless. In general, the true structure may never be known in detail.

An alternative is the use of canonical models, as discussed in Chapter 4. In a population setting, one might begin with Lotka–Volterra models. For instance, in the case of (5.18), one could try composing different sums and differences of terms like c_1N_1, c_2N_2, $c_3N_1^2$, $c_4N_2^2$, and $c_5N_1N_2$, which in our situation would actually include the structure that we assumed for the example. For enzyme kinetic models, one would prefer **generalized mass action (GMA)** or **S-systems**. These canonical models are relatively general and simple, and they are often great default models for starting an analysis. However, they are local approximations, which may or may not be able to capture the full dynamics of the observed data.

There are no easy solutions to the structure identification problem, and much more research will be required to develop promising solution strategies. However, it is likely that dissecting the task into three steps might be beneficial. In the first, one would merely try to establish which variables are important. This **feature selection** step ideally reduces the number of variables to be included in the model. Second, one tries to characterize which variable affects which other variables, and, possibly, whether the effects are likely be augmenting or diminishing. Once the likely topology of the system has been established, at least in a coarse manner, one might select mechanistic or canonical models for developing a specific symbolic and numeric representation [37].

Other very difficult issues arise in structure identification tasks. First, it may not even be known whether the right number of variables is included in the model, which can obviously lead to problems. Second, it might happen that the data are simply insufficient in quantity and/or quality for identifying the true model structure, and many groups have started analyzing this topic of **identifiability** with different approaches (for example, [19, 38–41]). An often effective approach toward this issue is the computation of the Fisher information matrix [42]. Finally, it is possible that the estimated model contains more parameters than can be validly estimated from the data, so that predictions toward new data become very unreliable. Methods of cross-validation can help to some degree with this **overfitting** issue [43].

As we saw in one of the examples, a search algorithm may suggest a negative parameter value where we would have expected a positive value. More generically, a single solution computed with a search algorithm sometimes turns out to be unreliable. To address this issue, it has become popular to search not for the uniquely optimal solution but rather for an entire ensemble of parameter settings that all, more or less, have the same SSE [39, 40, 44]. Expressed differently, the argument of unidentifiability and sloppiness is being turned around, and the optimal solution now consists of an entire set of acceptable parameter settings.

Finally, if several comprehensive and representative time series datasets are available, it might even be possible to establish a model in a nonparametric manner, that is, without specification of functions or parameter values. In this case, the data are computationally converted into a library that contains information about how each process in the study is affected by the various state variables [45]. This new method has not been tested enough to allow any final judgment to be made, but it does seem to have the potential to serve as a rather unbiased alternative to traditional modeling.

EXERCISES

Naturally, many of the exercises in this chapter require computational efforts and access to software, which is available, for instance, in Microsoft Excel®, MATLAB® and Mathematica®. Specific techniques needed in some of the exercises include solutions to ODEs, linear and nonlinear regression, genetic algorithms, and splines. The data tables are available on the website supporting the book (http://www.garlandscience.com/product/isbn/9780815345688).

5.1. Perform linear regression with the bacterial growth data in **Table 5.10**. Plot the results. Repeat the analysis after a logarithmic transformation of the data. Compare and interpret the results.

TABLE 5.10

t	Size	t	Size
0	2.2	1.1	22
0.1	2.9	1.2	21
0.2	3.2	1.3	30
0.3	6	1.4	35
0.4	6	1.5	38
0.5	6.2	1.6	47
0.6	8	1.7	58
0.7	11	1.8	70
0.8	14	1.9	81
0.9	13	2	90
1	18		

5.2. Perform linear regression with the bacterial growth data in **Table 5.11**. Plot the results. Repeat the analysis after a logarithmic transformation of the data. Compare and interpret the results.

TABLE 5.11

t	Size	t	Size
0	3	1.1	18
0.1	3	1.2	26
0.2	2.8	1.3	31
0.3	3.7	1.4	36
0.4	3.9	1.5	44
0.5	4.4	1.6	50
0.6	5.6	1.7	64
0.7	7.4	1.8	72
0.8	11	1.9	88
0.9	14	2	96
1	16		

5.3. Confirm that the Michaelis–Menten function becomes linear if one plots $1/v$ against $1/S$. Is the same true for a Michaelis–Menten process with competitive inhibition, which is given by the function

$$v_I = \frac{V_{\max} S}{K_M\left(1+\frac{I}{K_I}\right)+S},$$

where I is the inhibitor concentration and K_I is the inhibition constant? Support your answer with mathematical arguments.

5.4. Use linear regression to confirm the results ($V_{\max} = 2.2$, $K_M = 6.4$) in the text (see (5.4) and Table 5.2).

5.5. Apply a nonlinear regression algorithm directly to the data in Tables 5.2 and 5.3. Compare the results with the linear estimation upon transformation.

5.6. Apply a nonlinear regression algorithm to the differential equation (5.4), using the data given in **Table 5.12** and plotted in Figure 5.6. Furthermore, assume $S = 1.2$. Look at the code on the book's website for inspiration.

TABLE 5.12

t	M
0	2.000
1	1.061
2	0.671
3	0.509
4	0.442
5	0.414
6	0.403
7	0.398
8	0.396
9	0.395

5.7. Lineweaver and Burk [46] suggested plotting $1/v$ against $1/S$, thereby reducing the nonlinear estimation problem for Michaelis–Menten rate laws to one of simple linear regressions. Woolf [47] had another idea. If one plots v/S against v, the Michaelis–Menten function also becomes linear (Confirm this!). Even though Woolf came up with this trick, the resulting plot is usually called a Scatchard plot [48]. While the nonlinear method, the Lineweaver–Burk method, and the Woolf method, *should* all give the same results, they do so only if the data are error-free. As soon as there is noise in the data, equivalence is not guaranteed; in fact, the results can be quite different. To analyze this situation, use the data in **Table 5.13**. For all four cases, perform nonlinear regression and the two types of linear regression mentioned above.

Summarize your findings, along with graphs and regression results, in a brief report.

TABLE 5.13

S	v_0 (no error)	v_1 (errors distributed)	v_2 (errors at low end)	v_3 (errors at high end)
0.2	0.5454552	0.5	1	0.5
0.4	1	1.2	1.2	1.2
0.6	1.384616	1.25	1.5	1.25
0.8	1.714287	1.8	1.35	1.8
1	2	2.2	2.2	2.2
2	3	2.75	2.75	2.75
3	3.6	3.55	3.55	3
4	4	4.2	4.2	3.8
5	4.285715	4.2	4.2	5
10	5	4.85	4.85	5.6
20	5.454546	5.6	5.6	4.9

5.8. Imagine an experiment where a process V is governed by a substrate S and an inhibitor I. In the system, S decreases and I increases over time. Use the data in **Table 5.14** and multiple linear regression to compute the parameters for process V, which has the form $V = \gamma S^a I^b$. Plot the results in a pseudo-three-dimensional plot. Explain why it seems that the results lie on a slightly nonlinear curve rather than forming a (linear) regression plane.

TABLE 5.14

t	S	I	V
0	10	1.2	6.3
1	7.5	1.7	4.5
2	5.5	2.3	3.2
3	4	2.9	2.4
4	2.5	3.5	1.8
5	1.5	4.0	1.4
6	1.0	4.5	1.1
7	0.7	4.5	0.8
8	0.4	4.8	0.6
9	0.2	4.8	0.5
10	0.1	4.9	0.3

5.9. (a) The Michaelis–Menten rate law (MMRL) with competitive inhibition has the formula shown in Exercise 5.3. In addition to the parameters of the MMRL, it contains the inhibition parameter K_I and the concentration of inhibitor, which we assume can be controlled in the experiment. Suppose two datasets (**Table 5.15**) were measured with different inhibitor concentrations I. Using grid searches, determine the parameters from the two datasets separately and from both together. Discuss your findings.

TABLE 5.15

S	v (I = 10)	v (I = 20)	S	v (I = 10)	v (I = 20)
1	1.8	2	11	11.7	10.7
2	2.2	2.2	12	11.5	10.1
3	6.1	3.2	13	11.2	10.2
4	5.5	4.8	14	13	10.2
5	7.1	6.6	15	12.2	12.3
6	9.3	7.9	16	12.4	12.7
7	9	7.5	17	12.6	11.4
8	9.4	8	18	14.8	12.6
9	11.2	8.2	19	14.2	12.3
10	10.6	9.4	20	16.5	15.5

(b) Use a gradient method to perform the parameter estimation. Start with different guesstimates. Run the estimation for different numbers of iterations.

5.10. Visualize the error landscape for the nonlinear regression with data v_1 in Exercise 5.7.

5.11. Estimate the 10 parameters of the function

$$F(t) = \sin(at + b)[ct + d(t - e)^2 + f(t - g)^3]e^{h + i(t - j)},$$

which is plotted in **Figure 5.24** from the noise-free data in **Table 5.16**. Compare the requirements, procedures and results for different methods.

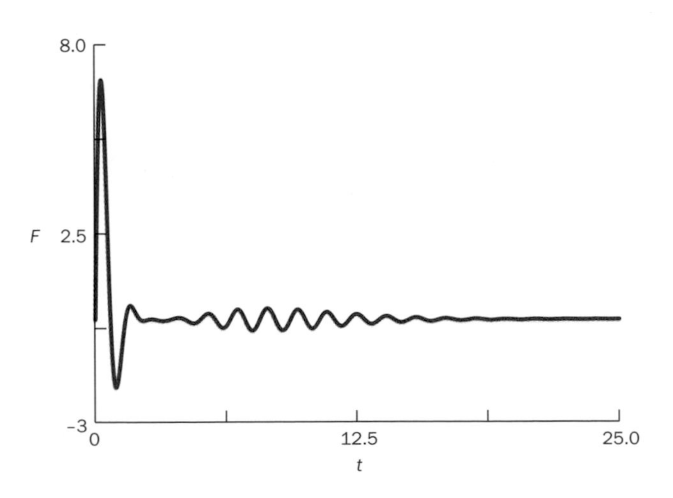

Figure 5.24 The function $F(t)$ with 10 parameters. This strange function is used for estimation purposes.

5.12. Use a genetic algorithm to estimate parameters for the MMRL with inhibition from the data in Table 5.15 (see Exercise 5.3). Look at the website code for inspiration on how to set up the genetic algorithm. Run the algorithm with different bounds on parameter values.

5.13. Estimate the parameters of an MMRL with inhibition from the data in **Table 5.17**, assuming that I = 20. Redo the analysis with the

TABLE 5.16

t	F	t	F	t	F	t	F	t	F
0.0	0.000	5.000	0.001	10.000	−0.005	15.000	0.002	20.000	0.000
0.1	4.184	5.100	0.064	10.100	−0.119	15.100	0.031	20.100	−0.004
0.2	6.471	5.200	0.122	10.200	−0.207	15.200	0.052	20.200	−0.007
0.3	6.929	5.300	0.162	10.300	−0.253	15.300	0.062	20.300	−0.009
0.4	5.964	5.400	0.174	10.400	−0.250	15.400	0.060	20.400	−0.008
0.5	4.152	5.500	0.150	10.500	−0.200	15.500	0.047	20.500	−0.006
0.6	2.080	5.600	0.093	10.600	−0.114	15.600	0.026	20.600	−0.004
0.7	0.223	5.700	0.011	10.700	−0.011	15.700	0.002	20.700	0.000
0.8	−1.119	5.800	−0.082	10.800	0.091	15.800	−0.021	20.800	0.003
0.9	−1.836	5.900	−0.166	10.900	0.171	15.900	−0.038	20.900	0.005
1.0	−1.976	6.000	−0.224	11.000	0.216	16.000	−0.047	21.000	0.006
1.1	−1.690	6.100	−0.242	11.100	0.219	16.100	−0.047	21.100	0.006
1.2	−1.170	6.200	−0.211	11.200	0.180	16.200	−0.038	21.200	0.005
1.3	−0.599	6.300	−0.137	11.300	0.110	16.300	−0.022	21.300	0.003
1.4	−0.110	6.400	−0.031	11.400	0.022	16.400	−0.004	21.400	0.000
1.5	0.222	6.500	0.088	11.500	−0.066	16.500	0.014	21.500	−0.002
1.6	0.383	6.600	0.195	11.600	−0.137	16.600	0.028	21.600	−0.004
1.7	0.401	6.700	0.269	11.700	−0.180	16.700	0.036	21.700	−0.005
1.8	0.326	6.800	0.293	11.800	−0.186	16.800	0.036	21.800	−0.005
1.9	0.213	6.900	0.260	11.900	−0.158	16.900	0.030	21.900	−0.004
2.0	0.103	7.000	0.176	12.000	−0.101	17.000	0.019	22.000	−0.002
2.1	0.022	7.100	0.054	12.100	−0.029	17.100	0.005	22.100	−0.001
2.2	−0.020	7.200	−0.082	12.200	0.045	17.200	−0.009	22.200	0.001
2.3	−0.030	7.300	−0.204	12.300	0.107	17.300	−0.020	22.300	0.002
2.4	−0.020	7.400	−0.290	12.400	0.146	17.400	−0.027	22.400	0.003
2.5	−0.004	7.500	−0.321	12.500	0.155	17.500	−0.028	22.500	0.003
2.6	0.009	7.600	−0.291	12.600	0.135	17.600	−0.024	22.600	0.003
2.7	0.012	7.700	−0.204	12.700	0.091	17.700	−0.016	22.700	0.002
2.8	0.006	7.800	−0.077	12.800	0.032	17.800	−0.005	22.800	0.001
2.9	−0.006	7.900	0.066	12.900	−0.029	17.900	0.005	22.900	−0.001
3.0	−0.020	8.000	0.196	13.000	−0.081	18.000	0.014	23.000	−0.002
3.1	−0.031	8.100	0.289	13.100	−0.115	18.100	0.020	23.100	−0.002
3.2	−0.036	8.200	0.327	13.200	−0.126	18.200	0.021	23.200	−0.002
3.3	−0.035	8.300	0.302	13.300	−0.113	18.300	0.019	23.300	−0.002
3.4	−0.026	8.400	0.219	13.400	−0.079	18.400	0.013	23.400	−0.001
3.5	−0.012	8.500	0.095	13.500	−0.033	18.500	0.005	23.500	−0.001
3.6	0.006	8.600	−0.045	13.600	0.017	18.600	−0.003	23.600	0.000
3.7	0.024	8.700	−0.176	13.700	0.060	18.700	−0.010	23.700	0.001
3.8	0.041	8.800	−0.271	13.800	0.090	18.800	−0.014	23.800	0.002
3.9	0.052	8.900	−0.314	13.900	0.101	18.900	−0.016	23.900	0.002
4.0	0.054	9.000	−0.296	14.000	0.093	19.000	−0.014	24.000	0.002
4.1	0.045	9.100	−0.222	14.100	0.068	19.100	−0.010	24.100	0.001
4.2	0.025	9.200	−0.108	14.200	0.032	19.200	−0.005	24.200	0.001
4.3	−0.005	9.300	0.024	14.300	−0.008	19.300	0.001	24.300	0.000
4.4	−0.040	9.400	0.149	14.400	−0.043	19.400	0.007	24.400	−0.001
4.5	−0.073	9.500	0.242	14.500	−0.069	19.500	0.010	24.500	−0.001
4.6	−0.097	9.600	0.287	14.600	−0.080	19.600	0.012	24.600	−0.001
4.7	−0.104	9.700	0.277	14.700	−0.075	19.700	0.011	24.700	−0.001
4.8	−0.089	9.800	0.215	14.800	−0.057	19.800	0.008	24.800	−0.001
4.9	−0.053	9.900	0.114	14.900	−0.029	19.900	0.004	24.900	0.000
								25.000	0.000

bootstrapping and jackknifing methods. Compare and interpret results. Also, average the parameter values obtained from (A) different bootstrapping runs and (B) different jackknifing runs and assess the fits.

TABLE 5.17

S	$v(I = 20)$
1	2
4	4.8
8	6
10	9.4
12	12
14	10.2
16	13.3
18	12.6
20	12.2

5.14. (a) Use a nonlinear regression algorithm to estimate the parameters of the function f in (5.9) from the data in Table 5.4. Redo the estimation several times with different start guesses. Study the quality of solutions with averaged parameter values.

(b) Transform the function f in (5.9) so that it becomes linear. Try to use linear regression to estimate the parameter values. Report your findings.

(c) Generate more equivalent representations of the function f in (5.9), similar to that in (5.10), using the relationship $Y = 1.75X^{0.8}$. Formulate a functional description that encompasses all possible representations of this type.

(d) Create a different situation of an estimation problem with multiple solutions.

5.15. Construct a function that has at least two minima with the same value (recall the situation in Figure 5.15). Is it possible to have infinitely many, unconnected minima with the same error? If you think yes, provide an example. If not, provide a proof, counterexample, or at least compelling arguments.

5.16. Use a spline function in MATLAB® to connect the data on population sizes N_i in Table 5.7, use this function to compute slopes at all time points, and estimate parameter values.

5.17. Estimate the parameter values in (5.16) upon dividing by N_1, which makes the problem linear. Compare the results with those in the text. Discuss the advantages and potential problems of each method.

5.18. (a) Use the data with and without noise (see Tables 5.7 and 5.8) and estimate parameter values with and without slope substitution, starting with different sets of initial guesses for the parameter values.

(b) Estimate parameter values with and without slope substitution, using subsets of the data with and without noise. Use different subsets with 10 and 15 time points. Compare results. Test whether averaging of the parameter values from different runs improves the fits.

5.19. Construct the diagram of a small pathway system that would be described by the following equations:

$$\dot{X}_1 = \alpha_1 X_3^{g_{13}} - \beta_1 X_1^{h_{11}},$$
$$\dot{X}_2 = \alpha_2 X_1^{g_{21}} - \beta_2 X_2^{h_{22}},$$
$$\dot{X}_3 = \alpha_3 X_2^{g_{32}} - \beta_3 X_3^{h_{33}} X_4^{h_{34}},$$
$$\dot{X}_4 = \alpha_4 X_1^{g_{41}} - \beta_4 X_4^{h_{44}}.$$

Is the diagram unique? Estimate parameter values for the system, using the datasets in **Table 5.18**, either one at a time or combined.

TABLE 5.18

t	X_1	X_2	X_3	X_4
DATASET 1				
0	1.400	2.700	1.200	0.400
0.2	1.039	3.122	1.591	0.317
0.4	0.698	3.176	1.980	0.250
0.6	0.482	2.992	2.298	0.195
0.8	0.372	2.711	2.519	0.157
1	0.322	2.431	2.643	0.135
1.2	0.302	2.198	2.682	0.125
1.4	0.300	2.025	2.657	0.120
1.6	0.307	1.911	2.591	0.120
1.8	0.322	1.846	2.506	0.123
2	0.340	1.822	2.416	0.126
2.2	0.359	1.826	2.335	0.130
2.4	0.377	1.851	2.269	0.135
2.6	0.393	1.886	2.222	0.139
2.8	0.404	1.924	2.192	0.142
3	0.412	1.959	2.177	0.144
3.2	0.415	1.988	2.174	0.145
3.4	0.415	2.010	2.180	0.146
3.6	0.413	2.024	2.189	0.146
3.8	0.410	2.030	2.201	0.145
4	0.406	2.031	2.212	0.145
DATASET 2				
0	0.200	0.300	2.200	0.010
0.2	0.439	0.833	1.789	0.107
0.4	0.597	1.347	1.569	0.156
0.6	0.685	1.791	1.540	0.185
0.8	0.686	2.125	1.634	0.197
1	0.627	2.327	1.787	0.195
1.2	0.548	2.406	1.954	0.184
1.4	0.479	2.392	2.105	0.170
1.6	0.428	2.324	2.223	0.157
1.8	0.395	2.235	2.304	0.147
2	0.376	2.146	2.349	0.141
2.2	0.367	2.071	2.363	0.137

TABLE 5.18 (CONTINUED)

t	X_1	X_2	X_3	X_4
DATASET 2				
2.4	0.365	2.013	2.356	0.135
2.6	0.368	1.975	2.336	0.135
2.8	0.373	1.954	2.309	0.136
3	0.380	1.946	2.281	0.137
3.2	0.387	1.949	2.256	0.139
3.4	0.394	1.958	2.237	0.140
3.6	0.399	1.970	2.223	0.142
3.8	0.402	1.982	2.214	0.143
4	0.404	1.993	2.211	0.143
DATASET 3				
0	4.000	1.000	3.000	4.000
0.2	1.864	2.667	2.076	1.909
0.4	0.952	3.151	1.952	0.872
0.6	0.598	3.111	2.094	0.413
0.8	0.447	2.886	2.296	0.232
1	0.371	2.622	2.468	0.165
1.2	0.332	2.376	2.572	0.139
1.4	0.316	2.175	2.607	0.128
1.6	0.313	2.026	2.588	0.124
1.8	0.320	1.926	2.533	0.124
2	0.332	1.870	2.460	0.125
2.2	0.348	1.849	2.385	0.129
2.4	0.365	1.853	2.317	0.132
2.6	0.381	1.874	2.261	0.136
2.8	0.394	1.904	2.221	0.139
3	0.404	1.937	2.196	0.142
3.2	0.410	1.967	2.184	0.144
3.4	0.413	1.992	2.182	0.145
3.6	0.412	2.010	2.187	0.146
3.8	0.411	2.021	2.195	0.145
4	0.408	2.027	2.205	0.145

5.20. Assume again the model structure of Exercise 5.19, but with new parameter values:

$$\dot{X}_1 = \alpha_1 X_3^{g_{13}} - \beta_1 X_1^{h_{11}},$$
$$\dot{X}_2 = \alpha_2 X_1^{g_{21}} - \beta_2 X_2^{h_{22}},$$
$$\dot{X}_3 = \alpha_3 X_2^{g_{32}} - \beta_3 X_3^{h_{33}} X_4^{h_{34}},$$
$$\dot{X}_4 = \alpha_4 X_1^{g_{41}} - \beta_4 X_4^{h_{44}}.$$

Use a gradient-based method, such as **`lsqcurvefit`** in MATLAB®, to estimate parameter values for the system, using the noisy datasets in **Table 5.19**, either one at a time or combined. Use as bounds for the parameters $0 \le \alpha_i, \beta_i \le 20$; and $-2 \le g_{ij}, h_{ij} \le 2$.

TABLE 5.19

t	X_1	X_2	X_3	X_4
DATASET 1				
0	0.109	3.145	1.304	0.817
0.2	0.468	2.336	2.290	0.539
0.4	0.430	1.998	2.778	0.445
0.6	0.354	1.499	3.216	0.329
0.8	0.340	1.131	3.162	0.259
1	0.352	0.871	3.455	0.244
1.2	0.326	0.673	3.223	0.222
1.4	0.353	0.590	3.061	0.233
1.6	0.362	0.513	2.733	0.252
1.8	0.406	0.481	2.620	0.285
2	0.451	0.455	2.473	0.311
2.2	0.460	0.473	2.262	0.336
2.4	0.508	0.508	1.977	0.383
2.6	0.525	0.552	1.951	0.466
2.8	0.559	0.595	1.800	0.492
3	0.576	0.676	1.760	0.547
3.2	0.607	0.707	1.744	0.557
3.4	0.613	0.788	1.674	0.587
3.6	0.616	0.782	1.627	0.646
3.8	0.609	0.794	1.624	0.630
4	0.611	0.848	1.736	0.618
DATASET 2				
0	2.017	3.046	1.143	0.115
0.2	0.711	2.962	2.320	0.966
0.4	0.443	2.652	3.225	0.754
0.6	0.381	2.052	3.249	0.485
0.8	0.385	1.494	3.636	0.312
1	0.338	1.206	3.769	0.275
1.2	0.352	0.835	3.608	0.246
1.4	0.342	0.630	3.448	0.234
1.6	0.353	0.568	3.332	0.245
1.8	0.400	0.541	3.276	0.270
2	0.409	0.472	2.524	0.295
2.2	0.469	0.514	2.580	0.342
2.4	0.517	0.485	2.414	0.329
2.6	0.530	0.544	2.081	0.424
2.8	0.614	0.560	1.848	0.451
3	0.636	0.606	1.927	0.506
3.2	0.644	0.685	1.899	0.585
3.4	0.687	0.724	1.668	0.575
3.6	0.615	0.772	1.748	0.664
3.8	0.650	0.821	1.871	0.626
4	0.602	0.753	1.500	0.619

5.21. For the two-population example in the text (see Table 5.9 and Figure 5.22), what is the absolute deviation of the optimized solution from the true solution at each time point if SSE ≈ 7 or SSE ≈ 10,000?

5.22. Find out from the literature what sloppiness means in the context of parameter estimation. Write a brief report.

5.23. Find out details regarding Fisher's information matrix. Write a brief report.

5.24. Write a brief report on overfitting and cross-validation.

5.25. Discuss the advantages and drawbacks of canonical models for structure identification.

REFERENCES

[1] Kutner MH, Nachtsheim CJ, Neter J & Li W. Applied Linear Statistical Models, 5th ed. McGraw-Hill/Irwin, 2004.

[2] Montgomery DC. Design and Analysis of Experiments, 9th ed. Wiley, 2017.

[3] Donahue MM, Buzzard GT & Rundell AE. Parameter identification with adaptive sparse grid-based optimization for models of cellular processes. In Methods in Bioengineering: Systems Analysis of Biological Networks (A Jayaraman & J Hahn, eds), pp 211–232. Artech House, 2009.

[4] Land AH & Doig AG. An automatic method of solving discrete programming problems. *Econometrica* 28 (1960) 497–520.

[5] Falk JE & Soland RM. An algorithm for separable nonconvex programming problems. *Manage. Sci.* 15 (1969) 550–569.

[6] Glantz SA, Slinker BK & Neilands TB. Primer of Applied Regression and Analysis of Variance, 3rd ed. McGraw-Hill, 2016.

[7] Goldberg DE. Genetic Algorithms in Search, Optimization and Machine Learning. Addison-Wesley, 1989.

[8] Holland JH. Adaptation in Natural and Artificial Systems: An Introductory Analysis with Applications to Biology, Control, and Artificial Intelligence. University of Michigan Press, 1975.

[9] Fogel DB. Evolutionary Computation: Toward a New Philosophy of Machine Intelligence, 3rd ed. IEEE Press/Wiley-Interscience, 2006.

[10] Dorigo M & Stützle T. Ant Colony Optimization. MIT Press, 2004.

[11] Catanzaro D, Pesenti R & Milinkovitch MC. An ant colony optimization algorithm for phylogenetic estimation under the minimum evolution principle. *BMC Evol. Biol.* 7 (2007) 228.

[12] Kennedy J & Eberhart RC. Particle swarm optimization. In Proceedings of IEEE International Conference on Neural Networks, Perth, Australia, 27 November–1 December 1995. IEEE, 1995.

[13] Poli R. Analysis of the publications on the applications of particle swarm optimisation. *J. Artif. Evol. Appl.* (2008) Article ID 685175.

[14] Metropolis N, Rosenbluth AW, Rosenbluth MN, et al. Equations of state calculations by fast computing machines. *J. Chem. Phys.* 21 (1953) 1087–1092.

[15] Salamon P, Sibani P & Frost R. Facts, Conjectures, and Improvements for Simulated Annealing. SIAM, 2002.

[16] Li S, Liu Y & Yu H. Parameter estimation approach in groundwater hydrology using hybrid ant colony system. In Computational Intelligence and Bioinformatics: ICIC 2006 (D-S Huang, K Li & GW Irwin, eds), pp 182–191. Lecture Notes in Computer Science, Volume 4115, Springer, 2006.

[17] Voit EO. What if the fit is unfit? Criteria for biological systems estimation beyond residual errors. In Applied Statistics for Biological Networks (M Dehmer, F Emmert-Streib & A Salvador, eds), pp 183–200. Wiley, 2011.

[18] Bertalanffy L von. Principles and theory of growth. In Fundamental Aspects of Normal and Malignant Growth (WW Nowinski, ed.), pp 137–259. Elsevier, 1960.

[19] Gutenkunst RN, Casey FP, Waterfall JJ, et al. Extracting falsifiable predictions from sloppy models. *Ann. NY Acad. Sci.* 1115 (2007) 203–211.

[20] Raue A, Kreutz C, Maiwald T, et al. Structural and practical identifiability analysis of partially observed dynamical models by exploiting the profile likelihood. *Bioinformatics* 25 (2009) 1923–1929.

[21] Efron B & Tibshirani RJ. An Introduction to the Bootstrap. Chapman & Hall/CRC, 1993.

[22] Good P. Resampling Methods: A Practical Guide to Data Analysis, 3rd ed. Birkhäuser, 2006.

[23] Manly BFJ. Randomization, Bootstrap and Monte Carlo Methods in Biology, 3rd ed. Chapman & Hall/CRC, 2006.

[24] Neves AR, Ramos A, Nunes MC, et al. *In vivo* nuclear magnetic resonance studies of glycolytic kinetics in *Lactococcus lactis*. *Biotechnol. Bioeng.* 64 (1999) 200–212.

[25] Voit EO & Almeida J. Decoupling dynamical systems for pathway identification from metabolic profiles. *Bioinformatics* 20 (2004) 1670–1681.

[26] Chou I-C & Voit EO. Recent developments in parameter estimation and structure identification of biochemical and genomic systems. *Math. Biosci.* 219 (2009) 57–83.

[27] Gennemark P & Wedelin D. Benchmarks for identification of ordinary differential equations from time series data. *Bioinformatics* 25 (2009) 780–786.

[28] Lillacci G & Khammash M. Parameter estimation and model selection in computational biology. *PLoS Comput. Biol.* 6(3) (2010) e1000696.

[29] Gengjie J, Stephanopoulos GN, & Gunawan R. Parameter estimation of kinetic models from metabolic profiles: two-phase dynamic decoupling method. *Bioinformatics* 27 (2011) 1964–1970.

[30] Sun J, Garibaldi JM & Hodgman C. Parameter estimation using meta-heuristics in systems biology: a comprehensive review. *IEEE/ACM Trans. Comput. Biol. Bioinform.* 9 (2012) 185–202.

[31] Verhulst PF. Notice sur la loi que la population suit dans son accroissement. *Corr. Math. Phys.* 10 (1838) 113–121.

[32] de Boor C. A Practical Guide to Splines, revised edition. Springer, 2001.

[33] Vilela M, Borges CC, Vinga S, Vasconcelos AT, et al. Automated smoother for the numerical decoupling of dynamics models. *BMC Bioinformatics* 8 (2007) 305.

[34] Voit EO & Chou I-C. Parameter estimation in canonical biological systems models. *Int. J. Syst. Synth. Biol.* 1 (2010) 1–19.

[35] Dam P, Fonseca LL, Konstantinidis KT & Voit EO. Dynamic models of the complex microbial metapopulation of Lake Mendota. *npj Syst. Biol. Appl.* 2 (2016) 16007.

[36] Schulz AR. Enzyme Kinetics: From Diastase to Multi-Enzyme Systems. Cambridge University Press, 1994.

[37] Goel G, Chou I-C & Voit EO. System estimation from metabolic time-series data. *Bioinformatics* 24 (2008) 2505–2511.

[38] Vilela M, Vinga S, Maia MA, et al. Identification of neutral biochemical network models from time series data. *BMC Syst. Biol.* 3 (2009) 47.

[39] Battogtokh D, Asch DK, Case ME, et al. An ensemble method for identifying regulatory circuits with special reference to the qa gene cluster of *Neurospora crassa*. *Proc. Natl Acad. Sci.* USA 99 (2002) 16904–16909.

[40] Lee Y & Voit EO. Mathematical modeling of monolignol biosynthesis in *Populus*. *Math. Biosci.* 228 (2010) 78–89.
[41] Gennemark P & Wedelin D. ODEion—a software module for structural identification of ordinary differential equations *J. Bioinform. Comput. Biol.* 12 (2014) 1350015.
[42] Srinath S & Gunawan R. Parameter identifiability of power-law biochemical system models. *J. Biotechnol.* 149 (2010) 132–140.
[43] Almeida JS. Predictive non-linear modeling of complex data by artificial neural networks. *Curr. Opin. Biotechnol.* 13 (2002) 72–76.
[44] Tan Y & Liao JC. Metabolic ensemble modeling for strain engineers. *Biotechnol. J.* 7 (2012) 343–353.
[45] Faraji M & Voit EO. Nonparametric dynamic modeling. *Math. Biosci.* (2016) S0025-5564(16)30113-4. doi: 10.1016/j.mbs.2016.08.004. (Epub ahead of print.)
[46] Lineweaver H & Burk D. The determination of enzyme dissociation constants. *J. Am. Chem. Soc.* 56 (1934) 658–666.
[47] Woolf B. Quoted in Haldane JBS & Stern KG. Allgemeine Chemie der Enzyme, p 19. Steinkopf, 1932.
[48] Scatchard G. The attractions of proteins for small molecules and ions. *Ann. NY Acad. Sci.* 51 (1949) 660–672.

FURTHER READING

Chou IC & Voit EO. Recent developments in parameter estimation and structure identification of biochemical and genomic systems. *Math. Biosci.* 219 (2009) 57–83.

Eiben AE & Smith JE. Introduction to Evolutionary Computing, 2nd ed. Springer, 2015.

Glantz SA, Slinker BK & Neilands TB. Primer of Applied Regression and Analysis of Variance, 3rd ed. McGraw-Hill, 2016.

Kutner MH, Nachtsheim CJ, Neter J & Li W. Applied Linear Statistical Models, 5th ed. McGraw-Hill/Irwin, 2004.

Montgomery DC. Design and Analysis of Experiments, 9th ed. Wiley, 2017.

Voit EO. What if the fit is unfit? Criteria for biological systems estimation beyond residual errors. In Applied Statistics for Biological Networks (M Dehmer, F Emmert-Streib & A Salvador, eds), pp 183–200. Wiley, 2011.

Gene Systems

When you have read this chapter, you should be able to:

- Discuss the Central Dogma of molecular biology, as well as modern amendments and refinements
- Describe the key features of DNA and RNA
- Identify the roles of different types of RNAs
- Outline the principles of gene regulation
- Set up simple models of gene regulation
- Summarize current methods for assessing gene expression and its location, along with their advantages and drawbacks

THE CENTRAL DOGMA

Genes are the main carriers of biological information from one generation to the next. They make us unique, are responsible for many of our good features and the bad features of others, contribute to a number of chronic diseases, and are the prime targets of evolution. In the olden days, which in this context means about half a century ago, the role of genes was simple, namely:

> Genes consist of self-replicating deoxyribonucleic acid (DNA), DNA is transcribed into matching ribonucleic acid (RNA), RNA is translated into proteins, and proteins manage the processes of daily life.

This unidirectional **Central Dogma**, proposed by Nobel Laureate Francis Crick [1, 2], became the centerpiece of modern biology.

The Central Dogma is still mostly true today, although we have learned that everything surrounding it is much, much more complicated: some DNA does not code for genes; some organisms inherit their information through RNA; molecular modifications such as DNA methylation can control **transcription**; and alternative splicing in eukaryotic cells may lead to many different RNAs and hence many different proteins from the same DNA. Maybe most significant of all, and in stark contrast to Crick's original ideas, proteins have a huge impact on DNA through their roles as **transcription factors**. In addition, most organisms employ various types of RNAs that affect which gene is transcribed when, where, and how much, and when a gene is to be silenced. Moreover, large and small metabolites can control gene expression. Large signaling metabolites include sphingolipids, which can "sense" environmental stresses and control the up- or down-regulation of appropriate genes in response [3]. An example of a small metabolite is glucose, which indirectly affects the transcription of many genes in organisms from bacteria to humans and, for instance, represses a wide variety of genes in the β-cells of the pancreas, which can be an especially serious problem in people with diabetes [4]. In bacteria, any excess of the rare amino acid tryptophan in the environment shuts down the expression of genes that are responsible for tryptophan production, because this production is metabolically expensive and would be wasteful. These strategies of avoiding unnecessary action are implemented through regulatory mechanisms that interact with the responsible section of the chromosome. In the case of tryptophan, the amino acid directly activates a repressor of the DNA segment that contains the genes that code for the enzymes of the tryptophan pathway [5]. Furthermore, there is evidence suggesting that metabolites can affect translation (see, for example, [6]). To make matters even more complicated, we have started to discover that about 98% of our DNA does not code for proteins, that genes can overlap each other, that the same stretches of DNA can lead to several alternative transcripts, and that regulatory elements are not necessarily located adjacent to the genes whose transcription they control [7].

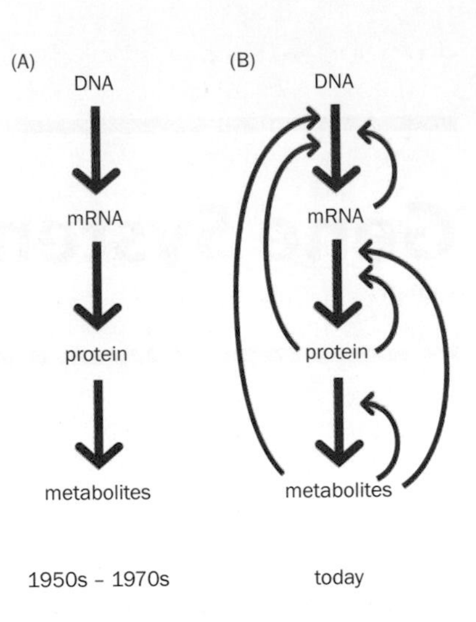

Figure 6.1 The original Central Dogma compared with modern understanding. In the original concept of the Central Dogma, transcription, translation, and enzymatic catalysis were proposed to form a linear chain of processes, although nobody doubted the role of regulation. We know now that a complex feedback structure at every level is crucial for appropriate functioning.

Thus, it is still true that genes are responsible for information transfer between generations, but they are clearly not the sole agents (**Figure 6.1**). In fact, genes only determine which proteins can be made by a cell, but they contain no direct information about the quantities and the timing with which proteins are made [8]. This information is obviously critical, because otherwise all cells of an organism would continuously produce the same proteins in the same quantities. Instead, genes, together with proteins and metabolites, act as the components of a well-organized system: genes provide the template for the synthesis of RNAs, which lead to proteins and metabolites, which in turn feed back to affect and control gene expression in multiple ways. This shared manner of control is a compromise between the original, unidirectional Central Dogma, where genes were seen as the true drivers of life, and the notion of genes as mere databases, or as "prisoners" of the physiological control of the organism, where they have no autonomy or independent control [8].

It is worth noting that genetic information is not exclusively transmitted through the organism's chromosomes. For instance, we have learned in recent years that epigenetic modifications in the form of methyl or acetyl groups attached to specific locations on the DNA can be inherited (see later in this chapter). Plants contain additional DNA in their chloroplasts, and the mitochondria of eukaryotes contain DNA that is transmitted from the mother to her offspring. Inheritance of male mitochondrial DNA has been reported in some species, although it is rare [9]. The contributions of chloroplast and mitochondrial DNA are generally small, but significant nevertheless. In humans, the mitochondrial DNA codes for 13 proteins, 22 **transfer RNAs**, and two subunits of ribosomal RNA.

Other nonchromosomal genetic elements include plasmids and transposable elements. Plasmids are typically circular, double-stranded DNA molecules that are natural in bacteria, although they have also been found in Archaea and in eukaryotes [10]. A bacterium may contain hundreds of plasmids at the same time and can share them with other bacteria, even from different species, in a process called horizontal gene transfer, which is crucial for the evolution of new species [11]. Plasmids can replicate independently of the main chromosome. They often contain genes that permit bacteria to survive under hostile conditions, such as exposure to antibiotics, which in hospitals can result in bacterial strains that are resistant to multiple drugs and very difficult to treat [12]. Perhaps the best-known example is methicillin-resistant *Staphylococcus aureus* (MRSA), which is a formidable culprit in many human infections [13]. On the positive side, plasmids have become extremely important research tools in biotechnology and synthetic biology, where they are used as vectors for the transport of genes into foreign species (see Chapter 14).

Transposable elements, or **transposons**, are stretches of DNA that can move from one location in the **genome** to another [14]. In some instances, the transposon is cut out of its original location and moved, whereas in other cases, the transposon is first copied and only the copy is relocated. In bacteria, transposons can carry several genes, which might be inserted into a plasmid and make the organism resistant to environmental stresses. Transposons can lead to important mutations and also to the expansion of the genome. It has been estimated that 60% of the maize genome consists of transposons. As one might expect, some transposons have been associated with human diseases. The *Alu* sequence can be found over a million times within the human genome and has been associated with diabetes, leukemia, breast cancer, and a number of other diseases [15].

The body of information on genes and on their structure, function, regulation, and evolution is overwhelmingly huge and growing at breakneck speed, primarily owing to the rapid evolution of sequencing techniques. Some of this information is of great and direct pertinence to systems biology, while other aspects are less so. Arguably most important for systems analyses is an understanding of which genes are expressed when, and how this complicated process is regulated by proteins, RNAs, and even metabolites. To hone this understanding in a fashion that is as brief as feasible, this chapter begins with a barebones discussion of the key features of DNA and RNA and afterwards devotes more attention to issues of the control and regulation of gene expression. The chapter ends with a brief description of standard methods for measuring gene expression. This description is certainly not sufficient for learning how to execute these methods, and the intent is rather to provide a feel for the different types of data that are of the greatest value to systems biology and for their qualitative and quantitative reliability. Thus—as with the following chapters on proteins and metabolites—this chapter is bound to disappoint many biologists, because it will only scratch the surface of the fascinating world of genes, genomes, and gene regulatory systems. In particular, this chapter keeps many aspects of bioinformatics to a minimum, and methodological issues of sequence analysis, **gene annotation**, evolution, and **phylogeny**, as well as experimental methods of genome analysis and manipulation, will be discussed rather cursorily or not at all. Excellent texts and Internet resources, including [16–20], cover these topics very well.

KEY PROPERTIES OF DNA AND RNA

6.1 Chemical and Physical Features

The basic building blocks of DNA are **nucleotides**, which form two long polymer strands. These strands are oriented in opposite (anti-parallel) directions and are held together by hydrogen bonds. Because of the physical and chemical features of the nucleotides, the two strands curl slightly, and the result is the famous double helix, which James Watson and Francis Crick discovered in 1953, based in part on Rosalind Franklin's X-ray photographs of DNA and her ideas of two nucleotide strands, and for which Watson and Crick were awarded the 1962 Nobel Prize in Physiology or Medicine [21, 22].

DNA makes use of four different nucleotides. Each consists of a nitrogen-containing nucleo-base, a five-carbon sugar called 2-deoxyribose, and a phosphate group. The nucleo-bases fall into the biochemical classes of purines and pyrimidines and are called adenine, cytosine, guanine, and thymine (abbreviated as A, C, G, and T). A combination of a nucleo-base and a sugar is called a **nucleoside**, and if a phosphate group is attached to the sugar as well, the resulting molecule is called a nucleotide. RNA uses very similar components, but the sugar is ribose rather than 2-deoxyribose and the nucleo-base thymine is replaced with uracil (U). The building blocks are visualized in **Figure 6.2**.

The chemical structure of the nucleo-bases has important consequences, since it is hydrogen bonding between them that holds the DNA strands together. Namely, guanine on one strand of DNA always pairs with cytosine on the anti-parallel strand, and adenine always pairs with thymine (**Figures 6.3** and **6.4**). Thus, one speaks of GC and AT **base pairs**, and this notion of a pair has become the most widely used size unit of DNA. Because of the strict pairing, one strand is in some sense the mirror image of the other, and this fact is of utmost importance for two fundamental processes. The first is DNA replication, which precedes cell division: double-stranded DNA opens up and both strands serve as templates for newly created DNA strands that bind to them. As a result of this **hybridization** process, the double-stranded DNA is now available in two copies, which

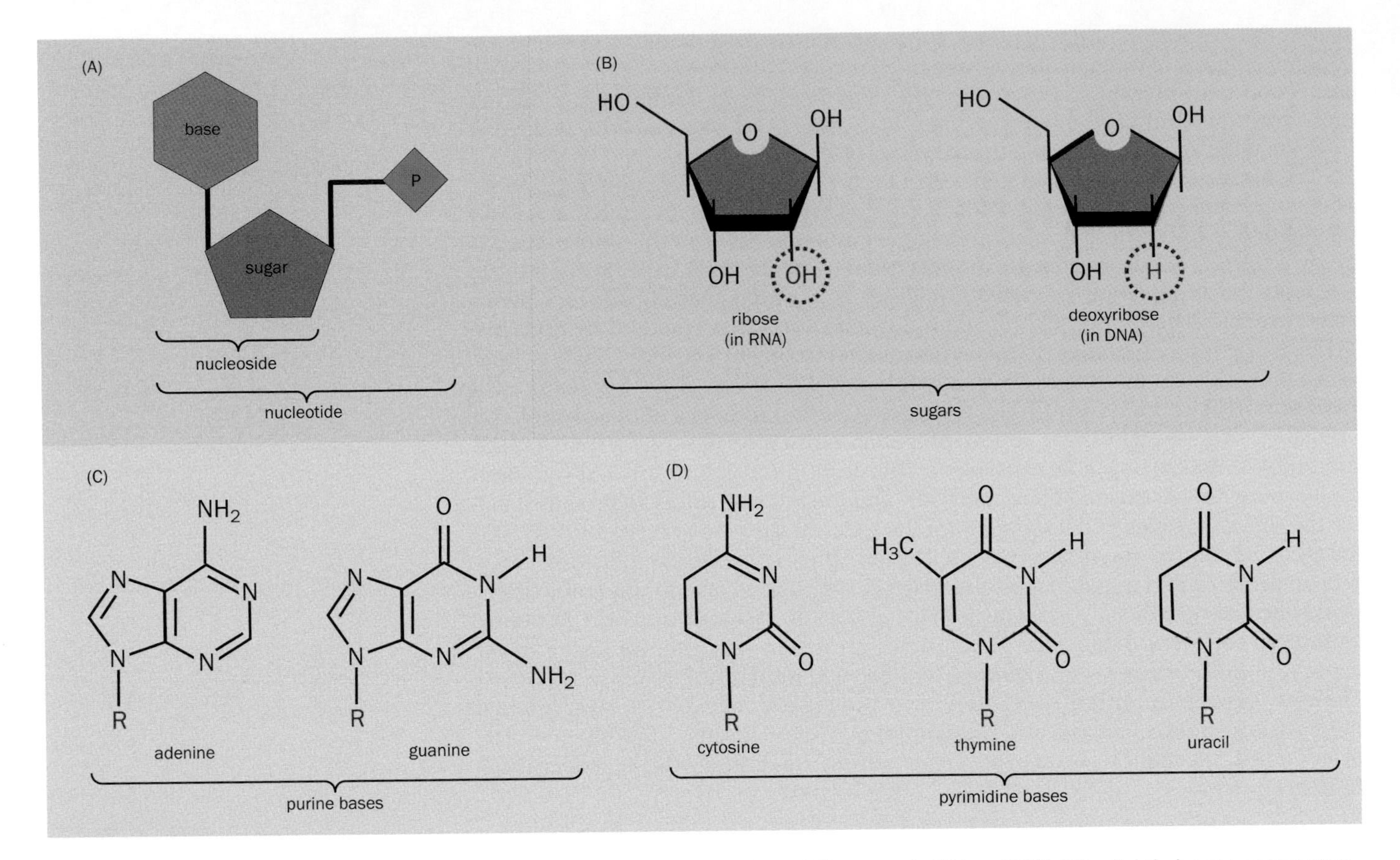

Figure 6.2 Building blocks of DNA and RNA. (A) Generic composition of nucleotides. (B) The sugars in RNA and DNA differ slightly, by one oxygen atom. (C, D) Both DNA and RNA contain purine and pyrimidine bases. Adenine, guanine, and cytosine are common to DNA and RNA, but the fourth base in DNA is thymine, while in RNA it is uracil.

eventually end up in the two daughter cells. The replication process is facilitated by specific proteins (**Figure 6.5**). The second fundamental process is the transcription of DNA into RNA, which follows the same principles, except that uracil replaces thymine.

The phosphate–deoxyribose backbone of each DNA strand has a definite orientation, with a 3-prime (3′) end and a 5-prime (5′) end, which are defined by the location of the bonds on the final sugar of the strand. When the strands form a double helix, they do so in an anti-parallel manner, with one strand in the 3′-to-5′ direction binding to the other in the 5′-to-3′ direction, as shown for example in Figure 6.4. In Figure 6.3, the backbone molecules are located at positions indicated by R. A sequence of three base pairs GC-AT-GC is illustrated in Figure 6.4. DNA consists of millions of base pairs that are connected in this fashion to form two complementary strands. The chemical and physical properties of these strands

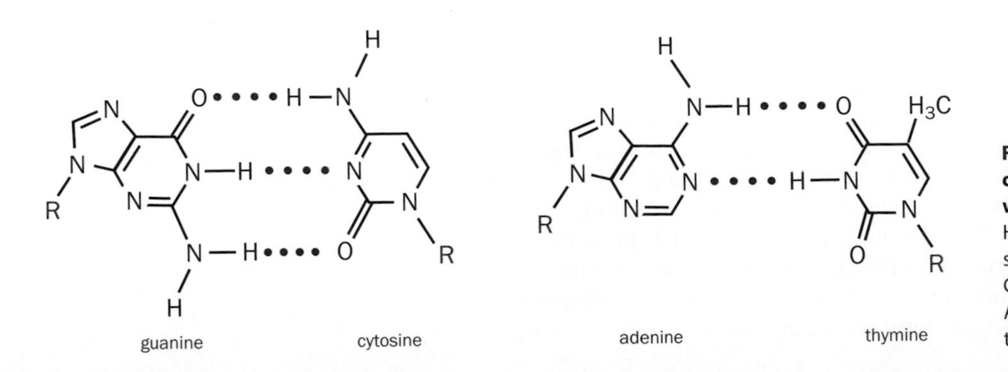

Figure 6.3 Guanine always pairs with cytosine on the anti-parallel strand, while adenine always pairs with thymine. Hydrogen bonds between the partners are shown as dotted red lines. Note that the GC pair contains three bonds, whereas the AT pair contains two. The GC bonding is therefore stronger.

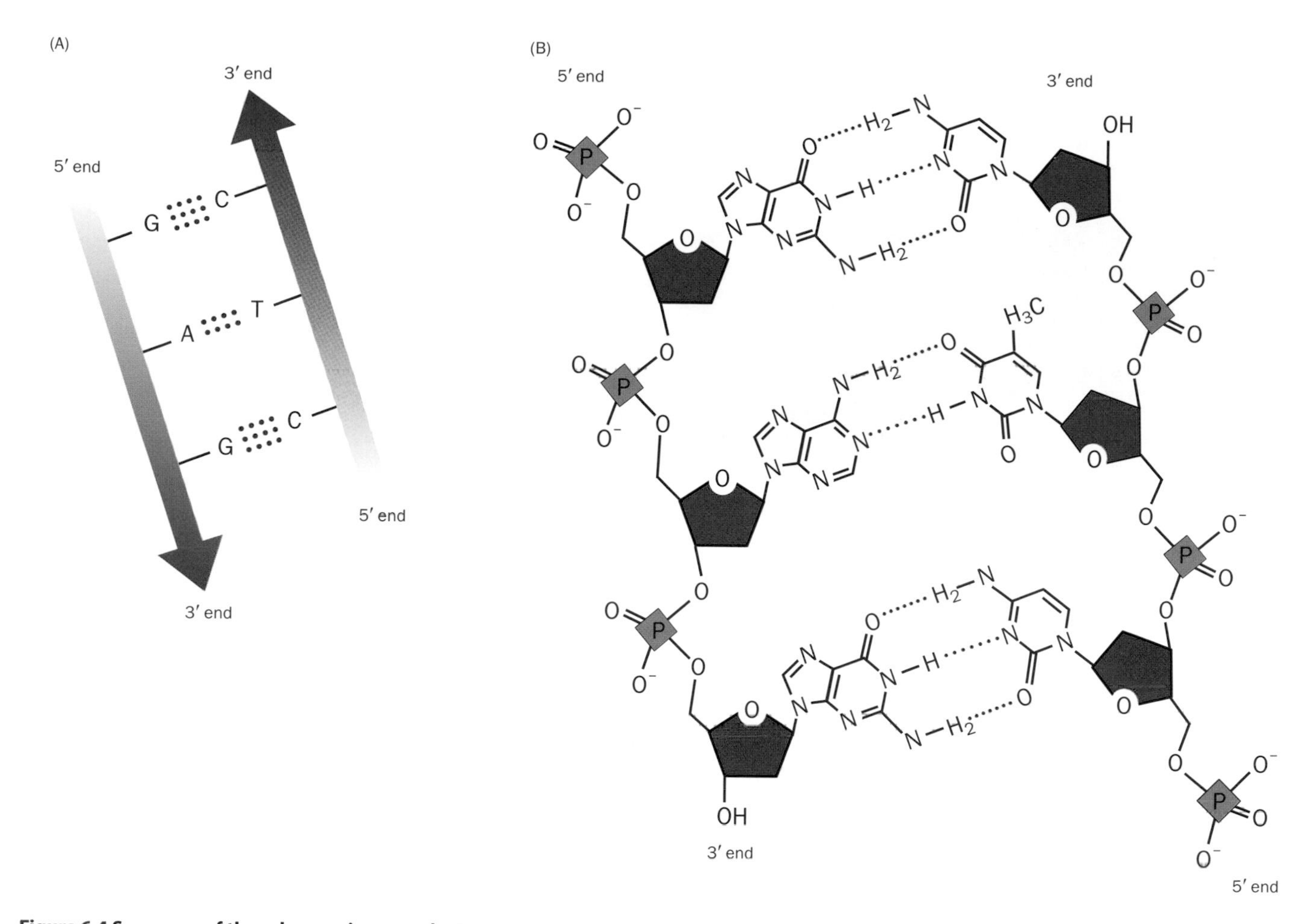

Figure 6.4 Sequence of three base pairs, namely GC, AT, GC, as shown in Figure 6.3. The connection between two neighboring bases is made by sugars and phosphates, which form one phosphate–deoxyribose backbone per strand. The backbones run in an anti-parallel fashion. (A) Diagram of backbones (blue and red) and bases with hydrogen bonds. (B) Molecular details.

Figure 6.5 Two simian virus SV40 protein domains interact with DNA just before replication. The high-resolution crystal structure shows the origin binding domains (OBDs) of a large T antigen (T-Ag) complexed with the DNA fragment that contains the origin for replication. The core origin contains four binding sites, each consisting of five nucleotides, which are organized as pairs of inverted repeats. In the co-structure, two T-Ag OBDs are oriented in a head-to-head fashion on the same face of the DNA, and each T-Ag OBD engages the major groove. Although the OBDs are very close to each other when bound to their DNA targets, they do not contact each other. (Courtesy of Huiling Chen and Ying Xu, University of Georgia.)

its encoded information and organization. Indeed, even the concept of a gene has changed over time.

In his revolutionary work on the inheritance of certain features of peas, the Austrian monk Gregor Mendel (1822–1884) noticed that biological traits are inherited from one generation to the next as discrete units. The discrete nature of this process allowed the separation of traits and explained the old observation that siblings can have different features. Mendel's findings were more or less ignored for a long time, until DNA was identified as the carrier of inheritance in the 1940s. At that point, genes were seen as the DNA code for traits such as eye color and body height, and possibly for characteristics like personality and intelligence.

Modern biology uses a more specific definition for a gene, namely, as a hereditary unit or as a specifically localized region of DNA that codes for a protein. The direct function of a coding gene is therefore the protein that results from its transcription and **translation**. By contrast, the assignment of higher-order function is often questionable, because such functions usually involve many processes that are executed by different proteins. It is therefore too ambiguous and vague to talk about an "eye-color gene," an "intelligence gene," or an "Alzheimer gene." Indeed, the **Gene Ontology (GO)** Consortium [30], which assigns functions to genes, states that "oncogenesis is not a valid GO term because causing cancer is not the normal function of any gene." Instead, the GO Consortium uses three categories that characterize different molecular properties of gene products:

- *The cellular component*: this refers to parts of a cell or its extracellular environment where the gene product is located.
- *Molecular function*: this covers fundamental activities of the gene product, such as binding or a specific enzymatic activity.
- *Biological process*: this refers to well-definable operations or molecular events, but not to higher functions such as renal blood purification.

It is rare that a single gene is solely responsible for some macroscopic feature or phenotype. Instead, the inheritance of complex, multifactorial traits such as body mass, skin color, or predisposition to diabetes depends on the contribution of many genes passed from parents to their children. This polygenic inheritance makes genetic studies challenging, because it involves the co-regulation of several genes and the control of possibly complex interactions between genes and the environment through epigenetic effects (see later). These issues are currently being investigated with analyses of quantitative trait loci (QTL), which are isolated stretches of DNA that are associated with a phenotypic trait and may be distributed over several chromosomes [31] (see also Chapter 13). Many successes of QTL analysis have been reported in plant breeding and crop development [32, 33], and also with respect to human diseases (see, for example, [34, 35]). However, while originally hailed as a new horizon in human genetics [36], QTL pose the severe problem that they often explain only a modest percentage of phenotypic variability. In an investigation of human body size, 20 loci explained only 3% of the variability in height among about 30,000 individuals [37]. It seems that many traits, including chronic diseases, involve a large number of genes, whose individual contributions are apparently very modest and therefore statistically elusive [38].

If a gene is a specifically localized region of DNA that codes for a protein, is there DNA that is not a gene? The answer is a resounding "Yes." It is currently assumed that humans possess between 20,000 and 25,000 genes. That sounds like a lot, but only comprises an estimated 2% of our DNA! Much of the remaining DNA has unknown functions, although one should be very careful calling it "junk DNA," as it was termed in the 1970s [39], because it is likely that the "junk" turns out to be good for something. Indeed, the more we learn about this noncoding DNA, the more we find that it has intriguing and very substantial roles in living cells, including regulatory roles at the transcriptional and translational levels. That these long stretches of DNA are not at all useless is also supported by the observation that some of them are remarkably stable throughout evolution and are even conserved between diverse species such as humans and pufferfish, which implies that they are under strong selective pressure and therefore most probably rather important [7]. At the same time, other stretches of DNA are not well conserved between species, which simply implies that we do not really understand the hidden signals in some noncoding DNA and their role in gene regulation. Current research therefore focuses on identifying markers for locations of noncoding elements and on characterizing their specific functions. In some cases, we have a vague idea of their roles, as reflected in the names of such DNA stretches—promoters, enhancers, repressors, silencers, insulators, and locus control regions, for example—but we have yet to develop a complete inventory.

Our current understanding of the organization of DNA is as follows [19]. The vast majority of DNA is noncoding and only a few percent of DNA codes for the synthesis of proteins in eukaryotes; this percentage increases in less advanced organisms. The average gene size in humans is about 3000 base pairs, but varies widely; the largest known human gene is dystrophin, with 2.4 million base pairs. Coding regions are usually rich in pairs of guanine and cytosine (GC pairs), which are more stable than AT pairs owing to their higher numbers of hydrogen bonds (see Figure 6.3). The pattern of distribution of genes within DNA cannot so far be explained and appears to be random in humans. Other species have more regulatory patterns of genes and noncoding regions. Long stretches of DNA containing lots of G's and C's apparently flank genes and form a barrier between coding and noncoding regions. The role of noncoding DNA is so far insufficiently understood, but some of this DNA is suspected to serve regulatory roles.

The prediction of genes from sequence data is still a challenge. One piece of evidence for identifying a gene is the open reading frame (ORF), which is a stretch of DNA that does not contain an in-frame stop codon. The presence of a start codon and a following sufficiently long ORF are often used as an initial screening tool for potential coding regions, but they do not prove that these regions will be transcribed and translated into protein. The challenges of gene prediction and annotation are greatly magnified in the new field of metagenomics, in which all genomes of large metapopulations are analyzed simultaneously. Such metapopulations may consist of hundreds if not thousands of microbial species, which coexist in environments such as the human gut, the soil, and the oceans (see Chapter 15).

Even within a eukaryotic gene, the DNA often contains regions that are not translated into protein. These so-called **introns** are typically transcribed, but later removed in a process called splicing, whereas the remaining regions of the gene, the **exons**, are merged into one piece of message and ultimately translated into protein. The splicing process, which is controlled by a variety of signaling molecules, permits the translation from different combinations of RNA pieces and thereby increases the possible number of resulting proteins (alternative splice variants) considerably (**Figure 6.8**). It is currently believed that some introns are transcribed as RNA genes such as **microRNAs**, which can regulate protein-encoding gene expression (see later).

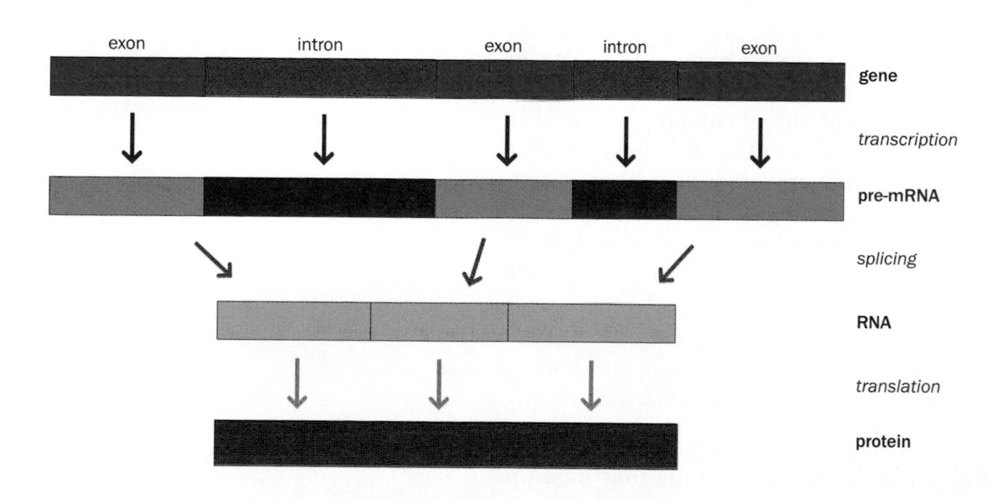

Figure 6.8 Exons and introns. Many eukaryotic genes contain introns, which ultimately do not correspond to amino acid sequences in the protein for which the gene codes. The splicing mechanism at the mRNA level permits alternative combinations and greatly increases the possible number and variability of proteins.

Over the past decades, molecular biology and bioinformatics have developed an enormous repertoire of laboratory techniques and computational tools for determining DNA sequences for prokaryotic and eukaryotic species, and even for individual humans [40], as well as for comparing DNAs from different species. These sequencing capabilities have allowed us to establish gene inventories for many species, along with their variations; notable, very comprehensive databases with this information include [19, 30, 41–43]. Furthermore, the journal *Nucleic Acids Research* annually publishes a special issue on biological databases. The relative ease of DNA sequencing has also led to much more accurate phylogenetic trees and other evolutionary insights than had been possible before. The methods for sequencing and sequence comparisons are fascinating; they are detailed in many books (for example, [16, 17, 28, 44]) and uncounted technical papers and review articles. An ongoing major challenge is the annotation of genes, that is, the identification of the function of a coding region. The Gene Oncology Consortium [30] collects methods and results of such efforts.

6.4 Eukaryotic DNA Packing

In prokaryotes, most of the DNA is arranged in a single chromosome, but some may be located separately, forming plasmids, as we discussed before. These plasmids can be shared among members of the same species or even other species, which leads to a potentially wide distribution of DNA and is a key mechanism of evolution. In higher organisms, DNA is organized in a species-specific number of chromosomes, which consist of DNA together with proteins that help package DNA efficiently and contribute to the control of gene transcription. Of particular importance are histones, which are spherical proteins around which DNA is wound (**Figure 6.9**), and protein scaffolds that are used for the DNA packaging process (**Figure 6.10**) [45]. A basic unit of about 150 DNA base pairs, wound around a histone core, is called a nucleosome (**Figure 6.11**; see also Figure 6.9). The nucleosomes are connected by stretches of about 80 base pairs of free DNA and coiled into higher-order structures, which eventually form chromatin, which makes up the chromosomes. The component *chrom* in these terms is Greek for color and reflects the fact that chromatin, when stained, becomes visible under a light microscope. The repeating pattern of nucleosomes and free DNA, which can be seen with an electron microscope, permits the astonishing packaging of about 2 m of eukaryotic DNA into a nucleus with a diameter of only 10 μ. At the same time, this packaging allows selective access to the DNA for transcription, which is controlled by chromatin remodeling motors that consist of proteins and use ATP for energy [46].

6.5 Epigenetics

We are just at the beginning of understanding how the environment, diet, and diseases can affect transcription through **epigenetic** processes that influence where and when genetic information is to be used (see Figure 6.9) [47]. In these processes, small molecules such as methyl groups can be attached to DNA as adducts and lead to heritable modifications of the chromatin. This means that it is possible that environmental factors affecting an individual can still be found in children and even grandchildren. The DNA adducts can allow the opening DNA for transcription or silence gene expression, and such processes have been associated with various diseases and all stages of tumor development, from initiation to invasion and metastasis. Fortunately, it seems possible, at least in principle, to reverse adverse epigenetic effects through clinical treatment and healthy living (see Figure 6.9C) [48].

RNA

RNA and DNA differ only slightly in their basic biochemical properties: the sugars in their backbone differ by just one oxygen atom, and one of the four bases is

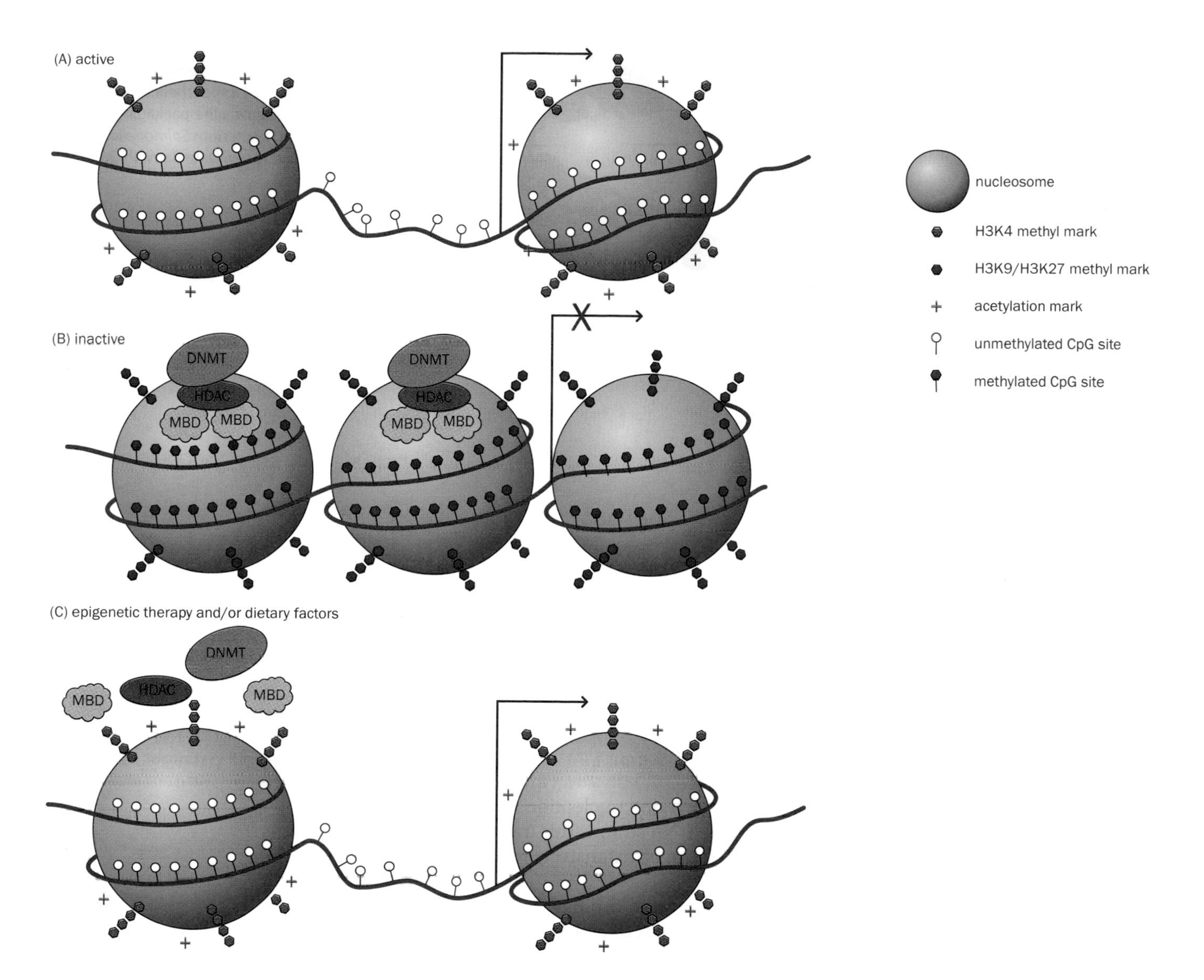

Figure 6.9 Diagram illustrating DNA organization around histones and epigenetic effects influencing the expression of genes close to a nucleosome (blue spheres). (A) Promoters of active genes are often associated with CG pairs that are not methylated (white circles), with acetylated histones (green crosses), and with methylated histone H3 (yellow hexagons); this configuration is conducive to transcription. (B) In some diseases such as cancer, CG sites in the promoter region of tumor suppressor genes are often methylated (red circles). Methyl-CG-binding domains (MBDs), histone deacetylases (HDACs), and DNA methyltransferases (DNMTs) make chromatin inaccessible and lead to repression of transcription. (C) Epigenetic therapy could potentially decrease MBDs, HDACs, and DNMTs and reverse their deleterious effects. (From Huang Y.-W, Kuo C.-T, Stoner K, et al. *FEBS Lett.* 585 [2011] 2129–2136. With permission from Elsevier.)

thymine in DNA and uracil in RNA (see Figure 6.2). However, in spite of their close similarities, DNA and RNA have very distinct roles in biology. DNA is the typical carrier of genetic information, whereas different types of RNA have diverse roles as intermediate information transducers and as regulators. One key difference is that, unlike DNA, most RNA molecules are single-stranded. Because these single RNA strands are not matched up with partners, as in the case of the DNA double helix, they are small in size, or short-lived, or exhibit complex two- and three-dimensional structures, collectively called **secondary structure**, where some of the nucleotides form bonds with other nucleotides of the same RNA strand (**Figures 6.12** and **6.13**).

6.6 Messenger RNA (mRNA)

For most organisms, mRNA provides the means of converting the DNA blueprint into proteins. In this transcription process, DNA opens up and serves as a template

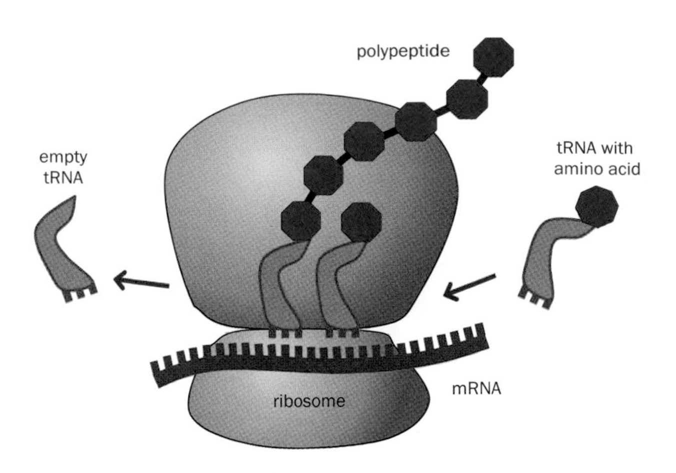

Figure 6.14 Schematic representation of the translation of messenger RNA (mRNA) into a polypeptide. A ribosome moves along the mRNA (green). Corresponding to each codon on the mRNA, a transfer RNA (tRNA; yellow) with the appropriate amino acid (blue octagons) is recruited. It enters the ribosome and permits the binding of its attached amino acid to the growing polypeptide chain (blue chain).

6.7 Transfer RNA (tRNA)

This type of RNA is relatively short, consisting of only about 80–100 nucleotides. It is synthesized from so-called RNA genes, which are usually located within stretches of noncoding DNA. The role of each tRNA is to transfer a specific amino acid to the growing peptide chain that is being constructed by a ribosome (**Figure 6.14**). Each tRNA contains a triplet of nucleotides that corresponds to a sequence of three DNA nucleotides, that is, a codon, which "codes" for the target amino acid (see Chapter 7 for a more detailed discussion of codons). This amino acid attaches to the 3′ end of the tRNA and, once associated with the ribosome, is covalently bound to the already formed polypeptide, a process that is catalyzed by the enzyme aminoacyl tRNA synthetase.

Each tRNA is single-stranded, but forms bonds between its own nucleotides that result in a secondary cloverleaf structure. This structure consists of three stems with loops, a variable loop, and an acceptor arm that permits the amino acid to attach. The center arm contains the anticodon region, whose three nucleotides match the codon on the mRNA and ensure that the correct amino acid is selected. To fit into the ribosome for translation, the tRNA must bend at its variable loop into an L-shape (**Figure 6.15**).

6.8 Ribosomal RNA (rRNA)

Ribosomes facilitate the process of translating mRNAs into chains of amino acids. They consist of proteins and rRNAs and are composed of two subunits (see Figure 6.14). The smaller one attaches to the mRNA and the larger one associates with the tRNAs and the amino acids they carry. When reading of an mRNA is completed, the two subunits separate from each other. Functionally, ribosomes are considered

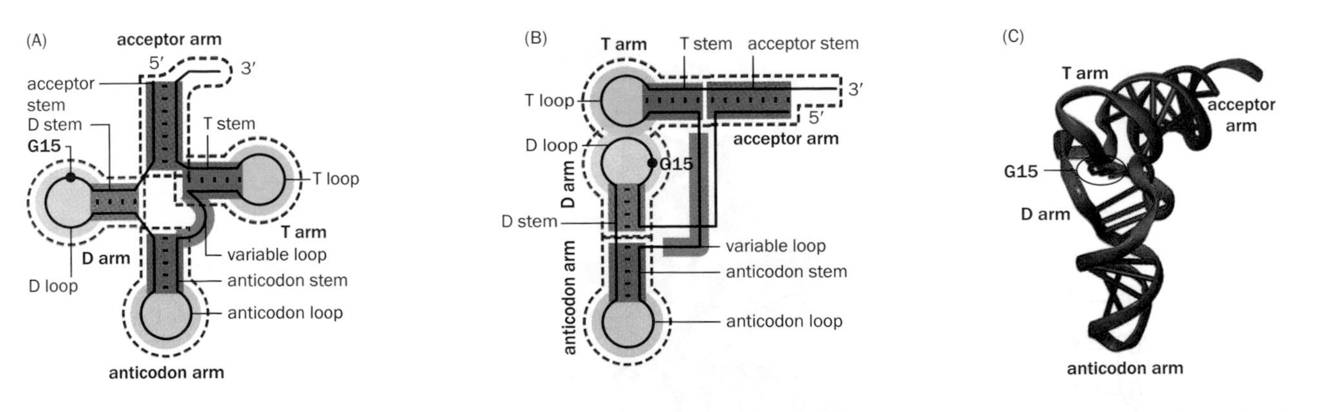

Figure 6.15 Cloverleaf structure of a transfer RNA (tRNA). (A) This secondary structure consists of four arms. The acceptor arm allows the attachment of a specific amino acid, whereas the opposing arm contains the matching anticodon, which mirrors the corresponding codon on the mRNA. (B) To serve its role within the ribosome, the tRNA has to bend at its variable loop into an L-shape. (C) The tRNA's three-dimensional structure. At position G15, this particular tRNA has been modified. (From Ishitani R, Nureki O, Nameki N, et al. *Cell* 113 [2003] 383–394. With permission from Elsevier.)

Figure 6.16 Crystal structure of a rather simple "hammerhead" ribozyme from the satellite RNA of the tobacco ringspot virus (sTRSV). This molecule was the first hammerhead ribozyme discovered. It uses a guanine residue as a general base during RNA catalysis. The structure shows how the single-stranded RNA folds and binds to itself. Also seen are guanosine triphosphate (balls and sticks) as a modified residue and magnesium ions (spheres) as ligands. (Protein Data Bank: PDB 2QUS [49].)

Figure 6.17 Structure of a ribozyme. The atomic structure of the large ribosomal subunit from the Dead Sea archaeon *Haloarcula marismortui* shows that the ribosome is a ribozyme, which catalyzes the formation of peptide bonds during translation. (Protein Data Bank: PDB 1FFZ [50].)

enzymes, belonging to the class of **ribozymes**, because they are catalysts for the linking of amino acids through peptide bonds (**Figures 6.16** and **6.17**).

6.9 Small RNAs

Several types of small, double-stranded RNAs have been discovered and characterized in recent years [51]. Because of their small size, they had been overlooked for a long time, but it is now clear that they play important roles in numerous fundamental biological processes and diseases [52]. Small RNAs fall into several classes, the most prominent of which are small-interfering RNA (siRNA) and micro-RNA (miRNA). Interestingly, RNA interference (RNAi) pathways involving these types of

RNAs are highly conserved throughout evolution, which underscores their central importance from a different angle. The first discovery of RNAi was actually made in an effort to enhance the flower patterns of petunias [53], demonstrating once again that one can never foretell where some seemingly obscure scientific endeavor may eventually lead.

Small interfering RNA goes by different names, including **silencing** RNA and short interfering RNA, both of which luckily permit the same abbreviation siRNA. These short, double-stranded RNA molecules consist of 20–30 nucleotides per strand, of which usually 21–25 form base pairs, while the remaining two or three stick out at either side and are not bound (**Figure 6.18**). The importance of siRNAs derives from the fact that they can interfere with the expression of genes and even silence it. This interference constitutes a potent regulatory mechanism that allows the cell to slow down or shut down translation at appropriate times [54].

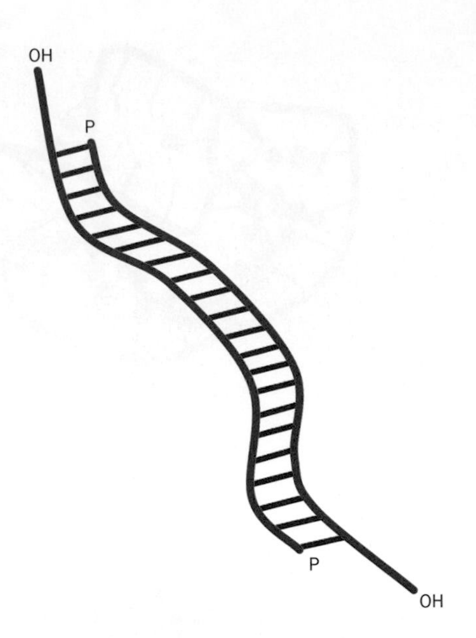

Figure 6.18 Schematic representation of a small interfering (siRNA) molecule. A typical siRNA consists of 21 base pairs and overhangs at both ends. P indicates the phosphate group at the 5′ end, while OH represents the hydroxyl group at the 3′ end.

Once introduced into cells, siRNAs permit the silencing of precisely targeted gene transcripts. The main mechanism seems to be the degradation of mRNA. This RNAi process begins with long double-stranded RNA, which is processed by an enzyme called dicer into small, sequence-specific siRNA. This siRNA triggers an RNA-induced silencing complex (RISC), which is an enzyme that catalyzes the cleavage of the target mRNA. Because siRNAs can be constructed artificially, the discovery and experimental mastery of RNAi has led to an enormous repertoire of tools for gene expression studies, including so-called loss-of-function techniques, and to a new class of potentially powerful therapeutics.

Other small RNAs have been discovered in recent times. Similar to siRNAs are micro-RNAs (miRNAs), which have also been called small temporal RNAs (stRNAs). Like siRNAs, miRNAs are produced from double-stranded RNA by a dicer enzyme. Their function is not quite clear, but it seems that they negatively regulate the expression of target transcripts. Apparently, miRNAs modulate protein expression through (a) a slicer mechanism for high-homology mRNA targets, (b) slicer-independent cleavage or degradation of mRNA, or (c) inhibition of translation without appreciable changes in mRNA levels [51]. It has been estimated that the human genome might encode over 1000 miRNAs, which potentially target half of all genes. This widespread influence is apparently possible because the homology between miRNAs and mRNAs does not have to be perfect. This imperfect matching is a significant difference between siRNAs and miRNAs in animals [55].

The discovery of miRNAs has led to exciting new insights and options for manipulating biological systems and possibly the treatment of cancer [56, 57]. However, one must be careful not to jump to conclusions regarding potential drug treatments too quickly, because the function of miRNAs in complex systems is not predictable from silencing studies executed in simple experimental systems. The reason is that an miRNA may silence some gene x whose normal function is down-regulation of another gene y, so that the effect of the miRNA on gene y may be activating rather than inhibiting. In fact, to understand the full impact of an miRNA it is necessary to consider the entire system of direct and indirect influences of the miRNA on the genome.

Double-stranded RNAs can also activate gene expression, which has led to the term small activating RNA (saRNA). Finally, so-called piwi-interacting RNAs (piRNAs) play an important role in germ-line development. It seems that these RNAs may silence transcription by regulating DNA methylation [52]. The characterization of small RNAs and their functional pathways is the subject of intense ongoing research.

Recently, Lai and collaborators reviewed numerous articles demonstrating how our understanding of the roles of miRNAs for gene regulation can be enhanced through mathematical modeling [58, 59].

6.10 RNA Viruses

As an exception to the Central Dogma, many viruses use single-stranded RNA instead of DNA as their inheritable material; some even use double-stranded RNA (**Figure 6.19**). The single strand is arranged either in a positive sense, which means that it acts like a messenger RNA that had been transcribed from DNA, or in a negative sense, in which case the strand is first transcribed into a matching positive strand. RNA viruses mutate very easily, because their polymerases do not possess the proof-reading ability of most DNA polymerases. Thus, transcription errors go

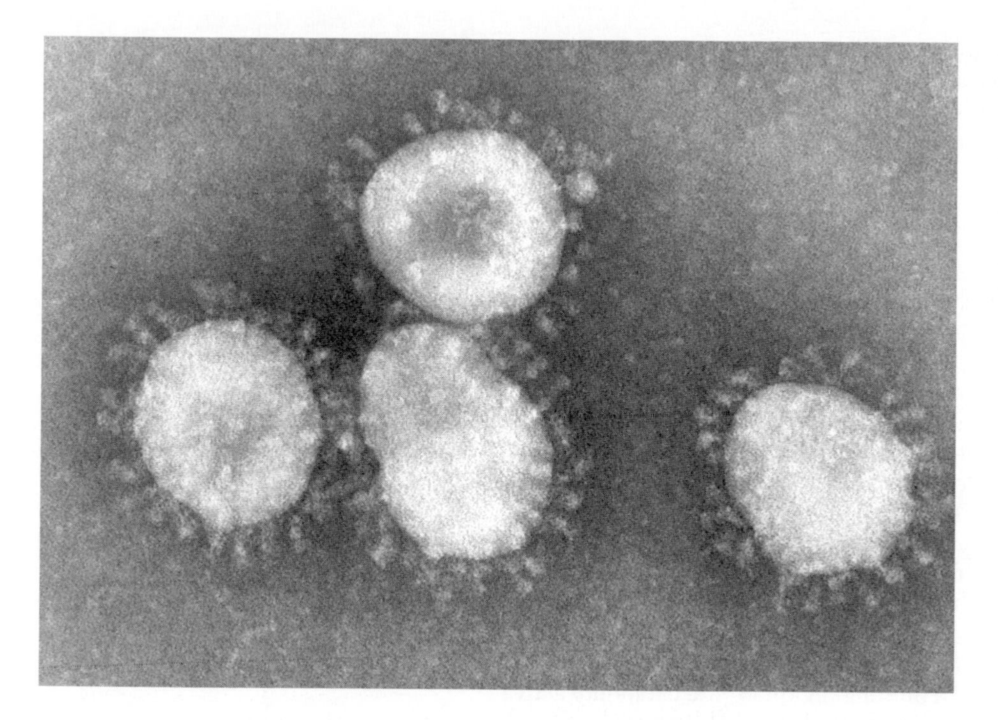

Figure 6.19 Electron micrograph of an RNA-containing coronavirus. The virus causes severe acute respiratory syndrome (SARS), a disease that rapidly spread around the world in 2003. Each virus is about 100 nm in diameter. (From Murray CL, Oha TS & Rice CM. Keeping track of viruses. In Microbial Forensics [B Budowle, SE Schutzer, RG Breeze, et al. eds], pp 137–153. Academic Press, 2011. With permission from Elsevier.)

undetected, but are advantageous in a sense that some of them allow the virus to develop resistance to environmental or pharmacological stressors. The positive strand of RNA is directly translated into a single protein, which the host cell cuts and modifies so that all proteins for virus replication are obtained.

GENE REGULATION

While essentially all cells of a multicellular organism contain the same genes, it is obvious that not all cells execute the same functions. The key to this variability is the coordinated expression of different sets of cell-type-specific genes at the appropriate times. But even unicellular organisms do not continuously transcribe all genes, and instead carefully regulate expression in order to satisfy the current demands for metabolites and for the enzymes that are needed for their production. Because most systems are simpler in microorganisms, very many investigations characterizing the expression of genes have targeted model species such as the bacterium *Escherichia coli* and the baker's yeast *Saccharomyces cerevisiae*, and a vast body of information has been assembled. And indeed, many of the fundamental concepts of gene regulation can be learned best from simpler organisms, as long as one keeps in mind that higher, multicellular organisms use additional means of control and regulation.

A major breakthrough in our understanding of the regulation of gene expression was the discovery of **operons** in bacteria in the 1950s. According to the original concept, each operon consisted of a defined stretch of DNA that contained at its beginning at least one regulatory gene that allowed the bacterium to control the expression of all other genes in the operon [60]. This view has been slightly modified in recent years, because it turned out that an operon does not necessarily contain a regulatory gene. Instead, the promoter region of an operon generally has multiple ***cis*-regulatory** motifs that can bind with either repressors or inducers.

The regulation of an operon can occur through repressors or inducers. Repressors are DNA-binding proteins that prevent the transcription of genes. These repressors are themselves the products of genes, and this fact creates hierarchies of regulation that we will discuss later. Repressors often have binding sites for specific non-protein cofactors. If these are present, the repressor is effective in preventing transcription; if they are absent, the repressor dissociates from the operator site, and the genes in the operon are transcribed. Inducers are molecules that initiate gene expression. An example is the sugar lactose, which functions as an inducer for the so-called *lac* operon, which is responsible for lactose uptake and utilization.

Lactose inactivates the repressor and thereby induces transcription. Other examples include certain antibiotics, which can induce the expression of genes that make the organism resistant [61, 62].

6.11 The *lac* Operon

It is illustrative to study the unit structure of an operon within the genome of an organism in more detail. The first well-characterized example of this structure was indeed the *lac* operon in *E. coli* [63]. Its discovery was a true breakthrough in genetics and molecular biology, and François Jacob, André Michel Lwoff, and Jacques Monod were awarded the 1965 Nobel Prize in Physiology or Medicine for their discoveries associated with this and other operons.

The *lac* operon contains three structural genes that code for enzymes associated with the uptake and utilization of the disaccharide (double-sugar) lactose, which chemically is a β-galactoside and prevalent in mammalian milk. These enzymes are β-galactoside permease, which is a transport protein responsible for the uptake of lactose from the medium (gene *22acy*), galactose transacetylase, which transfers an acetyl group from acetyl coenzyme A (acetyl-CoA) to β-galactosides (gene *lacA*), and β-galactosidase, an enzyme that converts lactose into the simpler sugars glucose and galactose (gene *lacZ*). In addition to these structural genes, the *lac* operon contains a promoter region P and an operator region O (**Figure 6.20**). Also of great importance are a regulator gene *lacI* and a binding region for the catabolite activator protein (CAP). The names of these components are mainly historical.

The bacterium only transcribes the structural genes if their products are needed. The cell manages the decision of whether to transcribe or not to transcribe with a multicomponent regulatory mechanism. If lactose is available in the environment, it serves as an inducer. It binds to the repressor, which is a protein that is coded for by a regulatory gene. In the case of the *lac* operon, the regulatory gene is directly adjacent to the operon, but in other cases it can be located quite distant from it. The repressor is normally bound to the operator region. However, binding of the inducer to the repressor causes the repressor to dissociate from the operator region. Without the repressor bound to the operator, the enzyme RNA polymerase, which synthesizes RNA, can in principle attach to the promoter region and transcribe the structural genes into mRNA. However, the promoter region often also contains response elements in the form of short DNA sequences that permit the binding of transcription factors under the appropriate conditions. Transcription factors recruit RNA polymerase and are of utmost importance for the regulation of gene expression. They will be discussed later in more detail.

In the case of the *lac* operon, the response elements react indirectly to high concentrations of glucose in the medium. Thus, glucose can repress the transcription of

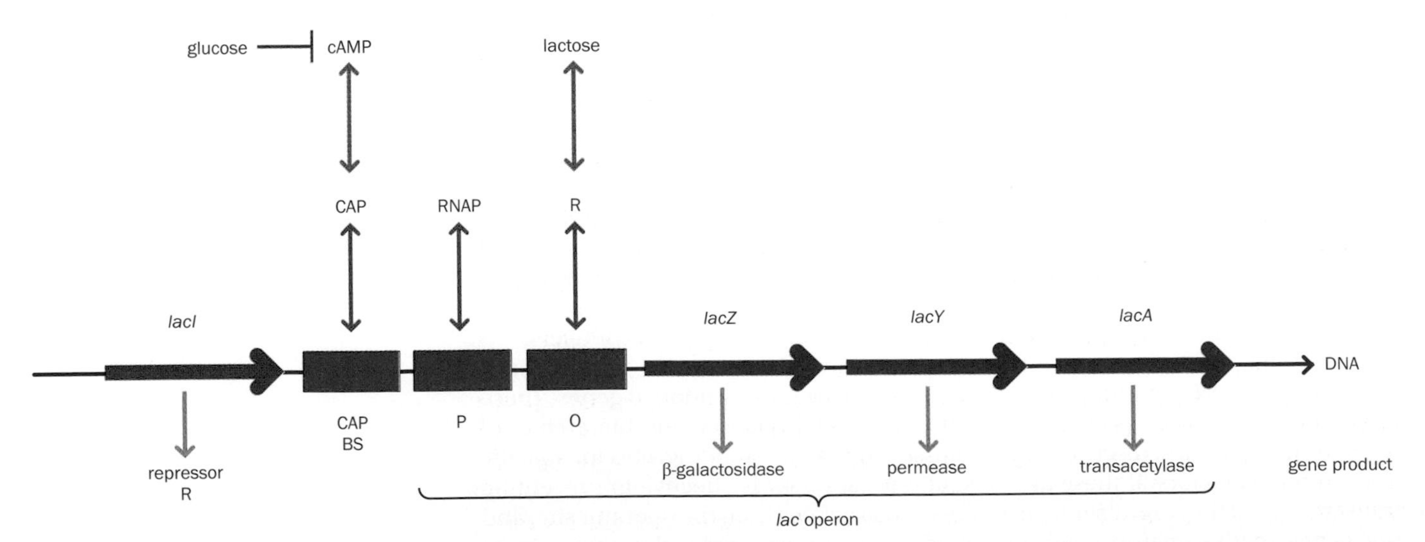

Figure 6.20 Diagram of the structure and function of the *lac* operon. The coding genes lead to the production of a repressor and three enzymes. Small molecules such as glucose and lactose affect expression. See the text and Table 6.1 for details.

TABLE 6.1: FUNCTION OF THE *lac* OPERON UNDER DIFFERENT ENVIRONMENTAL CONDITIONS

Lactose	Glucose	Repressor	CAP BS	Promoter	Operator	Transcription
⇓	⇓	U	B	F	R	off
⇑	⇓	I	B	B	F	on
⇑	⇑	I	F	F	F	off
⇓	⇑	U	F	BF	R	off

Key: Arrows indicate high (⇑) and low (⇓) concentrations in the environment. I, induced; U, uninduced; B, bound (in the case of the CAP-binding site, CAP BS, by CAP facilitated by cAMP, whose concentration is inversely related to the glucose concentration in the medium; in the case of the promoter P by RNA polymerase); F, free of binding partner; R, repressed. Transcription occurs only under high-lactose and low-glucose conditions.

the *lac* genes. This repression is advantageous because the bacterium saves energy by consuming the more easily digestible glucose rather than lactose. This repression is accomplished in the following fashion. As a consequence of complex metabolic and signaling mechanisms, the cell responds to environmental glucose levels by regulating its internal level of cyclic adenosine monophosphate (cAMP). For low glucose levels, cAMP is high, and for high glucose levels, cAMP is low. The signaling molecule cAMP binds to CAP, which attaches to a specific binding site right next to the promoter region (CAP BS in Figure 6.20). This binding permits the attachment of RNA polymerase to the promoter and transcription of the structural *lac* genes, if lactose is present. By contrast, if glucose in the medium is high, the bacterium does not bother with lactose and waste energy synthesizing enzymes, and rather uses the glucose. The high glucose concentration results in low cAMP, RNA polymerase does not attach to the promoter region, and the organism does not transcribe the *lac* genes. Taken together, the system expresses the structural genes only if lactose is available but glucose is not (**Table 6.1**).

Several groups have developed detailed dynamic models of the *lac* operon (see, for example, [64–67]). Each model focuses on a particular question, is correspondingly structured, and has a different level of complexity, which again shows that biological questions and data determine the type, structure, and implementation of a model (Chapter 2).

6.12 Modes of Regulation

In general, the regulation of an operon may occur through induction or repression, and it may be positive or negative, so that there are four base forms of controlling the initiation of gene transcription.

Positively inducible operons are controlled by activator proteins, which under normal conditions do not bind to the regulatory DNA region. Transcription is initiated only if a specific inducer binds to the activator. The inducer changes the **conformation** (three-dimensional shape; see Chapter 7) of the activator, which allows its interaction with the operator and triggers transcription.

Positively repressible operons are also controlled by activator proteins, but these are normally bound to the regulatory DNA region. When a co-repressor protein binds to the activator, the activator no longer binds to the DNA and transcription is stopped.

Negatively inducible operons are controlled by repressor proteins, which are normally bound to the operator of the operon and thereby prevent transcription. Specific inducers can bind to repressors, change their conformation so that binding to DNA is no longer effective, and thus trigger transcription.

Negatively repressible operons are normally transcribed. They have binding sites for repressor proteins, which, however, only prevent transcription if specific co-repressors are present. Binding of the repressor and co-repressor leads to attachment to the operator and prevents transcription.

The type of regulation found in each particular case appeared to be random at first, but a later, careful analysis led to clear design principles dictating the mode of control (see also Chapter 14). The discovery and interpretation of these design principles resulted in the demand theory of gene regulation, according to which there are strict rules for when a particular mode of regulation is beneficial and therefore evolutionarily advantageous [61, 68]. As one example, a repressor mode is advantageous if the regulated gene codes for a function that is in low demand within the organism's natural environment. By contrast, if the function is in high demand, the corresponding gene is under activator control.

Much of what we know about operons and their functions was determined through uncounted laboratory experiments, beginning even before the work of Jacob and Monod [63]. By contrast, much of what we are now beginning to understand about the higher-order organization of bacterial genomes has been deduced with the help of sophisticated computational approaches. These approaches fall mostly within the realm of bioinformatics, and we will only discuss some pertinent results without describing the methods themselves in detail. The interested reader is encouraged to consult [44].

In prokaryotes, all genes are located on a single chromosome, which consists of a closed circle of double-stranded DNA, and on much smaller plasmids, which can be transferred from one bacterium to another, as we discussed before. The chromosome usually contains a few thousand genes, while a typical plasmid contains between a few and several hundred genes. In higher organisms, nonchromosomal DNA is furthermore found in mitochondria and chloroplasts.

6.13 Transcription Factors

The most prominent regulators of gene expression are transcription factors (TFs), which include the repressors and activators we discussed before and which must attach to a binding site in the vicinity of the promoter region to execute their function. TFs may up- or down-regulate the operon; some dual-regulator TFs can do both, depending on the conditions. A TF is often activated through phosphorylation by a kinase, which can "sense" specific changes in the intra- or extracellular environment [69]. The kinase and the TF therefore constitute a two-component signaling system that permits the cell to respond adequately to changes in the environment. Chapter 9 discusses such systems in greater detail.

Typical bacterial genomes contain relatively few genes for TFs in comparison with their number of operons, and while some genes can be expressed without TFs, the mismatch between the numbers of TFs and operons has led to the inference that at least some TFs must be responsible for several operons [69]. The collection of all operons under the control of the same TF is called the regulon of the TF. Because an operon may be regulated by several TFs, they may belong to different regulons as well. Evidence suggests that bacteria tend to keep operons encoding the same biological pathway, or even systems of related pathways, in nearby genomic locations [70].

The prediction of TF-binding sites and regulons from sequence information is a difficult problem, but many algorithms have been developed for this purpose, and numerous databases are available that contain information on known TF-binding sites and regulons (see, for example, [69, 71]). In addition to regulons, bioinformaticists have discussed über-operons (*über* is German for "above"), which are sets of genes or operons that are regulated together even after many rearrangements of the DNA of the organism [72]. Other concepts include stimulons [73], which are sets of genes that respond to the same stimulus, and modulons, which are regulons associated with several pathways or functions that are under the control of the same regulator [74].

Responses to environmental stresses often require more than the up- or down-regulation of a single operon. In fact, a complex response to a stress such as starvation may consist of changes in the expression of dozens of genes. The coordination of such responses therefore mandates means for co-regulating genes in different locations of the chromosome. The premier mechanism for this coordination is a hierarchy of TFs. Conceptually, such a hierarchy is not difficult to understand. Let us suppose that six operons $O_1, \ldots, O_6$ need to be called up for a proper response and that they are under the control of one transcription factor each, which we just call $TF_1, \ldots, TF_6$ for simplicity. These TFs are proteins, which are coded for by their own genes $G_1, \ldots, G_6$. If these genes are under the control of a TF at the next level of the hierarchy, say TF_A, then a single manipulation of TF_A will result in expression changes of all six target operons.

A complex example of this hierarchy is the control of processes converting sugars into the amino acids lysine and glutamate in *Corynebacterium glutamicum* [75] (**Figure 6.21**). Another impressive example is the gene regulatory network controlling metabolism in *E. coli*, which consists of about 100 TF genes that control about 40 individual regulatory modules. These modules can be triggered by various external conditions and organize proper responses and control the expression of about 500 metabolic genes that code for enzymes [76, 77].

The hierarchical and context-specific mode of TF control is not restricted to bacteria, but can be found throughout evolution. An intriguing example is the analysis of transcriptional networks that regulate transitions between physiological and pathological states. For instance, the activation of genes in malignant brain tumors was found to be governed by about 50 TFs, of which five regulate three-quarters of the brain tumor genes as activators and one as repressor. Two TFs serve as master regulators of the system [78]. Algorithms have been developed to study these types of hierarchical TF networks [79].

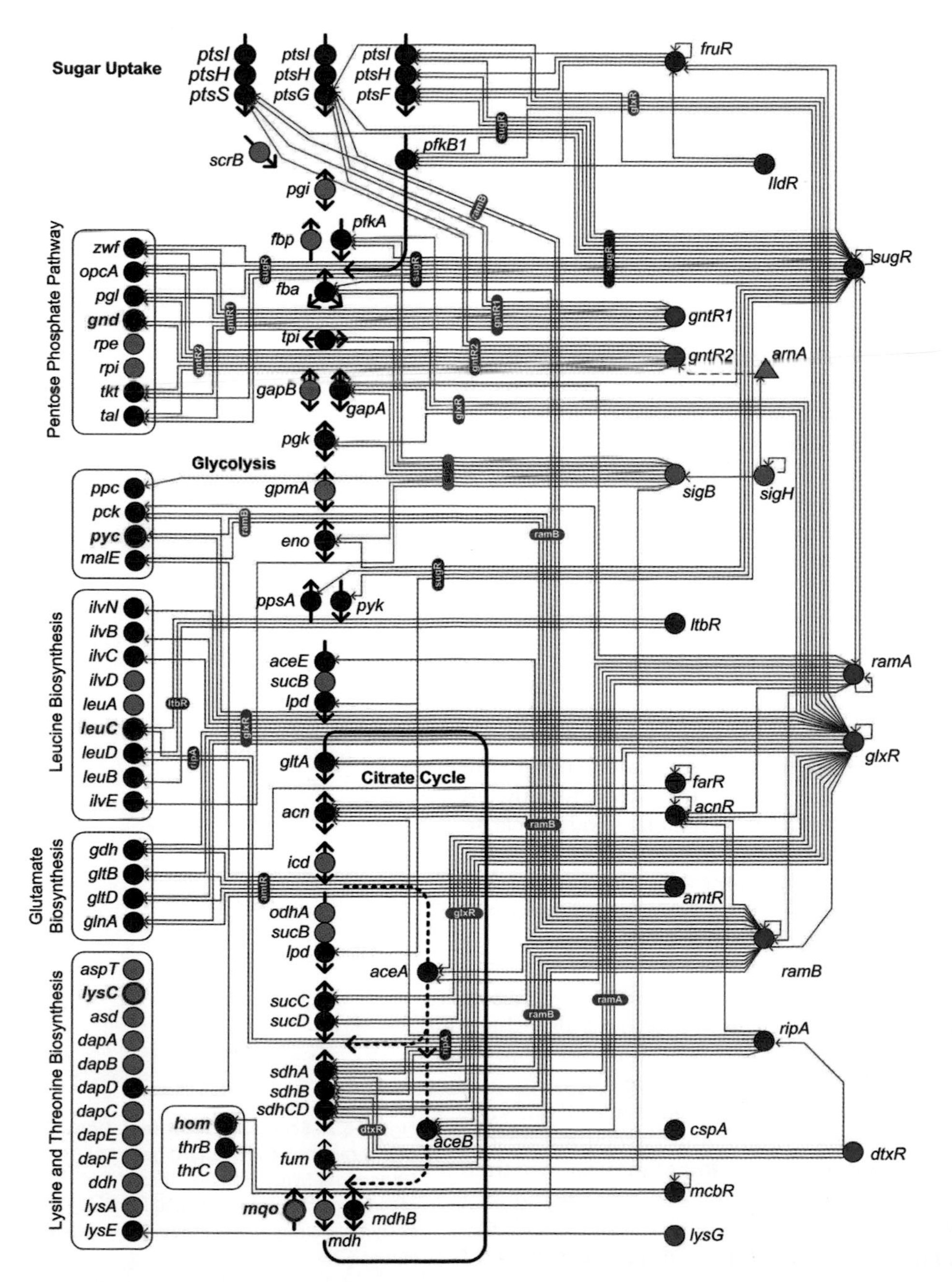

Figure 6.21 Transcription is often hierarchically controlled. Shown here is the hierarchy of transcription regulators involved in the conversion of glucose, fructose, and sucrose into the amino acids lysine and glutamate in *C. glutamicum*. Pertinent metabolic pathways are indicated as black arrows with corresponding gene names. (From Brinkrolf K, Schröder J, Pühler A & Tauch A. *J. Biotechnol.* 149 [2010] 173–182. With permission from Elsevier.)

The responses of TFs to environmental stimuli are typically achieved through small molecules that ultimately act as inducers for the TF genes. For example, a particular sphingolipid in yeast, which otherwise has no recognized function, was found to regulate several dozen genes associated with the response to heat stress by influencing 11 TFs positively and 2 negatively [3]. Moreover, groups of sphingolipids were associated with the expression of clusters of genes with coherently related functions.

The alteration of TFs in response to environmental stimuli is not a one-time, all-or-nothing event. Instead, TFs can be altered differentially with respect both to magnitude and to time periods following a stimulus. For instance, in the sphingolipid-induced change in TFs, clear time domains for different groups were distinguishable, with some groups being turned on and others turned off about 10–15 minutes after the beginning of heat stress [3].

Numerous other mechanisms can regulate the activity of TFs; for a detailed description, see [80]. Cells can sequester the regulator of some physiological function if this function is not needed at the time. For instance, if a sugar like maltose is not available in the medium, the regulator controlling the utilization of maltose is disabled. Mechanisms such as phosphorylation can control the activity of TFs spatially and temporally. Similarly, it is possible to fine-tune the cellular concentration of a TF through its expression and degradation. Finally, a TF or regulator may be translocated within the cell, for instance from the cytosol to the membrane in prokaryotes or between the nucleus and the cytosol in eukaryotes, thereby changing its activity state.

6.14 Models of Gene Regulation

Gene regulation may be approached with distinct computational modeling techniques. Different types of examples can be found in [79, 81–85] and excellent detailed reviews are provided in [86–88]. Generically, the approaches fall into four categories.

First, one may look at the structure of a gene interaction network, which is assessed as static, that is, the focus is on a fixed connectivity among genes under defined conditions. We discussed graph methods and methods such as Bayesian network inference in Chapter 3; see also the discussion of Boolean methods in Chapter 9. In this type of approach, the roles of TFs and their interactions are only implicit, although they are crucial, because genes do not interact directly, but by means of their products, including proteins that serve as TFs or their modulators and small signaling molecules, such as calcium, cAMP, and sphingolipids. Large-scale gene interaction networks of thousands of genes and their interactions have been reconstructed with experimental and computational means for model species such as yeast and *E. coli* [89, 90].

A second approach is currently possible only with a much smaller scope. Here the goal is to analyze the dynamics of gene regulatory networks. The typical techniques involve various types of **differential equations**, including nonlinear, piecewise-linear, or power-law-based **ordinary differential equations (ODEs)**, as well as qualitative differential equations, which are abstractions of ODEs, and delay differential equations, which permit the modeling of processes that are not instantaneous. In the forward mode, the differential equation models are typically designed with mRNAs and proteins as variables. Sometimes, metabolites are represented as well, because they may affect gene expression as co-inhibitors or co-activators, and these modulating effects are often formalized as Hill or power-law functions. It is quite evident that the sky is the limit, and that one quickly arrives at complex systems as we discussed them in Chapter 4. Specific examples of models with different levels of sophistication have already been mentioned in the context of the *lac* operon.

The biological literature on gene regulatory networks often represents genes as nodes and gene–gene interactions as edges, as shown in **Figure 6.22** (see also Chapter 3). Although such representations might seem to depict clearly how one gene (or gene transcript) affects another, they do not provide a satisfactory basis for the design of mathematical models, since, as we have seen, gene–gene interactions are mediated through transcription factors and other modulators, which are absent from these diagrams. In particular, these representations do not indicate whether, for instance, a process reduces the transcription into mRNA or hastens the removal of an mRNA. A clearer representation of the two-gene regulatory system in Figure 6.22B uses the instructions from Chapter 2. It is shown in **Figure 6.23** and

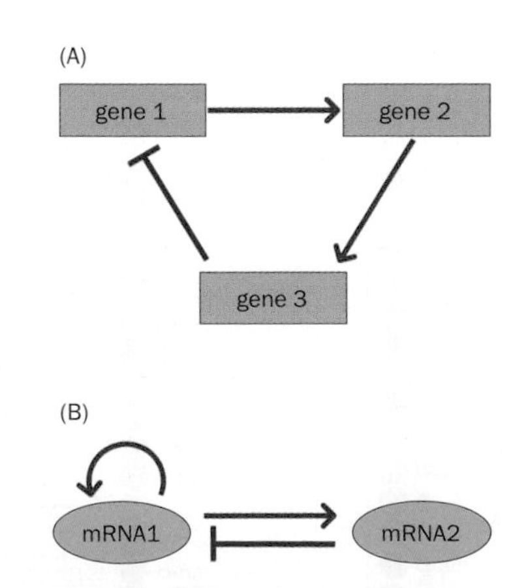

Figure 6.22 Schematic representations of interactions between (A) genes or (B) transcripts. While intuitive, these types of diagrams are not well suited for the design of mathematical models. See instead Figure 6.23.

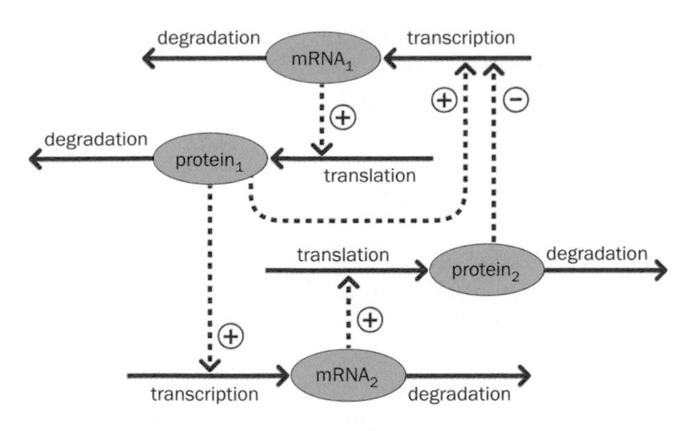

Figure 6.23 A representation of the diagram in Figure 6.22B that is better suited for designing a mathematical model. This representation appears to be more complicated, but it makes the processes involved explicit and immediately prescribes the set-up of model equations such as (6.1).

immediately suggests how to design a corresponding model. For example, equations in the following power-law format can be obtained directly from this diagram, and the theory behind these functions tells us that all parameter values are positive except for g_2:

$$
\begin{aligned}
(m\dot{R}NA_2) &= \alpha_1 P_1^{g_1} P_2^{g_2} - \beta_1 (mRNA_1)^{h_1}, \\
(m\dot{R}NA_2) &= \alpha_2 P_1^{g_3} - \beta_2 (mRNA_2)^{h_2}, \\
\dot{P}_1 &= \alpha_3 (mRNA_1)^{g_4} - \beta_3 P_1^{h_3}, \\
\dot{P}_2 &= \alpha_4 (mRNA_2)^{g_5} - \beta_4 P_2^{h_4},
\end{aligned}
\tag{6.1}
$$

In the reverse mode, ODE models have been used to infer gene interaction networks from expression data that had been recorded as time series (see, for example, [91] and Chapter 5). **Partial differential equation (PDE)** models have been used to account for reaction and diffusion processes within the cell [92].

The third class of models consists of **stochastic** formulations, which permit the modeling of variability in gene expression among individuals and scenarios [93] and can, for instance, model outcomes in systems where gene expression can toggle on and off in an apparently random fashion [94].

Finally, it is possible to use rule-based or **agent-based** models, which permit simulations of very diverse and complex scenarios, but are difficult to analyze mathematically [95]. We will discuss such approaches in Chapter 15.

MEASURING GENE EXPRESSION

When a gene is expressed, it is transcribed into mRNA, which may later be translated into protein. Thus, measuring gene expression means measuring the emergence of mRNA or protein. If protein is actually produced, its amount can be assessed with a variety of methods, including a **Western blot**. This name was chosen as a play on an earlier DNA test, which had been developed by Edwin Southern. In the Western blot technique, a cell culture or tissue sample is homogenized and proteins are extracted. The protein mixture is separated through gel electrophoresis based on molecular weight or other criteria. The most common variant, SDS-PAGE, uses buffers loaded with the detergent sodium dodecyl sulfate (SDS) for a polyacrylamide (PA) gel electrophoresis (GE). The SDS is negatively charged and covers the proteins, which therefore move through the gel toward a positively charged electrode. Furthermore, SDS-PAGE prevents polypeptides and proteins from forming three-dimensional structures (see Chapter 7), and therefore allows a separation of the proteins by molecular weight, because smaller proteins move through the acrylamide gel faster. The proteins in each band of the gel are transferred (blotted) to a membrane, where they are probed by the binding of specific antibodies.

Thousands of antibodies are available for this purpose. These antibodies can be detected in various ways. The most common detection method uses antibodies conjugated to enzymes (often horseradish peroxidase, HRP), which convert a substrate to a light-emitting product that can be detected and quantified. Technically, the primary antibody recognizes the protein of interest and is produced by a known species (for example, mouse). A secondary antibody, which is enzyme-linked, recognizes the constant region of the mouse antibody; an example is the goat anti-mouse IgG-HRP, which is a goat antibody that recognizes a mouse antibody and is linked to HRP. This construct allows for amplification of signal and the requirement for relatively few of the more expensive enzyme-linked antibodies. As an alternative to enzyme linkage, the antibodies can carry a fluorescent or radioactive marker that makes them detectable.

The antibody methods are semiquantitative, but calibration experiments can make them quantitative and therefore allow inferences regarding the expression of the genes that code for the measured proteins. The implicit assumption with respect to gene expression is that the target genes actually code for proteins and that the amounts of protein produced are proportional to the degree of gene expression.

An alternative to protein-based methods is the measurement of the relative abundance of the transcripts (that is, the mRNA molecules) that correspond to the target gene. The assumption in this case is that the number of mRNA molecules is (linearly) proportional to the degree of gene expression. For about a quarter century, the base technique for this purpose has been the **Northern blot** (**Figure 6.24A, B**). Again, this name was chosen in contrast to Southern's DNA test. The Northern blot procedure also uses gel electrophoresis, which in this case separates the mRNA molecules in a tissue sample based on their sizes. Subsequently, complementary RNA or DNA is applied. This RNA or DNA is labeled with a radioactive isotope such as ^{32}P or with a chemiluminescent marker. The probe is conjugated to an enzyme, which metabolizes the inactive chemiluminescent substrate to a product that emits light. If the labeled RNA or DNA finds a matching partner on the gel, it binds (hybridizes), thereby forming a double-stranded RNA or an RNA–DNA pair. After washing off unbound RNA or DNA, the chemiluminescence is detected. Alternatively, a radioactive label can be detected with autoradiography by exposing an X-ray film, which turns black when it receives radiation from the label. The degree of blackening is directly correlated with the amount of RNA, which makes the method quantitative. A database contains hundreds of blots for comparisons, searching, and identification [96].

A newer method of mRNA quantification is reverse transcription followed by a quantitative polymerase chain reaction (**RT-PCR**) (Figure 6.24C). In the reverse-transcription step, the mRNA is used to create a complementary DNA strand (**cDNA**) with an enzyme called reverse transcriptase. This cDNA does not correspond to the entire mRNA but consists only of a relatively short segment. The cDNA template is now amplified with quantitative real-time PCR (which unfortunately is also

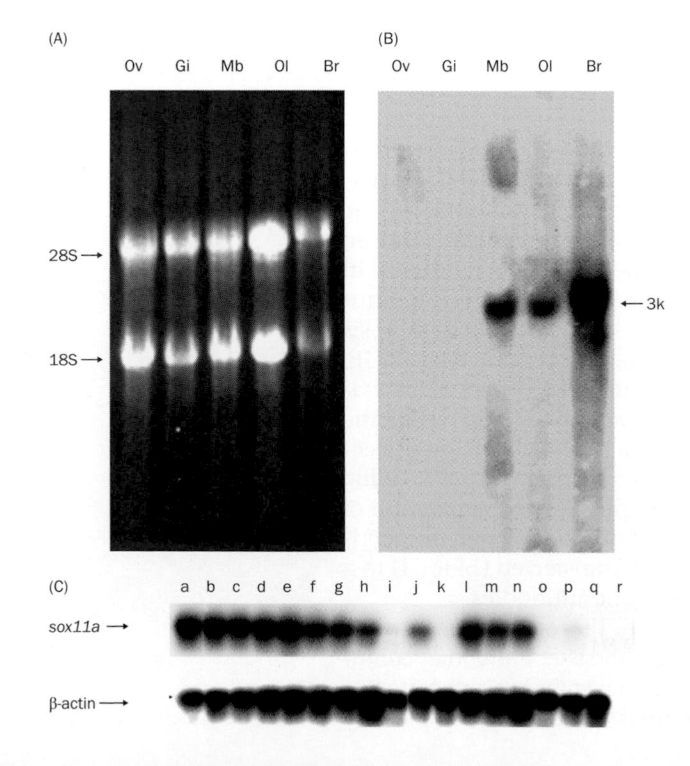

Figure 6.24 Blotting techniques for assessing gene expression. (A, B) Northern blots of the gene expression (mRNA) of the *sox11a* gene in ovary (Ov), gill filament (Gi), midbrain (Mb), olfactory bulb (Ol), and brain (Br) of the grouper: (A) separation on an agarose gel; (B) result of hybridization with labeled *sox11a* cDNA. (C) RT-PCR coupled with Southern blot of the gene expression (mRNA) of the *sox11a* gene in different tissues (a–q; r as a negative control) of the grouper. (From Zhang L, Zhu T, Lin D, et al. *Comput. Biochem. Physiol. B Biochem. Mol. Biol.* 157 [2010] 415–422. With permission from Elsevier.)

abbreviated RT-PCR, unless one includes "q" for "quantitative": RT-qPCR). The terminology "real-time" stems from the fact that one can use a fluorescent dye, which attaches only to double-stranded DNA, to monitor in real time how much DNA has been polymerized so far. qPCR is very sensitive and, once standards are established, very accurate. The overall result of RT-qPCR is the number of mRNAs per cell or in some small, defined volume. Alternative methods are described in [16].

Northern blots and RT-PCR are useful for accurate measurements of a few mRNAs. For larger tasks, other methods have been developed. One is serial analysis of gene expression (SAGE) [16, 97, 98]. This method permits rather exact measurements of both known and unknown transcripts. It begins with the isolation of mRNA from an input sample, such as some tissue. The mRNA is immobilized in a chromatography column and converted into cDNA, which is cut with restriction enzymes into short chains of 10–15 nucleotides (the so-called **expressed sequence tag, EST**). The small cDNA chains are assembled to form long DNA molecules, called concatamers, which are replicated by insertion into bacteria and subsequently sequenced. Computational analysis identifies which mRNAs were available in the original sample and how abundant they were. In contrast to most other methods, SAGE does not include an artificial hybridization step. The method is rather accurate and does not require prior knowledge of gene sequences or their transcripts.

The second class of methods for large-scale gene expression analysis uses **microarrays** or **DNA chips** (**Figure 6.25**). These are cheaper than SAGE and permit larger numbers of genes to be tested. However, quantification of the results is more difficult and less accurate than in SAGE. Microarrays and DNA chips both use hybridization and are similar in concept, but differ significantly in the specific type of DNA they employ. In both cases, thousands of tiny amounts of different DNA sequences, called probes, are spotted onto defined positions on a glass, plastic, or silicon wafer or a nylon membrane, where they are immobilized. The sample of mRNAs, characterizing the **transcriptome** of the cells of interest, is converted into a mixture of cDNAs (called the targets), labeled with a fluorescent dye, and applied to the chip or microarray. Wherever the cDNAs find matching partners among the spotted DNA probes, they hybridize, forming double-stranded DNA. After washing, which removes unbound or loosely bound cDNAs, the chip or microarray is scanned with a laser that detects the fluorescence of the bound cDNAs. The more cDNA is bound to a spot, the higher will be the intensity of fluorescence. The intensities are computationally converted into colors that reflect the up- and down-regulation states of genes [17]. The truly attractive feature of microarrays and DNA chips is that thousands of genes can be tested simultaneously.

In the case of a microarray, the DNA sequences are either synthetic short stretches of DNA, called oligonucleotides, or other short DNAs, such as cDNAs or products of a polymerase chain reaction. Because the amounts of DNA involved are

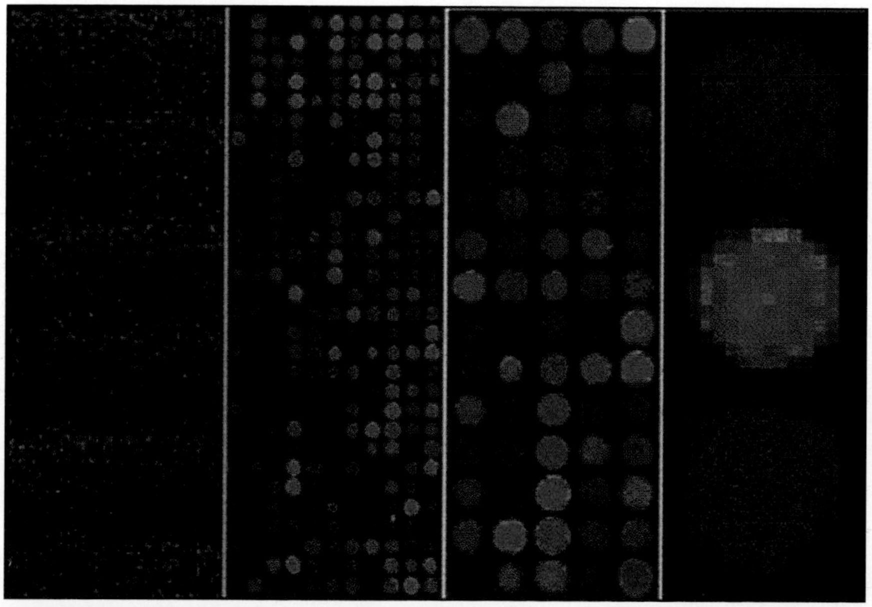

Figure 6.25 Different magnifications of a cDNA microarray slide. The microarray allows the profiling of changes in gene expression, here associated with chemotherapy of a human brain tumor. Each spot is associated with a particular gene. Different colors represent differences in gene expression between healthy and tumor cells. Green represents healthy control DNA, red represents sample DNA associated with disease, while yellow represents a combination of the two. Black areas correspond to spots where neither the control nor the sample hybridized to the target DNA. (From Bredel M, Bredel C & Sikic BI. *Lancet Oncol.* 5 [2004] 89–100. With permission from Elsevier.)

very small, the spotting of DNA on the chip is done by a robot, the intensity of fluorescence is measured with a laser scanner, and the results are interpreted with computer software and with specialized statistics. Microarrays allow the characterization of RNA populations through very large numbers of spots.

A particularly intriguing option is the two-color microarray, which is used for the comparison of DNA content in two samples, such as healthy cells and corresponding cancer cells. For this method, cDNAs are prepared from the two cell types and labeled with one of two variants of fluorescent dyes that emit light at different wavelengths. The two cDNA samples are mixed and hybridized to the same microarray. Laser scanning measures the relative intensities of the two types of fluorescence, and these intensities can be computationally converted into colors that indicate the up- and down-regulation states of genes [17].

In the case of DNA chips, the probes are constructed artificially directly on the chip by adding nucleotides, one by one, to growing chains of oligonucleotides in the different spots of the chip. Using sophisticated methods of photolithography and nucleotides that are light-activated, the resulting oligonucleotides in the different spots have ultimately genuine, defined sequences. DNA chips permit an astonishing density of up to 300,000 oligonucleotides per square centimeter, which allows very detailed analyses, for instance, of single nucleotide changes in a DNA sequence.

Many variations on these types of methods have been developed; the interested reader is referred to [16] and a rich body of literature. Each new method faces its own technical challenges and has advantages and drawbacks. The technical challenges and complications sometimes correlate with the quality of the output data. For instance, gene expression levels measured with the first microarrays only had an accuracy of a few fold, and apparent changes in expression of 20% or 50% had to be considered with caution. The accuracy of these methods has been improved quite a bit and is continuing to do so. Large, rapidly growing collections of gene expression data include the Gene Expression Omnibus [99], the Gene Expression Atlas [100], and numerous organism- and tissue-specific databases. Collectively, these databases contain information on several hundred thousand genes.

Newer alternatives for measuring gene-expression levels, which may eventually replace microarrays, include emerging sequencing-based technique such as deep sequencing, which records the number of times a particular base is read. This deep sequencing of a transcriptome, also called **RNA-Seq** (pronounced RNA-seek), reveals the sequence as well as the frequency with which particular RNAs are found. RNA-Seq has several advantages over hybridization-based techniques, including independence from the requirement of annotation, improved sensitivity, a wider dynamic range, and an assessment of the frequencies of particular RNA segments and of the percentage of the genome that was actually covered [101].

LOCALIZATION OF GENE EXPRESSION

In some systems, not only the degree of gene regulation, but also its localization within an organism or even within a single cell is of interest. To determine localization patterns of gene expression in space and time, one can again focus either on mRNAs or on proteins that correspond to the target genes. Proteins can be detected with labeled antibodies or by inserting the gene for the green fluorescent protein (**GFP**) into the genome of the host organism in such a manner that it is under the same control as the target gene. Not only that, but the DNA sequence of GFP can be inserted in frame with another protein-coding region, resulting in a GFP-tagged protein that renders it possible to follow the localization and dynamics of a protein (see Chapter 7). Both genes are expressed together, and GFP permits visualization of their location. mRNAs can also be visualized, because they contain specific localization elements or zip codes that are recognized by corresponding RNA-binding proteins, which can be tagged with a fluorescent marker such as tyramide.

As a particularly impressive example, different sets of genes are expressed in specific locations and to specific degrees during development. Beautiful demonstrations are presented in **Figures 6.26** and **6.27**, which show the localization of different mRNAs and the collocation of mRNAs and proteins during the early embryonic development of the fruit fly *Drosophila* [102, 103].

Figure 6.26 Many mRNAs in *Drosophila* display striking subcellular localization patterns during development. In this high-resolution fluorescence image, nuclei are shown in red and mRNAs in green. The head of the embryo will form on the left side of each image and the top will become its back. (From Martin KC & Ephrussi A. *Cell. Mol. Life Sci.* 136 [2009] 719–730. With permission from Elsevier.)

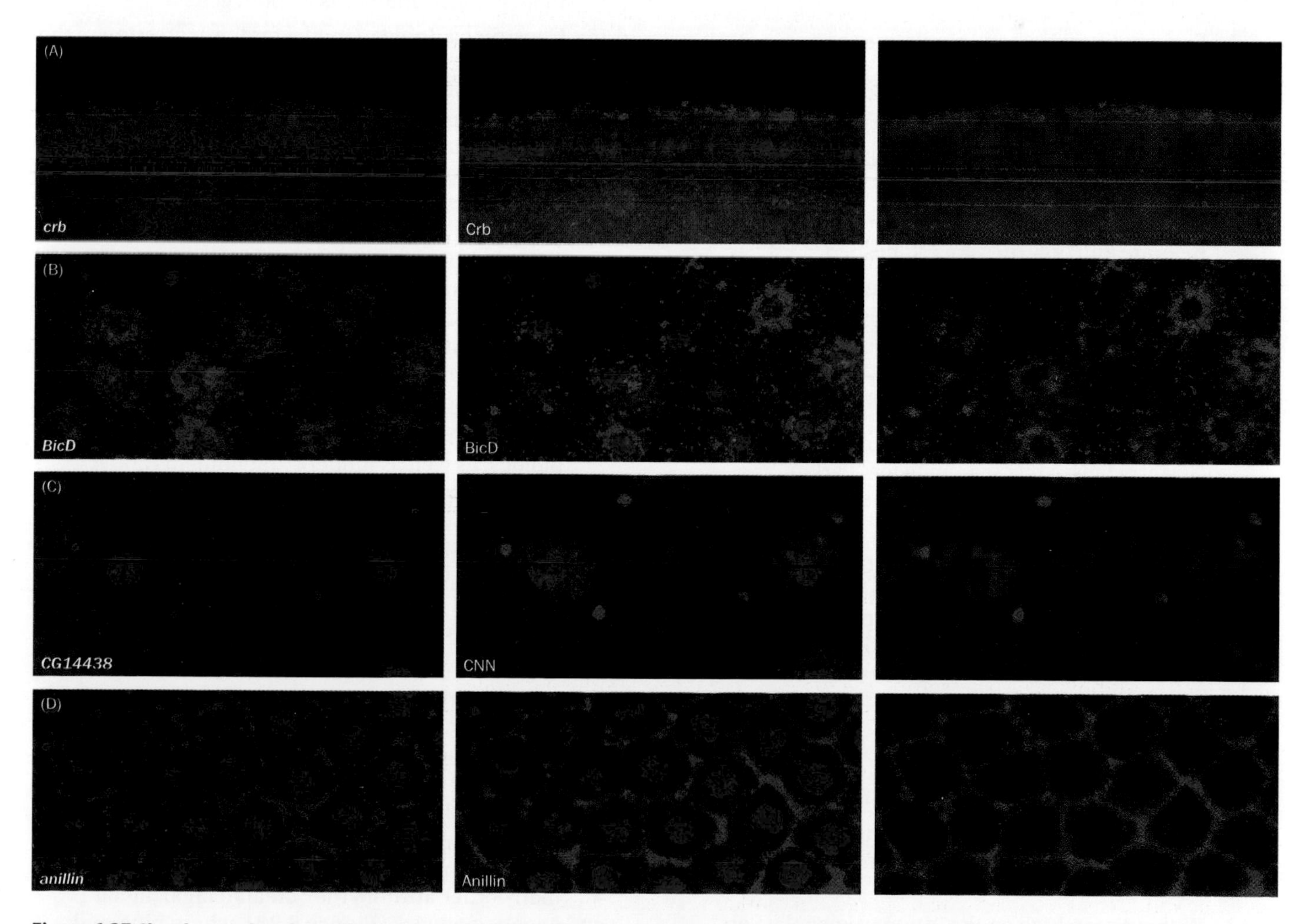

Figure 6.27 Closely correlated distribution patterns (right column) of mRNAs (left column) and proteins (center column) during early embryogenesis of the fruit fly *Drosophila*. The images were obtained using double-staining for selected mRNAs, proteins, or relevant markers. (A) mRNAs/proteins located at the apex. (B, C) mRNAs/proteins associated with the cell division apparatus. (D) mRNAs/proteins residing at cell junctions. Nuclei in the left and middle panels are shown in blue. Closer analysis demonstrated that mRNAs precede the appearance of the corresponding proteins. (From Lécuyer E, Yoshida H, Parthasarathy N, et al. *Cell. Mol. Life Sci.* 131 [2007] 174–187. With permission from Elsevier.)

OUTLOOK

Genomes have been and will continue to be a very exciting area of biological research. The developments in the field are often mind-boggling in their scope and speed, and yet one might expect that we have so far only scratched the surface. In the past, most gene studies have focused either on individual genes or on static gene regulatory networks. The field has now begun to move toward the roles of small RNAs, which comprise much more crucial and sophisticated regulatory systems than anyone had expected, and toward other means of regulating and controlling gene expression throughout the genome, so that the right gene is transcribed to the right degree, at the right time, and in the right location. The regulation and dynamics of these networks, and the localization and integration of their sub-networks, will without doubt be a cornerstone of experimental and computational systems biology throughout the foreseeable future.

EXERCISES

6.1. Identify and describe examples of overlapping genes.

6.2. Collect information on biological materials other than DNA that can be passed from one generation to the next.

6.3. Explore differences and similarities between DNA transposons and retrotransposons. Summarize your findings in a report.

6.4. GC base pairs contain three hydrogen bonds and AT pairs only two. Find out how much stronger GC base pairs are than AT base pairs.

6.5. Discuss in a two-page report what phylogenetic trees are and how they are established.

6.6. Gregor Mendel is sometimes called the "Father of Genetics." Summarize his major accomplishments in a one-page report.

6.7. Use Gene Ontology or some other web resource to determine the function of gene Hs.376209.

6.8. Explain why a search for "Parkinson's disease" as a GO term fails, while searching for "Parkin" as a "gene or protein" leads to genes associated with the disease.

6.9. Obtain information on several variations on QTL, called eQTL (expression QTL) and pQTL (which unfortunately can mean phenotypic, physiological, or protein QTL). Describe their advantages and drawbacks over traditional QTL.

6.10. How many DNA base pairs in humans are constituents of coding genes?

6.11. Compare genetic and epigenetic inheritance. Summarize your findings in a report.

6.12. Find out how cells attempt to ensure that genes are correctly translated into mRNA and proteins. Write a brief report.

6.13. Without searching on the Internet, speculate on how it is possible that only 1000 miRNAs can specifically target half of all human genes. Afterwards, check in the literature and the web how close your speculations were to reality.

6.14. Establish a list of diseases that are associated with RNA viruses.

6.15. Determine which cells in the human body do not contain the same genes as most other cells. Are there living human cells without genes?

6.16. Construct a Boolean model for the expression of the structural genes of the *lac* operon. For this purpose, look in the literature or on the Internet for the basics of Boolean logic. Then assign to lactose and glucose each a value of 0 if they are absent or 1 if they are present in the environment and use a combination of multiplication and addition to construct a formula for the "on" (1) and "off" (0) states of gene expression.

6.17. Determine the implications of mitochondrial DNA for studies of the evolution of the human race. Explain why studies of human evolution pay particular attention to genes located on the Y chromosome.

6.18. Discuss the original Central Dogma of molecular biology, as well as modern amendments and refinements.

6.19. Search the literature for a study where the connectivity of a gene network was inferred with statistical methods from observation (or artificial) data. Write a one-page summary of methods, results, and shortcomings.

6.20. Search the literature for a study where the connectivity of a gene network was inferred from observation (or artificial) data with methods based on differential equations. Write a one-page summary of methods, results, and shortcomings.

6.21. Kimura and collaborators described a method for inferring the structure of genetic networks. Like many other authors they created an artificial genetic network (**Figure 6.28**), where blue and red arrows are supposed to indicate "positive" and "negative regulations." What does "regulation" in this case mean? What would be its biological basis?

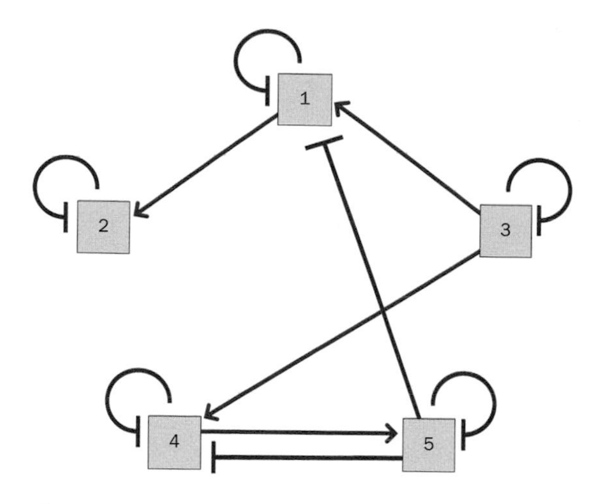

Figure 6.28 Artificial genetic network, used for testing an inference algorithm. (Adapted from Kimura S, Sonoda K, Yamane S, et al. *BMC Bioinformatics* 9 [2008] 23. With permission from Biomed Central.)

6.22. Discuss different options for developing a mathematical model for the gene network in Figure 6.28. Set up a mathematical model in symbolic form. What kinds of data would you need to estimate parameter values for a mathematical model? What type of algorithm would you need for the estimation?

6.23. Read the review of models for gene regulation by Schlitt and Brazma [87] and write a brief report about the use of difference models in the field.

6.24. Suppose two genes indirectly regulate each other's expression in the fashion shown in **Figure 6.29**, where green arrows indicate direct or indirect activating effects and the red arrow indicates repression. A possible model for this system has the form

$$\dot{X}_1 = \alpha_1 X_1^{0.4} X_2^{-0.15} - X_1^{0.2},$$
$$\dot{X}_2 = X_1 - X_2^{0.2}.$$

Begin by setting α_1 and the initial values equal to 1. Study the responses of this system for different values of the parameter α_1. Argue whether α_1 may be positive or negative. Alter the kinetic orders and study the effects. Summarize your findings in a report.

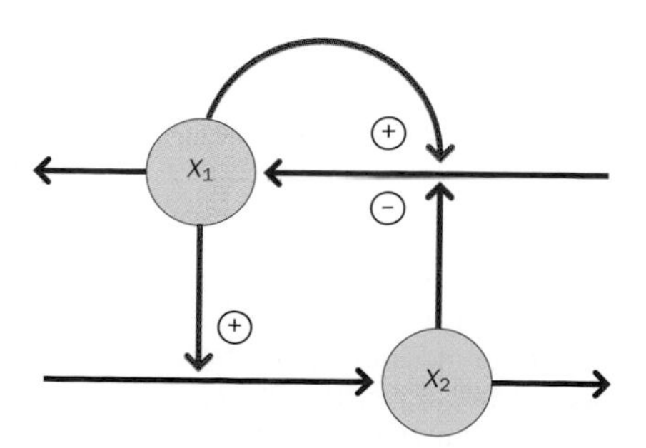

Figure 6.29 Artificial system in which two genes indirectly regulate each other's expression.

6.25. Implement a model for the artificial gene network in **Figure 6.30**. Begin with the model

$$\dot{G}_1 = \alpha_1 TF\, G_2 - \beta_1 G_1^{-0.5},$$
$$\dot{G}_2 = \alpha_2 G_1^{-1} - \beta_2 G_2^{0.5},$$

and set all rate constants, as well as the initial values and the value of the transcription factor *TF*, equal to 1. Initiate the system with different initial values. Second, study the effects of changes in the value of *TF*. Finally, study how the systems' dynamics changes with alterations in parameter values, including the negative kinetic orders associated with G_1.

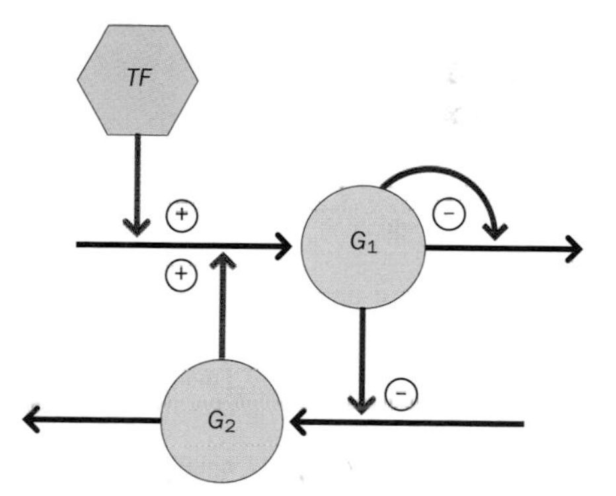

Figure 6.30 Artificial gene network. Although the system consists of just two genes and one transcription factor, it can generate interesting dynamics.

6.26. Set up an ODE model for the simple metabolic pathway shown in **Figure 6.31A**, using power-law functions (see Chapter 4). Choose kinetic orders between 0.2 and 0.8 and adjust the rate constants to your liking. Study the responses of the pathway to changes in the input. In the second phase of the exercise, use the same system, but augment it to reflect the system in Figure 6.31B, where the end product *Z* affects the production of a transcription factor *TF*, which turns on a gene *G*, which codes for an enzyme *E*, which catalyzes the conversion of *X* into *Y*. Choose appropriate parameter values to your liking. Before you simulate, make predictions of the effects of the addition of *TF*, *G*, and *E*.

6.27. Many studies explicitly or implicitly assume that a gene is transcribed into mRNA and that the mRNA is translated into protein, so that there is a direct linear correlation between gene expression and protein prevalence. Search the literature and the Internet for information that supports or refutes the biological validity of this assumption. Discuss the implications of this assumption for the design of models describing transcription–translation processes.

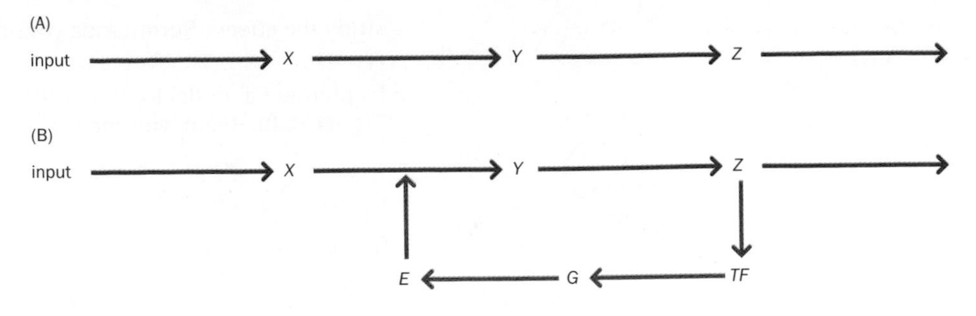

Figure 6.31 Pathways with and without genomic feedback. The responses of the system in (A) to changes in the input are easy to predict, but the same is not true for the system in (B).

6.28. Explore the details of a Southern blot for DNA analysis and summarize the highlights in a one-page report.

6.29. Find out how one typically prevents an RNA from forming a secondary structure, which would hinder hybridization.

6.30. Explore which steps in a SAGE, microarray, or DNA chip experiment cause the greatest uncertainties in results. Compare the degrees of accuracy attainable with the three methods.

REFERENCES

[1] Crick FHC. On protein synthesis. *Symp. Soc. Exp. Biol.* 12 (1956) 139–163. [An early draft of an article describing the Central Dogma. Also available at http://profiles.nlm.nih.gov/ps/access/SCBBFT.pdf.]

[2] Crick FHC Central dogma of molecular biology. *Nature* 227 (1970) 561–563.

[3] Cowart LA, Shotwell M, Worley ML, et al. Revealing a signaling role of phytosphingosine-1-phosphate in yeast. *Mol. Syst. Biol.* 6 (2010) 349.

[4] Schuit F, Flamez D, De Vos A & Pipeleers D. Glucose-regulated gene expression maintaining the glucose-responsive state of β-cells. *Diabetes* 51 (2002) S326–S332.

[5] Merino E, Jensen RA & Yanofsky C. Evolution of bacterial *trp* operons and their regulation. *Curr. Opin. Microbiol.* 11 (2008) 78–86.

[6] Rock F, Gerrits N, Kortstee A, van Kampen M, Borrias M, Weisbeek P & Smeekens S. Sucrose-specific signalling represses translation of the *Arabidopsis ATB2* bZIP transcription factor gene. *Plant J.* 15 (1988) 253–263.

[7] Elgar G & Vavouri T. Tuning in to the signals: noncoding sequence conservation in vertebrate genomes. *Trends Genet.* 24 (2008) 344–352.

[8] Noble D. Claude Bernard, the first systems biologist, and the future of physiology. *Exp. Physiol.* 93 (2008) 16–26.

[9] Breton S, Beaupre HD, Stewart DT, et al. The unusual system of doubly uniparental inheritance of mtDNA: Isn't one enough? *Trends Genet.* 23 (2007) 465–474.

[10] Siefert JL. Defining the mobilome. *Methods Mol. Biol.* 532 (2009) 13–27.

[11] Keeling PJ. Functional and ecological impacts of horizontal gene transfer in eukaryotes. *Curr. Opin. Genet. Dev.* 19 (2009) 613–619.

[12] Nikaido H. Multidrug resistance in bacteria. *Annu. Rev. Biochem.* 78 (2009) 119–146.

[13] Tang YW & Stratton CW. *Staphylococcus aureus*: an old pathogen with new weapons. *Clin. Lab. Med.* 30 (2010) 179–208.

[14] Sinzelle L, Izsvak Z & Ivics Z. Molecular domestication of transposable elements: from detrimental parasites to useful host genes. *Cell. Mol. Life Sci.* 66 (2009) 1073–1093.

[15] Batzer MA & Deininger PL. *Alu* repeats and human genomic diversity. *Nat. Rev. Genet.* 3 (2002) 370–379.

[16] Brown TA. Genomes, 3rd ed. Garland Science, 2006.

[17] Zvelebil M & Baum JO. Understanding BioInformatics. Garland Science, 2007.

[18] Alberts B, Johnson A, Lewis J, et al. Molecular Biology of the Cell, 6th ed. Garland Science, 2015.

[19] Human Genome Project (HGP). Insights Learned from the Human DNA Sequence. https://www.genome.gov/12011238/an-overview-of-the-human-genome-project/.

[20] Buehler LK & Rashidi HH (eds). Bioinformatics Basics: Applications in Biological Science and Medicine, 2nd ed. CRC Press, 2005.

[21] Maddox B. The double helix and the wronged heroine. *Nature* 421 (2003) 407–408.

[22] Watson JD & Crick FHC. A structure for deoxyribose nucleic acid. *Nature* 171 (1953) 737–738.

[23] Gregory TR & Hebert PD. The modulation of DNA content: proximate causes and ultimate consequences. *Genome Res.* 9 (1999) 317–324.

[24] Fugu Genome Project. http://www.fugu-sg.org/.

[25] Black DL. Mechanisms of alternative pre-messenger RNA splicing. *Annu. Rev. Biochem.* 72 (2003) 291–336.

[26] Wu H & Moore E. Association analysis of the general environmental conditions and prokaryotes' gene distributions in various functional groups. *Genomics* 96 (2010) 27–38.

[27] Vaidyanathan S & Goodacre R. Metabolome and proteome profiling for microbial characterization. In Metabolic Profiling: Its Role in Biomarker Discovery and Gene Function Analysis (GG Harrigan & R Goodacre, eds), pp 9–38. Kluwer, 2003.

[28] Durban R, Eddy S, Krogh A & Mitchison G. Biological Sequence Analysis: Probabilistic Models of Proteins and Nucleic Acids. Cambridge University Press, 1998.

[29] Zhou F, Olman V & Xu Y. Barcodes for genomes and applications. *BMC Bioinformatics* 9 (2008) 546.

[30] The Gene Ontology Project. http://www.geneontology.org/.

[31] Da Costa e Silva L & Zeng ZB. Current progress on statistical methods for mapping quantitative trait loci from inbred line crosses. *J. Biopharm. Stat.* 20 (2010) 454–481.

[32] Mittler R & Blumwald E. Genetic engineering for modern agriculture: challenges and perspectives. *Annu. Rev. Plant Biol.* 61 (2010) 443–462.

[33] van Eeuwijk FA, Bink MC, Chenu K & Chapman SC. Detection and use of QTL for complex traits in multiple environments. *Curr. Opin. Plant Biol.* 13 (2010) 193–205.

[34] Singh V, Tiwari RL, Dikshit M & Barthwal MK. Models to study atherosclerosis: a mechanistic insight. *Curr. Vasc. Pharmacol.* 7 (2009) 75–109.

[35] Zhong H, Beaulaurier J, Lum PY, et al. Liver and adipose expression associated SNPs are enriched for association to type 2 diabetes. *PLoS Genet.* 6 (2010) e1000932.

[36] Botstein D, White RL, Skolnick M & Davis RW. Construction of a genetic linkage map in man using restriction fragment length polymorphisms. *Am. J. Hum. Genet.* 32 (1980) 314–331.

[37] Weedon MN, Lango H, Lindgren CM, et al. Genome-wide association analysis identifies 20 loci that influence adult height. *Nat. Genet.* 40 (2008) 575–583.

[38] Weiss KM. Tilting at quixotic trait loci (QTL): an evolutionary perspective on genetic causation. *Genetics* 179 (2008) 1741–1756.

[39] Ohno S. So much "junk" DNA in our genome. *Brookhaven Symp. Biol.* 23 (1972) 366–370.

[40] Wheeler DA, Srinivasan M, Egholm M, et al. The complete genome of an individual by massively parallel DNA sequencing. *Nature* 452 (2008) 872–876.

[41] DNA Data Bank of Japan (DDBJ). http://www.ddbj.nig.ac.jp/.

[42] National Center for Biotechnology Information (NCBI). Databases. http://www.ncbi.nlm.nih.gov/guide/all/.

[43] NCBI. GenBank. http://www.ncbi.nlm.nih.gov/genbank/.

[44] Xu Y & Gogarten JP (eds). Computational Methods for Understanding Bacterial and Archaeal Genomes. Imperial College Press, 2008.

[45] Arold ST, Leonard PG, Parkinson GN & Ladbury JE. H-NS forms a superhelical protein scaffold for DNA condensation. *Proc. Natl Acad. Sci. USA* 107 (2010) 15728–15732.

[46] Narlikar GJ. A proposal for kinetic proof reading by ISWI family chromatin remodeling motors. *Curr. Opin. Chem. Biol.* 14 (2010) 660–665.

[47] Huang Y-W, Kuo C-T, Stoner K, et al. An overview of epigenetics and chemoprevention. *FEBS Lett.* 585 (2011) 2129–2136.

[48] Yoo CB & Jones PA. Epigenetic therapy of cancer: past, present and future. *Nat. Rev. Drug Discov.* 5 (2006) 37–50.

[49] Chi YI, Martick M, Lares M, et al. Capturing hammerhead ribozyme structures in action by modulating general base catalysis. *PLoS Biol.* 6 (2008) e234.

[50] Nissen P, Hansen J, Ban N, et al. The structural basis of ribosome activity in peptide bond synthesis. *Science* 289 (2000) 920–930.

[51] Liu Q & Paroo Z. Biochemical principles of small RNA pathways. *Annu. Rev. Biochem.* 79 (2010) 295–319.

[52] Siomi H & Siomi MC. On the road to reading the RNA-interference code. *Nature* 457 (2009) 396–404.

[53] Napoli C, Lemieux C & Jorgensen R. Introduction of a chimeric chalcone synthase gene into petunia results in reversible co-suppression of homologous genes *in trans*. *Plant Cell* 2 (1990) 279–289.

[54] Deng W, Zhu X, Skogerbø G, et al. Organization of the *Caenorhabditis elegans* small noncoding transcriptome: genomic features, biogenesis, and expression. *Genome Res.* 16 (2006) 20–29.

[55] Pillai RS, Bhattacharyya SN & Filipowicz W. Repression of protein synthesis by miRNAs: How many mechanisms? *Trends Cell Biol.* 17 (2007) 118–126.

[56] Chen J, Wang L, Matyunina LV, et al. Overexpression of miR-429 induces mesenchymal-to-epithelial transition (MET) in metastatic ovarian cancer cells. *Gynecol. Oncol.* 121 (2011) 200–205.

[57] Gregory PA, Bracken CP, Bert AG & Goodall GJ. MicroRNAs as regulators of epithelial–mesenchymal transition. *Cell Cycle* 7 (2008) 3112–3118.

[58] Lai X, Bhattacharya A, Schmitz U, Kunz M, Vera J. & Wolkenhauer O. A systems' biology approach to study microRNA-mediated gene regulatory networks. *Biomed. Res. Int.* 2013 (2013) Article ID 703849.

[59] Lai X, Wolkenhauer O & Vera J. Understanding microRNA-mediated gene regulatory networks through mathematical modelling. *Nucleic Acids Res.* 44 (2016) 6019–6035.

[60] Mayer G. Bacteriology—Chapter Nine: Genetic regulatory mechanism. In Microbiology and Immunology On-line. University of South Carolina School of Medicine. http://www.microbiologybook.org/mayer/geneticreg.htm.

[61] Savageau MA. Design of molecular control mechanisms and the demand for gene expression. *Proc. Natl Acad. Sci. USA* 74 (1977) 5647–5651.

[62] Adam M, Murali B, Glenn NO & Potter SS. Epigenetic inheritance based evolution of antibiotic resistance in bacteria. *BMC Evol. Biol.* 8 (2008) 52.

[63] Jacob F, Perrin D, Sanchez C & Monod J. Operon: a group of genes with the expression coordinated by an operator. *C. R. Hebd. Séances Acad. Sci.* 250 (1960) 1727–1729.

[64] Wong P, Gladney S & Keasling J. Mathematical model of the *lac* operon: Inducer exclusion, catabolite repression, and diauxic growth on glucose and lactose. *Biotechnol. Prog.* 13 (1997) 132–143.

[65] Vilar JMG, Guet CC & Leibler S. Modeling network dynamics: the *lac* operon, a case study. *J. Cell. Biol.* 161 (2003) 471–476.

[66] Yildrim N & Mackey MC. Feedback regulation in the lactose operon: a mathematical modeling study and comparison with experimental data. *Biophys. J.* 84 (2003) 2841–2851.

[67] Tian T & Burrage K. A mathematical model for genetic regulation of the lactose operon. In Computational Science and its Applications—ICCSA 2005. Lecture Notes in Computer Science, Volume 3481 (O Gervasi, ML Gavrilova, V Kumar, et al., eds). pp 1245–1253. Springer-Verlag, 2005.

[68] Savageau MA. Demand theory of gene regulation. I. Quantitative development of the theory. *Genetics* 149 (1998) 1665–1676.

[69] Su Z, Li G & Xu Y. Prediction of regulons through comparative genome analyses. In Computational Methods for Understanding Bacterial and Archaeal Genomes (Y Xu & JP Gogarten, eds), pp 259–279. Imperial College Press, 2008.

[70] Yin Y, Zhang H, Olman V & Xu Y. Genomic arrangement of bacterial operons is constrained by biological pathways encoded in the genome. *Proc. Natl Acad. Sci. USA* 107 (2010) 6310–6315.

[71] RegulonDB Database. http://regulondb.ccg.unam.mx/.

[72] Che D, Li G, Mao F, et al. Detecting uber-operons in prokaryotic genomes. *Nucleic Acids Res.* 34 (2006) 2418–2427.

[73] Godon C, Lagniel G, Lee J, et al. The H_2O_2 stimulon in *Saccharomyces cerevisiae*. *J. Biol. Chem.* 273 (1998) 22480–22489.

[74] Igarashi K & Kashiwagi K. Polyamine modulon in *Escherichia coli*: genes involved in the stimulation of cell growth by polyamines. *J. Biochem.* 139 (2006) 11–16.

[75] Brinkrolf K, Schröder J, Pühler A & Tauch A. The transcriptional regulatory repertoire of *Corynebacterium glutamicum*: reconstruction of the network controlling pathways involved in lysine and glutamate production. *J. Biotechnol.* 149 (2010) 173–182.

[76] Covert MW, Knight EM, Reed JL, et al. Integrating high-throughput and computational data elucidates bacterial networks. *Nature* 429 (2004) 92–96.

[77] Samal A & Jain S. The regulatory network of *E. coli* metabolism as a Boolean dynamical system exhibits both homeostasis and flexibility of response. *BMC Syst. Biol.* 2 (2008) 21.
[78] Carro MS, Lim WK, Alvarez MJ, et al. The transcriptional network for mesenchymal transformation of brain tumours. *Nature* 463 (2010) 318–325.
[79] Basso K, Margolin AA, Stolovitzky G, et al. Reverse engineering of regulatory networks in human B cells. *Nat. Genet.* 37 (2005) 382–390.
[80] Martínez-Antonio A & Collado-Vides J. Comparative mechanisms for transcription and regulatory signals in archaea and bacteria. In Computational Methods for Understanding Bacterial and Archaeal Genomes (Y Xu & JP Gogarten, eds), pp 185–208. Imperial College Press, 2008.
[81] Plahte E, Mestl T & Omholt SW. A methodological basis for description and analysis of systems with complex switch-like interactions. *J. Math. Biol.* 36 (1998) 321–348.
[82] Gardner TS, di Bernardo D, Lorenz D & Collins JJ. Inferring genetic networks and identifying compound mode of action via expression profiling. *Science* 301 (2003) 102–105.
[83] Geier F, Timmer J & Fleck C. Reconstructing gene-regulatory networks from time series, knock-out data, and prior knowledge. *BMC Syst. Biol.* 1 (2007) 11.
[84] Nakatsui M, Ueda T, Maki Y, et al. Method for inferring and extracting reliable genetic interactions from time-series profile of gene expression. *Math. Biosci.* 215 (2008) 105–114.
[85] Yeung MK, Tegner J & Collins JJ. Reverse engineering gene networks using singular value decomposition and robust regression. *Proc. Natl Acad. Sci. USA* 99 (2002) 6163–6168.
[86] de Jong H. Modeling and simulation of genetic regulatory systems: a literature review. *J. Comput. Biol.* 9 (2002) 67–103.
[87] Schlitt T & Brazma A. Current approaches to gene regulatory network modelling. *BMC Bioinformatics* 8 (Suppl 6) (2007) S9.
[88] Le Novère N. Quantitative and logic modelling of molecular and gene networks. *Nat. Rev. Genet.* 16 (2015) 146–158.
[89] Tong AH, Lesage G, Bader GD, et al. Global mapping of the yeast genetic interaction network. *Science* 303 (2004) 808–813.
[90] Babu M, Musso G, Diaz-Mejia JJ, et al. Systems-level approaches for identifying and analyzing genetic interaction networks in *Escherichia coli* and extensions to other prokaryotes. *Mol. Biosyst.* 5 (2009) 1439–1455.
[91] Gasch AP, Spellman PT, Kao CM, et al. Genomic expression programs in the response of yeast cells to environmental changes. *Mol. Biol. Cell* 11 (2000) 4241–4257.
[92] Reinitz J & Sharp DH. Gene circuits and their uses. In Integrative Approaches to Molecular Biology (J Collado-Vides, B Magasanik & TF Smith, eds), pp. 253–272. MIT Press, 1996.
[93] McAdams HH & Arkin A. It's a noisy business! Genetic regulation at the nanomolar scale. *Trends Genet.* 15 (1999) 65–69.
[94] Gardner TS, Cantor CR & Collins JJ. Construction of a genetic toggle switch in *Escherichia coli. Nature* 403 (2000) 339–342.
[95] Zhang L, Wang Z, Sagotsky JA & Deisboeck TS. Multiscale agent-based cancer modeling. *J. Math. Biol.* 58 (2009) 545–559.
[96] Medicalgenomics. BlotBase. http://www.medicalgenomics.org/blotbase/.
[97] Sagenet. SAGE: Serial Analysis of Gene Expression. http://www.sagenet.org/.
[98] Velculescu VE, Zhang L, Vogelstein B & Kinzler KW. Serial analysis of gene expression. *Science* 270 (1995) 484–487.
[99] NCBI. GEO: Gene Expression Omnibus. http://www.ncbi.nlm.nih.gov/geo/.
[100] European Molecular Biology Laboratory–European Bioinformatics Institute (EMBL–EBI). ATLAS: Gene Expression Atlas. http://www.ebi.ac.uk/gxa/.
[101] Croucher NJ & Thomson NR. Studying bacterial transcriptomes using RNA-seq. *Curr. Opin. Microbiol.* 13 (2010) 619–624.
[102] Martin KC & Ephrussi A. mRNA localization: gene expression in the spatial dimension. *Cell. Mol. Life Sci.* 136 (2009) 719–730.
[103] Lécuyer E, Yoshida H, Parthasarathy N, et al. Global analysis of mRNA localization reveals a prominent role in organizing cellular architecture and function. *Cell. Mol. Life Sci.* 131 (2007) 174–187.

FURTHER READING

Brown TA. Genomes, 3rd ed. Garland Science, 2006.
Carro MS, Lim WK, Alvarez MJ, et al. The transcriptional network for mesenchymal transformation of brain tumours. *Nature* 463 (2010) 318–325.
de Jong H. Modeling and simulation of genetic regulatory systems: a literature review. *J. Comput. Biol.* 9 (2002) 67–103.
Durban R, Eddy S, Krogh A & Mitchison G. Biological Sequence Analysis: Probabilistic Models of Proteins and Nucleic Acids. Cambridge University Press, 1998.
Huang Y-W, Kuo C-T, Stoner K, et al. An overview of epigenetics and chemoprevention. *FEBS Lett.* 585 (2011) 2129–2136.
Noble D. The Music of Life; Biology Beyond Genes. Oxford University Press, 2006.
Schlitt T & Brazma A. Current approaches to gene regulatory network modelling. *BMC Bioinformatics* 8(Suppl 6) (2007) S9.
Zvelebil M & Baum JO. Understanding Bioinformatics. Garland Science, 2007.

Protein Systems

When you have read this chapter, you should be able to:

- Discuss types of proteins and their roles
- Describe the basic chemical properties of proteins
- Explain the four hierarchical levels of protein structure
- Retrieve from databases information about the crystal structure of proteins
- Outline concepts of protein separation and proteomic techniques
- Discuss the basic concepts of protein structure prediction and protein localization
- Describe interactions among proteins and between proteins and other molecules

Biological systems are complicated machines, and proteins supply most of their gears, gaskets, valves, sensors, and amplifiers. Indeed, proteins are the most versatile and fascinating molecules in living cells. Although all proteins have generically the same molecular composition, their sizes, functions, and roles could not be more diverse, and their importance for life is hard to overestimate. Proteins are constantly generated and destroyed, with some existing just for minutes or less, and others, such as the proteins in the lens of the human eye, persisting for an entire human life. Owing to their central importance and versatility, an astounding amount of information has been accumulated over the past two centuries, since the Swedish chemist Jöns Jacob Berzelius coined the term protein (from the Greek for "primary") in 1838, and uncounted databases for different aspects of proteins are now available (see, for example, [1–4]). Proteins have been the subject of intense investigation in biochemistry and protein chemistry; molecular, structural, and computational biology; metabolic engineering; bioinformatics; proteomics; systems biology; and synthetic biology. Yet, what we know today might just be the tip of the iceberg. Thus, this chapter will not come close to portraying all aspects of proteins in a just manner, and it will not even present much detail about relevant topics as far as they are covered comprehensively in texts on computational biology and bioinformatics (for example, [5]). Instead, it will focus on some of the main features of proteins and discuss those roles that are of particular importance for the dynamic functioning of biological systems. We begin with some chemical properties and the assembly of proteins according to the building instructions provided in the genome. A good starting point for exploring the known properties of a protein is the Human Protein Atlas [6].

CHEMICAL AND PHYSICAL FEATURES OF PROTEINS

Proteins are polymers composed of amino acids, which are joined into a linear string by **peptide bonds** (**Figure 7.1**). Each amino acid unit (or residue of the polymer) has a particular molecular side chain, denoted by R in Figure 7.1. The chemical properties of the side chains determine the features of the amino acids and collectively of the protein. Importantly, the side chains can interact with each other through different noncovalent mechanisms, such as hydrogen bonding and van der Waals forces. Furthermore, the spatial arrangements of amino acid residues are affected by hydrophobic effects, which tend to maximize the packing density of the protein, as well as by covalent links and ionic bonds, called salt bridges. The interactions between the side chains of distant amino acids of the same or a different protein lead to cross-linking and the formation of the **tertiary** (three-dimensional) **structure** of the protein (see later). They are also responsible for the stability of a protein. Of note is the most prevalent covalent link, namely, a disulfide bond between two cysteine residues, which is so important that the resulting dimeric (two-component) amino acid has its own name: cystine. Almost all of nature uses 20 L-amino acids (**Table 7.1**). Alas, as so often in nature, there are exceptions: a very few organisms make use of two additional amino acids, namely, selenocysteine and pyrrolysine [7, 8].

As we discussed in Chapter 6, proteins are generated through transcription, translation, and possible subsequent modifications from stretches of DNA within the organism's genome. Specifically, each amino acid within a protein corresponds to a DNA codon, consisting of a triplet of nucleotides within the genetic code, which is transcribed into a matching RNA codon. Because DNA consists of four nucleotides, $4 \times 4 \times 4 = 64$ different codons are possible. A few of these serve as start or stop signals for transcription, but more codons than necessary are left to code for the 20 amino acids, with the consequence that most amino acids are encoded by more than one codon (**Table 7.2**). In fact, in cases such as alanine, the third nucleotide of the codon is virtually, although not totally, immaterial. While different codons can thus represent the same amino acid, a phenomenon called wobble, the reverse is not true: every codon codes for one and only one amino acid, or, as a mathematician might say, the transcription–translation system implements a "many-to-one mapping." Nonetheless, the choice of a codon is apparently not random. For instance, different species often prefer particular codons over alternatives, a phenomenon

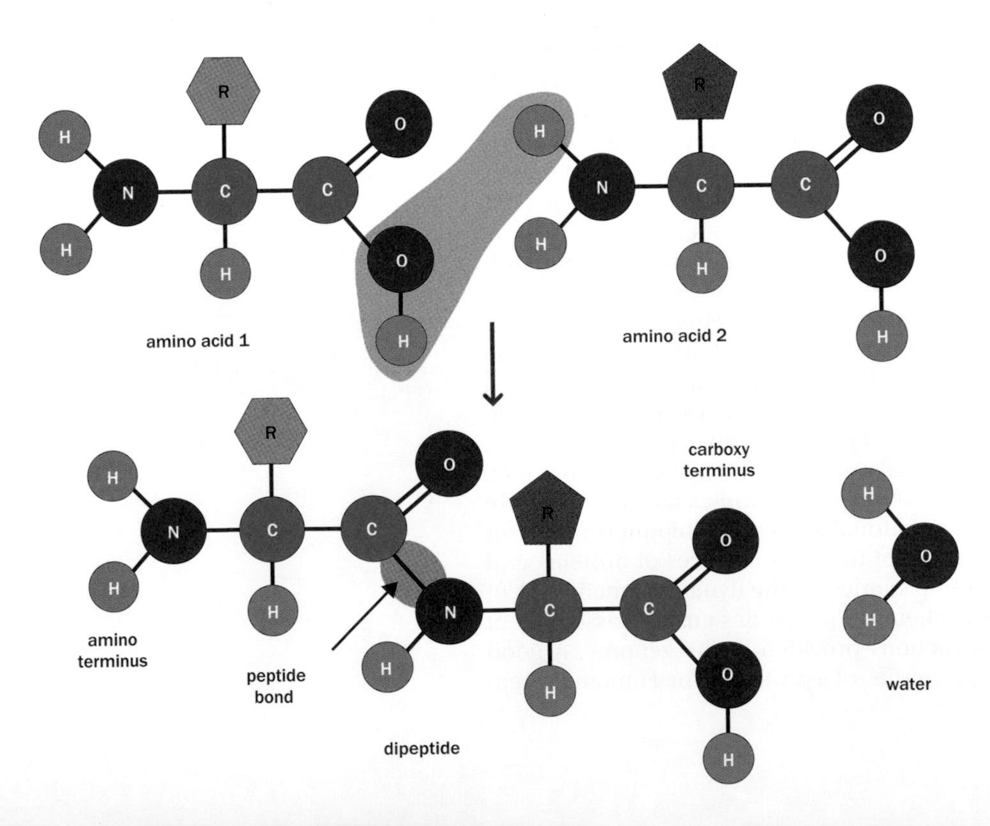

Figure 7.1 Generic representation of a peptide bond between two amino acids. R represents the specific side chain of each amino acid.

TABLE 7.1: AMINO ACIDS IN ALPHABETICAL ORDER WITH THEIR ABBREVIATIONS

The generic backbone is shown in blue. Color coding of the amino acid's name represents types of side chains: orange, hydrophobic; green, positively charged; red, negatively charged; yellow, polar uncharged; turquoise, special cases; purple, nonstandard amino acids.

Generic Amino Acid	Generic Amino Acid (simplified notation)	Alanine (A/Ala)	Arginine (R/Arg)	Asparagine (N/Asn)	Aspartic Acid (D/Asp)
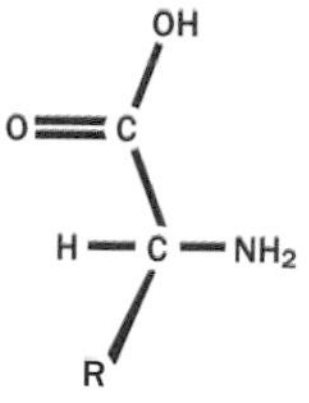 OH O=C H—C—NH_2 R	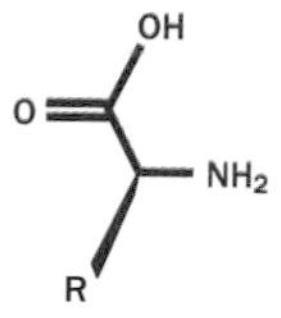 OH O NH_2 R	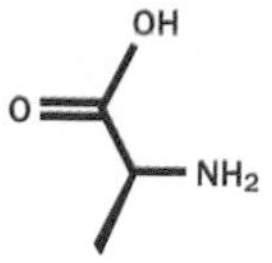 OH O NH_2	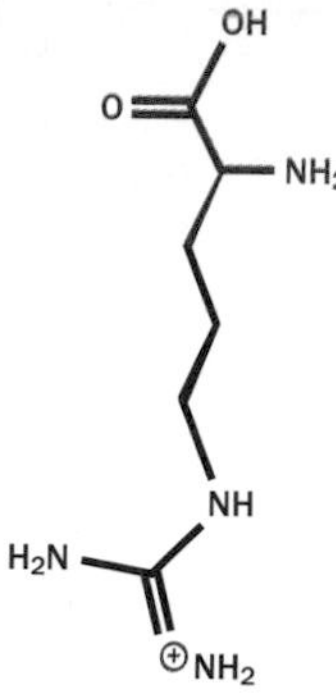 OH O NH_2 NH H_2N $^{\oplus}NH_2$	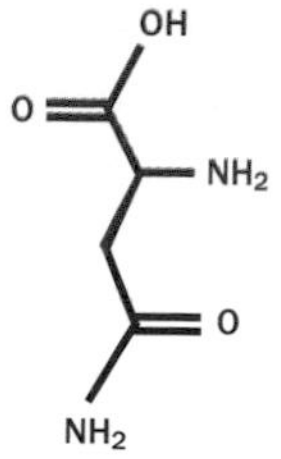 OH O NH_2 O NH_2	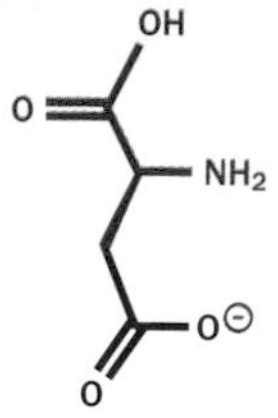 OH O NH_2 $O^{\ominus}$ O
Cysteine (C/Cys)	Glutamic Acid (E/Glu)	Glutamine (Q/Gln)	Glycine (G/Gly)	Histidine (H/His)	Isoleucine (I/Ile)
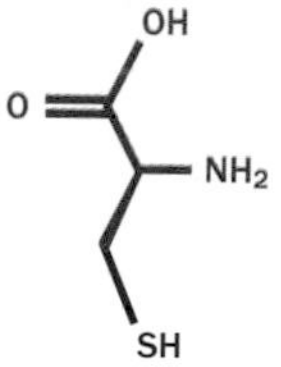 OH O NH_2 SH	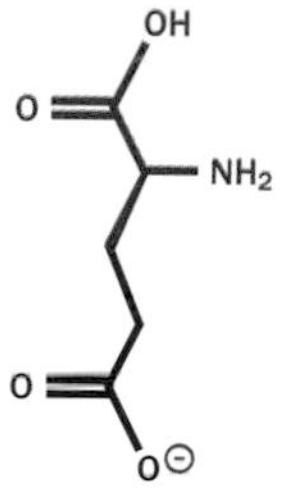 OH O NH_2 O $O^{\ominus}$	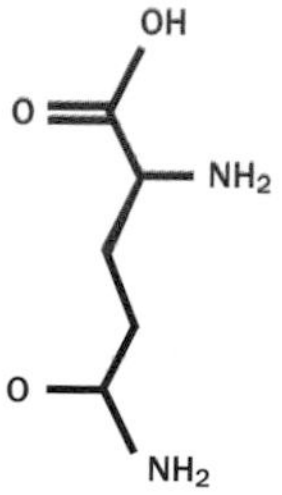 OH O NH_2 O NH_2	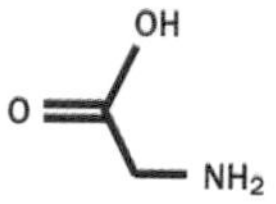 OH O NH_2	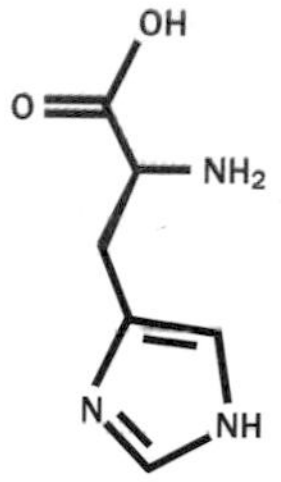 OH O NH_2 N NH	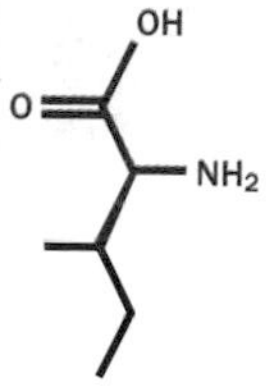 OH O NH_2
Leucine (L/Leu)	Lysine (K/Lys)	Methionine (M/Met)	Phenylalanine (F/Phe)	Proline (P/Pro)	Pyrrolysine (O/Pyl)
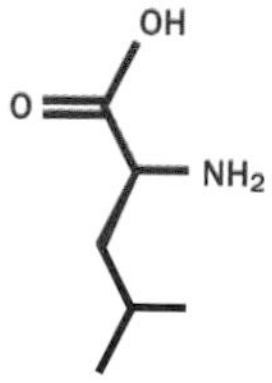 OH O NH_2	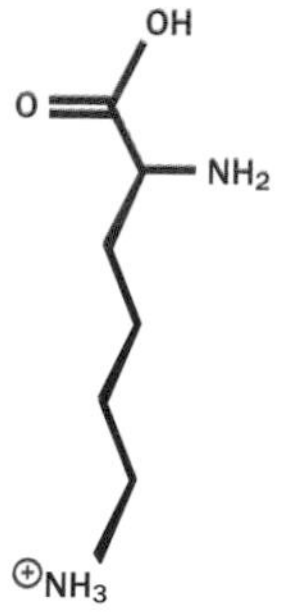 OH O NH_2 $^{\oplus}NH_3$	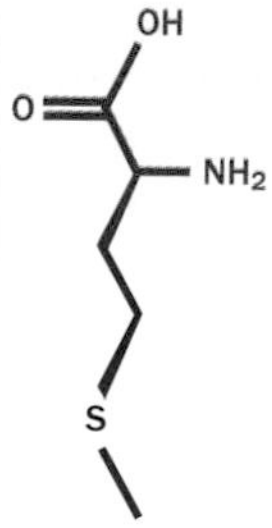 OH O NH_2 S	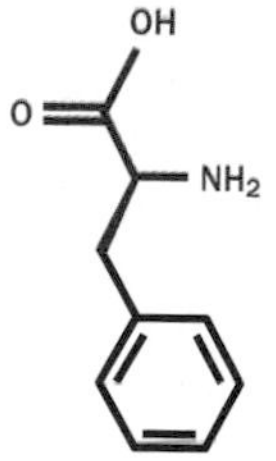 OH O NH_2	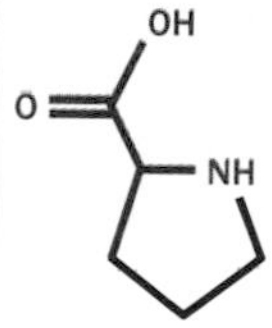 OH O NH	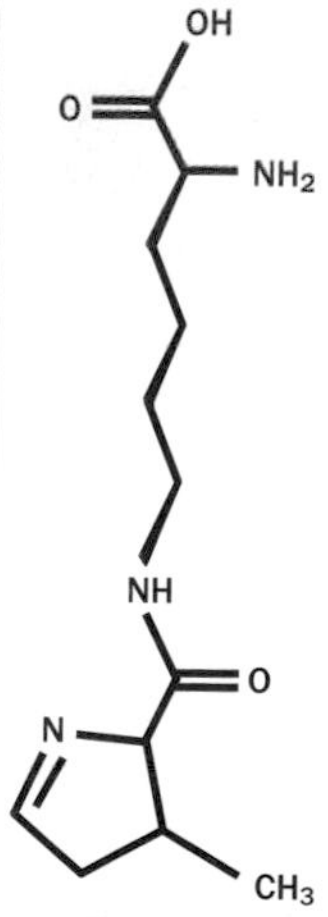 OH O NH_2 NH O N CH_3
Selenocysteine (U/Sec)	Serine (S/Ser)	Threonine (T/Thr)	Tryptophan (W/Trp)	Tyrosine (Y/Tyr)	Valine (V/Val)
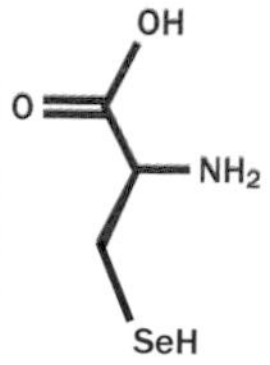 OH O NH_2 SeH	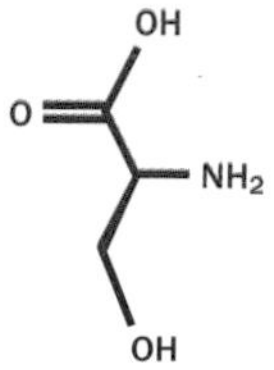 OH O NH_2 OH	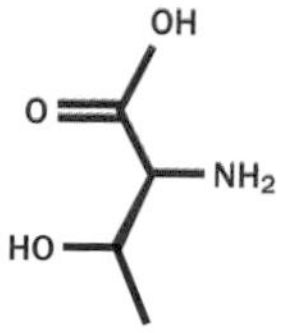 OH O NH_2 HO	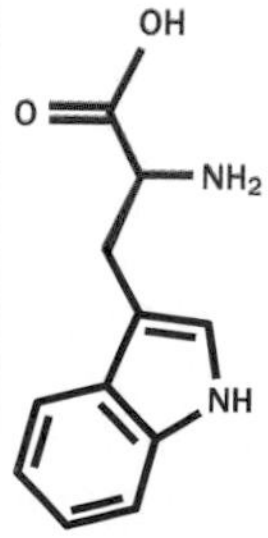 OH O NH_2 NH	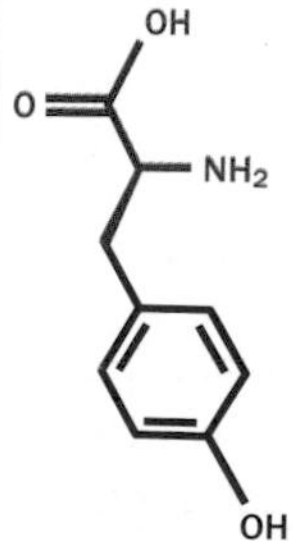 OH O NH_2 OH	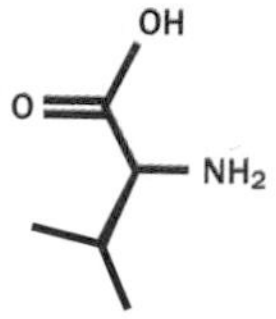 OH O NH_2

TABLE 7.2: AMINO ACIDS AND THEIR CORRESPONDING RNA CODONS

Ala	GCA, GCC, GCG, GCU	**Leu**	CUA, CUG, CUC, CUU, UUA, UUG
Arg	AGA, AGG, CGA, CGC, CGG, CGU	**Lys**	AAA, AAG
Asn	AAC, AAU	**Met**	AUG
Asp	GAC, GAU	**Phe**	UUC, UUU
Cys	UGC, UGU	**Pro**	CCA, CCC, CCG, CCU
Gln	CAA, CAG	**Ser**	UCA, UCC, UCG, UCU, AGC, AGU
Glu	GAA, GAG	**Thr**	ACA, ACC, ACG, ACU
Gly	GGA, GGC, GGG, GGU	**Trp**	UGG
His	CAC, CAU	**Tyr**	UAC, UAU
Ile	AUA, AUC, AUU	**Val**	GUA, GUC, GUG, GUU
START	AUG	**STOP**	UAA, UAG, UGA

called codon usage bias (see Table 3 in [9]). In prokaryotes, this bias is so pronounced that species can be barcoded based on codons [10] (see Chapter 6). Also intriguing, research seems to suggest that alternative codons, while resulting in the same amino acid, can nevertheless affect the function of a protein in some cases [11].

Each side chain makes an amino acid chemically unique. Because there are 20 standard amino acids, and because proteins contain many of them, the potential variability among proteins is enormous. It might sound exaggerated to consider the possible effects of single amino acid substitutions, but it is well known that even a single change can indeed alter the structure of a protein drastically [12]. For instance, sickle cell disease is caused by a single DNA mutation that results in one amino acid substitution and has drastic physiological effects.

Proteins vary tremendously in size. At the low end of the size scale are **peptides**, which consist of up to about 30 amino acids; this boundary between peptides and proteins is not clearly defined. On the high end, the largest protein that has so far been characterized is titin, which is a constituent of muscle tissue and consists of about 27,000 amino acids. To gauge the degree of variability among proteins, compute how many different proteins would be possible if they consisted of exactly (or at most) 1000 or 27,000 amino acids. The size of a protein is usually not given in terms of the number of amino acids though, but in atomic mass units or daltons. Because of their large sizes, proteins are actually measured in kilodaltons (kDa). As an example, titin has a molecular mass of close to 3000 kDa.

The physical and chemical characteristics of two neighboring amino acid side chains slightly bend their peptide bond in a predictable fashion, and all these bends collectively cause the protein to fold into a specific three-dimensional (3D) structure or **conformation**. This structure can contain coils, sheets, hairpins, barrels, and other features, which we will discuss later. Rearranging the same amino acids in another order would almost certainly result in a different 3D structure. One should note in this context that some proteins contain unstructured peptide regions of various lengths that allow different bending and folding, which in turn increases the repertoire of roles a protein can assume [13].

Over 40 years ago, Nobel Laureate Christian Anfinsen postulated the so-called thermodynamic hypothesis stating that, at least for the class of small globular proteins, the structure of a protein is determined exclusively by the amino acid sequence, so that under normal environmental conditions the protein structure is unique, stable, and such that it minimizes the total Gibbs free energy [14]. This assertion was based on *in vitro* experiments where he and his colleagues showed that the protein ribonuclease, which had been unfolded with urea, spontaneously refolded into the right conformation once the urea was removed. One might think that disulfide bonds could be a driving force in this process, but it is our current understanding that rather they are the result of correct folding and serve to stabilize a protein, whether it is folded correctly or incorrectly.

The automatic folding posited by Anfinsen's Dogma poses an interesting puzzle, called Levinthal's Paradox. Cyrus Levinthal computed that even if a protein consisted of only 100 amino acids, it would allow so many conformations that correct folding would take 100 billion years [15]. Of course, that is out of the question, and in reality the process is remarkably quick: it takes only a bit more than one minute to fold a protein of 300 amino acids! While we do not completely understand how this is possible, we do know that the folding of a newly created amino acid chain into the proper 3D structure is facilitated by two means. First, a special class of cytosolic proteins, called chaperonins, supports the folding task as the newly formed chain of amino acids exits the **ribosome**. A prominent example is the GroEL/ES protein complex. GroEL consists of 14 subunits that form a barrel, the inside of which is hydrophobic, while its partner GroES resembles a lid. The unfolded amino acid chain and adenosine triphosphate (ATP) bind to the interior rim of a cavity in the barrel-shaped GroEL and trigger a change in conformation, as well as an association of the growing protein with GroEL. Binding between the two proteins causes the subunits of GroEL to rotate and to eject the folded protein, and GroES is released. Proteins often undergo several rounds of refolding until the appropriate conformation is reached. Modern methods of cryo-electron microscopy permit a glimpse into the structure of proteins like GroEL/ES (see Figure 7.3 below). Chaperonins belong to a larger class of molecules, called molecular **chaperones**. These not only facilitate folding, they also protect a protein's 3D structure in situations of stress, such as elevated heat. In reference to this role, chaperones belong to the class of heat-shock proteins (HSPs).

The second mechanism facilitating proper folding is accomplished through special enzymes in the endoplasmic reticulum of eukaryotes. These enzymes, called protein disulfide isomerases (PDIs), catalyze the formation and breakage of disulfide bonds between cysteine residues within the protein [16]. This mechanism helps proteins find the correct arrangement of disulfide bonds. PDIs also serve as chaperones that aid the refolding of wrongly arranged proteins. In the process of forming a disulfide bond, the PDI is chemically reduced. The protein Ero1 subsequently oxidizes PDI and thereby refreshes its active state.

While the chain of amino acids in a protein corresponds to the nucleotide sequence in the corresponding mRNA and DNA, a protein may be chemically altered once the transcription and translation processes are completed. Collectively these alterations are called **post-translational modifications (PTMs)**. They consist of the attachment of biochemical groups, for example, phosphate, acetate, or larger molecules such as carbohydrates or lipids. Furthermore, amino acids may be removed from the amino end of the protein, enzymes may cut a peptide chain into two, and disulfide bridges may form between two cysteine residues. It even happens that the amino acid arginine is changed into a different amino acid, citrulline, which has no corresponding codon. Hundreds of PTMs are listed in a document file within the UniProt Knowledgebase [3], and proteome-wide statistics on PTMs are available at [17]. PTMs are of enormous importance, because they expand the repertoire of protein functions manifold and permit fine-tuned, dynamic control of the specific roles of proteins. A prominent example of the functionality of a PTM is phosphorylation or dephosphorylation in a specific location of a protein, which in the case of an enzyme or signaling protein can lead to activation or deactivation (see Chapters 8 and 9).

The turnover of proteins can be very fast or very slow, depending on the particular protein. If a protein is no longer needed, it is marked for degradation by the attachment of arginine [11] or of multiple copies of a specific regulatory protein called **ubiquitin**. This marking is recognized by a cellular protein recycling organelle, called the **proteasome**, which disassembles the protein into peptides and amino acids [18]. Aaron Ciechanover, Avram Hershko, and Irwin Rose received the 2004 Nobel Prize in Chemistry for the discovery of this process. Interestingly, ubiquitin and ubiquitin-like proteins are directly involved in the assembly of ribosomes [19]. This coupling of the nuclear ubiquitin–proteasome system with the assembly of new ribosomal subunits ensures that these two central processes of protein production and destruction are well coordinated.

Thus, the genetic code determines the amino acid sequence, the amino acid residues and the protein's environment determine the base structure of a protein, and the structure and regulatory modifications determine its role. Special proteins help other proteins fold into the right conformation, affect their activity, and mark them for destruction in an ongoing dynamic process. During their short or long lifetimes, the proteins in a functioning cell or organism have roles that are amazingly diverse, as we will see throughout this and other chapters.

7.1 Experimental Protein Structure Determination and Visualization

If a protein can be crystallized, its 3D structure can be determined with **X-ray crystallography** [20]. In this very well-established technique, which has generated tens of thousands of protein structures over the past decades, an X-ray beam is shot at the protein crystal and diffracts (scatters) into many directions, which are determined by the atomic arrangements within the protein (**Figure 7.2A**). The angles and intensities of the diffracted beams allow the reconstruction of a 3D picture of the density of the electrons within the crystal, which in turn provides information on the mean positions of the atoms and their chemical bonds. Another popular method for 3D reconstructions of proteins is **NMR** (nuclear magnetic resonance) **spectroscopy** [21], which also has produced thousands of images of protein structures (Figure 7.2B).

A newer and very powerful alternative to X-ray crystallography is cryo-electron microscopy (cryo-EM) (**Figures 7.3** and **7.4**). This method can be useful, for instance, for the analysis of large proteins and protein complexes [26], as well as of membrane proteins, which are often fragile and do not always yield sufficient amounts of material for X-ray crystallographic methods [27, 28]. Cryo-EM is based on freezing a thin suspension of molecules in vitreous ice on an electron microscope grid and subsequently imaging the molecules.

The different structural techniques are complementary in providing information on proteins. For example, cryo-EM may be used in conjunction with X-ray crystallography for single-particle analysis. In this case, X-ray crystal structures of one or more subunits are docked into the cryo-EM structure of a large complex that is not amenable to 3D crystallization in its entirety. Some proteins are highly flexible, which necessitates the study of each conformation in parallel.

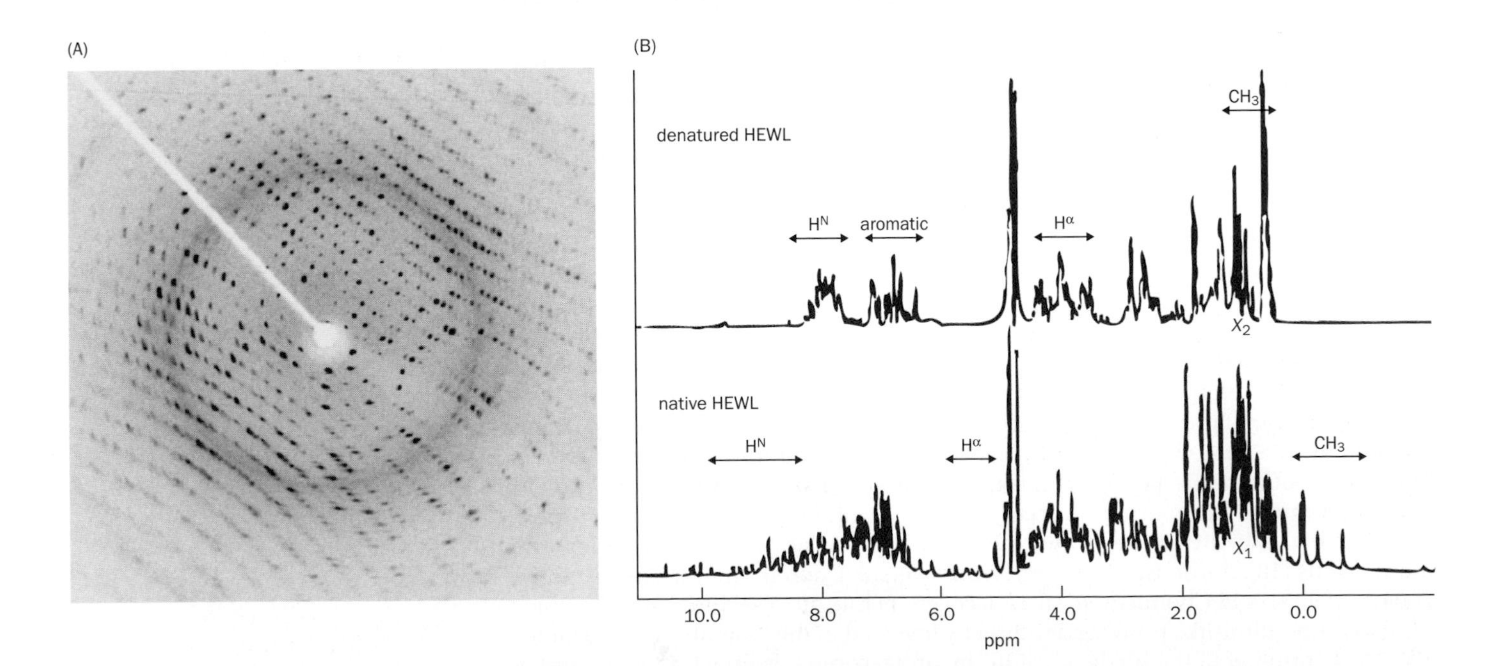

Figure 7.2 X-ray crystallography and NMR spectroscopy have led to insights into thousands of protein structures. (A) Diffraction image of a crystal of hen egg lysozyme. The protein structure is inferred from the angles and intensities of the diffracted X-ray beam. (B) One-dimensional ^{1}H-NMR spectra of denatured and native hen egg lysozyme. The presence (or absence) of protein structure can be assessed from the distribution of methyl, H^α, and H^N signals, where the latter identify the molecular positions of the hydrogen ions. In the native (structured) protein (bottom), the methyl signals are shifted upfield and the H^α and H^N signals downfield. In the denatured (unstructured) protein (top), these signals are absent from the regions of interest. (From Redfield C. Proteins studied by NMR. In Encyclopedia of Spectroscopy and Spectrometry, 2nd ed. [J Lindon, GE Tranter & DW Koppenaal, eds], pp. 2272–2279. Elsevier, 2010. With permission from Elsevier.)

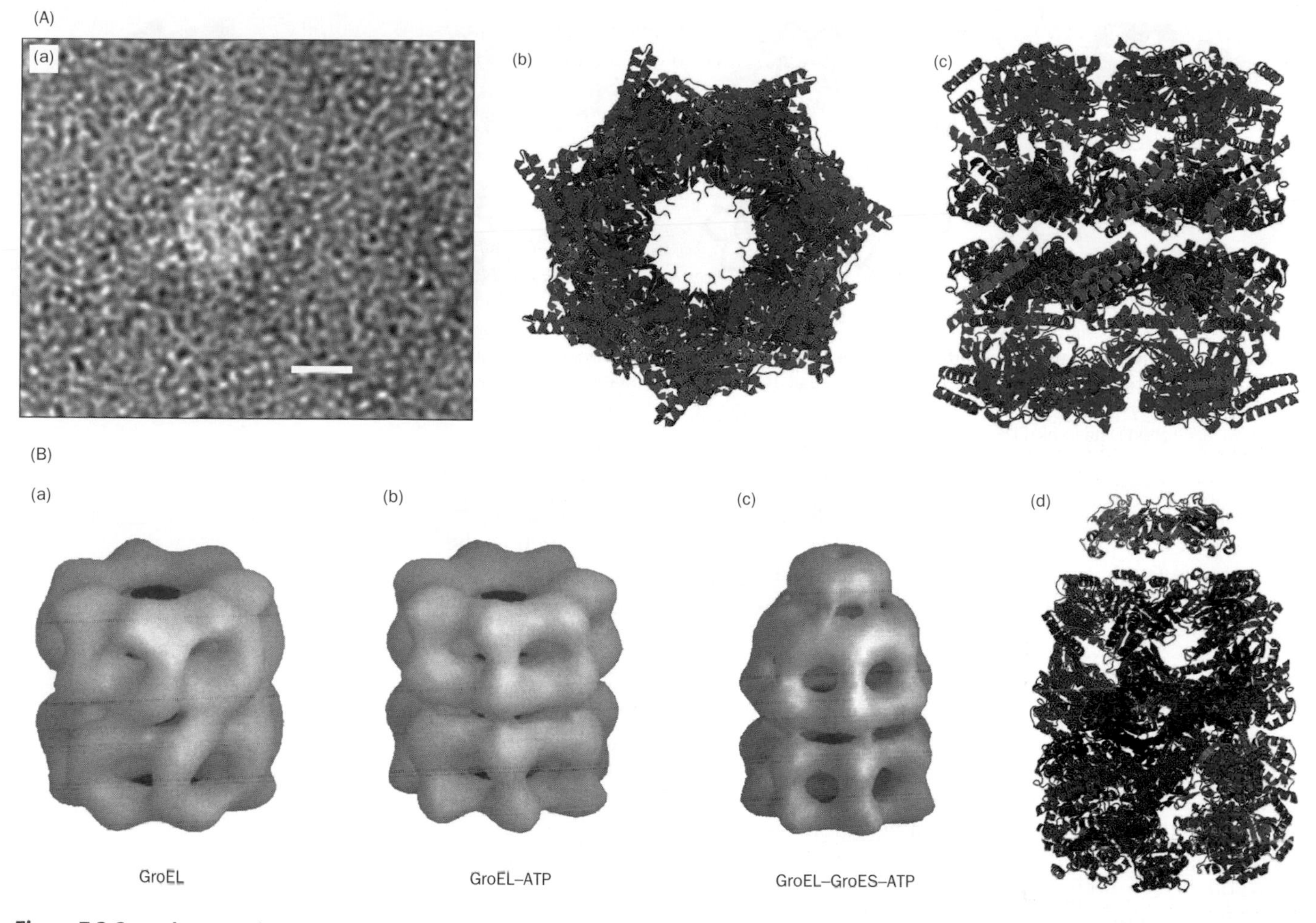

Figure 7.3 Cryo-electron microscopy (cryo-EM) results for the molecular chaperone GroEL. (A) Electron micrograph (a) and the inferred ribbon structure of the protein GroEL in end-view orientation (b) and in a side view (c) (Protein Data Bank: PDB 2C7E [22]). GroEL measures 140 Å in diameter in the end-on orientation; the scale bar represents 100 nm. Courtesy of Inga Schmidt-Krey, Georgia Institute of Technology. (B) Cryo-EM-based, surface-rendered views of 3D reconstructions of (a) GroEL, (b) GroEL–ATP, and (c) GroEL–GroES–ATP. The GroES ring is seen as a disk above the GroEL. (d) Corresponding ribbon structure. It consists of a β-barrel structure with a β-hairpin forming the roof of the dome-like heptamer (Protein Data Bank: PDB 2CGT [23]). ([a–c] from Roseman AM, Chen S, White H, et al. *Cell* 87 [1996] 241–251. With permission from Elsevier.)

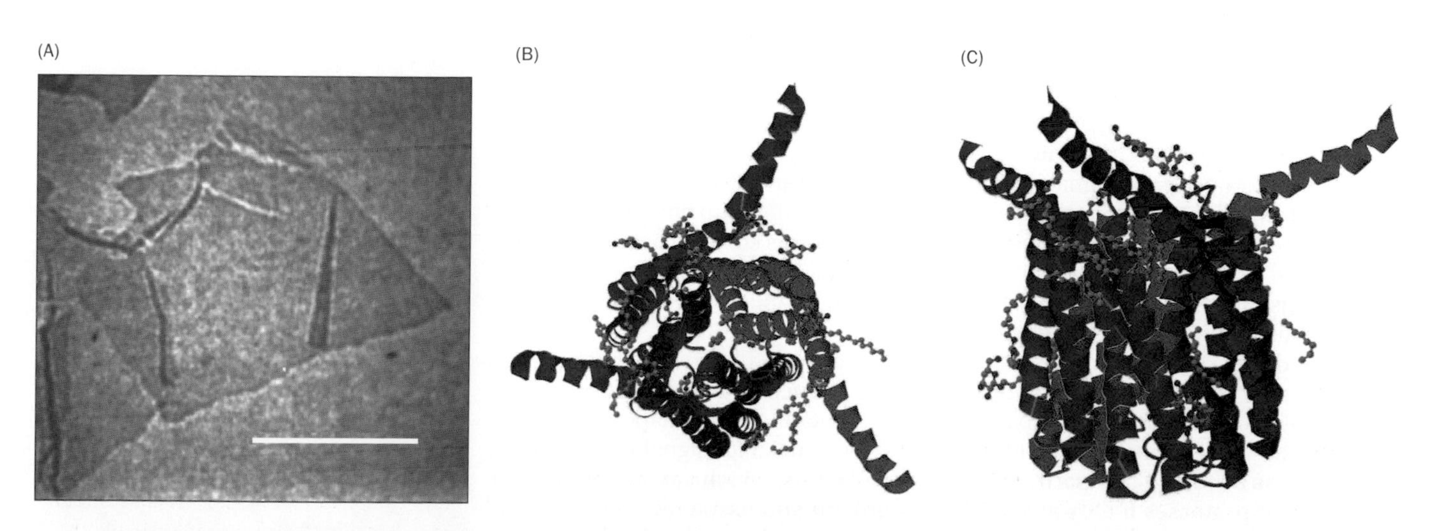

Figure 7.4 Cryo-EM is a valuable tool for elucidating structure–function questions in membrane proteins. Here, two-dimensional crystallization (A) and electron crystallography are used to study the human membrane protein leukotriene C4 synthase, which has a size of only 42 Å; the scale bar represents 1 μm. (B, C) Inferred ribbon structures (Protein Data Bank: PDB 2PNO [24, 25]). (Courtesy of Inga Schmidt-Krey, Georgia Institute of Technology.)

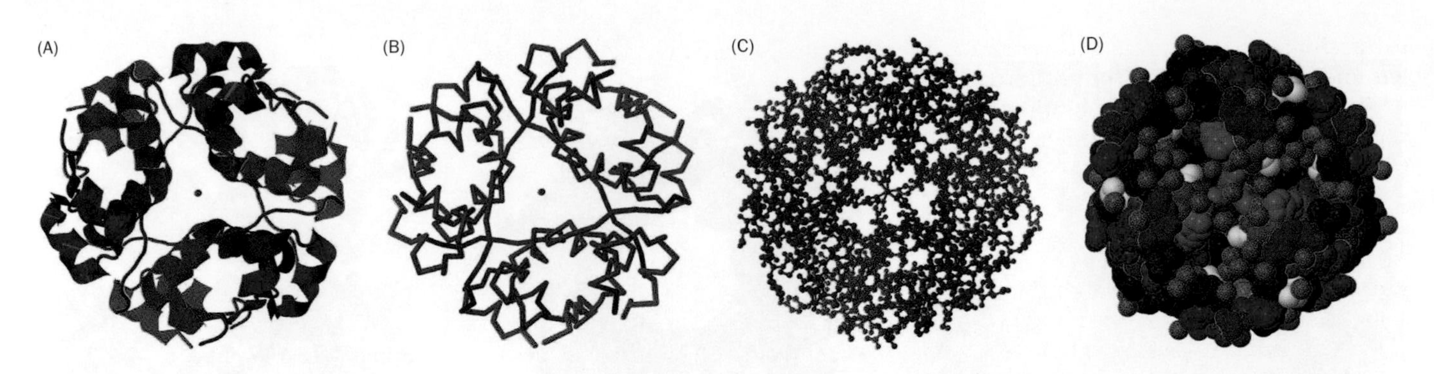

Figure 7.5 Different PDB-Jmol representations of the crystal structure of human insulin, complexed with a copper ion (in the center). (A) Ribbon cartoon representation. (B) Backbone representation. (C) Ball-and-stick representation. (D) Sphere representation. The last of these, also called the space-filling or Corey–Pauling–Koltun (CPK) representation, uses a convention showing atoms in defined colors (for example, C, gray; O, red; N, blue; Cu, light brown). (Protein Data Bank: PDB 3IR0 [29, 31].)

Because of the large number of amino acids in a protein, there is a huge repertoire of potential protein shapes and functions. Good software tools, such as the open-source Java viewer Jmol in the Protein Data Bank (PDB) [1, 29], have been developed to visualize the 3D shapes of protein crystal structures (**Figure 7.5**). One should, however, keep in mind that a crystal structure cannot always be obtained. For instance, some proteins contain unstructured peptide regions of various lengths, which permit different folding, so that the global 3D structure and the functional role of a protein are not always fixed but depend on its milieu [13]. Furthermore, other molecules, such as sugars and phosphate groups, may be attached to proteins at certain times and thereby change their conformation and activity state. Finally, proteins have of course experienced changes in their structures throughout evolution, so that the "same" protein from different species may have different structures. Interestingly, the most important features, such as the active sites of enzymes, have been conserved much more strictly than other domains, leading to the terminology of anchor sites and variable sites of protein structures [30].

AN INCOMPLETE SURVEY OF THE ROLES AND FUNCTIONS OF PROTEINS

It is impossible to rank proteins by importance, because removal of any class or type of proteins would be disastrous. Nonetheless, if one were to sort all proteins purely by their abundance, the clear winner would be RuBisCO, a protein that most people have never heard of. If you are surprised, consider that RuBisCO (ribulose 1,5-bisphosphate carboxylase oxygenase) is the first enzyme in the Calvin cycle, which is the metabolic pathway responsible for photosynthesis. It has been estimated that the total photosynthetic biomass production on Earth is over 100 billion tons of carbon per year [32], and RuBisCO is instrumental in all of it. Photosynthetic cyanobacteria (blue algae) alone comprise an estimated biomass between 40 and 50 billion tons. This is an enormous amount, considering that all humans together make up only about 250 million tons. The most abundant protein in the human body is collagen, which accounts for between a quarter and a third of the body's total protein content [33]. Within human cells, actin is the winner [34]. We can see that even the most abundant proteins are distributed among different functional classes.

When discussing the function of a protein, one should be aware that proteins often serve more than one role. For instance, the extracellular matrix consists of a number of proteins and other substances that maintain a cell's shape. The matrix also aids the movement of cells and molecules, it is crucial for signal transduction, and it acts as a storage compartment for growth factors, which can be released by the action of proteases if they are needed for growth and tissue regeneration. The same release processes are probably also involved in the dynamics of tumor formation and cancer metastasis. Specific examples of matrix proteins are fibronectins, which attach to collagen and other proteins called integrins on the surfaces of cells. Fibronectins allow cells to reorganize their cytoskeleton and to move through the extracellular space. They are also critical in blood clotting, where they bind to platelets and thereby assist in wound closure and healing.

The following will highlight the main function of some representative proteins, but many of these have secondary, yet important, roles—for instance, by serving as enzymes and as structural proteins that have a strong effect on signal transduction.

7.2 Enzymes

Enzymes are proteins that catalyze chemical reactions; that is, they convert organic substrates into other organic products. For instance, the first enzyme of glycolysis, hexokinase, takes the substrate glucose and converts it into the product glucose 6-phosphate (see Chapter 8 for an illustration). This conversion would not occur in any appreciable amount without the enzyme, because glucose 6-phosphate has a higher energy state than glucose. The enzyme makes the conversion possible by coupling the reaction to a second reaction. Namely, the required energy cost is paid for by the energy-rich molecule adenosine triphosphate (ATP), which donates one of its phosphates to the product and is thereby dephosphorylated to adenosine diphosphate (ADP), which contains less energy. A key element of this coupling of processes is that the enzyme itself is unchanged in the reaction and directly available for converting further substrate molecules. The vast majority of all metabolic conversions in a living cell are catalyzed by enzymes in this fashion. We will discuss details of such conversions in Chapter 8, which addresses metabolism.

With respect to the enzyme itself, it is of interest to study how a protein is capable of facilitating a metabolic reaction. The details of such a process depend of course on the specific enzyme and its substrate(s), but one can generically say that the enzyme contains an active site to which the substrate molecule binds. This active site is a physical 3D structure, which typically consists of a groove or pocket in the surface of the enzyme molecule and whose shape very specifically matches the

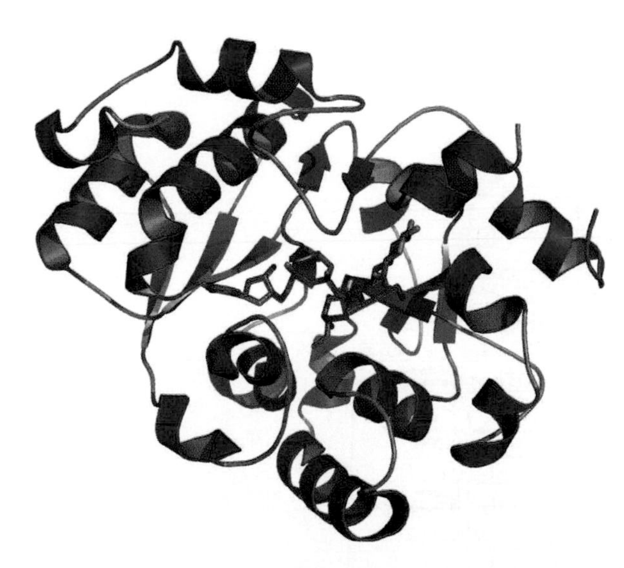

Figure 7.6 Many pathogenic bacteria produce lipo-oligosaccharides, which are composed of lipid and sugar components and resemble molecules on human cell surfaces. This molecular mimicry can deceive membrane receptors in the host, which do not recognize the invaders as foreign. As a consequence, the bacteria attach to the unsuspecting receptors and manage to evade the host's immune response. An example is the meningitis-causing bacterium *Neisseria meningitidis*, and a key tool of its success is the enzyme galactosyltransferase LgtC, whose crystal structure is shown here. The enzyme catalyzes an important step in the biosynthesis of a specific lipo-oligosaccharide; namely; it transfers α-D-galactose from UDP-galactose to a terminal lactose. The enzyme (turquoise ribbons) is shown in complex with sugar analogs (UDP 2-deoxy-2-fluoro-galactose and 4′-deoxylactose; stick molecules), which respectively resemble the true donor and acceptor sugars encountered in nature. Investigations of such complexes, combined with methods of directed evolution and kinetic analysis, offer important insights into the mechanisms with which bacteria outsmart the mammalian immune system and provide a starting point for targeted drug design. (Courtesy of Huiling Chen and Ying Xu, University of Georgia.)

substrate (**Figure 7.6**). Because of this specificity, enzymes typically catalyze just one substrate, or one class of substrates, with high efficiency. Initially, the active site was assumed to be rather rigid, but the current understanding is that it is more flexible, thereby allowing the transient formation of a complex between enzyme and substrate. Chemical residues near to the active site act as donors of small molecular groupings or protons, which are attached to the substrate during the reaction. Once attached, the modified molecule falls off the enzyme as the product of the reaction. Other classes of enzymes act on substrates by removing molecular groups, which are then accepted by residues close to the active site. Once the product is formed, the residues are returned to their original state.

The involvement of an enzyme in a metabolic reaction is of enormous importance for several reasons. First, very many reactions have to be executed against a thermodynamic gradient. In other words, such a reaction would not be possible without an enzyme. Second, even if the product has a lower energy state than the substrate, the energy state of the substrate is still lower than a transition state that needs to be overcome before the product can be formed. An enzyme is usually needed to surmount this energy barrier. Specifically, the enzyme lowers the energy requirement of the reaction by stabilizing the transition state. Third, the cell can regulate the enzyme and thereby affect its activity. Through this mechanism, the cell is able to steer substrates into alternative pathways and toward products that are in highest demand. For instance, pyruvate may be used as the starting substrate for the generation of oxaloacetate, which then enters the citric acid cycle, or it may be used to create acetyl-CoA, which can subsequently be used in fatty acid biosynthesis or different amino acid pathways. Pyruvate can also be converted into lactate and used by some organisms for the generation of ethanol (**Figure 7.7**).

The "decision" of how much substrate should be channeled toward alternative pathways and products is the result of complex regulation. The most immediate controllers are the enzymes that catalyze the various possible conversions of the substrate. Indirectly, the decision is much more distributed among factors that regulate the activities of these enzymes. This regulation is implemented in various, distinct ways. First, metabolites within or outside the target pathways can enhance or inhibit the activity of an enzyme. This modulation occurs essentially immediately.

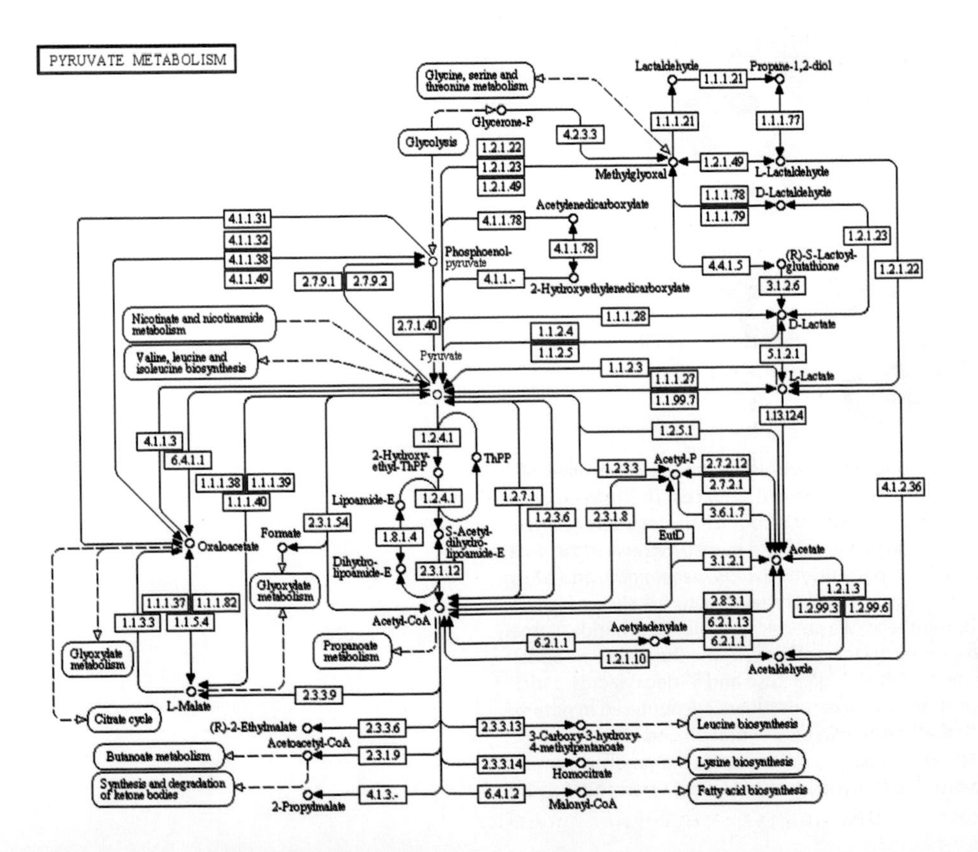

Figure 7.7 The metabolite pyruvate is involved in numerous biochemical reactions. These reactions are catalyzed by enzymes (identified by their Enzyme Commission (EC) numbers in boxes). Such pathways are discussed further in Chapter 8. (Courtesy of Kanehisa Laboratories.)

In some cases, the product of a pathway directly binds to the active site on the enzyme, thereby competing with the substrate and slowing down the conversion of substrate to product. An alternative is **allosteric** inhibition or activation, where the product attaches to the enzyme in a location different from the active site. This phenomenon leads to a physical deformation of the active site that alters the efficiency of substrate binding and is called **cooperativity**. The second mode of regulation is exerted by other proteins, which can quickly activate or deactivate an enzyme through the attachment or removal of a key molecular grouping, such as a phosphate or glycosyl group. More permanently, the regulatory protein ubiquitin can bind to the enzyme and mark it for degradation by the proteasome, as we discussed before. Finally, the activity of enzymes can be changed over much longer time horizons by means of gene regulation. Namely, the genes coding for the enzyme or its modulators may be up- or down-regulated, which ultimately affects the amount of enzyme present in the cell and therefore its throughput. As one might expect, the different modes of regulation are not mutually exclusive, and situations such as environmental stresses can easily evoke some or all of them simultaneously (see Chapter 11).

7.3 Transporters and Carriers

In addition to enzymes, transport or carrier proteins are crucially important for metabolism. Transport is needed between the inside and outside of a cell, within a cell, and outside cells, for instance, in the bloodstream. Transport across the cell membrane, in either direction, is often facilitated by **transmembrane proteins**, which accomplish their task either through active transport or through facilitated diffusion in the case of small molecules. A very important class of these proteins carry ions across the membrane; this ion transport is central to a vast number of processes, including the action and deactivation of nerve and muscle cells. We will discuss this in greater detail in Chapter 12.

Transport of small molecules within cells often occurs via diffusion. However, larger molecules are too big for this process to be efficient. Instead, they are often packed into vesicles, where they can be stored and eventually expelled somewhere else inside the cell or to the outside. Vesicles typically consist of lipid bilayers, but it is the current understanding that proteins are often involved in their formation and their docking to a membrane [35]. An important example of vesicle dynamics is the movement of neurotransmitters, such as dopamine, from the nerve cell into the synaptic cleft between cells, a process that is the basis for signal transduction in the brain. Another example of vesicle transport is the uptake of lipids from the intestines into the bloodstream via the lymphatic system [36].

A different, very interesting, type of movement within cells is facilitated by kinesin and dynein proteins, which use chemical energy from the hydrolysis of ATP to move along microtubule cables [37]. This movement is involved in a number of processes, including mitosis and the transport of molecules within the axons of nerve cells. The microtubules themselves also consist of proteins.

Transport between neighboring cells can also occur through different mechanisms. A particularly prevalent task is cell-to-cell communication, which is often accomplished by sending ions or other small molecules through physical channels. These channels, called **gap junctions**, are formed by proteins that span the intracellular space between neighboring cells. Specifically, these connectors consist of two half-channels that each span the membrane of one of the neighboring cells and are composed of four helices. This arrangement constitutes a positively charged entrance within the cytoplasm, a funnel leading into a negatively charged tunnel, and an extracellular cavity. The funnel itself is formed by six helices that determine the maximal size of molecules entering the channel [19]. Gap junctions can be opened or closed through rotating and sliding motion of their component proteins, called connexins, which thereby create or close a center opening (**Figure 7.8**). Important examples of gap junctions are found in the heart, where they help the cardiac myocytes contract in a coordinated fashion, making up the heartbeat (see Chapter 12).

Transport outside the cell includes the delivery of oxygen from the arterial blood to all tissues of the body. Responsible for this process is the iron-containing

Figure 7.8 Gap junctions are composed of proteins called connexins. (A) Side view of a connexin complex. The red domains span the membranes of two neighboring cells, while the yellow domains bridge the extracellular space. (B) Top view of the same connexin complex. (C) Symbolic representation of the opening and closing of a gap junction by means of rotating connexin subunits. (Protein Data Bank: PDB 2ZW3 [38].)

metalloprotein hemoglobin, which accounts for about 97% of a red blood cell's dry mass (**Figure 7.9**). Like many enzymes, this protein is subject to potential competitive and essentially irreversible inhibition of its binding of oxygen, for instance by carbon monoxide, which is therefore very poisonous [39]. Another carrier protein is serum albumin, which transports water-insoluble lipids through the bloodstream. Albumin is the most abundant protein in blood plasma. It is a tightly packed globular protein with lipophilic groups on the inside and hydrophilic side groups on the

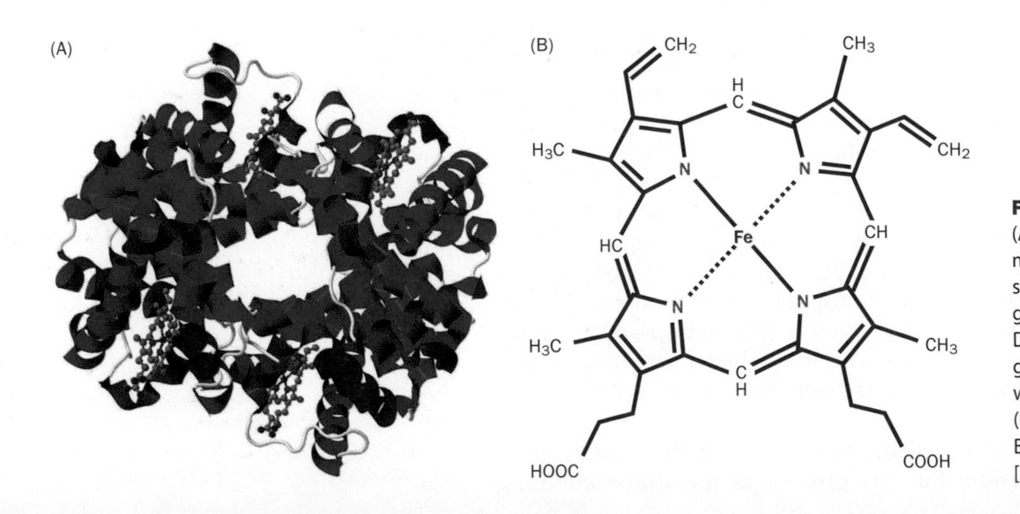

Figure 7.9 Heme and hemoglobin. (A) Human hemoglobin A is a globular metalloprotein (red) composed of four subunits that anchor the planar heme groups (the ball-and-stick assembly) (Protein Data Bank: PDB 3MMM). (B) Each heme group consists of a complex porphyrin ring with four nitrogen atoms that holds an iron (Fe^{2+}) ion, to which oxygen can bind. (From Brucker EA. *Acta Crystallogr*, Sect. D 560 [2000] 812–816.)

outside. The primary task of serum albumin is the regulation of osmotic pressure in the plasma, which is needed for an adequate distribution of fluids between the blood and extracellular spaces.

Another class of carrier proteins contains cytochromes, which operate in the electron transport chain and thus in the respiration process. They receive hydrogen ions from the citric acid cycle and combine them with oxygen to form water. This process releases energy, which is used to generate ATP from ADP. Cytochromes can be inhibited by the poison cyanide, which therefore suppresses cellular respiration.

Proteins in the extracellular matrix, such as fibronectins, are also involved with the coordinated transport of materials between blood capillaries and cells. This interstitial transport is important for the delivery of substrates. It is also the key mechanism for exposing cancer cells to chemotherapy [40]. We will discuss fibronectins later.

A particularly fascinating transport protein complex is the flagellar motor, which allows many bacteria and archaea to move (**Figure 7.10**) [41]. The flagellum itself is a hollow tube that is made from the protein flagellin and contains an internal cytoskeleton composed of microtubules and other proteins that render sliding and rotating motions possible [37]. Its shaft runs through an arrangement of protein bearings in the cell membrane. It is turned by a rotor that is located on the inside of the cell membrane and consists of several distinct proteins. This flagellar motor converts electrochemical energy into torque and is powered by a proton pump in the cell membrane, which moves protons, or in some cases sodium, across the membrane. The flow of protons is possible because of a metabolically generated concentration gradient. The motor can achieve several hundred rotations of the flagellum per minute, which translates into an amazing bacterial movement speed of about 50 cell lengths per second. Furthermore, by slightly changing the positioning of a specific type of motor protein, the organism can switch direction.

Flagella and similarly structured cilia are not only found in prokaryotes but are also crucial in eukaryotes, where they often have a more complex microtubular structure [42]. Prominent examples are sperm cells, cells of the fallopian tube, which move the egg from the ovary to the uterus, and endothelial cells in the respiratory tract, which use cilia to clear the airways of mucus and debris. Mammals often develop immune responses to flagellar antigens such as flagellin, with which they fight bacterial infections. Key components in these immune responses are proteins

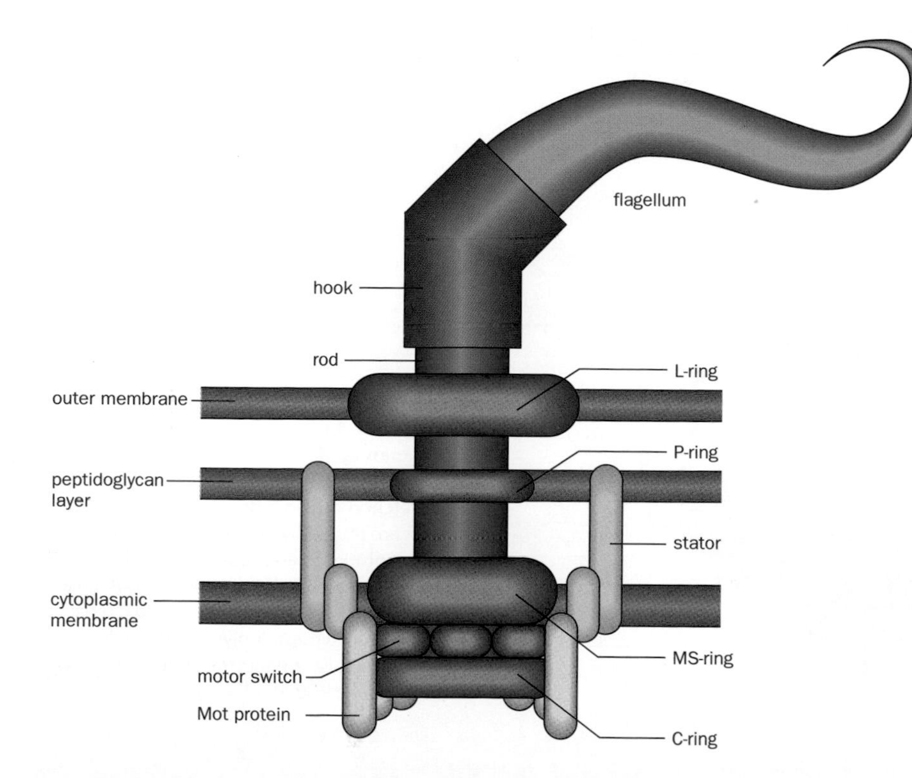

Figure 7.10 Diagram of a flagellar motor from a bacterium. This complex protein arrangement allows bacteria to swim in different directions. Similar assemblies are found in sperm cells, in the fallopian tubes (where eggs need to be moved from the ovary to the uterus), and in the respiratory tract (where cilia clear the airways of mucus and debris).

that recognize certain molecular patterns and are called Toll-like receptors (TLRs) [43], which we will discuss in Section 7.5.

7.4 Signaling and Messenger Proteins

Cells in higher organisms communicate with other cells and with the outside on a regular basis. They send and receive signals, which can be chemical, electrical, or mechanical, and must respond to these signals in a coordinated manner. This process, which is generically called **signal transduction**, is of obvious and utmost importance, and the whole of Chapter 9 is devoted to it.

As just one example, let us consider a particularly important class of membrane-spanning signaling proteins, called **receptor tyrosine kinases (RTKs)**. Generically, kinases are enzymes that transfer a phosphate group from a donor such as ATP to a substrate. In the case of RTKs, this enzymatic action amounts to the initiation of signal transduction. Namely, when a ligand, such as an epidermal growth factor, cytokine, or hormone, binds to the receptor on the outside of the cell, its protein domain on the inside (that is, in the cytoplasm) initiates autophosphorylation. This process leads to the recruitment of signaling proteins and thereby starts the intracellular signal transduction process. RTK-based processes are crucial in healthy cells and also seem to be important in the development and progression of cancers, as well as neurodegenerative diseases such as Alzheimer's disease and multiple sclerosis. An example of an epidermal growth factor receptor is shown in **Figure 7.11**.

In addition to receptors and signaling proteins, the signal transduction process is supported by scaffold proteins, which interact with signaling proteins in a fashion that is not entirely understood. It seems that scaffold proteins assist in the appropriate localization of pathway components to the cell membrane, the nucleus, the Golgi apparatus, and other organelles. Scaffold proteins furthermore assist the condensation of DNA and are involved with the recognition of surface structures of viruses by antibodies [45].

It should be noted that some, although not all, hormones are proteins or peptides, which act over long distances. Hormones are produced by animal cells in glands, such as the pituitary, thyroid, adrenal glands, and the pancreas, are transported through the bloodstream, and bind to specific **receptors** in distant locations of the body, where they signal the need for some action. Examples are the blood-sugar-regulating protein insulin, growth hormone, the blood-vessel-control hormones angiotensin and vasopressin, and the exhilaration-causing endorphins, as well as the more recently identified "hunger hormones" orexin and ghrelin. Plants also produce hormones.

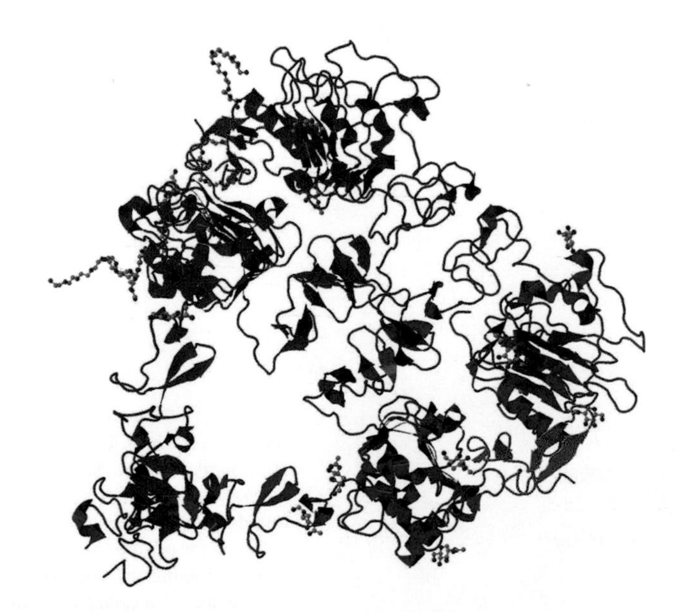

Figure 7.11 Crystal structure of the human epidermal growth factor receptor. Two types of ligands are shown. The stretched-out ligands at the top and center left are nonaethylene glycol molecules. On the right, one can see several *N*-acetyl-D-glucosamine (GlcNAc) molecules, which are components of a biopolymer in bacterial cell walls, called peptidoglycan. Interestingly, GlcNAc is also a neurotransmitter. (Protein Data Bank: PDB 3NJP [44].)

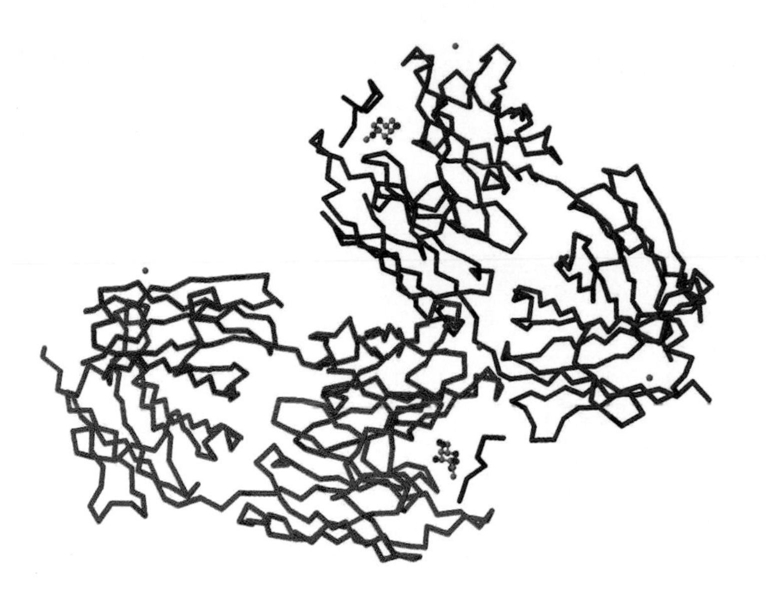

Figure 7.12 A mouse antibody recognizes a so-called Tn antigen (center top and bottom). Individual spheres are zinc molecules. Tn antigens have been associated with many cancers and made them targets for the development of cancer vaccines. Tn stands for threonine/*N*-acetyl-D-galactosamine α-*O*-serine. (Protein Data Bank: PDB 3IET [46].)

7.5 Proteins of the Immune System

Proteins are crucial players at many levels of the vertebrate immune system. Antibodies, or **immunoglobulins**, are globular plasma proteins that are produced by white blood cells (**Figure 7.12**). They are either attached to the surface of immune cells, called effector B cells, or circulate with the blood and other fluids. A typical B cell carries tens of thousands of antibodies on its surface. Antibodies bind to neutralize foreign molecules (antigens) as they are found on the surfaces of bacteria, viruses, and cancer cells, as well as nonliving chemicals, such as drugs and toxins. Owing to this binding, the invaders lose their ability to enter their target cells in the host. The antibodies also initiate the destruction of pathogens by macrophages.

Antibodies typically consist of a complex of two heavy-chain and two light-chain proteins, and while these always form the same generic structure, specific genetic mechanisms have evolved to generate tips at the ends of these proteins that contain extremely variable antigen-binding sites [47]. When B cells proliferate in response to an insult, high mutation rates increase the variability of these sites. As a consequence, the antigen-binding sites are capable of distinguishing millions of antigens by their structural features, or epitopes. In autoimmune disease, the same antibody responses are directed against the body's own cells.

A second group of proteins that are crucial for immune responses is the large family of **cytokines**. These signaling molecules are used for cell-to-cell communication and drive the progression of inflammatory processes. Cytokines are produced by T helper cells, which can respond to antigen peptides within a few seconds, yet require hours of signal exposure to become fully activated [48, 49]. Cytokines help recruit further immune cells and macrophages to the site of injury or a bacterial or viral infection. Such an infection can trigger 1000-fold increases in cytokines, and slight imbalances in the numbers and types of cytokines can lead to "cytokine storms," sepsis, multi-organ failure, and death. Well-known examples of cytokines include interleukins and interferons. We discuss some recent modeling approaches to understanding the exceedingly complicated dynamic system of cytokines in Chapter 15. A final group of "defense proteins" consists of enzymes in tears and the oils of the skin that guard against invaders and thus contribute to the innate immunity of the body.

Prominent proteins at the intersection of signaling and immune response are the **Toll-like receptors (TLRs)**—*toll* is German for "Wow!"—which function as sentinels that recognize numerous epitopes, including peptidoglycans and lipopolysaccharides, which are typical components of bacterial cell walls (**Figure 7.13**). Upon recognizing a specific epitope, the TLR activates pro-inflammatory cytokines and signaling pathways and thereby initiates a proper immune response. TLRs come in a number of variants and form a very complex functional interaction network [52].

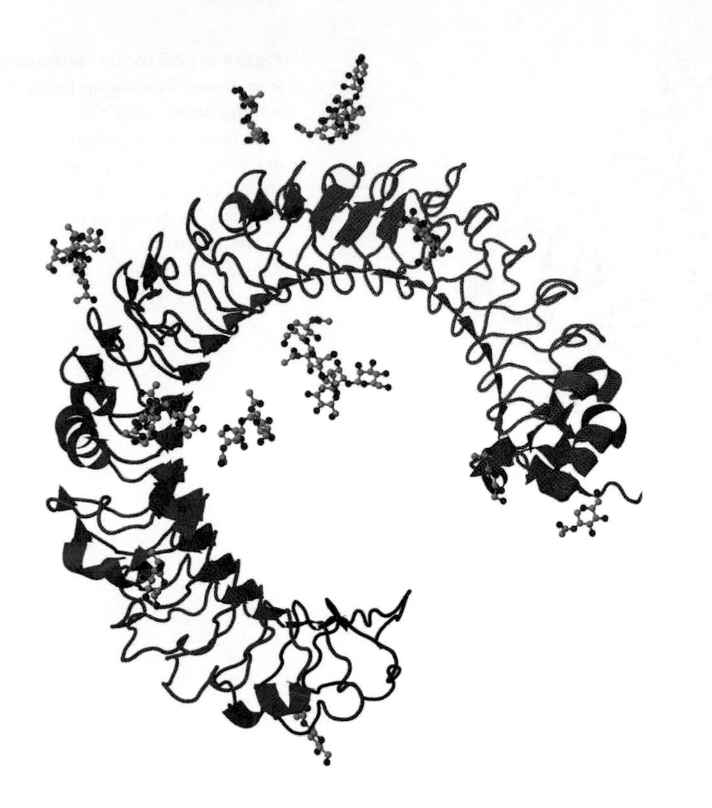

Figure 7.13 Toll-like receptors (TLRs) sense and bind various ligands, a process that initiates a signaling cascade and leads to the production of inflammation mediators. Shown here is the outer domain of the human TLR3, which recognizes double-stranded RNA, a molecular signature of viruses. The horseshoe shape is due to 23 repeated leucine-rich domains and establishes a distinct receptor for pathogen recognition, such as the cell surface sugars fucose, mannose, and *N*-acetyl-D-glucosamine (GlcNAc), which can be seen inside and outside the receptor. (Protein Data Bank: PDB 3CIG [50]; visualization with methods from [51].)

7.6 Structure Proteins

Collagen is the most abundant protein in humans. It is found in many tissues that provide structural support to the body, including bone, cartilage, connective tissue, tendons, ligaments, and the discs in the spine. It also constitutes between 2% and 5% of muscle tissue and is a component of blood vessels and the cornea of the eye. Collagen supports the structure of the extracellular matrix that provides support to cells and is involved in cell adhesion and the transport of molecules within the space between cells.

Collagen consists of three polypeptide strands that contain a large number of repeated patterns of specific amino acids, such as glycine–proline–AA or glycine–AA–hydroxyproline, where "AA" represents some other amino acid. Owing to the presence of these patterns, the amount of glycine is unusually high—the only other protein with this feature seems to be a protein in silk fibers. The abundance of glycine, combined with the relatively high content of proline and hydroxyproline rings, causes collagen polypeptides to curl into left-handed helices. These helical strands in turn are twisted into a right-handed triple helix, which is thereby a prime example of a quaternary structure that is stabilized by numerous hydrogen bonds. Each triple helix can form a super-coiled collagen microfibril, and many of these microfibrils make up a strong yet flexible macro-structure. When hydrolyzed, collagen becomes gelatin, which has multiple uses in the food industry.

Like other proteins, collagen is involved in control tasks as well, especially in the context of cell growth and differentiation. For instance, it induces the receptor tyrosine kinase DDR2 (member 2 of the discoidin domain receptor family), which triggers autophosphorylation of targets in the cytosol (**Figure 7.14**). To permit these control tasks, some sections of the collagen polypeptide do not show the regular curling patterns described above.

Complementing the load-bearing strength of collagen, elastins equip tissues with shape flexibility, which is particularly important in muscle cells, blood vessels, the lungs, and skin. Like collagen, elastins contain high amounts of glycine, but also similar amounts of proline, valine, and alanine. Fibroblasts secrete the glycoprotein fibrillin into the extracellular matrix, where it is incorporated into microfibrils that are thought to provide a scaffold for the deposition of elastin.

Another structural protein is keratin, which is the base material for skin, hair, nails, horns, feathers, beaks, and scales. Keratin contains many α-helices, stabilized with disulfide bonds between cysteine residues. Similar to collagen, three

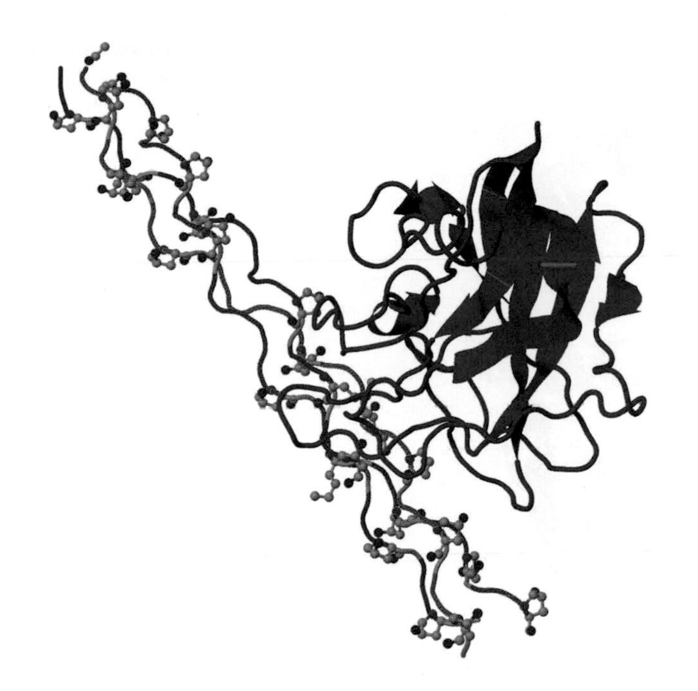

Figure 7.14 Human collagen (diagonal strands) complexed with the discoidin domain receptor DDR2 (to the right). DDR2 is a common receptor tyrosine kinase that is activated by triple-helical collagen. The binding of collagen to DDR2 causes structural changes in the surface loops of the receptor and has been linked to its activation. The receptors control cell behavior and are incorrectly regulated in several human diseases. (Protein Data Bank: PDB 2WUH [53].)

α-helices are twisted around each other into a protofibril. Eleven of these form a tough microfibril, in which nine protofibrils are wound around two center protofibrils. Hundreds of microfibrils form a macrofibril. Interestingly, the toughness of keratin is only matched by chitin, which is not a protein but a long-chain polymer of *N*-acetylglucosamine that provides strength to the protective armor of crustaceans like crabs and lobsters, and is also found on the surfaces of bacteria.

Actin is a globular protein in essentially all eukaryotic cells. It has two important roles. First, it interacts in muscle cells with the protein myosin. Actin forms thin filaments, while myosin forms thick filaments. Together they form the contractile unit of the muscle cell (see Chapter 12). Second, actin interacts with cellular membranes. In this role, actin and the proteins cadherin and vinculin form microfilaments. These microfilaments are components of the cytoskeleton and participate in numerous fundamental cellular functions, including cell division, cell motility, vesicle and organelle movement, and the creation and maintenance of cell shape. They are also directly involved in processes such as the phagocytosis of bacteria by macrophages.

The lens of the mammalian eye is a complicated structure that contains fibers composed of long transparent cells. More than a third of its weight is due to proteins. These proteins, which have fascinated scientists for over 100 years, include crystallins [54], collagen, and the membrane protein aquaporin [55], which maintains an unusually tight packing between neighboring cells (**Figure 7.15**). Intriguingly, there is very little turnover in the lens, and humans retain their embryonic lens proteins throughout life. Age-related changes in the lens, such as cataracts, are therefore difficult to repair.

The versatility of proteins has resulted in uncounted roles associated with the maintenance of shape and structure. In many cases, these roles are performed with "mixed" molecules that consist of proteins and something else. For instance, apolipoproteins, together with several types of lipids, can form lipoproteins, which serve as carriers for lipids in the blood, and also as enzymes and structural proteins. Well-known examples are high-density and low-density lipoproteins (HDLs and LDLs), which, respectively, carry cholesterol from the body's tissues to the liver and vice versa. They are sometimes referred to as "good" and "bad" cholesterols, respectively.

Glycoproteins contain sugars (glycans) that are bound to the side chains of the polypeptides. Glycans are often attached during or after translation of the mRNA into the amino acid chain, in a process called glycosylation. Glycoproteins are often integrated into membranes, where they are used for cell-to-cell communication. Prominent examples are various hormones and mucins, which are contained in the

Figure 7.15 The vertebrate eye lens protein aquaporin-0 (AQP0). AQP0 consists of a complex of four subunits, which are arranged in a three-dimensional lattice. According to current understanding, AQP0 proteins mediate the connections among the very tightly packed fiber cells of the lens. (Protein Data Bank: PDB 2C32 [55].)

mucus of the respiratory and digestive tracts. Glycoproteins are important for two aspects of immune responses. First, some of these proteins are mammalian immunoglobulins or surface proteins on T cells and platelets. Second, glycoproteins form structures on the surfaces of bacteria such as streptococci, which are recognized by the host's immune system.

CURRENT CHALLENGES IN PROTEIN RESEARCH

Given the enormous distribution and versatility of proteins, it is not surprising that intense research efforts are ongoing at all levels, from protein chemistry to the dynamics of interacting protein systems. Outside pure chemistry and biochemistry, current challenges fall into the categories of proteomics, structure and function prediction, localization, and systems dynamics. Of course, there is not always a clear boundary between these activities.

7.7 Proteomics

The **proteome** is the totality of proteins in a system, and proteomics attempts to characterize it. This characterization typically includes an account of the inventory and the abundance of many or all proteins at a given point in time. As is to be expected, both inventory and abundance can significantly differ among the tissues of an organism. Extensive research has addressed the spatial distribution and localization of target proteins, as well as temporal changes due to physiological trends, such as differentiation and aging, or in response to infection or disease. It is hoped that early identification of these spatial and temporal changes might suggest diagnostic biomarkers that show up before a disease manifests.

Proteomics is indeed a grand challenge, because it is not known how many proteins exist in model organisms. It is not even clear how to count them. For instance, should the same protein with two different post-translational modifications count as one or two proteins? Many signaling proteins and enzymes can be phosphorylated or dephosphorylated, and these processes change the activity state of the protein. Should both forms be counted separately? Methylation, glycosylation, arginylation, ubiquitination, oxidation, and other modifications raise similar questions. Finally, newly formed transcripts can be spliced in different ways, thereby leading to different proteins. These splice variants are presumably the reason that humans can manage with only about 20,000–25,000 genes; some trees are thought to possess twice as many.

A good start to identifying proteins, and to comparing and discerning two similar tissues or protein systems, for instance in the characterization of cancer cells, is one- or **two-dimensional (2D) gel electrophoresis** [56]. In the one-dimensional case, proteins are first coated with a negatively charged molecule, called SDS (sodium dodecyl sulfate). In this coating, the number of negative charges is proportional to the size of the protein. Thus, the proteins can be separated according to their molecular size, which is reflected by their mobility through a gel that is placed into an electric field. In the typical 2D case, the proteins are first separated with respect to their isoelectric point, which is the pH at which the protein has a neutral charge. In the second dimension, they are separated with respect to size. In addition to these traditional features, other separation criteria have been developed (see, for example, [57]). As the result of the separation in two coordinates, every protein migrates to a specific spot on the gel. These spots can be visualized with various stains, such as a silver stain, which binds to cysteine groups in the protein, or with a specific dye from the wool industry, called Coomassie Brilliant Blue. The degree of staining indicates to a reasonable approximation the amount of protein in a given spot. The 2D separation is effective, because it seldom happens that two proteins have the same mass as well as the same isoelectric point. If a protein is absent in one of the two systems being compared, its spot remains empty. If it is phosphorylated or otherwise modified, it migrates in a predictable fashion to a different spot. One important advantage of this method is that one does not even have to know which protein has migrated to a particular spot. Thus, the method can be used as a discovery tool. Many software packages are available for supporting the analysis of 2D gel electrophoreses. A variation is DIGE (difference gel electrophoresis), which shows the differences in proteomic profiles between comparable cells in different functional states (**Figure 7.16**) [58].

Numerous other methods for separating proteins have been developed in recent years [59, 60]. These include MudPIT (multidimensional protein identification technology), which uses two-dimensional liquid chromatography, and the so-called iTRAQ technique (isobaric tags for relative and absolute quantitation). iTRAQ is used to compare protein mixtures from different sources in the same experiment. The key is that the proteins in each sample are "tagged" with specific markers that contain different isotopes of the same atom and therefore have slightly different masses, which can be distinguished with mass spectrometry (see below).

Two-dimensional gel electrophoresis identifies spots where proteins are present. In favorable cases, these same spots have been observed in other studies, and a comparison with corresponding databases suggests what the so-far unknown protein might be [61]. If this comparison is unsuccessful, one can cut the protein spot out of the gel or use the liquid from MudPIT and analyze it in order to deduce its amino acid sequence. Once the sequence is known, comparisons with protein databases provide clues about the nature and function of the target protein [2].

Figure 7.16 Two-dimensional difference gel electrophoresis (DIGE) image of lung alveolar type II cells from mouse. The globin transcription factor GATA1 was knocked down in these cells. Each spot indicates the presence of a protein with a specific charge and mass. Different colors identify differences between knock-downs and normal mice. (Courtesy of John Baatz, Medical University of South Carolina.)

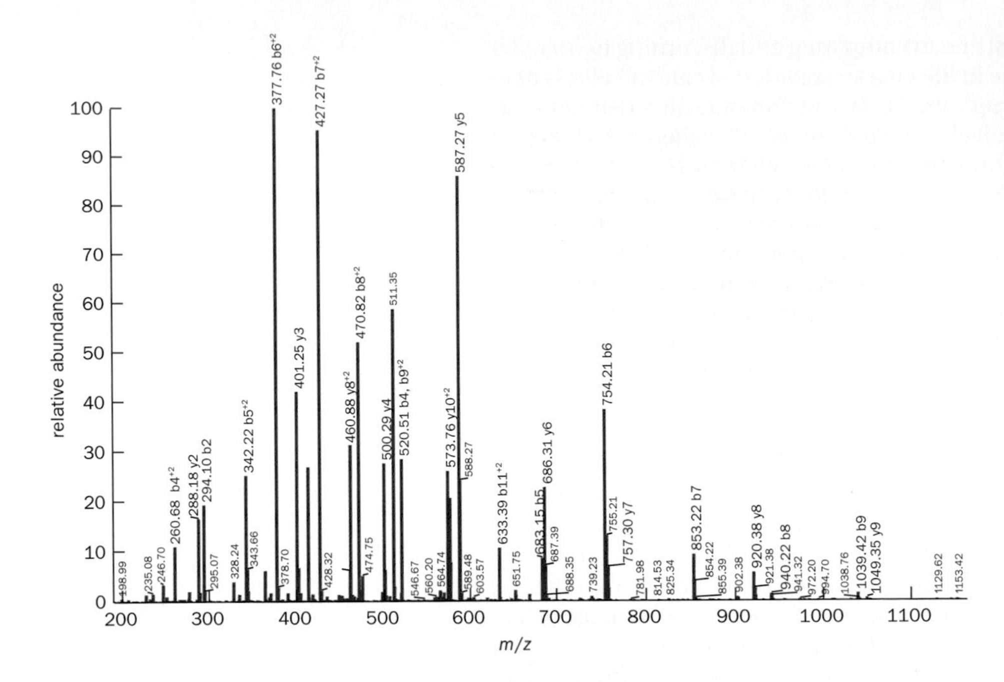

Figure 7.17 Tandem mass spectrum of the tryptic peptide RHPEYAVSVLLR from bovine serum albumin. The plot shows, in the form of peaks, the fragments of the molecule as they were detected in the mass spectrometer. The height of each peak indicates the abundance of the observed fragment, while the location indicates the mass-to-charge ratio (m/z). Since peptides tend to fragment at the linkages between specific amino acids, it is possible to deduce the amino acid sequence of the peptide from the ionized fragments. Designations like b4, b7, and y5 are codes within a specific nomenclature for labeling peptide fragments. (Courtesy of Dan Knapp and Jennifer Bethard, Mass Spectrometry Facility of the Medical University of South Carolina.)

The determination of the protein sequence from an isolated spot of a 2D electrophoresis or another sample of protein is typically accomplished with **mass spectrometry (MS)**. One of the most widely used variants of this technique is MALDI–TOF, which stands for matrix-assisted laser desorption/ionization–time of flight. In this method, a sample of peptide solution is deposited on a substrate (the matrix). A laser is beamed onto the matrix and desorbs (ejects) and ionizes (electrically charges) the peptides. The ionized peptides are shot through an electric field toward a detector, which measures for each peptide the time in flight. This time depends on the mass of the peptide and the applied charge. Correspondingly, the output of a MALDI–TOF experiment shows the amounts of material at the detector, which are exhibited in the form of peaks, as a function of mass divided by charge (m/z). There is ample literature on this and other separation techniques.

It is difficult to analyze large intact proteins by MS. A necessary preparatory step is therefore the cutting of a protein into smaller peptides. This step is achieved with enzymes such as trypsin that cut peptide bonds between predictable pairs of amino acids. While this step solves the problem of large masses, it creates a new problem. Namely, instead of one protein, one now has to address thousands of peptide fragments that are mixed up. Thus, in a second round of a tandem MS, or MS/MS, analysis, these peptides are typically forced to collide with an inert gas, which randomly breaks one of the remaining peptide bonds in the fragment, thereby replacing it with a positive charge. The result is a collection of "b" and "y" peaks, which correspond to the left or right pieces of the broken-up peptide. The fragment y1 represents the first amino acid in the peptide, the m/z difference between y1 and y2 represents the second amino acid, and so on (**Figure 7.17**). Computer algorithms have been developed to reconstitute the peptides from these fragment mixtures. Current research is focused on, among other things, miniaturization of the process, which would permit proteomic analysis with smaller protein samples than is possible now [62], and the development of methods that permit larger molecules to be analyzed without fragmentation [63].

7.8 Structure and Function Prediction

The amino acid sequence of a protein can be deduced from the nucleotide sequence of its gene, if it is known, or determined with methods of gel electrophoresis and mass spectrometry. Once the amino acid sequence is known, each residue is known, and it may seem a small step to derive what the protein looks like from the chemical and physical features of all the residues. Alas, this intuitive deduction is utterly wrong, and the identification or prediction of the 3D structure of a protein from its amino acids is in reality an extremely difficult challenge that has attracted enormous experimental and computational attention. Nonetheless, addressing this challenge is very important, because one rightly expects that the 3D structure of a protein is instrumental for its function. As a consequence, an entire branch of biology, called structural biology, focuses on macromolecules and, in particular, proteins.

In order to dissect the big challenge of protein structure and function prediction into smaller steps, the field has defined four types of structures. The primary structure of a protein refers to its amino acid sequence. The secondary structure is a localized feature involving regular 3D shapes or motifs. The best understood of these motifs are α-helices, β-sheets, and turns. Each α-helix is the consequence of regularly spaced hydrogen bonds between backbone oxygen and amide hydrogen atoms, while a β-sheet occurs when two strands are joined by hydrogen bonds and each strand contains alternating residues. Turns and bends are due to three or four specific sequences of amino acids that permit tight folds that are stabilized with hydrogen bonds.

The local motifs are usually interrupted by less well-defined stretches within the protein, and all defined and undefined portions combine to form the tertiary structure, that is, the overall folding of the protein. The tertiary structure of a protein is maintained by a hydrophobic core, which prevents the easy dissolution of the structure in the aqueous medium of the cell, and furthermore through chemical bonds, such as hydrogen bonds, disulfide bonds, and salt bridges. One should note that this tertiary structure is not ironclad, and post-translational modifications, changes in environmental conditions, and various cellular control mechanisms of protein activation and deactivation can alter it significantly [11, 13]. This flexibility in folding is crucial, because it is directly related to the function of the protein. A premier example is the activation of an enzyme by a metabolic modulator, which we discussed before. Finally, many proteins consist of subunits and are functional only as complexes. The structure formed by two or more protein subunits is called the quaternary structure of a protein. The functional rearrangements, which may affect both the tertiary and quaternary structure, are often called conformations, and transitions between such states are called conformational changes.

Two distinct approaches can be employed to identify the 3D structure of a protein. On the experimental side, crystallography, cryo-EM, and NMR are among the most widely used methods. On the computational side, very large computers and customized, fast algorithms are being used to predict the 3D structure of proteins and their binding sites from their amino acid sequences, the chemical and physical features of their amino acid residues, and possibly other available information [64–67]. While some successes have been reported, this protein structure prediction task is in general an unsolved problem. Reliable structure prediction is a very important and ultimately rewarding problem, because it has immediate implications for an understanding of disease and the development of specific drugs.

As a first step toward full predictability, powerful computational methods have been developed to identify smaller, recurring 3D structures within proteins, such as α-helices and β-strands. α-helices (**Figure 7.18**) cause the protein locally to coil, while β-strands form stretched-out sheets, which are often connected with U-turns or hairpins. In some channel proteins, these sheets are arranged into barrels (**Figure 7.19**). β-sheets are quite common and have been implicated in a number of human diseases, such as Alzheimer's disease. There is a rich literature on these and other structural motifs in proteins.

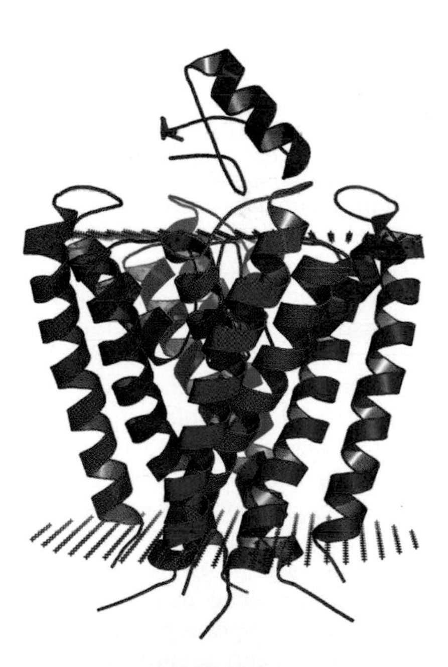

Figure 7.18 Helical bundles form a transmembrane potassium channel. The channel is shown in complex with charybdotoxin. These types of ion channels play a critical role in signaling processes and are attractive targets for treating various diseases. The structure reveals how charybdotoxin binds to the closed form of KcsA and thereby makes specific contact with the extracellular surface of the ion channel, resulting in pore blockage. This blockage yields direct structural information about an ion channel complexed to a peptide antagonist. (Courtesy of Huiling Chen and Ying Xu, University of Georgia.)

Prediction of the function of a protein from knowledge of its amino acid sequence is very difficult. One must remember that while the nucleotide sequence provides the initial code for peptides and proteins, many modifications can occur between the translation of mRNA and the availability of the fully functional protein. These modifications can affect many physical and chemical properties of the protein and thereby alter its folding, stability, activity, and function. Furthermore, peptides may be spliced together in different ways, thereby forming distinct proteins. Presently available methods for predicting the so-far

Figure 7.19 Example of a β-barrel. The outer membrane enzyme phospholipase A of *Escherichia coli* participates in the bacterium's secretion of colicins, which are toxins that are released into the environment to reduce the competition from other bacterial species. The protein forms a β-barrel, whose activity is regulated by reversible dimerization of the enzyme, that is, by the formation of a complex of two subunits (green and red). As a result of the dimerization, the barrels become functional channels for oxyanions. (Courtesy of Huiling Chen and Ying Xu, University of Georgia.)

uncharacterized function of a protein include a variety of similarity, homology, and structure comparisons with proteins of known sequence and function [61]. Also helpful in many cases are phylogenetic approaches and information from the Gene Ontology Project [68] (see Chapter 6). All these methods fall within the realm of bioinformatics and structural biology, and readers interested in the topic are encouraged to consult texts such as [5]. An extension of these computational methods is the field of **molecular dynamics**, which attempts to describe how molecules, for instance in proteins, move in order to achieve a specific function.

7.9 Localization

To serve their specific functions, proteins must be in the correct location within a cell or tissue. It is therefore of great interest to study which proteins are located where. An intriguing method is the attachment of the small **green fluorescent protein (GFP)** to a target protein [69, 70]. This attachment can be achieved through genetic engineering by expressing in the cell of interest a fusion protein consisting of the natural target protein with a GFP reporter linked to it (**Figure 7.20**). GFP was first isolated from a jellyfish, but many variants in different colors have since been engineered. GFP consists of a β-barrel containing a chromophore, which is a molecule that absorbs light of a certain wavelength and emits another wavelength. GFP is not limited to locating proteins but is widely used in molecular biology for reporting the presence of a variety of cellular components and the expression of genes that are of particular interest. In fact, the discovery of GFP has truly transformed cell biology, especially in conjunction with automated real-time systems for fluorescence microscopy in living cells. For example, Singleton and colleagues used enhanced GFP-signaling sensors to investigate the spatial and temporal processes of 30 signaling intermediates, including receptors, kinases, and adaptor proteins, within individual T cells during their activation [71]. Because of the importance of protein localization, a dedicated website has been established [72]. Martin Chalfie, Osamu Shimomura, and Roger Y. Tsien were awarded the 2008 Nobel Prize in Chemistry for the discovery and development of GFP.

A distinctly different methodology was developed by Richard Caprioli's laboratory at Vanderbilt University. This methodology scans tissue slices and applies MALDI individually to very many small areas in a tight grid. The result is a peptide spectrum for every grid point, and visualization reveals a very detailed picture of peptide abundances within the tissue section (**Figure 7.21**). This MALDI imaging method even permits the creation of 3D pictures of peptide abundances in tissues (**Figure 7.22**). For such an analysis, the tissue is thinly sliced and every slice is analyzed with 2D MALDI imaging. Computer software is used to stack up all these images to create a 3D impression [73].

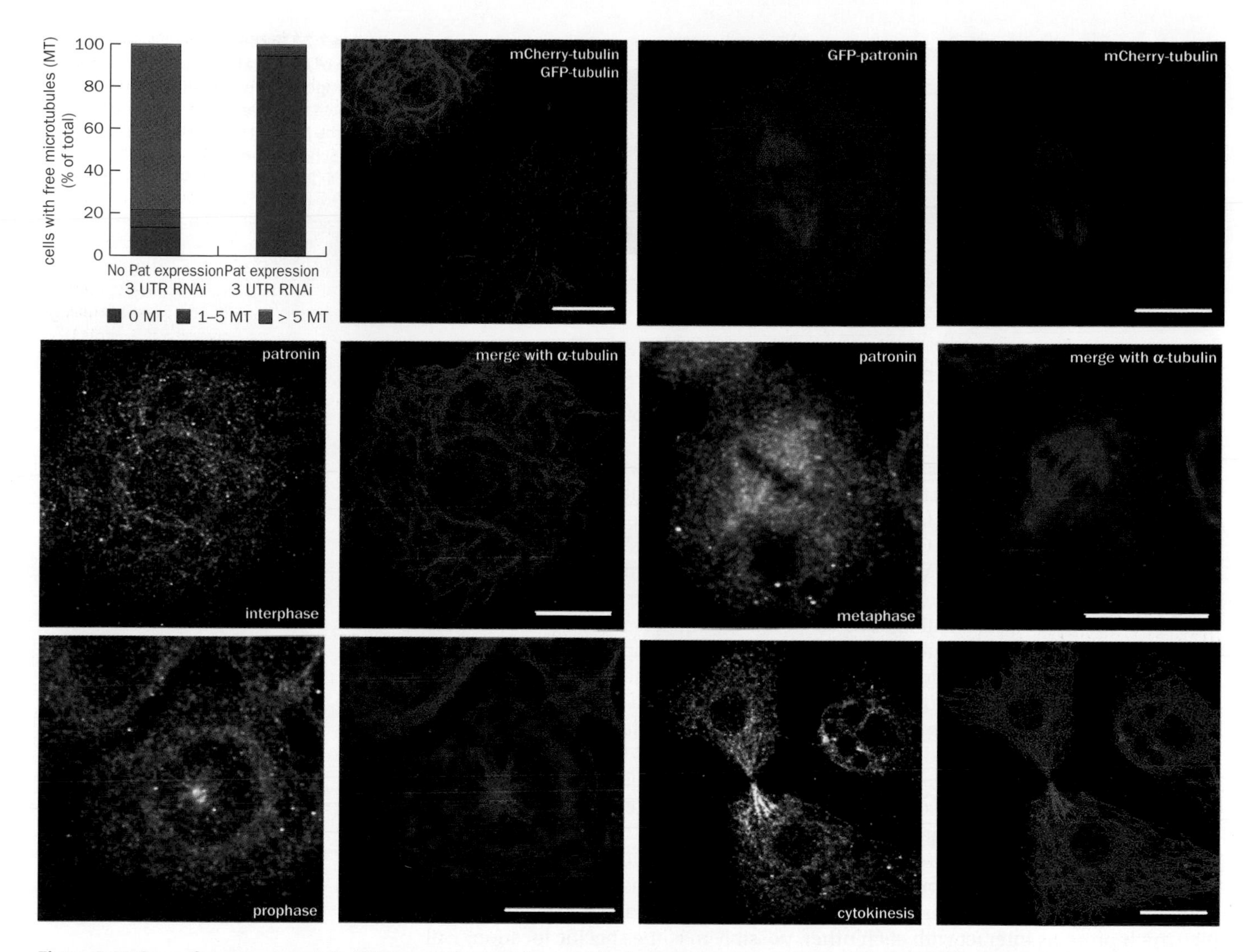

Figure 7.20 Green fluorescent protein (GFP) and other tagging proteins have become powerful tools for localization studies. Shown here are the specific intracellular locations of the proteins patronin and tubulin in different phases of the cell cycle. (From supplements to Goodwin SS & Vale RD. *Cell* 143 [2010] 263–274. With permission from Elsevier.)

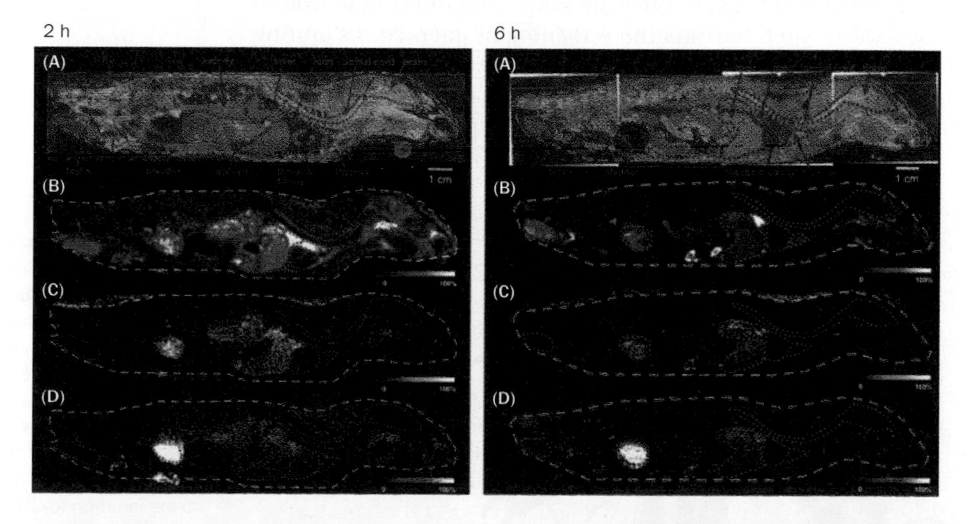

Figure 7.21 Mass spectrometry can be used to image the spatial distribution of a drug and related metabolites *in vivo*. Shown are cross-sectional images of a rat 2 and 6 h after dosing with the antipsychotic olanzapine. Organs are outlined in red. (A) Whole-body section across four gold-coated MALDI target plates. (B) MS/MS ion image of olanzapine within the cross-section. (C) MS/MS ion image of the first-pass derivative *N*-desmethylolanzapine. (D) MS/MS ion image of another first-pass derivative, 2-hydroxymethylolanzapine. The original drug, olanzapine, is visible throughout the body at 2 h, but shows decreased intensity in the brain and spinal cord after 6 h. The first-pass derivatives accumulate in the liver and bladder, rather than the brain. (From Reyzer ML & Caprioli RM. *Curr. Opin. Chem. Biol.* 11 [2006] 29–35. With permission from Elsevier.)

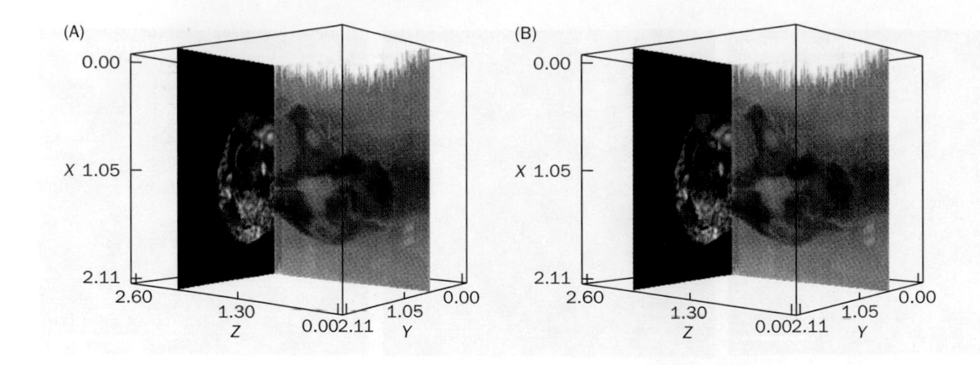

Figure 7.22 Reconstructed 3D localization image of a rat brain tumor. The images are composed of numerous 2D slices, each evaluated with mass spectrometry. (A) and (B) show the localization of different proteins. (From Schwamborn K & Caprioli RM. *Mol. Oncol.* 4 [2010] 529–538. With permission from Elsevier.)

7.10 Protein Activity and Dynamics

It should have become clear by now that proteins are drivers of dynamic activities at many temporal and spatial scales. Some proteins interact with other molecules on a timescale of seconds or even faster, as is the case with enzymes, whereas the dynamic changes in some proteins may take many years, as we can see in the very slow bonding of sugars to proteins in the lens of the eye, which clouds the lens in advanced age and can lead to cataracts [74]. Correspondingly, studying proteins very often means studying protein interactions. As is to be expected, these investigations can take very different forms. At the low extreme of the time and space spectrum, the main topic of interest is the docking of small ligands on proteins, which is the basis for enzyme action and most drugs; these actions will be discussed in Chapter 8. Slightly further up on the scale is the entire field of transcription factors and their crucial role in gene regulation, which we discussed in Chapter 6. A further step up, we find protein–protein interactions, which, owing to their importance and prevalence, are often abbreviated as PPI, and can be found in numerous manifestations [75].

PPIs exhibit many aspects worth investigating. First, there is great interest in complexes of a small number of peptides or proteins in quaternary structures. The relevant methods of investigation require the techniques of theoretical chemistry and algorithmic optimization, similar to those used for the analysis of ligand docking (**Figure 7.23**). Second, enormous effort in recent times has been devoted to PPI networks of medium to very large sizes (**Figure 7.24**). The rationale is that deciphering which proteins interact with each other, possibly in some specific location, will provide insights into the functioning of a cell as a complex system.

Several approaches have been applied to these types of investigations. Arguably the most prominent has been the yeast two-hybrid assay, which uses artificial fusion proteins inside the cell nucleus of a yeast to infer the binding partners of a protein [76, 77]. The assay is scalable, which permits the screening of interactions among very many proteins. A significant drawback of this method is the often very high number of false positives, which means that the test suggests many more interactions than truly exist in a living cell. For relatively small interaction networks, it is

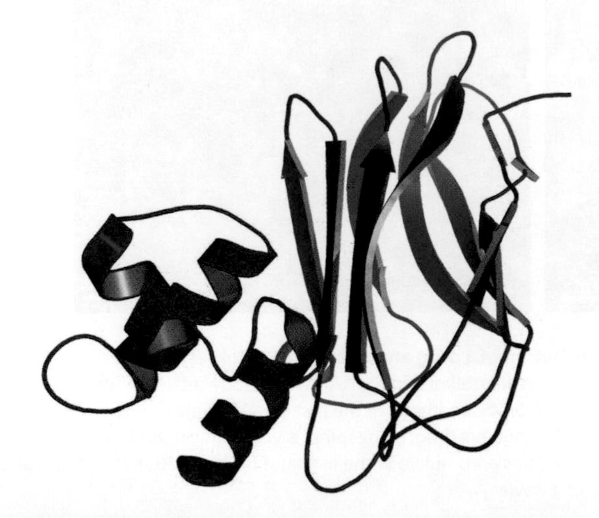

Figure 7.23 Interaction between two proteins forming the cohesin–dockerin complex of the cellulosome in the anaerobic bacterium *Clostridium thermocellum*. The cellulosome is a multiprotein complex for the efficient degradation of plant cell walls. The β-sheet domain of cohesin (blue) interacts predominantly with one of the helices of the dockerin (red). The structure provides an explanation for the lack of cross-species recognition between cohesin–dockerin pairs and thus provides a blueprint for the rational design, construction, and exploitation of these catalytic assemblies. (Courtesy of Huiling Chen and Ying Xu, University of Georgia.)

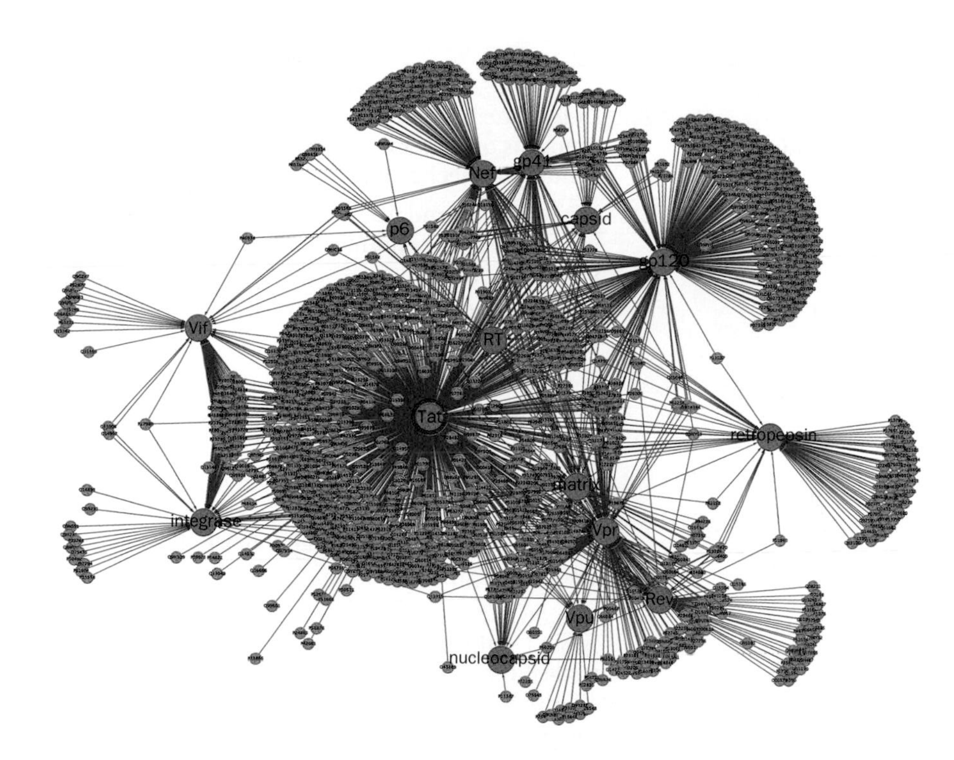

Figure 7.24 Network of 1785 unique protein–protein interactions between HIV and the human host. The graph was assembled from a database of the National Institute of Allergy and Infectious Diseases, Division of AIDS. Red nodes represent HIV proteins and light orange nodes represent host factors. It is unclear at present which interactions are direct and which are functionally relevant. (From Jäger S, Gulbahce N, Cimermancic P, et al. *Methods* 53 [2011] 13–19. With permission from Elsevier.)

possible to verify or refute these interactions with co-localization studies or through **immunoprecipitation** with the help of specific antibodies [78]. It is also possible, at least to some degree, to use computational validation methods [79]. Once PPIs have been identified and validated, graph methods of static network analysis are used to characterize the topological features of the interaction network, such as types of nodes and connectivity patterns, which we discussed in Chapter 3. The effective visualization of such networks is a challenging topic of ongoing research.

The analysis of PPIs becomes much more complicated when their dynamics is of interest, or even if one just needs to know the directionality of the edges in the network. Typical examples are signal transduction systems, where it is necessary to identify which component is triggering changes in other components downstream (**Figure 7.25**). A different example of a short-term PPI is the transport of a protein by another protein, for example between the nucleus and the cytoplasm. Similarly, we have already mentioned the process of ubiquitination, which marks a protein for degradation. PPIs with directionality and dynamics can be analyzed with differential equation models. As a prominent class of examples, namely, signaling systems, are discussed in Chapter 9.

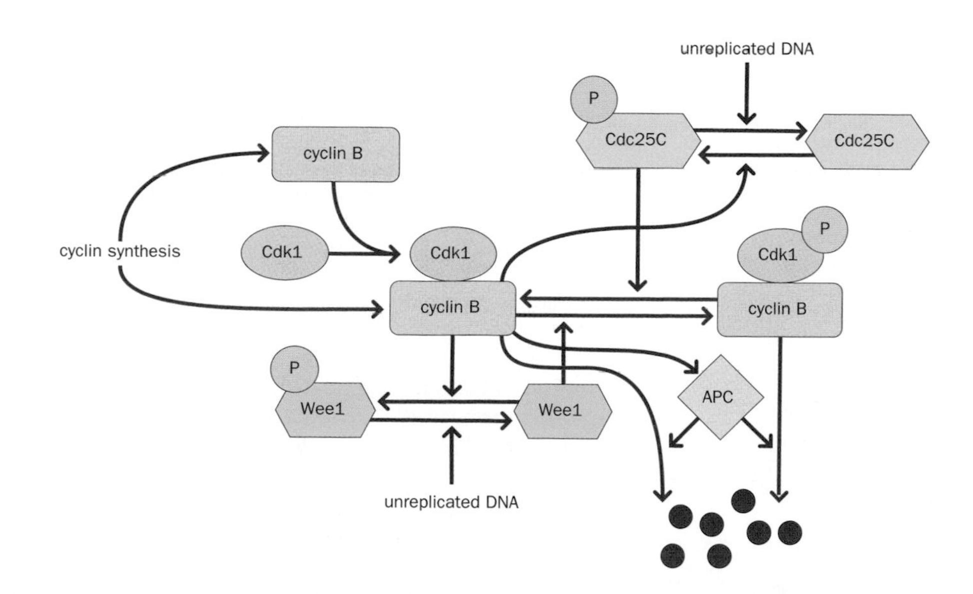

Figure 7.25 Protein–protein interaction network controlling the entry of cells into mitosis. Leland Hartwell, Tim Hunt, and Paul Nurse received the 2001 Nobel Prize in Physiology or Medicine for their discoveries of key regulators of this process. While the details are not critical here, the diagram illustrates that there is clear directionality in some signals. Unreplicated DNA blocks the entry of cells into mitosis. The process is governed by positive feedback between the Cdk1–cyclin B complex and Cdc25 and by double-negative feedback between Cdk1 and Wee1. This arrangement renders it possible for gradual mitotic cyclin synthesis to lead to a switch-like increase in Cdk1 activation. Cdk1 is inactivated owing to a negative feedback in which Cdk1 indirectly targets cyclin B for degradation through APC. (Adapted from Zwolak J, Adjerid N, Bagci EZ, et al. *J. Theor. Biol.* 260 [2009] 110–120. With permission from Elsevier.)

EXERCISES

7.1. Search the literature to determine the relative frequency of amino acids in different biological systems.

7.2. Compute the theoretical number of different proteins consisting of 100, 1000, or 27,000 amino acids.

7.3. Make intuitive predictions of how much the total variability of proteins of a given length would increase if pyrrolysine and selenocysteine were found as often as standard amino acids. Check your predictions with calculations.

7.4. Is the number of possible proteins with a fixed number of amino acids affected by the relative frequencies with which different amino acids are found? Argue for or against a significant effect and/or do some computations.

7.5. Explore the visualization options of the PDB and summarize the highlights in a report.

7.6. Search the literature and the Internet for important physical, chemical, and biological properties of the widely found proteins RuBisCO, collagen, and actin.

7.7. Explore whether plants have proteins, or even classes of proteins, that are not found in animals.

7.8. Search the literature for a visualization of the allosteric inhibition site of an enzyme.

7.9. Retrieve from the literature a picture of an enzyme as it binds to its substrate.

7.10. Discuss differences and similarities between prokaryotic flagella and eukaryotic cilia and their molecular motors.

7.11. In a short report, explain differences between hormones and pheromones. Include a discussion of broad chemical categories that contain hormones and pheromones.

7.12. Discuss in a one-page report how nature accomplishes the enormous variety and adaptability among antibodies.

7.13. Discuss the terms "pro-inflammatory" and "anti-inflammatory" in the context of cytokines.

7.14. Discuss the concepts behind difference gel electrophoresis (DIGE).

7.15. Why is it necessary to cut proteins into peptides before using mass spectrometry? Find literature to back up your answers.

7.16. Discuss the concepts on which the iTRAQ method is based. Weigh its advantages and disadvantages in comparison with MALDI-TOF.

7.17. Discuss the concepts behind the MudPIT method, along with advantages and drawbacks.

7.18. Discuss in a report advantages and disadvantages of different separation techniques for proteins.

7.19. Explore the literature and the Internet and describe the concepts of computer algorithms that reconstitute peptides from digestion fragments.

7.20. Determine from the literature the minimal amounts of protein samples needed for different proteomics techniques.

7.21. List chemical and physical properties of amino acids and their residues that are useful for protein structure prediction.

7.22. Determine three reasons why protein structure prediction is challenging.

7.23. Survey the literature and the Internet for different types of membrane proteins. Write a report describing their properties and functions.

7.24. Explain in more detail the "positive feedback between the Cdk1–cyclin B complex and Cdc25 and the double-negative feedback between Cdk1 and Wee1" in Figure 7.25.

7.25. What would be the steps for setting up a dynamic model of the system in Figure 7.25? Once you have determined a strategy, study the article by Zwolak and collaborators [80] and compare your strategy and theirs.

7.26. According to the Central Dogma of molecular biology (see Chapter 6), DNA is transcribed into RNA, which is translated into protein, which in this case is an enzyme that catalyzes a metabolic reaction. Develop a simple cascaded model to describe these processes, as shown in **Figure 7.26**, where X_1, X_2, and X_3 are mRNA, protein, and a metabolite, respectively, X_4, X_5, and X_6 are precursors (explain what these should be), and X_7 is a transcription factor or activating metabolite. Set up equations with power-law functions (Chapter 4) and select reasonable parameter values. Analyze what happens if a gene is expressed. Study what happens if X_3 and X_7 are the same metabolite.

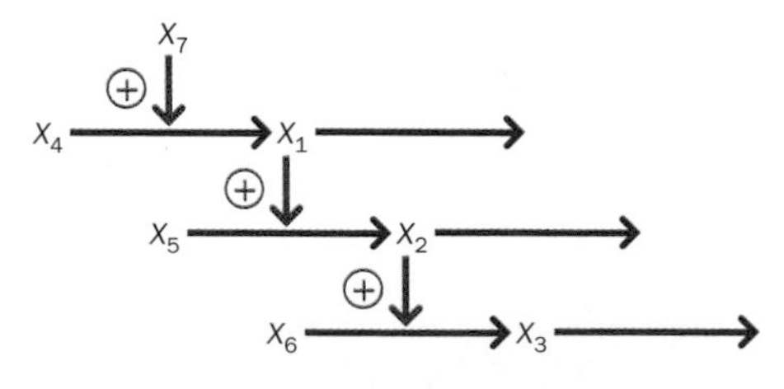

Figure 7.26 Schematic of the Central Dogma. X_1 represents a transcribed mRNA, X_2 a translated protein, which here is an enzyme, and X_3 a metabolite produced in the reaction catalyzed by X_2. X_4, X_5, and X_6 represent sources or precursors of each respective step. X_7 is a transcription factor or activating metabolite.

7.27. How would you model post-translational modifications and alternative splice variants in the model in Exercise 7.26?

7.28. Assume that metabolite X_3 in Exercise 7.26 is the substrate for a linear pathway beginning with Z_1, leading to Z_2, Z_3, Z_4, and so on, until the final product Z_{10} is synthesized. Suppose Z_{10} strongly represses transcription at the top level of the cascade. Compare simulation results with those in Exercise 7.26.

7.29. Suppose a protein can be unphosphorylated, phosphorylated in some position A, phosphorylated in some position B, or phosphorylated in positions A and B. Sketch a diagram of the different phosphorylation states and transitions between them. Design an ordinary differential equation (ODE) model describing the transitions and states. How do you assure that the total amount of protein does not change over time? Suppose an external regulator inhibits one or two particular transitions. What is its effect?

7.30. Consider the same phosphorylation states as in Exercise 7.29. Design a Markov model describing the transitions and states. Compare features of the model and of simulations with those in Exercise 7.29. Is it possible to account for external regulators?

7.31. Many signaling cascades are based on the phosphorylation and dephosphorylation of proteins and have a structure as shown in simplified form in **Figure 7.27**. Using methods of systems analysis with differential equations, set up a model for the cascade, select reasonable parameter values, and perform sufficiently many simulations to write a confident report about the dynamical features of the cascade.

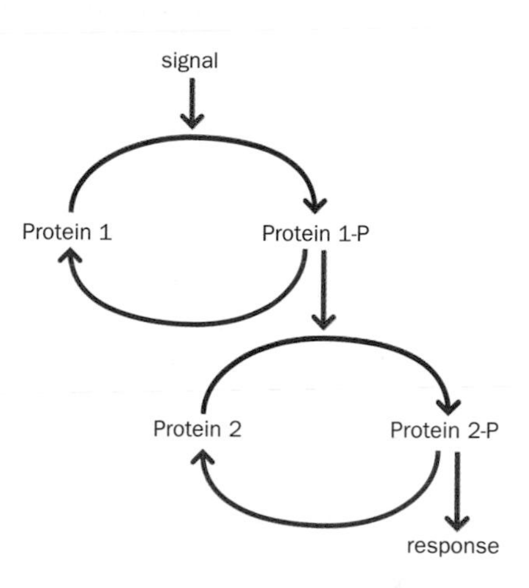

Figure 7.27 Simplified signaling cascade. The protein at either level may be phosphorylated ("-P") or dephosphorylated. Only the phosphorylated form of Protein 1 triggers the second level of the cascade.

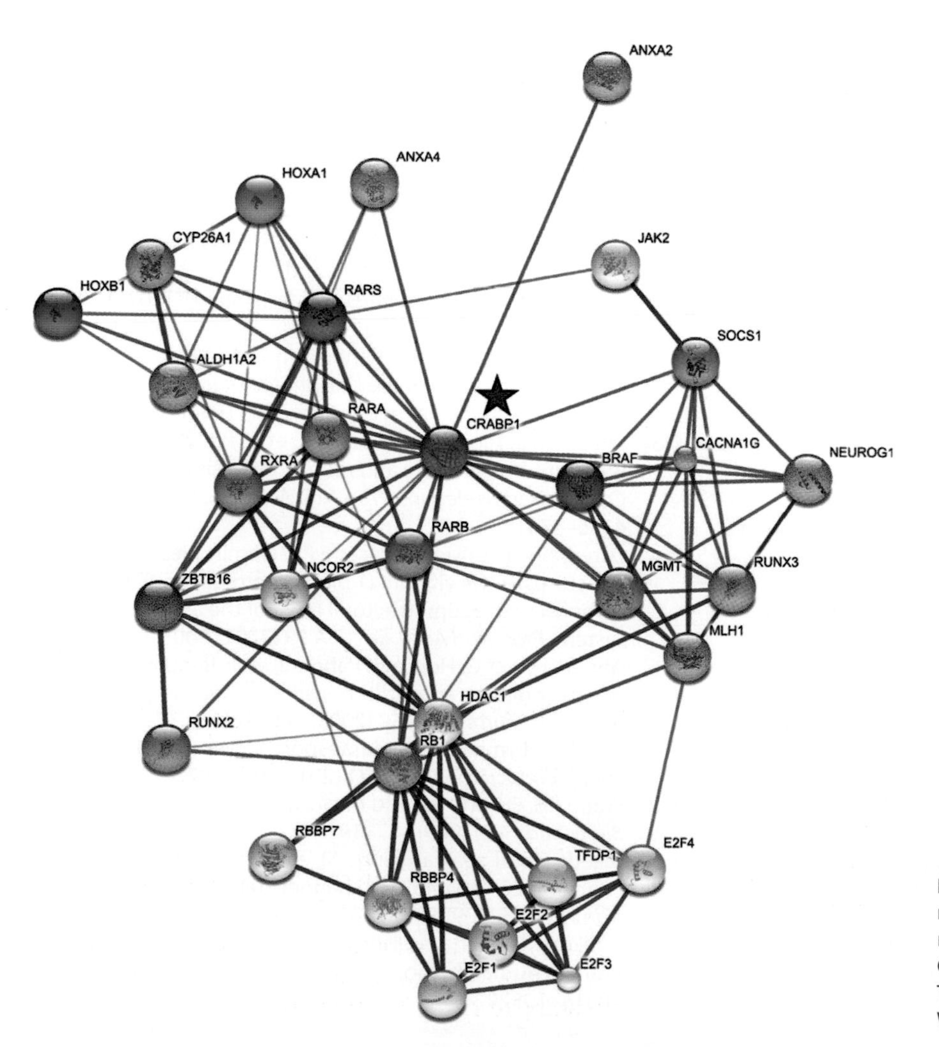

Figure 7.28 Protein–protein interaction network of CRABP1. It is evident that different nodes exhibit distinct connection patterns (see Chapter 3). (From Akama K, Horikoshi T, Nakayama T, et al. *Biochim. Biophys. Acta* 1814 [2011] 265–276. With permission from Elsevier.)

7.32. Search the literature and the Internet for a protein–protein interaction network whose structure was inferred with computational means. Write a one-page summary of the study.

7.33. Akama et al. [81] describe the protein–protein interaction network of the cellular retinoic acid binding protein CRABP1, which is involved in the development of the nervous system (**Figure 7.28**). What standard types of numerical features that characterize this network can you extract?

REFERENCES

[1] Protein Data Bank (PDB). http://www.pdb.org/pdb/home/home.do.
[2] Swiss-Prot. http://ca.expasy.org/sprot/.
[3] UniProt. http://www.uniprot.org/.
[4] BioGRID. http://www.thebiogrid.org/.
[5] Zvelebil M & Baum JO. Understanding Bioinformatics. Garland Science, 2008.
[6] The Human Protein Atlas. http://www.proteinatlas.org.
[7] Stadtman TC. Selenocysteine. *Annu. Rev. Biochem.* 65 (1996) 83–100.
[8] Atkins JF & Gesteland R. The 22nd amino acid. *Science* 296 (2002) 1409–1410.
[9] Brown TA. Genomes, 3rd ed. Garland Science, 2006.
[10] Zhou F, Olman V & Xu Y. Barcodes for genomes and applications. *BMC Bioinformatics* 9 (2008) 546.
[11] Zhang F, Saha S, Shabalina SA & Kashina A. Differential arginylation of actin isoforms is regulated by coding sequence-dependent degradation. *Science* 329 (2010) 1534–1537.
[12] Seppälä S, Slusky JS, Lloris-Garcerá P, et al. Control of membrane protein topology by a single C-terminal residue. *Science* 328 (2010) 1698–1700.
[13] Gsponer J, Futschik ME, Teichmann SA & Babu MM. Tight regulation of unstructured proteins: from transcript synthesis to protein degradation. *Science* 322 (2008) 1365–1368.
[14] Anfinsen CB. Principles that govern the folding of protein chains. *Science* 181 (1973) 223–330.
[15] Levinthal C. How to fold graciously. In Mossbauer Spectroscopy in Biological Systems: Proceedings of a Meeting held at Allerton House, Monticello, Illinois (JTP DeBrunner & E Munck, eds), pp 22–24. University of Illinois Press, 1969.
[16] Hatahet F & Ruddock LW. Protein disulfide isomerase: a critical evaluation of its function in disulfide bond formation. *Antioxid. Redox Signal.* 11 (2009) 2807–2850.
[17] PTM Statistics Curator: Automated Curation and Population of PTM Statistics from the Swiss-Prot Knowledgebase. http://selene.princeton.edu/PTMCuration/.
[18] Vellai T & Takacs-Vellai K. Regulation of protein turnover by longevity pathways. *Adv. Exp. Med. Biol.* 694 (2010) 69–80.
[19] Shcherbik N & Pestov DG. Ubiquitin and ubiquitin-like proteins in the nucleolus: multitasking tools for a ribosome factory. *Genes Cancer* 1 (2010) 681–689.
[20] Sherwood D & Cooper J. Crystals, X-rays and Proteins: Comprehensive Protein Crystallography. Oxford University Press, 2011.
[21] Jacobsen NE. NMR Spectroscopy Explained: Simplified Theory, Applications and Examples for Organic Chemistry and Structural Biology. Wiley, 2007.
[22] Ranson NA, Farr GW, Roseman AM, et al. ATP-bound states of GroEL captured by cryo-electron microscopy. *Cell* 107 (2001) 869–879.
[23] Clare DK, Bakkes PJ, van Heerikhuizen H, et al. An expanded protein folding cage in the GroEL–gp31 complex. *J. Mol. Biol.* 358 (2006) 905–911.
[24] Ago H, Kanaoka Y, Irikura D, et al. Crystal structure of a human membrane protein involved in cysteinyl leukotriene biosynthesis. *Nature* 448 (2007) 609–612.
[25] Zhao G, Johnson MC, Schnell JR, et al. Two-dimensional crystallization conditions of human leukotriene C4 synthase requiring a particularly large combination of specific parameters. *J. Struct. Biol.* 169 (2010) 450–454.
[26] Ranson NA, Clare DK, Farr GW, et al. Allosteric signaling of ATP hydrolysis in GroEL–GroES complexes. *Nat. Struct. Mol. Biol.* 13 (2006) 147–152.
[27] Schmidt-Krey I. Electron crystallography of membrane proteins: two-dimensional crystallization and screening by electron microscopy. *Methods* 41 (2007) 417–426.
[28] Schmidt-Krey I & Rubinstein JL. Electron cryomicroscopy of membrane proteins: specimen preparation for two-dimensional crystals and single particles. *Micron* 42 (2011) 107–116.
[29] Jmol: An Open-Source Java Viewer for Chemical Structures in 3D. http://jmol.sourceforge.net/.
[30] Brylinski M & Skolnick J. FINDSITE^{LHM}: a threading-based approach to ligand homology modeling. *PLoS Comput. Biol.* 5 (2009) e1000405.
[31] Raghavendra SK, Nagampalli P, Vasantha S, Rajan S. Metal induced conformational changes in human insulin: Crystal structures of Sr^{2+}, Ni^{2+} and Cu^{2+} complexes of human insulin. *Protein and Peptide Letters* 18 (2011) 457–466.
[32] Field CB, Behrenfeld MJ, Randerson JT & Falkowski P. Primary production of the biosphere: integrating terrestrial and oceanic components. *Science* 281 (1998) 237–240.
[33] Friedman L, Higgin JJ, Moulder G, et al. Prolyl 4-hydroxylase is required for viability and morphogenesis in *Caenorhabditis elegans*. *Proc. Natl Acad. Sci. USA* 97 (2000) 4736–4741.
[34] Otterbein LR, Cosio C, Graceffa P & Dominguez R. Crystal structures of the vitamin D-binding protein and its complex with actin: structural basis of the actin-scavenger system. *Proc. Natl Acad. Sci. USA* 99 (2002) 8003–8008.
[35] Ahnert-Hilger G, Holtje M, Pahner I, et al. Regulation of vesicular neurotransmitter transporters. *Rev. Physiol. Biochem. Pharmacol.* 150 (2003) 140–160.
[36] Dixon JB. Lymphatic lipid transport: sewer or subway? *Trends Endocrinol. Metab.* 21 (2010) 480–487.
[37] Wade RH. On and around microtubules: an overview. *Mol. Biotechnol.* 43 (2009) 177–191.
[38] Maeda S, Nakagawa S, Suga M, et al. Structure of the connexin 26 gap junction channel at 3.5 Å resolution. *Nature* 458 (2009) 597–602.
[39] Birukou I, Soman J & Olson JS. Blocking the gate to ligand entry in human hemoglobin. *J. Biochem.* 286 (2011) 10515–10529

[40] Pietras K. Increasing tumor uptake of anticancer drugs with imatinib. *Semin. Oncol.* 31 (2004) 18–23.
[41] Kearns DB. A field guide to bacterial swarming motility. *Nat. Rev. Microbiol.* 8 (2010) 634–644.
[42] Lindemann CB & Lesich KA. Flagellar and ciliary beating: the proven and the possible. *J. Cell Sci.* 123 (2010) 519–528.
[43] Bessa J & Bachmann MF. T cell-dependent and -independent IgA responses: role of TLR signalling. *Immunol. Invest.* 39 (2010) 407–428.
[44] Lu C, Mi LZ, Grey MJ, et al. Structural evidence for loose linkage between ligand binding and kinase activation in the epidermal growth factor receptor. *Mol. Cell. Biol.* 30 (2010) 5432–5443.
[45] Holmes MA. Computational design of epitope-scaffolds allows induction of antibodies specific for a poorly immunogenic HIV vaccine epitope. *Structure* 18 (2010) 1116–1126.
[46] Nemazee D. Receptor editing in lymphocyte development and central tolerance. *Nat. Rev. Immunol.* 6 (2006) 728–740.
[47] Brooks CL, Schietinger A, Borisova SN, et al. Antibody recognition of a unique tumor-specific glycopeptide antigen. *Proc. Natl Acad. Sci. USA* 107 (2010) 10056–10061.
[48] Weiss A, Shields R, Newton M, et al. Ligand–receptor interactions required for commitment to the activation of the interleukin 2 gene. *J. Immunol.* 138 (1987) 2169–2176.
[49] Davis MM, Krogsgaard M, Huse M, et al. T cells as a self-referential, sensory organ. *Annu. Rev. Immunol.* 25 (2007) 681–695.
[50] Bell JK, Botos I, Hall PR, et al. The molecular structure of the Toll-like receptor 3 ligand-binding domain. *Proc. Natl Acad. Sci. USA* 102 (2005) 10976–10980.
[51] Moreland JL, Gramada A, Buzko OV, et al. The molecular biology toolkit (MBT): a modular platform for developing molecular visualization applications. *BMC Bioinformatics* 6 (2005) 21.
[52] Oda K & Kitano H. A comprehensive map of the Toll-like receptor signaling network. *Mol. Syst. Biol.* 2 (2006) 2006.0015.
[53] Carafoli F, Bihan D, Stathopoulos S, et al. Crystallographic insight into collagen recognition by discoidin domain receptor 2. *Structure* 17 (2009) 1573–1581.
[54] Bloemendal H. Lens proteins. *CRC Crit. Rev. Biochem.* 12 (1982) 1–38.
[55] Palanivelu DV, Kozono DE, Engel A, et al. Co-axial association of recombinant eye lens aquaporin-0 observed in loosely packed 3D crystals. *J. Mol. Biol.* 355 (2006) 605–611.
[56] Rabilloud T. Variations on a theme: changes to electrophoretic separations that can make a difference. *J. Proteomics* 73 (2010) 1562–1572.
[57] Waller LN, Shores K & Knapp DR. Shotgun proteomic analysis of cerebrospinal fluid using off-gel electrophoresis as the first-dimension separation. *J. Proteome Res.* 7 (2008) 4577–4584.
[58] Minden JS, Dowd SR, Meyer HE & Stühler K. Difference gel electrophoresis. *Electrophoresis* 30(Suppl 1) (2009) S156–S161.
[59] Latterich M, Abramovitz M & Leyland-Jones B. Proteomics: new technologies and clinical applications. *Eur. J. Cancer* 44 (2008) 2737–2741.
[60] Yates JR, Ruse CI & Nakorchevsky A. Proteomics by mass spectrometry: approaches, advances, and applications. *Annu. Rev. Biomed. Eng.* 11 (2009) 49–79.
[61] ExPASy. http://www.expasy.ch/.
[62] Sen AK, Darabi J & Knapp DR. Design, fabrication and test of a microfluidic nebulizer chip for desorption electrospray ionization mass spectrometry. *Sens. Actuators B Chem.* 137 (2009) 789–796.
[63] Grey AC, Chaurand P, Caprioli RM & Schey KL. MALDI imaging mass spectrometry of integral membrane proteins from ocular lens and retinal tissue. *J. Proteome Res.* 8 (2009) 3278–3283.
[64] Das R & Baker D. Macromolecular modeling with Rosetta. *Annu. Rev. Biochem.* 77 (2008) 363–382.
[65] Skolnick J. In quest of an empirical potential for protein structure prediction. *Curr. Opin. Struct. Biol.* 16 (2006) 166–171.
[66] Skolnick J & Brylinski M. FINDSITE: a combined evolution/structure-based approach to protein function prediction. *Brief Bioinform.* 10 (2009) 378–391.
[67] Lee EH, Hsin J, Sotomayor M, et al. Discovery through the computational microscope. *Structure* 17 (2009) 1295–1306.
[68] The Gene Ontology Project. http://www.geneontology.org/.
[69] Phillips GJ. Green fluorescent protein—a bright idea for the study of bacterial protein localization. *FEMS Microbiol. Lett.* 204 (2001) 9–18.
[70] Stepanenko OV, Verkhusha VV, Kuznetsova IM, et al. Fluorescent proteins as biomarkers and biosensors: throwing color lights on molecular and cellular processes. *Curr. Protein Pept. Sci.* 9 (2008) 338–369.
[71] Singleton KL, Roybal KT, Sun Y, et al. Spatiotemporal patterning during T cell activation is highly diverse. *Science Signaling* 2 (2009) ra15.
[72] PSLID—Protein Subcellular Location Image Database. http://pslid.org/start.html.
[73] Sinha TK, Khatib-Shahidi S, Yankeelov TE, et al. Integrating spatially resolved three-dimensional MALDI IMS with *in vivo* magnetic resonance imaging. *Nat. Methods* 5 (2008) 57–59.
[74] Ulrich P & Cerami A. Protein glycation, diabetes, and aging. *Recent Prog. Horm. Res.* 56 (2001) 1–21.
[75] Nibbe RK, Chowdhury SA, Koyuturk M, et al. Protein–protein interaction networks and subnetworks in the biology of disease. *Wiley Interdiscip. Rev. Syst. Biol. Med.* 3 (2011) 357–367.
[76] Chautard E, Thierry-Mieg N & Ricard-Blum S. Interaction networks: from protein functions to drug discovery. A review. *Pathol. Biol. (Paris)* 57 (2009) 324–333.
[77] Fields S. Interactive learning: lessons from two hybrids over two decades. *Proteomics* 9 (2009) 5209–5213.
[78] Monti M, Cozzolino M, Cozzolino F, et al. Puzzle of protein complexes *in vivo*: a present and future challenge for functional proteomics. *Expert Rev. Proteomics* 6 (2009) 159–169.
[79] Tong AH, Drees B, Nardelli G, et al. A combined experimental and computational strategy to define protein interaction networks for peptide recognition modules. *Science* 295 (2002) 321–324.
[80] Zwolak J, Adjerid N, Bagci EZ, et al. A quantitative model of the effect of unreplicated DNA on cell cycle progression in frog egg extracts. *J. Theor. Biol.* 260 (2009) 110–120.
[81] Akama K, Horikoshi T, Nakayama T, et al. Proteomic identification of differentially expressed genes in neural stem cells and neurons differentiated from embryonic stem cells of cynomolgus monkey (*Macaca fascicularis*) *in vitro*. *Biochim. Biophys. Acta* 1814 (2011) 265–276.

FURTHER READING

Chautard E, Thierry-Mieg N & Ricard-Blum S. Interaction networks: from protein functions to drug discovery. A review. *Pathol. Biol. (Paris)* 57 (2009) 324–333.

Latterich M, Abramovitz M & Leyland-Jones B. Proteomics: new technologies and clinical applications. *Eur. J. Cancer* 44 (2008) 2737–2741.

Nibbe RK, Chowdhury SA, Koyuturk M, et al. Protein–protein interaction networks and subnetworks in the biology of disease. *Wiley Interdiscip. Rev. Syst. Biol. Med.* 3 (2011) 357–367.

Shenoy SR & Jayaram B. Proteins: sequence to structure and function—current status. *Curr. Protein Pept. Sci.* 11 (2010) 498–514.

Twyman R. Principles of Proteomics, 2nd ed. Garland Science, 2014.

Yates JR, Ruse CI & Nakorchevsky A. Proteomics by mass spectrometry: approaches, advances, and applications. *Annu. Rev. Biomed. Eng.* 11 (2009) 49–79.

Zvelebil M & Baum JO. Understanding Bioinformatics. Garland Science, 2008.

Metabolic Systems

8

When you have read this chapter, you should be able to:

- Identify and characterize the components of metabolic systems
- Describe conceptually how the components of metabolic systems interact
- Understand the basics of mass action and enzyme kinetics
- Be aware of the various data sources supporting metabolic analyses
- Associate different types of metabolic data with different modeling tasks
- Explain different purposes of metabolic analyses

Our genome is sometimes called the blueprint of who we are. We have seen in previous chapters that this notion is somewhat simplified, because the macro- and micro-environments have significant roles throughout the biological hierarchy, including the expression of genes. Nonetheless, our genes do contain the code that ultimately determines the machinery of life. The lion's share of this machinery is present in the form of proteins, and, like most machines, they take input and convert it into output. In the case of living systems, the vast majority of both inputs and outputs consist of metabolites. Most prominently, proteins convert metabolites, including food or substrate, into energy and into a slew of other metabolites ranging from sugars and amino acids to vitamins and neurotransmitters. Signal transduction is sometimes seen as distinct from metabolic action, but it is also often accomplished through metabolic changes, such as the phosphorylation or glycosylation of proteins or the synthesis or degradation of signaling lipids. Frequently, it is at the level of metabolites that a normal physiological process becomes pathological and health turns into disease. A gene may be mutated but, on its own, this change does not cause problems. However, if the mutation leads to an abnormal protein or regulatory mechanism and thereby affects the dynamics of a metabolic pathway, it may ultimately cause an excess or dearth of specific metabolites, with consequences that in some cases might be tolerable, but in other cases can compromise the affected cell and possibly even kill it.

DNA, proteins, and metabolites are composed of the same "organic" atoms (mainly C, O, N, and H), and there is really no essential biochemical difference between them, except that DNA and proteins are large molecules with particular chemical structures, while metabolites in the strict sense are much smaller by comparison. For instance, the paradigm metabolite glucose ($C_6H_{12}O_6$) has a molecular mass of $(6 \times 12 + 12 \times 1 + 6 \times 16) = 180$, whereas hexokinase, the enzyme that converts glucose into glucose 6-phosphate, has a molecular mass of roughly 100,000 daltons (Da) [1] (**Figure 8.1**). The average molecular mass of a nucleotide in DNA is about 330 Da, which for the average-sized human gene of roughly 3000 bases corresponds to a molecular mass of about 1,000,000 Da. The largest known human

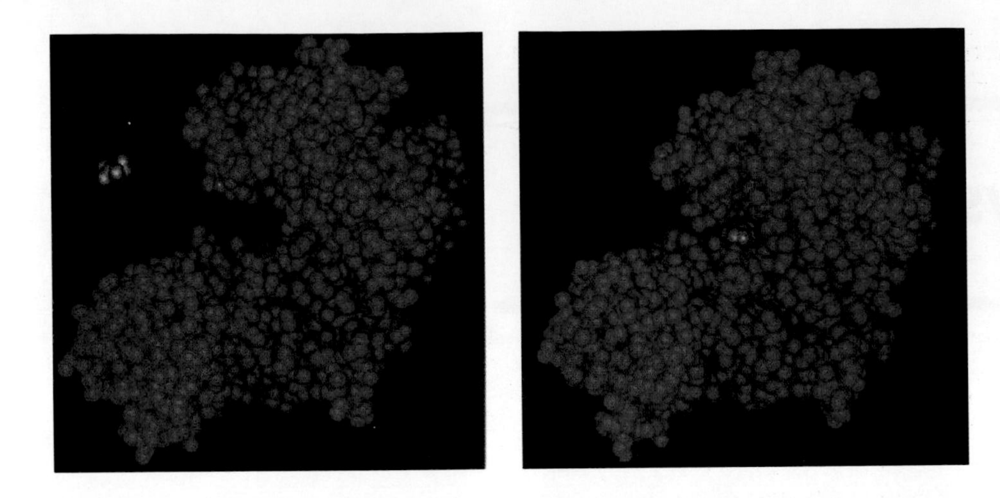

Figure 8.1 Typical metabolites are small in comparison with proteins and DNA. The enzyme hexokinase (blue) binds a molecule of glucose (red). Note the huge difference in sizes of the two molecules. (Courtesy of Juan Cui and Ying Xu, University of Georgia.)

gene codes for dystrophin, a key protein in muscle, which, if altered, is one of the root causes of muscular dystrophy. This gene consists of 2.4 million bases and thus has a molecular mass of almost 800 million daltons [2]. Indeed, a typical metabolite is tiny in comparison with proteins and nucleic acids.

The conversion of metabolites into other metabolites has been the bread and butter of traditional biochemistry. This conversion typically occurs in several steps that form metabolic pathways. Many biochemistry books are organized according to these pathways, containing chapters on central metabolism, energy metabolism, amino acid metabolism, fatty acid metabolism, and so on, sometimes including less prevalent pathways such as the synthesis of poisons in insects and spiders or pathways leading to secondary metabolites, such as color pigments, which may not be absolutely required for survival. In reality, metabolism is not neatly arranged in pathways, but pathways are merely the most obvious contributors to complex regulated pathway systems, just as interstate highways are the most prominent features of a much larger and highly connected network of roads, streets, and alleyways.

Traditionally, biochemistry has mostly focused on the individual steps a cell employs to convert different metabolites into each other. Of particular importance in these discussions are enzymes, which we discussed in Chapter 7. Each enzyme facilitates one or maybe a few of the specific steps in a string of metabolic conversions. In recent years, much of the emphasis on single reaction steps has shifted toward high-throughput methodologies of **metabolomics** that try to characterize the entire metabolic state of a living system (or subsystem) at a given point in time. A good example is a mass spectrogram, which is capable of quantitatively assessing the relative amounts of thousands of metabolites in a biological sample (see later).

This chapter is very clearly not an attempt to review biochemistry, and it will mention specific pathways only in the context of illustrative examples. Rather, it tries to describe some of the basic and general concepts of biochemistry and metabolomics that are particularly pertinent for analyses in computational systems biology. It begins with the characterization of enzyme-catalyzed reactions, discusses means of controlling the rates of these reactions, establishes the foundation for mathematical methods that capture the dynamics of metabolic systems, and lists some of the methods of data generation and the comprehensive resources of information regarding metabolites that are presently available.

BIOCHEMICAL REACTIONS

8.1 Background

Metabolism is the sequential conversion of organic compounds into other compounds. As chemical reactions, these conversions must of course obey the laws of chemistry and physics. Intriguingly, thermodynamics tells us that for the unaided conversion of one chemical compound into another to take place, the latter must be in a lower-energy state than the former. Are biochemistry and metabolism at odds with one of the fundamental principles of physics by converting food into metabolites of high energy content, such as amino acids or lipids? The situation is particularly puzzling in plants, where the "food" consists merely of carbon dioxide, CO_2, which has very low energy. How is the production of high-energy compounds possible?

The answer, which of course is fully compatible with the laws of thermodynamics, consists of three components:

- Plants are able to utilize sunlight as energy for converting CO_2 into metabolites of higher energy content.
- In all organisms, the metabolic production of a desired high-energy molecule is "paid for" by the conversion of another highly energetic molecule into a lower-energy molecule. For instance, the conversion of glucose into glucose 6-phosphate is coupled with the conversion of the high-energy molecule adenosine triphosphate (ATP) into the lower-energy molecule adenosine diphosphate (ADP).
- Highly energetic molecules in cells do not simply degrade into lower-energy molecules, as thermodynamics might suggest. Instead, they can only move to the lower-energy state if they transit an intermediate state of even higher energy. This transition is therefore not automatic, but must be facilitated, which is accomplished with enzymes (see Chapter 7 and **Box 8.1**). Indeed, the vast majority of metabolic reactions require an enzyme, and they very often involve a secondary substrate or co-factor that provides the energy for the creation of a higher-energy molecule from a lower-energy molecule. The involvement of an enzyme speeds up the reaction rate enormously and furthermore affords the possibility of regulation, in which one or more metabolites influence the degree of enzyme activity and thereby the rate of the metabolic reaction.

For the purposes of metabolic systems analysis, we often do not need to know much more about the molecular mechanisms associated with an enzyme than that the enzyme enables the transition from substrate to product and that its activity can be regulated with various mechanisms that act on different timescales

BOX 8.1 THE FUNCTION OF AN ENZYME DEMONSTRATED WITH AN ANALOGY

The function of an enzyme is illustrated in **Figure 1A**, where the original metabolite (typically called the substrate) with a high-energy state (red) must be briefly elevated to a transition state of even higher energy (blue), before it moves to the product state with lower energy. The enzyme temporarily provides the necessary activation energy E_A. An intuitive analogy is popcorn (Figure 1B). In its original state, consisting of cold corn kernels, its energy is constrained: although the popcorn is in a pan high above the floor, it cannot fall down to the lower-energy level of the floor. Now we turn on the heat (activation energy), which releases the stored energy and allows each kernel to jump during its pop (the transition state), and it is indeed possible that some kernels will jump over the rim of the pan onto the floor.

Figure 1 Processes overcoming thresholds of high energy. (A) The conversion of a metabolic substrate with energy E_S into a product with lower energy E_P requires activation energy E_A that exceeds a threshold (blue). (B) The situation in (A) is similar to popping corn, where individual kernels may receive enough (heat) energy to jump out of the pan and onto the floor.

(see Chapter 7 and later in this chapter). What we do need to explore in detail is how enzymatic processes are translated into mathematical models.

8.2 Mathematical Formulation of Elementary Reactions

To ease our way into this translation task, we begin with the simpler situation in which a chemical compound degrades over time without the involvement of an enzyme or co-factor. This situation is actually not very interesting biochemically, but a good start for model development. It has two prominent applications. One is radioactive decay, in which radionuclides spontaneously disintegrate and the collective disintegration process is proportional to the presently existing amount. The second is a diffusion process, in which the transported amount of material is proportional to the current amount.

The proportionality in both cases gives us a direct hint for how to set up a describing equation: namely, the change at time t is proportional to the amount at time t. Recalling Chapters 2 and 4, change is expressed mathematically as the derivative with respect to time, the amount is a function of time, and proportional implies a linear function. Putting the pieces together into an equation yields

$$\frac{dX}{dt} = \dot{X} = -k \cdot X. \tag{8.1}$$

The right-hand side of this **ordinary differential equation (ODE)** contains three items that are reminiscent of the SIR model in Chapter 2. X is the amount (for example, of the radionuclide), and one often refers to X generically as a pool (of molecules). The second component is the rate constant k, which is always positive (or zero) and constant over time. It quantifies how many units of X are changing per time unit. Finally, the minus sign indicates that the change ($\dot{X}$) is in the negative direction. In other words, material disappears rather than accumulates. The formulation in (8.1) as a description of a chemical process is actually based on considerations of **statistical mechanics** and thermodynamics and was proposed more than one hundred years ago by Svante Arrhenius (1859–1927).

As we discussed in Chapter 4, (8.1) is a linear differential equation that describes exponential behavior. Suppose now that X is converted into Y and that the disappearance of X is captured well by (8.1). Because all material leaving pool X moves into pool Y, it is easy to see that the change in Y must be equal to the change in X, except that the two have opposite signs. The dynamics of the two variables is therefore easily described as

$$\begin{aligned} \dot{X} &= -kX, \\ \dot{Y} &= kX. \end{aligned} \tag{8.2}$$

We are allowed to add the two differential equations, which yields

$$\dot{X} + \dot{Y} = 0 \tag{8.3}$$

and has the following interpretation. The total change in the system, consisting of the change in X plus the change in Y, equals zero. There is no overall change. This makes sense, because material is just flowing from one pool to the other, while no material is added or lost.

Many metabolic reactions involve two substrates and are therefore called bimolecular. Their mathematical description is constructed in analogy to the one-substrate case and leads to a differential equation where the right-hand side involves the product of the two substrates. Specifically, suppose X_1 and X_2 are the substrates of a bimolecular reaction that generates product X_3. Then the increase in the concentration of X_3 is given as

$$\dot{X}_3 = k_3 X_1 X_2. \tag{8.4}$$

Note that the substrates enter the equation as a product and not a sum, even though one might speak of adding a second substrate or formulate the reaction as $X_1 + X_2 \rightarrow X_3$. The reason for this formulation can be traced back to thermodynamics and to the fact that the two molecules have to come into physical contact within the physical space where the reaction happens, which is a matter of probability and leads to the product.

Because X_3 is the recipient of material and does not affect its own synthesis, the right-hand side is positive and does not depend on X_3, and its concentration continues to increase as long as X_1 and X_2 are available. For every molecule of X_3 that is produced, one molecule of X_1 and one molecule of X_2 are used up. Therefore, the loss in either one substrate is

$$\dot{X}_1 = \dot{X}_2 = -\dot{X}_3 = -k_3 X_1 X_2. \tag{8.5}$$

It is also possible that X_3 is produced from two molecules of type X_1 rather than from X_1 and X_2. In this case, the describing equations are

$$\begin{aligned} \dot{X}_1 &= -2k_3 X_1^2, \\ \dot{X}_3 &= k_3 X_1^2. \end{aligned} \tag{8.6}$$

The product of X_1 and X_2 in (8.5) becomes X_1^2 in both equations of (8.6), and one says that the process is of second order with respect to X_1. The first equation in (8.6) furthermore contains the stoichiometric factor 2, which does not enter the second equation. The reason is that X_1 is used up twice as fast as X_3 is produced, because two molecules of X_1 are needed to generate one molecule of X_3.

The mathematical formulations discussed here are the foundation of **mass action kinetics**, which was introduced about 150 years ago [3–5]. According to this widely used analytical framework, all substrates enter a reaction term as factors in a product, where powers reflect the number of contributing molecules of each type. The term also contains a rate constant, which is positive (or possibly zero) and does not change over time.

Mass action formulations implicitly assume that many substrate molecules are available and that they are freely moving about in a homogeneous medium. In many cases, and especially in living cells, these assumptions are not really true, but they do provide good approximations to realistic metabolic systems. They are frequently used because more accurate formulations are incomparably more complicated. For instance, one could more realistically consider the reaction between X_1 and X_2 as a random encounter (stochastic) process. While intuitively plausible, a model of such a process requires heavy-duty mathematics for further analysis [6, 7].

8.3 Rate Laws

Most biochemical reactions in a metabolic system are catalyzed by enzymes. About a century ago, the biochemists Henri, Michaelis, and Menten proposed a mechanism and a mathematical formula describing this process [8, 9]. They postulated that the substrate S and the catalyzing enzyme E reversibly form an intermediary complex (ES), which subsequently breaks apart and irreversibly yields the reaction product P while releasing the enzyme unchanged and ready for recycling. The diagram of the mechanism with typical rate constants for all steps is shown in **Figure 8.2**. Modern methods of transient and single-molecule kinetics have shown that this simple scheme is somewhat simplistic and that many reactions in fact

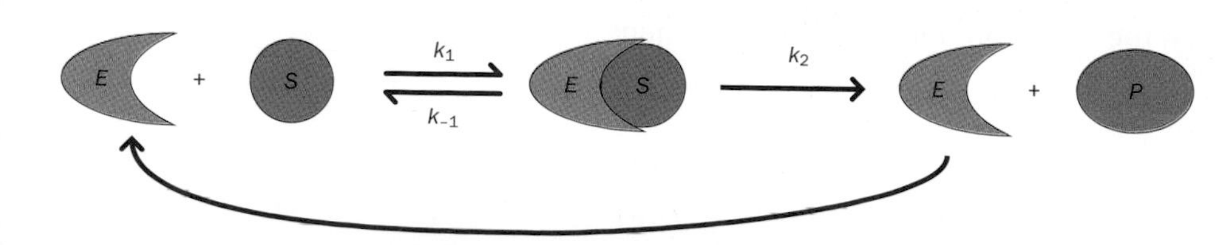

Figure 8.2 Conceptual model of an enzyme-catalyzed reaction mechanism, as proposed by Michaelis and Menten. A substrate S and an enzyme E reversibly form a complex (ES), which may either return S and E or lead to the product P, thereby releasing the free enzyme, which may be used again.

consist of entire networks of fast sub-reactions with different intermediary complexes that form a multidimensional free-energy surface [10]. Nonetheless, the Michaelis–Menten mechanism is a very useful conceptual framework, and an enormous number of studies have measured its characteristic parameters.

A mathematical model for the mechanism is obtained straightforwardly by setting up mass action functions (see Chapter 2). Specifically, each process term is formulated by including all variables that are directly involved, as well as the appropriate rate constant. The result is

$$\dot{S} = -k_1 \cdot S \cdot E + k_{-1} \cdot (ES), \tag{8.7}$$

$$\dot{(ES)} = k_1 \cdot S \cdot E - (k_{-1} + k_2) \cdot (ES), \tag{8.8}$$

$$\dot{P} = k_2 \cdot (ES). \tag{8.9}$$

These equations are easy to interpret. For instance, (8.7) says that the change in substrate is governed by two processes. In the first, $-k_1SE$, the substrate and enzyme enter a bimolecular reaction with rate k_1. The process carries a minus sign, because substrate is lost. The second process, $k_{-1}(ES)$, describes that some of the complex (ES) reverts to S and E, and this happens with rate k_{-1}. The sign here is positive, because the process augments the pool S.

An equation for E is not formulated, because the sum of free enzyme E and enzyme bound in the complex (ES) is assumed to remain constant; this total enzyme concentration is denoted E_{tot}. Thus, once we know the dynamics of (ES), we can infer the dynamics of E. It is usually assumed that the reversible formation of the complex (ES) occurs much faster than the production of P, and that therefore $k_1 \gg k_2$ and $k_{-1} \gg k_2$. The parameter k_2 is sometimes also called the catalytic constant and denoted by k_{cat}. The production of P is assumed to be irreversible.

The differential equations in (8.7)–(8.9) do not have an explicit analytical solution but can of course be solved computationally with a numerical integrator, which is available in many software packages. Since Henri, Michaelis, and Menten did not have computers, they proposed reasonable assumptions that yielded a significant simplification from the ODE system to an algebraic formulation. The assumptions are that substrate is present in excess, relative to the enzyme ($S \gg E$), and that the enzyme–substrate complex (ES) is in equilibrium with the free enzyme and substrate. As a more realistic variation of the latter, Briggs and Haldane [11] proposed the **quasi-steady-state assumption (QSSA)**, which entails that (ES) does not appreciably change in concentration over time and that the total enzyme concentration E_{tot} is constant. These assumptions were made for *in vitro* experiments, where indeed the solution can be stirred well and substrate easily exceeds the substrate concentration many times.

With regard to metabolism in living cells, the QSSA is often wrong at the very beginning of an experiment, when substrate and enzyme encounter each other for the first time, and it becomes inaccurate at the end, when almost all substrate is used up (see Figure 8.3; however, also see Exercise 8.7). In addition, the implicit assumption of a well-stirred homogeneous reaction medium, which is a prerequisite for mass action processes, is of course not satisfied in living cells. In spite of these issues, the QSSA is sufficiently close to being satisfied in many realistic situations, and a lot of experience has demonstrated that its benefits often outweigh its problems (for further discussion, see [12, 13]).

If one accepts QSSA, life becomes much simpler. The left-hand side of (8.8) is now equal to zero, because (ES) is assumed not to change. Furthermore, one defines the rate of product formation as $v_P = \dot{P}$ and introduces the maximum velocity of the reaction as $V_{max} = k_2 \ E_{tot}$ and the Michaelis constant as $K_M = (k_{-1} + k_2)/k_1$. Using straightforward algebra, one can show that the rate of product formation can then be expressed as

$$v_P = \frac{V_{max} S}{K_M + S} \tag{8.10}$$

(See Exercise 8.5). The function in (8.10) is referred to as the **Michaelis–Menten rate law (MMRL)** or the Henri–Michaelis–Menten rate law (HMMRL). A plot of the MMRL shows a monotonically increasing function that approaches $V_{\max}$ (**Figure 8.3**), which means that $V_{\max}$ quantifies the fastest possible speed with which the reaction can proceed. K_M corresponds to the substrate concentration for which the reaction speed is exactly half the maximum velocity. Its numerical value can vary widely among different enzymes. In many cases, it is close to the natural substrate concentration *in vivo*. Recall from Chapter 5 how the parameter values of the MMRL can be obtained from *in vitro* measurements with simple linear regression.

Instead of using algebra, the MMRL can also be obtained from the following biochemical arguments. The reversible reaction between substrate, enzyme, and intermediate complex on the left-hand side of Figure 8.2 is governed by the mass action formulation

$$\frac{k_{-1}}{k_1} = k_d = \frac{E \cdot S}{(ES)}, \tag{8.11}$$

where k_d is called the equilibrium or dissociation constant. Because k_2 is traditionally assumed to be small in comparison to k_1 and k_{-1}, k_d is approximately equal to $(k_{-1} + k_2)/k_1$ and thus to K_M.

The occupancy of an enzyme reflects how much of the enzyme is bound to the substrate. In other words, it is the ratio $(ES)/E_{\text{tot}}$. We can use (8.11) to rearrange this ratio as

$$\frac{(ES)}{E_{\text{tot}}} = \frac{(ES)}{E + (ES)} = \frac{E \cdot S / k_d}{E + E \cdot S / k_d} = \frac{S / k_d}{1 + S / k_d} = \frac{S}{k_d + S} \approx \frac{S}{K_M + S}. \tag{8.12}$$

Using this result, we immediately obtain the MMRL as

$$v_P = \dot{P} = k_2 \cdot (ES) \approx \frac{k_2 \cdot E_{\text{tot}} \cdot S}{K_M + S} = \frac{V_{\max} \cdot S}{K_M + S}. \tag{8.13}$$

Note that the solution of an ODE is typically presented as a plot of the dependent variables as functions of time t. In (8.7)–(8.9), these are S, (ES), and P. By contrast, (8.10), which is the form almost always found in biochemistry books and articles,

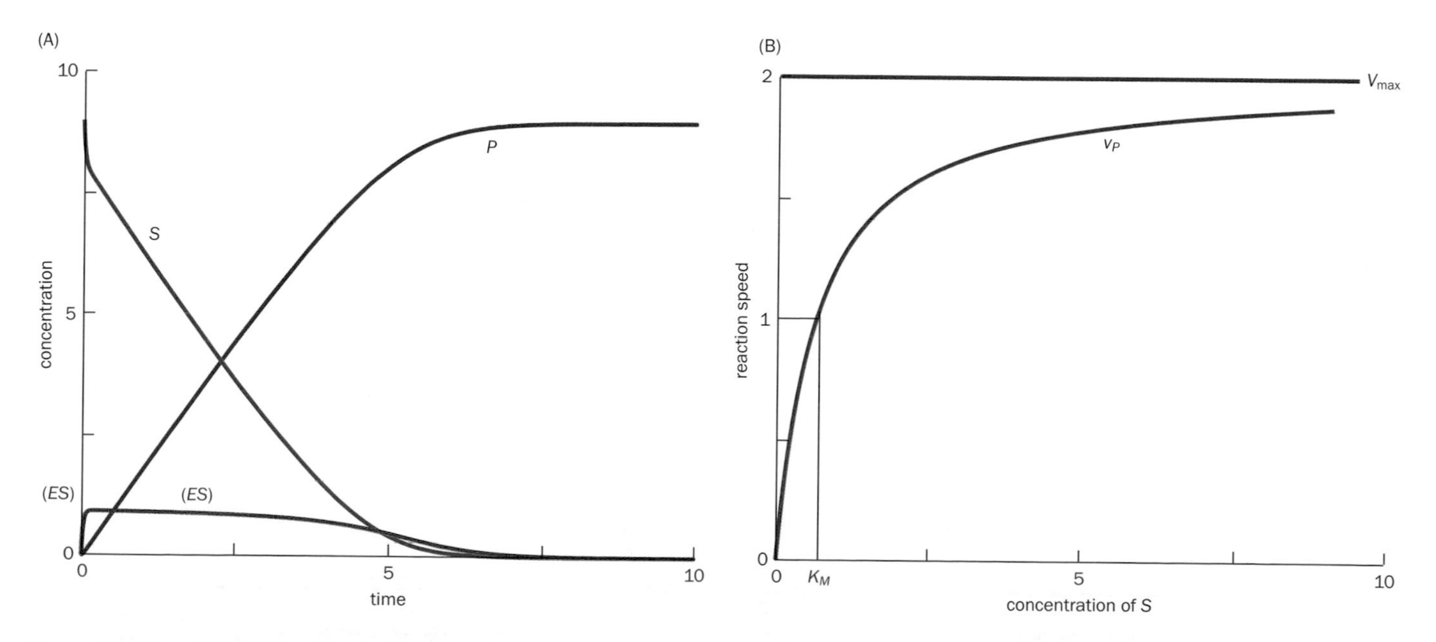

Figure 8.3 Plots of the Michaelis–Menten rate law. (A) Solution of the system of equations (8.7)–(8.9) in the form of concentrations versus time. (B) The more familiar plot of reaction speed versus substrate concentration, (8.10). Parameters are $k_1 = 4$, $k_{-1} = 0.5$, $k_2 = 2$, $K_M = 0.625$, $E_{\text{tot}} = 1$, $V_{\max} = 2$, $S(0) = 9$, and $(ES)(0) = P(0) = 0$.

formulates the speed of product formation v_P as a function of the substrate concentration S. The two plots are distinctly different (see Figure 8.3).

Thousands of investigations in biochemistry have used this formulation, and the MMRL is even used as a "black-box" model in other contexts that have nothing to do with biochemical kinetics, just because its shape fits many observed data. Of interest is that K_M is considered a property of an enzyme–substrate pair, so that K_M values from the literature or a database may be useful for an analysis even if they were measured by different groups and under slightly different conditions. By contrast, experimental measurements of V_{max} can vary substantially, depending on the experimental conditions and can seldom be assumed to have the same value under different conditions.

A biochemical and mathematical extension of the MMRL is the **Hill rate function** [14], which describes biochemical processes involving enzymes with several subunits. The mathematical consequence is that the substrate appears in the rate law with a Hill coefficient, that is, with an integer power that reflects the number of subunits, which is typically between 2 and 4. With Hill coefficient n, the rate law reads

$$v_{P,\text{Hill}} = \frac{V_{max} S^n}{K_M^n + S^n}. \tag{8.14}$$

In contrast to the MMRL, the Hill rate law with $n > 1$ is sigmoidal (**Figure 8.4**). For $n = 1$, the Hill rate law is the same as the MMRL, and the S-shape becomes a simple shoulder curve. The derivation of this rate law follows exactly the same steps as in the case of the MMRL, except that one accounts for n substrate molecules and n product molecules. For $n = 2$, (8.7) becomes

$$\dot{S}_H = -2k_1 S_H^2 E_H + 2k_{-1}(ES)_H, \tag{8.15}$$

where the index H simply distinguishes the equation from (8.7) [15]. The algebraic form (8.14) of the rate law can be derived again from the corresponding differential equations, with arguments from either algebra or biochemistry, as we discussed for the MMRL.

As described so far, the speed of an enzyme-catalyzed reaction depends exclusively on the substrate concentration, the parameters K_M and V_{max}, and possibly n. In reality, many reactions are controllable by inhibitors and activators that sense the demand for a product and adjust the turnover rate of the reaction. Modulation of the

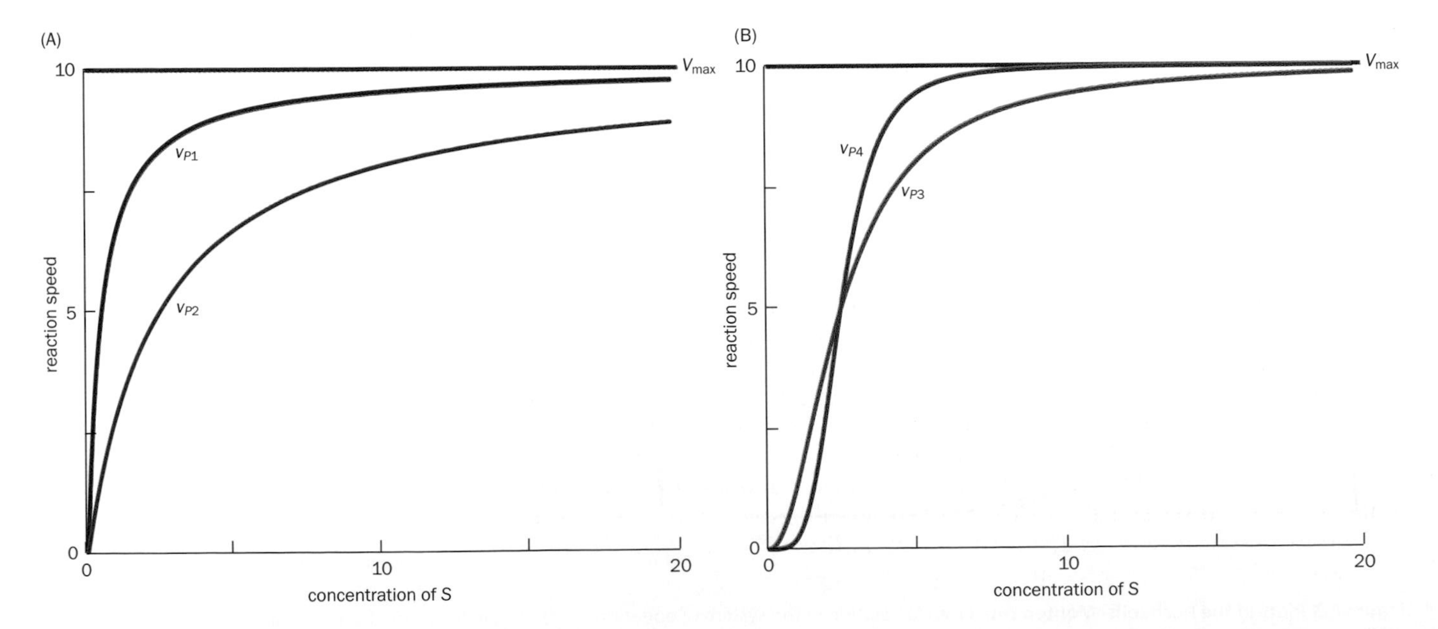

Figure 8.4 Plots of Michaelis-Menten (MMRL) and Hill rate laws (HRL). (A) Two MMRLs, v_{P1} and v_{P2}, with the same $V_{max} = 10$ and $K_{M1} = 0.5$ and $K_{M2} = 2.5$, respectively. (B) Two HRLs, v_{P3} and v_{P4}, with the same $V_{max} = 10$ and $K_M = 2.5$ and with Hill coefficients $n = 2$ and $n = 4$, respectively. Note that v_{P2} in (A) corresponds to an HRL with $V_{max} = 10$, $K_M = 2.5$, and $n = 1$.

reaction speed by an inhibitor or activator may occur at several locations in the Michaelis–Menten mechanism shown in Figure 8.2. For instance, an inhibitor may compete with the substrate molecules for an active site on the enzyme. This may happen if the inhibitor is structurally similar to the substrate and therefore fits into the same binding site on the enzyme. It may also happen that the inhibitor forms a rather stable complex with the enzyme–substrate complex (ES), which partially prevents the reaction toward the product from occurring. If the inhibitor and the substrate are structurally different, inhibition of the enzyme activity can be accomplished if the inhibitor binds to a different site than the active site on the enzyme. In the process of binding, the catalytic activity of the enzyme at the active site is reduced. This type of **allosteric** inhibition is quite frequent. It is also possible that the allosteric effect is activating rather than inhibiting.

Mathematical descriptions differ for the various modes of inhibition and may in effect increase the K_M or decrease the V_{max} value, or lead to a Hill-type function as in (8.14). Thousands of articles and many books have discussed enzyme regulation and described biochemical techniques for characterizing the regulatory mechanisms. Since these investigations are firmly in the realm of biochemistry, we will not discuss them much further. Good representations of mathematical models of enzyme modulation can be found in [16–18].

Three types of inhibition are called competitive, uncompetitive, and noncompetitive. The terminology of the latter two is somewhat confusing, but has become the standard. The three types are ultimately associated with the K_M, the substrate concentration in the denominator of the MMRL, and V_{max}, respectively. Specifically, in competitive inhibition, the inhibitor with concentration I competes with the substrate for the binding site on the enzyme. The corresponding rate law is

$$v_{P,CI} = \frac{V_{max}S}{K_M\left(1+\frac{I}{K_I}\right)+S}. \tag{8.16}$$

The new parameter K_I is the dissociation constant of the inhibitor.

In uncompetitive inhibition, the inhibitor binds to the complex (ES). The corresponding rate law is

$$v_{P,UI} = \frac{V_{max}S}{K_M+S\left(1+\frac{I}{K_I}\right)}. \tag{8.17}$$

In noncompetitive inhibition, the inhibitor reduces the binding affinity between enzyme and substrate without directly affecting the substrate-binding site. The corresponding rate law is therefore

$$v_{P,NI} = \frac{V_{max}S}{K_M+S}\left(1+\frac{I}{K_I}\right)^{-1}. \tag{8.18}$$

A different type of inhibition is called allosteric regulation. In this case, the inhibitor binds to the enzyme, but not at the substrate-binding site. Nonetheless, the docking of the inhibitor alters the substrate-binding site and thereby reduces the affinity between enzyme and inhibitor. The general mathematical formulation is much more complicated and is not based on the MMRL; a detailed derivation can be found in [18].

In reality, the inhibition may be a mixture of several mechanisms, which increases the complexity of a mathematical representation. Circumventing this issue, biochemical systems theory (BST) treats an inhibitor like any other variable and represents it with a power-law approximation. Thus, if the degradation reaction of X_1 is inhibited by X_j, one sets up the model for X_1 as

$$\dot{X}_1 = production - \beta_1 X_1^{h_{11}} X_j^{h_{1j}}, \tag{8.19}$$

and theory says that h_{1j} is a negative quantity. If an inhibitor I is small in quantity, it is sometimes replaced with $(1 + I)$, and (8.19) then reads [19]

$$\dot{X}_1 = production - \beta_1 X_1^{h_{11}} \left(1 + \frac{I}{K_I}\right)^{h_{1I}}. \tag{8.20}$$

Thus, if the inhibitor concentration is reduced to 0, the MMRL is regained. Note that this representation is similar to the noncompetitive rate law, although it is derived purely through linearization in logarithmic space (Chapter 4).

Independent of the details of inhibition, data directly related to a traditional enzyme kinetic investigation are typically presented as quantitative features of enzymes that are characterized by parameters such as K_M, inhibition and dissociation constants K_I and K_D, and sometimes V_{max} values. Often, co-factors and modulators of reactions are listed as well. Originally, these features were described in a huge body of literature, but nowadays databases such as BRENDA [20] (see Box 8.2 below) and ENZYME [21] offer collections of many properties of enzymes and reactions, along with pertinent literature references.

While few functions in biology have received as much attention as the MMRL and Hill functions, one must caution that the two are approximations that are often not really justified (see, for example, [12, 13, 16]). First and foremost, these functions, just like mass action kinetics, assume reactions occurring in well-stirred, homogeneous media, which obviously is not the case inside a living cell that is full of materials and organelles and where reactions often happen at surfaces. Second, the quasi-steady-state assumption is seldom truly valid, and neither is the amount of enzyme always constant and much smaller than the substrate concentration, as the MMRL assumes it to be. Finally, the kinetic functions fail for small numbers of reactants, and ODEs or continuous functions have to be replaced with much more complicated stochastic processes [6].

PATHWAYS AND PATHWAY SYSTEMS

8.4 Biochemistry and Metabolomics

If one had to describe the difference between traditional enzyme kinetics and metabolomics, one might simplistically say that the former focuses primarily on individual reactions and pathways, along with factors influencing them, while the latter is primarily interested in (large) assemblies of reactions that collectively form pathway systems. Expressed differently, the main focus of traditional biochemistry and enzyme kinetics can be seen as targeted and local, while the focus of metabolomics is comprehensive and global. This simplified distinction may suggest that biochemistry is a subset of metabolomics, but this is not entirely so, because metabolomics looks at so many reactions simultaneously that details at the local level are often only of secondary interest. As a case in point, an important component of enzyme kinetic investigations is a detailed protocol for the purification of an enzyme and the characterization of its affinity to different substrates. By contrast, typical metabolomic studies attempt to determine two features. The first is the **metabolic profile** of a biological sample, which consists of the relative (or absolute) amounts or concentrations of all metabolites. The second feature is the characterization and distribution of the relative (or absolute) magnitudes of all fluxes throughout the system. Here, each flux is defined as the rate of turnover of molecules in a reaction step or pathway or, expressed differently, as the amount of material flowing through a metabolite pool or pathway at a given time. Some authors reserve the term flux for pathways, while calling essentially the same quantity a rate when a single reaction is being studied [22].

Studying hundreds of metabolites or fluxes at the same time usually mandates that details of individual reaction steps or mechanisms of inhibition or activation be ignored. This situation is comparable to the different foci in zoology and ecology, where the former primarily targets the unique features of specific animals or animal species, while the latter tries to understand the interactions among (many) species, including animals, plants, and sometimes even bacteria and viruses.

BOX 8.2 METABOLIC PATHWAY INFORMATION DISPLAYED IN BRENDA

The main focus of BRENDA [20] is quantitative information on enzymes, which can be searched by name or Enzyme Commission (EC) number, as well as by species. **Figure 1** is a screenshot showing the results of a typical search. BRENDA offers values for important quantities such as K_M values, substrates, co-factors, inhibitors, and other features that can be very important for modeling purposes. All information is backed up by references.

BRENDA

The Comprehensive Enzyme Information System

EC 3.2.1.28 - alpha,alpha-trehalase

MgCl2	Saccharomyces cerevisiae	50 mM, C-trehalase, complete inhibition	136150	2D-image
phosphate	Saccharomyces cerevisiae	-	136165	2D-image
ZnCl2	Saccharomyces cerevisiae	0.1 mM, C-trehalase, complete inhibition	136150	2D-image

ACTIVATING COMPOUND	ORGANISM	COMMENTARY	LITERATURE	IMAGE
3',5'-cAMP	Saccharomyces cerevisiae	activation is dependent on presence of ATP and a divalent cation such as Mg2+, Mn2+ or Co2+	136162	2D-image
MgATP2-	Saccharomyces cerevisiae	C-trehalase can be activated by MgATP2- in presence of cAMP	136150	2D-image
additional information	Saccharomyces cerevisiae	the cryptic enzyme form is completely activated at protein concentrations higher than 60 mg/ml	136162	-

KM VALUE [mM]	KM VALUE [mM] Maximum	SUBSTRATE	ORGANISM	COMMENTARY	LITERATURE	IMAGE
11.78	-	alpha,alpha-Trehalose	Saccharomyces cerevisiae	-	656417	2D-image
0.5	-	Trehalose	Saccharomyces cerevisiae	-	136158	2D-image
1.4	-	Trehalose	Saccharomyces cerevisiae	V-trehalase	136150	2D-image
4.7	-	Trehalose	Saccharomyces cerevisiae	-	136141	2D-image
4.79	-	Trehalose	Saccharomyces cerevisiae	in absence of trehalose-c	136144	2D-image
5	-	Trehalose	Saccharomyces cerevisiae	-	136165	2D-image
5.28	-	Trehalose	Saccharomyces cerevisiae	in presence of trehalose-c	136144	2D-image
5.7	-	Trehalose	Saccharomyces cerevisiae	C-trehalase	136150	2D-image
10	-	Trehalose	Saccharomyces cerevisiae	apparent value, at pH 7.1	677711	2D-image

Figure 1 Typical information provided in BRENDA. Specifying the EC number or name of trehalase offers a lot of quantitative information that can be very useful for pathway modeling (see Chapter 11). The screenshot here shows only a selection of results related to K_M values obtained under various conditions. BRENDA also has direct links to KEGG (see Box 8.3), BioCyc (see Box 8.4), and other databases. (Courtesy of Dietmar Schomburg, Technische Universität Braunschweig, Germany.)

Of course, there is plenty of overlap between enzyme kinetics and metabolomics. For instance, the flux through a pathway system is related to the enzyme activities and V_{max} values of the individual reaction steps. One could surmise that the analysis of the dynamics of a metabolic system might not require very detailed information regarding each and every reaction step, as long as we know the turnover rate of the reaction and its modulators. However, once we start extrapolating results from the analysis of one set of conditions to another, these details can become crucially important. Thus, traditional biochemistry and metabolomics have their own foci but also much common ground.

While a little naive, the above distinction between traditional enzyme kinetics and metabolomics is useful for now, because it provides rationales and explanations for the different classes of analytical methods in the two fields, and for the types of data that are associated with them. Using painstaking analysis, traditional biochemistry has yielded a solid understanding of most reactions in all major and many minor metabolic pathways, and huge amounts of data related to enzyme kinetics, such as K_M and V_{max} values, have been measured and are available under their Enzyme Commission (EC) number in databases such as BRENDA [20] (**Box 8.2**). In stark contrast, typical metabolomic data consist of comprehensive metabolic concentration or flux profiles, either at a steady state or dynamically changing in response to a stimulus.

8.5 Resources for Computational Pathway Analysis

The analysis of metabolic pathway systems typically follows one of two routes, which are distinct and complement each other. The first is stoichiometric analysis. It focuses on the connectivity within the pathway system and addresses which metabolite can be converted directly into which other metabolite. The first step of any stoichiometric analysis is therefore the collection of all known or alleged reactions producing or degrading the metabolites in the pathway system of interest. These reactions form a stoichiometric map consisting of nodes (metabolites) and vertices (arrows: reactions). In mathematical terms, this map is a directed graph, for which many methods of analysis are available. Chapter 3 discussed the most pertinent features of stoichiometric networks, along with analyses based on graphs and linear algebra. That chapter also discussed possible transitions from these static networks to dynamics networks, for instance, with methods of **metabolic control analysis (MCA)** [22] or by taking into account that static networks can be considered steady states of fully dynamic systems, for example, within the framework of **biochemical systems theory (BST)** (see, for example, [18, 23]).

As an example, consider glucose 6-phosphate, which is typically isomerized into fructose 6-phosphate during glycolysis, but may also be converted into glucose 1-phosphate via the phosphoglucomutase reaction, reverted to glucose by glucose 6-phosphatase, or enter the pentose phosphate pathway through the glucose 6-phosphate dehydrogenase reaction. In a stoichiometric map, glucose 6-phosphate is represented as a node and the reactions are represented as arrows pointing from glucose 6-phosphate to other nodes, representing the products of the reactions.

Of enormous help for the construction of a stoichiometric map are freely accessible databases such as KEGG (the Kyoto Encyclopedia of Genes and Genomes) [24] and BioCyc [25] (**Boxes 8.3** and **8.4**), which contain collections of pathways, along with information regarding their enzymes, genes, and plenty of useful comments, references, and other pieces of information. The databases furthermore provide numerous tools for exploring pathways in well-characterized organisms such as *Escherichia coli*, yeast, and mouse, and also for inferring the composition of ill-characterized pathways, for instance, in lesser-known microbial organisms.

In addition to KEGG and BioCyc, other tools are available for metabolic pathway analysis. For instance, the commercial software package ERGO [26] offers a bioinformatics suite with tools supporting comparative analyses of genomes and metabolic pathways. Users of ERGO can combine various sources of information, including sequence similarity, protein and gene context, and regulatory and expression data, for optimal functional predictions and for computational reconstructions of large parts of metabolism. Pathway Studio [27] allows automated information mining, the identification and interpretation of relationships among genes, proteins, metabolites, and cellular processes, as well as the construction and analysis of pathways. KInfer (Kinetics Inference) [28] is a tool for estimating rate constants of systems of reactions from experimental time-series data on metabolite concentrations. Another interesting database of biological pathways is Reactome [29, 30], which contains not only metabolic pathways, but also information regarding DNA replication, transcription, translation, the cell cycle, and signaling cascades.

The Lipid Maps project [31] provides information and databases specifically for lipids and lipidomics, a term that was chosen to indicate that the totality of lipids may be comparable to genomics and proteomics in scope and importance. Indeed, the more we learn about lipids, the more we must marvel at the variety of functions that they serve—from energy storage to membranes and to genuine mechanisms of signal transduction. And not to forget: 60% of the human brain is fat.

Metabolic concentration and flux profiles are usually, but not always, measured at a steady state. In this condition, material is flowing through the system, but all metabolite pools and flux rates remain constant. At a steady state, all fluxes and metabolite concentrations in the system are constant. Moreover, all fluxes entering any given metabolite pool must in overall quantity be equal to all fluxes leaving this pool. These are strong conditions that permit the inference of some fluxes from knowledge of other fluxes. Indeed, if sufficiently many fluxes in the pathway system could be measured, it would be possible to infer all other fluxes. In reality, the number of measured fluxes is almost always too small for a complete inference, but the method of flux balance analysis (see [32] and Chapter 3) renders the inference possible, if certain assumptions are made.

The second type of representation of a pathway system also includes the stoichiometric connectivity map, but in addition describes how the fluxes functionally depend on metabolites, enzymes, and modulators (see Chapter 4). Mathematical flux descriptions may consist of Michaelis–Menten or Hill rate laws, as discussed earlier, but alternatives, such as power-law functions [23], linear–logarithmic functions [33], and a variety of other formulations, are available and have their

BOX 8.3 METABOLIC PATHWAY REPRESENTATION IN KEGG

Chapter 11 will analyze a small metabolic pathway in yeast that is very important for the organism's response to stresses such as heat. The main compound of interest within this pathway is the sugar trehalose. **Figures 1–3** are screenshots from KEGG—the Kyoto Encyclopedia of Genes and Genomes [24]. KEGG consists of 16 main databases with information on genomes, pathways, functional aspects of biological systems, chemicals, drugs, and enzymes.

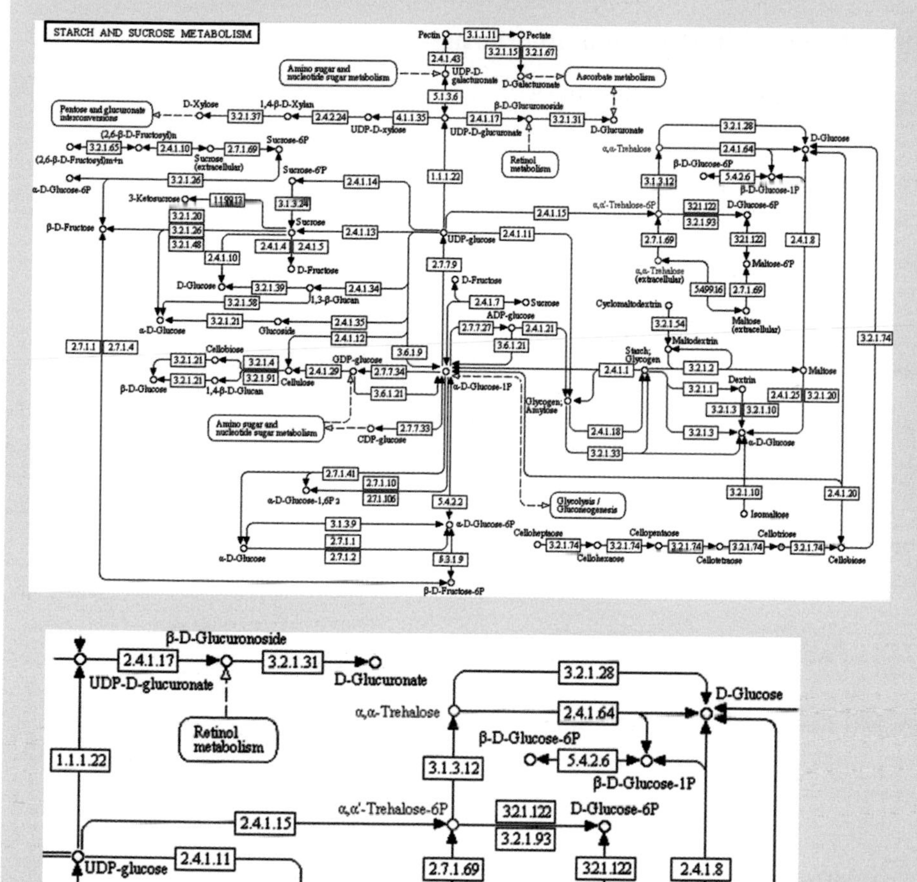

Figure 1 Metabolic network associated with trehalose dynamics. Trehalose is a carbohydrate and can be found in KEGG's starch and sucrose pathway system. In this representation, the metabolites are the nodes. Associated with the reactions (edges) are numbers in boxes, which are the EC numbers that uniquely identify the catalyzing enzymes. (Courtesy of Kanehisa Laboratories.)

Figure 2 A more detailed view of the synthetic and degradation pathways of trehalose in KEGG. Clicking on an item in a box leads to further information (see Figure 3). (Courtesy of Kanehisa Laboratories.)

KEGG Saccharomyces cerevisiae (budding yeast): YBR001C

Help

Entry	YBR001C CDS S.cerevisiae
Gene name	NTH2
Definition	Nth2p (EC:3.2.1.28)
Orthology	K01194 alpha,alpha-trehalase [EC:3.2.1.28]
Pathway	sce00500 Starch and sucrose metabolism
Class	Metabolism; Carbohydrate Metabolism; Starch and sucrose metabolism [PATH:sce00500] BRITE hierarchy
SSDB	Ortholog Paralog Gene cluster GFIT
Motif	Pfam: Trehalase Trehalase_Ca-bi PROSITE: TREHALASE_1 TREHALASE_2 Motif
Other DBs	NCBI-GI: 6319473 NCBI-GeneID: 852286 SGD: S000000205 MIPS: YBR001C UniProt: P35172
Position	II:complement(238941..241283) Genome map
AA seq	780 aa AA seq DB search MVDFLPKVTEINPPSEGNDGEDNIKPLSSGSEQRPLKEEGQQGGRRHHRRLSSMHEYFDP FSNAEVYYGPITDPRKQSKIHRLNRTRTMSVFNKVSDFKNGMKDYTLKRRGSEDDSFLSS QGNRRFYIDNVDLALDELLASEDTDKNHQITIEDTGPKVIKVGTANSNGFKHVNVRGTYM LSNLLQELTIAKSFGRHQIFLDEARINENPVDRLSRLITTQFWTSLTRRVDLYNIAEIAR DSKIDTPGAKNPRIYVPYNCPEQYEFYIQASQMNPSLKLEVEYLPKDITAEYVKSLNDTP

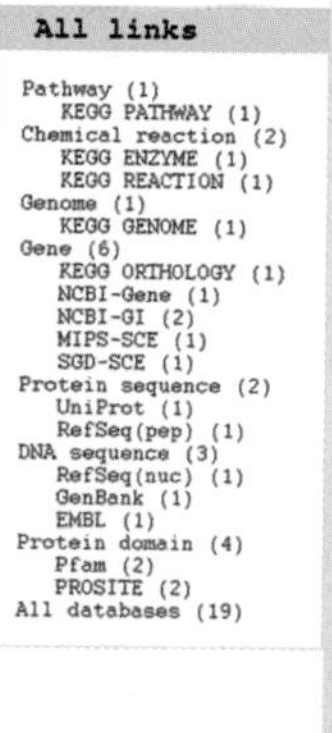

Figure 3 Details provided by KEGG. Clicking on an EC number in a pathway representation as in Figure 2 leads to biochemical and genome-related features of the corresponding enzyme. The information displayed here describes the enzyme trehalase, which degrades trehalose into glucose (see Chapter 11). (Courtesy of Kanehisa Laboratories.)

BOX 8.4 METABOLIC PATHWAY REPRESENTATION IN BIOCYC

As mentioned in Box 8.3, Chapter 11 will require information on the sugar trehalose in yeast. **Figures 1** and **2** are screenshots from BioCyc. Running the cursor over any of the entities shown on the website generates names, synonyms, reactions, and other useful information. BioCyc also provides information on pertinent genes, a summary of pathway features, references, and comparisons with other organisms. BioCyc furthermore offers sophisticated tools for mining **curated** databases.

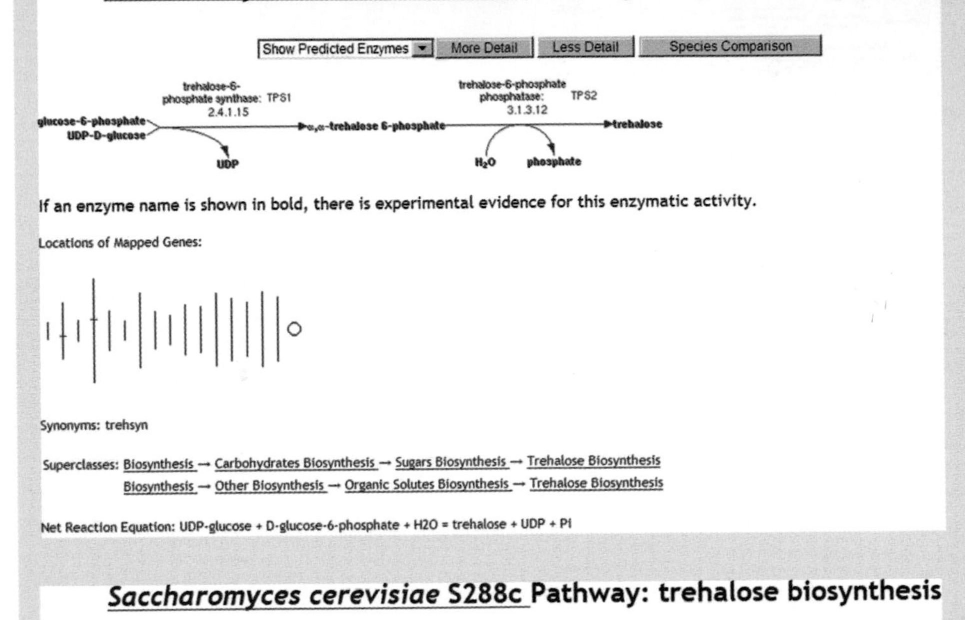

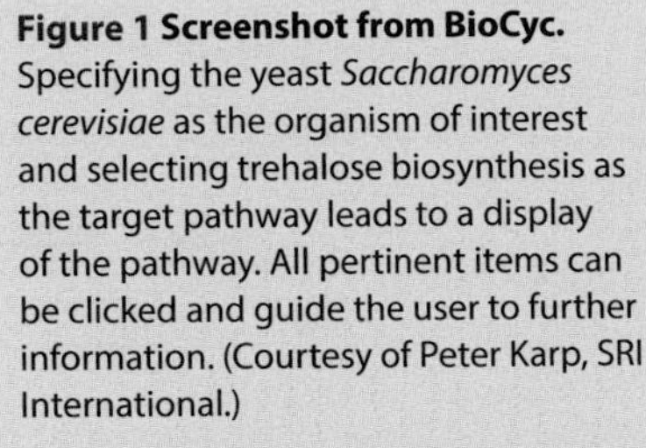

Figure 1 Screenshot from BioCyc. Specifying the yeast *Saccharomyces cerevisiae* as the organism of interest and selecting trehalose biosynthesis as the target pathway leads to a display of the pathway. All pertinent items can be clicked and guide the user to further information. (Courtesy of Peter Karp, SRI International.)

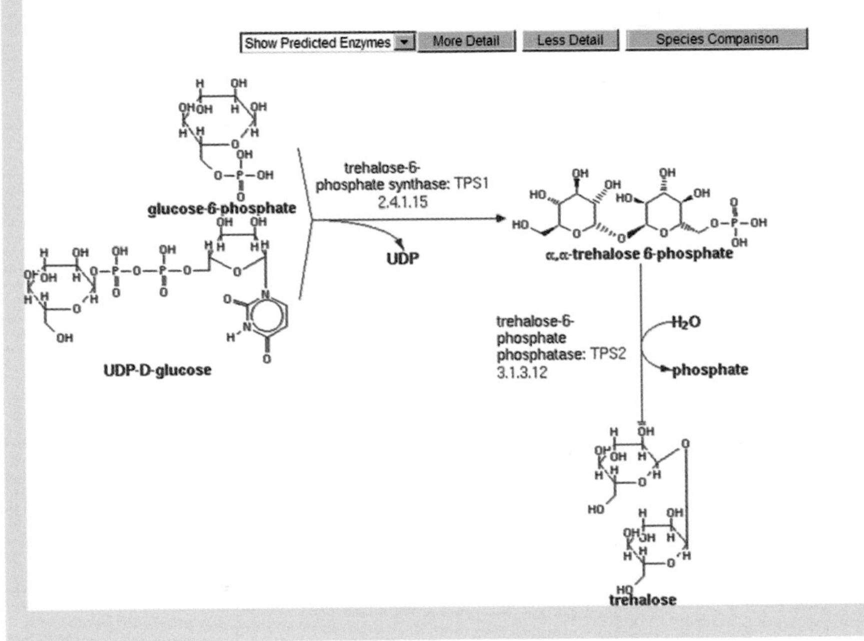

Figure 2 Details available in Biocyc. The simplified pathway representation in Figure 1 allows the user to request more detail, which is shown here. (Courtesy of Peter Karp, SRI International.)

advantages and drawbacks. Because these dynamic–kinetic representations contain much more information than simple stoichiometric representations, their construction is more complicated and requires considerably more input data in the form of kinetic information and/or metabolic time series.

8.6 Control of Pathway Systems

A key feature that is not addressed in stoichiometric models, but is present in dynamic models, is the regulation and control of the flow of material throughout the metabolic pathway system. Regulation allows the system to respond to changing environments and changing demands where some metabolite is needed in greater amounts than normal. For instance, by changing the activity of an enzyme at a branch point where two pathways diverge, more material can be channeled into the one or other pathway, depending on demand. We can easily see the importance of controllability by studying what would happen without control. Imagine a water distribution system with pipes of fixed widths, where at each node the collective cross section of all input pipes equals the collective cross section of all output pipes. If the system is running at maximal capacity, the amount of water at each point is proportional to the pipe size. If less water is entering the system, the amounts decrease correspondingly, but without means of intervention it is impossible to steer the water flow selectively into a direction where it is most needed for a particular purpose.

The control of metabolic pathway systems occurs at several levels. The fastest is a (temporary) change in enzyme activity, which is achieved at the metabolic level itself. A premier example is feedback inhibition by the end product of a pathway: the more product is generated the stronger becomes the inhibition of the enzyme at the beginning of the pathway (**Figure 8.5**). Similarly, by inhibiting one pathway at a branch point, the other branch receives more substrate. In many cases, the end product not only inhibits its own branch, but also activates the competing branch and/or inhibits a reaction step upstream of the branch. One possibility is a nested feedback design where either end product inhibits its own branch and also the initial substrate production (**Figure 8.6**). These types of feedback patterns have been analyzed in detail with regard to the efficiency of alternative designs (see, for example, [18, 34]). The methods described in Chapters 2 and 4 allow us to set up models for these types of pathways, and Chapter 5 provides recipes for finding appropriate parameter values, if the right data are available.

The second manner of control happens at the proteomic level, where enzymes can be activated or de-activated, just like other proteins—for instance, by covalent modification in the form of phosphorylation, ubiquitination, or glycosylation, by S–S bonding, or through protein–protein interactions. Of course, it is also possible to degrade enzymes permanently by means of proteases.

The third option for controlling metabolism occurs at the level of gene expression, where transcription, translation, or both can be affected. At the level of transcription, the gene coding for an enzyme may be up-regulated and lead to a larger amount of enzyme and therefore to an increased turnover capacity. This mode of control, which often relies on the mobilization or translocation of transcription factors, is very effective, but much slower than metabolic regulation such as end-product inhibition, because it involves transcription and translation, which may

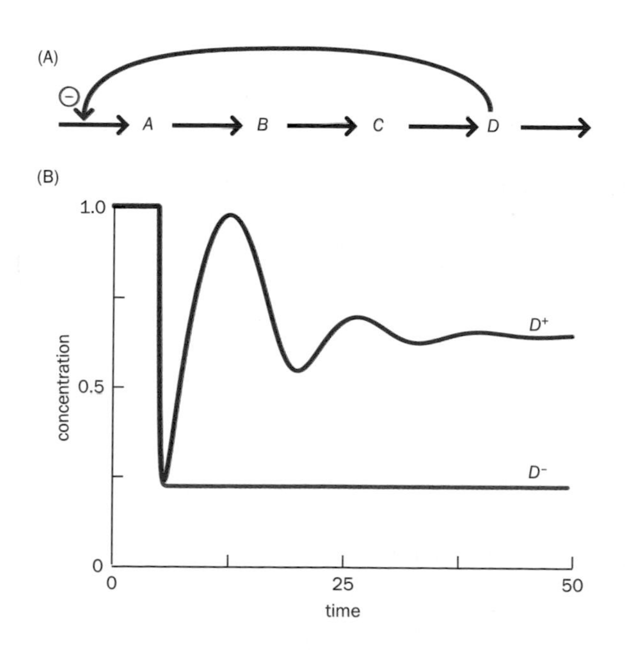

Figure 8.5 Linear pathway with feedback. (A) Reaction scheme with feedback inhibition of the initial step by the end product. (B) Comparison of responses to a sudden and persistent demand of metabolite D, starting at time $t = 8$. With feedback (D^+), the concentration of D oscillates and converges to a level of about two-thirds of the initial value. Without feedback (D^-), the concentration of D sinks to less than one-quarter.

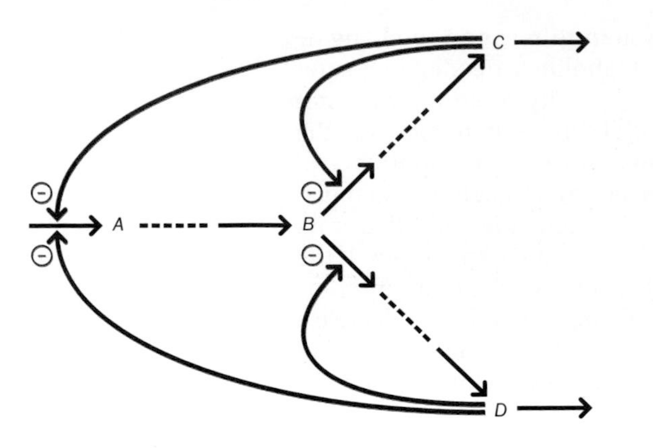

Figure 8.6 Branched pathway with several feedback inhibition signals. Depending on the strengths of the feedback signals, material may be channeled preferentially into either one of the two branches. (Adapted from Savageau MA. Biochemical Systems Analysis: A Study of Function and Design in Molecular Biology. Addison-Wesley, 1976. With permission from John Wiley & Sons.)

take on a timescale of the order of tens of minutes if not hours. It is also possible to exert regulation at the level of translating RNA into proteins.

Finally, metabolic control can be implemented through the design of the pathway and, in particular, with isozymes, that is, with distinct but similar enzymes catalyzing the same reaction. For instance, in a situation as shown in Figure 8.6, it is not uncommon to find two isozymes for the production of A, each of which is inhibited by only one of the end products, C or D. In addition to feedback inhibition, these isozymes may also be differently regulated at the genetic level. The two isozymes may furthermore be physically separated by forming different supramolecular complexes or through their location in different compartments.

In complicated situations such as stress responses, the regulation of metabolism is the result of a combination of several modes of action, including direct metabolic adjustments, the activation of transcriptional regulators, post-translational modifications, and various signaling events [35]. The immediate response may also occur directly at the molecular level. For instance, it is known that temperature can change the activity of enzymes that are involved in heat-stress response [36], and it has been shown that this mechanism is sufficient for mounting an immediate cellular response, whereas heat-induced changes in gene expression govern heat-stress responses over longer time horizons (Chapter 11 and [37]).

METHODS OF METABOLOMIC DATA GENERATION

Throughout the history of biochemistry, measurement of concentrations and fluxes has been as important as characterization of enzymes, modulators, and turnover rates. The distinguishing feature of modern high-throughput metabolomics is the simultaneous generation of large metabolite or flux profiles, which sometimes consist of hundreds of data points obtained in a single experiment [38, 39]. Most of these methods measure systems in a steady state, but some techniques also allow characterization of changing profiles over a sequence of time points, following some stimulus. It is worth noting in this context that there is a genuine difference in complexity between metabolomics on one hand and genomics and proteomics on the other. The genome consists of only four bases and the proteome of 20 amino acids, and these are rather similar. By contrast, the metabolome consists of thousands of metabolites, which are chemically very diverse and come in all sizes, chemical families, and affinities to water or lipids.

Uncounted methods have been developed for measuring metabolites. Some of these are quite generic, whereas others are specific for the metabolites in question. Generically, metabolite analysis normally entails a sequence of processes [39, 40]:

1. Sample preparation and quenching of metabolic activity
2. An extraction procedure
3. Separation of metabolites
4. Quantification and/or profiling with methods such as nuclear magnetic resonance (NMR) and mass spectrometry (MS)
5. Data analysis

8.7 Sampling, Extraction, and Separation Methods

The key step in sample preparation is efficient rapid sampling [41]. This step is crucial, because any changes induced by a stimulus or any intended or unintended variation in the environment directly affect the metabolite concentrations in the cell. For instance, intracellular metabolites in yeast typically have a turnover rate at the order of one mole per second, and many metabolites respond to new conditions within the millisecond range [39]. Thus, to yield valid snapshots of a metabolic system, any rapid-sampling method must instantaneously quench all metabolic activity. This quenching is typically achieved through very fast changes in temperature, for instance, by injecting the sample into liquid nitrogen, methanol, or a solution containing ethanol and sodium chloride [42].

Once the sample has been quenched, intracellular metabolites must be separated from metabolites of the external medium and extracted with a reagent such as perchloric acid that disrupts the cell, releases the intracellular metabolites, and denatures the enzymes, thereby preventing further degradation and biochemical conversions. These initial steps of metabolite analysis are quite complex, and it is virtually impossible to execute them without losing some of the metabolites [40]. As a consequence, and because of the natural variability among cells, the reproducibility of metabolite extraction is often compromised. In some cases, it is possible to obtain direct measurements in living cells, for instance, with fluorescence techniques that allow visualization of specific metabolites such as glucose and NADH *in vivo*, but it is not always clear how precise such direct measurements are. A promising alternative is sometimes *in vivo* NMR, which we will discuss in a moment.

Many methods have been developed for the quantitative measurement of intracellular metabolites in extracts resulting from the sampling steps mentioned above. The traditional option has been an enzymatic assay, which is specific and often very precise. The disadvantages of this approach include relatively large sample volumes and the need for separate assays for most metabolites [39]. A distinct alternative is chromatography, which is one of the oldest and best-documented separation techniques in biochemistry [40]. High-throughput metabolic studies often use gas chromatography (GC), combined with mass spectrometry (GC-MS), or high-performance liquid chromatography (HPLC). GC-MS has the advantage that it offers high resolution—but only if the analyzed chemicals are volatile or can be vaporized without decomposition, which is not always possible, in particular, if the metabolites are large. In HPLC, a biological sample is pumped through a column, separated, and measured with a detector. HPLC permits a wider range of compounds to be measured than GC, but has a lower resolution.

Capillary electrophoresis (CE) encompasses a family of methods for separating ionized molecules based on their charge and has become a valuable tool for metabolic fingerprinting [43]. The separation takes place inside a small capillary connecting a source vial with the biological sample in an aqueous buffer solution and a destination vial. The sample enters the capillary owing to capillary action, and the analytes migrate and separate when a strong electric field is applied. A detector close to the end of the capillary measures the amount of each analyte and sends the data to a computer.

8.8 Detection Methods

Among the detection methods, **mass spectrometry (MS)** and **nuclear magnetic resonance (NMR) spectroscopy** stand out. The main concepts of MS were discussed in Chapter 7. MS may be used as a stand-alone technique without prior separation, or in combination with a separation technique such as GC, HPLC, or CE [44]. In the very powerful combination of CE and MS, the efflux from the capillary is subjected to electrospray ionization, and the resulting ions are analyzed, for instance, with a time-of-flight mass spectrometer.

The power of MS methods in metabolomics becomes especially evident when hundreds of spectrograms are stacked up into metabolic landscapes. As a beautiful example, Johnson and collaborators [45] combined liquid chromatography (LC) with a specific high-resolution variant of MS, which was accurate and sensitive enough to allow the discrimination of thousands of mass-to-charge (m/z) peaks (see Chapter 7). Furthermore, because the method can be automated and is relatively cost-effective,

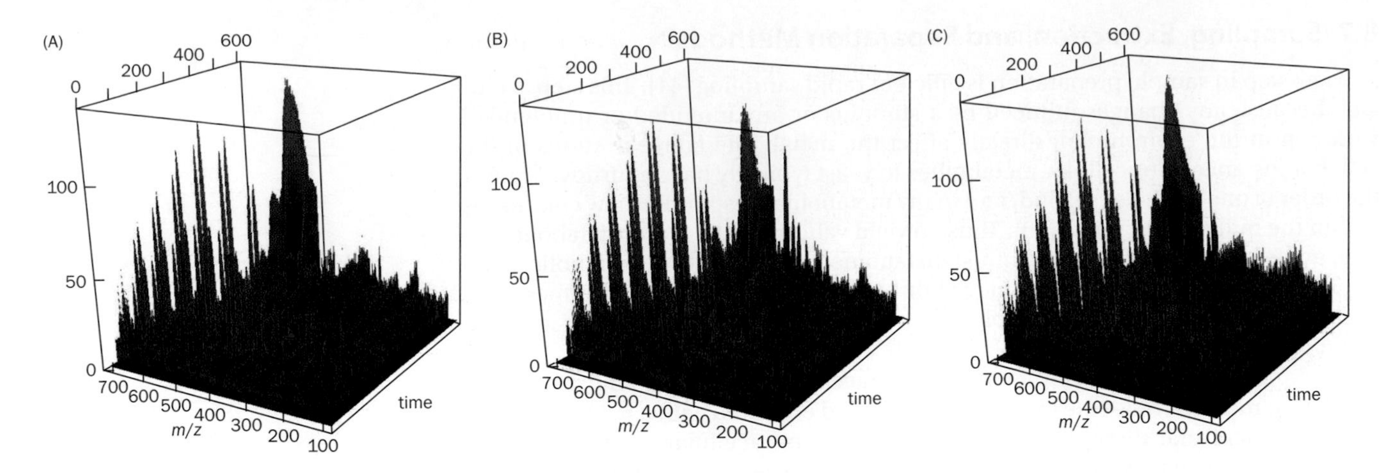

Figure 8.7 Metabolic landscapes of plasma samples from three human subjects (A–C). A typical MS spectrogram shows just one slice within these three-dimensional plots, with mass over charge (*m/z*) as horizontal axis and ion intensity as vertical axis. The third axis (time) here indicates the retention time during the sample separation with liquid chromatography. Analysis of the data in the figure shows that many features are common among the three individuals, but also that subtle differences in metabolic states exist and can be detected. (Courtesy of Tianwei Yu, Kichun Lee, and Dean Jones, Emory University.)

it can be used to compare the metabolic profiles of individual subjects and thereby provides a powerful tool for personalized medicine (see Chapter 13). As an illustration of the resolution power of the method, **Figure 8.7** shows the metabolic profiles of plasma samples from three individuals, as they change after stimulation with cystine. Many of the peaks in such profiles are at first unknown and require further chemical analysis.

NMR spectroscopy exploits a physical feature of matter called nuclear spin. This spin is a property of all atomic nuclei possessing an odd number of protons and neutrons, and can be regarded as a spinning motion of the nucleus on its own axis. Associated with this spinning motion is a magnetic moment, which makes the atoms behave like small magnets. When a static magnetic field is applied, the atoms align parallel to this field. Atoms possessing a spin quantum number of ½, which is the case for most biologically relevant nuclei, separate into two populations that are characterized by states with different energy levels and orientations. If an oscillating magnetic field is applied in the plane perpendicular to the static magnetic field, transitions between the two states can be induced. The resulting magnetization in the atoms can be recorded and mathematically transformed into a frequency spectrum. The nucleus is to some degree protected from the full force of the applied field by its surrounding cloud of electrons, and this "shielding" slightly changes the transition frequency of the nucleus. This effect on the frequency enables NMR to distinguish not only different molecules but also the nuclei within each molecule. Because both excitation and detection are due to magnetic fields, the sample is preserved and does not require prior treatment or separation; thus, NMR is non-invasive and nondestructive (**Figures 8.8** and **8.9**).

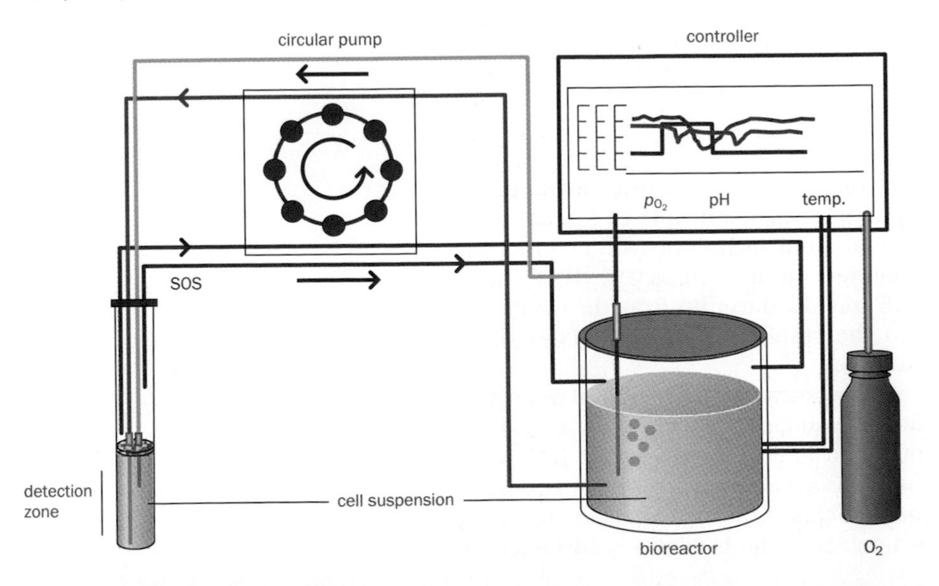

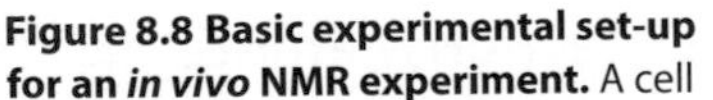

Figure 8.8 Basic experimental set-up for an *in vivo* NMR experiment. A cell solution is kept in a bioreactor, where pH, temperature, and other physicochemical parameters are controlled. A circulation pump is used to move small quantities of cell solution, via the blue tube, into the detection zone, where a magnetic field is applied and the spectrogram is measured. The cell solution is returned to the bioreactor through the black tube. Should the black tube be clogged up with cells, the blue SOS tube serves as a back-up to "save our system" from spillage. The experiments can be done under aerobic or anaerobic conditions. (Courtesy of Luis Fonseca and Ana Rute Neves, ITQB, Portugal.)

The most relevant nuclei detectable with NMR are ^{1}H, ^{13}C, ^{15}N, ^{19}F, and ^{31}P. The nuclei ^{1}H, ^{19}F, and ^{31}P are the predominant isotopes in nature, and NMR is useful for analyses of mixtures, extracts, and biofluids. By contrast, ^{13}C and ^{15}N are rare. ^{13}C, for instance, has a natural abundance of only about 1%, which makes it particularly attractive for metabolic studies, where a ^{13}C-enriched substrate is given to cells and the fate of the labeled carbon atom is traced over time, thereby allowing the identification of metabolic pathways and the measurement of carbon fluxes (see Case Study 3 in Section 8.12). In some cases, it is useful to couple ^{13}C-NMR with ^{31}P-NMR, which permits additional measurements of time series of metabolites such as ATP, ADP, phosphocreatine, and inorganic phosphate, and provides an indication of the energy state and the intracellular pH of the cells. Like MS, NMR can also be used to analyze metabolites *in vitro*. The advantages are less sample preparation and better structural details, while the main disadvantage is lower sensitivity.

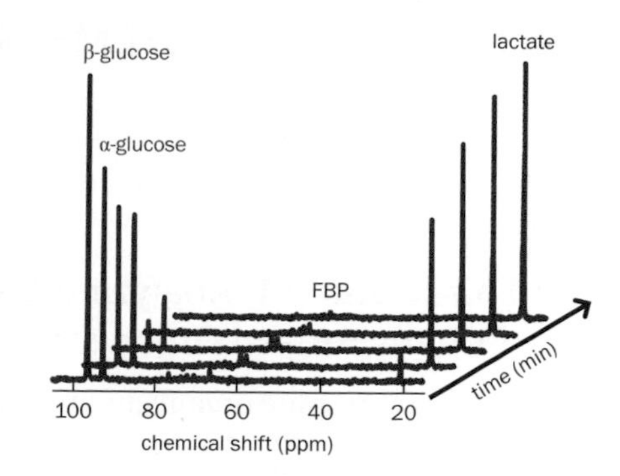

Figure 8.9 Result of an *in vivo* NMR experiment. The raw data consist of frequency spectra, which are subsequently converted into metabolite concentrations at different time points. Here, glucose in the medium decreases over time, while lactate accumulates. In addition to external substrates and end products, internal metabolites such as fructose bisphosphate can be measured. (Courtesy of Ana Rute Neves and Luis Fonseca, ITQB, Portugal.)

8.9 Flux Analysis

Stoichiometric and flux balance analysis (Chapter 3) require the experimental determination of metabolic fluxes. The fluxes fall in two categories. The first consists of fluxes entering or leaving the biological system. These are comparatively easy to measure with methods of analytical chemistry, for example, as uptake (disappearance) of substrate from the medium or the generation of waste materials, such as lactate, acetate, or CO_2. In contrast to these influxes and effluxes, the majority of interesting processes happen inside cells, where they are difficult to measure. It is not feasible to kill the cells for these measurements, because many internal fluxes would presumably stop. In some cases, the analysis of external fluxes is sufficient to infer the distribution of internal fluxes, but usually additional assumptions are required [46]. Furthermore, this inference fails if the system contains parallel pathways, bidirectional reaction steps, or metabolic cycles. Finally, one has to assume that energy-producing and energy-consuming reactions are known in some detail.

One relatively recent method that addresses some of these issues is called **metabolic flux analysis** and is based on isotopomers, which are isomers with isotopic atoms in different positions [47]. For instance, in a typical isotopomer experiment with ^{13}C, a metabolite with three carbons can at most have 2^3 isotopomers, because each carbon could be labeled or unlabeled. The various isotopomers are usually not equally likely and instead are produced with a frequency distribution that contains clues regarding the internal fluxes of a system. Specifically, one assumes that the cellular system is in an open steady state, where material is fed to and metabolized by the system, but where all metabolite concentrations and fluxes are constant. Once the labeled material has been given to the cells, one waits until the label has equilibrated throughout the metabolic system. Subsequently, the cells are harvested and metabolism is stopped, a cell extract is prepared, and the isotopomer distribution for many or all metabolites is determined with NMR or MS techniques. Evaluation of the results requires a mathematical model of the pathway that describes how the labeled material is distributed throughout the system. The parameters of this model are initially unknown and must be estimated from the isotopomer distributions. Once the model has been estimated, the internal fluxes can be quantified. Isotopomer analysis can be very powerful, but the generation and analysis of isotopomer data is not trivial. It requires solid skills in NMR spectroscopy or MS, as well as in computational modeling and statistics. Also, there are many challenges in the details. For instance, the time for the isotopomer equilibration to stabilize is sometimes difficult to determine.

Proteome and genome data may be used as coarse substitutes for indirect assessments of flux distributions, but caution is required. The abundance of enzymes obtained with the methods of proteomics may be assumed to correlate (linearly) with the overall turnover rates of the corresponding reaction steps, and changes in gene expression may be assumed to correlate with amounts of proteins and enzyme activities, but this is not always true. As is to be expected, these assumptions sometimes lead to good, but sometimes to rather unreliable, approximations of the actual processes in living cells [48].

FROM DATA TO SYSTEMS MODELS

The variety and diversity of biochemical and metabolomic data offer distinct approaches for the development of models. These have the potential for insights of different types and are exemplified below with three case studies.

8.10 Case Study 1: Analyzing Metabolism in an Incompletely Characterized Organism

It is obvious that some organisms are better understood than others. Much research has focused on model organisms such as the bacterium *Escherichia coli*, the fruit fly *Drosophila melanogaster*, and the mouse *Mus musculus*, with the expectation that insights from these representatives are applicable to lesser-known organisms. In many cases, such inferences are valid, but nature also has a way of varying details so that, for instance, high percentages of genes still have unknown functions, even in moderately well-characterized organisms.

A beautiful example of characterizing metabolism with a variety of modern methods has been published in *Science* [35]. In this study, an international group of researchers established a complete metabolic map of the small bacterium *Mycoplasma pneumoniae*, which is the culprit in some types of human pneumonia. They selected this particular organism mainly because of its comparatively small size and simple organization: *M. pneumoniae* contains fewer than 700 protein-coding genes, of which the authors, however, found only 231 to be annotated. Earlier studies had characterized a number of metabolic aspects in this bacterium with biochemical and computational methods [49, 50], and the results of these studies were integrated here with new analyses.

The authors of [35] began by assembling a catalog of known reactions, using KEGG as a starting point and assessing the activities of the various reactions with information from the literature and some new annotations. Complementing this information, they collected genome information, such as the co-occurrence of genes in the same operon and the homology of sequences with known enzymes in related organisms, and evaluated the likely functionality of inferred enzymes using structural information regarding pertinent catalytic residues. This combined metabolomic and genomic **data mining** led to apparently disconnected pathways, but also to suggestions for completing the metabolic map. As an example, the authors used methods discussed in Chapter 3 to fill a critical gap in the seemingly disconnected ascorbate pathway. They succeeded by inferring from sequence homology, predicted activity, and the position within the responsible operon that the organism is likely to possess the critical enzyme L-ascorbate 6-phosphate lactonase, without which the ascorbate pathway would not be functional. In other cases, the structural analysis permitted the elimination of putative enzymes that lacked the catalytic residues necessary for substrate turnover. Finally, putative enzyme-catalyzed reactions were eliminated from the metabolic map if their substrates were not produced by any of the other pathways. The result of this information mining and manual curation was a metabolic map without gaps, isolated reactions, or open metabolic loops. The map was then further refined with targeted laboratory experiments in order to validate the presence of alternative pathways and to determine the primary direction in reversible pathways. These experiments included growth on different substrates and ^{13}C-labeled glucose.

Further analysis of the refined metabolic map, supported by additional wet experiments, led to interesting insights. One finding confirmed the intuitive assumption that genes are up- or down-regulated in functional, pathway-associated clusters and in a dynamic fashion. Moreover, there appeared to be good concordance between gene expression, protein abundance, and metabolic turnover. For instance, the authors observed concomitant increases in mRNA, protein expression, and turnover rate associated with glycolysis, following medium acidification. Closer inspection of the results actually indicated, not surprisingly, a definite time delay between gene regulation and changes in metabolite concentrations.

Experiments with ^{13}C-labeled glucose showed that 95% of all glucose in *M. pneumoniae* ultimately becomes lactate and acetate, which implies that the lion's share of glycolysis is used for energy generation. Another interesting finding was that this simple organism contains many more linear pathways, and fewer diverging and converging branch points, than more complicated organisms, and is therefore less redundant. This decreased redundancy has direct implications for the robustness of the organism, because some metabolites can only be produced in a single manner. As a consequence, 60% of the metabolic enzymes in *M. pneumoniae* were found to be essential, which is about four times as many as in *E. coli*. Finally, the authors concluded that *M. pneumoniae* is not optimized for fast growth and biomass production, which is often assumed in flux balance analyses as the main objective of a microbe, but possibly for survival strategies that suppress the growth of competitor populations and interactions with the host organism.

8.11 Case Study 2: Metabolic Network Analysis

Microorganisms can produce valuable organic compounds in high quantities and often exquisite purity. Examples include yeast cells converting sugars into alcohol, fungi producing citric acid, and bacteria producing insulin. However, many wild-type organisms produce these compounds only in small quantities, and it is the task of metabolic engineers to improve yield. Traditionally, they pursued this goal with long sequences of genetic manipulations, repeated fine-tuning of the growth medium, and selection of the best strains [51]. As an alternative, modern methods of metabolic systems analysis may be pursued (see [12, 32, 52] and Chapter 14).

A representative example is the industrial production of the amino acid lysine, using the microorganism *Corynebacterium glutamicum*. Lysine is a valuable additive for animal feeds, and many studies have therefore worked on improving its yield. Most of these studies used stoichiometric and flux balance analysis (see [53–55] and Chapter 3), but some groups also designed fully kinetic models of the pathway [56, 57]. This option was feasible because the pathway structure of lysine biosynthesis is well known, and databases such as KEGG, BioCyc, and BRENDA contain information on all the enzymes involved.

Generally, the development of kinetic models is much more challenging than a stoichiometric analysis, because the material balance at all metabolite nodes at the steady state must be augmented with detailed rate equations. This process is complicated for two reasons. First, the best-suited mathematical representations of specific enzyme reactions are usually not known and any assumptions regarding their appropriateness are difficult to validate. One may circumvent this challenge by selecting a canonical representation, but it is still not trivial to determine to what degree such a representation is valid (see the discussions in Chapter 4 and [12, 58]). The second challenge is the determination of numerical values for all parameters in the selected rate equations, such as the turnover rate of a process or the Michaelis constant of an enzyme. Plenty of pertinent information regarding numerous enzymes can be found in BRENDA and ExPASy, but this information was often inferred from experiments *in vitro*, and it is unclear whether the presented kinetic characteristics are numerically valid *in vivo* (see, for example, [59]).

8.12 Case Study 3: Extraction of Dynamic Models from Experimental Data

The vast majority of dynamic models of metabolic systems have been constructed from the bottom up. In the first step, each reaction or process within the system is assigned a mathematical representation, such as a mass action term, the MMRL, or a power-law function. Second, data from the literature are used to estimate the parameters of each rate function. Third, all representations are merged into a model of the entire system. Fourth, with high hopes, one tests the integrated model against validation data. Unfortunately, this test fails more often than not, and several rounds of refinements and searches for better parameter values ensue. Even for moderately sized systems, the process often takes months, if not years. However, the result is a fully dynamic model that has a much higher potential than a static model. Examples are manifold and include [60–62].

A distinct alternative is model development from the top down, using time-series data. The group of Santos and Neves used *in vivo* ^{13}C- and ^{31}P-NMR spectroscopy to measure metabolite concentrations in living populations of the bacterium *Lactococcus lactis* [63, 64]. Because NMR is nondestructive, the group was able to measure dense time series of the decreasing concentration of labeled substrate, as well as the time-dependent concentrations of metabolic intermediates and end products, all within the same cell culture. By fine-tuning their methods, they were able to measure the key metabolites of glycolysis in intervals of 30 seconds or less, thereby creating hour-long time series of metabolite profiles. The resulting data were successfully converted into computational models (see, for example, [65, 66]).

Kinoshita and co-workers [67] measured time-course profiles of over 100 metabolites in human red blood cells, using capillary electrophoresis with subsequent mass spectrometry (CE-MS). Using these data, the authors developed a mathematical pathway model, consisting of about 50 enzyme kinetic rate equations, whose formats were extracted from the literature. The model allowed manipulations mimicking the condition of hypoxia, where the organism is deprived of oxygen.

EXERCISES

8.1. Screen the literature or Internet for large metabolites that are not proteins, peptides, RNA, or DNA. Compare their sizes with those of proteins.

8.2. Describe (8.8) and (8.9) in words.

8.3. Confirm by mathematical means that no material is lost or gained in the Michaelis–Menten process.

8.4. Formulate a differential equation for E in the Michaelis–Menten process. Compare this equation with (8.8) and interpret the result.

8.5. Derive (8.10) from the ODE system (8.7)–(8.9) or find the derivation in the literature or Internet. For the derivation, make use of the quasi-steady-state assumption and of the fact that the sum $E + (ES)$ is constant.

8.6. Simulate (8.7)–(8.10) and compare $\dot{P}$ with v_p in (8.10) for different initial values and parameter values. Why are no initial values needed in (8.10)?

8.7. Redo the analysis of Exercise 8.6, but assume that S is 1000 times larger than E, which might mimic *in vitro* conditions. Evaluate the validity of QSSA in comparison with Figure 8.3.

8.8. Explain the details of the dynamics of the substrate in the Hill model (8.15). Formulate equations for (ES) and P.

8.9. For a reaction with competitive inhibition, plot $v_{P,CI}$ against S, and $1/v_{P,CI}$ against $1/S$, for several inhibitor concentrations and compare the plots with the corresponding plot from the uninhibited reaction.

8.10. For a reaction with uncompetitive inhibition, plot $v_{P,UI}$ against S, and $1/v_{P,UI}$ against $1/S$, for several inhibitor concentrations and compare the plots with the corresponding plot from the uninhibited reaction. If you have done Exercise (8.9), compare the two types of inhibition.

8.11. For a reaction with noncompetitive inhibition, plot $v_{P,NI}$ against S, and $1/v_{P,NI}$ against $1/S$, for several inhibitor concentrations and compare the plots with the corresponding plot from the uninhibited reaction. If you have done Exercise (8.9) or (8.10), compare your results.

8.12. Extract from KEGG and BioCyc all pathways using glucose 6-phosphate as a substrate. Are these pathways the same in *E. coli*, baker's yeast, and humans?

8.13. Mine databases and the literature to reconstruct a metabolic pathway system describing the production of lysine. Search for kinetic parameter values for as many steps as possible. *Note*: This is an open-ended problem with many possible solutions.

8.14. Given the known influxes (blue) and effluxes (green) in **Figure 8.10**, compute the internal fluxes (red) of the system. Is the solution unique? Discuss

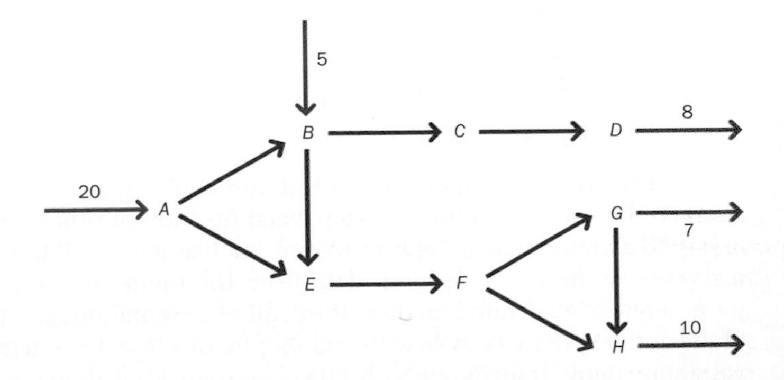

Figure 8.10 Generic stoichiometric map. The influxes (blue) and effluxes (green) are given in units of mmol L^{-1} min^{-1}.

issues of uniqueness generically and provide examples and counterexamples. Is the system in Figure 8.10 necessarily in a steady state? Is the flux information sufficient to determine the concentrations of the metabolites *A*, ..., *H*?

8.15. Design models for the pathway in Figure 8.5. Represent the production of *A* as $1 \cdot D^{-0.5}$. For all simulations, start at the steady state (which you have to compute first) and double the efflux from *D* at $t = 10$. In the first set of simulations, use the MMRL for the reactions that use *A*, *B*, *C*, or *D* as substrate. Choose different combinations of V_{max} and K_M values as you like. Repeat the simulations with power-law rate laws. Summarize your findings in a report.

8.16. Design a power-law model for the pathway in Figure 8.6. Ignore intermediates that are suggested by dotted lines and set all rate constants equal to 1. For all substrate degradation processes, use kinetic orders of 0.5. For inhibition effects, start with kinetic orders of –0.5. Subsequently, change the inhibition effects one by one, or in combination, to other negative values. Always start your simulations at the steady state. Summarize your findings in a report.

8.17. How could one model the effect of altered gene expression on a metabolic pathway? Discuss different options.

8.18. Compare the ascorbate pathway in *M. pneumoniae* with that in *E. coli*. Report your findings.

8.19. Find five pathways in *E. coli* that do not exist in *M. pneumoniae*.

8.20. Explore how much citric acid is being produced annually and which microorganism is primarily used for its large-scale industrial production. Search for models that have addressed citric acid production and review their key features.

8.21. Compare kinetic parameters of the fermentation pathway in yeast that were presented in the work of Galazzo and Bailey [68] and Curto et al. [60] with those that are listed in BRENDA and ExPASy. Summarize your findings in a spreadsheet or a report.

8.22. Consider the model in **Figure 8.11**, which we have already studied in Chapter 3 (Exercise 3.28). It describes the pentose pathway, which exchanges carbohydrates with different numbers of carbon atoms, as shown in **Table 8.1**. For instance, two units of X_1 are needed to produce one unit of X_2 and X_3, and each reaction $V_{5,3}$ uses one unit of X_5 to generate two units of X_3. Note that X_3 appears in two locations, but represents the same pool. Assume that all processes follow mass action kinetics and that the influx ($V_{7,1}$) and the effluxes ($V_{3,0}$ and CO_2) are 20 mmol L^{-1} min^{-1} and that all other fluxes have a value of 10 mmol L^{-1} min^{-1}. Confirm that these settings correspond to a consistent steady-state flux distribution. Assume the following steady-state concentrations: $X_1 = 10$, $X_2 = 6$, $X_3 = 12$, $X_4 = 8$, $X_5 = 8$, $X_6 = 4$, $X_7 = 20$. Compute the rate constants of all processes.

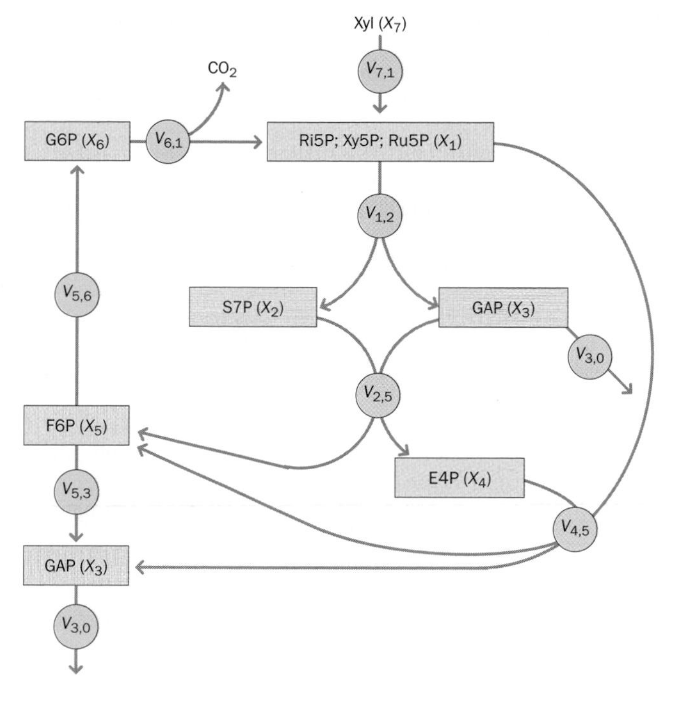

Figure 8.11 Pentose phosphate pathway. Information on influxes and effluxes permits the computation of internal fluxes. (From Wiechert W & de Graaf AA. *Biotechnol. Bioeng.* 55 [1997] 101–117. With permission from John Wiley & Sons.)

TABLE 8.1: NUMBER OF CARBON ATOMS IN EACH METABOLITE IN THE PATHWAY IN FIGURE 8.11

Compound Pool	Number of Carbon Atoms
X_1	5
X_2	7
X_3	3
X_4	4
X_5	6
X_6	6
X_7	5

8.23. Using the model in Exercise 8.22 with the same flux distribution, is it possible that the system can have different steady-state concentrations? If not, explain. If the answer is yes, construct one or more examples. Write a brief report on your findings.

8.24. Using the results from Exercise 8.22, perform a simulation that starts at the steady state and models a bolus perturbation to X_1. Make a prediction of the model responses before you simulate and summarize your findings in a brief report.

8.25. Using the results from Exercises 8.22–8.24, perform two simulations that start at the steady state and model either a persistent or a bolus perturbation to X_7. Make a prediction of the model responses before you simulate and write a brief report on your findings.

REFERENCES

[1] Derechin M, Ramel AH, Lazarus NR & Barnard EA. Yeast hexokinase. II. Molecular weight and dissociation behavior. *Biochemistry* 5 (1966) 4017–4025.
[2] Human Genome Project Information. http://www.ornl.gov/sci/techresources/Human_Genome/home.shtml.
[3] Guldberg CM & Waage P. Studier i Affiniteten. *Forhandlinger i Videnskabs-Selskabet i Christiania* (1864) 35.
[4] Guldberg CM & Waage P. Études sur les affinites chimiques. Brøgger & Christie, 1867.
[5] Voit EO, Martens HA & Omholt SW. 150 years of the mass action law. *PLoS Comput. Biol.* 11 (2015) e1004012.
[6] Gillespie DT. Stochastic simulation of chemical kinetics. *Annu. Rev. Phys. Chem.* 58 (2007) 35–55.
[7] Wolkenhauer O, Ullah M, Kolch W & Cho K-H. Modelling and simulation of intracellular dynamics: choosing an appropriate framework. *IEEE Trans. Nanobiosci.* 3 (2004) 200–207.
[8] Henri MV. Lois générales de l'action des diastases. Hermann, 1903.
[9] Michaelis L & Menten ML. Die Kinetik der Invertinwirkung. *Biochem. Z.* 49 (1913) 333–369.
[10] Benkovic SJ, Hammes GG & Hammes-Schiffer S. Free-energy landscape of enzyme catalysis. *Biochemistry* 47 (2008) 3317–3321.
[11] Briggs GE & Haldane JBS. A note on the kinetics of enzyme action. *Biochem. J.* 19 (1925) 338–339.
[12] Torres NV & Voit EO. Pathway Analysis and Optimization in Metabolic Engineering. Cambridge University Press, 2002.
[13] Savageau MA. Enzyme kinetics *in vitro* and *in vivo*: Michaelis–Menten revisited. In Principles of Medical Biology (EE Bittar, ed.), pp 93–146. JAI Press, 1995.
[14] Hill AV. The possible effects of the aggregation of the molecules of hæmoglobin on its dissociation curves. *J. Physiol.* 40 (1910) iv–vii.
[15] Edelstein-Keshet L. Mathematical Models in Biology. McGraw-Hill, 1988.
[16] Schultz AR. Enzyme Kinetics: From Diastase to Multi-Enzyme Systems. Cambridge University Press, 1994.
[17] Cornish-Bowden A. Fundamentals of Enzyme Kinetics, 4th ed. Wiley, 2012.
[18] Savageau MA. Biochemical Systems Analysis: A Study of Function and Design in Molecular Biology. Addison-Wesley, 1976.
[19] Berg PH, Voit EO & White R. A pharmacodynamic model for the action of the antibiotic imipenem on *Pseudomonas in vitro*. *Bull. Math. Biol.* 58 (1966) 923–938.
[20] BRENDA: The Comprehensive Enzyme Information System. http://www.brenda-enzymes.org/.
[21] ENZYME: Enzyme Nomenclature Database. http://enzyme.expasy.org/.
[22] Fell DA. Understanding the Control of Metabolism. Portland Press, 1997.
[23] Voit EO. Computational Analysis of Biochemical Systems: A Practical Guide for Biochemists and Molecular Biologists. Cambridge University Press, 2000.
[24] KEGG (Kyoto Encyclopedia of Genes and Genomes) Pathway Database. http://www.genome.jp/kegg/pathway.html.
[25] BioCyc Database Collection. http://biocyc.org/.
[26] ERGO: Genome Analysis and Discovery System. Integrated Genomics. http://ergo.integratedgenomics.com/ERGO/.
[27] Pathway Studio. Ariadne Genomics. http://ariadnegenomics.com/products/pathway-studio.
[28] KInfer. COSBI, The Microsoft Research–University of Toronto Centre for Computational and Systems Biology. www.cosbi.eu/index.php/research/prototypes/kinfer.
[29] Reactome. http://www.reactome.org.
[30] Joshi-Tope G, Gillespie M, Vastrik I, et al. Reactome: a knowledgebase of biological pathways. *Nucleic Acids Res.* 33 (2005) D428–D432.
[31] Lipid Maps: Lipidomics Gateway. http://www.lipidmaps.org/.
[32] Palsson BØ. Systems Biology: Properties of Reconstructed Networks. Cambridge University Press, 2006.
[33] Visser D & Heijnen JJ. Dynamic simulation and metabolic re-design of a branched pathway using linlog kinetics. *Metab. Eng.* 5 (2003) 164–176.
[34] Alves R & Savageau MA. Effect of overall feedback inhibition in unbranched biosynthetic pathways. *Biophys. J.* 79 (2000) 2290–2304.
[35] Yus E, Maier T, Michalodimitrakis K, et al. Impact of genome reduction on bacterial metabolism and its regulation. *Science* 326 (2009) 1263–1268.
[36] Neves M-J & François J. On the mechanism by which a heat shock induces trehalose accumulation in *Saccharomyces cerevisiae*. *Biochem. J.* 288 (1992) 859–864.
[37] Fonseca LL, Sánchez C, Santos H & Voit EO. Complex coordination of multi-scale cellular responses to environmental stress. *Mol. BioSyst.* 7 (2011) 731–741.
[38] Harrigan GG & Goodacre R (eds). Metabolic Profiling. Kluwer, 2003.
[39] Oldiges M & Takors R. Applying metabolic profiling techniques for stimulus–response experiments: chances and pitfalls. *Adv. Biochem. Eng. Biotechnol.* 92 (2005) 173–196.
[40] Villas-Bôas SG, Mas S, Åkesson M, et al. Mass spectrometry in metabolome analysis. *Mass Spectrom. Rev.* 24 (2005) 613–646.
[41] Theobald U, Mailinger W, Baltes M, et al. *In vivo* analysis of metabolic dynamics in *Saccharomyces cerevisiae*: I. Experimental observations. *Biotechnol. Bioeng.* 55 (1997) 305–316.
[42] Spura J, Reimer LC, Wieloch P, et al. A method for enzyme quenching in microbial metabolome analysis successfully applied to Gram-positive and Gram-negative bacteria and yeast. *Anal. Biochem.* 394 (2009) 192–201.
[43] Garcia-Perez I, Vallejo M, Garcia A, et al. Metabolic fingerprinting with capillary electrophoresis. *J. Chromatogr. A* 1204 (2008) 130–139.
[44] Scriba GK. Nonaqueous capillary electrophoresis–mass spectrometry. *J. Chromatogr. A* 1159 (2007) 28–41.
[45] Johnson JM, Yu T, Strobel FH & Jones DP. A practical approach to detect unique metabolic patterns for personalized medicine. *Analyst* 135 (2010) 2864–2870.
[46] Varma A & Palsson BØ. Metabolic flux balancing: basic concepts, scientific and practical use. *Biotechnology (NY)* 12 (1994) 994–998.
[47] Wiechert W. ^{13}C metabolic flux analysis. *Metab. Eng.* 3 (2001) 195–206.
[48] Griffin TJ, Gygi SP, Ideker T, et al. Complementary profiling of gene expression at the transcriptome and proteome levels in *Saccharomyces cerevisiae*. *Mol. Cell. Proteomics* 1 (2002) 323–333.
[49] Pachkov M, Dandekar T, Korbel J, et al. Use of pathway analysis and genome context methods for functional genomics of *Mycoplasma pneumoniae* nucleotide metabolism. *Gene* 396 (2007) 215–225.
[50] Pollack JD. *Mycoplasma* genes: a case for reflective annotation. *Trends Microbiol.* 5 (1997) 413–419.
[51] Ratledge C & Kristiansen B (eds). Basic Biotechnology, 3rd ed. Cambridge University Press, 2006.
[52] Stephanopoulos GN, Aristidou AA & Nielsen J. Metabolic Engineering: Principles and Methodologies. Academic Press, 1998.

[53] Koffas M & Stephanopoulos G. Strain improvement by metabolic engineering: lysine production as a case study for systems biology. *Curr. Opin. Biotechnol.* 16 (2005) 361–366.
[54] Kjeldsen KR & Nielsen J. *In silico* genome-scale reconstruction and validation of the *Corynebacterium glutamicum* metabolic network. *Biotechnol. Bioeng.* 102 (2009) 583–597.
[55] Wittmann C, Kim HM & Heinzle E. Metabolic network analysis of lysine producing *Corynebacterium glutamicum* at a miniaturized scale. *Biotechnol. Bioeng.* 87 (2004) 1–6.
[56] Dräger A, Kronfeld M, Ziller MJ, et al. Modeling metabolic networks in *C. glutamicum*: a comparison of rate laws in combination with various parameter optimization strategies. *BMC Syst. Biol.* 3 (2009) 5.
[57] Yang C, Hua Q & Shimizu K. Development of a kinetic model for L-lysine biosynthesis in *Corynebacterium glutamicum* and its application to metabolic control analysis. *J. Biosci. Bioeng.* 88 (1999) 393–403.
[58] Voit EO. Modelling metabolic networks using power-laws and S-systems. *Essays Biochem.* 45 (2008) 29–40.
[59] Albe KR, Butler MH & Wright BE. Cellular concentrations of enzymes and their substrates. *J. Theor. Biol.* 143 (1989) 163–195.
[60] Curto R, Sorribas A & Cascante M. Comparative characterization of the fermentation pathway of *Saccharomyces cerevisiae* using biochemical systems theory and metabolic control analysis. Model definition and nomenclature. *Math. Biosci.* 130 (1995) 25–50.
[61] Hynne F, Danø S & Sørensen PG. Full-scale model of glycolysis in *Saccharomyces cerevisiae. Biophys. Chem.* 94 (2001) 121–163.
[62] Alvarez-Vasquez F, Sims KJ, Hannun YA & Voit EO. Integration of kinetic information on yeast sphingolipid metabolism in dynamical pathway models. *J. Theor. Biol.* 226 (2004) 265–291.
[63] Neves AR, Pool WA, Kok J, et al. Overview on sugar metabolism and its control in *Lactococcus lactis*—the input from *in vivo* NMR. *FEMS Microbiol. Rev.* 29 (2005) 531–554.
[64] Neves AR, Ventura R, Mansour N, et al. Is the glycolytic flux in *Lactococcus lactis* primarily controlled by the redox charge? Kinetics of NAD^+ and NADH pools determined *in vivo* by ^{13}C NMR. *J. Biol. Chem.* 277 (2002) 28088–28098.
[65] Voit EO, Almeida JO, Marino S, Lall R. et al. Regulation of glycolysis in *Lactococcus lactis*: an unfinished systems biological case study. *Syst. Biol. (Stevenage)* 153 (2006) 286–298.
[66] Chou IC & Voit EO. Recent developments in parameter estimation and structure identification of biochemical and genomic systems. *Math. Biosci.* 219 (2009) 57–83.
[67] Kinoshita A, Tsukada K, Soga T, et al. Roles of hemoglobin allostery in hypoxia-induced metabolic alterations in erythrocytes: simulation and its verification by metabolome analysis. *J. Biol. Chem.* 282 (2007) 10731–10741.
[68] Galazzo JL & Bailey JE. Fermentation pathway kinetics and metabolic flux control in suspended and immobilized *S. cerevisiae. Enzyme Microbiol. Technol.* 12 (1990) 162–172.

FURTHER READING

Cornish-Bowden A. Fundamentals of Enzyme Kinetics, 4th ed. Wiley, 2012.
Edelstein-Keshet L. Mathematical Models in Biology. McGraw-Hill, 1988.
Fell DA. Understanding the Control of Metabolism. Portland Press, 1997.
Harrigan GG & Goodacre R (eds). Metabolic Profiling. Kluwer, 2003.
Koffas M & Stephanopoulos G. Strain improvement by metabolic engineering: lysine production as a case study for systems biology. *Curr. Opin. Biotechnol.* 16 (2005) 361–366.
Savageau MA. Biochemical Systems Analysis: A Study of Function and Design in Molecular Biology. Addison-Wesley, 1976.
Schultz AR. Enzyme Kinetics: From Diastase to Multi-Enzyme Systems. Cambridge University Press, 1994.
Torres NV & Voit EO. Pathway Analysis and Optimization in Metabolic Engineering. Cambridge University Press, 2002.
Voit EO. Computational Analysis of Biochemical Systems: A Practical Guide for Biochemists and Molecular Biologists. Cambridge University Press, 2000.
Voit EO. Biochemical systems theory: a review. *ISRN Biomath.* 2013 (2013) Article ID 897658.

Signaling Systems

9

When you have read this chapter, you should be able to:

- Understand the fundamentals of signal transduction systems
- Discuss the advantages and disadvantages of discrete and continuous models of signal transduction
- Describe the role of G-proteins
- Solve problems involving Boolean models
- Understand bistability and hysteresis
- Discuss quorum sensing
- Analyze differential equation models of two-component signal transduction systems
- Analyze differential equation models of the MAPK signal transduction system

In cell biology, **signal transduction** describes a class of mechanisms through which a cell receives internal or external **signals**, processes them, and triggers appropriate responses. In many cases, the signals are chemical, but cells can also respond to different kinds of physical signals, such as light, electrical stimuli, and mechanical stresses. In fact, many different signals are often processed at the same time [1]. If the cell receives an external chemical signal, the transduction process usually begins with the binding of specific ligands to receptors on the outer surface of the cell membrane. These receptors are transmembrane proteins, and binding of a ligand to the extracellular domain of the receptor triggers a change in the conformation or function of the receptor that propagates to the intracellular domain of the receptor. This change mediates other changes within the cell that ultimately result in an appropriate response to the original signal. As a specific example, integrins are transmembrane proteins that transduce information from the extracellular matrix or the surrounding tissue to the inside of the cell and are involved in cell survival, apoptosis, proliferation, and differentiation [2].

Good examples of signal transduction circuits can be found in an important family of signaling molecules called **G-proteins**. These are located on the inside of the cell membrane and are coupled to specific receptors that reach from the inside to the outer surface of the cell and whose conformation changes when a ligand binds to them. Specifically, upon binding to the extracellular domain of such a **G-protein-coupled receptor (GPCR)** (**Figure 9.1**), the change in the intracellular domain of the GPCR causes a subunit of the G-protein to be released and to bind to a molecule of guanosine triphosphate (GTP), which leads to the dissociation of two molecular subunits. The importance of this process led to the prefix "G," which

Figure 9.1 Example of a membrane-spanning G-protein-coupled receptor (GPCR_3ny8). The human β_2-adrenergic receptor (β_2-AR) consists of a seven-helical-bundle membrane protein. The two planes represent the lipid bilayer. GPCRs represent a large fraction of current pharmaceutical targets, and β_2-AR is one of the most extensively studied. (Courtesy of Huiling Chen and Ying Xu, University of Georgia.)

stands for "guanine-nucleotide-binding." The activated G-protein subunits detach from the receptor and initiate internal signaling processes that are specific to the particular type of G-protein. For instance, the process may trigger the production of cyclic adenosine monophosphate (cAMP), which can activate protein kinase A, which in turn can phosphorylate and thereby activate many possible downstream targets. Other G-proteins stimulate the generation of signaling compounds such as diacylglycerol and inositol trisphosphate, which control the release of intracellular calcium from storage compartments into the cytoplasm. After some recovery period, hydrolysis removes the third phosphate group from GTP and readies the G-protein for transduction of a new signal. The functionality of very many drugs is associated with GPCRs. For their "discovery of G-proteins and the role of these proteins in signal transduction in cells," Alfred Gilman and Martin Rodbell received the 1994 Nobel Prize in Physiology or Medicine.

The signal transduction processes inside the cell are mediated by chains of biochemical reactions that are catalyzed by enzymes and often modulated by **second messengers**, which are water-soluble or -insoluble molecules or gases such as nitric oxide. Typical water-soluble, hydrophilic messengers are calcium and cAMP. They are usually located in the cytosol and may be activated or deactivated by specific enzymes. Hydrophobic signaling molecules are often associated with the plasma membrane. Examples include different phosphatidylinositols, inositol trisphosphate, and the sphingolipid ceramide. The steps between receiving a signal at the membrane and triggering a response in the cytosol or nucleus are often organized as a cascade of events that amplify the incoming signal and filter out noise. The speed of signal transduction depends strongly on the mechanism. Calcium-triggered electrical signals occur at the order of milliseconds, while protein- and lipid-based cascades respond at the order of minutes. Signals involving genomic responses happen within tens of minutes, hours, or even days. Chapter 12 discusses the complex, calcium-based events occurring during every beat of the heart.

Because signaling processes are ubiquitous in biology, an enormous literature has been amassed over the past decades, and even a cursory review of all signal transduction systems is far beyond the scope of this chapter. In contrast to the wealth of experimental information, modeling studies are much fewer, although their number has also been rising quickly in recent years. In this chapter, we will discuss representative examples of models that illustrate certain aspects of signal transduction systems. These models fall into two distinct classes with only scant overlap. The first class considers signaling networks as **graphs** and tries (a) to understand their functionality and (b) to deduce their causal connectivity from experimental observations with methods of discrete mathematics and statistics. The second class attempts to capture the dynamics of signaling networks with **ordinary or stochastic differential equation models**.

STATIC MODELS OF SIGNAL TRANSDUCTION NETWORKS

9.1 Boolean Networks

An intuitive approach toward assessing signaling systems is a network with hardwired connections and signaling events that happen on a discrete timescale. Suppose the signaling system consists of 25 components. To be less vague, let us imagine that these components are genes, and a that response is triggered if these genes show a particular combination of expression, where certain genes are *on* (meaning that they are expressed), whereas others are *off* (for instance, because they are presently repressed). Furthermore, suppose that an expressed gene ultimately leads to the production of a transcription factor or repressor that may affect its own expression and/or that of other genes in the system.

Let us at once abstract and simplify the set-up by assembling the 25 genes in a 5×5 grid of boxes (**Figure 9.2A**) and let us name the genes according to their position, such that G_{11} refers to the top left corner, G_{12} lies just to the right of G_{11}, G_{21} is just below G_{11}, and so on. An arbitrary decision says that a white box means the gene is *on* and that black means *off*. Since we like numbers, we translate *off* into 0 and *on* into 1. Again, this setting is rather arbitrary. We can now formulate statements such as the following:

for $i, j = 1, \ldots, 5$,

$$G_{ij} = \begin{cases} 1 & \text{if} \quad i = j, \\ 0 & \text{if} \quad i \neq j, \end{cases} \tag{9.1}$$

which means that the genes on the northwest–southeast diagonal of the grid are *on* (Figure 9.2B).

The beauty of these types of networks lies in the simplicity of defining their dynamics. Time moves forward in a discrete fashion, which we easily formulate as $t = 0, 1, 2, \ldots$ (see Chapter 4). Each G_{ij} is now a recursive function of time, $G_{ij} = G_{ij}(t)$, and changes in the states of the network are determined by rules that state what happens to each state during the transition from one time point to the next. As an example of a very simple rule, consider the following situation:

for $i, j = 1, \ldots, 5$,

$$G_{ij}(t+1) = \begin{cases} 1 & \text{if} \quad G_{ij}(t) = 0, \\ 0 & \text{if} \quad G_{ij}(t) = 1. \end{cases} \tag{9.2}$$

Put into words, this rule says that 0 switches to 1 and 1 switches to 0 at every time transition (Figure 9.2C). In other words, the rule defines a blinking pattern, switching back and forth between Figures 9.2B and C. This dynamics is independent of the initial state of the network. In other words, we can define $G_{ij}(0)$ any way we want, as long as all settings are either 0 or 1.

This type of network with *on*/*off* states and transition rules is often called a **Boolean** network, honoring the nineteenth-century British mathematician and philosopher George Boole, who is credited with inventing the logical foundation of modern computer science. Much has been written about Boolean networks, and an

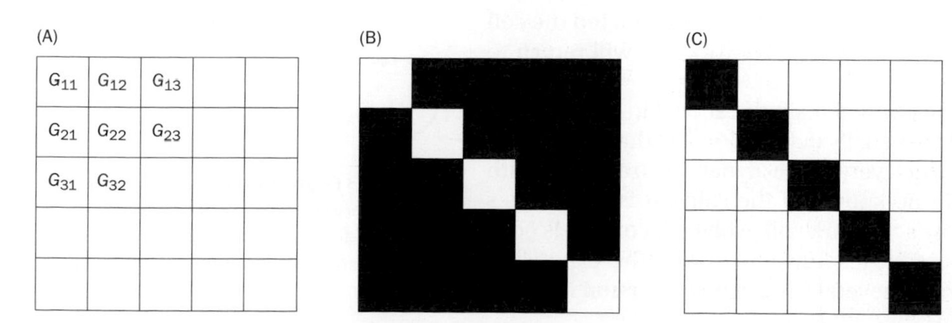

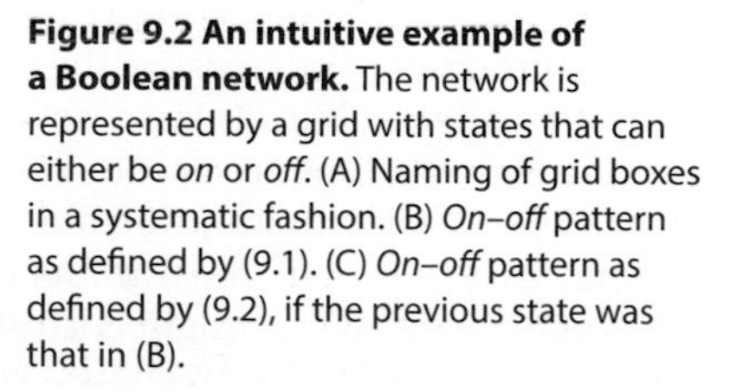

Figure 9.2 An intuitive example of a Boolean network. The network is represented by a grid with states that can either be *on* or *off*. (A) Naming of grid boxes in a systematic fashion. (B) *On–off* pattern as defined by (9.1). (C) *On–off* pattern as defined by (9.2), if the previous state was that in (B).

excellent starting point for further exploration is Stuart Kauffman's widely recognized treatise *The Origins of Order* [3].

The rules in a Boolean network may be much more complicated than in the illustration above. For instance, a rule could say that G_{ij} switches if the majority of its neighbors are 1. How could we formulate such a rule? Let us start with a node that is not located at one of the sides of the grid. Thus, it has eight neighbors, which are

$$G_{i-1,j-1},\; G_{i-1,j},\; G_{i-1,j+1},\; G_{i,j-1},\; G_{i,j+1},\; G_{i+1,j-1},\; G_{i+1,j},\; G_{i+1,j+1}.$$

Because each neighbor can only take values of 0 or 1, we may formulate the rule as

$$G_{ij}(t+1)=\begin{cases}1 & \text{if} \quad G_{ij}(t)=0 \;\; \text{and sum of all neighbors } >4,\\ 0 & \text{if} \quad G_{ij}(t)=0 \;\; \text{and sum of all neighbors } \leq 4,\\ 0 & \text{if} \quad G_{ij}(t)=1 \;\; \text{and sum of all neighbors } >4,\\ 1 & \text{if} \quad G_{ij}(t)=1 \;\; \text{and sum of all neighbors } \leq 4.\end{cases} \tag{9.3}$$

The rule is applied iteratively at the transition from any time t to time $t+1$, during which the values of the neighbors of each node at time t determine the value of this node at time $t+1$. It should be noted that the rule in (9.3) may be written in different, equivalent, ways, some of which will certainly be more compact and elegant.

For intriguing entertainment with a more complex network that has many conceptual similarities with a Boolean network, explore the *Game of Life*, which was created by John Conway in 1970 and attracted interest from an almost Zen-like following of biologists, mathematicians, computer scientists, philosophers, and many others. The game is still available on the Internet and even has its own Wikipedia page.

Boolean networks can easily be interpreted as signaling systems, for instance, in genomes. Instead of using a fixed grid, it is more common to locate the genes G_i as nodes in a directed graph (see Chapter 3), where incoming arrows show the influences of other genes (G_j, G_k, ...), while outgoing arrows indicate which genes are affected by the gene G_i. A typical illustrative example is given in **Figure 9.3**. Because of evolutionary advantages with respect to robustness, gene networks appear to be only sparsely connected, and most genes have only one or two upstream regulators [4], which greatly facilitates the construction of gene regulatory network models.

The signaling function is accomplished by defining a specific rule for each gene G_i at the transition from time t to $t+1$, which is determined by the expression state of some or all other genes (or gene products) at time t. Under well-defined constellations of the system, G_i will "fire" (become 1), while it is otherwise silent ($G_i = 0$). As an example, suppose that at $t = 0$ only G_1 in Figure 9.3 is *on* and all other genes are *off*. The activation arrows and inhibition symbols suggest that, at $t = 1$, G_2 will still be *off*, while G_3 and G_4 will turn *on*.

To capture the full dynamics of the network, it is necessary to define for each gene whether it will be *on* or *off* at time $t+1$, given an *on–off* pattern of all genes at time t. These complete definitions are needed, because, for instance, the graphical representation in Figure 9.3 does not prescribe what will happen to G_3 if both G_1 and G_2 are *on*. Will the activation trump the inhibition or will the opposite be true? The state of G_3 at $t+1$ may depend on its own state at time t. In general, the rules may be very complicated, which implies that rather complex signaling systems may be constructed and analyzed. As an example, Davidich and Bornholdt [5] modeled the cell cycle regulatory network in fission yeast with a Boolean network. We will return to this example at the end of this chapter.

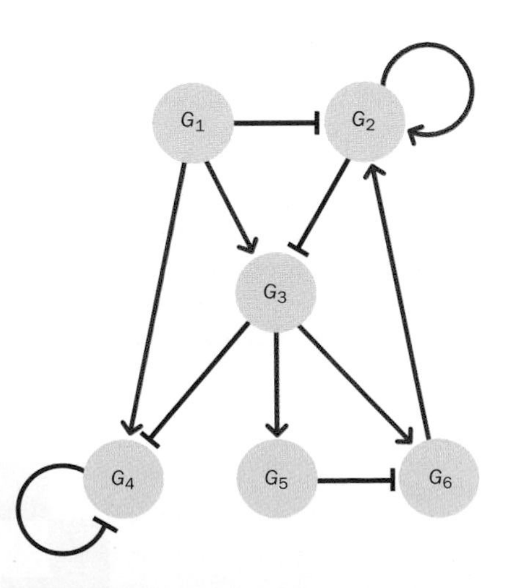

Figure 9.3 Representation of a generic gene interaction network as a graph. Arrows indicate activation while lines with terminal bars represent inhibition or repression.

Two features of Boolean networks are of particular significance. First, there is no real size limitation, and because the rules are usually defined locally, that is, for each node, it is often relatively easy to construct very large dynamic processes with complicated *on–off* patterns. But even if one knows all the rules, it is sometimes difficult to predict with intuition alone how a network will evolve. Second, it is conceptually easy to expand the number of possible responses per state. This expansion allows gray tones, which might correspond to several (discrete) expression levels of genes. The rules become somewhat more complicated, but the principles remain the same. It is even possible to allow for a continuous range of each expression state. With this expansion, the formal description becomes much more complicated and resembles that of an artificial neural network except that an artificial neural network allows modifications of the strengths of connections between nodes (see [6] and Chapter 13). An expansion that is far from trivial is the permission to make the rules themselves time-dependent or adaptive in a sense that they change in response to the overall state of the network.

A limitation of traditional Boolean networks is indeed that typically they are not adaptive and do not allow time-dependent rules, and also they have no memory: as soon as a new state is reached, the previous state of the network is forgotten. In many biological systems, such memory is important. Another limitation is the strictly deterministic nature of these networks: given the initial state and the transition rules, the future of the network is completely determined. In order to overcome this particular limitation and to allow for biological variability and stochasticity, Shmulevich and colleagues proposed probabilistic Boolean networks, where the transitions between states occur with predefined probabilities [7, 8]. This set-up may remind us of the Markov models that we discussed in Chapter 4.

9.2 Network Inference

Natural signal transduction networks have the ability to function robustly in notoriously noisy environments. Whether it is gene expression, which can show significant stochastic fluctuations, or a system of neurons, where action potentials may fire spontaneously, a signal transduction system ultimately responds with a high degree of reliability to the correct inputs while generally filtering out spurious inputs.

The genuinely stochastic environments and internal features of signal transduction systems create many challenges for modeling and analysis. The biggest challenge is actually not the determination of whether a signal, corrupted by noise, will trigger a response—usually, this merely requires a comparatively simple application and evaluation of the rules governing conditional probabilities. The true challenge consists of the opposite task, as we have already discussed in Chapter 3. Namely, given only a set of input signals and the corresponding set of responses, is it possible to infer or reconstruct the structure and numerical features of the inner workings of the signal transduction network? Is it possible to conclude, with any degree of reliability, that component X only fires if components Y and Z are both active? The answers to such questions obviously depend on a number of factors, including the amount of data and their degree of certainty, the expected noise, and, last but not least, the size of the network: it does not take much imagination to expect that large networks are much more difficult to infer than systems with just a handful of components.

The topic of network inference has traditionally been addressed with two methods: either one designs a static graph model and uses methods of Bayesian network inference, which we discussed in Chapter 3, or one constructs a system of differential equations and attempts to identify the connectivity and regulation of the signaling system from sets of time-series data with inverse (parameter estimation) methods (see [9, 10] and Chapter 5).

SIGNAL TRANSDUCTION SYSTEMS MODELED WITH DIFFERENTIAL EQUATIONS

9.3 Bistability and Hysteresis

The construction of systems of differential equations that are geared toward capturing the dynamics of signal transduction systems can take many forms. While the sky is the limit in this approach, the key component is usually a module that permits **bistability**. As the name suggests, such a module possesses two stable states, which are separated by an unstable state. If a signal is low, the system assumes one stable steady state, and if the signal exceeds some threshold, which is related to the unstable state, the system moves to the other stable steady state. The simplest example of such a system is a switch that can be triggered by a signal. If the signal is present, the system is *on*, and when the signal disappears, the system turns *off*. Bistable systems of different types are found in many areas of biology, examples including classical genetic toggle switches [11, 12], cell cycle control, cellular signal transduction pathways, switches to new programs during development, and the switch between lysis and lysogeny in bacteriophages [13].

Bistable systems are easy to construct. For instance, in Chapter 4 we used a polynomial of the form

$$\dot{X} = c(SS_1 - X)(SS_2 - X)(SS_3 - X). \tag{9.4}$$

Here, the parameter c just determines the speed of a response, while SS_1, SS_2, and SS_3 are three steady states. It is easy to confirm the latter, because setting X to any of the steady-state values makes $\dot{X}$ zero. For instance, for $(SS_1, SS_2, SS_3) = (10, 40, 100)$, and assuming that c is positive, $SS_1 = 10$ and $SS_3 = 100$ are stable steady states, while $SS_2 = 40$ is unstable.

A different approach to designing a system with a bistable switch is to combine a sigmoidal trigger with a linear or power-law relaxation function that lets the response variable return to its *off* state. A minimalistic model consists of a single ODE, such as

$$\dot{X} = 10 + \frac{20X^4}{100^4 + X^4} - 2X^{0.5}. \tag{9.5}$$

This barely qualifies as a "system," since it only has one variable but, notwithstanding semantics, we may interpret the equation as the description of a quantity X that is produced with a constant input of magnitude 10 and replicates or activates its own production by means of a Hill term. Furthermore, X is consumed or degraded according to a power-law function.

As always, the steady state(s) of the system can be assessed by setting (9.5) equal to zero. We could try to solve the equation algebraically, but that is a bit messy. However, we can easily convince ourselves that a steady state is achieved if the two positive terms exactly balance the negative term. Thus, we can plot the two functions and study where they intersect. The graph (**Figure 9.4**) indicates that there are actually three intersection points, which correspond directly to the steady states. One occurs for $X \approx 25$, the second for $X \approx 100$, and the third for $X \approx 210$. Implementing the system in PLAS and starting the solution with different initial values quickly shows that the true steady states for $X \approx 25.41736$ and $X \approx 210.7623$ are stable. The third steady state is actually at exactly $X = 100$, and it is unstable: starting slightly higher, even for 100.1, the system diverges toward the higher stable steady state, whereas it approaches the lower state for smaller initial values, even if they are as close as 99.9. An interpretation could be that the system is in either a healthy (low steady) state or a diseased (high steady) state. Both are stable, and small perturbations are easily tolerated in each state, whereas large perturbations may shift the system from health to disease or vice versa. Biologically, the unstable state is never realized, at least not for long, because it does not tolerate even the smallest disturbances.

As an example, suppose the system resides at the low steady state ($X \approx 25$), but external signals reset the value of X at a series of time points, as indicated in **Table 9.1**. **Figure 9.5** illustrates the responses of the system. Upon each resetting event, the system approaches the stable steady state in whose **basin of attraction** it lies. The unstable state does not have such a basin, because it is repelling: solutions starting even slightly higher than $X = 100$ wander toward the high steady state, and solutions starting lower than the unstable state approach the low steady state.

As an actual biological example of a bistable system, consider an auto-activation mechanism for the expression of a mutant *T7 RNA* polymerase (*T7R*), which was investigated by Tan and collaborators [14]. A priori, the most straightforward representation seemed to require merely a positive activation loop (**Figure 9.6**). However, closer observations indicated that the natural system exhibited bistability, which the simple activation motif is unable to do. The authors therefore combined two positive

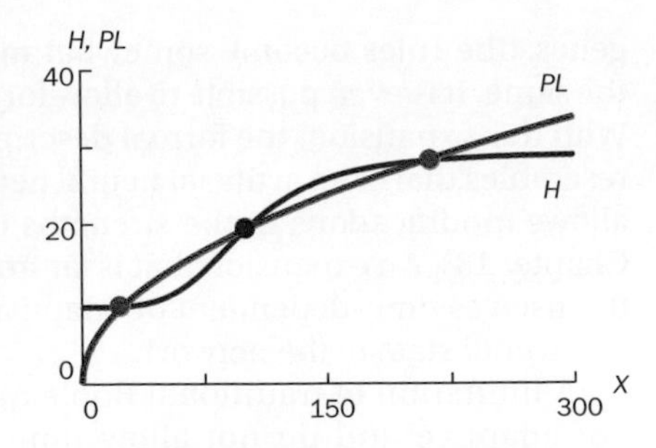

Figure 9.4 Bistability is often implemented as a differential equation containing the difference between an S-shaped function and a linear or power-law term. Shown here are the graphs of a typical signal response relationship in the form of a shifted sigmoidal Hill function H (blue; the first two terms on the right-hand side of (9.5)) and a power-law degradation function (green; the last term on the right-hand side of (9.5)). Their intersections indicate steady-state points. The most pertinent cases contain two stable (yellow) and one unstable state (red).

TABLE 9.1: SCHEDULE OF RESETTING X IN (9.5) AT SEVERAL TIME POINTS t (SEE FIGURE 9.5)

t	X
0	25
20	10
40	90
100	60
180	120
300	280
400	150
480	50
520	80
550	10

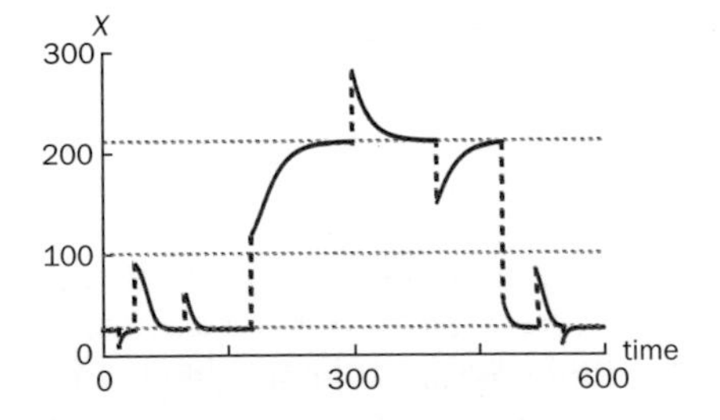

Figure 9.5 Response of X in (9.5) to external signals that reset the value of X at the time points identified with dashed red lines (see Table 9.1). The three dotted lines indicate the three steady states. The central steady state at $X = 100$ is unstable and separates the two basins of attraction of the stable steady states at about 25 and 211. If reset above 100, the system approaches the higher steady state, whereas resetting below 100 causes the system to approach the low steady state.

activation cycles, one for the auto-activation and the second representing the metabolic cost of the additional mechanism and subsequent growth retardation, as well as dilution of *T7R* due to growth (**Figure 9.7**). Indeed, the merging of the two positive loops generated bistability that was qualitatively similar to actual cellular responses.

With suitable scaling, Tan et al. [14] proposed the following equation to represent the system:

$$(\dot{T7R}) = \frac{\delta + \alpha \cdot (T7R)}{1 + (T7R)} - \frac{\phi \cdot (T7R)}{1 + \gamma \cdot (T7R)} - (T7R). \tag{9.6}$$

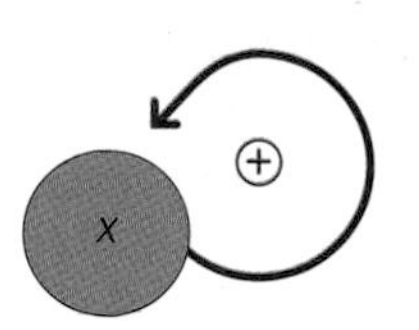

Figure 9.6 A simple positive activation loop. "Motifs" of this type are discussed in Chapter 14.

The first term on the right-hand side describes the production of *T7R*, including self-activation, the second term represents utilization of *T7R* for growth, metabolic costs, and dilution, while the last term accounts for intrinsic decay. With clever numerical scaling, the authors were able to keep the number of parameters to a minimum, and the remaining parameter values were set as $\delta = 0.01$, $\alpha = 10$, $\phi = 20$, and $\gamma = 10$.

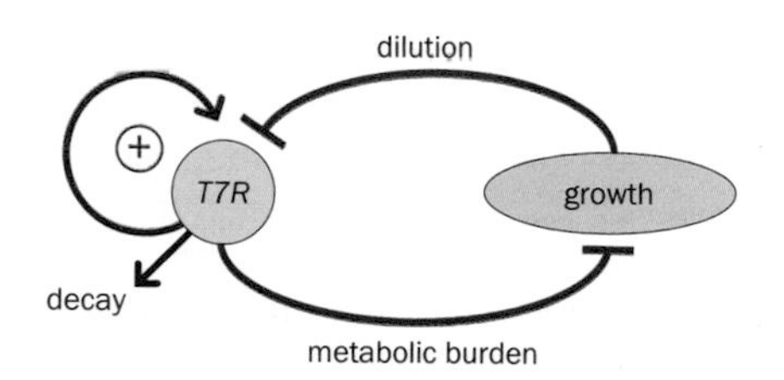

Figure 9.7 The combination of two auto-activation loops can lead to bistable behavior. Here, a mutant *T7 RNA* polymerase (*T7R*) is auto-active, but causes growth retardation. Growth, in turn, dilutes *T7R*. All effects together cause bitable behavior. See [14] for details.

Of course we can easily solve the differential equation (9.6), but of primary importance is whether the system has the potential for bistability. Thus, the aspect of major interest is the characterization of fixed points, which requires setting the differential equation equal to zero. The result is an algebraic equation that we can again evaluate by plotting the production term and the two degradation terms against *T7R* (**Figure 9.8**). This plot indicates that production and degradation intersect three times, and a little bit more analysis shows that the system has two stable steady states (for very high and very low values of *T7R*) and one unstable steady state for an intermediate value of *T7R*.

With a little bit more effort, ODEs with sigmoidal components can be made more interesting. For instance, consider a small two-variable system, which could be interpreted as consisting of two genes that affect each other's expression. A high signal *S* triggers the expression of *Y*, *Y* affects *X*, and *X* affects *Y*. The system is shown in **Figure 9.9**.

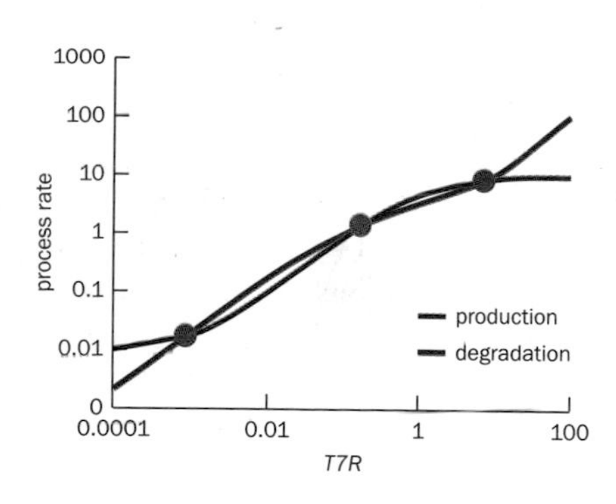

Figure 9.8 Visualization of bistability. Plotting the production and degradation rates of *T7R* in the system in Figure 9.7 and (9.6) against *T7R* shows that the system has three steady states. Two of these (green) are stable, whereas the center state (yellow) is unstable, resulting in bistability of the system.

As we have seen in Chapters 2 and 4, it is not too difficult to translate a diagram like that in Figure 9.9 into differential equations. Of course there are many choices for an implementation, and we select one that contains a sigmoidal function representing the effect of *Y* on the expression of *X*. Specifically, let us consider the following system:

$$\begin{aligned} \dot{X} &= 200 + \frac{800Y^4}{16^4 + Y^4} - 50X^{0.5}, \\ \dot{Y} &= 50X^{0.5}S - 100Y. \end{aligned} \tag{9.7}$$

The signal is an independent variable that affects the system from the outside. Suppose the signal changes during our simulation experiment as shown in **Table 9.2**. Starting close to the steady state $(X, Y) \approx (20.114, 6.727)$, and following the signal resetting schedule in the table, we obtain the response in **Figure 9.10**.

Of greatest interest in comparison with the bistable system is that something truly new and intriguing is happening here. The system starts with $S = 3$, which corresponds to the low steady-state value $X \approx 20$ and is to be interpreted as *off*. When the signal is increased to 10, *X* shoots up to about 400 and is *on*. Fifteen time units later, the signal goes back to $S = 3$, and we would expect the response to shut off again. However, *X* does not return to its *off* state of about 20, but instead drops only modestly to a value of about 337, which is still to be interpreted as *on*. At the end of the experiment, the signal is back to $S = 3$ again, and *X* is *off* again, but this happens only after the signal had been reduced temporarily to an even smaller value ($S = 2.5$). Thus, for a signal strength of $S = 3$, the system can be either *on* or *off*! Further analysis of this curious behavior implies is that the system has built-in memory: the state of the response variable depends not only on the current value of the external signal, but also on the recent history of the signal and the system. Mathematically, this history is stored in the dependent variables. Indeed, if we look at signal *S* and response *X*, it appears that there is memory, or **hysteresis**, as it is called in the jargon of the field. In conveniently structured models, conditions can be derived for the existence or absence of bistability and hysteretic effects (see, for example, [14–16]). An

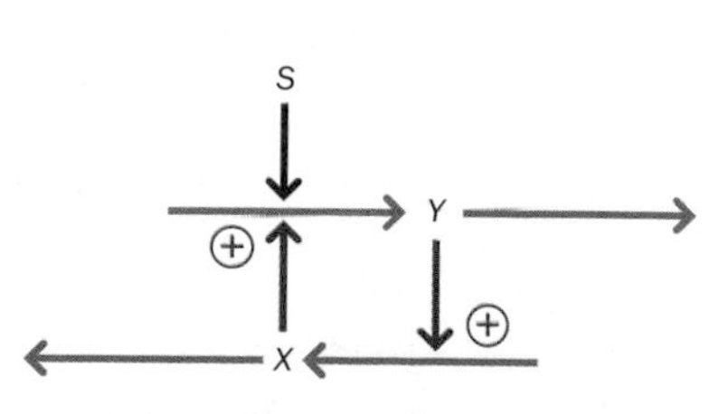

Figure 9.9 Artificial system with two components and an external signal. *X* and *Y* might be interpreted as the expression states of two genes. The system is affected by signal *S*, which could be interpreted as a stressor or a transcription factor.

example of a relatively simple real-world system is a so-called two-component signaling system in bacteria, which we will discuss in the next section of this chapter.

Hysteresis can be explored in greater detail with a two-phase simulation study (**Figure 9.11**). In the first phase, we start with a steady state resulting from a low value of S. We increase S slowly, in a stepwise fashion, let the system come sufficiently close to the steady state, and then increase S again. For this illustration, let us concentrate on the variable X. For values of S between 0 and 2.5, the steady-state value of X, X_{ss}, is close to 16 and changes very little, and on further raising S to about 3.3, X_{ss} increases gradually to about 30. On raising S further, to about 3.4, X_{ss} suddenly jumps to about 365. Checking signal strength in the vicinity of 3.4 demonstrates that this is a real jump and not a very fast, but gradual, increase in X_{ss}. Further rises in S, above 3.4, increase X_{ss} a little more, but not much.

Now we enter the second phase, where we start with a high value of S and slowly lower it in a stepwise manner, each time recording X_{ss}. At first, the steady-state values exactly trace those obtained from the first phase, and we would probably expect a jump back for $S \approx 3.4$. Surprisingly, this jump does not happen, and X_{ss} decreases only slowly. As a second surprise, X_{ss} crashes from about 230 to about 17 once S reaches a value close to 2.5. Again, this is not a fast, gradual decrease, but a real jump. Lowering S further causes X_{ss} to trace the early results of the first phase of the simulation.

The important conclusion here is the following: for a signal with a strength S in the range between about 2.6 and 3.3, the value of X_{ss} depends on history of the system. The situation is depicted in Figure 9.11B, which shows the different steady states of X for S in this range, which depend on whether S is increased from low values or decreased from high values. Note that there are no steady-state values on the dashed lines. The response of Y is similar, although of lower magnitude, and we will return to it later in the chapter, when we discuss clocks.

A simple bistable system models a "hard" switch: a signal just a tiny bit above the threshold will always trigger an *on* response, while a signal just below results in an *off* response. These response patterns are entirely deterministic. Suppose now that the system operates close to the threshold value for X_{ss}, that is, the unstable steady-state point. This is a reasonable assumption, because otherwise the capability of bistability of the system would be wasted. Let us suppose further that the environment is mildly stochastic. If so, then every change in S across the threshold, up or down, will make the system jump. Hysteresis softens this situation. In the range of S between 2.6 and 3.3, the illustration system tends to remain *on* or *off*, depending where it was just before, and noise in the signal is therefore easily tolerated.

TABLE 9.2: SUSTAINED SIGNAL STRENGTHS S BEGINNING AT TIME POINTS t

t	S
0	3
5	10
20	3
40	2.5
50	3

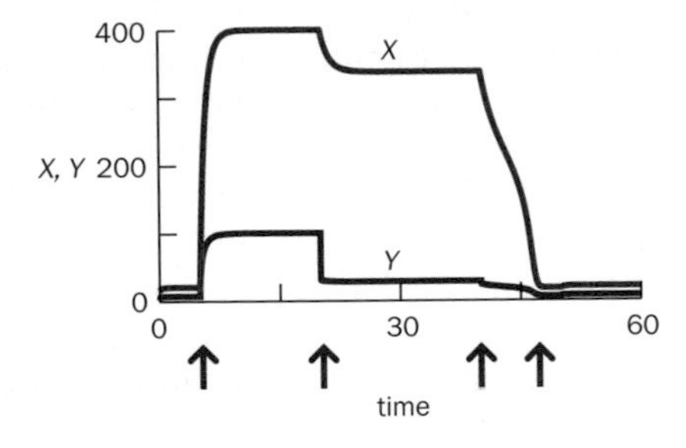

Figure 9.10 The two-variable signaling system in (9.7) shows hysteresis. The response variable assumes different values for the same input signal, depending on the history of the system. See the text for values of the signal, which changes at the times indicated by the arrows.

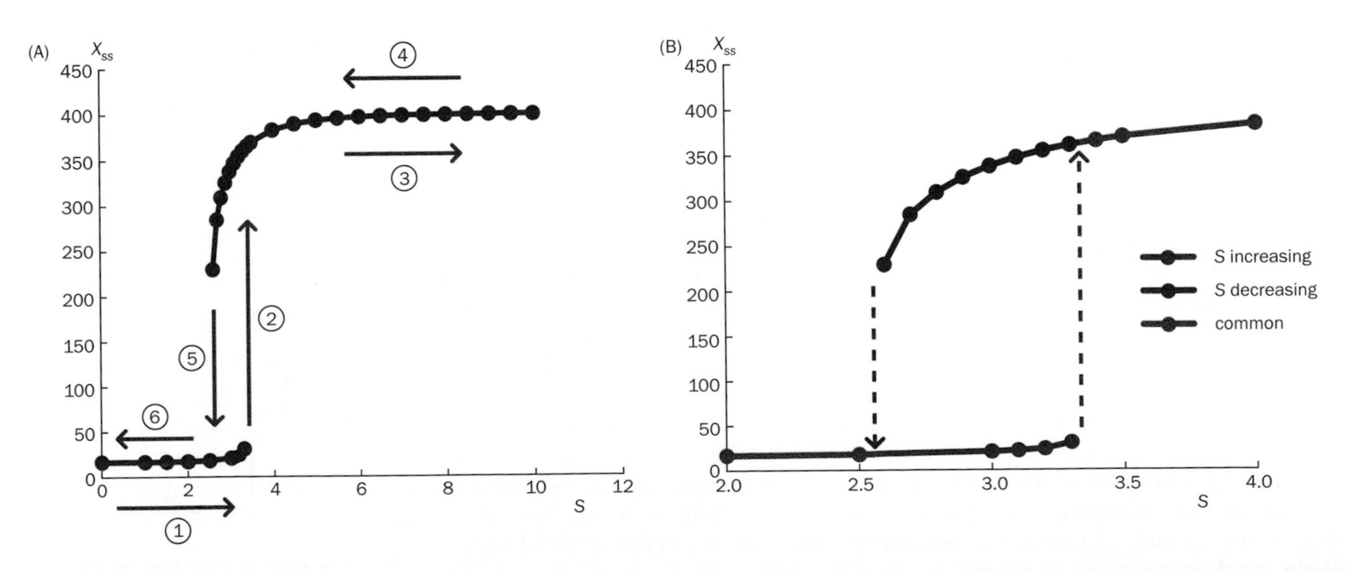

Figure 9.11 Demonstration of hysteresis. (A) indicates a series of alterations in S and the corresponding responses in X_{ss}. Increasing S from 0 to about 3 affects X_{ss} only slightly. However, changing S from about 3.3 to 3.4 results in an enormous jump in X_{ss}. Further increases in S do not change the new steady state much. Decreasing S from 10 to 3.4 traces the same steady states. However, there is no jump at 3.4. Surprisingly, the jump happens between about 2.6 and 2.5. (B) shows a magnification of the system response plot for signal strengths around 3, where X_{ss} depends on the recent history of the system. The dashed arrows indicate jumps and do not reflect steady states.

Expressed differently, it takes a true, substantial change in signal to make the system jump. Section 9.7 and **Box 9.1** discuss further analyses and variants of this basic hysteretic system.

In recent years, many authors have studied what happens if a signal and/or a system are corrupted by noise. Such investigations require methods that allow stochastic input, such as stochastic modeling, differential equations with stochastic components, or stochastic **Petri nets** [13, 19, 20] (see Exercise 9.18).

BOX 9.1: ONE-WAY HYSTERESIS

An important step in the mammalian immune response is the differentiation of B cells into antibody-secreting plasma cells. This step is triggered by an antigen that the host has not encountered before. The signaling pathway controlling this differentiation process is governed by a gene regulatory network, where several genes and transcription factors play critical roles. Focusing on the three key transcription factors Bcl-6, Blimp-1, and Pax5, Bhattacharya and collaborators [17] developed a mathematical model capturing the dynamics of this differentiation system. They were specifically interested in understanding how dioxin (TCDD; 2,3,7,8-tetrachlorodibenzo-*p*-dioxin) can interrupt the healthy functioning of this signaling pathway. Dioxln Is a toxic environmental contaminant that is produced in incineration processes. It is also infamous as Agent Orange, which was used in the Vietnam War to defoliate crops and forests. A simplified diagram describing the functional interactions among the genes and transcription factors in the model is shown in **Figure 1A**. The model that Bhattacharya et al. proposed is actually more complicated, since they explicitly considered gene expression, transcription, and translation. Of interest is that the multiple repression signals form a dynamic switch that, upon stimulation with an antigen, ultimately directs a B cell to differentiate into a plasma cell. The antigen considered here is a lipopolysaccharide (LPS), a compound found on the outer membranes of many bacteria and a known trigger of strong immune responses in animals. Bhattacharya and collaborators analyzed in detail how this pathway responds to deterministic and stochastic noise in the LPS signal [17, 18].

To study the switching behavior of the deterministic system, we make further drastic simplifications. In particular, we omit TCDD and its effect on the pathway. Furthermore, we leave out Bcl-6 and replace the associated dual inhibition with an auto-activation process for Blimp-1, because two serial inhibitions correspond to activation. Finally, we create a differentiation marker DM by combining Pax5 with IgM, with the understanding that IgM is high when Pax5 is low, and vice versa. The simplified signaling pathway is shown in Figure 1B. Defining X as Blimp-1, Y as the DM, and S as LPS, it is easy to transcribe the system into our typical notation (Figure 1C). Note that this diagram is actually quite similar to the small system in Figure 9.9 that we studied earlier.

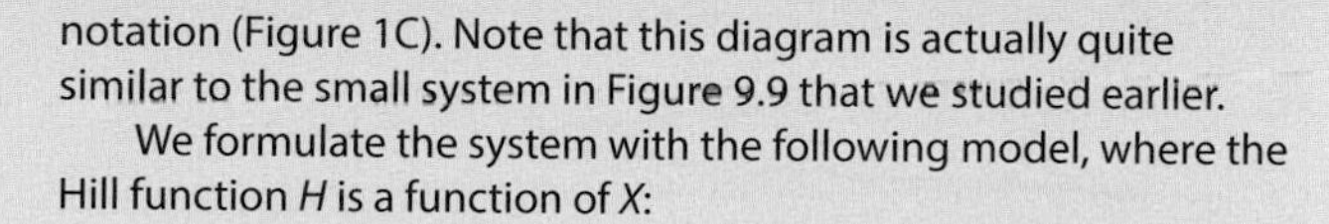

We formulate the system with the following model, where the Hill function H is a function of X:

$$\begin{aligned} H &= 1 + \frac{4X^4}{16^4 + X^4}, \\ \dot{X} &= 3.2H^2 + Y^{0.5}S - 2X, \\ \dot{Y} &= 4H - Y^{0.5}. \end{aligned} \tag{1}$$

To examine the responses of the system to different magnitudes of the stimulus, we perform exactly the same analysis as for the system in (9.7). Namely, we increase S in repeated small steps from a very low value to a high value, every time computing the stable steady state of the system and focusing in particular on the steady-state values X_{ss}. The results are shown in **Figure 2**.

For small signal strengths, X_{ss} is only mildly affected. For $S = 0$, it has a value of about 1.6, which slowly increases to about 8 with higher values of S. Because we have some experience with these types of systems, we are not too surprised that there is a critical point, here for $S \approx 2.25$, where the response jumps to a high level of about 62 (dashed arrow). For higher signal strengths, the response increases further, but at a relatively slow pace. Now we go backwards, decreasing S at every step a little bit. Again, we are not surprised that the system does not jump back down to the lower branch at $S \approx 2.25$, because we expect it to make this jump at a smaller value. However, it never does. Even reducing the signal to 0 does not move the system from the upper branch to the lower branch. Mathematically, such a jump would happen for about $S \approx -0.8$, but of course that is biologically not possible. In other words, the system has similar *off* steady states for low signal strengths, jumps to *on* steady states at about 2.25, but now is stuck with these *on* steady states, even if the signal disappears entirely. Interpreting this observation for a B cell, we are witnessing its irreversible differentiation into an antibody-secreting plasma cell. The unstable steady states (red) cannot be simulated. However, they can be computed if we express S as a function of X_{ss} (see Exercise 9.34).

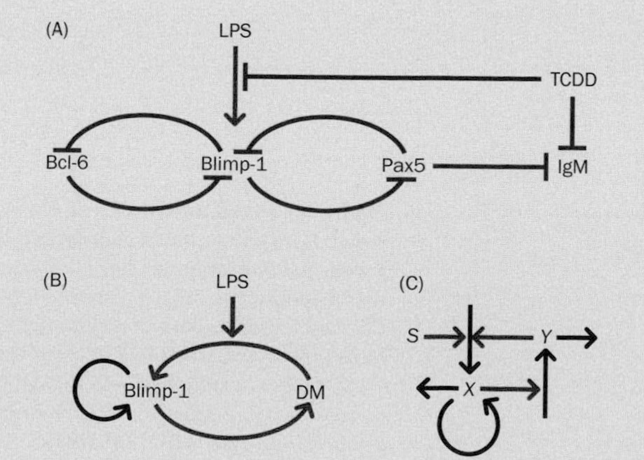

Figure 1 Simplified diagrams of the B-cell differentiation pathway. (A) A slightly simplified version of the model proposed by Bhattacharya et al. [17]. Lipopolysaccharide (LPS) stimulates this system. Blimp-1 is a master regulator that represses the transcription factors Bcl-6 and Pax5, which are typical for the B-cell phenotype. Pax5 inhibits the expression of immunoglobulin M (IgM), which mediates the immune response and can lead to the differentiation of the B cell into an antigen-secreting plasma cell. The environmental pollutant dioxin (TCDD) interferes with the signaling pathway. (B) A drastic reduction of the pathway, where TCDD is left out and Pax5 and IgM are combined into a differentiation marker, DM. (C) The same diagram in our typical notation, with X representing Blimp-1, Y the differentiation marker, and S the lipopolysaccharide.

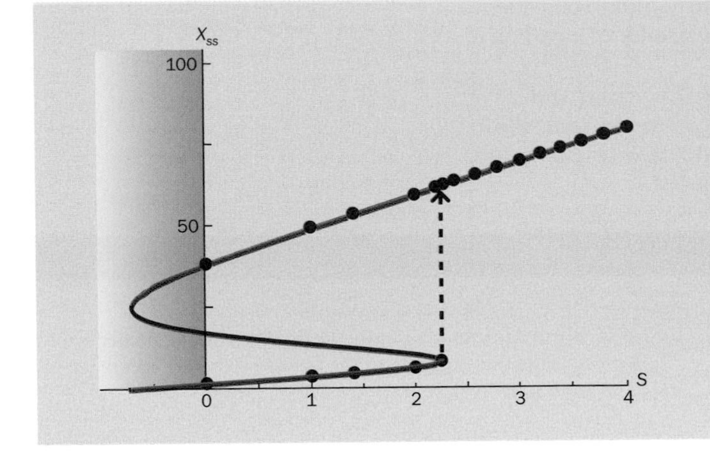

Figure 2 The B-cell differentiation pathway exhibits "one-way hysteresis." For small signals, the system has a steady state on the lower branch of the green hysteresis curve, which corresponds to the phenotype of a B cell. When the signal exceeds 2.25 in magnitude, the system jumps (dashed arrow) to a steady state on the upper branch, which corresponds to the phenotype of an antibody-secreting plasma cell. However, once the system has entered a steady state on the upper branch, it can never return to a steady state on the lower branch, even if the signal disappears entirely. Thus, the differentiation step is irreversible. The red component depicts unstable steady states. The dots indicate some simulation results.

9.4 Two-Component Signaling Systems

Two-component signaling (TCS) systems occur widely in archaea and bacteria, as well as in some fungi and plants. They are absent in animals, which rather transduce signals with three-component systems, protein kinase cascades, and/or lipid-based systems. TCS systems allow organisms to sense changes in environmental factors such as temperature and osmolarity, and to discern the direction of chemical gradients. Once the signal has been received, it is processed and ultimately allows the organism to respond in an appropriate fashion. In many cases, the TCS system is integrated with other regulatory mechanisms, such as allosteric enzyme regulation in the nitrogen assimilation of enteric bacteria [21]. For a recent collection of articles on TCS systems, see [22].

A relatively well-understood variant of a bacterial TCS system is the signal transduction protein CheY in *Escherichia coli*, which reacts to nutrients and other chemical attractants in the environment and affects the direction of the rotation of the organism's flagellar motor [23] (**Figure 9.12**). Binding of an attractant lowers the activity of the receptor-coupled kinase CheA, which decreases the level of phosphorylated CheY. This CheY-P in turn binds to a component of the motor (see Chapter 7). The motor usually rotates counterclockwise, but changes direction when CheY is phosphorylated. As a result, the bacterium tumbles and changes direction. Thus, mediated by CheY, the cell controls its swimming behavior.

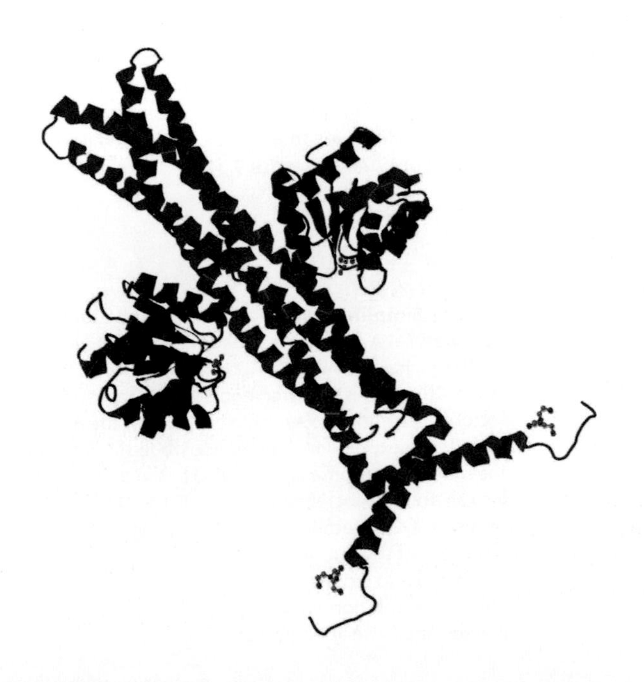

Figure 9.12 Structure of a CheZ dimer, which is instrumental in chemotaxis in *E. coli*. The phosphatase CheZ stimulates the dephosphorylation of the response regulator CheY by a so-far unknown mechanism. The most prominent structural feature of the CheZ dimer is a long four-helix bundle that consists of two helices from each monomer. (Protein Data Bank: PDB 1KMI [24].)

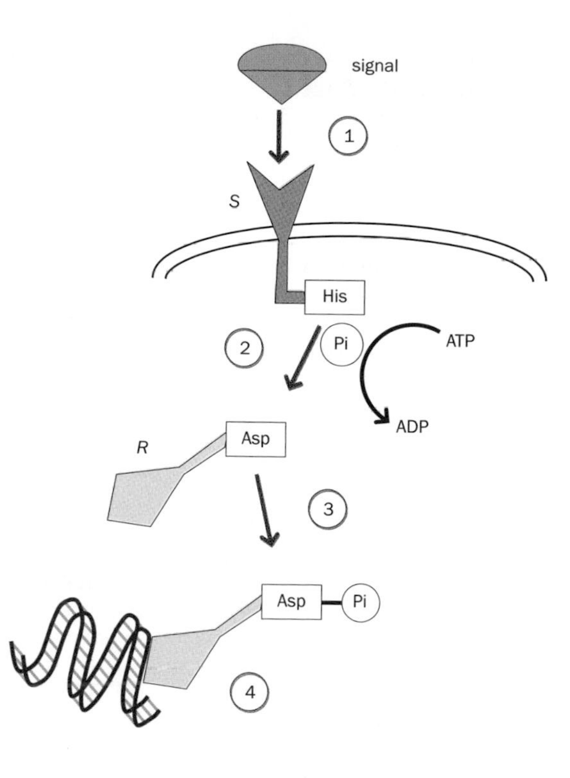

Figure 9.13 Signal transduction in a two-component system. The environmental signal binds to the sensor *S*. A histidine residue (His) on *S* receives a phosphate group (Pi), which is quickly transferred to an aspartic acid (Asp) residue on the response regulator *R*, which binds to DNA and effects a change in gene expression.

The first component of a TCS system acts as the sensor or transmitter and consists of an unphosphorylated membrane-bound histidine kinase (**Figure 9.13**). If a suitable signal is present, a phosphate group is transferred from ATP to a histidine residue on the sensor, resulting in ADP and a phosphohistidine residue. Because all this happens on the same molecule, the kinase is sometimes called an autokinase and the process is called autophosphorylation. The bond between the residue and the phosphate group is rather unstable, which allows the histidine kinase to transfer the phosphate group to an aspartic acid residue on the second component, the response regulator or receiver. The phosphate transfer causes a conformational change in the receiver, which changes its affinity to a target protein, or more typically to a specific DNA sequence, causing differential expression of one or more target genes. The ultimate physiological effects can be manifold.

Once the task of the TCS system has been accomplished, the response regulator must be dephosphorylated in order to return it to its receptive state. The rate at which the aspartyl phosphate is released as inorganic phosphate is well controlled and results in half-lives of the phosphorylated form of the regulator that may vary from seconds to hours. Interestingly, two slightly different variants have been observed in the dephosphorylation mechanism of the response regulator. In the simpler case, the mechanism is independent of the state of the sensor, and the sensor is called monofunctional. In the more complicated variant, the dephosphorylation of the response regulator is enhanced by the unphosphorylated form of the sensor protein, which implies that the sensor has two roles and is therefore bifunctional. The two variants are shown in **Figure 9.14**. In the following, we perform simulations with both of them and refer the reader to the literature [15, 25] for a further discussion of their comparative relevance.

It should be noted that the phosphorylation state of the regulator in both variants of the design may change in the absence of a signal, which introduces a certain level of noise into the system [15, 25]. Other variations of TCS are coupled TCS systems and analogous systems containing more components. An example of the latter is the sporulation system Spo in organisms like *Bacillus anthracis*, which consists of nine possible histidine sensors and two response regulators [26]. Other systems of a similar type are found in phosphorelays that regulate virulence [22]. Signaling systems in plants also often contain more than two components.

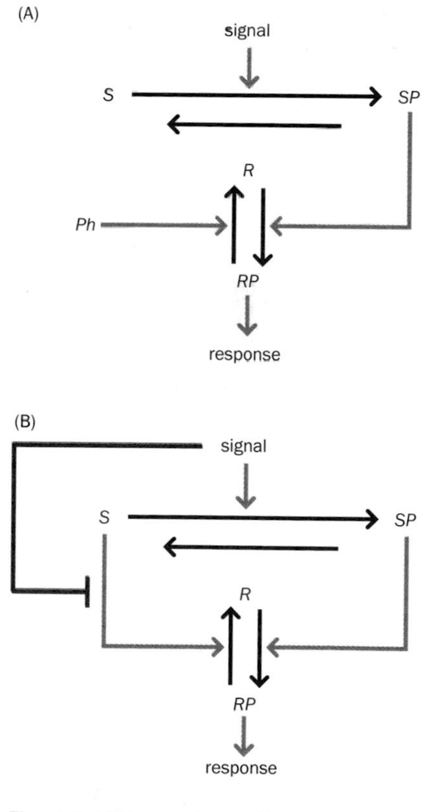

Figure 9.14 Two variants of a two-component signaling (TCS) system. The signal triggers the phosphorylation of a sensor *S*, which leads to phosphorylation of a response regulator or receiver *R*, leading to a response. (A) The monofunctional variant includes a phosphatase (*Ph*), which removes the phosphate from the phosphorylated receiver *RP*. (B) In the bifunctional variant, the sensor facilitates dephosphorylation of the phosphorylated receiver, a process that is inhibited by the signal.

For a model analysis that is representative of these types of studies, we follow the work of Igoshin et al. [15], which built upon earlier analyses by Alves and Savageau [25] and Batchelor and Goulian [27]. Of the many questions that could be analyzed, we focus on the shape of the signal–response curves and ask whether the TCS system can be bistable and show hysteresis. **Figure 9.15** shows a detailed diagram that contains both the mono- and bifunctional variants (Figure 9.14A and B, respectively) as special cases. It also contains intermediate complexes and shows the rate constants used in the model; their names are taken directly from Igoshin et al. [15]. By setting some parameters equal either to zero or to a different value, the mono- or bifunctional models are obtained.

Following Igoshin and collaborators, we set the model up as a mass-action model, so that the only parameters are rate constants, while all kinetic orders are 1 or 0. Translation of the diagram in Figure 9.15 leads directly to the following model:

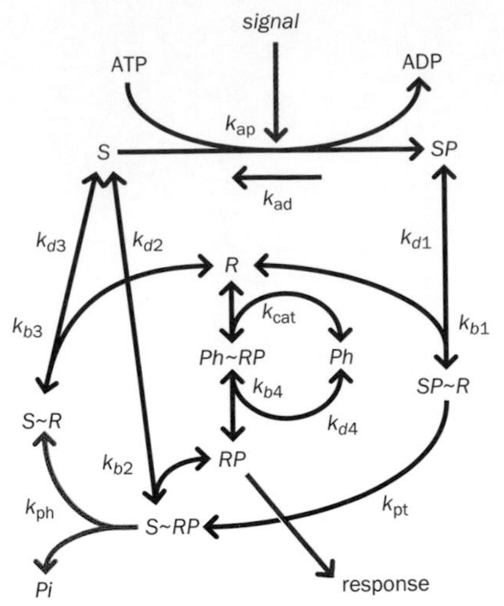

$$
\begin{aligned}
\dot{S} &= -k_{\text{ap}}(signal)\cdot S + k_{\text{ad}}(SP) + k_{d3}(S{\sim}R) - k_{b3}S\cdot R + k_{d2}(S{\sim}RP) - k_{b2}S\cdot(RP),\\
\dot{(SP)} &= k_{\text{ap}}S - k_{\text{ad}}(SP) - k_{b1}(SP)\cdot R + k_{d1}(SP{\sim}R),\\
\dot{(SP{\sim}R)} &= k_{b1}(SP)\cdot R - k_{d1}(SP{\sim}R) - k_{\text{pt}}(SP{\sim}R),\\
\dot{(S{\sim}RP)} &= k_{\text{pt}}(SP{\sim}R) + k_{b2}S\cdot(RP) - k_{d2}(S{\sim}RP) - k_{\text{ph}}(S{\sim}RP),\\
\dot{(S{\sim}R)} &= k_{b3}S\cdot R - k_{d3}(S{\sim}R) + k_{\text{ph}}(S{\sim}RP),\\
\dot{R} &= -k_{b1}(SP)\cdot R + k_{d1}(SP{\sim}R) + k_{d3}(S{\sim}R) - k_{b3}S\cdot R + k_{\text{cat}}(Ph{\sim}RP),\\
\dot{(RP)} &= k_{d2}(S{\sim}RP) - k_{b2}S\cdot(RP) + k_{d4}(Ph{\sim}RP) - k_{b4}(Ph)\cdot(RP),\\
\dot{(Ph{\sim}RP)} &= k_{b4}(Ph)\cdot(RP) - k_{d4}(Ph{\sim}RP) - k_{\text{cat}}(Ph{\sim}RP),\\
\dot{(Ph)} &= k_{d4}(Ph{\sim}RP) - k_{b4}(Ph)\cdot(RP) + k_{\text{cat}}(Ph{\sim}RP),
\end{aligned}
\tag{9.8}
$$

Figure 9.15 Detailed model diagram of the TCS system. By setting $k_{\text{ph}} = 0$, one obtains the monofunctional variant.

Igoshin and co-workers obtained most of the equations and parameter values from Batchelor and Goulian [27], but added some reactions and the corresponding parameters. A list of rate constants and initial values is given in **Table 9.3**; other initial values are zero. These parameter values were originally determined for the EnvZ/OmpR TCS system. EnvZ is an inner membrane histidine kinase/phosphatase, for instance in *E. coli*, that senses changes in the osmolarity of the surrounding medium. EnvZ thus corresponds to *S* in our model. It regulates the phosphorylation state of the transcription factor OmpR, which corresponds to *R* in our notation. OmpR controls the expression of genes that code for membrane channel proteins, called porins, that are relatively abundant and permit the diffusion of molecules across the membrane.

We begin our analysis with a typical simulation experiment, namely, we start the system at its steady state and send a signal at time $t = 100$. Mechanistically, we just change the independent variable *signal* in (9.8) from 1 to 3. Since we are not really interested in intermediate complexes, we show only the responses in *S*, *SP*, *R*, and *RP*, among which *RP* is the most important output to be recorded, because it is an indicator of the response. The mono- and bifunctional systems show similar, yet distinctly different, responses. In particular, the bifunctional system responds noticeably faster (**Figure 9.16**). If the signal is stronger, (for example, *signal* = 20), the two systems show very similar responses. Note that the two systems start at different values, which correspond to their (different) steady states.

As a second test, we model a short signal. Specifically, we set *signal* = 20 at time $t = 100$ and stop the signal at time $t = 200$ by resetting it to 1. Interestingly, the two TCS systems now respond in a qualitatively different fashion (**Figure 9.17**).

TABLE 9.3: RATE CONSTANTS AND INITIAL VALUES FOR MONOFUNCTIONAL (MF) AND BIFUNCTIONAL (BF) TCS SYSTEMS (DIFFERENCES ARE SHADED)

	k_{ap}	k_{ad}	k_{b1}	k_{d1}	k_{b2}	k_{d2}	k_{b3}	k_{d3}	k_{b4}	k_{d4}	k_{pt}	k_{ph}	k_{cat}	S(0)	R(0)	Ph(0)
MF	0.1	0.001	0.5	0.5	0.05	0.5	0.5	0.5	0.5	0.5	1.5	0	0.005	0.17	6.0	0.17
BF	0.1	0.001	0.5	0.5	0.05	0.5	0.5	0.5	0.5	0.5	1.5	0.05	0.025	0.17	6.0	0.17

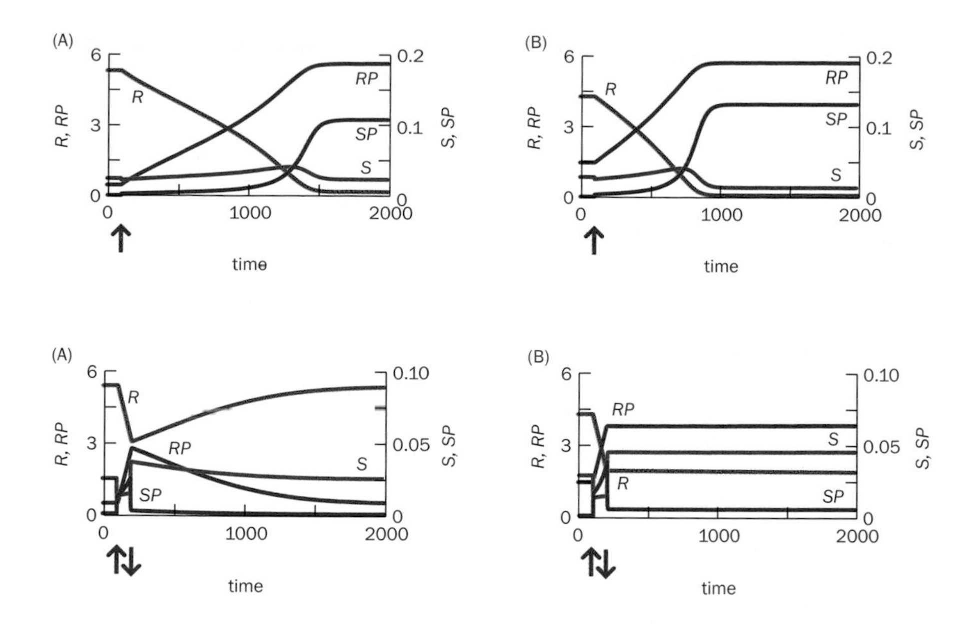

Figure 9.16 Simulations with the two variants of the TCS system. (A) Responses of the monofunctional variant to a signal at time $t = 100$ (arrow). (B) Corresponding responses of the bifunctional design.

Figure 9.17 Differences in responses of the two variants of the TCS system. The monofunctional (A) and bifunctional (B) TCS systems show qualitatively different responses to a short, transient signal (arrows), with the former returning to its steady state and the latter reaching a new state that appears to be a steady state; however, see Figure 9.18.

The monofunctional system responds to the signal and returns to its steady state as soon as the signal stops. By contrast, the bifunctional system seems to assume a new steady state. Will it eventually return to the initial state? The answer is difficult to predict but easy to check with a simulation: we simply extend the simulation to something like $t = 12{,}000$. Surprisingly, the system does not stay at this apparent steady state (which really is none), nor does it return to the initial state. Instead, after a while, it starts rising again (without any intervention from the outside) and assumes a new steady state (**Figure 9.18**).

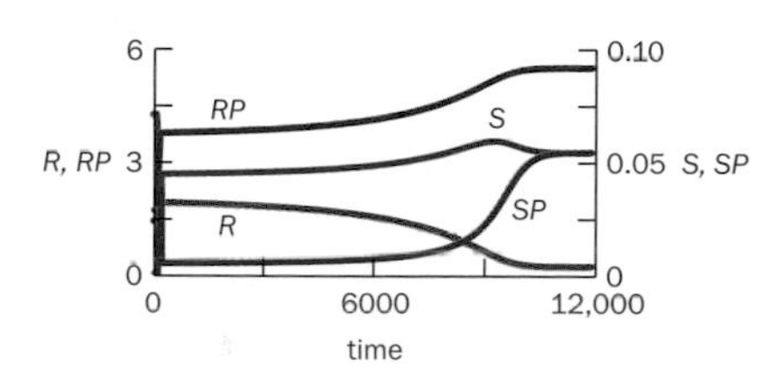

Figure 9.18 Continuation of the simulation of the bifunctional system in Figure 9.17. The presumed steady state is only a transient state, and the system shows a late response to the signal by approaching the true steady state.

The short-signal experiment suggests at the very least that the bifunctional TCS system permits bistability. But does it also permit hysteresis? Moreover, can we test whether the monofunctional system exhibits similar responses? To answer these questions, we execute a series of experiments with varying signal strengths and study the response in *RP*. We begin with the monofunctional variant, by running it from the initial values in Table 9.3 (with all other initial values at zero) with a low signal of 1 and computing its steady state. Now the real series of experiments begins. We initiate the system at the steady state and reset the signal to a different value between 0 and 4 at time $t = 100$. For signal strengths between 0 and about 1.795, the corresponding values of the response variable *RP* increase slowly from 0 to about 2. Intriguingly, a tiny further increase in the signal to 1.8 evokes a response of 5.57! Further increases in signal do not change this response much. It is pretty clear that the system is bistable.

Does the system also exhibit hysteresis? To find out, we perform a second set of experiments, this time starting with the high steady state, which is characterized by values that we obtain by running the original system with a high signal towards its steady state. Beginning with a high signal of 4, we again see a response of about 5.6. Now we slowly lower the signal from experiment to experiment, every time recording the value of *RP*. Nothing much happens until 1.8, and the results are essentially identical to those in the previous series of experiments. However, lowering the signal to 1.795 does not cause a jump. In fact, the response is high until the signal strength is reduced to about 1.23, where the response value suddenly drops to about 0.65. These results are a clear indication of hysteresis. The signal–response relationship is shown in **Figure 9.19**.

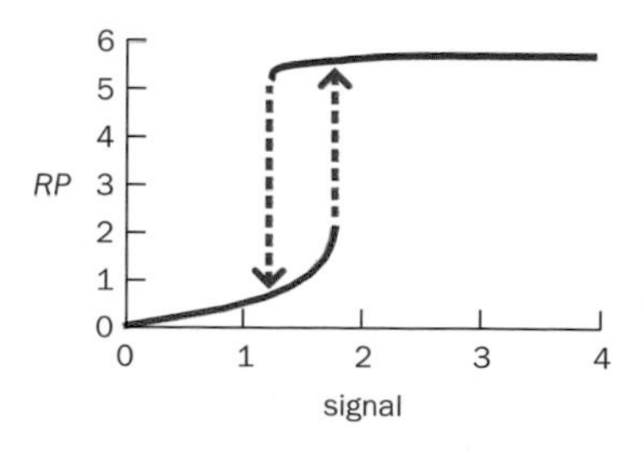

Figure 9.19 The signal–response relationship of the monofunctional TCS model shows clear hysteresis. For signal strengths between about 1.23 and 1.8, the response depends on whether the system was in a low or high state when the signal hit. This type of phenomenon shows that the monofunctional TCS system has memory. The response is similar to that of system (9.5) (see Figure 9.11).

Igoshin et al. [15] and Alves and Savageau [25] derived conditions for when the different TCS systems exhibit bistability and hysteresis and studied the effect of noise in the signal, which in the case of TCS systems appears to be of minor significance. They also discussed the advantages and limitations of the mono- and bifunctional variants in detail. As in many cases of this nature, one variant is more efficient with respect to some physiologically relevant aspects of the response but less efficient with respect to other aspects, and the overall superiority of one over the other depends on the environment in which the organism lives. We will discuss some methods for the analysis of such variants in Chapter 14.

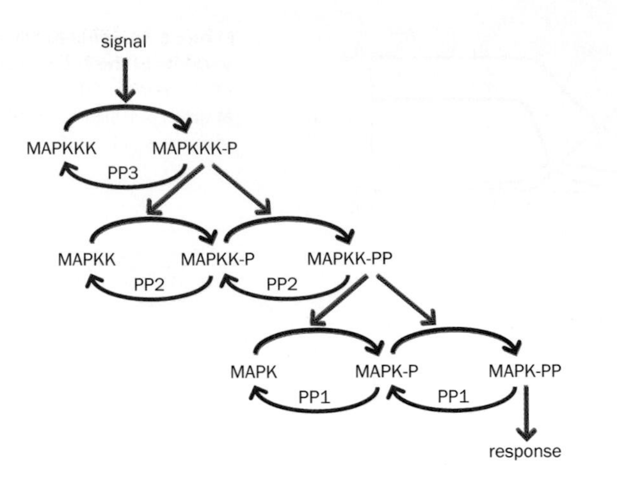

Figure 9.20 Typical three-layer MAPK cascade. At each level, the inactive form on the left is phosphorylated (once or twice) into the active form, which serves as enzyme at the next layer. Phosphatases (PP1, PP2, and PP3) return the active forms to their inactive states by removing phosphate.

9.5 Mitogen-Activated Protein Kinase Cascades

In contrast to the TCS system in bacteria, most higher organisms use a more complicated signal transduction system whose key component is the mitogen-activated protein kinase (**MAPK**) **signaling cascade**. MAPK (usually pronounced "map-kinase") systems appear to be present in all eukaryotes and also in a few prokaryotes, such as the biofilm-forming bacterium *Myxococcus xanthus*. Intriguingly, MAPK cascades have a highly conserved architecture. The MAPK signaling cascade receives signals from cell surface receptors or cytosolic events, processes them by filtering out noise, usually amplifies them, and ultimately affects various downstream targets, such as cytosolic proteins and nuclear transcription factors. The external signal may consist of a mitogen or inflammatory cytokine, a growth factor, or some physiological stress. This signal is transduced, for example, by a G-protein-coupled receptor that activates the MAPK cascade. The ultimate result of the MAPK system may be as different as inflammation, differentiation, apoptosis, or initiation of the cell cycle.

The typical MAPK cascade is shown in **Figure 9.20**. It consists of three layers where proteins are phosphorylated or dephosphorylated. The kinase nearest to the signal source is generically called MAP kinase kinase kinase (MAPKKK). If it is activated by a cytosolic or external signal, MAPKKK is phosphorylated to MAPKKK-P. Being a kinase itself, MAPKKK-P activates phosphorylation of the MAP kinase kinase (MAPKK) in the second layer. In fact, full activation of MAPKK requires sequential phosphorylation at two sites: a tyrosine site and a threonine site. The resulting MAPKK-PP in turn phosphorylates and thereby activates the kinase of the third layer, the MAP kinase (MAPK). Again, full activation requires two phosphorylation steps. The activated MAPK-PP can phosphorylate cytosolic targets or can translocate to the nucleus, where it activates specific transcriptional programs. At each layer, a phosphatase can dephosphorylate the active forms into the corresponding inactive forms. Two prominent examples of MAP kinases are the extracellular-signal-regulated kinase (ERK) and the c-Jun N-terminal kinase (JNK), where Jun refers to a family of transcription factors. The responses of the cascades are quite fast: within the first 5 minutes of stimulation, ERK is activated up to 70% [28], and within 10 minutes, significant amounts of activated ERK are translocated to the nucleus [29]. A mutated form of MAPKKK in the ERK pathway has often been found in malignant melanomas and other cancers.

An obvious question is this: The three-layer cascade has been conserved through evolution, so it is fair to assume that this design has advantages. But what are these, for instance, in comparison with a simple or double phosphorylation at a single layer? We use a model analysis to shed light on this intriguing question.

The key module of the cascade is the sequential phosphorylation at two sites, which is catalyzed by the same enzyme from the next higher layer. Simplified and detailed diagrams of this module are presented in **Figures 9.21** and **9.22**, where X stands for MAPK or MAPKK, E is the catalyzing enzyme, XP and XPP are singly or doubly phosphorylated forms, and XE and XPE are complexes of X and XP with the enzyme.

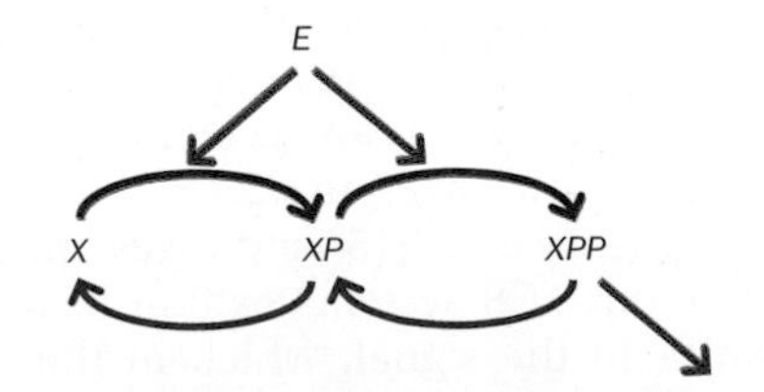

Figure 9.21 Key module of the three-layer MAPK cascade in Figure 9.20. It is tempting to use Michaelis–Menten functions to model the two phosphorylation steps, but the two steps compete for the same enzyme, whose concentration is not constant.

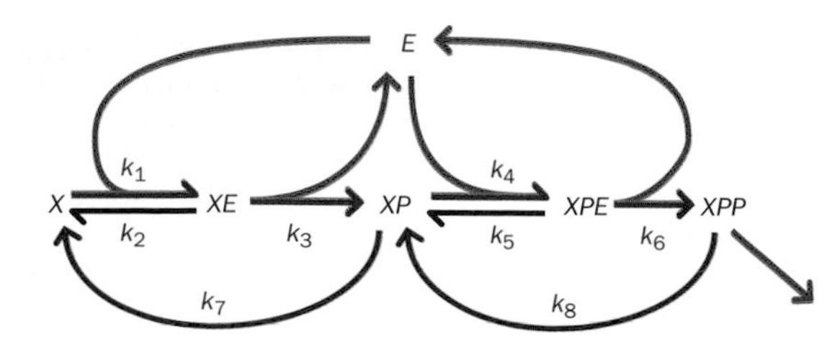

Figure 9.22 Detailed representation of the module in Figure 9.21. Accounting for the two substrate–enzyme complexes *XE* and *XPE* shows the dual role of enzyme *E*.

It is tempting to set up the two phosphorylation steps with Michaelis–Menten rate functions, but such a strategy is not the best option, because (1) the enzyme concentration is not constant, (2) the enzyme concentration is not necessarily smaller than the substrate concentration, and (3) the two reaction steps are competing for the same enzyme. Instead, it is useful to retain the mechanistic ideas of the Michaelis–Menten mechanism, which postulates the formation of a substrate–enzyme complex, and to formulate this mechanism in the basic format of mass-action kinetics (see Chapters 2, 4, and 8). What we do *not* want to do is to rely on the quasi-steady-state assumption that would simplify this system toward the well-known Michaelis–Menten function, because in this format the enzyme concentration is no longer explicit, let alone dynamic. A direct translation of the diagram in Figure 9.22 into mass-action equations is straightforward:

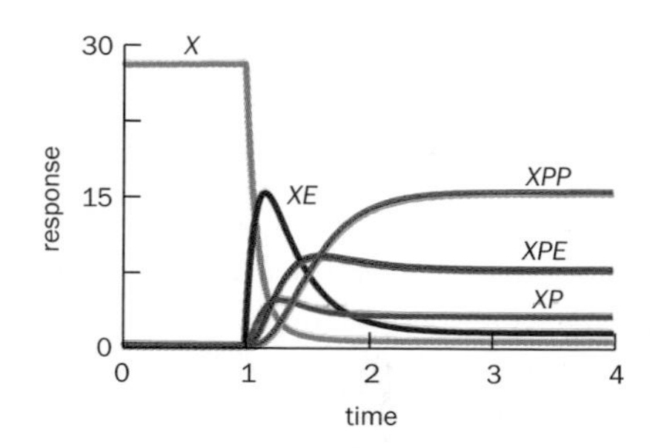

Figure 9.23 Response of the MAPK module in Figure 9.22. At time $t = 1$, the signal is increased. The result is single and double phosphorylation of *X*.

$$
\begin{aligned}
\dot{X} &= -k_1 X \cdot E + k_2(XE) + k_7(XP), \\
(\dot{XE}) &= k_1 X \cdot E - (k_2 + k_3)(XE), \\
(\dot{XP}) &= k_3(XE) - k_7(XP) - k_4(XP) \cdot E + k_5(XPE) + k_8(XPP), \\
(\dot{XPE}) &= k_4(XP) \cdot E - (k_5 + k_6)(XPE), \\
(\dot{XPP}) &= k_6(XPE) - k_8(XPP).
\end{aligned}
\tag{9.9}
$$

To perform simulations, we specify more or less arbitrarily chosen values for the parameters and initial concentrations in **Table 9.4**. It is now easy to study the effects of changes in the enzyme *E* on the variables of the module. At the beginning of the simulation, $E = 0.01$, and the initial conditions for the various forms of *X* are set such that the system is more or less in a steady state. Suppose now that at time $t = 1$ the signal *E* is increased to 10. This is a strong signal, and the system responds with phosphorylation on both sites. After a brief transition, the balance between *X*, *XP*, and *XPP* is entirely switched, with very little unphosphorylated *X* left (**Figure 9.23**). Notably, the amount of *XPE* is relatively large, because quite a lot of enzyme is available. If the signal is reset to 0.01, the module returns to its initial state (**Figure 9.24**).

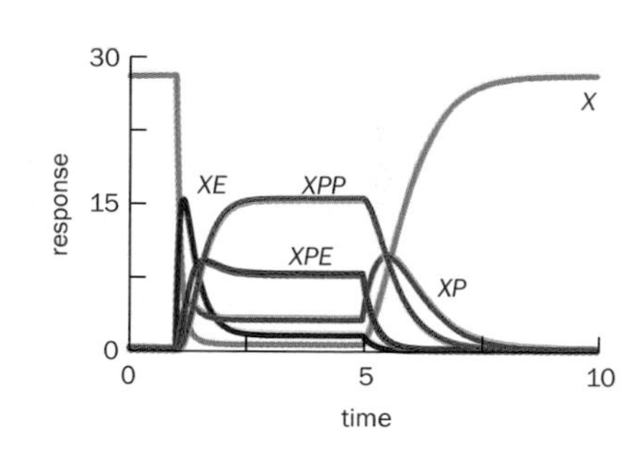

Figure 9.24 Continuation of the response in Figure 9.23. At time $t = 5$, the signal is decreased to its original low state, and *X* returns to its unphosphorylated state.

It is also easy to study how strong the signal must be to trigger a response. If *E* is set to 3, rather than 10, the response is rather similar to the previous scenario, although the response is not as pronounced (results not shown). If *E* is set to 1 instead, the response is much slower. *X* is reduced only to about 13.6, while *XPP* assumes a value of about 3.2. In other words, no real switch is triggered. For even weaker signals, the response is insignificant (**Figure 9.25**).

Now that we have a bit of a feel for the module, we can use it to implement a complete MAPK cascade (see Figure 9.20). Specifically, we set up the module in triplicate, with additional variable names *Y*, *YP*, *YPP*, *Z*, *ZP*, and *ZPP*, which represent the middle and bottom layers of MAPK and MAPKK phosphorylation, respectively, and use the same parameter values and initial conditions as before. Regarding the first layer, which involves only one phosphorylation of MAPKKK,

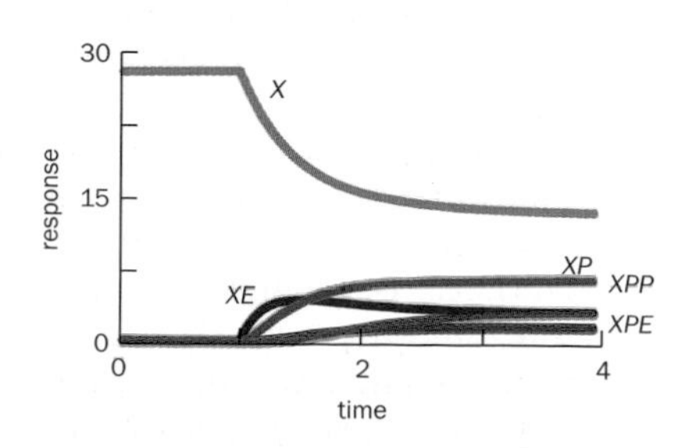

Figure 9.25 Response of the MAPK module in Figure 9.22 to a weak signal at $t = 1$. While *X* responds to the signal with a decrease, the doubly phosphorylated form *XPP* remains low.

TABLE 9.4: KINETIC PARAMETER VALUES AND INITIAL CONDITIONS FOR THE MAPK SYSTEM (9.9)

$k_1 = k_4$	$k_2 = k_5$	$k_3 = k_6$	$k_7 = k_8$	*X*(0)	*XE*(0)	*XP*(0)	*XPE*(0)	*XPP*(0)
1	0.1	4	2	28	0.1	0.2	0.01	0.02

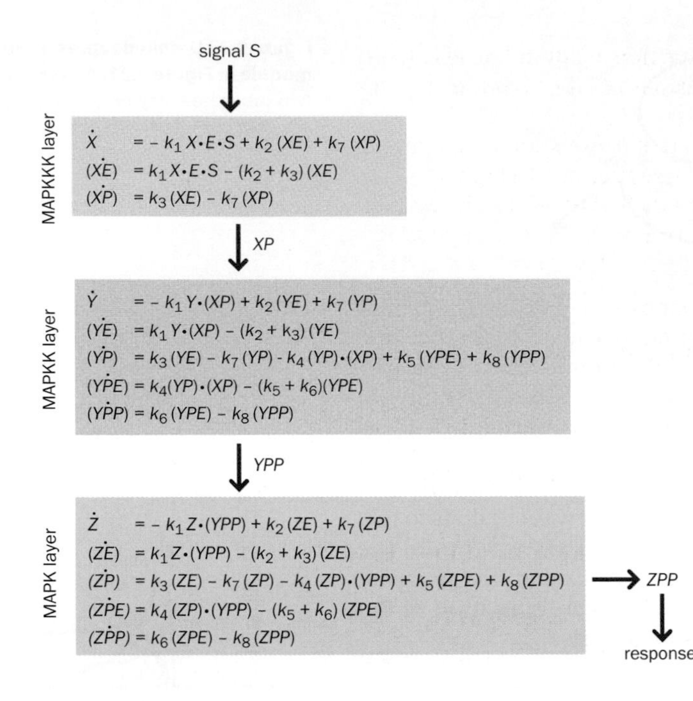

Figure 9.26 Implementation of a model of the MAPK cascade in Figure 9.20. The MAPK and MAPKK layers are analogous, while the MAPKKK layer is simpler, since it contains only one phosphorylation step. The kinetic rate constants may be different for each layer. See the text for a description of simplifications in the equations for *XP* and *YPP*.

we us a single Michaelis–Menten mechanism, again formulated in mass-action representation, with $X(0) = 28$ and $XE(0) = XP(0) = 0.1$. The input to the first module is signal *S*, and the output of this module is the singly phosphorylated form *XP*, which catalyzes the two phosphorylation steps $Y \rightarrow YP$ and $YP \rightarrow YPP$ at the next layer. Similarly, the output of the second layer, *YPP*, catalyzes the reactions $Z \rightarrow ZP$ and $ZP \rightarrow ZPP$ at the bottom layer. *ZPP* triggers the actual physiological response, but we simply use *ZPP* as output indicator of the cascaded system. The model implementation is shown in **Figure 9.26**, and we set the kinetic parameters to exactly the same values as before (see Table 9.4).

Strictly speaking, this implementation is a simplification that assumes that the "enzymes" *XP* and *YPP*, as well as *ZPP*, remain constant, which is not really the case, since *XP* undergoes reactions with *Y* and *YP*, and *YPP* reacts with *Z* and *ZP*; the role of *ZPP* is harder to quantify in general. Thus, for a more accurate representation, all reactions utilizing or releasing *XP*, *YPP*, or *ZPP* should be included in the equations governing these variables. For instance, the equation for *XP* should read

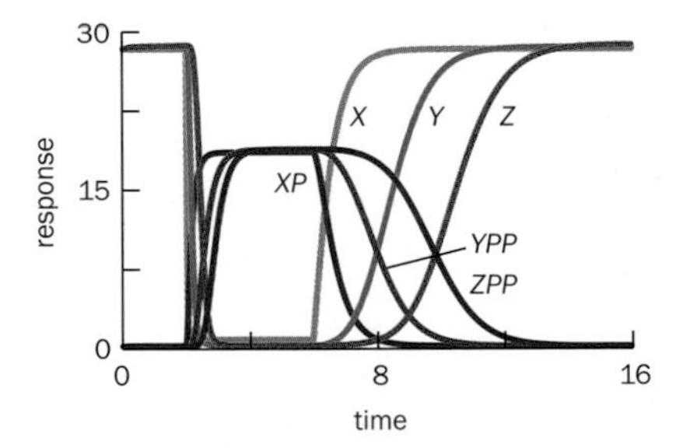

Figure 9.27 Responses of the main variables of the MAPK model to switches in signal. The *X*, *Y*, and *Z* variables respond to an increase and decrease in signal in a staggered manner (see the text for details).

$$(\dot{XP}) = k_3(XE) - k_7(XP) - k_1 Y \cdot (XP) + (k_2 + k_3)(YE) \\ - k_4(YP) \cdot (XP) + (k_5 + k_6)(YPE). \tag{9.10}$$

If the signal is off, *XP* and *YPP* have rather small values, but in the case of greater interest, namely if the signal is on and stays on, *XP* and *YPP* are constant, with values of about 14.3 and 12.7 for the given parameters k_i, and simulations show that accounting explicitly for interactions with the lower levels slows down the responses of the cascade a little bit, but that ignoring them does not affect the results qualitatively. We therefore ignore these processes for simplicity. Exercise 9.30 assesses the differences between models.

As a first simulation, suppose the external signal *S* starts at 0.01, switches to 50 at $t = 2$, and switches back to 0.01 at $t = 6$. The responses of the most important variables are shown in **Figure 9.27**. The strong signal causes rapid phosphorylation at all layers. If the signal is not as strong ($S = 5$ at $t = 2$), *X* does not dip as close to zero as before, but only to about 6. Otherwise the responses are similar (results are not shown). For a relatively weak signal ($S = 0.5$ at $t = 2$), we see the benefit of the cascaded system. While *XP* rises only to about 5, the ultimate response variable *ZPP* shows a robust response (**Figure 9.28**). In other words, we have solid signal amplification in addition to a steeper response and clear signal transduction. If the signal rises to only two or three times its initial value ($S = 0.02$ or $S = 0.03$ at $t = 2$), the response in *ZPP* is very weak. The results give us a first answer to our earlier question of why there are three layers in the cascade: three layers permit improved signal

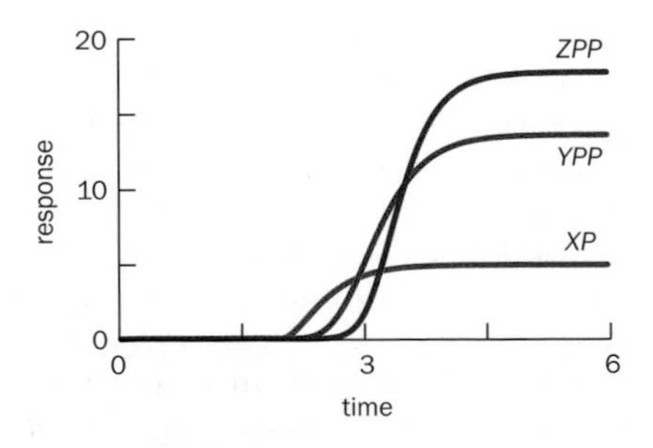

Figure 9.28 Signal amplification by the MAPK cascade. Even for a relatively weak signal, the responses at the three layers of the MAPK cascade are successively stronger, indicating amplification of the signal by the cascaded structure.

amplification. At the same time, they effectively reduce noise [30]. It has also been shown that the three-layer design improves the robustness of the signal transduction system against variations in parameter values [31].

It is easy to investigate how the cascade responds to brief repeated signals. We perform a simulation as before, but create an *off-on* sequence of signals. As an example, if the signal switches between 0.01 and 0.5 every half time unit, beginning with $S = 0.5$ at $t = 2$, we obtain the result shown in **Figure 9.29**. We can see that the responses at the different layers are quite different. The first layer exhibits fast, but very weak, responses in *XP*, whereas the ultimate, amplified response *ZPP* at the third layer in some sense smoothes over the individual short signals.

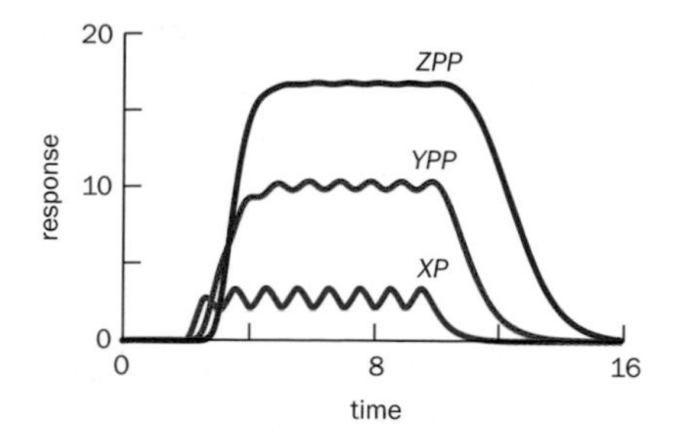

Figure 9.29 Responses of different layers of the MAPK cascade to brief, repeated signals. The signal at the output layer smoothes the responses at the MAPKK and MAPKKK layers into a solid and sustained, amplified signal.

It should be clear that no attempt was made in the implementation of the cascade to estimate biologically relevant parameter values; our purpose was simply to demonstrate the concepts of the dual-phosphorylation module and its role in the MAPK cascade. Huang and Ferrell [32] and Bhalla and Iyengar [33] discuss parameter values and features of realistic MAPK cascades in some detail, and Schwacke and Voit [34] and Schwacke [35] provide a computational rationale for parameter values of the cascade that are actually observed in nature and that lead to stronger amplification than our ad hoc parameter set.

It should be noted that all signaling systems obviously have a spatial component, which the typical ODE models ignore. In fact, it seems that some cascades are organized along protein scaffolds, which can significantly affect their efficiency [36–38]. Furthermore, MAPK cascades typically do not work in isolation. Instead, the cell contains several parallel cascades, which communicate with each other via **crosstalk** [39–41]. For instance, the output of the top layer in one cascade might activate or inhibit the center layer in another. Crosstalk is thought to increase the reliability and fidelity of signal transduction. It furthermore creates options for very complex signaling tasks, such as an in-band detector, where the cascade only fires if the signal is within a certain range, but neither weaker nor stronger. Crosstalk also renders all logic functions (*AND, OR, NOT*) including the so-called *EXCLUSIVE OR* (*XOR*) mechanism possible. In the latter case, the cascade fires only if either one of two signals is present, but it does not fire if both signals are *on* or if both signals are *off* [33] (see also the discussion of bi-fans in Chapter 14). Detailed experimental and theoretical work seems to indicate that normal and cancer cells may differ not so much in their signaling components, but rather in the way in which these components are causally connected within complex signaling networks [42].

It is interesting to note that cells often use the same or similar components of a signaling cascade for different purposes. For instance, yeast cells employ Ste20p, Ste11p, and Ste7p, for a number of different stress responses, including pheromone exposure, osmotic stress, and starvation [38].

9.6 Adaptation

The human body responds to many signals from the outside world on a daily basis. Our pupils constrict in bright sunlight and dilate when it is dark, we sweat to cool the body, and we pull our hands back when we touch something hot. Many of these responses are driven by the body itself, independent of our conscious control. Intriguingly, we often get used to a signal, if it is repeated in short order or if it persists for some time. We start breathing normally again a few days after moving to high altitudes, and after a while we no longer really notice some noises, unless we specifically focus on them. This phenomenon of a diminished biological response to the same repeated signal is called **adaptation**. In the language of systems biology, the organism is initially in a homeostatic steady state and responds to occasional signals by moving from an *off* state to an *on* state, but then returns to the *off* state when the signal has vanished. However, if the same type of signal persists for some while, the system returns from the *on* state to the *off* state anyway, even though the *off* state normally corresponds to the absence of the signal.

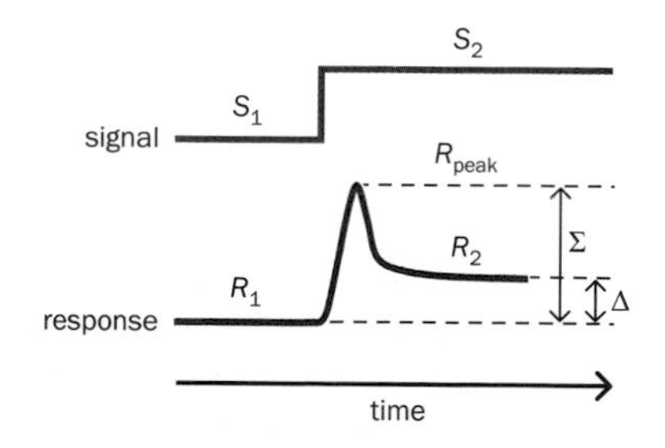

Figure 9.30 Features of adaptation. For a low signal S_1, the typical response is an *off* state, characterized by the response R_1. For short spikes in the signal, the response is transient, and the system returns to R_1 (not shown). In the case of adaptation, the system responds to the signal S_2 by reaching R_{peak}, but then approaches a value R_2 that is close to R_1, even though the signal S_2 persists. The response can be characterized by its sensitivity Σ and precision Π; the latter is inversely related to the relative difference Δ between R_1 and R_2.

According to Ma and co-workers [43], adaptation can be characterized by two features, which they call *sensitivity* and *precision* (**Figure 9.30**). The sensitivity Σ measures the normal strength of the response to a signal, that is, the relative difference between the output values of the *on* and *off* states in relation to the relative signal strength. The precision Π is defined as the inverse of the difference Δ between the original homeostatic state and the state the system attains through adaptation. Specifically,

$$\Sigma = \frac{|R_{\text{peak}} - R_1| / |R_1|}{|S_2 - S_1| / |S_1|}, \tag{9.11}$$

$$\Pi = \Delta^{-1} = \left(\frac{|R_2 - R_1| / |R_1|}{|S_2 - S_1| / |S_1|} \right)^{-1}. \tag{9.12}$$

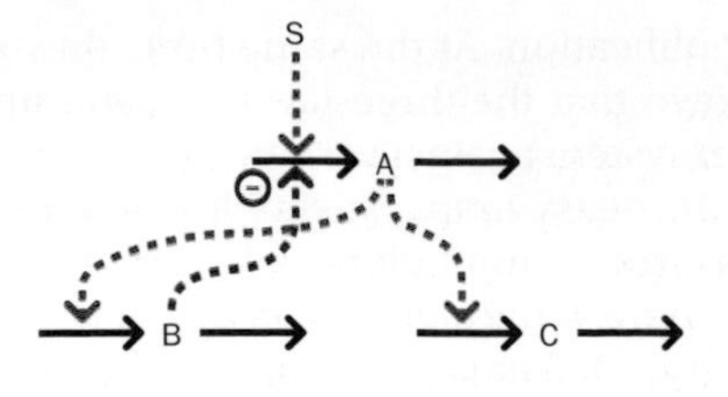

Figure 9.31 Three-node system for studying adaptation. The system is adapted from Ma et al. [43], but represented in the format we have been using throughout the book.

Ma's group explored the question of what types of systems with three nodes are capable of displaying adaptation. One such motif is shown in **Figure 9.31** in our notation. Whereas Ma used a Michaelis–Menten representation, we use an S-system model with simplified notation, which here has the intriguing advantage that one can actually compute conditions under which the system displays adaptation (Exercise 9.31). A symbolic S-system model with positive variables *A*, *B*, and *C* can be written down immediately:

$$\begin{aligned} \dot{A} &= \alpha_1 S B^{g_1} - \beta_1 A^{h_1}, \\ \dot{B} &= \alpha_2 A^{g_2} - \beta_2 B^{h_2}, \\ \dot{C} &= \alpha_3 A^{g_3} - \beta_3 C^{h_3}. \end{aligned} \tag{9.13}$$

TABLE 9.5: PARAMETER VALUES FOR THE ADAPTATION SYSTEM (9.13)

i	$\alpha_i = \beta_i$	g_i	h_i
1	2	–1	0.4
2	0.1	1	0.05
3	4	0.75	0.5

Without loss of generality, we may suppose that the normal steady state is (1, 1, 1) for a signal $S = 1$, which immediately means that $\alpha_i = \beta_i$ for $i = 1, 2, 3$; convince yourself that this is generally true. **Table 9.5** provides typical parameter values.

To explore the responses of the system, we start at the steady state and send signals as brief "boluses" at times $t = 5$, 30 and 45. In PLAS, this is easily done with commands like @ 5 $S = 4$, @ 5.1 $S = 1$. In all three cases, the output variable *C* shoots up, then undershoots, before returning toward the homeostatic value of 1 (**Figure 9.32**). The variables *A* and *B* are not of prime interest, but they return to 1 as well. These responses are not surprising, because the steady state (1, 1, 1) is stable. Now the real experiment begins. At $t = 75$, we reset the signal permanently to $S = 1.2$, which corresponds to the relative change Ma and collaborators used in their paper. Although the signal stays *on* at this value throughout the rest of the experiment, the system adapts, and *C* approaches a value of 1.013, which corresponds to a deviation of 1.3% from the original value. Ma and colleagues consider a difference below 2% as sufficiently precise, so that our model is indeed both sensitive and precise. Note that the internal variable *B*, which is of no real interest, shows a larger deviation.

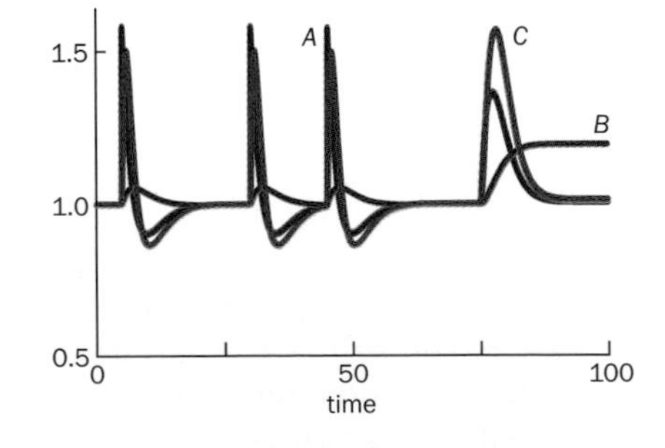

Figure 9.32 Responses of the three-node system. Following transient signals at times $t = 5$, 30, and 45, the variable of interest, *C*, shoots up and then returns to homeostasis. This response is to be expected, because the homeostatic steady state is stable. However, in response to a persistent signal, here starting at $t = 75$, the system adapts, and *C* approaches a new state that is very close to its homeostatic value. The time trends of variables *A* and *B* are shown, but are not of primary interest here.

9.7 Other Signaling Systems

The differential equation models discussed so far are paradigmatic signal transduction systems in the narrowest sense. Many other model approaches have been proposed under the same rubric of signaling systems. For instance, gene regulatory networks are sometimes considered signaling systems, because the expression of one gene provides the signal for other genes to be expressed or repressed, which is mediated through transcription factors, repressors, or small RNAs (see [10, 44] and Chapter 6).

A specific example is the genomic regulation of the cell cycle, which we have already discussed in the context of Boolean network analysis [5]. For many years, Tyson, Novak, and others have been using ordinary differential equations to study cell cycles and their control. In the case of yeast, these models have become so good that they capture essentially all alterations in cell cycle dynamics that stem from mutants in any of the involved genes [45, 46]. Other approaches have been based on stochastic differential equations [47], recursive models [48], hybrid Petri nets [49], or machine learning methods [50]. The juxtaposition of these diverse models of gene regulatory networks demonstrates again that modeling can take many shapes and forms, and that the choice of a particular model depends greatly on the purpose of the model and the questions that are to be answered.

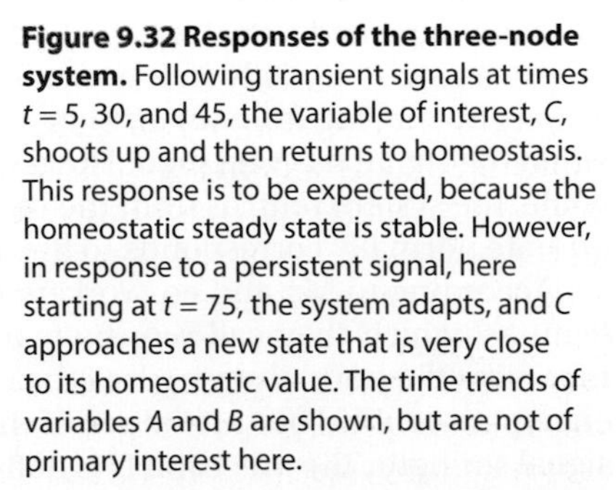

Another fascinating class of signaling systems concerns circadian clocks. These molecular clocks sense and interpret light or other environmental conditions and allow organisms, ranging from prokaryotes to humans, to be active at appropriate times during the day–night cycle. The oscillations are essentially self-sustained, because they continue even if the organism is held in continuous light or continuous darkness. However, they usually run out of phase after a while, indicating that humans, for instance, have an intrinsic cycle duration that is usually slightly longer than 24 hours. Because the clocks are based on biochemical events, one might surmise that they would depend on the environmental temperature, but, interestingly, this is not the case. An immense body of literature documents the experimental and computational research that has been done on circadian rhythms over the past 100 years, since J. S. Szymanski observed that animals could maintain a 24-hour rhythm without external cues [51]. The relatively new field of chronobiology focuses on issues of circadian clocks and, for instance, tries to understand the effects on night shifts and jet lags on human health [52, 53].

It is surprisingly simple to construct clocks from systems with hysteresis. As a demonstration, let us return to the system in (9.7) and study the variable Y. Its hysteresis plot is similar to that of X, although its numerical values are an order of magnitude smaller. Importantly, we see the same type of jumping behavior for the same values of S. This is no coincidence, because X and Y are tightly interconnected.

In the earlier demonstration of hysteresis, we artificially raised S from low values below 2 to high values above 4 and then lowered it back. This procedure suggests a thought experiment where some force within or outside the system would do this raising and lowering repeatedly. If so, we can quite easily convince ourselves that X and Y should start to oscillate in response.

But we can do even better, namely, by formulating a model. So far, the steady states of X and Y are functions of the independent variable S. Now, by making S a dependent variable, we can have its dynamics driven by X or Y so that S becomes high if X or Y is low, and low if X or Y is high.

To implement this plan, let us express S as a function of Y_{ss}, which here is feasible with a little bit of algebra. For Y to be at a steady state, the second equation of (9.7) must equal 0. Thus, we obtain

$$S = \frac{Y_{ss}}{0.5X_{ss}^{0.5}}. \tag{9.14}$$

For X to be at a steady state, the first equation of (9.7) must equal 0, which yields

$$0.5X_{ss}^{0.5} = 2 + \frac{8Y_{ss}^4}{16^4 + Y_{ss}^4}. \tag{9.15}$$

Plugging (9.15) into (9.14), we obtain S as a function of Y_{ss}:

$$S = Y_{ss} \bigg/ \left(2 + \frac{8Y_{ss}^4}{16^4 + Y_{ss}^4}\right). \tag{9.16}$$

This function is shown in **Figure 9.33A**. Figure 9.33B flips this plot over, so that Y_{ss} depends on S. Note, however, that Y_{ss} is not a function of S, because for values around $S = 3$, Y_{ss} has three values. Nonetheless, this mapping is very useful, because it shows the relationship between Y_{ss} and S. In fact, we can superimpose this plot on **Figure 9.34A** and obtain Figure 9.34B. Two critical observations can be made. First, the low and high branches of the plot in Figure 9.34A are exactly matched by the mapping between Y_{ss} and S. Second, this mapping adds a piece to the curve, connecting the jump-off points of Y_{ss} as a function of S. This new piece, curving from $(S, Y_{ss}) \approx (3.3, 9.1)$ upward to $(S, Y_{ss}) \approx (2.6, 19.7)$, exhibits unstable steady states between the high and low stable steady states. All values on this piece are repelling, and unless one starts exactly at such a point, the solution will go either to the upper or to the lower steady state, both of which are stable.

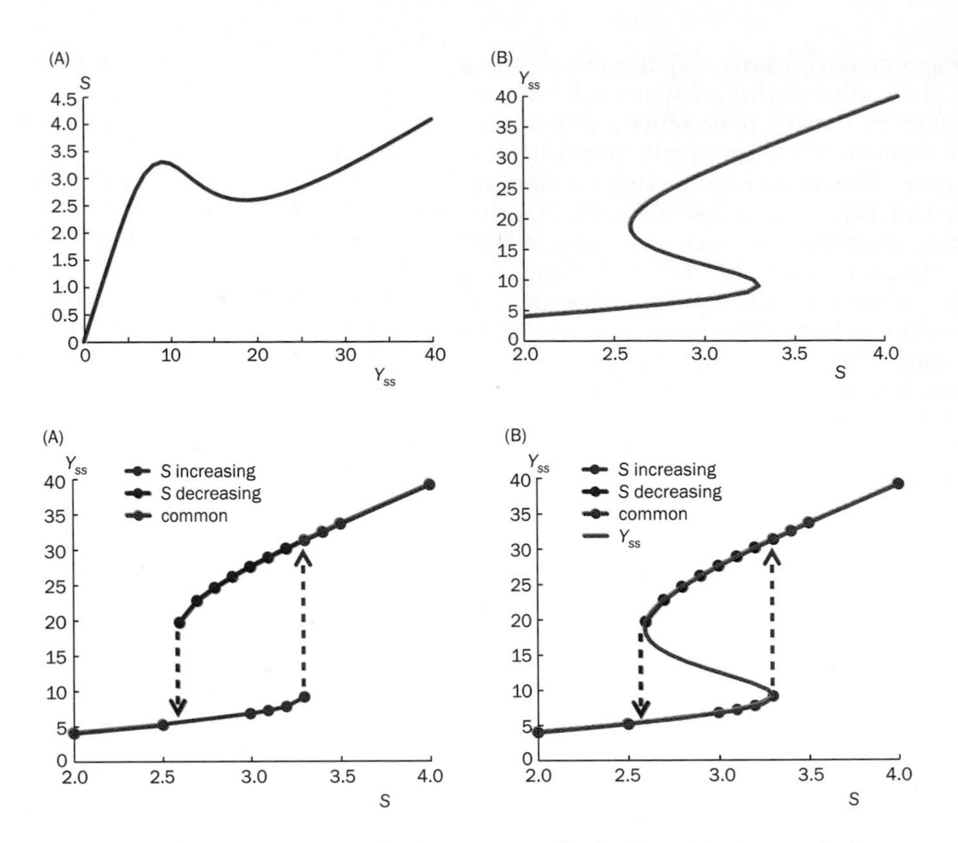

Figure 9.33 Relationship between Y_{ss} and the signal *S*. (A) A plot of (9.16), which results from computing *S* from the steady-state equations of (9.7). (B) The same plot flipped over for *S* between 2 and 4. Although *S* is a function of Y_{ss}, Y_{ss} is not a function of *S*, because, for a value of *S* around 3, Y_{ss} can have three different values.

Figure 9.34 Hysteretic behavior of Y_{ss} in response to the signal *S*. (A) The hysteresis diagram of (9.7) with respect to Y_{ss}, which is qualitatively similar to the diagram for X_{ss} (Figure 9.11). (B) Superimposing Figure 9.33B on this plot "fills in the blanks" by showing the unstable steady states for signals between about 2.6 and 3.3.

Now let us return to the task of creating a clock. To minimize confusion, we replace *S* with the dependent variable *Z*, which we define as

$$\dot{Z} = 5Y^{-1} - 0.25Z^{0.5}. \tag{9.17}$$

Other than exchanging *Z* for *S*, we retain the system in (9.7) without change. The exact form of *Z* is actually not all that critical, but it does affect the shape of the oscillations. The important feature on which we must insist is that large values of *Y* strongly reduce the new signal and that low values strongly increase it. This requirement is met by *Y*'s exponent of –1. Also, *Z* should not wander off to infinity or zero, and should instead have the potential for a steady state, which is accomplished through the degradation term. With these settings, the signal becomes small as soon as *Y* becomes big, and vice versa. A small signal causes *Y* to approach the low steady state, but as soon as *Y* becomes small enough, the signal increases, causing *Y* to approach the high steady state. As a consequence, *X*, *Y*, and *Z* never actually reach their steady states and instead oscillate in the manner of a limit cycle (**Figure 9.35**), which we discussed in Chapter 4, and which is a good model for a clock. It is easy to check that the system is indeed a limit cycle by perturbing one of the variables. For instance, even if *X* is externally reduced a lot (here, from about 270 to 150 at time 20, the system recovers very quickly (**Figure 9.36**).

This transition from bistability to limit cycles has been studied in a model of an embryonic cell cycle oscillator in the frog *Xenopus* [54] and also with the so-called repressilator system [55], which is discussed in Chapter 14.

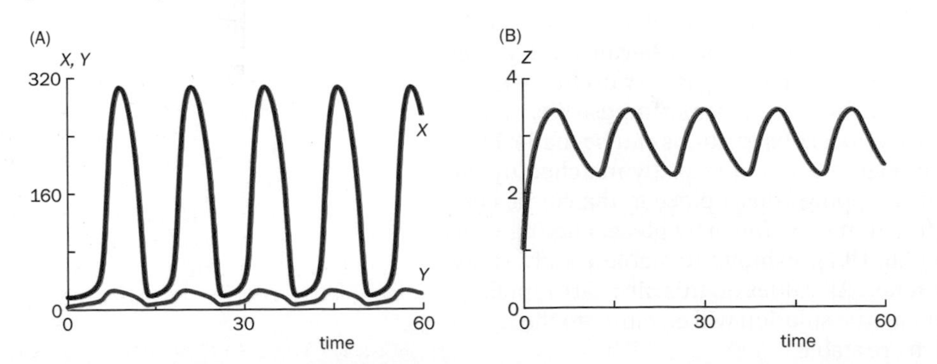

Figure 9.35 Oscillations resulting from making the signal dependent on *Y*. Owing to the construction where high values of *Y* lead to low values of *Z*, and low values of *Y* lead to high values of *Z*, the system never reaches a steady state and instead oscillates in a sustained manner.

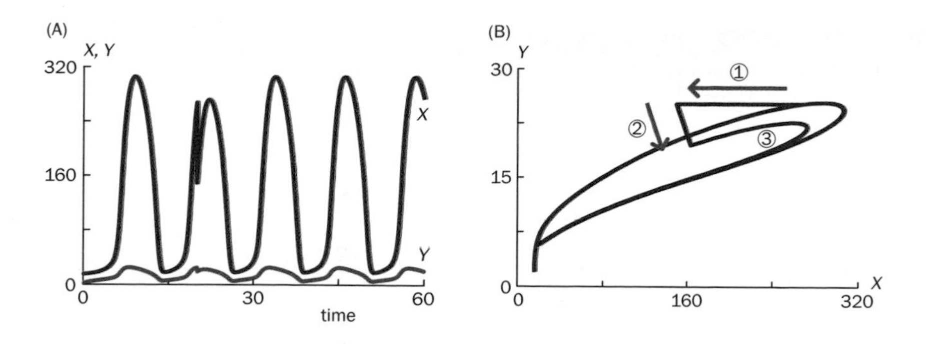

Figure 9.36 The oscillations resulting from making the signal dependent on Y form a limit cycle. If one of the variables is perturbed, the system quickly regains its limit-cycle behavior. Here, X was reduced at time $t = 20$ from about 270 to 150. The perturbation leads to a slight phase shift, but the original oscillation is regained. (A) shows the oscillation in the time domain, while (B) shows the corresponding phase-plane plot of Y versus X. Note that Z is too small to be recognizable.

While it is not surprising that clocks in mammals are quite complex, it is intriguing that it has been possible to create simple circadian rhythms even *in vitro*, without transcription or translation [56]. All the self-sustained oscillator requires is a set of three proteins, called KaiA, KaiB, and KaiC, from the cyanobacterium *Synechococcus elongatus*, and an energy source in the form of ATP. The simple system is even temperature-compensated, exhibiting the same oscillation period under temperature variations. *In vivo*, the Kai system is of course more intricate, and it has been demonstrated how very slow, but ordered and fine-tuned, enzymatic phosphorylation of the Kai proteins allows the cyanobacterium not only to sustain oscillations, but also to respond appropriately to *zeitgebers* (timing signals) in the environment [57]. Specifically, KaiC has two phosphorylation sites and can therefore be present in four forms: unphosphorylated (U), phosphorylated only on a serine residue (S), phosphorylated only on a threonine residue (T), or phosphorylated on both sites (ST). These different forms cycle in different phases. KaiA enhances the autophosphorylation of KaiC, and KaiC dephosphorylates in the absence of KaiA. KaiB in turn impedes the activity of KaiA (**Figure 9.37**).

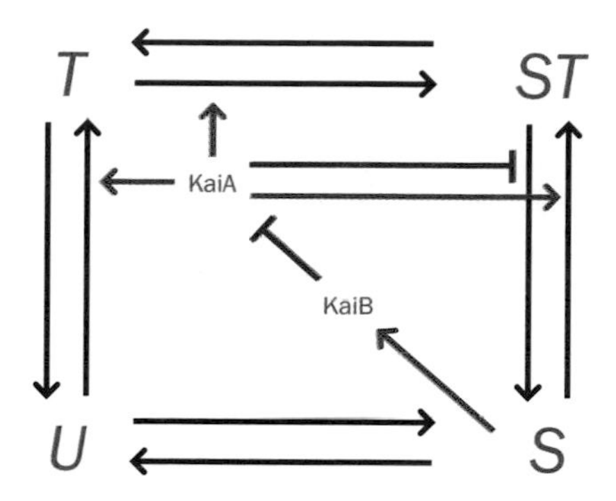

Figure 9.37 Schematic of the Kai oscillation system associated with circadian rhythms. Green arrows indicate activation and red lines with bar show inhibition. See the text for details. (Data from Rust MJ, Markson JS, Lane WS, et al. *Science* 318 [2007] 809–812.)

A conversion of the diagram into a kinetic model allows further analyses of the oscillator and of various perturbations and mutations [57]. Under baseline conditions, the model oscillates as shown in **Figure 9.38**. Using first-order kinetics, but with a nonlinear influence of the KaiA concentration, the model of Rust and collaborators [57] reads

$$\begin{aligned}
\dot{T} &= k_{UT}(S)U + k_{DT}(S)D - k_{TU}(S)T - k_{TD}(S)T, \quad T(0)=0.68,\\
\dot{D} &= k_{TD}(S)T + k_{SD}(S)S - k_{DT}(S)D - k_{DS}(S)D, \quad D(0)=1.36,\\
\dot{S} &= k_{US}(S)U + k_{DS}(S)D - k_{SU}(S)S - k_{SD}(S)S, \quad S(0)=0.34,\\
A &= \max\{0, [\text{KaiA}] - 2S\},\\
k_{XY}(S) &= k_{XY}^{0} + \frac{k_{XY}^{A} A(S)}{K + A(S)}.
\end{aligned} \tag{9.18}$$

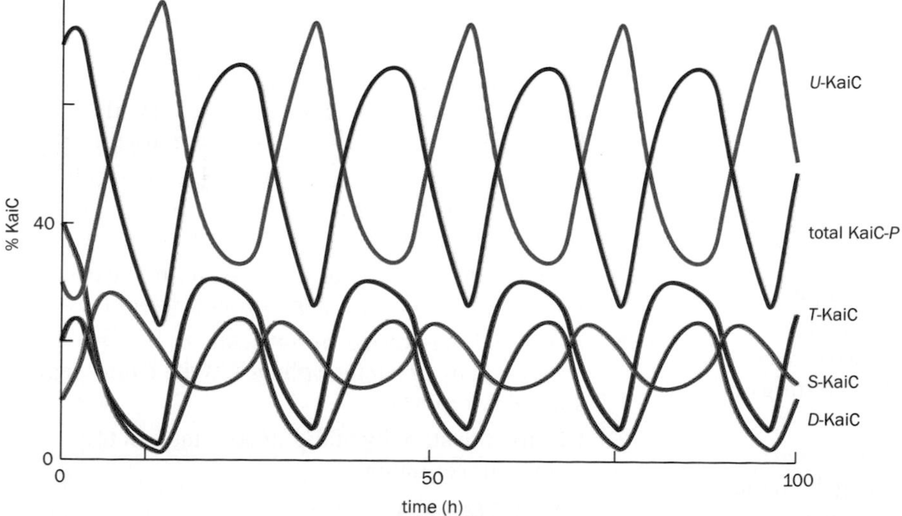

Figure 9.38 Oscillations of the Kai system in Figure 9.37. (From Rust MJ, Markson JS, Lane WS, et al. *Science* 318 [2007] 809–812. With permission from the American Association for the Advancement of Science.)

TABLE 9.6: RATE PARAMETER VALUES FOR THE CLOCK SYSTEM (9.18)

	UT	*TD*	*SD*	*US*	*TU*	*DT*	*DS*	*SU*
k^0	0	0	0	0	0.21	0	0.31	0.11
k^A	0.479077	0.212923	0.505692	0.0532308	0.0798462	0.173	−0.319385	−0.133077

Here, D refers to the double-phosphorylation state ST. Suitable kinetic parameter values are given in **Table 9.6**. The total concentration was taken as 3.4, and the amount of unphosphorylated KaiC was thus $U = 3.4 - T - D - S$. Furthermore, [KaiA] = 1.3 and $K = 0.43$ [57].

Maybe most intriguing is the precision of this relatively simple oscillator in the actual organism. In spite of asynchronous cell division, the clocks of a cell and its offspring run with very little deviation over several weeks, even in the absence of external cues. Many other processes in the organism are driven by the Kai oscillations, by virtue of rhythmic changes in the supercoiling state of the bacterial DNA and subsequent global changes in gene expression [58].

As a final example of a widespread signal transduction system, consider cell-to-cell signaling among microbes, which leads to **quorum sensing** [59]. This phenomenon allows groups of cells or organisms to coordinate their responses, such as swarming or aggregation, based on the current population density and the emission of specific signaling molecules. Once emitted, these molecules diffuse and bind to dedicated receptors on the surfaces of other microbes, which leads to changes in the expression of target genes, including those responsible for the production of the signaling molecule. Thus, triggering the signal transduction event requires a certain concentration of the molecule, which is only achieved if a quorum of bacteria is sending out the signal. Quorum sensing has been observed in bioluminescence and biofilms and is of great importance and concern for the virulence of pathogens, such as *Pseudomonas aeruginosa* and *Staphylococcus aureus*, some of which have become multidrug-resistant. Understanding the mechanistic, molecular details of quorum sensing may point to means of controlling these pathogens and offer new avenues of synthetic biology, with the goal of engineering cells with rewired metabolic pathways or other desirable properties [60] (see also Chapter 14).

EXERCISES

9.1. Formulate rules, corresponding to (9.3), for nodes with fewer than eight neighbors.

9.2. Create a 5 × 5 grid with 25 nodes. Starting with an arbitrary combination of *on–off* states, we established a simple rule producing a blinking dynamics in which every state switched from *on* to *off* or from *off* to *on* at every transition. Is this rule unique, or can other rule sets produce blinking? Is it possible for each state to stay *on* (*off*) for two time units, then switch to *off* (*on*) and stay *off* for two time units, and then to switch back and forth every two time units? Explain your answer.

9.3. For a 5 × 5 grid with 25 nodes and initially arbitrary *on–off* states, create rules such that eventually all states are zero except for the center state, which should be constantly *on*. Are the rules unique?

9.4. For a 5 × 5 grid with 25 nodes, define rules and initial conditions such that an *on* column moves from left to right, then wraps around by starting on the left again.

9.5. Establish rules that make an *on* diagonal move in the southwest direction and then start again in the northeast corner for the next diagonal wave.

9.6. For a 5 × 5 grid with 25 nodes and initially arbitrary *on–off* states, is it possible to create rules such that no *on–off* pattern is ever repeated? Either create such rules or prove that this is not possible.

9.7. Suppose each node may have one of three states, such as red, blue, or green. Construct a blinking pattern, in which each state cycles through the three colors.

9.8. Search the literature or the Internet for at least three biological examples that were modeled with Boolean methods. Write a brief summary report.

9.9. How could one set up a Boolean network model that remembers the two previous states? Sketch out a plan and implement a small network.

9.10. Review the principles and challenges of Bayesian inference. Write a one-page summary report.

9.11. Review what is known about switches between lysis and lysogeny in bacteriophages. Write a one-page summary report.

9.12. Discuss the stability of the steady states of the differential equation

$$\dot{X} = 2X(4 - X^2)$$

from a mathematical and a biological point of view. Discuss what happens if the multiplier 2 is replaced with −2.

9.13. Perform simulations with the bistable system in (9.5) to find out in which range(s) the system is most sensitive to noise. Interpret your findings.

9.14. Construct a new "bi-unstable" system with two unstable states and a stable state in between. Demonstrate the responses of the system to perturbations similar to those shown in Table 9.1 and Figure 9.5.

9.15. Demonstrate with simulations that the model in (9.6) is bistable.

9.16. Determine whether the model in (9.4) can exhibit hysteresis.

9.17. Implement the two-variable hysteretic system in (9.7) and explore how strong a perturbation must be to move the response incorrectly from *off* to *on*. Study persistent and temporary perturbations. Explain your findings.

9.18. Create a sequence of signals that are corrupted by noise. Study its effects on the bistable system in (9.7).

9.19. Test whether the bifunctional variant of (9.8) also exhibits bistability and hysteresis.

9.20. Igoshin and collaborators [15] emphasized the importance of the phosphatase *Ph* in the bifunctional model. Explore bistability and hysteresis if there is no phosphatase *Ph*. For this purpose, use the same model as in the text, but set $k_{b2} = 0.5$ and set the initial value of *Ph* equal to zero. Repeat the experiment with values of *Ph* between 0 and 0.17. Summarize your findings in a report.

9.21. Study the responses of the two TCS systems (mono- and bifunctional) to repeated brief signals.

9.22. Implement the MAPK cascade shown in Figure 9.26 and study the effects of noise on the signal–response relationship.

9.23. Explore the responses of a MAPK cascade with parameter values that are closer to reality (cf. [32, 34, 35]), namely, $k_1 = k_4 = 0.034$, $k_2 = k_5 = 7.75$, $k_3 = k_6 = 2.5$, $k_7 = k_8 = 1$; $X(0) = 800$, $Y(0) = 1000$, $Z(0) = 10{,}000$, and $E = 1$; all other initial values are 0.

9.24. Explore the effects of increasing the rate of phosphorylation at the three layers. Use either the model in the text or the model in Exercise 9.23.

9.25. Study bistability and hysteresis in the MAPK cascade. Can a single module, as shown in Figure 9.22, be bistable and/or hysteretic? Is the complete MAPK cascade bistable? Is it hysteretic? Investigate these questions with simulations. Summarize your answers in a report.

9.26. Create a small pathway model with two sequential reactions that are each modeled as regular Michaelis–Menten rate laws. Compare its responses with those of the double-phosphorylation system at the center and bottom layers of the MAPK cascade, where the two phosphorylation steps compete for the same, limited amount of enzyme.

9.27. Investigate what happens if the MAPK cascade receives repeated brief signals. Discuss your findings.

9.28. Investigate what happens if the phosphatases in the MAPK cascade have very low activity.

9.29. Kholodenko [61] discussed the consequences of potential negative feedback, in which MAPK inhibits the activation of MAPKKK. Implement such a feedback mechanism in the model of the MAPK cascade discussed in the text and study the implications of different strengths of this feedback.

9.30. The MAPK model, as used in the text, assumes that *XP* and *YPP* are constant. Study the consequences of making them dependent on their next lower level in the cascade.

9.31. Derive conditions on the parameter values in (9.13) under which the system exhibits adaptation. Assume that the initial steady state is (1, 1, 1), that the signal is permanently increased from 1 to 1.2, and that the output variable *C* approaches the steady-state value of 1.02. Solve the system for the steady state and derive the conditions.

9.32. Explore the numerical features and functions in (9.17) and their effect on the limit-cycle clock.

9.33. Use (9.18) to analyze biologically relevant scenarios. To get started, consult the original literature. Document your findings in a report.

9.34. Compute the red portion of the hysteresis curve in Figure 2 of Box 9.1 by formulating *S* as a function of X_{ss}.

REFERENCES

[1] Ma'ayan A, Blitzer RD & Iyengar R. Toward predictive models of mammalian cells. *Annu. Rev. Biophys. Biomol. Struct.* 34 (2005) 319–349.

[2] Cheresh DE (ed.). Integrins (Methods in Enzymology, Volume 426). Academic Press, 2007.

[3] Kauffman SA. Origins of Order: Self-Organization and Selection in Evolution. Oxford University Press, 1993.

[4] Leclerc R. Survival of the sparsest: robust gene networks are parsimonious. *Mol. Syst. Biol.* 4 (2008) 213.

[5] Davidich MI & Bornholdt S. Boolean network model predicts cell cycle sequence of fission yeast. *PLoS One* 3 (2008) e1672.

[6] Russell S & Norvig P. Artificial Intelligence: A Modern Approach, 3rd ed. Prentice Hall, 2010.

[7] Shmulevich I, Dougherty ER, Kim S & Zhang W. Probabilistic Boolean networks: a rule-based uncertainty model for gene regulatory networks. *Bioinformatics* 18 (2002) 261–274.

[8] Shmulevich I, Dougherty ER & Zhang W. From Boolean to probabilistic Boolean networks as models of genetic regulatory networks. *Proc. IEEE* 90 (2002) 1778–1792.

[9] Chou IC & Voit EO. Recent developments in parameter estimation and structure identification of biochemical and genomic systems. *Math. Biosci.* 219 (2009) 57–83.

[10] Nakatsui M, Ueda T, Maki Y, et al. Method for inferring and extracting reliable genetic interactions from time-series profile of gene expression. *Math. Biosci.* 215 (2008) 105–114.
[11] Monod J & Jacob F. General conclusions—teleonomic mechanisms in cellular metabolism, growth, and differentiation. *Cold Spring Harbor Symp. Quant. Biol.* 28 (1961) 389–401.
[12] Novick A & Weiner M. Enzyme induction as an all-or-none phenomenon. *Proc. Natl Acad. Sci. USA* 43 (1957) 553–566.
[13] Tian T & Burrage K. Stochastic models for regulatory networks of the genetic toggle switch. *Proc. Natl Acad. Sci. USA* 103 (2006) 8372–8377.
[14] Tan C, Marguet P, &You L, Emergent bistability by a growth-modulating positive feedback circuit. *Nat. Chem. Biol.* 5 (2009) 942–948.
[15] Igoshin OA, Alves R & Savageau MA. Hysteretic and graded responses in bacterial two-component signal transduction. *Mol. Microbiol.* 68 (2008) 1196–1215.
[16] Savageau MA. Design of gene circuitry by natural selection: analysis of the lactose catabolic system in *Escherichia coli. Biochem. Soc. Trans.* 27 (1999) 264–270.
[17] Bhattacharya S, Conolly RB, Kaminski NE, et al. A bistable switch underlying B cell differentiation and its disruption by the environmental contaminant 2,3,7,8-tetrachlorodibenzo-*p*-dioxin. *Toxicol. Sci.* 115 (2010) 51–65.
[18] Zhang Q, Bhattacharya S, Kline DE, et al. Stochastic modeling of B lymphocyte terminal differentiation and its suppression by dioxin. *BMC Syst. Biol.* 4 (2010) 40.
[19] Hasty J, Pradines J, Dolnik M & Collins JJ. Noise-based switches and amplifiers for gene expression. *Proc. Natl Acad. Sci. USA* 97 (2000) 2075–2080.
[20] Wu J & Voit EO. Hybrid modeling in biochemical systems theory by means of functional Petri nets. *J. Bioinf. Comp. Biol.* 7 (2009) 107–134.
[21] Jiang P & Ninfa AJ. Regulation of autophosphorylation of *Escherichia coli* nitrogen regulator II by the PII signal transduction protein. *J. Bacteriol.* 181 (1999) 1906–1911.
[22] Utsumi RE. Bacterial Signal Transduction: Networks and Drug Targets. Landes Bioscience, 2008.
[23] Bourret RB & Stock AM. Molecular Information processing: lessons from bacterial chemotaxis. *J. Biol. Chem.* 277 (2002) 9625–9628.
[24] Zhao R, Collins EJ, Bourret RB & Silversmith RE. Structure and catalytic mechanism of the *E. coli* chemotaxis phosphatase CheZ. *Nat. Struct. Biol.* 9 (2002) 570–575.
[25] Alves R & Savageau MA. Comparative analysis of prototype two-component systems with either bifunctional or monofunctional sensors: differences in molecular structure and physiological function. *Mol. Microbiol.* 48 (2003) 25–51.
[26] Bongiorni C, Stoessel R & Perego M. Negative regulation of *Bacillus anthracis* sporulation by the Spo0E family of phosphatases. *J. Bacteriol.* 189 (2007) 2637–2645.
[27] Batchelor E & Goulian M. Robustness and the cycle of phosphorylation and dephosphorylation in a two-component system. *Proc. Natl Acad. Sci. USA* 100 (2003) 691–696.
[28] Schoeberl B, Eichler-Jonsson C, Gilles ED & Müller G. Computational modeling of the dynamics of the MAP kinase cascade activated by surface and internalized EGF receptors. *Nat. Biotechnol.* 20 (2002) 370–375.
[29] Pouyssegur J, Volmat V & Lenormand P. Fidelity and spatio-temporal control in MAP kinase (ERKs) signalling. *Biochem. Pharmacol.* 64 (2002) 755–763.
[30] Thattai M & van Oudenaarden A. Attenuation of noise in ultrasensitive signaling cascades. *Biophys. J.* 82 (2002) 2943–2950.
[31] Bluthgen N & Herzel H. How robust are switches in intracellular signaling cascades? *J. Theor. Biol.* 225 (2003) 293–300.
[32] Huang C-YF & Ferrell JEJ. Ultrasensitivity in the mitogen-activated protein kinase cascade. *Proc. Natl Acad. Sci. USA* 93 (1996) 10078–10083.
[33] Bhalla US & Iyengar R. Emergent properties of networks of biological signaling pathways. *Science* 283 (1999) 381–387.
[34] Schwacke JH & Voit EO. Concentration-dependent effects on the rapid and efficient activation of MAPK. *Proteomics* 7 (2007) 890–899.
[35] Schwacke JH. The Potential for Signal Processing in Interacting Mitogen Activated Protein Kinase Cascades. Doctoral Dissertation, Medical University of South Carolina, Charleston, 2004.
[36] Kholodenko BN & Birtwistle MR. Four-dimensional dynamics of MAPK information processing systems. *Wiley Interdiscip. Rev. Syst. Biol. Med.* 1 (2009) 28–44.
[37] Muñoz-García J & Kholodenko BN. Signalling over a distance: gradient patterns and phosphorylation waves within single cells. *Biochem. Soc. Trans.* 38 (2010) 1235–1241.
[38] Whitmarsh AJ & Davis RJ. Structural organization of MAP-kinase signaling modules by scaffold proteins in yeast and mammals. *Trends Biochem. Sci.* 23 (1998) 481–485.
[39] Jordan JD, Landau EM & Iyengar R. Signaling networks: the origins of cellular multitasking. *Cell* 103 (2000) 193–200.
[40] Hanahan D & Weinberg RA. The hallmarks of cancer. *Cell* 100 (2000) 57–70.
[41] Kumar N, Afeyan R, Kim HD & Lauffenburger DA. Multipathway model enables prediction of kinase inhibitor cross-talk effects on migration of Her2-overexpressing mammary epithelial cells. *Mol. Pharmacol.* 73 (2008) 1668–1678.
[42] Pritchard JR, Cosgrove BD, Hemann MT, et al. Three-kinase inhibitor combination recreates multipathway effects of a geldanamycin analogue on hepatocellular carcinoma cell death. *Mol. Cancer Ther.* 8 (2009) 2183–2192.
[43] Ma W, Trusina A, El-Samad H, et al. Defining network topologies that can achieve biochemical adaptation. *Cell* 138 (2009) 760–773.
[44] Beisel CL & Smolke CD. Design principles for riboswitch function. *PLoS Comput. Biol.* 5 (2009) e1000363.
[45] Tyson JJ, Csikasz-Nagy A & Novak B. The dynamics of cell cycle regulation. *Bioessays* 24 (2002) 1095–1109.
[46] Tyson JJ & Novak B. Temporal organization of the cell cycle. *Curr. Biol.* 18 (2008) R759–R768.
[47] Sveiczer A, Tyson JJ & Novak B. A stochastic, molecular model of the fission yeast cell cycle: role of the nucleocytoplasmic ratio in cycle time regulation. *Biophys. Chem.* 92 (2001) 1–15.
[48] Vu TT & Vohradsky J. Inference of active transcriptional networks by integration of gene expression kinetics modeling and multisource data. *Genomics* 93 (2009) 426–433.
[49] Matsuno H, Doi A, Nagasaki M & Miyano S. Hybrid Petri net representation of gene regulatory network. *Pac. Symp. Biocomput.* 5 (2000) 341–352.
[50] To CC & Vohradsky J. Supervised inference of gene-regulatory networks. *BMC Bioinformatics* 9 (2008).
[51] Szymanski JS, Aktivität und Ruhe bei Tieren und Menschen. *Z. Allg. Physiol.* 18 (1919) 509–517.
[52] Cermakian N & Bolvin DB. The regulation of central and peripheral circadian clocks in humans. *Obes. Rev.* 10(Suppl. 2) (2009) 25–36.
[53] Fuhr L, Abreu M, Pett P & Relógio A. Circadian systems biology: when time matters. *Comput. Struct. Biotechnol. J.* 13 (2015) 417–426.
[54] Ferrell JE Jr, Tsai TYC & Yang Q. Modeling the cell cycle: Why do certain circuits oscillate? *Cell* 144 (2011) 874 – 885.
[55] Elowitz MB & Leibler S. A synthetic oscillatory network of transcriptional regulators. *Nature* 403 (2000) 335–338.

[56] Nakajima M, Imai K, Ito H, et al. Reconstitution of circadian oscillation of cyanobacterial KaiC phosphorylation *in vitro*. *Science* 308 (2005) 414–415.
[57] Rust MJ, Markson JS, Lane WS, et al. Ordered phosphorylation governs oscillation of a three-protein circadian clock. *Science* 318 (2007) 809–812.
[58] Woelfle MA, Xu Y, Qin X & Johnson CH. Circadian rhythms of superhelical status of DNA in cyanobacteria. *Proc. Natl. Acad. Sci. USA* 104 (2007) 18819–18824.
[59] Fuqua WC, Winans SC & Greenberg EP. Quorum sensing in bacteria: the LuxR–LuxI family of cell density-responsive transcriptional regulators. *J. Bacteriol.* 176 (1994) 269–275.
[60] Hooshangi S & Bentley WE. From unicellular properties to multicellular behavior: bacteria quorum sensing circuitry and applications. *Curr. Opin. Biotechnol.* 19 (2008) 550–555.
[61] Kholodenko BN. Negative feedback and ultrasensitivity can bring about oscillations in the mitogen-activated protein kinase cascades. *Eur. J. Biochem.* 267 (2000) 1583–1588.

FURTHER READING

Bhalla US & Iyengar R. Emergent properties of networks of biological signaling pathways. *Science* 283 (1999) 381–387.

Hasty J, Pradines J, Dolnik M & Collins JJ. Noise-based switches and amplifiers for gene expression. *Proc. Natl Acad. Sci. USA* 97 (2000) 2075–2080.

Huang C-YF & Ferrell JEJ. Ultrasensitivity in the mitogen-activated protein kinase cascade. *Proc. Natl Acad. Sci. USA* 93 (1996) 10078–10083.

Igoshin OA, Alves R & Savageau MA. Hysteretic and graded responses in bacterial two-component signal transduction. *Mol. Microbiol.* 68 (2008) 1196–1215.

Jordan JD, Landau EM & Iyengar R. Signaling networks: the origins of cellular multitasking. *Cell* 103 (2000) 193–200.

Kauffman SA. Origins of Order: Self-Organization and Selection in Evolution. Oxford University Press, 1993.

Kholodenko BN, Hancock JF & Kolch W. Signalling ballet in space and time. *Nat. Rev. Mol. Cell Biol.* 11 (2010) 414–426.

Ma'ayan A, Blitzer RD & Iyengar R. Toward predictive models of mammalian cells. *Annu. Rev. Biophys. Biomol. Struct.* 34 (2005) 319–349.

Sachs K, Perez O, Pe'er S, et al. Causal protein-signaling networks derived from multiparameter single-cell data. *Science* 308 (2005) 523–529.

Schwacke JH & Voit EO. Concentration-dependent effects on the rapid and efficient activation of MAPK. *Proteomics* 7 (2007) 890–899.

Tyson JJ, Csikasz-Nagy A & Novak B. The dynamics of cell cycle regulation. *Bioessays* 24 (2002) 1095–1109.

Utsumi RE. Bacterial Signal Transduction: Networks and Drug Targets. Landes Bioscience, 2008.

Population Systems

10

When you have read this chapter, you should be able to:

- Identify and discuss standard approaches to studying populations
- Distinguish models for homogenous and stratified populations
- Design and analyze models for age-structured populations
- Develop and analyze models for competing populations
- Understand how to incorporate population growth into larger systems models

Trends in the sizes of populations have fascinated humans for a long time. Hunters and gatherers were not only interested in the size of their own population, but had a strongly vested interest in the populations of animals and plants in their surroundings, upon which their existence depended [1]. Kings, emperors, and other politicians throughout the ages found population sizes important for forecasting food needs and for levying taxes. Apparently, the art of quantifying population sizes can be traced back at least to the Paleolithic period of 30,000 years ago, when early humans in Central Europe used tally sticks to keep track of changes in populations; **exponential growth** was recorded as early as 4000 years ago by the Babylonians [1–3]. Thus, throughout recorded history, population growth has been measured, documented, predicted, and analyzed.

This chapter takes a very broad view of populations. While we might immediately think of the human world population or the population growth in our home town, populations may also consist of free-living bacteria, viruses, and healthy or tumor cells. One might even include unconventional populations, such as the alleles of a gene within a human or animal population [4] or a population of molecules that is converted into a population of different molecules by the action of an enzyme.

POPULATION GROWTH

The earliest rigorous mathematical descriptions of population sizes and their trends are often attributed to the British clergyman and economist Thomas Robert Malthus [5], who formulated the law of exponential growth, and to the Belgian mathematician Pierre-François Verhulst [6], who proposed the sigmoidal **logistic growth** function that we often encounter in microbial populations (see Chapter 4). An enormous number of other growth functions were subsequently proposed for diverse populations, ranging from humans, animals, and plants to bacteria, cells, and viruses, and the literature on growth functions is huge. Most of these are relatively simple nonlinear functions, some discrete, some continuous, while others can be represented as the solutions of differential equations [1]. One might find it odd that population sizes, which are clearly discrete integers, are often modeled with differential equations, which could predict something like 320.2 individuals. The simple rationale for this strategy is that differential equations are often easier to set up and analyze than the corresponding discrete models. The situation is in reality even more complicated, because processes like growth that evolve over a long time period are almost always affected by **stochastic** events. Yet differential equations that capture the average growth behavior are often more useful than fully stochastic models, because they are so much easier to analyze and understand.

Previous chapters have made it clear that biological systems, even at the subcellular level, are exceedingly complex. Since higher organisms consist of very many cells, one should therefore expect that the growth of organisms, and of populations of organisms, is several orders of magnitude more complex. In some sense, this would be true if we wanted to keep track of every cell or even every biological molecule in a population of organisms. However, while molecular and intracellular details are of course crucial, the study of populations focuses on a different level, where cells, individuals, or organisms are the units of choice, and processes occurring within these units are not considered.

One rationale supporting this simplification is the fact that intracellular processes are much, much faster than the growth **dynamics** of individuals and organisms. In fact, they are so fast that they are always essentially in a **steady state**, so that their fast dynamics is not of importance for the much slower dynamics of the growth and aging of individuals and organisms. As a result, there are often only a few processes that are occurring on just the right timescale to affect growth [1]. Furthermore, population studies by their nature consider large numbers of individuals, which leads to an averaging of individual variability. Thus, in simplified growth models, all individuals or members of the population are essentially the same, and one studies their collective behavior.

The logistic growth function, which we discussed in some detail in Chapter 4, is a great example of these approximation and averaging effects. By describing millions of cells, aspects of individuality and diversity, which can actually be quite significant even in bacteria, are averaged away, and the populations follow simple, smooth trends (as, for example, in **Figure 10.1**) that are characterized by just a very few population **parameters** such as the **growth rate** and the **carrying capacity** (see below). This approach might seem overly crude, but it is very powerful, as the analogy with an ideal gas shows: while it is impossible to keep track of every gas molecule and its interactions with other molecules or the wall of the container, physicists have developed surprisingly simple gas laws that relate volume, pressure, and temperature to each other in a very accurate manner.

10.1 Traditional Models of Population Growth

The exponential and logistic growth functions are not the only growth functions. In fact, very many **growth laws** have been proposed over the past 200 years, either as explicit functions or as **ordinary differential equations (ODEs)**. One example is the Richards function [7]

$$N_R(t) = K\{1 + Q\exp[-\alpha\nu(t - t_0)]\}^{-1/\nu}, \tag{10.1}$$

which is very flexible in shape and has been used widely in plant science, agriculture, and forestry. Clearly, N_R is a function of time, t. $N_0 = N_R(t_0) = N_R(0)$ is the population size at some time point declared as 0, K is the carrying capacity, α and ν are positive parameters and $Q = (K/N_0)^\nu - 1$. The various parameters permit shifting and stretching the sigmoid curve in different directions.

Another growth function, which is often used in actuarial sciences, is the Gompertz growth law [8]

$$N_G(t) = K\exp[-b\exp(-rt)], \tag{10.2}$$

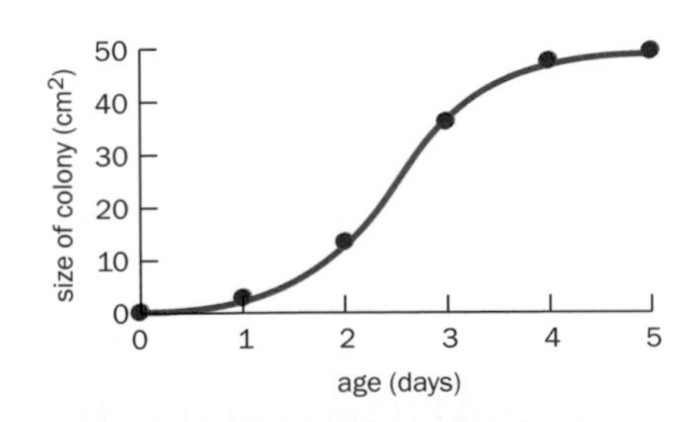

Figure 10.1 Growth of a bacterial colony. The colony, measured by its area on a Petri dish, was assessed over time (red dots). The logistic function $N(t) = 0.2524/(e^{-2.128t} + 0.005125)$ (blue line; see Chapter 4) models the data very well. (Data from Lotka A. Elements of Physical Biology. Williams & Wilkins, 1924.)

where K and r are again the carrying capacity and the growth rate, respectively, and b is an additional positive parameter.

It has turned out that it is beneficial for many purposes to represent growth functions as ODEs. For instance, the logistic growth function $N(t) = 0.2524/(e^{-2.128t} + 0.005125)$ (see Figure 10.1 and Chapter 4) may be written as

$$\dot{N} = rN - \frac{r}{K}N^2. \tag{10.3}$$

For the specific example in Figure 10.1, the corresponding parameter values are $r = 2.128$, $K = 49.25$, and $N_0 = 0.2511$.

Mathematically, this conversion is often not too difficult (Exercise 10.2), and from a modeling point of view, the ODE format facilitates the incorporation of the growth process into more complicated dynamic systems [9]. Specifically, the ODE formulation treats the growth process like any other process in a system, and (10.3) could be interpreted as a small system that describes the balance between one augmenting term, rN, and one diminishing term, $(r/K)N^2$. As a consequence, the growth process may be modulated by other components of the system. For instance, suppose that bacteria are growing in a human lung during pneumonia. Drug treatment with penicillin does not kill bacteria, but prevents them from proliferating. Thus, this effect would be incorporated into the positive growth term of the population model (10.3). By contrast, the body's immune system kills bacteria, and a pneumonia model would include this killing in the negative degradation term of (10.3). Another example with wide applicability involves ODEs that describe in a rather intuitive fashion how different populations living in the same environment interact and affect their growth characteristics. The typical approach for assessing this dynamics is the formulation of logistic functions with added interaction terms. We will discuss this example later in this chapter.

In many cases, ODE representations of growth processes are also more intuitive than an explicit growth function, because they can be interpreted in comparison with exponential growth, for which we have a good intuitive feel. In (10.3), the first term on the right-hand side represents unabated exponential growth, whereas the second term is sometimes interpreted as a diminishing **crowding** effect, which is proportional to the square of the number of individuals in a population. As an alternative, one may formulate (10.3) as

$$\dot{N} = r\left(\frac{K-N}{K}\right)N. \tag{10.4}$$

Mathematically, (10.3) and (10.4) are exactly equivalent, but (10.4) suggests the interpretation of a growth rate that depends on the population density. This growth rate is almost equal to r for very small populations, where N is much smaller than K, and decreases toward 0 if the population reaches the carrying capacity K.

Similarly, the Richards function (10.1) may be written as

$$\dot{N}_R = \alpha\left[1 - \left(\frac{N_R}{K}\right)^{\nu}\right]N_R, \tag{10.5}$$

which again can be interpreted as a variant of an exponential process with a population-density-dependent growth rate, which consists of α and the term in square brackets.

The Gompertz function (10.2) may also be written in ODE form with a density-dependent rate:

$$\dot{N}_G = r\ln\left(\frac{K}{N_G}\right)N_G. \tag{10.6}$$

Alternately, the same Gompertz function may be formulated as a set of two ODEs:

$$\dot{N}_G = RN_G,$$
$$\dot{R} = -rR. \tag{10.7}$$

While this representation may seem unnecessarily complicated, it is actually easier to interpret than (10.2) and (10.6). In the first equation, the new variable R is easily identified as the rate of an exponential growth process. The second equation shows that this growth rate is time-dependent; namely, it represents an exponential decay of the type R_0e^{-rt} (see Chapter 4). Thus, different representations may be mathematically equivalent but can have different biological interpretations. Furthermore, the embedding of functions into differential equations can be a tool for comparing and classifying growth laws [1, 9–11].

10.2 More Complex Growth Phenomena

Although many growth laws have been collected over time, there is still no guarantee that an observed dataset can be accurately modeled by any of them. As a case in point, it is still being debated what the maximal sustainable size of the human population is (**Box 10.1**). If we knew the growth function of the world population, we could easily determine the maximal size.

In some cases, the parameters of growth functions are relatively easy to assess. For instance, exponential growth is characterized by a growth rate parameter, which often is directly related to features such as the division time of cells (see [12, 13] and Chapters 2 and 4). However, other population parameters are not directly related to parameters that can be measured in individuals or even in small, growing populations. For instance, the bacterial colony in Figure 10.1 approaches a final size of about 50 cm^2, but this value cannot be inferred from studying individual bacteria or **subpopulations**, because it combines genetic, physiological, and environmental factors in some unknown combination. Trees in a planted plot of land grow according to a widely acknowledged "3/2 rule" that relates their size to the planting density. This relationship of a **power-law function** with negative exponent 3/2 has been observed uncounted times, but there is no true rationale for the relationship or the specific number [14, 15].

A different unsolved problem regarding population dynamics is the formulation and analysis of models for so-called **metapopulations**. These metapopulations sometimes consist of hundreds or thousands of different species of microbes that at once compete for space and nutrients and rely on each other for survival. Indeed, metapopulations are the rule rather than the exception, because we can find them in "environments" from our mouth to soil and sewage pipes. We will discuss some emerging approaches to studying such metapopulations in Chapter 15.

Typical population studies ignore spatial considerations, even though they may be very important. Intriguing examples are the growth of tumors, the incremental spread of introduced pests, such as the red fire ant *Solenopsis invicta* (see Exercise 10.32), and the spread of a disease throughout a country. A complicating aspect of human populations is that humans behave in complex and often unpredictable ways. For instance, imagine the task of predicting the trends in a population of people with diabetes under different proposed health-care schemes. A relatively recent approach to studying such cases, at least via computer **simulation**, is agent-based modeling, where individuals are modeled with heuristic rules that guide their decisions and interactions with others [16, 17]. Typically, all individuals have the same rule set, but it is also possible to account for stratified populations that consist of different subpopulations. Agent-based models are well suited for spatial phenomena, and the actions of individuals can easily be made to depend on their present states, their neighbors, and their local environment. While simulations with these types of models can be very powerful, mathematical analyses are often difficult. We will discuss this approach in Chapter 15.

BOX 10.1: TRENDS IN THE GROWTH OF THE HUMAN POPULATION

The human world population has been steadily growing throughout recorded history. The big question on many people's minds is whether or when this growth will slow down or even stop. Census bureaus around the world try to answer this enormously important question, using estimates of current population sizes (**Figure 1**) and birth rates (**Figure 2**). Even with the best data available, the resulting predictions vary tremendously (**Figure 3**).

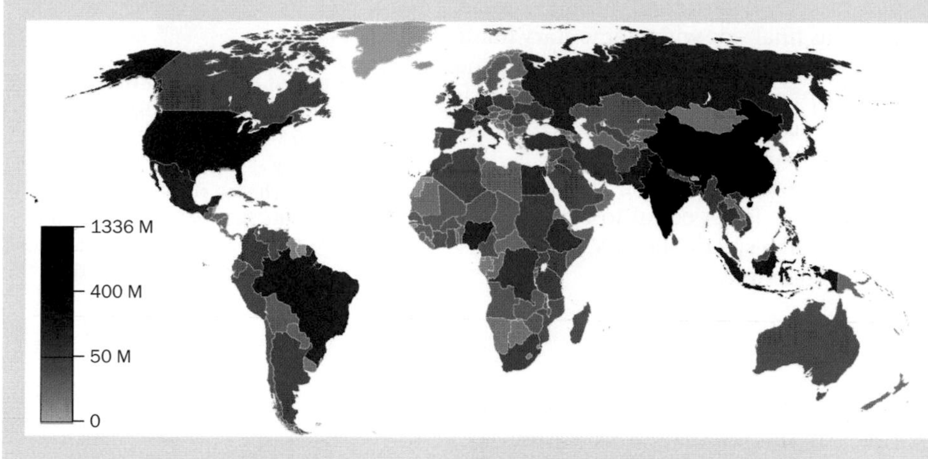

Figure 1 Estimated human world population in 2009. Current population distributions, combined with country-specific birth rates, are used to predict future population trends. (Courtesy of Roke under the Creative Commons Attribution–Share Alike 3.0 Unported license.)

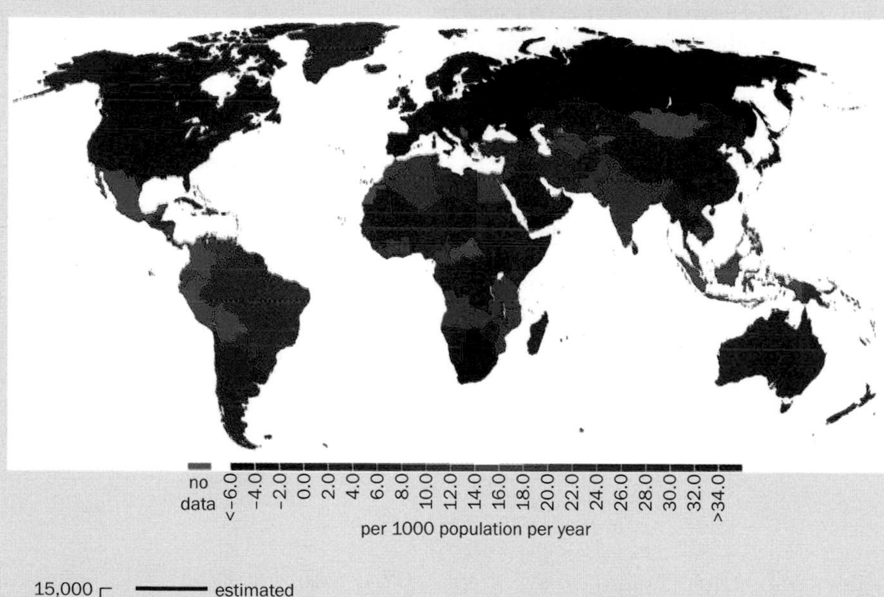

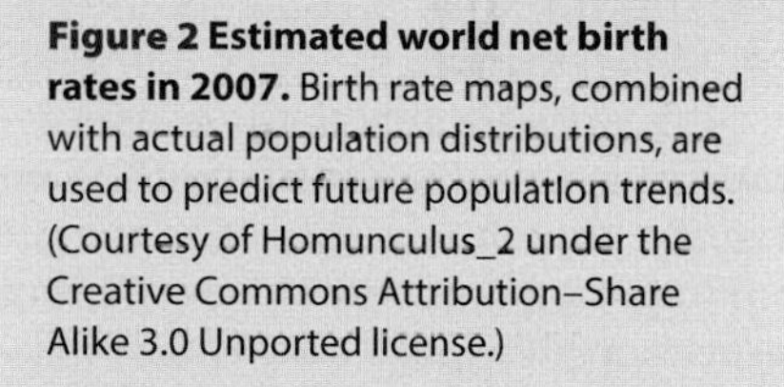

Figure 2 Estimated world net birth rates in 2007. Birth rate maps, combined with actual population distributions, are used to predict future population trends. (Courtesy of Homunculus_2 under the Creative Commons Attribution–Share Alike 3.0 Unported license.)

Figure 3 Recent history and differing forecasts for the growth of the human world population. The red, orange, and green curves reflect the United Nations' high, medium, and low estimates. (Courtesy of Loren Cobb Creative Commons Attribution–Share Alike 3.0 Unported license.)

POPULATION DYNAMICS UNDER EXTERNAL PERTURBATIONS

The dynamics of a single unperturbed population is usually not very complicated, as we have seen in the previous section. For instance, it does not really matter all that much how small the initial population size is: in a realistic model, the population increases toward some final, stable size, which is easily determined through simulation (**Figure 10.2**). We can also often compute this final size with elementary mathematical methods. If the growth process is formulated as an explicit function of time, we consider the growth function for time going to infinity. For example, in the case of the Gompertz function (10.2), the exponential term $\exp(-rt)$ approaches zero, which causes the term $\exp[-b \exp(-rt)]$ to approach 1, so that $N_G(t)$ approaches K.

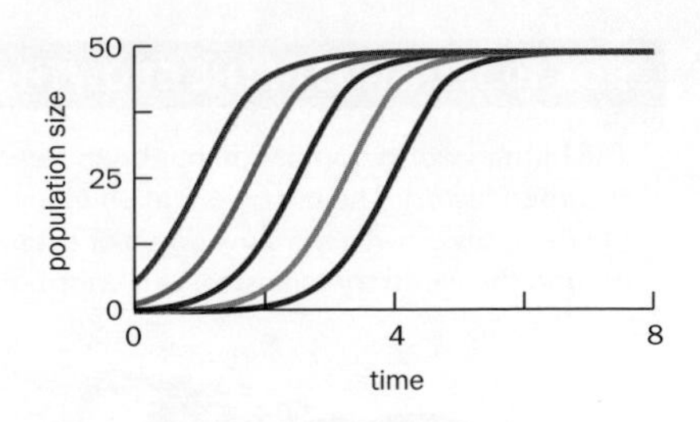

Figure 10.2 Shifted logistic functions. Different initial values shift the logistic function (10.3) in the horizontal direction, but eventually all curves reach the same final value, K. The red curve $N(0) = 0.2511$ corresponds to the process in Figure 10.1. Initial values for the curves, from left to right, are 20×0.2511; 5×0.2511; 0.2511; $0.2511/5$; $0.2511/20$.

If the growth process is represented by a differential equation, we know from earlier chapters how to proceed: the time derivative is set equal to zero and the resulting algebraic equation is solved. For instance, the steady states N_{ss} of the logistic function (10.3) are determined by solving the equation

$$0 = rN_{ss} - \frac{r}{K} N_{ss}^2. \tag{10.8}$$

The problem has two solutions. First, for $N_{ss} = 0$, the population is "empty" and remains that way. Second, for the case $N_{ss} \neq 0$, we are allowed to divide by r and by N_{ss}, which leads to $1 - N_{ss}/K = 0$, which gives $N_{ss} = K$ as the carrying capacity.

Is it possible that the population ever exceeds this value? Sure. It could be that individuals immigrate from other areas, thus creating a higher value. It might also happen that the conditions in the previous years were very favorable, thereby permitting the population to grow beyond its typical carrying capacity. In contrast to growth from a small size, which is S-shaped, the decrease in population size looks like an exponential function, which approaches the same steady state, K, as before (**Figure 10.3**).

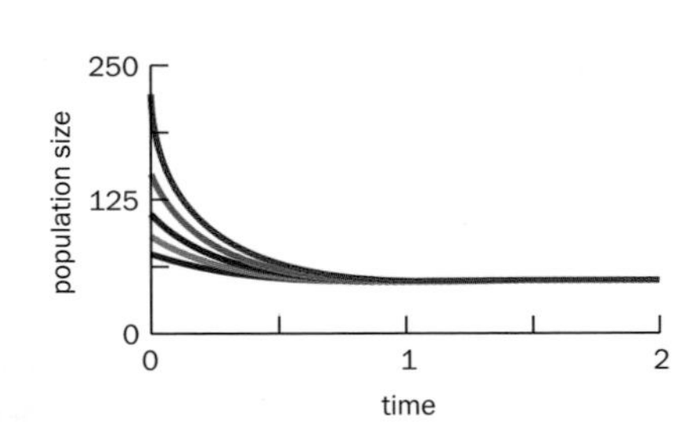

Figure 10.3 Logistic functions above their carrying capacity. If the logistic function is initialized above K, it decreases toward K in a monotone fashion that does not resemble the S-shaped growth from low initial values. Initial values, from bottom to top, are 75; 1.2×75; 1.5×75; 2×75; 3×75.

In nature, many organisms are subjected to predation, and bacteria are no exception. For instance, some ameba and other protozoans eat bacteria. As Richter [18] pointed out in the context of whale hunting, the dynamics of a species depends critically on the type of predation. In many models, predation is assumed to be proportional to the current number of individuals, while it occurs with a constant rate in another typical case. It is easy to explore the consequences. If predation is proportional to population size, we formulate the dynamics as

$$\dot{N}_1 = rN_1 - \frac{r}{K} N_1^2 - pN_1, \tag{10.9}$$

where we have added the index 1 merely for easy discussion and comparisons. In the case of constant predation, the formulation is

$$\dot{N}_2 = rN_2 - \frac{r}{K} N_2^2 - P. \tag{10.10}$$

Let us suppose that the common parameters in the two models are the same as before and that both populations start with an initial size of 50. When we more or less arbitrarily set the first predation parameter as $p = 0.75$ and select $P = 25$, the two populations have very similar dynamics, including roughly the same steady state (**Figure 10.4A**), and one might think that the type of predation does not really matter. However, the two populations exhibit distinctly different responses to perturbations. For example, if the two populations are at some point reduced to 15, maybe owing to other predators, disease, or some environmental stress, the first population recovers quite quickly, while the second population never recovers, and eventually dies out (Figure 10.4B). Stability analysis allows the computation of the perturbation threshold above which the second population recovers and below which it goes extinct.

In typical models of population growth, all parameters are constant, and the dynamics is therefore fully determined. In reality, the growth rate may depend on

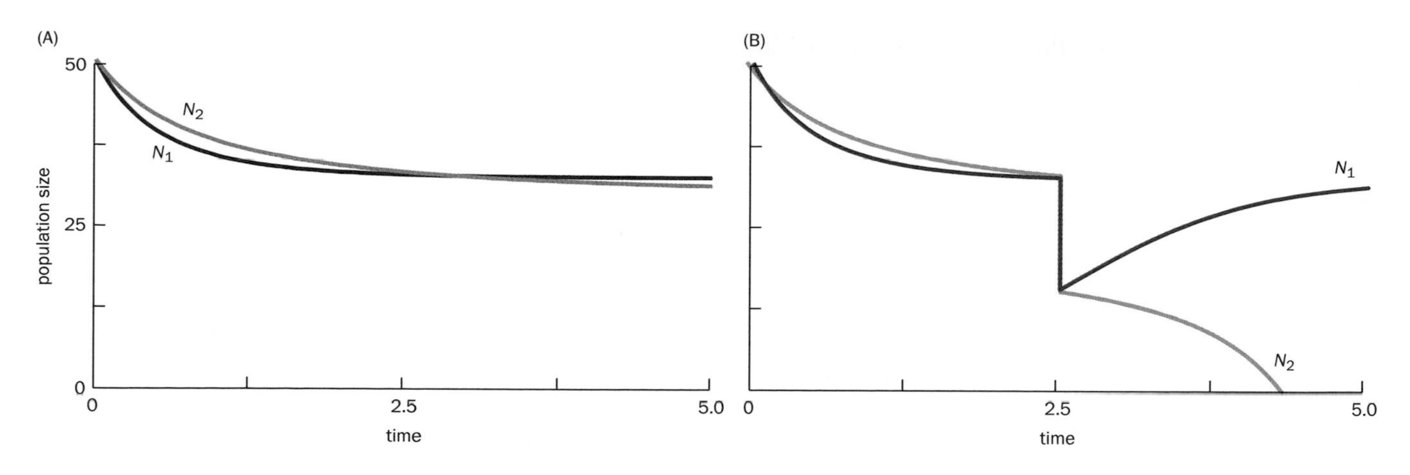

Figure 10.4 Comparison of two logistic growth processes exposed to different types of predation. Without perturbations (A), the two functions (10.9) and (10.10) are quite similar. However, if some external factor reduces both populations to a sufficiently small value, the population with proportional predation (N_1) recovers, while the population exposed to constant predation (N_2) becomes extinct. In (B), both populations are reduced to 15 at time 2.5.

environmental factors, such as temperature, which of course is a function of time, or it may depend on the current population size [18]. Furthermore, the parameters usually fluctuate stochastically within some range. For very large populations, such variations are not problematic, but they can become detrimental if a population is very small.

ANALYSIS OF SUBPOPULATIONS

While the growth of a **homogeneous** population is interesting in many respects, the range of likely dynamics is somewhat limited. More interesting are the dynamic behaviors of stratified populations containing different categories of individuals and the interactions between populations. An example of the dynamics of subpopulations was rather extensively discussed in Chapter 2, where models described how susceptible individuals become infected, recover, and possibly become susceptible again. Uncounted variations of these **SIR models** have appeared in the mathematical modeling literature, and we will not discuss them here again. A variation of such subpopulation models accounts for the different ages of individuals that are exposed to a disease vector. Specific questions in such a study ask whether a disease outbreak can be controlled and at what age it is most efficacious to vaccinate individuals. If many age classes, along with different health and treatment states, are considered, these models can consist of hundreds of differential equations [19]. Interestingly, the dynamics of the actual disease-causing agents, such as viruses, is often not explicitly accounted for in these models.

An interesting example of a simple, yet fairly realistic, growth model with age classes was proposed by Leslie [20] many decades ago. It is framed as a linear recursive model of the type we discussed in Chapter 4. The population is categorized into subpopulations of individuals with different ages and different proliferation rates. All individuals have the same potential lifespan, but they may of course die earlier. For humans with a maximum age of 120, one could select $m = 12$ age classes of $\tau = 10$ years each and set a corresponding recording interval of τ years. In other words, population sizes are exclusively documented only every τ years, but not in between.

The Leslie model is most easily formulated as a **matrix** equation, which for four age classes has the form

$$\begin{pmatrix} P_1 \\ P_2 \\ P_3 \\ P_4 \end{pmatrix}_{t+\tau} = \begin{pmatrix} \alpha_1 & \alpha_2 & \alpha_3 & \alpha_4 \\ \sigma_1 & 0 & 0 & 0 \\ 0 & \sigma_2 & 0 & 0 \\ 0 & 0 & \sigma_3 & 0 \end{pmatrix} \begin{pmatrix} P_1 \\ P_2 \\ P_3 \\ P_4 \end{pmatrix}_t . \tag{10.11}$$

Let us dissect this equation. The vector on the left contains $P_1, \ldots, P_4$ at time $t + \tau$. These quantities are the sizes of the subpopulations in age classes 1, 2, 3, 4, respectively. Almost the same vector appears on the far right, except that it captures the population sizes at time t. Thus, the recursive character of the equation is the same as in the simple case of exponential growth (Chapter 4), and writing (10.11) in vector and matrix notation yields

$$\mathbf{P}_{t+\tau} = \mathbf{L}\mathbf{P}_t. \tag{10.12}$$

Connecting the two vectors is the Leslie matrix **L**. It consists mainly of zeros, except for quantities α in the first row and quantities σ in the subdiagonal. The α's represent the reproduction rate for each age class, given as the number of offspring produced per individual in this age class within the time period τ. The parameters σ represent the fractions of individuals surviving from one age class into the next and have values between 0 and 1. Writing out the matrix equation shows these features more intuitively. For instance, the equation for P_2 is obtained by multiplying the elements in the second row of the matrix with the vector on the right-hand side, which yields

$$P_{2,t+\tau} = \sigma_1 P_{1,t} + 0 \cdot P_{2,t} + 0 \cdot P_{3,t} + 0 \cdot P_{4,t} = \sigma_1 P_{1,t}. \tag{10.13}$$

Expressed in words, age class 2 at time $t + \tau$ contains exactly those individuals that were in age class 1 at time t and survived time period τ. Age classes 3 and 4 have a similar structure. (Please write down the equations to convince yourself.) Now let us look at age class 1. There is no class 0, so that there cannot be survival σ_0 into class 1. Instead, all individuals in class 1 come exclusively from reproduction, and the rate of this process differs for the various age classes. For instance, individuals in class 3 produce offspring at a rate of α_3 per time period τ. These birth processes are reflected in the first equation, describing $P_{1,t+\tau}$, which is

$$P_{1,t+\tau} = \alpha_1 P_{1,t} + \alpha_2 P_{2,t} + \alpha_3 P_{3,t} + \alpha_4 P_{4,t}. \tag{10.14}$$

How does this equation account for the fact that only a certain percentage (quantified by σ_1) of individuals of class 1 survive? Only indirectly. First, according to the set-up of the model, no individual can stay in class 1 (although this class can have offspring that again would be in the same class). Furthermore, those surviving ($\sigma_1 P_{1,t}$ individuals) show up in class 2, and those dying (namely, $(1 - \sigma_1)P_{1,t}$ individuals) are no longer explicitly accounted for; they disappear from the population and from the model.

Because the process is so rigidly structured, we can write the Leslie equation for any numbers of age classes:

$$\begin{pmatrix} P_1 \\ P_2 \\ \vdots \\ P_m \end{pmatrix}_{t+\tau} = \begin{pmatrix} \alpha_1 & \alpha_2 & \cdots & \alpha_{m-1} & \alpha_m \\ \sigma_1 & 0 & \cdots & 0 & 0 \\ 0 & \sigma_2 & \cdots & 0 & 0 \\ \vdots & \vdots & \ddots & \vdots & \vdots \\ 0 & 0 & \cdots & \sigma_{m-1} & 0 \end{pmatrix} \begin{pmatrix} P_1 \\ P_2 \\ \vdots \\ P_m \end{pmatrix}_t. \tag{10.15}$$

This **recursion** again moves forward by τ time units at a time. It is also possible to jump $n\tau$ steps at once. Namely, we simply use the nth power of the matrix:

$$\mathbf{P}_{t+n\tau} = \mathbf{L}^n \mathbf{P}_t. \tag{10.16}$$

Leslie models of this type are interesting tools for exploring trends in subpopulation, even though they make many explicit and implicit assumptions (for examples, see **Box 10.2**). For instance, there is no mention of population sizes at time points inside the intervals, that is, between $t + k\tau$ and $t + (k + 1)\tau$ for $k = 0, 1, 2, \ldots$. For simplicity, males are usually not included, because they do not bear children. Importantly, the growth process is independent of the current population size. In other words, the parameters α and σ do not depend on time or on the population size—which is at odds with many real populations. As a consequence, the population either always keeps growing or always keeps decreasing, but it almost never reaches a stable size that is not zero (for an exception, see Exercise 10.11). Of course, it is mathematically possible to make α and σ dependent on time or population size, but then the model is no longer linear, and certain computations become mathematically much more difficult. For pure simulation purposes, these extensions are easy to implement and explore.

The Leslie model uses average survival and reproduction rates and therefore is **deterministic**. As a consequence, every simulation with the same parameters and initial sizes of subpopulations yields exactly the same results. In a variation that is conceptually simple but not trivial in its implementation and analysis, one could consider randomness in survival and reproduction [12, 13]. For instance,

BOX 10.2: TYPES OF AGE STRUCTURES WITHIN POPULATIONS

In the Leslie model, the birth rates within a population typically differ by the age of the mother but they are constant over time within each age class. Similarly, death rates are usually higher for older individuals, but remain constant over time. In reality, neither birth nor death rates are constant. As an illustration, *National Geographic* (January 2011) contrasted typical population pyramids resulting from trends in birth and death rates. Representative pyramids are shown in **Figure 1**. China's actual population structure is shown in **Figure 2**. It is a mixture of the pyramids in Figure 1.

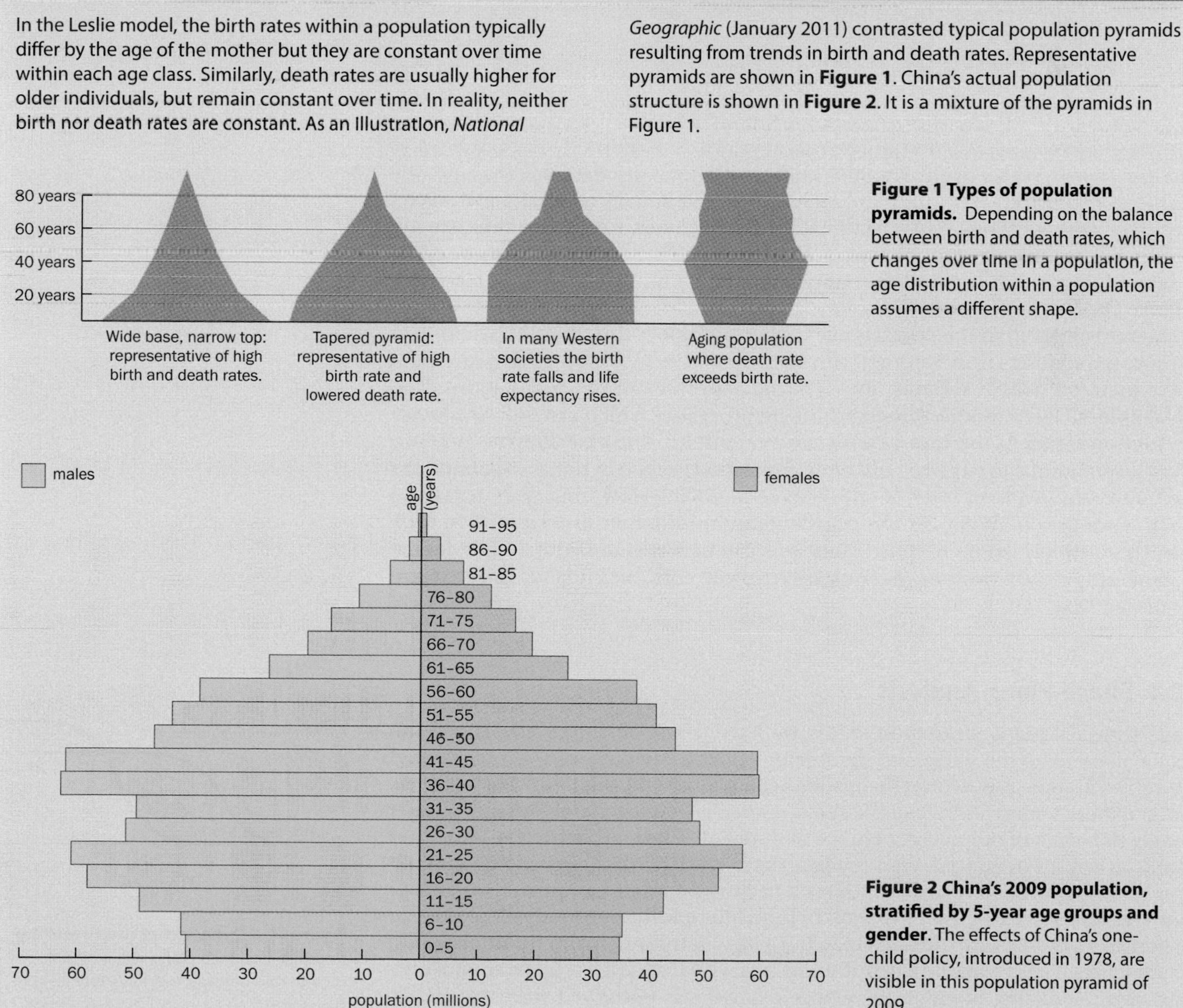

Figure 1 Types of population pyramids. Depending on the balance between birth and death rates, which changes over time in a population, the age distribution within a population assumes a different shape.

Figure 2 China's 2009 population, stratified by 5-year age groups and gender. The effects of China's one-child policy, introduced in 1978, are visible in this population pyramid of 2009.

one might imagine cells that are moving through the **cell cycle** and enter its different phases in a probabilistic fashion. Typical quantities to be assessed would include the numbers of cells in each phase and the distribution of transition times throughout the cell cycle. This seemingly harmless change from deterministic to **random** events moves the model into the realm of stochastic processes, which are much more complicated than the deterministic Leslie model. Good texts include [21–23].

INTERACTING POPULATIONS

10.3 General Modeling Strategy

In the simplest case of an interaction system, the individuals of two populations do not influence each other much, except that they might compete for the same resources. The simplest example may be a mixture of two exponentially growing bacterial populations with different growth rates. A more interesting example is a tumor growing within the tissue where it originated. Often the growth rate of tumor cells is not necessarily faster than that of healthy cells, but tumor cells undergo cell death at a much lower rate. We can easily model this situation with a system of healthy (H) and tumor (T) cells of the form

$$\begin{aligned} \dot{H} &= r_H H - m_H H^2, \\ \dot{T} &= r_T T - m_T T^2. \end{aligned} \tag{10.17}$$

The parameters r_H, r_T, m_H, and m_T represent growth and mortality rates of healthy and tumor cells, respectively. Suppose both growth rates are the same ($r_H = r_T = 1$), the mortality rates are $m_H = 0.01$ and $m_T = 0.0075$, and initially only 0.1% of the cells are tumor cells ($H(0) = 1$, $T(0) = 0.001$). With these settings, it is easy to simulate the system. Even though the growth rates are the same and the number of tumor cells is at first miniscule, the lowered mortality allows the tumor eventually to take over (**Figure 10.5**).

More realistic than the coexistence in this example is the situation where two or more populations truly interact with each other. The most straightforward case is competition, which reduces the carrying capacities of the populations. The usual approach is a model of logistic growth processes, which are expanded with interaction terms. If the interactions are very strong, one can imagine that not every population can survive. The typical questions asked in these situations are therefore: Who will survive? Is coexistence possible? How long does it take to reach equilibrium? What do the transients from initiation to termination look like? Interestingly, some of these questions can be answered with a high degree of generality if only two populations are involved. The methods for this task are discussed next.

10.4 Phase-Plane Analysis

Every time we run a simulation study, we have to ask ourselves how much the results depend on the particular parameter values we had chosen. If we changed the values in some way, would the results be fundamentally different? For instance, if population A survives in a simulated scenario, is that a general outcome or is it merely the result of our current choice of initial numbers, growth rates, and other characteristics? For complicated models, these types of questions can be difficult to answer. However, if we are dealing with just two populations, qualitative **phase-plane analysis** is a very effective tool. In principle, **qualitative analysis** could also be done for systems with more species, but a key contributor to our intuition would no longer be of help, namely a two-dimensional visualization. So, let us suppose we have two interacting populations whose dynamics is described with two ODEs. These equations consist of logistic growth functions of the type in (10.3) and (10.4), which furthermore contain one interaction term each. These interaction terms are typically of the same form but with different parameter values, because the two

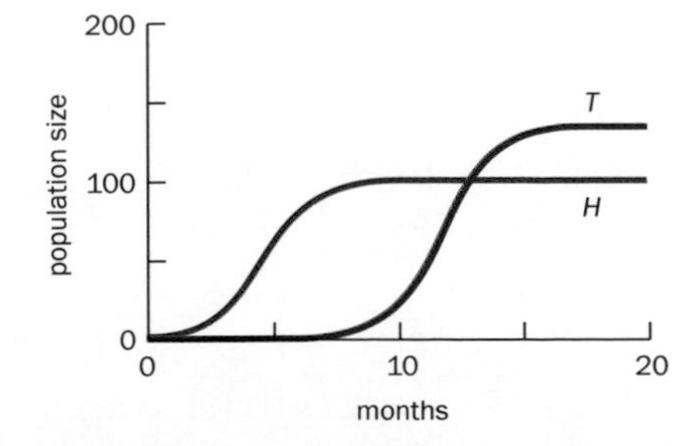

Figure 10.5 Dynamics of two coexisting populations. In the long run, a higher growth rate or lower mortality rate is more influential than a larger initial population size, as demonstrated here with a simplified model of healthy (H) and tumor (T) cells growing in the same tissue (see (10.17)).

species respond differently to the effect of the interaction. Specifically, we use the following pair of equations as an example:

$$\dot{N}_1 = r_1 N_1 \left(\frac{K_1 - N_1 - aN_2}{K_1} \right),$$
$$\dot{N}_2 = r_2 N_2 \left(\frac{K_2 - N_2 - bN_1}{K_2} \right). \tag{10.18}$$

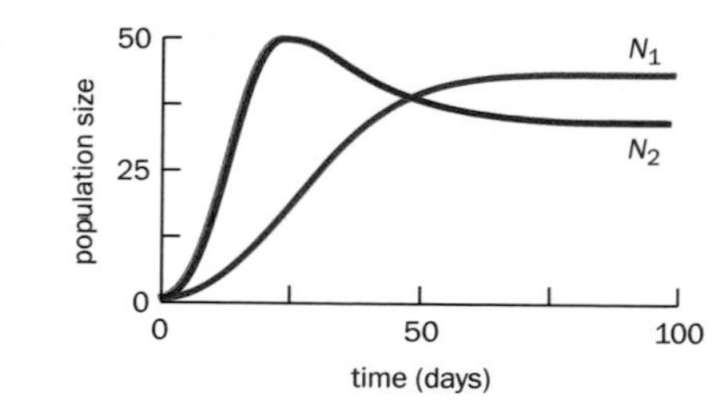

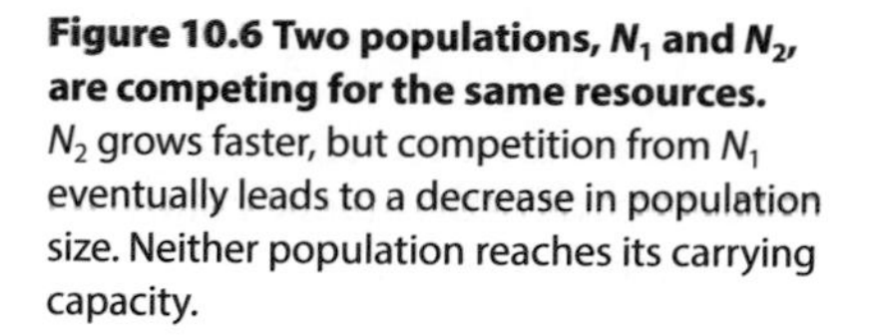
Figure 10.6 Two populations, N_1 and N_2, are competing for the same resources. N_2 grows faster, but competition from N_1 eventually leads to a decrease in population size. Neither population reaches its carrying capacity.

In order to facilitate both mathematical and computational analyses, we select the following, more or less arbitrary, parameter values: $r_1 = 0.15$, $K_1 = 50$, $a = 0.2$, $r_2 = 0.3$, $K_2 = 60$, $b = 0.6$, $N_1(0) = 1$, $N_2(0) = 1$.

It is easy to solve this system numerically for N_1 and N_2 as functions of time; a plot of the solution is shown in **Figure 10.6**. Some details are probably expected, others may be surprising. First, the dynamics of N_1 is similar to that of a logistic function that is not affected by outside forces. N_2 grows faster, owing to its larger growth rate, but begins to fade as population N_1 grows. Owing to the competition, neither population reaches its carrying capacity. In particular, N_2 only reaches a final size of about 34, while its carrying capacity is 60.

The key concept of the following type of qualitative analysis is that we do not take explicit account of time and instead plot the behavior of one population against the behavior of the other. Thus, we look at a coordinate system where the axes represent N_1 and N_2 (**Figure 10.7**). If we kept a record of N_1 and N_2 over time, the dataset would have the form shown in **Table 10.1**. We can easily see from Figure 10.7 that time is represented in a drastically different way than before, namely merely as labels along a curve that describes how N_2 depends on N_1. In fact, we could double the speed of the system, by multiplying all rate constants in (10.18) by 2, and the plot would look exactly the same, except that the time labels would be different. Expressed differently, time has moved to the background, and the focus is on the relationship between N_1 and N_2. Even though time no longer appears explicitly, this phase plot indicates the speed of the processes indirectly through the distance between consecutive time points. In Figure 10.7, the speed is greatest between $t = 10$ and $t = 20$, and the process almost comes to a halt beyond $t = 60$.

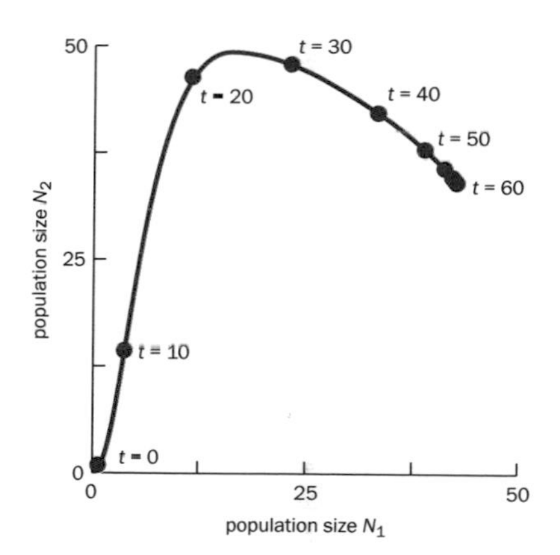

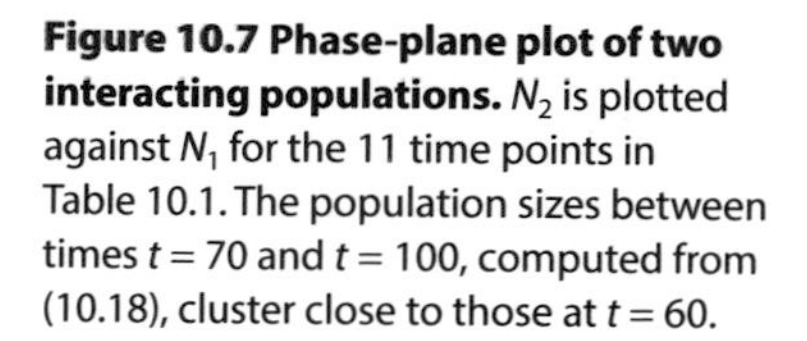
Figure 10.7 Phase-plane plot of two interacting populations. N_2 is plotted against N_1 for the 11 time points in Table 10.1. The population sizes between times $t = 70$ and $t = 100$, computed from (10.18), cluster close to those at $t = 60$.

Intriguingly, we can analyze this relationship generically with minimal information about the parameter values of the system. We begin with the steady states of the system. These are determined by setting both time derivatives in (10.18) equal to zero and solving for N_2 as a function of N_1 (or the other way around). The specific form of the equations makes this very easy. First, we can convince ourselves rather easily that $(N_1, N_2) = (0, 0)$ is a steady state. This makes sense, because if no individuals exist, there is no dynamics. This case is important to remember, but let us assume for the following that N_1 and N_2 are strictly positive. If we divide the two

TABLE 10.1: SIZES OF TWO COMPETING POPULATIONS, N_1 AND N_2, OVER TIME

Time	N_1	N_2
0	1	1
10	4.06	14.40
20	11.96	46.38
30	23.58	47.96
40	33.76	42.27
50	39.34	38.04
60	41.70	35.81
70	42.61	34.80
80	42.96	34.37
90	43.10	34.20
100	43.15	34.13

equations by N_1 and N_2, respectively, which we are allowed to do because we assume that these variables are not zero, we obtain two simpler equations in two unknowns, which we can solve directly:

$$\begin{aligned}0 &= r_1 N_1\left(\frac{K_1 - N_1 - aN_2}{K_1}\right) \Rightarrow N_2 = \frac{K_1 - N_1}{a},\\ 0 &= r_2 N_2\left(\frac{K_2 - N_2 - bN_1}{K_2}\right) \Rightarrow N_2 = K_2 - bN_1.\end{aligned} \tag{10.19}$$

These equations are valid for any parameter values. Using the numerical values from before, the nonzero steady-state solution is $N_1 = 43.18$, $N_2 = 34.09$. The system contains two further steady states, where either N_1 or N_2 is zero and the other variable is not.

In addition to the steady states, we can explore situations where only one of the time derivatives is zero. In fact, these situations are modeled by the two equations in (10.19) when they are considered separately. In the first case, the equation given by $\dot{N}_1 = 0$ is called the **N_1-nullcline**, where "null" refers to zero and "cline" comes from the Greek root for slope. In general, this function is nonlinear, but the N_1-nullcline in our example is represented by a linear function, namely $N_2 = (K_1 - N_1)/a$. It has a negative slope, and intersects the vertical axis at K_1/a, and the horizontal axis at K_1. Similarly, the N_2-nullcline is determined from setting the time derivative of N_2 equal to zero, and the result is $N_2 = K_2 - bN_1$, which is again a straight line with negative slope. It intersects the vertical axis at K_2, and the horizontal axis at K_2/b. The two nullclines intersect at the nonzero steady state, because at this point both time derivatives are zero. **Figure 10.8**, which is intentionally not drawn to scale to indicate the generic nature of the plot, shows the two nullclines and the two steady-state points.

Even without specifying parameter values, we can find out more about the system by simply looking at the phase plot with two nullclines. For instance, for every point exactly on the N_1-nullcline, the derivative of N_1 is by definition zero. Therefore, any **trajectory** of the ODE system must cross this nullcline vertically, because the value of N_1 does not change. Furthermore, $\dot{N}_1$ is less than zero on the right side of the nullcline and greater than zero on the left side. We can confirm this by moving any point on the nullcline a tiny bit to the right: the value of N_1 increases very slightly, which makes the right-hand side of the first differential equation slightly negative. Thus, all derivatives $\dot{N}_1$ on the right of the N_1-nullcline are negative, and all those on the left are positive. Analogous arguments hold for the N_2-nullcline, where all trajectories must cross horizontally. Taking these pieces of information together, the plot thus consists of four areas that are characterized by the signs of the two derivatives (see Figure 10.8). Every

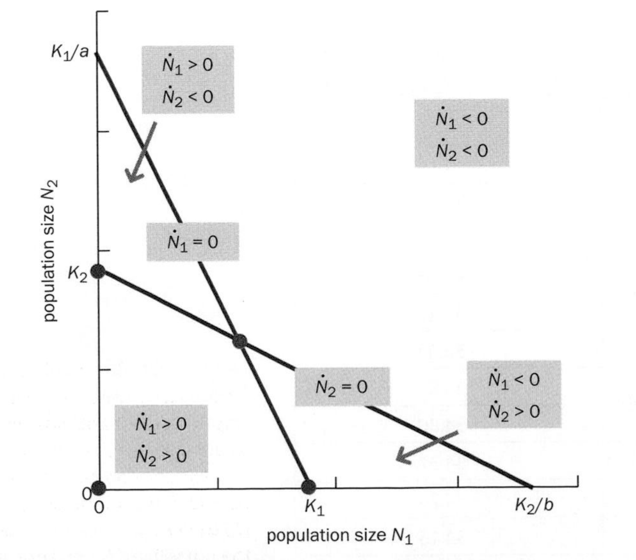

Figure 10.8 Phase-plane plot for two competing populations N_1 and N_2. The nullclines (red and blue lines) intersect at the nonzero steady state (yellow circle). At other steady states (green circles), N_1, N_2, or both populations are zero.

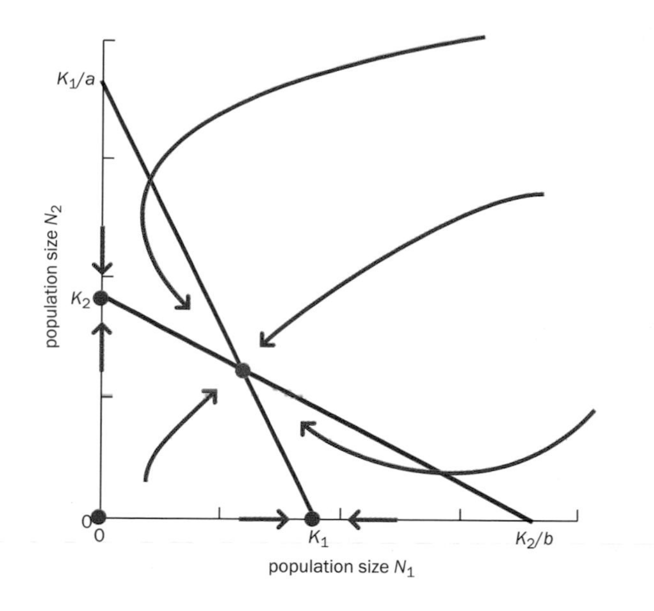

Figure 10.9 Coarse visualization of the dynamics of two competing populations. Sketching "generic" trajectories in the phase-plane plot shows that the two competing populations N_1 and N_2 approach the nonzero steady state (yellow circle), where both survive and coexist. Trivial mathematical exceptions are cases where N_1, N_2, or both are zero at the beginning.

trajectory that satisfies the pair of differential equations (10.18) must move according to these derivatives. This allows us to sketch all possible trajectories on the phase plane. We do not know exactly what these trajectories look like, but we do know their most significant features, namely, the direction in which they move and whether they cross a nullcline in a vertical or horizontal direction (**Figure 10.9**). Two special cases arise if one of the two variables is zero (see Exercises 10.19 and 10.20). Finally, the phase plot indicates that all trajectories move away from (0, 0): this trivial steady state is unstable.

In the example we have just discussed, the two nullclines intersect at a single point, the nontrivial steady state. It may also happen that they do not intersect in the positive quadrant, which is the only area of interest. For instance, suppose $K_1 > K_2/b$ and $K_1/a > K_2$. The situation is shown in **Figure 10.10**. Now there are only three areas, with different combinations of signs of the two derivatives. Exactly the same arguments hold as before, but now all trajectories converge to $N_1 = K_1$, $N_2 = 0$, with the exception of trajectories that start exactly on the vertical axis (that is, $N_1(0) = 0$). The interpretation is that, unless there are no individuals in population N_1 at all, N_1 will eventually survive and grow to its carrying capacity, while N_2 will become extinct. Examples are invasive species, which start with a few organisms but, if unchecked, lead to the extinction of the corresponding native species. The slopes and intersections of the two nullclines permit two additional situations, which are left for Exercises 10.21 and 10.22.

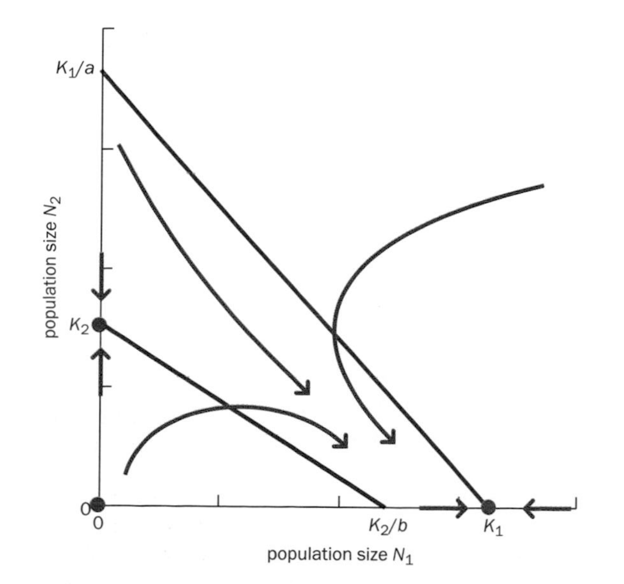

Figure 10.10 Depending on the parameter values, the two populations cannot coexist and one will eventually go extinct. If the parameter values are such that the nullclines do not intersect in the positive quadrant, one of the two populations (here N_1) survives at its carrying capacity, whereas the other (here N_2) disappears.

Arguably the most famous class of examples where this type of analysis has been used concerns predator–prey systems. These systems were first discussed with mathematical rigor almost one hundred years ago by Alfred Lotka, a physical chemist and biologist, and Vito Volterra, a mathematician. Lotka used this approach originally for the analysis of chemical reactions, but it is better remembered for population studies in ecology (see also Chapter 4). In its simplest form, the Lotka–Volterra (LV) model describes the interactions between one prey species N and a predatory species P. The equations are nonlinear and quite similar to those we discussed above:

$$\begin{aligned} \dot{N} &= \alpha N - \beta NP, \\ \dot{P} &= \gamma NP - \delta P. \end{aligned} \tag{10.20}$$

All parameters are positive. In the absence of predators, N grows exponentially with growth rate α, whereas the presence of predators limits this growth through the term βNP. The form of a product of the two variables with a rate constant is based on arguments similar to those discussed in the context of chemical mass action kinetics (Chapter 8), where the product is used to describe the probability of two (chemical) species encountering each other in a three-dimensional space. The predator population P is assumed to rely exclusively on N for food; without N, the predator population decreases exponentially toward extinction. Again, the growth term is formulated as a product, but the rate γ is usually different from the rate β in the prey equation, because the magnitude of the effect of the same process (predation) is significantly different for prey and predators.

With typical parameter values ($\alpha = 1.2$, $\beta = 0.2$, $\gamma = 0.05$, $\delta = 0.3$, $N_0 = 10$, $P_0 = 1$), N and P exhibit ongoing oscillations, in which P follows N with some delay and typically with a much smaller magnitude (**Figure 10.11**).

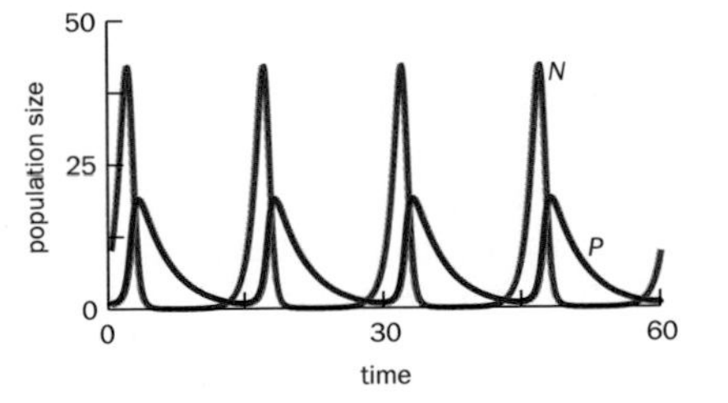

Figure 10.11 Idealized dynamics of predator (*P*) and prey (*N*) populations. The dynamics of the Lotka–Volterra system (10.20) consists of ongoing oscillations. In reality, such oscillations are only seen when the two species coexist in a complex environment.

Linearization and stability analysis demonstrates that (0, 0) is a saddle point and that the interior, nontrivial steady state is a center (see Chapter 4). Moreover, one can show that the general solution outside (0, 0) consists of closed orbits around the interior steady-state point, which correspond to sustained oscillations with the same amplitude [24, 25]. However, these oscillations are not limit cycles, and every perturbation, even if it is very small, moves the system to a different orbit or causes a species to die.

Many important features of a phase-plane analysis remain the same if the nullclines are nonlinear. In particular, the slope of a variable right on its nullcline is zero, so that trajectories again transect nullclines either horizontally or vertically. Also, the intersections of nullclines are steady-state points. The main difference is that the nullclines may intersect more than once. Thus, the system may have two interior intersection points, where neither of the variables is zero, one of which is typically stable whereas the other is unstable. Of course, there could be more than two intersections, depending on the model structure.

As an example, let us consider the system

$$\begin{aligned} \dot{N}_1 &= N_1\,(10 + 1.5N_1 - 0.1N_1^2 - N_2)/150, \\ \dot{N}_2 &= N_2(50 + 0.3N_1^2 - 7N_1 - N_2)/150, \end{aligned} \tag{10.21}$$

where each population is augmented and reduced through two processes each. The nullclines are easily computed by setting the first or the second equation equal to zero and expressing N_2 as a function of N_1. They are shown in **Figure 10.12A**. As before, the axes are additional nullclines, and the system allows for the extinction of either N_1 or N_2. Also, (0, 0) is a steady-state point, and we will see that it is unstable. To explore the system behavior within the different areas created by the nullclines, it is useful to compute the vector field of the system. This representation consists of solving the equations many times for a very short time period, each time starting with initial values that are taken from a grid of relevant (N_1, N_2) pairs. Such a representation is shown in Figure 10.12B. Relatively long lines correspond to fast changes, whereas lines that almost look like dots represent very slow changes; rationalize why this is so. True dots are steady-state points. A sample of trajectories is shown in Figure 10.12C; the arrows have been added to indicate the direction of each

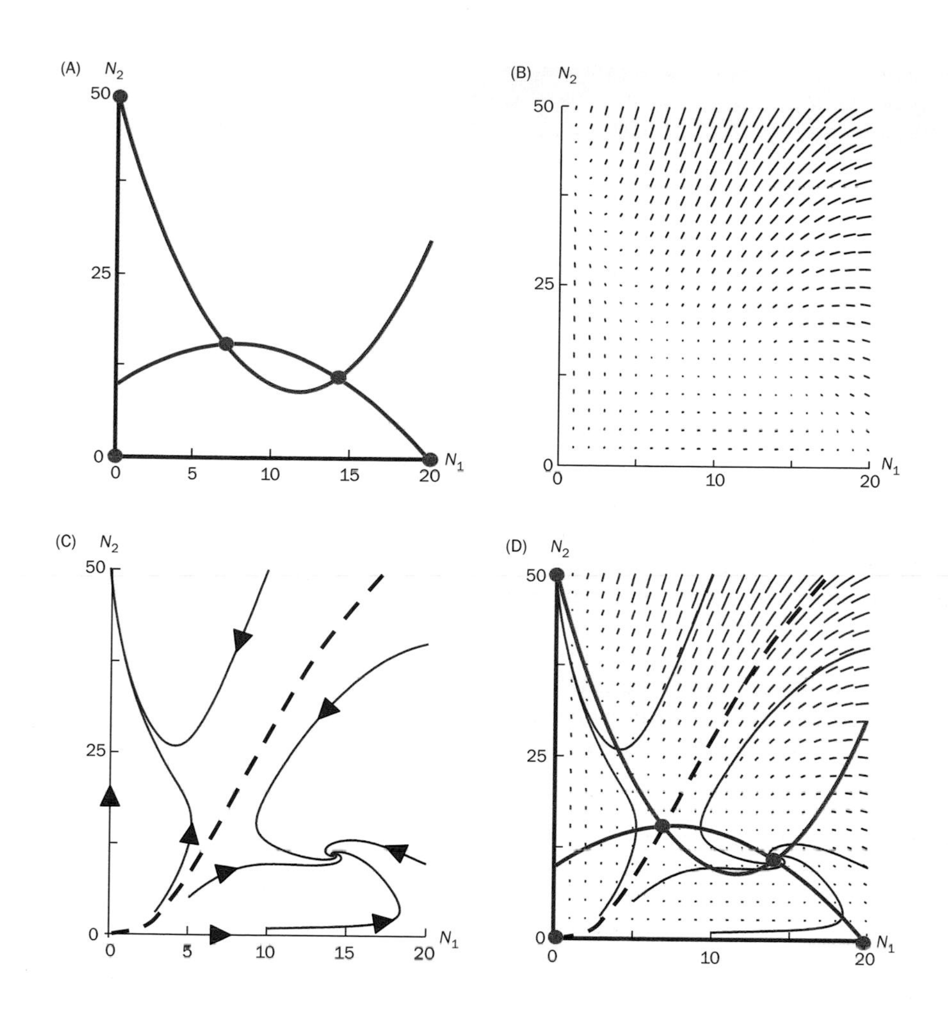

Figure 10.12 Phase-plane analysis of the system in (10.21). In contrast to the earlier systems, the nullclines in this case are nonlinear, thereby allowing two interior steady-state points. (A) shows the nullclines (including the axes) and the steady-state points (yellow). (B) depicts the vector field of the system, which summarizes the flow of all trajectories. A sample of trajectories is shown in (C), together with a separatrix (dashed), which separates trajectories ending either at (0, 50) or at (14.215, 11.115). (D) superimposes (A), (B), and (C) for a complete qualitative picture. Note that all trajectories transect the nullclines either in horizontal or vertical direction. The steady state at (7.035, 15.60) is a saddle point. Moving exactly on the separatrix toward this point would end in this point. However, deviating even slightly to the left or right of the separatrix pushes any trajectory further away.

trajectory. The dashed line is called the separatrix, which is Latin for "she who separates." Why this line is allegedly female may be pondered, but is not important here. What is important is that this line divides the phase plane into sections with different behaviors: Starting to the left of the separatrix, the system will eventually approach the steady state where N_1 becomes extinct and N_2 is 50. Starting to the right, the system will spiral into the coexistence state $(N_1, N_2) \approx (14.215, 11.115)$. Not surprisingly, the separatrix runs through the saddle-point steady state at (7.035, 15.60) and into the unstable point (0, 0). Of note is that every perturbation away from (0, 0), no matter how slight, can lead to the extinction of either N_1 or of N_2 or to coexistence. This observation highlights the existential risks faced by small populations.

10.5 More Complex Models of Population Dynamics

Obviously, all these models are very much simplified and sometimes distant from reality. For instance, predators very rarely rely on a single prey species and are instead participants in complex food webs where many predator species feed on many prey species. Other complications in population modeling result from complex dependences, such as the pollination of plants by specific insect species. Another issue of great import is the dynamics for small values. In the differential equation system, even 0.1 individuals of prey may recover and later provide food for the predators. In reality, the species would die out. Furthermore, all systems in nature are subject to numerous environmental influences, some of which can greatly stabilize or destabilize the system. For instance, refuges for prey can drastically change the simple dynamics exhibited in Figure 10.11. Finally, as soon as the spatial expansion of a population becomes important, as is prominent in the spread of **epidemics**, one has to resort to entirely different models, such as partial differential systems or agent-based models (Chapter 15).

TABLE 10.2: PARAMETER VALUES FOR THE LOTKA–VOLTERRA SYSTEM (10.22) PLOTTED IN FIGURE 10.12

a_1	1	b_{11}	−1	b_{21}	0	b_{31}	−2.33	b_{41}	−1.21
a_2	0.72	b_{12}	−1.09	b_{22}	−1	b_{32}	0	b_{42}	−0.51
a_3	1.53	b_{13}	−1.52	b_{23}	−0.44	b_{33}	−1	b_{43}	−0.35
a_4	1.27	b_{14}	0	b_{24}	−1.36	b_{34}	−0.47	b_{44}	−1

In spite of the numerous immediate and more indirect caveats, LV models have found much application in ecology and many other areas. In higher dimensions, a typical formulation for n variables is

$$\dot{X}_i = a_i X_i \times \left(1 + \sum_{j=1}^{n} b_{ij} X_j\right). \tag{10.22}$$

As in the lower-dimensional case of a predator–prey system, each right-hand side of this system of ODEs consists of a linear term and a sum of binary (second-degree) terms, which describe interactions between two species ($i \neq j$) or the crowding term within a species ($i = j$) that we encountered several times in logistic growth functions and competition models. Because of the linear structure inside the parentheses, this model allows steady-state analyses and parameter estimation for arbitrarily large systems (Chapters 4 and 5). It is even possible to include environmental factors in these types of models [26].

Higher-dimensional LV systems may have very interesting mathematical features. As a case in point, they can exhibit deterministic chaos. (see Chapter 4). As a reminder, this term indicates that the system is entirely deterministic and no randomness is included in the model structure. At the same time, the system oscillates in complicated patterns that are not predictable without the model and never repeat themselves. As a numerical example, consider a four-variable LV system of the type (10.22) with the parameters in **Table 10.2**, which was proposed by Sprott and colleagues [27, 28] and exhibits extremely complicated, **chaotic** dynamics (**Figure 10.13**); all four initial values were chosen as 1.

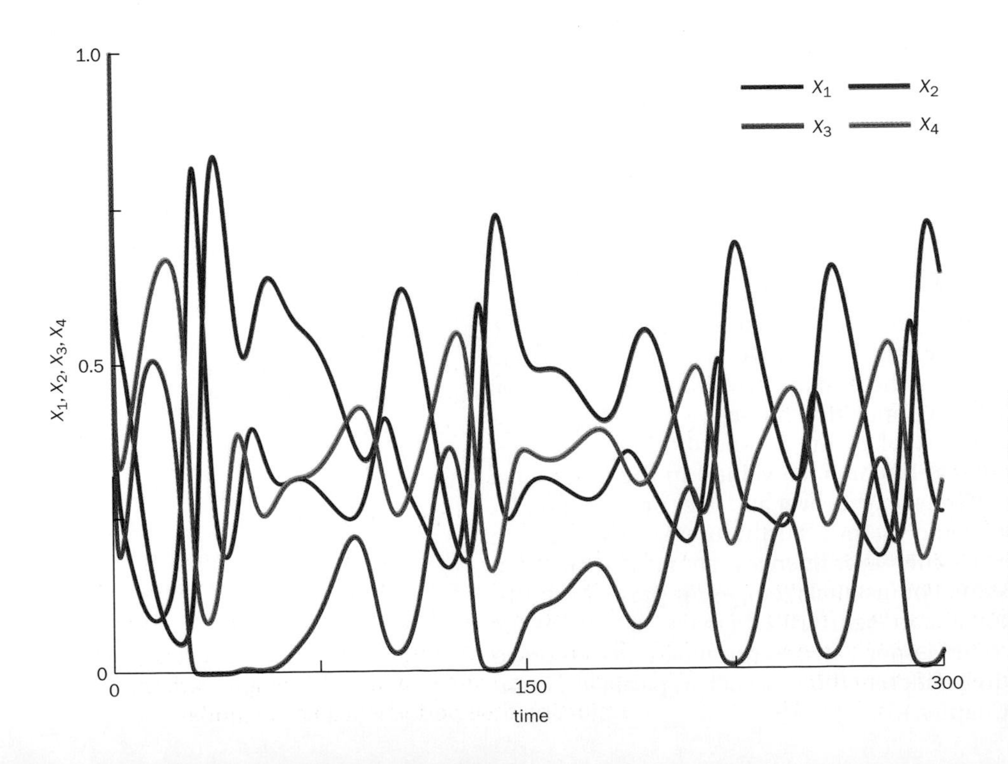

Figure 10.13 Even small dynamic systems may exhibit deterministic chaos. The form of a Lotka–Volterra model (10.21) is deceptively simple. In reality, such a system may exhibit very complicated dynamics, such as the chaotic behavior of the four-variable system shown here. (Adapted from Sprott JC, Vano JA, Wildenberg JC, et al. *Phys. Lett. A* 335 [2005] 207–212, with permission from Elsevier, and from Vano JA, Wildenberg JC, Anderson MB, et al. *Nonlinearity* 19 [2006] 2391–2404, with permission from IOP Publishing.)

While higher numbers of variables often make systems more complicated, other factors can contribute to the complexity of a model. For instance, some or all parameters could be made time-dependent. As so often, there is hardly any limit to the range of possibilities. A rather recent topic of strong interest is the characterization and analysis of metapopulations, consisting of hundreds or thousands of microbial species that live in the same space and simultaneously compete for resources and support each other through an intricate division of labor. We will discuss these complicated population systems in Chapter 15.

EXERCISES

10.1. Fit the data from **Table 10.3** (taken from Figure 10.1) with Richards and Gompertz functions, as well as with the generalized growth equation proposed by Savageau [17], which has the form

$$\dot{N}_S = \alpha N_S^g - \beta N_S^h,$$

with positive parameter values. Use fitting methods discussed in Chapter 5.

TABLE 10.3: GROWTH OF A BACTERIAL POPULATION*

Age of colony (days)	Colony size (area in cm^2)
0	0.24
1	2.78
2	13.53
3	36.3
4	47.5
5	49.4

*Data from Lotka A. Elements of Physical Biology. Williams & Wilkins, 1924.

10.2. Derive the ODE format of the logistic growth law (10.3) from the explicit form in Figure 10.1, including the appropriate initial value. Compute the derivative of the explicit function and express it in terms of N and suitable parameters.

10.3. For small ν, the Richards growth function can be represented approximately as the ODE

$$\dot{N}_R = \alpha N_R \ln\left(\frac{K}{N_R}\right)\nu.$$

Assess the quality of this approximation for different parameter values.

10.4. It is sometimes said that the world population is growing "super-exponentially." Use **Table 10.4**, which was assembled from US Census Bureau datasets, to analyze this claim. Is it possible to predict from these data what the carrying capacity of the human population on Earth might be? Try to use methods from Chapter 5 to represent the data with one of the growth models described in this chapter.

TABLE 10.4: GROWTH OF THE HUMAN WORLD POPULATION

Year	World population *P*	log *P*
1000	25,400,000	7.404833717
1100	30,100,000	7.478566496
1200	36,000,000	7.556302501
1300	36,000,000	7.556302501
1400	35,000,000	7.544068044
1500	42,500,000	7.62838893
1600	54,500,000	7.736396502
1700	60,000,000	7.77815125
1800	81,300,000	7.910090546
1900	155,000,000	8.190331698
1910	175,000,000	8.243038049
1920	186,000,000	8.269512944
1930	207,000,000	8.315970345
1940	230,000,000	8.361727836
1950	240,000,000	8.380211242
1960	3,042,389,609	9.483214829
1970	3,712,813,618	9.569703148
1980	4,452,686,744	9.648622143
1990	5,288,828,246	9.723359464
2000	6,088,683,554	9.784523403
2010	6,853,019,414	9.835881962
2020	7,597,238,738	9.880655774
2030	8,259,167,105	9.916936253

10.5. Use simulations or stability analysis to find the perturbation threshold above which the population with constant predation, (10.10), recovers and below which it goes extinct. Can you find a similar threshold for the population with proportional predation, (10.9)? Argue whether this is possible or not and/or execute simulations.

10.6. Provide arguments and/or simulation results for why small populations are more strongly affected by (random) perturbations than large populations.

10.7. Discuss why it is important to use age classes of the same length and to match this length with the reporting time period τ in a Leslie model.

10.8. For $n = 2$, $n = 3$, and $n = 5$, show that the matrix equation (10.16) generates the same numbers as using (10.15) twice, three times, or five times, respectively. Think before you execute the matrix equation five times.

10.9. Construct a Leslie model where the population size is constant but not zero. Execute simulations to show how robust this model is to changes in one of the age classes or in one of the parameters.

10.10. Select values for $\alpha_1, \ldots, \alpha_4$ and $\sigma_1, \ldots, \sigma_3$ and compute the total size of the population as a function of time.

10.11. Construct cases where the Leslie model shows either growth or decline. Construct a case where the population stays exactly the same over time. Is there only one solution to this problem?

10.12. In the Leslie model, make α and/or σ dependent on time. Perform simulations and study the consequences.

10.13. In the Leslie model, make α and/or σ dependent on population size. Perform simulations and study the consequences.

10.14. Population experts claim that delaying the first birth by a few years would have a significant impact on the human population growth explosion in developing countries. Set up a Leslie model to test this claim. State which assumptions (realistic or unrealistic) you made.

10.15. The roly poly or pillbug, *Armadillidium vulgare* (**Figure 10.14**), is a terrestrial crustacean in the grasslands of California. It can live up to 4 or sometimes even 5 years. The world is a dangerous place for the roly poly: of roughly 10,000 eggs hatched in a year, only about 11% survive to enjoy their first birthday. Of these, about 48.5% are female, of which only about 23% breed. As a consequence, only about 126 breeding females enter the second year to produce offspring. This number continues to decrease, as indicated in **Table 10.5**, which was compiled from data in [29]. Fortunately for the species, the breeding females lay between 25 and 150 eggs per year.

Use the information in Table 10.5 to construct a Leslie model. Study whether the population will increase, decrease, or remain constant with these settings. Explore what happens if the survival rate of eggs is 8% or 15%. Execute simulations to determine whether the order matters mathematically and/or biologically in which 10 years of 8% and 10 years of 20% survival occur? Discuss and interpret your simulation results.

Figure 10.14 Roly poly. A typical population of the Californian roly poly or pillbug, *Armadillidium vulgare*, exhibits complex birth and death patterns. (Courtesy of Franco Folini under the Creative Commons Attribution–Share Alike 3.0 Unported license.)

TABLE 10.5: LIFE STATISTICS OF THE ROLY POLY

Age class (years)	Number of eggs or breeding females*	Number of eggs produced
0	10,256	0
1	126	3,470
2	98	4,933
3	22	1,615
4	2	238

*The entry for age class 0 (<1 year old) represents the total number of eggs. Entries for later age classes are breeding females.

10.16. Make predictions regarding the dynamics of two coexisting bacterial populations that are growing exponentially with different growth rates. To what degree do the initial population sizes matter?

10.17. What happens if the mortality rate of tumor cells in (10.17) is 0? Predict the outcome and confirm or refute the predictions with simulations. Adjust other model parameters if the results appear to be unrealistic. Summarize your findings.

10.18. Study what happens to the phase-plane plot if one multiplies both equations in the population system (10.18) by the same positive factor. Explore the same with a negative factor. In all cases, make predictions, study the phase-plane plot, run simulations, and discuss biological relevance.

10.19. Compute steady-state points in the population system (10.18) when either N_1 or N_2 is zero and the other variable is not. Interpret these points in the context of logistic growth functions.

10.20. Use conceptual arguments, rather than a formal mathematical derivation, to discuss the stability of the two steady states in (10.19) when one variable is zero and the other is not.

10.21. Discuss the two nullcline situations not analyzed in the text. Sketch the nullclines and trajectories in each case and make predictions about survival and extinction.

10.22. Select parameter values for the four combinations of nullclines and explore different trajectories with simulations.

10.23. Study what happens in the predator–prey model (10.20) if an external factor alters the value of N or P at time $t = 20$. Discuss your findings.

10.24. Perform a qualitative analysis of the predator–prey model (10.20). Discuss coexistence and extinction.

10.25. Linearize the predator–prey model (10.20) at the trivial and nontrivial steady states and locate their stability patterns on the diagram in Figure 4.17.

10.26. Implement the chaotic Lotka–Volterra system and study the effects of changing the initial value of X_1. Summarize your findings in a report.

10.27. Perform a phase-plane analysis for the following system:

$$\dot{N}_1 = 0.1N_1\left(\frac{100-2N_2-N_1}{15}\right),$$
$$\dot{N}_2 = 0.7N_2\left(\frac{60-N_1-N_2}{45}\right).$$

10.28. Perform a phase-plane analysis for the following system:

$$\dot{N}_1 = 0.6N_1(50-N_1-0.833N_2),$$
$$\dot{N}_2 = 0.2N_2(30-0.75N_1-N_2).$$

10.29. Perform a phase-plane analysis for the following system:

$$\dot{N}_1 = 3N_1(10-N_2+1.5N_1-0.1N_1^2),$$
$$\dot{N}_2 = 0.7N_2(20-N_1-N_2).$$

10.30. Create a figure like Figure 10.12 for the following system:

$$\dot{N}_1 = 0.3N_1(500-10N_2+70N_1-3N_1^2),$$
$$\dot{N}_2 = 0.7N_2(80-2N_1-N_2).$$

10.31. Create a figure like Figure 10.12 for the following system:

$$\dot{N}_1 = 2N_1(10-N_2+1.5N_1-0.1N_1^2),$$
$$\dot{N}_2 = 2N_2(50-7N_1-N_2+0.3N_1^2).$$

10.32. Around the year 1918, the first black fire ant *Solenopsis richteri* reached the shores of the United States through the port of Mobile in Alabama. In the late 1930s, it was joined by the more aggressive red fire ant, *Solenopsis invicta*. Without significant predators, the two ant species spread quickly, honoring the red ant's Latin species name meaning "undefeated." In 1953, the United States Department of Agriculture found that 102 counties in 10 states had been invaded. By 1996, about 300,000 acres throughout the South were infected [30]. How would you design one or more models describing the past spread and predicting future invasion into new territories? What kinds of data would be needed for what kind of model? Search the literature for model studies of the phenomenon.

REFERENCES

[1] Savageau MA. Growth of complex systems can be related to the properties of their underlying determinants. *Proc. Natl Acad. Sci. USA* 76 (1979) 5413–5417.

[2] Neugebauer O & Sachs A (eds). Mathematical Cuneiform Texts, p 35. American Oriental Society/American Schools of Oriental Research, 1945.

[3] Bunt LNH, Jones PS & Bedient JD. The Historical Roots of Elementary Mathematics, p 2. Prentice-Hall, 1976.

[4] Segel LA. Modeling Dynamic Phenomena in Molecular and Cellular Biology. Cambridge University Press, 1984.

[5] Malthus TR. An Essay on the Principle of Population. Anonymously published, 1798 (reprinted by Cosimo Classics, 2007).

[6] Verhulst PF. Notice sur la loi que la population poursuit dans son accroissement. *Corr. Math. Phys.* 10 (1838) 113–121.

[7] Richards FJ. A flexible growth function for empirical use. *J. Exp. Bot.* 10 (1959) 290–130.

[8] Gompertz B. On the nature of the function expressive of the law of human mortality, and on a new mode of determining the value of life contingencies. *Philos. Trans. R. Soc. Lond.* 115 (1825) 513–585.

[9] Savageau MA. Growth equation: a general equation and a survey of special cases. *Math. Biosci.* 48 (1980) 267–278.

[10] Voit EO. Cell cycles and growth laws: the CCC model. *J. Theor. Biol.* 114 (1985) 589–599.

[11] Voit EO. Recasting nonlinear models as S-systems. *Math. Comput. Model.* 11 (1988) 140–145.

[12] Voit EO & Dick G. Growth of cell populations with arbitrarily distributed cycle durations. I. Basic model. *Math. Biosci.* 66 (1983) 229–246.

[13] Voit EO & Dick G. Growth of cell populations with arbitrarily distributed cycle durations. II. Extended model for correlated cycle durations of mother and daughter cells. *Math. Biosci.* 66 (1983) 247–262.

[14] White J. The allometric interpretation of the self-thinning rule. *J. Theor. Biol.* 89 (1981) 475–500.

[15] Voit EO. Dynamics of self-thinning plant stands. *Ann. Bot.* 62 (1988) 67–78.

[16] Bonabeau E. Agent-based modeling: methods and techniques for simulating human systems. *Proc. Natl Acad. Sci. USA* 14 (2002) 7280–7287.

[17] Macal CM & North M. Tutorial on agent-based modeling and simulation. Part 2: How to model with agents. In Proceedings of the Winter Simulation Conference, December 2006, Monterey, CA (LF Perrone, FP Wieland, J Liu, et al., eds), pp 73–83. IEEE, 2006.

[18] Richter O. Simulation des Verhaltens ökologischer Systems. Mathematische Methoden und Modelle. VCH, 1985.

[19] Hethcote HW, Horby P & McIntyre P. Using computer simulations to compare pertussis vaccination strategies in Australia. *Vaccine* 22 (2004) 2181–2191.

[20] Leslie PH. On the use of matrices in certain population mathematics. *Biometrika* 33 (1945) 183–212.
[21] Pinsky MA & Karlin S. An Introduction to Stochastic Modeling, 4th ed. Academic Press, 2011.
[22] Ross SM. Introduction to Probability Models, 10th ed. Academic Press, 2010.
[23] Wilkinson DJ. Stochastic Modelling for Systems Biology, 2nd ed. Chapman & Hall/CRC Press, 2012.
[24] Hirsch MW, Smale S & Devaney RL. Differential Equations, Dynamical Systems, and an Introduction to Chaos, 3rd ed. Academic Press, 2013.
[25] Kaplan D & Glass L. Understanding Nonlinear Dynamics. Springer-Verlag, 1995.
[26] Dam P, Fonseca LL, Konstantinidis KT & Voit EO. Dynamic models of the complex microbial metapopulation of Lake Mendota. *NPJ Syst. Biol. Appl.* 2 (2016) 16007.
[27] Sprott JC, Vano JA, Wildenberg JC, et al. Coexistence and chaos in complex ecologies. *Phys. Lett. A* 335 (2005) 207–212.
[28] Vano JA, Wildenberg JC, Anderson MB, et al. Chaos in low-dimensional Lotka–Volterra models of competition. *Nonlinearity* 19 (2006) 2391–2404.
[29] Paris OH & Pitelka FA. Population characteristics of the terrestrial isopod *Armadillidium vulgare* in California grassland. *Ecology* 43 (1962) 229–248.
[30] Lockley TC. Imported fire ants. In Radcliffe's IPM World Textbook. University of Minnesota. http://ipmworld.umn.edu/lockley.

FURTHER READING

Edelstein-Keshet L. Mathematical Models in Biology. McGraw-Hill, 1988.
Hassell MP. The Dynamics of Competition and Predation. Edward Arnold, 1976.
Hoppensteadt FC. Mathematical Methods of Population Biology. Cambridge University Press, 1982.
May RM. Stability and Complexity in Model Ecosystems. Princeton University Press, 1973 (reprinted, with a new introduction by the author, 2001).
Murray JD. Mathematical Biology I: Introduction, 3rd ed. Springer, 2002.
Murray JD. Mathematical Biology II: Spatial Models and Biomedical Applications, Springer, 2003.
Ptashne M. A Genetic Switch: Phage Lambda Revisited, 3rd ed. Cold Spring Harbor Laboratory Press, 2004.
Renshaw E. Modelling Biological Populations in Space and Time. Cambridge University Press, 1991.
Savageau MA. Growth of complex systems can be related to the properties of their underlying determinants. *Proc. Natl Acad. Sci. USA* 76 (1979) 5413–5417.

Integrative Analysis of Genome, Protein, and Metabolite Data: A Case Study in Yeast

11

When you have read this chapter, you should be able to:

- Discuss the steps for converting diverse experimental data into a computational model
- Describe how available data and research goals define the focus and scope of a model
- Match different data types with suitable modeling methods
- Identify the main components of the heat stress response in yeast
- Describe responses on different timescales and their impact on modeling analyses
- Explain the role of trehalose in heat stress responses
- Set up a metabolic model of the heat stress response from metabolic data
- Set up a metabolic model of the heat stress response based on gene expression data
- Discuss the advantages and limitations of time-series data for metabolic modeling

In the first part of the book, we learned the basics of the wondrous world of modeling in biology and gained a glimpse into the structure of networks and systems. Then, in the second part, we discussed various types of data at different levels of the hierarchy of biological organization, their information content, limitations, and peculiarities, and, in particular, their value for a deeper understanding of how systems are designed and how they operate. We saw that biological data can take many forms and that some are clearly more useful for systems analyses than others. It is now time to study some concrete applications of models, because these, more than any general discussions, reveal where the real challenges are and how they might be overcome. Also, as any modeler learns over time, actual biological systems have a way of surprising us with aspects that would have never occurred to us before we embarked on their analysis. Many hypotheses and draft theories have come crashing down when they were applied to real (or, as some theoreticians like to think, "dirty") data.

This chapter addresses some basic issues of multiscale analysis. The system we discuss is comparatively simple, and the multiple scales do not span the entire spectrum from molecules to the world oceans (see Chapter 15), but merely include genes, proteins, and metabolites. Nevertheless, the small case study is sufficient for reaching the accurate conclusion that it is often problematic to slice biological systems for analysis into isolated levels of organization. We will see that focusing solely on the genomic or the metabolic level masks important clues as to how the system as a whole is organized. The case study also shows that even apparently simple systems require a lot of information regarding their components, processes, and other aspects, and that it is difficult to discern *a priori* whether particular details can be omitted without much loss or whether they provide clues for important features of the function and regulation of the system. Putting the simplicity of the case study in this chapter into perspective, one will easily imagine how any multilevel system becomes much more complicated when the range of scales is widened, and we will discuss one such situation, the modeling of the heart, as a case study in Chapter 12.

ON THE ORIGIN OF MODELS

It is usually difficult to pinpoint where the ideas of a new research study originated. There are intriguing cases such as the dream of a snake catching its own tail, of monkeys chasing each other in a circle, or of dancing ballerinas that allegedly inspired the German chemist Friedrich August Kekulé von Stradonitz to propose a ring structure for the carbon atoms in a benzene molecule. Alas, in many other cases, new studies seem to have come from a spark of unknown origin or from ongoing discussions augmented by a loose collection of numerous tidbits of intellectual contributions. The situation is without doubt even more complicated in interdisciplinary studies, where one side usually does not fully understand the inner workings of the other. So, for the sake of argument, let us just imagine a fictitious water cooler encounter between an experimentalist E and a modeler M:

E: Have I ever told you about my heat stress stuff? I think it's pretty cool. Got some great data. I'm sure you could model them!

M: Hmm. I'm really sort of busy. What's so special about heat stress?

E: It's interesting.

M: I guess it would be nice to figure out how to respond to these recent heat waves.

E: No, no, we do yeast.

M: Yeast? Why is that interesting?

E: Well, because yeast is a good model.

M: Model for what?

E: Responses to stresses in all kinds of cells and organisms, including humans.

M: Yeah, it would be nice to know how to deal with stress! Let's have coffee, and you tell me more.

E: Well, let me start with an intriguing molecule: trehalose ...

As in a biochemical reaction that requires the collision of molecules, two scientists bump into each other, creating a spark that might quickly dissipate or lead to a long-term collaboration. Intriguingly, it is almost impossible to define conditions or predictors for either outcome, but it appears that every new collaborative project demands a certain leap of faith. And if the project is interdisciplinary, there must be two leaps and they must be rather large.

In our encounter, the modeler eventually becomes tempted to pursue the project, and the first order of business is now to gauge whether there is even enough common ground and what the prospects of success might be. The experimenter's statement "I'm sure you could model them!" is certainly not enough, because experimentalists often do not have a sufficient feel for modeling to judge the suitability of their data for a computational analysis. Furthermore, fathoming the potential and probability of success is not easy for anybody, and while experience certainly helps, it is a good idea early on to define what exactly the purpose of the model analysis might be. A crucial component of this assessment is limiting the scope, which is easier said than done, because the creation of a focus often requires extended discussions and hours of literature browsing. In our case, the modeler quickly learns from the experimentalist that any cellular **stress** response involves very many steps at various levels of biological organization: the response is clearly physiological, and dissecting it reveals that genes are up-regulated; the activities of existing proteins might be changed, new proteins are synthesized, and some proteins are deactivated; there are most certainly shifts in the cell's metabolic profile; and it is to be expected that the overall response is controlled by signal transduction.

It is impossible to capture and integrate all these components simultaneously and in detail with today's modeling methods. Even though teragrid and cloud computing have become a reality and computer scientists have produced very efficient algorithms for executing mega-size discrete-event **simulations**, biological systems are typically so ill-characterized in detail that the bottleneck for modeling analyses is often not computational power but the lack of enough quantitative biological information. It is therefore necessary for the experimentalist and the modeler to spend sufficient time to understand each other's intellectual context, milieu, concepts, and ways of thinking; the suitability, quality, and quantity of data; the potentially applicable modeling techniques; a realistic scope; and some targets that one might attain with reasonable likelihood of success. The following is a possible sample of lead questions:

- What is heat stress? What is heat shock? Does the study have broader significance?
- What is trehalose?
- Is the response natural? Does it occur under conditions that are typical for the organism?
- Which biological level is mainly responsible? Genomic, proteomic, metabolic, physiological?
- Is the heat response found in many/all creatures?
- What information is available?
- What types of data are available? How reliable are they?
- What is already known? What is not known, but significant?
- Has anyone already constructed pertinent mathematical models? Are they good?
- What aspects have been modeled?
- Which questions can be answered with the models?
- What is the main purpose of a new model analysis?
- Is it feasible to limit the scope without losing the big picture?
- Can the system be modularized without compromising its natural function?
- Is it possible to foresee whether a model would have a good chance of success?

An experienced experimentalist will be able to give at least partial answers to the biological questions, but the situation is complicated because there is so much known and it is quite evident that not all information is equally important for the modeling effort. What should the modeler be told, without being totally overwhelmed? There are also aspects that the experimentalist might not know in sufficient detail and that the modeler will have to find out independently. Let us fast-forward and suppose that the modeler is indeed interested in working with the experimentalist on one or more mathematical models that explain certain aspects of the heat stress response. The basis from which to start the new project is common knowledge of the heat response in yeast, and in particular the role of trehalose, which the experimentalist had identified as interesting for the ongoing investigation.

A BRIEF REVIEW OF THE HEAT STRESS RESPONSE IN YEAST

If the temperature of the medium is changed from 30 °C to 39 °C, several events are triggered within the yeast cells (for reviews, see [1–4]). These events are most certainly interconnected, but it is so far not clear what controls their coordination. A quick foray into the literature shows that there is a fine, although not always observed, distinction between heat stress and heat shock, with the first usually referring to increased heat conditions that can still be considered physiological and the latter referring to even higher temperatures, at which the cells barely survive. The experimentalist is most interested in heat stress, which thereby refines the scope of the study.

Maybe the most astounding response happens at the metabolic level: this is the production of the disaccharide trehalose, which essentially consists of two glucose molecules bound together (see later). Under normal physiological growth conditions, yeast cells contain merely trace concentrations of trehalose and its immediate precursor trehalose 6-phosphate (T6P). However, once the temperature rises above 37°C, the amount of trehalose may rise to an amazing concentration of up to 1 g of trehalose per 1 g of protein, or 40% of the entire cell biomass [5]. Mutants that lack the ability to form trehalose or that overproduce the trehalose-degrading enzyme trehalase are overly sensitive to environmental stresses, including heat, which indicates that the dramatic trehalose production is apparently vital for an effective response. Maybe of greatest importance, trehalose protects proteins against **denaturation** and reduces the formation of unnatural protein aggregates. In a similar role, it preserves the integrity of DNA, lipids, and membranes. Beyond protecting macromolecules and membranes, the trehalose pathway is involved in the control of glucose utilization by counteracting the accumulation of free cytoplasmic glucose and sugar phosphates, which would otherwise result in the depletion of cytoplasmic phosphate and adenosine triphosphate (ATP).

A different metabolic response is the dramatic activation of specific lipids, called **sphingolipids**, which have been implicated in a variety of fundamental decisions at the cellular level, including **differentiation, apoptosis**, and the response to various stresses, including heat [6, 7]. Within a few minutes of heat stress, key enzymes of the pathway of sphingolipid biosynthesis exhibit increased activity. Several intermediates of the pathway increase several-fold in quantity, and some remain elevated for two hours or more. The role of sphingolipids in the response to heat stress appears to be twofold. First, they are important components of plasma membranes, and because the stress response requires the transport and mobilization of proteins and metabolites, the additional quantities of sphingolipids might be required to sustain membrane integrity. Second, sphingolipids serve as second messengers in signal transduction processes. As one example, sphingolipid metabolites activate the transcription of a number of genes, including some of those associated with trehalose synthesis [8].

At the genome level, the response to heat is widespread, and many genes are up-regulated within a few minutes [9]. Not surprisingly, almost all genes coding for enzymes associated with the trehalose cycle are up-regulated, but two details are intriguing. First, the degree of up-regulation differs dramatically, spanning a range from essentially unchanged to over 100-fold. Second, the genes coding for trehalose-degrading enzymes, trehalases, are robustly up-regulated as well. Why would that happen, if the cell needs large amounts of trehalose? We will return to this question later. It should also be noted that some genes are strongly down-regulated. Examples are ribosomal protein transcripts. In fact, there is a **transient** decrease in overall **translation** initiation, coupled with transient **cell cycle** and growth arrest. It is possible that the combined effects of the decrease in transcript and protein synthesis reduce biosynthetic costs while the cell adapts to stress [1, 10].

Proteins are involved in stress responses as well [11]. Several dozen yeast proteins are transiently induced by heat stress. Roughly a third of them are **heat shock proteins**, which mediate protein–protein interactions and the transport of proteins through membranes. They are apparently involved in the recovery from stress and, in particular in the correct refolding of proteins that had unfolded during heat.

Some of these proteins respond very strongly. For instance, the heat shock protein Hsp90 is induced ten- to fifteenfold. Of particular importance is also Hsp104, which seems to be associated with accumulating trehalose and with the removal of denatured and unsalvageable protein fractions, for instance through the **proteasome** (see [12] and Chapter 7). In addition to heat shock proteins, several enzymes are activated. We will study later the manner in which this activation appears to be crucial to the metabolic heat stress response, and specifically to the huge increase in trehalose in response to strong rises in temperature.

Further aspects of the heat response might be called physiological. These include changes in potassium channels and the mobilization of **transcription factors**, such as the zinc finger proteins Msn2p and Msn4p, which control the expression of most of the enzymes of carbon metabolism, as well as antioxidant defense proteins of the heat stress response. These responses seem to be obligatory, because *MSN2/MSN4* double mutants are unable to accumulate trehalose under heat conditions. The importance of this mechanism is underscored by inbuilt redundancy: failure in one of the transcription factors is compensated by the other. Translocation of Msn2p and Msn4p from the cytosol to the nucleus occurs within minutes of heat stress and has been considered to be the rate-determining step in the activation of the heat stress response. In the nucleus, Msn2p and Msn4p interact with **stress response elements** (STREs), which are found in some of the genes associated with the trehalose cycle, among others. It appears that this interaction between Msn2/4p and the STREs is modulated by sphingolipids [8, 13].

This brief summary indicates that the heat stress response is very important to the cell, because its control spans all hierarchical levels of biological organization. These control mechanisms do not act in isolation, and there are checks and balances, as one would expect to find in a **robust** regulatory **design** (**Figure 11.1**). For instance, most of the mechanisms outlined above lead to an increase in trehalose. However, the cell also possesses mechanisms to slow down or block trehalose biosynthesis. Particularly important among them is the Ras–cAMP–PKA pathway, which in unstressed cells keeps the transcription of trehalose-associated enzymes and the concentration of trehalose at low levels. The pathway can also activate the trehalose-degrading enzyme trehalase. In direct opposition to the effects of heat, cyclic adenosine monophosphate/protein kinase A (cAMP/PKA) blocks STREs and causes movement of Msn2/4p from the nucleus back to the cytosol. Furthermore, it seems that PKA prevents the accumulation of Msn2/4 in the nucleus. These processes involve the enzyme Yak1 kinase. In addition, most genes that are induced by Msn2/4p after heat stress are repressed by an excess of cAMP. Thus, any heat stress response

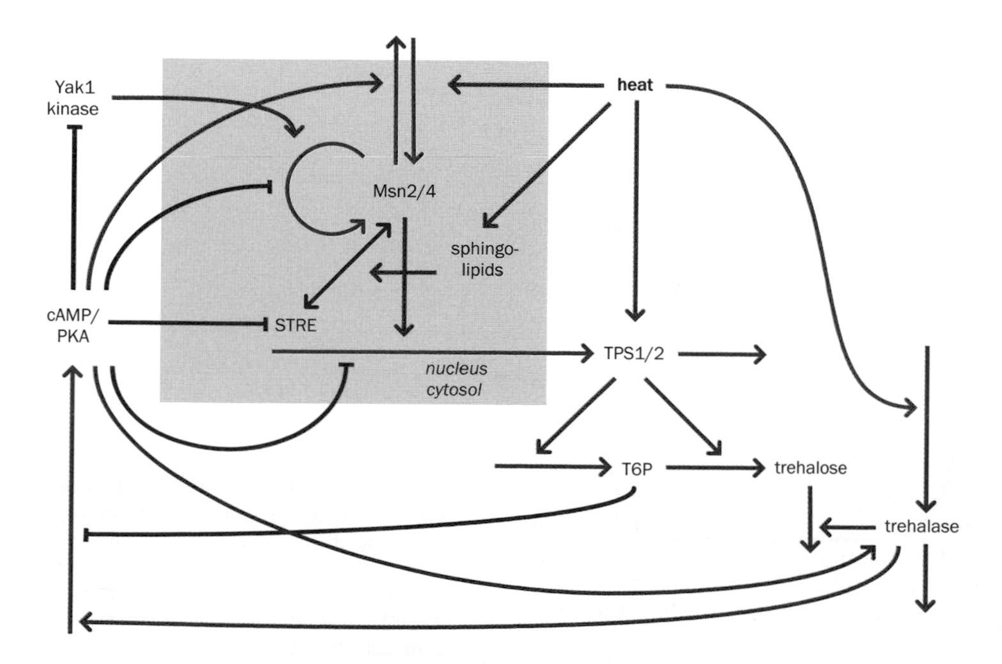

Figure 11.1 Tentative structure of the dynamic system of checks and balances regulating the heat stress response. The shaded yellow square represents the cell nucleus. The blue arrows indicate gene expression or metabolic reactions, the green arrows indicate activating effects, the blunted red lines indicate inhibitory effects, the orange arrow indicates genetic activation and enzyme deactivation, and the purple arrow indicates the interaction between transcription factors (Msn2/4) and stress response elements (STRE). See the text for further information.

must overcome or control intracellular cAMP levels. Interestingly, T6P inhibits one of the early glycolytic enzymes and indirectly affects the activation of the cAMP-dependent signaling pathway. Finally, trehalase is a substrate of PKA and therefore not only degrades trehalose but also induces the PKA pathway. While some of these details may not be crucial for the following analyses in this chapter, they serve to demonstrate that there is a fine-tuned balance between **antagonistic** pressures of the direct and indirect effects of heat on one hand and the activity of the Ras–cAMP–PKA signaling pathway on the other (for further details, see [2, 14, 15]).

11.1 The Trehalose Cycle

The experimentalist and modeler agree that the modeling effort should initially focus on the role of trehalose in the heat stress response. It is therefore necessary to explore the basic features of this metabolite and, in particular, to study its biosynthesis and degradation.

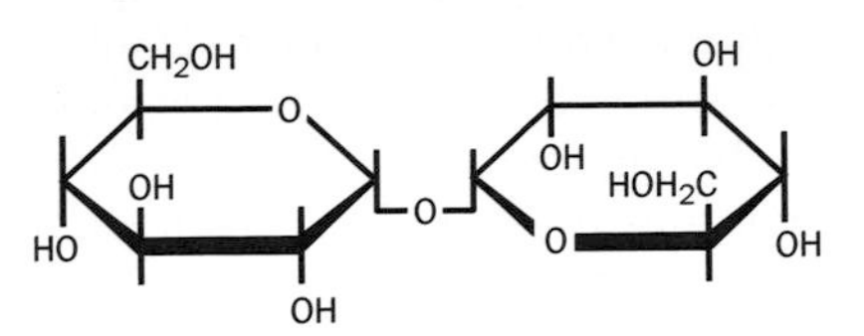

Figure 11.2 α,α′-trehalose. The molecule is a disaccharide; that is, it consists of two sugar (glucose) molecules.

Trehalose (α,α′-trehalose, α-D-glucopyranosyl α-D-glucopyranoside, or α,α-1,1-diglucose) (**Figure 11.2**) is a disaccharide (two sugars) that is produced from glucose in several enzymatic steps. Under normal conditions, most glucose enters glycolysis and the pentose pathway. Excess glucose is typically stored in the form of the highly branched **polysaccharide** glycogen. However, under heat conditions, a sufficient amount of glucose must be devoted to the synthesis of trehalose. These changing demands already imply that the cell has to solve a complex control task.

As in many cases of metabolic networks, information regarding pathway connectivity can be found in databases such as KEGG [16] and BioCyc [17], as well as in the original literature. A simplified diagram is presented in **Figure 11.3**. More difficult is the determination of the regulatory structure, which requires sophisticated **text mining** and a thorough study of the primary literature. A summary of the pathway with regulation is given in **Figure 11.4**. Furthermore, information on gene expression in response to heat stress can be visualized as in **Figure 11.5**.

Yeast cells can take up glucose from the medium with one or more of about 20 hexose transporters (HXTs), which are specialized for different conditions. The glucose is instantaneously converted into glucose 6-phosphate (G6P), and there is speculation that G6P might inhibit glucose transport into the cell (see, for example, [18]). The phosphorylation of glucose to G6P is catalyzed by three **isozymes**: hexokinase I, hexokinase II, and glucokinase; their corresponding genes are *HXK1*, *HXK2*, and *GLK1*. It is commonly assumed that this phosphorylation step is essentially **irreversible**. The three isozymes control each other through complex interactions that determine which kinase is most active under specific environmental conditions. For instance, during exponential growth on glucose, hexokinase

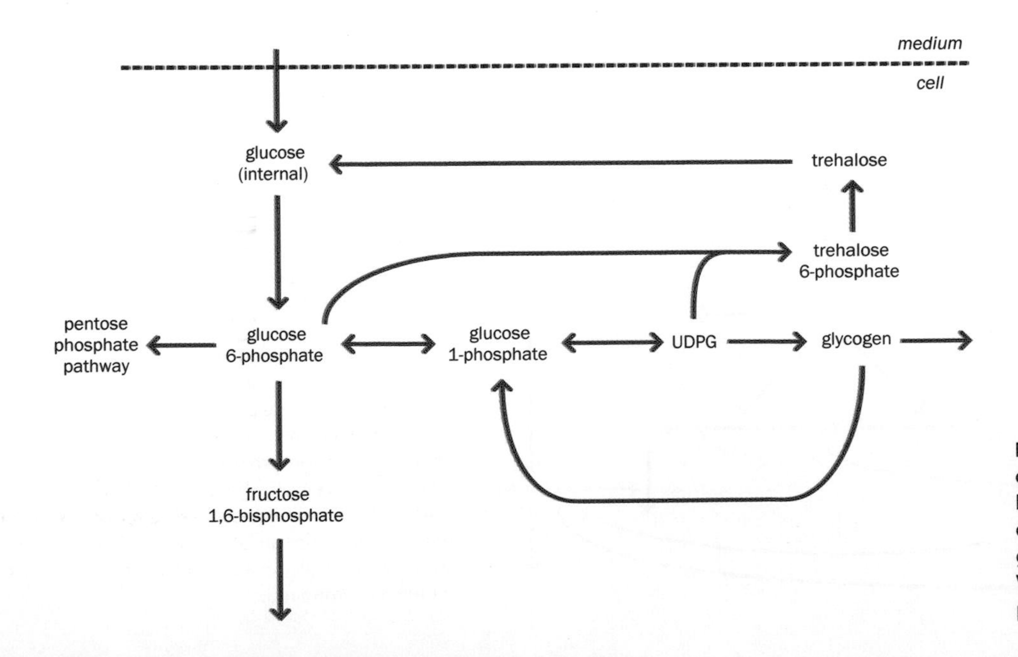

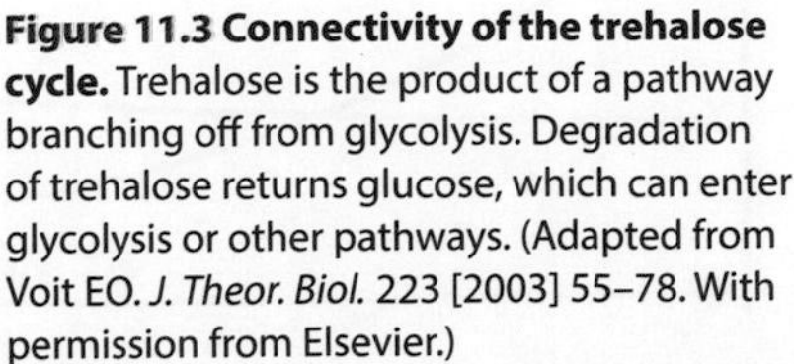

Figure 11.3 Connectivity of the trehalose cycle. Trehalose is the product of a pathway branching off from glycolysis. Degradation of trehalose returns glucose, which can enter glycolysis or other pathways. (Adapted from Voit EO. *J. Theor. Biol.* 223 [2003] 55–78. With permission from Elsevier.)

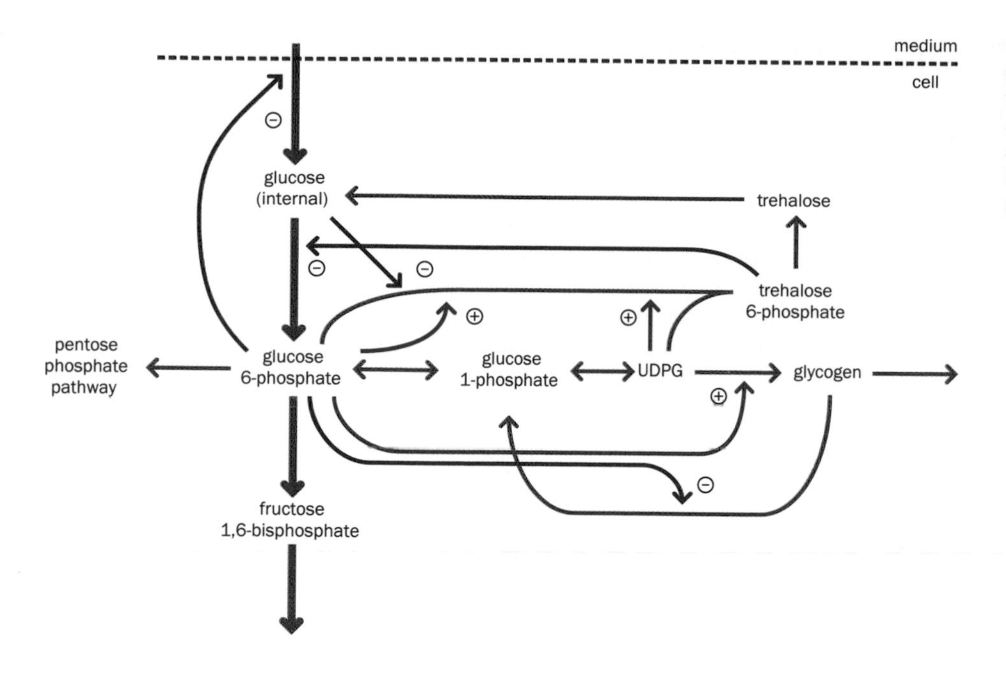

Figure 11.4 Regulation of the trehalose cycle. Green arrows show flow of material; thicker arrows indicate the main flux. Red and blue arrows indicate inhibition or activation, respectively. (Adapted from Voit EO. *J. Theor. Biol.* 223 [2003] 55–78. With permission from Elsevier.)

I and glucokinase are down-regulated, whereas hexokinase II is induced and becomes the dominant catalyst. The enzymes are rather stable and can remain active for several hours *in vivo*. Because our main focus here is on the trehalose cycle, it should be noted that T6P is a strong inhibitor of hexokinase II at physiological concentrations, but that it inhibits hexokinase I only weakly and glucokinase not at all. Again, things are much more complicated than they might appear at first.

G6P is arguably the most important branch point of carbohydrate metabolism, because it is here where material is channeled into glycolysis by isomerization to fructose 6-phosphate, into the pentose phosphate pathway by means of the G6P dehydrogenase reaction, or into other pathways, such as the trehalose cycle (see Figures 11.3–11.5). Under normal conditions, roughly two-thirds to three-quarters of G6P is used for glycolysis, 5–10% is channeled toward the pentose phosphate pathway (PPP), about 20% is reversibly converted into glucose 1-phosphate (G1P), and the remaining smaller quantities move into other pathways, for instance, toward glycogen and trehalose. Intriguingly, this distribution changes drastically under heat stress conditions. The gene for G6P dehydrogenase is up-regulated about four- to sixfold and the genes for the pathways toward trehalose and glycogen are up-regulated very strongly (ten- to twentyfold). By contrast, the genes

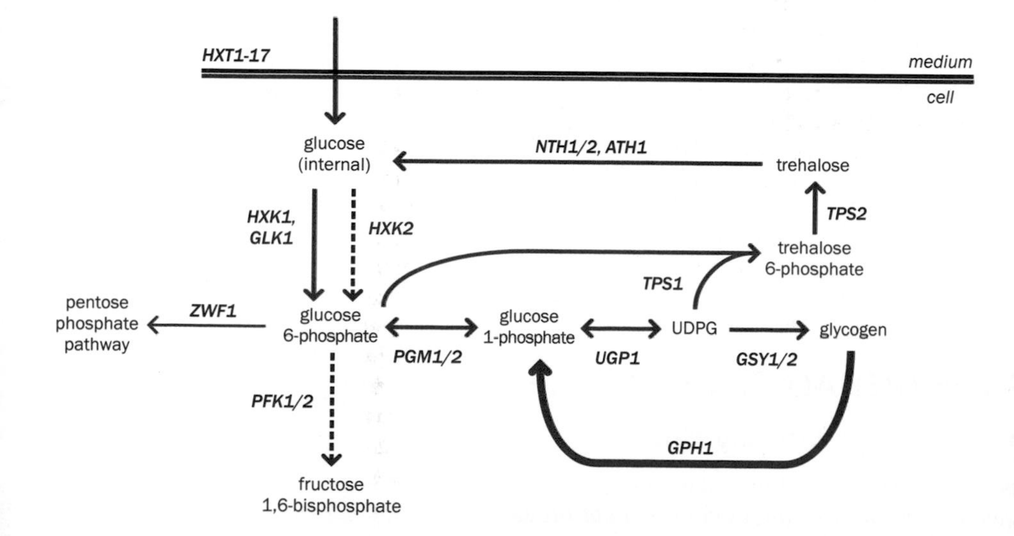

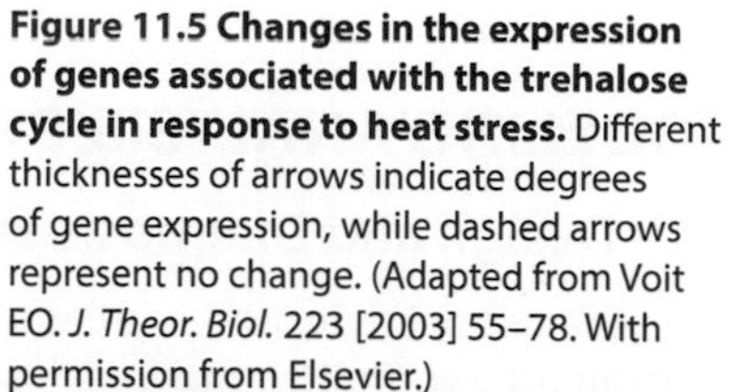
Figure 11.5 Changes in the expression of genes associated with the trehalose cycle in response to heat stress. Different thicknesses of arrows indicate degrees of gene expression, while dashed arrows represent no change. (Adapted from Voit EO. *J. Theor. Biol.* 223 [2003] 55–78. With permission from Elsevier.)

coding for phosphofructokinase (*PFK1/2*), the first committed step toward glycolysis, are apparently not affected at all.

The formation of trehalose occurs in several steps (see Figures 11.3–11.5). Phosphoglucose mutase reversibly converts G6P into G1P, which secondarily undergoes a reaction with uridine triphosphate to form uridine diphosphate glucose (UDPG). A subsequent reaction, which is catalyzed by T6P synthase (TPS1), synthesizes trehalose 6-phosphate (T6P) from G6P and UDPG. T6P phosphatase (TPS2) quickly converts T6P into trehalose through phosphoric ester hydrolysis. Indeed, TPS1 and TPS2 form a complex, together with two regulatory subunits, TPS3 and TSL1 (trehalose phosphate synthase and trehalose synthase long chain) that are co-regulated and induced by heat. At least in bacteria, and presumably in yeast, the T6P synthase reaction is irreversible. Its gene, *TPS1*, is repressed by glucose, and the degree of repression determines the concentration and activity of the trehalose production complex. In contrast, G6P and UDPG induce trehalose production. *TPS1* induction after heat stress is reflected in a corresponding production of T6P.

Through *O*-glycosyl bond hydrolysis, trehalose can be split into two molecules of glucose. This reaction is catalyzed by one or more of three trehalases, NTH1, NTH2, or ATH1, which are located in different cellular compartments and exhibit different activity profiles during growth. As discussed earlier, the trehalase step is important for trehalose degradation, and trehalase deficiency leads to trehalose accumulation. Furthermore, successful recovery from heat stress appears to require rather rapid removal of trehalose, because high concentrations of trehalose impede the refolding of proteins that had partially denatured during heat stress [19, 20]. Conversely, any overproduction of trehalase results in insufficient trehalose concentrations under stress and may increase mortality.

Taken together, the biosynthesis and degradation of trehalose begin and end with glucose, and thereby form the trehalose cycle:

$$2 \text{ glucose} \rightarrow 2 \text{ G6P}; 1 \text{ G6P} \rightarrow 1 \text{ UDPG};$$
$$1 \text{ G6P} + 1 \text{ UDPG} \rightarrow 1 \text{ T6P}; 1 \text{ trehalose} \rightarrow 2 \text{ glucose}$$

The trehalose cycle provides microorganisms with an amazing tolerance to heat, as well as to cold, osmotic pressure, and water shortage, and this tolerance is achieved effectively and cheaply. This combination of fortuitous features has piqued the interest of biotechnologists, who are beginning to replace some of the established preservation techniques for valuable organic materials, such as icing, liquid nitrogen, and vacuum drying, with much more subtle and less damaging metabolic methods involving trehalose [21, 22]. The most obvious and direct application is the optimization of freezing and freeze-drying of yeast cells, for instance, in the food industry. More generally, because of its unique ability to stabilize molecules, its mild sweetness, high solubility, and low hygroscopicity, and, last but not least, a price that has become affordable through genetic modifications of microorganisms, trehalose has become an important biotechnological target for manufacturing dried and processed food and making dry powder from foods and liquids. Currently, over 20,000 food products, made by 7000 companies around the world, contain trehalose [23]. For the same reasons, trehalose is of interest for the production of cosmetics, such as lipsticks. Furthermore, trehalose treatment permits efficient freeze-drying of platelets and other blood cells, which is of great importance because it could partially alleviate storage problems that are the cause of a chronic worldwide shortage of human blood platelets. Trehalose may also be used for the storage of vaccines, active recombinant **retroviruses**, and mammalian and invertebrate cells at low temperature, which is, for instance, needed for experiments in space laboratories.

MODELING ANALYSIS OF THE TREHALOSE CYCLE

11.2 Design and Diagnosis of a Metabolic Pathway Model

Given the connectivity and regulatory structure of the pathway, it is straightforward to formulate symbolic equations, especially if one uses the default of a **canonical**

modeling approach, such as **biochemical systems theory** (BST; see Chapter 4 and [24–26]). One begins by defining dependent variables for all metabolite pools of interest and constructs their differential equations such that all **fluxes** entering or exiting these pools are formulated as **power-law functions**. Each of these functions contains all variables that directly affect the corresponding flux, raised to an appropriate power, called the **kinetic order**, and in addition each flux term has a **rate constant** that is positive. If other types of functions are known (or can be assumed) to be appropriate representations, then they may also be used in the model construction. Functional forms such as **Michaelis–Menten** and **Hill** rate laws can often be found in the literature (see Chapter 8).

For the design of a model for the trehalose cycle, several starting points are possible. One could start from scratch, consulting pathway databases such as KEGG [16] and BioCyc [17] and extracting kinetic data from a database such as BRENDA [27]. However, it is of course easier to construct a model by adapting an already existing precursor. In this case, Galazzo and Bailey [18] performed targeted experiments on glycolysis in yeast and set up and parameterized a mathematical pathway model, using Michaelis–Menten rate laws and their generalizations. While they did not specifically consider the production of trehalose, owing to the low flux through this pathway under standard conditions, their model can directly serve as a baseline that is to be expanded toward the trehalose cycle. Further simplifying our task, Curto and co-workers [28] converted Galazzo and Bailey's model into BST models, which others later used for various purposes, such as illustrating model design and optimization methods [25, 29, 30]. Some of these models were even expanded to include the trehalose cycle [2, 31, 32], thereby making our life unusually easy. Furthermore, several datasets are available that characterize gene expression associated with heat stress in yeast [2, 9, 31–33]. As an additional data source, which we will discuss later in more detail, Fonseca et al. [34] measured **time series** of relevant metabolite concentrations.

A model of the trehalose cycle (see Figures 11.3–11.5) in **generalized mass-action** (GMA) form, as adapted from earlier studies, is as follows:

$$
\begin{aligned}
\text{Glucose:}\quad & \dot{X}_1 = 30X_2^{-0.2} - 90X_1^{0.75}X_6^{-0.4} + 2.5X_7^{0.3}, \quad X_1 = 0.03;\\
\text{G6P:}\quad & \dot{X}_2 = 90X_1^{0.75}X_6^{-0.4} + 54X_2^{-0.6}X_3^{0.6} - 23X_2^{0.75}\\
& \qquad - 3X_2^{0.2} - 7.2X_2^{0.4}X_3^{-0.4} - 0.2X_1^{-0.3}X_2^{0.3}X_4^{0.3}, \quad X_2 = 1;\\
\text{G1P:}\quad & \dot{X}_3 = 7.2X_2^{0.4}X_3^{-0.4} + 0.9X_2^{-0.2}X_4^{-0.2}X_5^{0.25} - 54X_2^{-0.6}X_3^{0.6}\\
& \qquad - 11X_3^{0.5} - 12X_2^{-0.2}X_3^{0.8}X_4^{-0.2}, \quad X_3 = 0.1;\\
\text{UDPG:}\quad & \dot{X}_4 = 11X_3^{0.5} - 0.2X_1^{-0.3}X_2^{0.3}X_4^{0.3} - 3.5X_2^{0.2}X_4^{0.4}, \quad X_4 = 0.66;\\
\text{Glycogen:}\quad & \dot{X}_5 = 3.5X_2^{0.2}X_4^{0.4} + 12X_2^{-0.2}X_3^{0.8}X_4^{-0.2} - 0.9X_2^{-0.2}X_4^{-0.2}X_5^{0.25}\\
& \qquad - 4X_5^{0.25}, \quad X_5 = 1.04;\\
\text{T6P:}\quad & \dot{X}_6 = 0.2X_1^{-0.3}X_2^{0.3}X_4^{0.3} - 1.1X_6^{0.2}, \quad X_6 = 0.02;\\
\text{Trehalose:}\quad & \dot{X}_7 = 1.1X_6^{0.2} - 1.25X_7^{0.3}, \quad X_7 = 0.05.
\end{aligned}
\tag{11.1}
$$

Here, the units are micromoles for concentrations and micromoles per minute for fluxes. Note that the model contains not only trehalose and T6P, in addition to the glycolytic metabolites, but also the pathway toward glycogen, which competes with the production of trehalose for the shared substrate UDPG. We will see later that the dynamics of glycogen is strongly affected by heat stress. It is at this point unclear to what degree the reaction between G1P and glycogen is reversible, and the model above includes both directions.

Before the model is used for further analyses, it should be subjected to the standard battery of diagnostic tests that, if successful, will increase our confidence in its reliability. Two typical tests explore the **steady state**, along with the **eigenvalues** and sensitivity values of the model (see Chapter 4). Indeed, with the given initial values, the system is very close to a steady state and all eigenvalues have negative real and zero imaginary parts, confirming local **stability**. The sensitivities and **gains** of metabolites and fluxes with respect to rate constants and kinetic orders are unremarkable, with the vast majority of them being smaller than 1 in magnitude, indicating that most perturbations in parameters or **independent variables** are readily attenuated. Only a few sensitivities are larger than 1 in magnitude, but none have values high enough to be worrisome. Because sensitivities and gains are characteristics associated with very small changes, it is also useful to spot-check the local and structural stability of the model with simulations of temporary or persistent increases in some of the variables or parameters. For instance, one might initiate the system at the steady state but with one of the initial values doubled or halved. Again, the results are unremarkable for our model, and the system very quickly recovers from such perturbations. Interestingly, the recovery from a bolus in UDPG takes the longest, but even here the system is essentially back to its steady state in less than 5 minutes.

As a specific example, suppose the influx of glucose into the system is augmented by a constant bolus of 10 units between times 1 and 5, with the system starting at time 0 in its steady state. As a comparison, the natural influx has about 30 units. As **Figure 11.6** indicates, all variables react to the bolus with increasing concentrations, which return to normal once the bolus is removed. Note that the variables are normalized with respect to their steady-state values, which moves the steady state to (1, 1, …, 1).

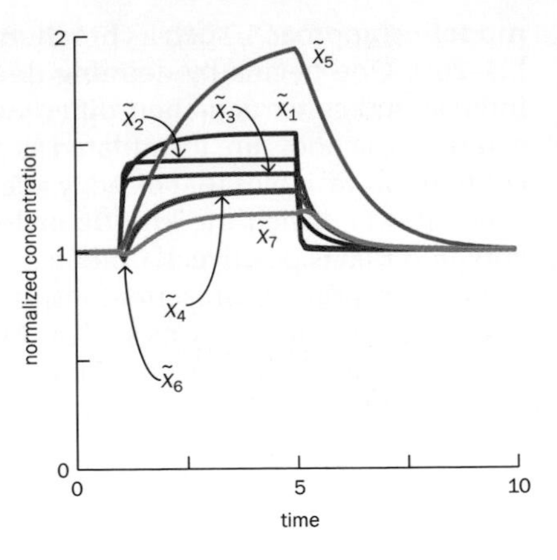

Figure 11.6 Responses of the trehalose pathway model to a glucose bolus between times 1 and 5. All variables are normalized with respect to their steady-state values. Variable names are the same as in (11.1), with the tilde (~) denoting normalization.

There is no limit to the kinds of simulations that one can perform with models of this type. One interesting case is a persistent excess or shortage of glucose. It has been speculated that trehalose is a carbohydrate storage molecule, but that glycogen is preferred because it is energetically superior. The simulation is simple: permanently increase or decrease the glucose influx to the system. As an example, if the influx is doubled, the glycolytic flux increases by 80% and the trehalose concentration by about 40%. In contrast, glycogen increases fivefold. Under glucose shortage, the glycogen storage pool becomes quickly depleted.

Another set of simulation experiments helps us explore the response of the system to enhanced or decreased enzyme activities, which could be due to natural or artificial genetic alterations. For instance, we can easily test and confirm that mutations in *TPS1* or *TPS2* lead to reduced trehalose production.

In addition to exploring the effects of temporary or persistent perturbations, one can use the model to study the specific roles of the various regulatory signals that are present in the trehalose cycle. The tool for these types of explorations of so-called design principles is the method of mathematically controlled comparison, which we will discuss in detail in Chapter 14. In a nutshell, one performs two analyses in parallel: one with the observed system and one with a system that is identical except for one feature of interest. Differences in responses between the two systems can then be attributed to the differing feature. An analysis of this type shows that each regulatory feature in the trehalose system offers the pathway a slight advantage, and that all signals together comprise a much more effective control system [2].

11.3 Analysis of Heat Stress

The model as formulated so far does not contain variables or parameters that represent heat. So, how can we study the effects of heat stress? Several strategies may be adopted, and we begin with the simplest, which focuses on metabolites and enzymes. The metabolites of interest are already in the model as dependent variables, but the enzymes of the pathway that catalyze the different reactions converting glucose eventually into trehalose and other products are not obvious. However, the model implicitly accounts for them, because they directly affect the rate constants. For instance, the term $1.1X_6^{0.2}$ in the equations for T6P and trehalose describes the conversion of T6P into trehalose, which is catalyzed by the enzyme T6P phosphatase. The rate of this reaction has a numerical value of 1.1, and this quantity is in truth the product of a rate constant and the enzyme activity. If the enzyme activity is doubled, the rate constant is doubled. This **linear** relationship is all we need in the given situation, and we do not really need to know how exactly the value 1.1 is split between the rate constant and the enzyme activity.

TABLE 11.1: INDEPENDENT VARIABLES IN MODEL (11.2)

Catalytic or transport step	Enzyme or transporter	Variable
Glucose uptake	HXT	X_8
Hexokinase/glucokinase	HXK1/2, GLK	X_9
Phosphofructokinase	PFK1/2	X_{10}
G6P dehydrogenase	ZWF1	X_{11}
Phosphoglucomutase	PGM1/2	X_{12}
UDPG pyrophosphorylase	UPG1	X_{13}
Glycogen synthase	GSY1/2	X_{14}
Glycogen phosphorylase	GPH	X_{15}
Glycogen use	GLC3	X_{16}
α,α-T6P synthase	TPS1	X_{17}
α,α-T6P phosphatase	TPS2	X_{18}
Trehalase	NTH	X_{19}

Nevertheless, when many enzyme activities are to be changed, it is convenient to display them explicitly in the equations. This is easily accomplished by defining an independent variable for each enzyme, include it in the appropriate reaction steps, which sometimes appear in two or more equations, and set its value equal to 1 for normal conditions. **Table 11.1** and a slightly expanded set of equations, (11.2), illustrate these changes. Note that these equations are exactly like (11.1), except that the independent variables for all pertinent enzymes are now explicitly shown. As long as these have their normal value of 1, the two sets of equations are numerically the same.

$$
\begin{aligned}
\text{Glucose:}\quad & \dot{X}_1 = 30X_2^{-0.2}X_8 - 90X_1^{0.75}X_6^{-0.4}X_9 + 2.5X_7^{0.3}X_{19}, \quad X_1 = 0.03;\\
\text{G6P:}\quad & \dot{X}_2 = 90X_1^{0.75}X_6^{-0.4}X_9 + 54X_2^{-0.6}X_3^{0.6}X_{12} - 23X_2^{0.75}X_{10}\\
& \quad -3X_2^{0.2}X_{11} - 7.2X_2^{0.4}X_3^{-0.4}X_{12} - 0.2X_1^{-0.3}X_2^{0.3}X_4^{0.3}X_{17}, \quad X_2 = 1;\\
\text{G1P:}\quad & \dot{X}_3 = 7.2X_2^{0.4}X_3^{-0.4}X_{12} + 0.9X_2^{-0.2}X_4^{-0.2}X_5^{0.25}X_{15} - 54X_2^{-0.6}X_3^{0.6}X_{12}\\
& \quad -11X_3^{0.5}X_{13} - 12X_2^{-0.2}X_3^{0.8}X_4^{-0.2}X_{15}, \quad X_3 = 0.1;\\
\text{UDPG:}\quad & \dot{X}_4 = 11X_3^{0.5}X_{13} - 0.2X_1^{-0.3}X_2^{0.3}X_4^{0.3}X_{17} - 3.5X_2^{0.2}X_4^{0.4}X_{14}, \quad X_4 = 0.66;\\
\text{Glycogen:}\quad & \dot{X}_5 = 3.5X_2^{0.2}X_4^{0.4}X_{14} + 12X_2^{-0.2}X_3^{0.8}X_4^{-0.2}X_{15} - 0.9X_2^{-0.2}X_4^{-0.2}X_5^{0.25}X_{15}\\
& \quad -4X_5^{0.25}X_{16}, \quad X_5 = 1.04;\\
\text{T6P:}\quad & \dot{X}_6 = 0.2X_1^{-0.3}X_2^{0.3}X_4^{0.3}X_{17} - 1.1X_6^{0.2}X_{18}, \quad X_6 = 0.02;\\
\text{Trehalose:}\quad & \dot{X}_7 = 1.1X_6^{0.2}X_{18} - 1.25X_7^{0.3}X_{19}, \quad X_7 = 0.05.
\end{aligned}
\tag{11.2}
$$

Interestingly, Neves and François [35] reported that the three enzymes directly associated with the trehalose dynamics exhibit activities that depend on the ambient temperature. In fact, they presented a temperature–activity plot for each enzyme (**Figure 11.7**). According to these plots, the activities of T6P synthase and T6P phosphatase increase roughly 2.5-fold, if the temperature in the medium is raised from a normal temperature (between 25°C and 30°C) to heat stress conditions of about 39°C. By contrast, the activity of trehalase falls to about half its original value. This information can be entered directly into our model. Namely, the corresponding enzyme terms (X_{17}, X_{18}, and X_{19}) are multiplied by 2.5, 2.5, and 0.5, respectively. As an illustration, the term $0.2X_1^{-0.3}X_2^{0.3}X_4^{0.3}X_{17}$ represents the T6P synthase reaction and appears in the equations for G6P, UDPG, and T6P; for the heat stress scenario, the enzyme activity X_{17} is changed from 1 to 2.5. The term $1.1X_6^{0.2}X_{18}$ appears in the equations for T6P and trehalose, and again the enzyme activity X_{18} is set to 2.5.

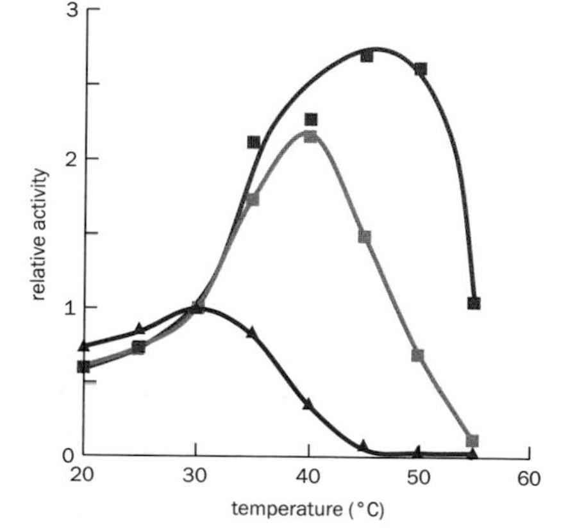

Figure 11.7 Effect of temperature on activities of enzymes producing (green) and degrading (red) trehalose. Light green, trehalose 6-phosphate synthase; dark green, trehalose 6-phosphate phosphatase; red, trehalase; lines are smoothed, interpolating trends. (Data from Neves MJ and François J. *Biochem. J.* 288 [1992] 859–864.)

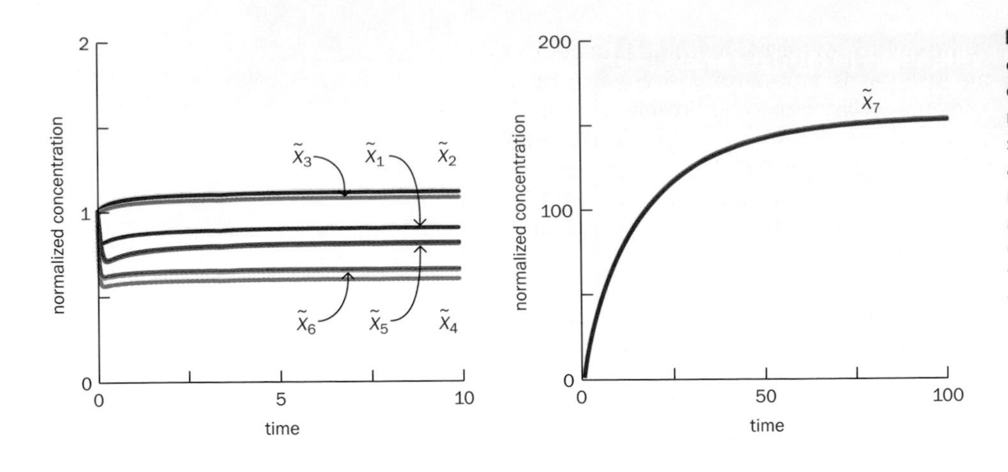

Figure 11.8 Differential response to changes in reported heat-stress-sensitive enzymes. When the activities of enzymes reported in [35] are adjusted to reflect heat stress conditions, trehalose (X_7) increases about 150-fold, while the other metabolites remain more or less unchanged or decrease. All quantities are presented as relative to the original steady-state values, as indicated by the tildes. Note the different scales along the axes.

Finally, the term $1.25X_7^{0.3}X_{19}$ is directly part of the equation for trehalose, and appears as $2.5X_7^{0.3}X_{19}$ in the equation for glucose; to represent heat stress conditions, the enzyme activity X_{19} is set to 0.5. Everything else in the model remains the same.

We can easily solve the model with the new parameter values, starting with the original steady state, and study the responses in metabolites. The simulation shows that trehalose increases tremendously to about 7.5 mM, which corresponds to roughly 150 times its level under normal temperature conditions. At the same time, G6P and G1P remain essentially unchanged, while the remaining metabolites decrease to between 50% and 75% of their original values (**Figure 11.8**). As before, the sky is the limit for variations on these types of simulations. As a final note on this set of studies, one could ask how quickly the enzyme activities change in response to heat stress. This question is not all that pertinent here, as long as the enzyme reacts with a similar speed, because a common time delay of δ minutes would simply move all transients by that much to the right.

11.4 Accounting for Glucose Dynamics

All simulations so far have assumed that the glucose concentration in the medium is so high that the cells can take up glucose at a constant rate, which in the glucose equation has the value 30. In a laboratory experiment, this situation may be feasible, but it might not be realistic in nature. We can adapt the model with reasonable effort to account for depletion of substrate in the medium. It is quite obvious that the constant rate of 30 must be replaced with a function that accounts for glucose utilization. As before, it is useful to replace the constant rate of 30 with the product of the concentration of the external glucose (GLU) and a rate constant α. For the sake of argument, let us assume the external concentration is $GLU = 100$, so that the new rate constant is $\alpha = 30/(GLU) = 0.3$, which numerically leads to the same model we had before, because $\alpha \cdot (GLU) = 30$. Thus, the glucose uptake term in the equation of X_1 becomes $0.3 \cdot (GLU) \cdot X_2^{-0.2}$. The advantage that we have gained is that we can now represent GLU as an independent or dependent variable, which may be constant or not. In particular, if we define GLU through a differential equation, we can model its change over time, while we still have the option of setting $(G\dot{L}U) = 0$, which corresponds to a constant glucose concentration in the medium.

The new option of interest is now the depletion of glucose by the cells. For instance, we could set

$$(G\dot{L}U) = -p \cdot 0.3 \cdot (GLU) \cdot X_2^{-0.2}. \tag{11.3}$$

In this formulation, glucose disappears from the medium, which is seen in the minus sign. Furthermore, the depletion of glucose is proportional to the uptake by the cells, with a proportionality constant p. This parameter p reflects the ratio between the volume of the medium and the volume of the cytosol, and is usually not equal to 1, because the external concentration is based on the volume of the medium, whereas the concentration inside the cells needs to reflect the intracellular volume. Even seemingly trivial aspects like concentrations can become complicated! To permit us to do some simulations, suppose $p = 0.025$, and we begin with normal temperature conditions and an initial concentration of external glucose of 100. The result is shown in **Figure 11.9**. All metabolites decrease in concentration, with glycogen (X_5) being affected most. This detail makes sense, because glycogen is a storage compartment for rainy days.

It is now easy to study what happens under heat stress and simultaneous glucose depletion. Namely, we use the most recent variant of the model, which accounts for changes in *GLU*, and multiply the trehalose enzymes by 2.5 and 0.5, respectively, as before. Some results are shown in **Figure 11.10**. The shape of the transients is somewhat different and, most prominently, trehalose reaches a maximum, before decreasing. If we were to extend the simulation to longer time windows, we would find that trehalose approaches 0 at about $t = 450$. However, we must be careful with such extensions, because cells under stress usually turn off all processes that are not absolutely essential and enter something like a state of suspended animation.

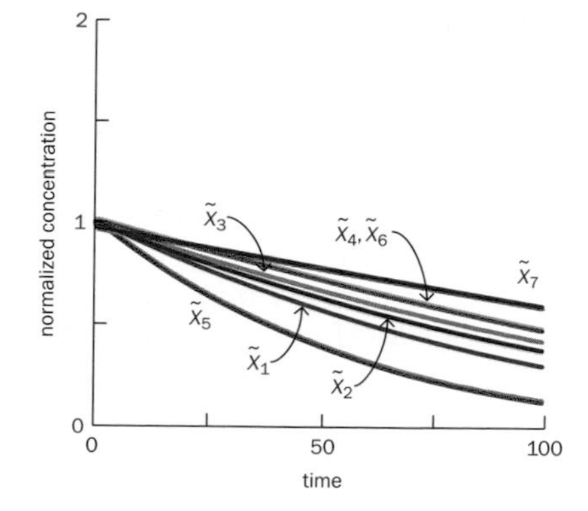

Figure 11.9 Under normal temperature conditions, a slow depletion of external glucose leads to decreases in all metabolites. Note that the storage compartment of glycogen (X_5), suffers the greatest loss. All quantities are normalized with respect to their steady-state values.

11.5 Gene Expression

Uncounted papers have been written about changes in gene expression under stress conditions, including heat stress and heat shock in yeast. Many of these papers address an assortment of heat shock proteins, whose genes are up-regulated under adverse conditions. However, changes in expression are also observed in a large number of genes coding for enzymes, and, indeed, several of the enzymes of the trehalose cycle are up-regulated in response to heat treatment. These changes in expression, which are usually measured with **microarray** and **DNA chip** experiments, are very interesting, because they can be obtained for many genes in single experiments, whereas metabolic data are often derived from different experiments and sometimes even from different types of experiments, so that one has to ask how compatible these diverse data really are.

Gene expression data are not without problems either, for two main reasons. First, technical challenges and experimental noise lead to uncertainties regarding the accuracy of the results, as we discussed in Chapter 6. More importantly, it is not clear whether gene expression is truly correlated with the observed numbers of mRNA molecules and the corresponding amounts of active proteins in a more or less linear fashion [36]. In spite of these uncertainties, gene expression data are very valuable owing to their large information content from single experiments, and they can be employed, at the very least, as coarse tools for providing clues regarding the coordinated responses that cells mount when under stress. A large collection of gene expression data from yeast under various stresses is available in

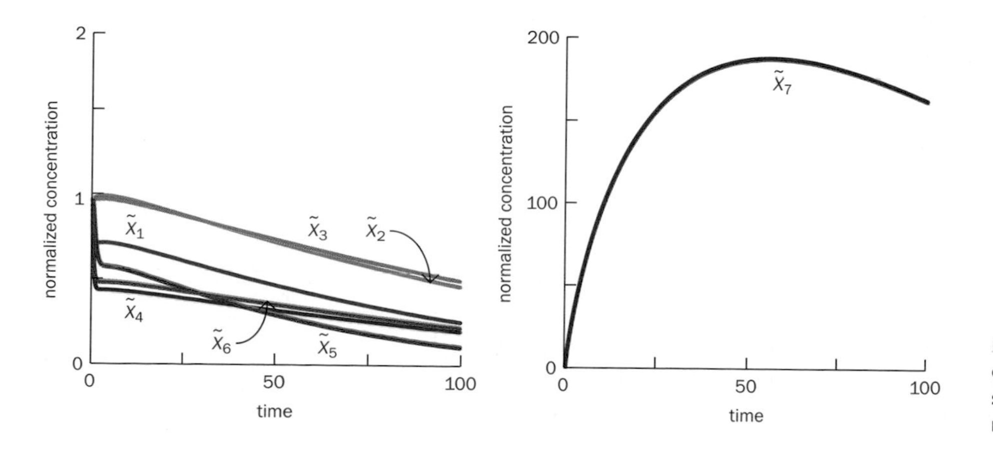

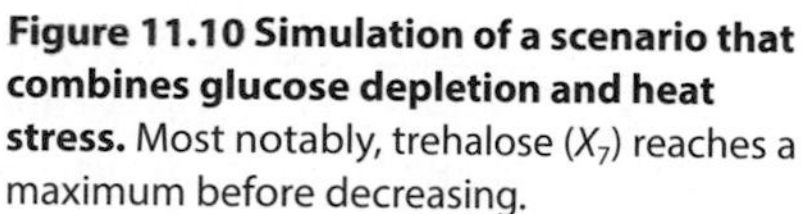

Figure 11.10 Simulation of a scenario that combines glucose depletion and heat stress. Most notably, trehalose (X_7) reaches a maximum before decreasing.

the Stanford Yeast Database [9, 33, 37]. Similar data, specifically for heat stress, were also presented in [31].

If one assumes, in first approximation, a direct proportionality between gene expression and protein prevalence, then it is possible to associate enzyme activities with gene expression and to make inferences from the genome level to the metabolic level. For instance, the gene *ZWF1* codes for the enzyme G6P dehydrogenase, which catalyzes the reaction that diverts carbon from glycolysis toward the pentose phosphate pathway. Suppose we wanted to investigate the effects of fourfold up-regulation of *ZWF1* on the trehalose cycle, under the assumption that none of the other reactions are affected. Presuming direct proportionality between gene expression and enzyme activity, we start with the original model in the form of (11.2), multiply the corresponding enzyme term X_{11} in the equation for G6P by 4, and solve the system. As one might expect, all concentrations rapidly fall, because material is irreversibly channeled out of the model system. Interestingly, the transients are all slightly different, and X_1 even shows a small, initial overshoot that is at first glance unexpected (**Figure 11.11**); see Exercise 11.10.

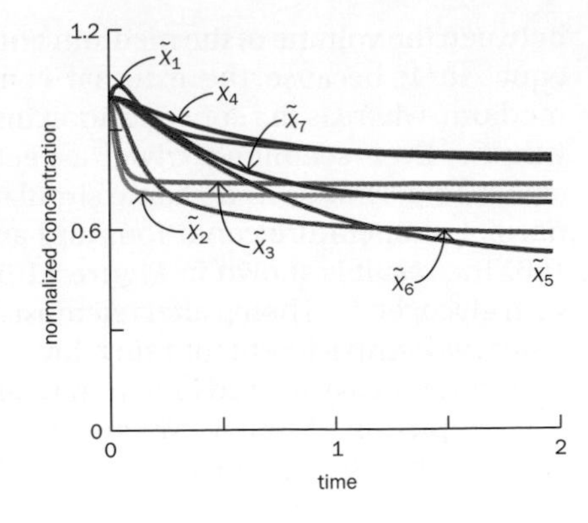

Figure 11.11 Effect of the pentose phosphate pathway (PPP). An increase in the activity of G6P dehydrogenase (X_{11}) channels material into the PPP and thereby leads to decreases in all system variables.

Using this association strategy, one can easily change enzyme activities according to observed changes in gene expression. We begin by extracting pertinent information from the Stanford database [37]. It is presented in **Table 11.2** in the form of the relative expression change in each gene of the trehalose cycle as a power of 2. Thus, a value of 3 indicates a 2^3 = eight-fold change, while −1 corresponds to half the expression level as under normal conditions. Blank entries indicate missing measurements.

The expression data clearly demonstrate that the yeast cells initiate a coordinated, large-scale response within minutes of the beginning of heat stress. In fact, other studies have demonstrated changes within 2 minutes. Most of the involved genes rise in expression and eventually return to the normal baseline. Furthermore, all genes seem to reach their maximal change in expression roughly between 15 and 20 minutes after heat stress.

TABLE 11.2: CHANGES IN SELECTED YEAST GENES DURING HEAT STRESS*

ORF	Gene	Enzyme or step	5	10	15	20	30	40	60	80
YPR026W	*ATH1*	Vacuolar trehalase			3.03		2.94		1.39	
YCL040W	*GLK1*	Glucokinase	3.21	4.38	5.32	5.12	4.59		2.19	1.99
YPR160W	*GPH1*	Glycogen phosphorylase	3.76	5.07	6.78	5.99		2.83		
YFR015C	*GSY1*	Glycogen synthase	2.97	2.96	4.02	3.29	1.76	1.05	0.12	0.4
YLR258W	*GSY2*	Glycogen synthase	2.96	4.04	4.42	4.24	3.45	2.67	1.59	1.61
YFR053C	*HXK1*	Hexokinase I	3.57	4.7	5.72	4.84	4.53	1.96	0.75	1.65
YGL253W	*HXK2*	Hexokinase II	0.5	0.4	0.64	0.15	−0.76	−1.09	−0.32	−0.09
YDR001C	*NTH1*	Neutral α, α-trehalase	2.26	2.79	3.52	3.24	3.33	2.22	1.56	1.43
YBR001C	*NTH2*	Neutral α, α-trehalase	1.45	1.75	2.17	2.5	2.12	1.66	1.45	1.19
YKL127W	*PGM1*	Phosphoglucomutase I	0.07	−0.12	0.06	0.08	−0.71	−0.32	0.12	−0.36
YMR105C	*PGM2*	Phosphoglucomutase II	4.6	6.05	7.22	6.83	6.59	4.8	2.45	2.28
YGR240C	*PFK1*	Phosphofructokinase I	−0.19	−0.9	−0.41	−0.68	−0.64	−0.15	−0.55	−0.45
YMR205C	*PFK2*	Phosphofructokinase II	−0.06	−1		0.29	−0.22	0.07	0.03	0.19
YBR126C	*TPS1*	T6P synthase	2.06	3.46	3.69	3.44	3.1	1.81	1.06	0.91
YDR074W	*TPS2*	T6P phosphatase	3.57	3.4	4.16	4.24	3.31	1.78	1.89	1.76
YMR261C	*TPS3*	T6P synthase regulator	−0.09	0.15	1.73			−0.09		
YML100W	*TSL1*	T6P synthase regulator	4.88	4.86	6.28	6.38	5.7	4.75	3.66	3.31
YKL035W	*UGP1*	UDPG pyrophosphorylase	2.23	3.05		4.25	3.36	1.92	0.99	0.99
YNL241C	*ZWF1*	G6P dehydrogenase	0.68	1.27	2.34	1.95	1.71	1.33	0.53	0.53

*Data extracted from [37].

Limiting our analysis to the trehalose cycle, we find that most genes are up-regulated within a few minutes and remain up-regulated for some while. In particular, the trehalose-producing and -degrading genes follow similar trends (**Figure 11.12**). Interestingly, the degree of up-regulation varies very widely (see Table 11.2), and some genes, such as the two coding for phosphofructokinase (*PFK1* and *PFK2*), are not up-regulated much in response to heat stress, while others are expressed very strongly. For instance, glycogen phosphorylase (*GPH1*), which uses glycogen to produce G1P, reaches an up-regulation level of almost 7, which really means $2^7 = 128$-fold! Also intriguing is that the gene coding for hexokinase 2 (*HXK2*) is not changed much at all, while its **isozymes** hexokinase 1 and glucokinase, which catalyze the same phosphorylation reaction of glucose to G6P, are strongly up-regulated. Another interesting detail is the fact that the expression patterns of *TPS1* and *TPS2* are very similar. As a consequence, the pool of T6P remains more or less constant throughout the heat stress response. This constancy is important, because T6P is toxic at high concentrations [38, 39].

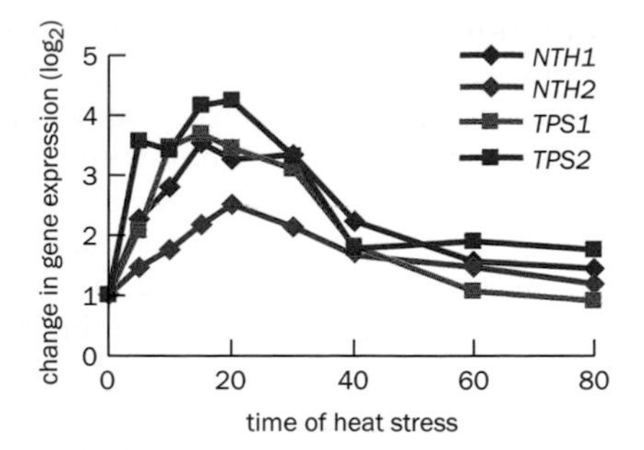

Figure 11.12 Heat-stress-induced changes in the expression of genes directly associated with trehalose. *NTH1/2*, trehalase; *TPS1*, T6P synthase; *TPS2*, T6P phosphatase. *TPS2* rises to a level of $2^{4.24} = 18.9$, which means an almost twenty-fold increase in gene expression over normal temperature conditions about 20 minutes after heat stress begins.

At first glance, many details of these patterns in expression may not make much sense. However, they can be explained to some degree by means of the metabolic model and the assumption that enzyme activities and gene expression levels are directly correlated. For simplicity of analysis, let us focus on the maximal deviation from the normal state, which occurs roughly 15–20 minutes into the heat stress and convert this maximum level of gene expression directly into changes in enzyme activity. Thus, if a gene is up-regulated eightfold, we multiply the activity of the corresponding enzyme by 8. Because most processes contribute to the degradation of a substrate as well as the synthesis of a product, two terms are often affected. For instance, the hexokinase reaction from glucose to G6P appears in the first two equations as $90X_1^{0.75}X_6^{-0.4}X_9$, once with a minus sign and once with a plus sign. One further complication comes from the fact that several reactions, such as the conversion of glucose into G6P, can be catalyzed by several isozymes. A particularly dramatic example is the uptake of external glucose by the cells, for which yeast uses roughly 20 different transporters. If several enzymes catalyze the same reaction, their involvement may be apportioned, for instance, by looking at their affinity under the most pertinent conditions or by weighing their importance by the number of corresponding mRNA molecules per cell [32].

Clearly, the conversion of gene expression into enzyme activity and the weighting of isozymes are drastic simplifications, and results from analyses based on such assumptions must be considered with caution. Nevertheless, when a comprehensive analysis of this type was executed [32], three insights were gained:

1. If one were to cluster all genes by their degree of up-regulation, as is sometimes done in bioinformatics for identifying shared control of ill-characterized pathway systems, it would appear that the glycolytic genes have nothing to do with each other, because they are expressed at such different degrees and would therefore be categorized in distinct classes. Therefore, **clustering** of genes by expression, while often beneficial, must be done with caution.
2. The up-regulation profile as observed after about 15–20 minutes of heat stress is not very intuitive. However, if this pattern is translated into enzyme activities, it appears that the response is well coordinated. The system produces plenty of trehalose, ATP (via glycolysis) for energy, and NADPH (within the pentose pathway) for an adequate **redox** balance. At the same time, the pathway system does not accumulate unnecessary or toxic intermediates, such as T6P.
3. It is possible to use the model for testing alternative hypothesized up-regulation patterns and assess their performance with respect to trehalose, ATP, and NADPH production, as well as the generation of intermediates. For instance, the cell would have to expend less effort by just up-regulating one or two key enzymes. Analysis of such hypothetical, alternative up-regulation profiles demonstrates that they do not seem to perform better than the observed profile and are in many respects inferior.

Many variations on this theme can be implemented. Foremost, the analysis described in the preceding paragraphs considered just one gene expression level (at 15–20 minutes) and only assessed the resulting steady state of the system under conditions of unlimited glucose in the medium. In expanded investigations, one could study transients and delays, change gene expression dynamically according to the published microarray data, compare constant glucose availability and glucose depletion, or combine some or all of these aspects. In addition, one could account for metabolic changes brought forth by temperature-dependent changes in enzyme activities, as we discussed before. The next section describes one option for such a combined analysis. The interested reader should also study [10] and [31] for different analyses of similar data on heat stress in yeast.

MULTISCALE ANALYSIS

In the first part of the analysis, we used the temperature-dependent changes in key enzymes to model the heat stress response. While the results looked reasonable, very few concrete data were available for comparison. Furthermore, this strategy entirely ignored the fact that a significant component of the heat stress response is controlled at the genomic level. In the second part of the analysis, we concentrated on gene expression and ignored direct temperature effects on enzymes. This strategy also led to some insights, but suffered from the uncertainties surrounding the assumption that gene expression can be converted directly and quantitatively into enzyme activities. This assumption was necessary because flux measurements over time were not available.

11.6 *In Vivo* NMR Profiles

New methodologies are capable of obtaining additional types of information. Specifically, it is possible to measure entire time series of concentrations *in vivo*, using **NMR spectroscopy** (see Chapter 8). Fonseca and collaborators [34] used this methodology to study heat stress in yeast. They grew the cells to a stable population size and then provided them with a bolus of ^{13}C-labeled glucose at a favorable temperature of 30°C. While the bolus was taken up by the cells, the investigators measured selected metabolites that were produced from the glucose substrate, including trehalose, G6P, fructose 1,6-bisphosphate (FBP) as an indicator of glycolysis, and final products such as ethanol, acetate, and glycerol (**Figure 11.13**). A second bolus of labeled glucose was given some while after the first bolus had been taken up, but at this time the yeast cells were under heat stress. Finally, the temperature was returned to its normal value of 30°C and another set of measurements was taken. Typical results are shown in **Figure 11.14**. Significant features of this experimental set-up are that glucose input is not constant but the external glucose is quickly depleted, and the cells are not growing under the given conditions.

Inspection of the results from this experiment leads to several insights. First, the cells quickly take up the glucose and essentially immediately convert it into G6P, FBP, trehalose, as well as end products like ethanol, acetate, and glycerol. G6P is not seen in Figure 11.14, because it requires labeling with glucose labeled on carbon 6, but it shows a time profile similar to that of FBP (**Figure 11.15**). Second, production of trehalose is much greater during heat stress. Instead of a peak of 4 mM under normal conditions, the cells generate new trehalose up to about 8 mM. While this doubling is significant, the final amount seems low in comparison with ample data from the literature that suggest much higher amounts. Under recovery conditions, trehalose production is similar to that under normal conditions during the first phase. Closer analysis also shows that the uptake of glucose is slightly faster under heat stress conditions.

Some of these and earlier results make intuitive sense, but some are rather surprising. We know that much of the heat stress response is controlled at the genome level. However, if genetic control were the driver, it would be impossible to see increased trehalose production within 2 minutes of glucose availability, because transcription and translation together take about 20 minutes in yeast. But if trehalose can be produced in large amounts *without* any specific action or intervention of genes, why do the cells expend the physiological costs of up-regulating the genes of the trehalose cycle at all? Furthermore, we have already noted that three of the trehalose enzymes are heat-dependent, with T6P synthase and T6P phosphatase about

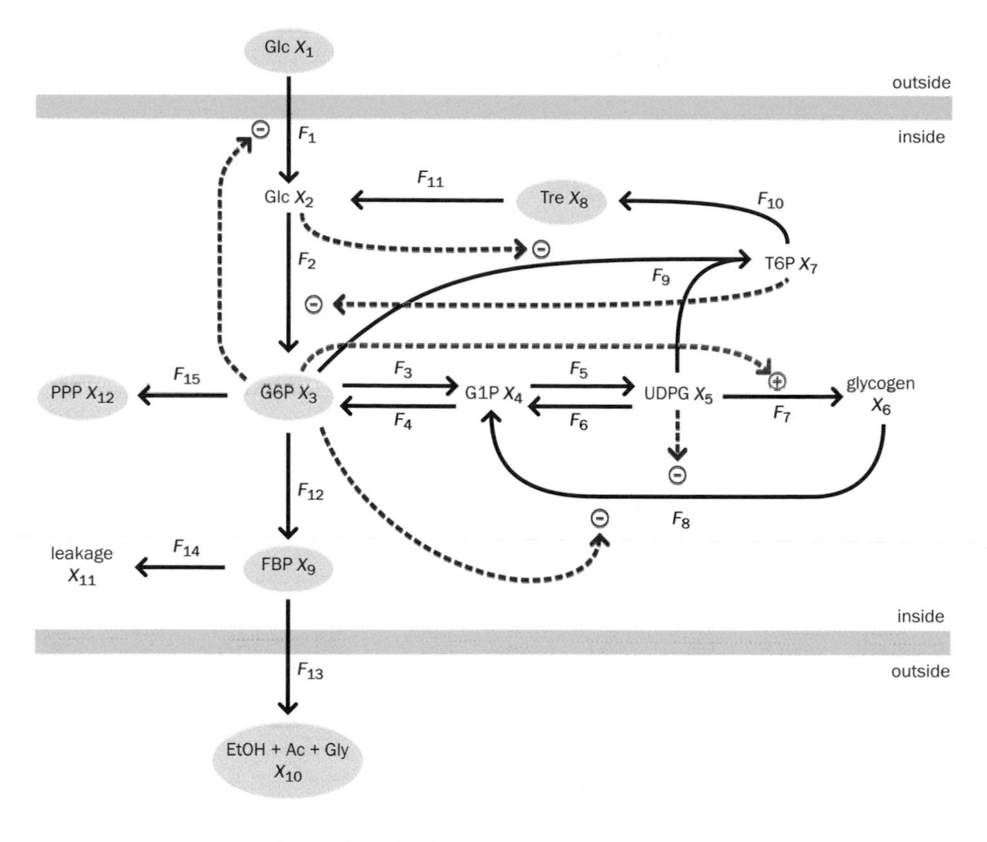

Figure 11.13 Model structure used to analyze time-series data from *in vivo* NMR experiments. Measured quantities are indicated by ellipses. (Adapted from Fonseca L, Sánchez C, Santos H & Voit EO. *Mol. BioSyst.* 7 [2011] 731–741. With permission from The Royal Society of Chemistry.)

2.5 times as active and trehalase half as active as under optimal temperature conditions (see Figure 11.7 and [35]). These effects of temperature make intuitive sense, because all three lead to increased trehalose production, which is a desirable goal. But are they sufficient to mount an effective response as we observe it in Figure 11.14? A question of this type is difficult to answer with hard thinking or even with wet experiments alone, which suggests the use of a modeling analysis.

In principle, such an analysis is straightforward. Once we have a model that fits the data under normal conditions, we can simply change the three enzyme activities in the model (by adjusting the corresponding rate constants) and solve the differential equations to see whether the heat response is modeled with sufficient accuracy. Furthermore, we can explore whether other processes might be directly affected by heat stress. For instance, closer inspection of Figure 11.14 seems to indicate that the uptake of glucose from the medium occurs slightly faster under

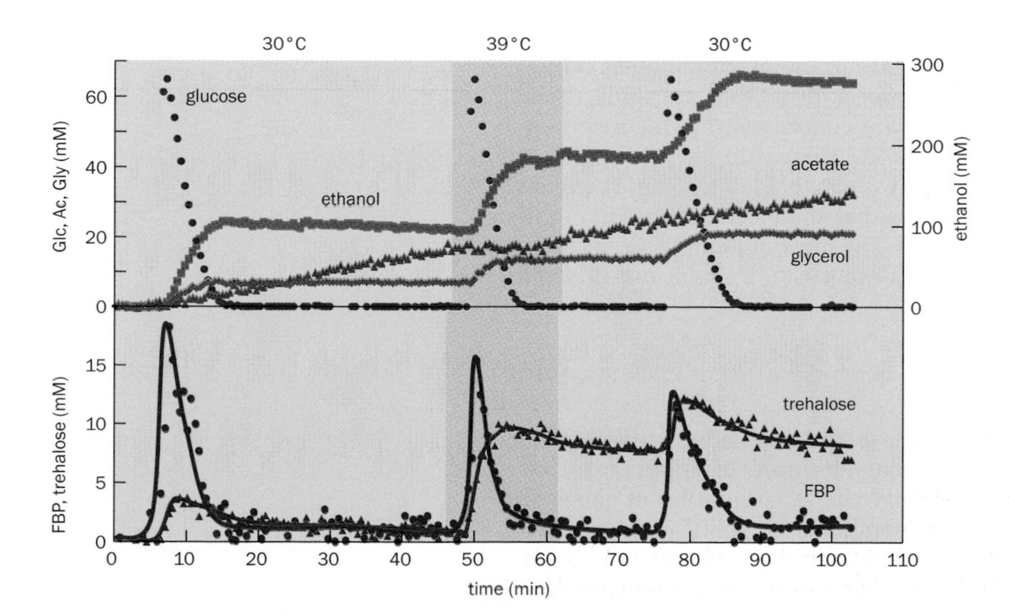

Figure 11.14 *In vivo* NMR profiles of metabolites during the consumption of three pulses of glucose (65 mM). This experiment measured glucose, ethanol, acetate, and glycerol in the medium, and FBP and trehalose inside the cells. Each pulse of glucose was supplied under normal (30ºC), heat stress (39ºC), and recovery (normal temperature) conditions. The pale orange band indicates the heat stress period. (Adapted from Fonseca L, Sánchez C, Santos H & Voit EO. *Mol. BioSyst.* 7 [2011] 731–741. With permission from The Royal Society of Chemistry.)

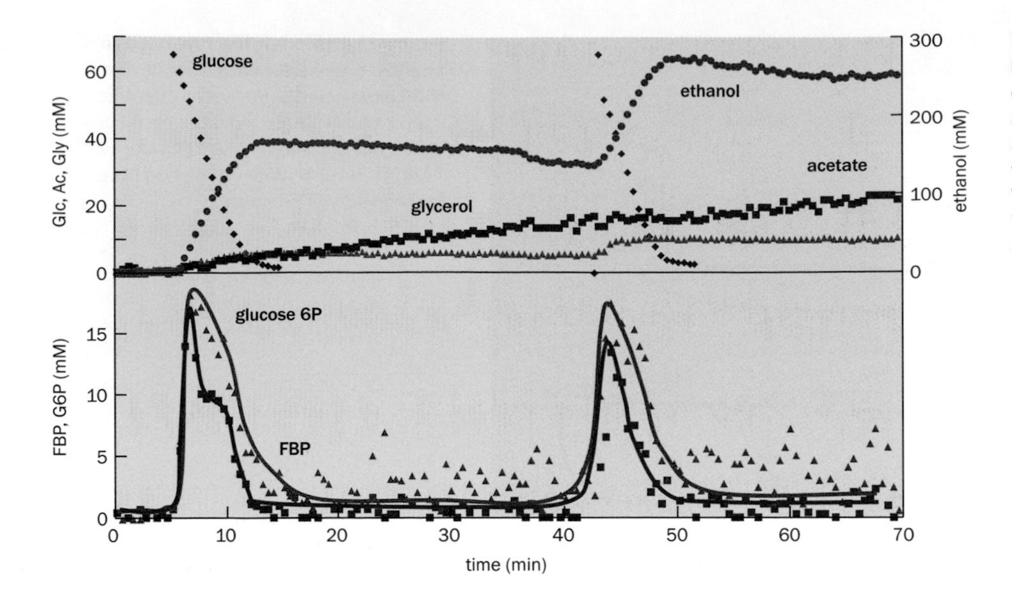

Figure 11.15 Metabolic profile of G6P, FBP, and end products obtained from an experiment with [6-^{13}C]glucose. The blue and pale orange blocks indicate optimal and heat stress conditions. (Adapted from Fonseca L, Sánchez C, Santos H & Voit EO. *Mol. BioSyst.* 7 [2011] 731–741. With permission from The Royal Society of Chemistry.)

heat stress conditions. In fact, it is not difficult to perform a comprehensive simulation study in which one, two, several, or all enzyme activities are made inducible by heat [34].

Before we embark on such modeling of the heat stress response, let us return to an earlier question, namely, why the cell might up-regulate any of the pertinent genes if a metabolic response alone is sufficient. An additional set of experiments, performed specifically to shed light on this question [34], may provide some clues. Because metabolic effects of changes in gene expression only begin to materialize about 20–30 minutes after the initiation of heat stress at the earliest, Fonseca grew cells and exposed them to 39°C heat stress during the last 40 minutes of growth, thereby allowing them time to adjust their gene expression profile. He then repeated the earlier experiment of glucose boluses under normal and then under heat conditions.

The results for these preconditioned cells are shown in **Figure 11.16** (right column) and should be compared with those from normally grown control cells (Figure 11.16, left column). The differences are striking. First, glucose uptake is much slower (note the different timescales). In the normally grown population, the glucose bolus at 30°C is depleted within about 8 minutes, while the preconditioned population needs almost twice as long. At 39°C, the uptake by the control cells is essentially the same as at 30°C, but in the preconditioned cells it has sped up from about 13 minutes to about 8 minutes. Second, the FBP profiles are distinctly different. For the control cells, similar peaks of about 18 mM are reached within 2 minutes at both temperatures. By contrast, the peaks in the preconditioned cells are barely recognizable. Third, and maybe most importantly for this case study, the trehalose profiles are dramatically different. In the control cells, trehalose reaches peaks of about 4 mM at 30°C and 10 mM at 39°C, whereas in the preconditioned cells, the peaks reach about 20 mM (30°C) and over 90 mM (39°C). These comparative results suggest that preconditioning by heat has significant effects later in life, when the cells are again exposed to heat stress. The preconditioning seems to prepare them for a redistribution of fluxes and, especially, a much more forceful response by the trehalose cycle.

11.7 Multiscale Model Design

Now let us design a model that integrates these findings and possibly explains them. The earlier models in (11.1) and (11.2) had the correct basic structure, but it is advantageous for our purposes here to make the separate roles of genes and enzymes explicit. Moreover, the NMR experiments do not include all components of the earlier model, such as UDPG and T6P, but do include others, such as ethanol and acetate (see Figure 11.13), so that it is best to set up a correspondingly

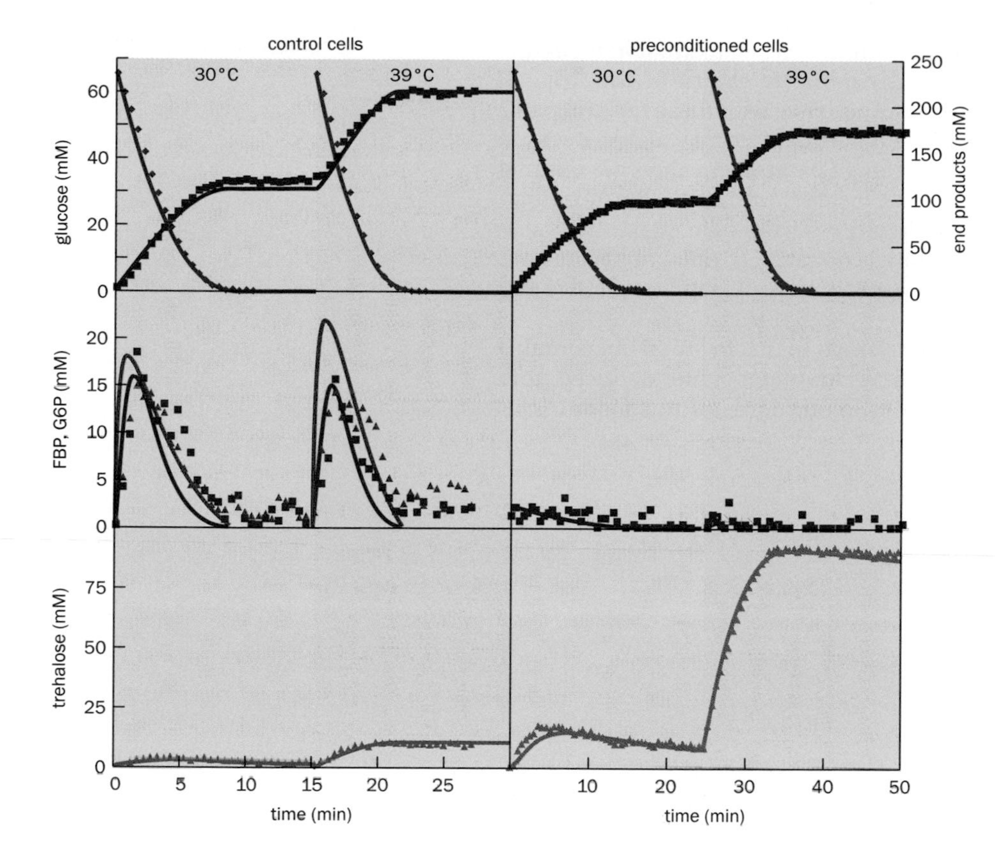

Figure 11.16 Experimental time courses (symbols) and model fits for key representatives of the trehalose cycle. The dark blue symbols and curves in the top row show glucose in the medium, while the dark red symbols and curves show all end products combined. Their concentration is much higher than that of glucose, because they are trioses (three-carbon sugars), while glucose is a hexose (six-carbon sugar). The profiles of FBP (lighter red curves and symbols in the center row) are dramatically different for control cells and preconditioned cells. G6P (green curves and symbols in the center row) was not measured for the preconditioned cells, but the profile is expected to mirror that of FBP. Note the striking differences in trehalose production between control and preconditioned cells (bottom row). Also note the different timescales for control and preconditioned cells, which clearly indicate that the control cells take up glucose more efficiently. (Adapted from Fonseca L, Sánchez C, Santos H & Voit EO. *Mol. BioSyst.* 7 [2011] 731–741. With permission from The Royal Society of Chemistry.)

adjusted model. This situation is quite common: as soon as the focus of an analysis shifts, it might be advisable to set up an amended model that contains exactly those components of the system for which we have data. This switch does not mean that one model is better than the other; it just means that it is often easier to develop models that more closely match the new focus. The new model design does not have to start from scratch, though, because information used for one model can often be used in the alternative model as well. Furthermore, comparisons of model results in overlapping aspects can serve as some sort of quality control. Ultimately, it would of course be desirable to have one model covering all different conditions, but such a high standard is not often achievable, at least not in the beginning.

The new model has to address four scenarios, namely control and preconditioning both at 30°C and at 39°C. Our hypothesis is that the switch from 30°C to 39°C for either control or preconditioned cells is due to heat-induced changes in the activities of some of the enzymes. A sub-hypothesis is that changes in the three enzymes identified by Neves and François [35] are sufficient, while the alternative sub-hypothesis postulates that more than these three enzymes must change in activity to launch the observed response.

The change from control to preconditioning is complicated, because the governing mechanisms have not been identified experimentally. Nonetheless, we can hypothesize that heat preconditioning affects the changes in the regulation of pertinent genes, which in turn have a longer-term effect on the amounts of corresponding enzymes. To test this hypothesis with the model, we can introduce into each power-law term of the model a factor representing the amount of enzyme. Under normal conditions, this factor is set to 1, while it takes a so-far unknown value for heat-preconditioned cells. Thus, each term in the model equations contains, as always, a rate constant and the contributing variables, raised to appropriate kinetic orders, and, in addition, a factor for heat-induced changes in enzyme activity and a factor for preconditioning.

The second issue with the original model is the fact that the parameter values had been collected from different sources and experiments executed under different conditions. In particular, the literature data corresponded to constant-glucose conditions, whereas the new time-series data are obtained during glucose utilization. Furthermore, several parameter values of the original model had to be assumed, based on collective experience with these types of pathways, but could not be tested further, since no suitable data were available. Because of these uncertainties, it is doubtful that the original parameter values are optimal for the new, rather different sets of experiments. Instead, we can use the new time-series data and deduce parameter values directly with the inverse methods discussed in Chapter 5. This estimation procedure is not at all trivial, but because we are at this point not interested in algorithmic aspects, we just skip these issues and focus on the results (see [34] for details).

The new model has a GMA structure that is similar to that in (11.2) but contains a slightly different set of variables, as well as new parameter values deduced from the new time-series data (**Table 11.3**). The symbolic form of the model is as follows:

$$
\dot{X}=\begin{cases}
-F_1/V_{\text{ext}}, \\
(F_1+2F_{11}-F_2)/V_{\text{int}}, \\
(F_2+F_4-F_3-F_9-F_{12}-F_{15})/V_{\text{int}}, \\
(F_3-F_4+F_6-F_5+F_8)/V_{\text{int}}, \\
(F_5-F_6-F_7-F_9)/V_{\text{int}}, \\
(F_7-F_8)/V_{\text{int}}, \\
(F_9-F_{10})/V_{\text{int}}, \\
(F_{10}-F_{11})/V_{\text{int}}, \\
(F_{12}-F_{13}-F_{14})/V_{\text{int}}, \\
2F_{13}/V_{\text{ext}}, \\
F_{14}/V_{\text{int}}, \\
F_{15}/V_{\text{int}},
\end{cases}
\qquad
F=\begin{cases}
B\gamma_1\tau^{P_1}X_1^{h_1}X_3^{hr_1}Q_1^{(T-30)/10}, \\
B\gamma_2\tau^{P_2}X_2^{h_2}X_7^{hr_2}, \\
B\gamma_3\tau^{P_3}X_3^{h_3}Q_3^{(T-30)/10}, \\
B\gamma_4\tau^{P_4}X_4^{h_4}, \\
B\gamma_5\tau^{P_5}X_4^{h_5}, \\
B\gamma_6\tau^{P_6}X_5^{h_6}, \\
B\gamma_7\tau^{P_7}X_5^{h_7}X_3^{hr_3}, \\
B\gamma_8\tau^{P_8}X_6^{h_8}X_3^{hr_4}X_5^{hr_5}, \\
B\gamma_9\tau^{P_9}X_3^{h_9}X_5^{h_{10}}X_2^{hr_6}Q_9^{(T-30)/10}, \\
B\gamma_{10}\tau^{P_{10}}X_7^{h_{11}}Q_{10}^{(T-30)/10}, \\
B\gamma_{11}\tau^{P_{11}}X_8^{h_{12}}Q_{11}^{(T-30)/10}, \\
B\gamma_{12}\tau^{P_{12}}X_3^{h_{13}}, \\
B\gamma_{13}\tau^{P_{13}}X_9^{h_{14}}Q_{11}^{(T-30)/10}, \\
B\gamma_{14}\tau^{P_{14}}X_9^{h_{15}}, \\
0.05F_{12}.
\end{cases}
\tag{11.4}
$$

The left set of equations in (11.4) are the differential equations, where, for instance, the first row is to be read as $\dot{X}_1=-F_1/V_{\text{ext}}$. The right set of equations are all the fluxes, sorted according to Figure 11.13. The fluxes contain the usual rate constants γ and the dependent variables X_i with their kinetic orders, h and hr, which characterize substrate dependence and regulatory influences, respectively. The equations contain several other quantities: V_{ext} is the volume of the cell suspension (50 mL), V_{int} is the total intracellular volume (7.17 mL), and B is the biomass in the reactor (3013 and 2410 mg of dry weight under normal temperature and heat, respectively). The factors τ^{P_i} represent changes in protein amounts due to altered gene expression. We set $\tau=1$ for cells grown under control conditions and $\tau=2$ for heat-preconditioned cells, so that all τ^{P_i} correspond to powers of 2, as is typical in gene expression studies. The values of the different P_i were estimated from the measured time series together with the other parameters. The temperature dependence of each affected enzyme is explicitly modeled as $Q_i^{(T-30)/10}$. Each Q_i is a typical temperature coefficient (Q_{10}) for enzymatic reaction i, which depends on the actual ambient temperature T (either 30 or 39°C) and represents the change in enzymatic activity brought about by a 10°C increase in temperature.

The model fits the data under normal temperature and heat preconditioning quite well (see Figure 11.16). In addition to confirming this fit, we can use the model

TABLE 11.3: MODEL PARAMETERS (RATE CONSTANTS AND KINETIC ORDERS) AND PROTEIN CHANGES OBTAINED FROM TIME-SERIES DATA BY MEANS OF OPTIMIZATION*

Flux	Model step	Rate constant γ	Kinetic orders for substrates†	Kinetic orders for regulators‡	Fold change (2^{P_i})	Q_i
1	HXT	2.87×10^{-5}	0.526 (30°C) 0.472 (39°C)	−0.002 (G6P)	0.7	1.57
2	HXK	1.90×10^{-4}	0.510	−0.209 (Tre6P)	9.2	
3	PGM^F	5.66×10^{-6}	0.400		20.7	1.48
4	PGM^R	3.13×10^{-5}	0.471		17.3	
5	UGP^F	3.58×10^{-5}	0.767		16.2	
6	UGP^R	1.31×10^{-5}	0.159		26.0	
7	GSY	9.43×10^{-7}	0.459	0.000 (G6P)	0.9	
8	GPH	6.94×10^{-8}	0.311	−0.002 (G6P) −0.001 (UDPG)	61.8	
9	TPS1	1.19×10^{-6}	0.659 (G1P) 0.625 (UDPG)	0.000 (Glc)	21.5	2.48
10	TPS2	3.24×10^{-6}	0.361		14.2	2.35
11	NTH	1.99×10^{-7}	0.082		4.9	0.42
12	PFK	2.89×10^{-5}	0.693		1.0	
13	"FBA"§	6.13×10^{-5}	0.369		1.2	1.26
14	Leakage	5.54×10^{-6}	0.672		4.1	
15	ZWF	1.45×10^{-7}	0.693		1.0	

*From Fonseca L, Sánchez C, Santos H & Voit EO. Complex coordination of multi-scale cellular responses to environmental stress. *Mol. BioSyst.* 7 (2011) 731–741. With permission from The Royal Society of Chemistry.
†The parentheses show the temperature associated with the kinetic order for glucose transport or the substrate associated with each kinetic order.
‡The parentheses show the regulating metabolite.
§"FBA" designates the collection of enzymatic steps between fructose 1,6-bisphosphate aldolase and the release of end products.

to explore the questions we posed earlier. First, are the heat-induced changes in enzyme activities, which we discussed earlier, sufficient for an adequate heat stress response? The answer is "Yes" and "No." It is "No" in a sense that the observed changes in the trehalose-associated enzymes (TPS1, TPS2, and NTH1/2) alone are not sufficient to explain the observed trehalose response, because the model indicates a much weaker response than that observed (**Figure 11.17**). However, the answer is a conditional "Yes" if the measured changes in these enzymes are accompanied by a slight reduction in glucose uptake (to about 60% hexose transporter activity; see also Figure 11.16), a 50% increase in phosphoglucomutase and a slighter (about 25%) collective increase in glycolytic activity (see Figure 11.17). Interestingly, *in vitro* activity assays of glycolytic enzymes show similar trends [34].

The data analysis with the model also makes predictions regarding alterations in protein amounts due to preconditioning. No direct data are available for comparisons. However, as we discussed earlier in the chapter, one may translate alterations in gene expression into amounts of protein, at least as a crude substitute. The comparison of the quantitative changes in protein levels with literature data is quite favorable (**Table 11.4**), even though they were obtained with entirely different methods.

Table 11.4 exhibits one striking difference, namely in glycogen synthase. In the time-series (bolus) experiments, the activity of this enzyme is slightly decreased, while it is strongly increased under constant-glucose conditions. The reasons for this difference are unclear, but it could be that the constant glucose supply was more than the cells needed at the time, and therefore allowed them to store some of the material as glycogen. Also, one must keep in mind that the constant-glucose experiments were performed with growing cells under steady-state conditions, while the dynamic experiments used resting cells. Finally, the inferred change in FBP used for non-glycolytic pathways, which in the time-series experiments increased about fourfold, has no analog in the mRNA study.

Summarizing the results from the time-series analysis, the model seems to capture the responses of the yeast cells under different conditions quite well. In particular, it shows that heat-induced changes in enzyme activities are sufficient to mount a

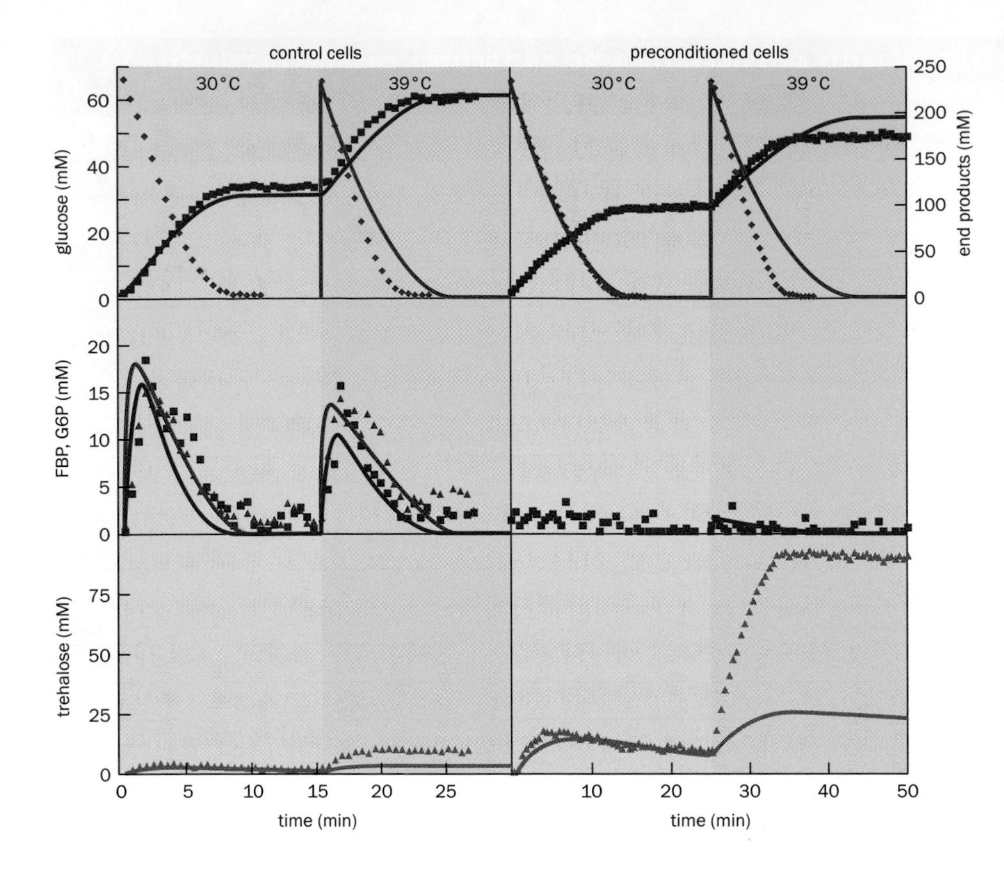

Figure 11.17 Testing the assumption that exclusively TPS1, TPS2, and NTH1/2 are affected by heat [35]. Experimental time courses (symbols) and model fits for key representatives of the trehalose cycle, under the assumption that only the three enzymes TPS1, TPS2, and NTH1/2 are affected by heat. Symbols and colors are the same as in Figure 11.16. (Adapted from Fonseca L, Sánchez C, Santos H & Voit EO. *Mol. BioSyst.* 7 [2011] 731–741. With permission from The Royal Society of Chemistry.)

short-term response, but preconditioning makes such a response much stronger and effective. Furthermore, the cellular adjustments between heat conditions and between normally grown and preconditioned cells seem to be reasonable.

11.8 The Trehalase Puzzle

Have all questions been addressed here? In biology, the answer to such a question is of course seldom “Yes.” Here, we still have an unresolved puzzle surrounding the role of trehalase in the heat stress response. There is clear evidence that

TABLE 11.4: CHANGES IN ENZYME AMOUNTS BETWEEN NORMALLY GROWN AND PRECONDITIONED CELLS, COMPARED WITH HEAT-INDUCED CHANGES IN mRNA LEVELS REPORTED IN THE LITERATURE

Enzyme	Change in amount inferred from time data*		Change in mRNA†
	Forward reaction	Reverse direction	
Hexokinase	9	—	8
Phosphoglucosemutase	21	17	16
UDPG pyrophosphorylase	16	26	16
Glycogen synthase	0.9	—	16
Glycogen phosphorylase	—	62	50
T6P synthase	21	—	12
T6P phosphatase	14	—	18
Neutral trehalase	5	—	6
Phosphofructokinase	1	—	1

*Expressed as ratio between protein amount after preconditioning and protein amount in normally grown cells.
†Data from [2].

the genes coding for trehalase (*NTH1/2* and *ATH1*) are immediately and strongly up-regulated under heat stress. This up-regulation seems at first counterintuitive, because it suggests an increase in the degradation of a metabolite that is clearly needed. We have also seen that the activity of trehalase is directly reduced by heat. Thus, there are concurrent but apparently counteracting changes in response to heat stress. One way to reconcile the opposing changes is to have a closer look at the timing of these processes. The direct reducing effect of heat on the activity of the enzyme very quickly allows the necessary accumulation of trehalose. After about 20 minutes for transcription and translation, the increased gene expression results in a higher amount of enzyme, which apparently compensates for the heat-induced reduction in enzyme activity and keeps trehalose production and degradation in balance. More importantly, as soon as the heat stress ceases, the heat-induced reduction in activity stops, while the amount of enzyme is still increased. This combination of a large amount of enzyme with uninhibited activity allows the cell to remove trehalose very quickly. This removal of unneeded trehalose is apparently crucial for a healthy reentry into normal physiological operation [19, 20].

While this seems reasonable, nature is once again not that simple. In both normally grown and preconditioned cells, the trehalose degradation profile does not seem to differ much under normal temperature, heat, and recovery conditions, although it might be slightly slower during heat stress. Intrigued by these observations, Fonseca and co-workers [34] performed an experiment in which first a bolus of [2-^{13}C]glucose (labeled on carbon position 2) and then a bolus of [1-^{13}C]glucose were given during heat stress. NMR spectroscopy can distinguish these two forms, and analysis of the data revealed the puzzling observation that the "new" trehalose appears to be degraded independently of the previously generated trehalose (**Figure 11.18**). This distinction also seems to occur during recovery, where it appears that the new trehalose is degraded quickly, while the old trehalose is degraded very slowly and in an essentially linear pattern that continues the dynamics that began during heat stress (see Figure 11.14). Our mathematical model assumes that trehalase does not differentiate between the two "types" of trehalose, and therefore it cannot offer an explanation. A possible reason for the differential degradation could be that, under heat conditions, trehalose binds to membranes [40] or associates with hydration cages that form around proteins during heat stress [41]. It could be that it is more difficult for trehalase to access the "old" trehalose, which was formed under heat stress. Further dedicated experiments will be necessary to provide definite answers to these questions. But would the effort of designing new experiments be worthwhile, or is the logic of the argument flawed from the onset?

Instead of waiting for further experimental evidence, let us formulate a specific hypothesis regarding the intriguing degradation patterns of different pulses of trehalose and test with a model analysis whether this hypothesis could be true, at least

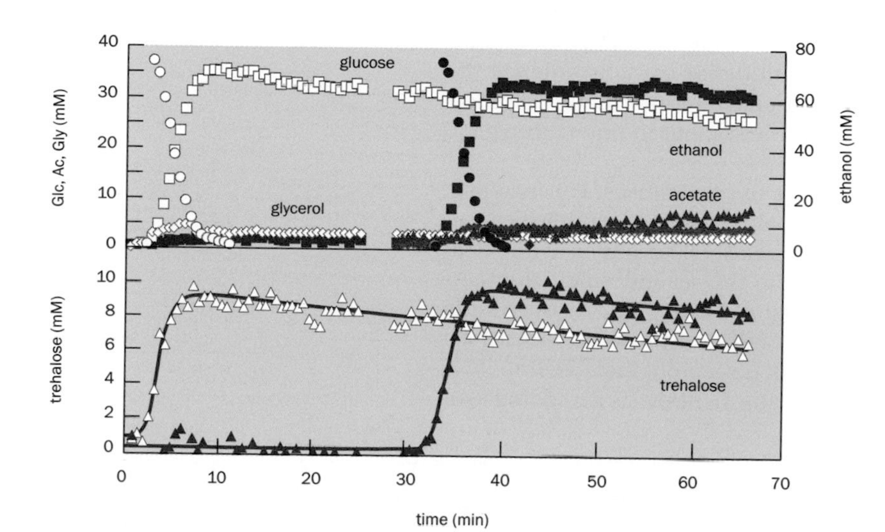

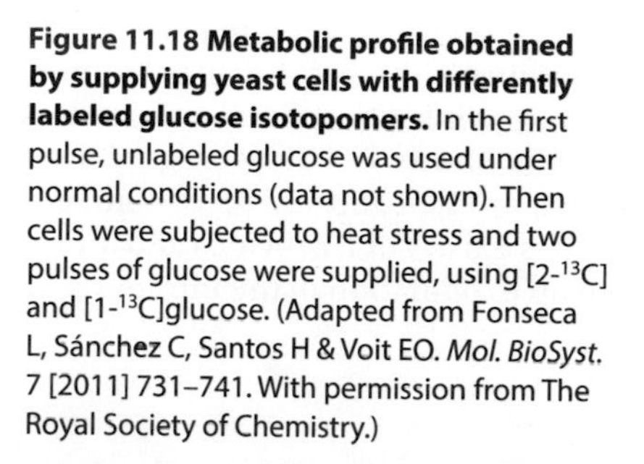

Figure 11.18 Metabolic profile obtained by supplying yeast cells with differently labeled glucose isotopomers. In the first pulse, unlabeled glucose was used under normal conditions (data not shown). Then cells were subjected to heat stress and two pulses of glucose were supplied, using [2-^{13}C] and [1-^{13}C]glucose. (Adapted from Fonseca L, Sánchez C, Santos H & Voit EO. *Mol. BioSyst.* 7 [2011] 731–741. With permission from The Royal Society of Chemistry.)

in principle. Thus, suppose the generated trehalose can be free (TF) or bound (TB) to intracellular structures, such as membranes, and that the enzyme trehalase (E) primarily (or exclusively) degrades TF (**Figure 11.19**). Suppose further that the association (binding) constant k_a between enzyme and substrate is normally much smaller than the dissociation constant k_d, which itself has a low value. If so, then most trehalose at optimal temperature is in the TF state, and its degradation follows something like a power-law or Michaelis–Menten process. Now let us make the assumption that heat stress increases k_a, so that it greatly exceeds k_d. This reversal of association and disassociation causes most trehalose to assume the TB state, which, similar to Styrofoam packaging, protects membranes and other structures from disintegration. The conversion of TB into TF is slow owing to the value of k_d, which is now relatively low in comparison with k_a. Because the concentration of TF is low, the degradation process is substrate-limited and essentially independent of E. Indeed, the concentration of TF is governed by the small k_d and therefore more or less constant, because there is a lot of TB. A second trehalose pulse under heat stress leads to twice as much TB and therefore to roughly a doubled amount of $k_d \times$ TB that becomes TF and can be degraded. Overall, this mechanism would lead to a process that appears to degrade the two trehalose pulses independently. By contrast, if the second trehalose pulse is given under cold conditions, it is immediately degraded according to some biochemical process, while the "old" trehalose is still mostly bound.

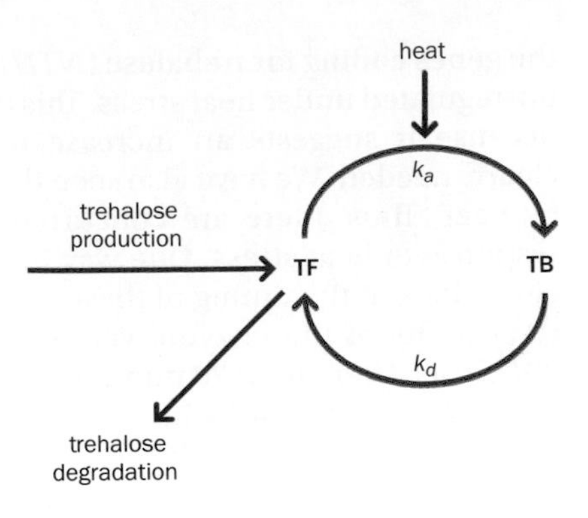

Figure 11.19 Simplified model diagram for exploring the trehalase puzzle. The specific hypothesis tested here is that trehalose may exist in two states, namely, as free (TF) or as bound (TB) trehalose.

An intuitive speculation like that in the previous paragraph might be attractive, but it is prone to logical or numerical inconsistencies. However, given the specificity of the hypothesis, it is not difficult to formulate it as a small mathematical model of the association/disassociation system. Reality is most likely much more complex, but simplicity in such a model analysis is key. After all, the analysis is only supposed to provide proof of concept that the distinction of TF and TB could possibly explain the intriguing trehalose degradation pattern in Figure 11.18. Thus, we translate the diagram of Figure 11.19 directly into a model, which may have the following simple form:

$$\begin{aligned} (\dot{TF}) &= k_d(TB) - k_a(TF) + Bolus - \beta E \cdot (TF)^h, \\ (\dot{TB}) &= k_a(TF) - k_d(TB). \end{aligned} \quad (11.5)$$

Here, the conversions between TF and TB are simple linear processes with rates k_a and k_d, the new production of trehalose is represented by the independent variable *Bolus*, and the degradation of TF is modeled as a power-law function with rate constant β, trehalase as enzyme E, and TF as substrate with a typical kinetic order h. Let us suppose the initial values are $TF(0) = TB(0) = 0.1$, that $k_d = 0.01$, and that $k_a = 0.002$ or $k_a = 5$ for optimal and heat stress conditions, respectively. For normal conditions, we set $E = 1$, but we are not even planning to change the enzyme amount or activity, so that we could really eliminate E from the system altogether. Furthermore, we define free trehalose degradation with $\beta = 0.15$ and $h = 0.1$ or $h = 0.8$ for optimal and heat stress conditions, respectively, and define for convenience the total trehalose concentration $TT = TF + TB$. This variable does not require a differential equation. Notice that all these settings are arbitrary, but in line with typical parameters in power-law modeling (see Chapter 4).

The model can be used to analyze the responses to two boluses. For instance, one might start the system under heat stress with a bolus of freshly synthesized trehalose. In the first simulation, a second bolus is given after 60 minutes, while heat continues. The second simulation is similar, but, with the second bolus, the temperature is set to optimal temperature conditions (**Figure 11.20**). Indeed, the simple model shows trehalose degradation patterns very similar to those observed (see Figures 11.14 and 11.18). Of course, this model output is no proof that two trehalose states exist, but it provides and affirms a possibly testable hypothesis that could lead to an explanation of the intriguing pattern.

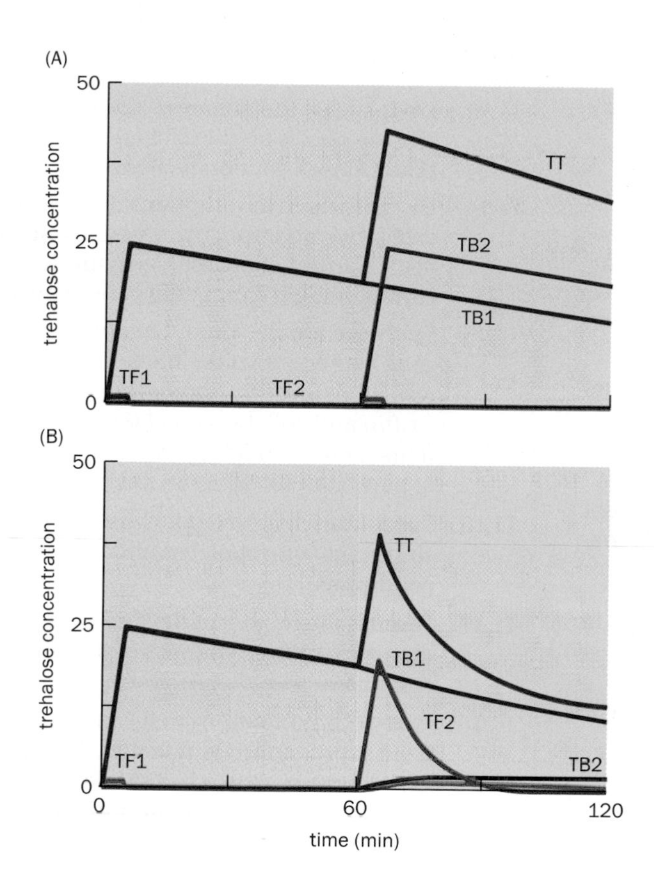

Figure 11.20 Results of model simulations of trehalose degradation under the hypothesis of free (TF) and bound (TB) trehalose (see Figure 11.19, (11.5), and the discussion in the text). Two boluses were administered (indicated by suffixes1 and 2), and TF, TB, and total trehalose (TT) were plotted. (A) corresponds to Figure 11.18 and represents the situation where both boluses are given under heat stress, while (B) corresponds to Figure 11.14, where the second bolus is given under cold conditions. (Adapted from Fonseca L, Sánchez C, Santos H & Voit EO. *Mol. BioSyst.* 7 [2011] 731–741. With permission from The Royal Society of Chemistry.)

CONCLUDING COMMENTS

Although heat stress has been studied extensively over the past few decades, many questions regarding the overall coordination of stress responses continue to puzzle the scientific community. One reason for the ongoing challenge is that typical stress responses are systemic. That is, they involve a wide repertoire of mechanisms at different levels of biological organization, and these mechanisms do not operate in isolation but are interdependent, functionally interwoven, and synergistic. Furthermore, the responses occur on different, yet overlapping, timescales.

We have discussed specific aspects of the heat stress response in yeast, along with small, simplified models that can aid our intuition and help us explain certain aspects, including sometimes confusing observations. Although this system is comparatively small and well understood, we have seen how challenging a detailed understanding of a coordinated response is. As a case in point, we have seen that even a narrow focus on the metabolic aspect of the response cannot be validly addressed without looking at processes in the realms of genes and proteins. Furthermore, the responses in the present case depend critically on the history of the cells. If these have been preconditioned by heat stress earlier in life, their metabolic responses turn out to be much stronger than if the cells were naive. This observation and its broader implications should give us reason to be very cautious with sweeping statements regarding cellular responses or diseases, because we often have no good information regarding the history of a cell or organism, although this history may have a critical role in its future fate.

As we continue to generate increasingly more comprehensive and sophisticated data, it is becoming ever clearer that the unaided human mind will soon reach its limits in integrating the numerous and diverse pieces of quantitative information into a functioning response and that only computational models will be able to organize this information to provide a fuller understanding of complex response systems.

EXERCISES

11.1. Implement the model (11.1) in a software program, such as MATLAB®, Mathematica®, or PLAS. Check whether the system has one or more steady states and whether (any of) these steady states are stable. Revisit the meaning of eigenvalues and their real and imaginary parts in this context. Investigate the sensitivities of the system and recall what exactly the magnitude and sign of a sensitivity or gain means.

11.2. Perform perturbation experiments with the model (11.1) in order to develop a feel for how it responds to slight alterations in numerical features. Investigate transient features of the model with simulations of temporary or persistent increases or decreases in some of the variables or parameters.

11.3. Expand your program to normalize all variables with respect to their steady-state values. What is the main advantage of this normalization?

11.4. Simulate excess and shortage of glucose with the model (11.1) and report your findings.

11.5. Simulate mutations in *TPS1* or *TPS2* with the model (11.1) and report your findings. Search the literature for data and compare them with your modeling results.

11.6. Simulate mutations leading to decreased activities (80%, 20%) of all enzymes, one at a time and two at a time. Discuss to what degree single mutations are predictive of double mutations.

11.7. How would you simulate alterations in enzymatic steps that are catalyzed by two or more enzymes, such as hexokinase I, hexokinase II, and glucokinase? Search the literature for responses to deletions of one of several enzymes catalyzing the same reaction. Write a brief summary report.

11.8. It has been observed that *TPS1* deletion mutants are unable to grow on glucose substrate. Search the literature for details. Discuss how this phenomenon can be demonstrated with the model or why such an analysis is not possible.

11.9. Why is the trehalase reaction in the equations of glucose and trehalose represented with different terms?

11.10. Explain with logic and with simulations why X_1 in (11.2) shows a small, initial overshoot when the amount of enzyme *ZWF1* in the equation for G6P is multiplied by 4 (see Figure 11.11).

11.11. Using the model (11.2), test whether changes in only one or two enzyme activities would result in appreciable increases in trehalose. Report how the metabolic profiles are affected by each change.

11.12. Using the model (11.2), explore different alterations in enzyme activities. Assemble a table showing which alterations lead to significant increases in trehalose. Add one sentence per simulation describing other effects on the system.

11.13 Predict what the influence of the size of the external glucose pool is on the dynamics of the model (11.2). Check your predictions with simulations.

11.14. Discuss which assumptions must be made, explicitly and implicitly, when one substitutes heat-induced changes in enzyme activities with corresponding changes in gene expression.

11.15. Study the article "Complementary profiling of gene expression at the transcriptome and proteome levels in *Saccharomyces cerevisiae*" by Griffin and collaborators [36] and discuss implications for the modeling efforts in this chapter. Look in particular for genes associated with glycolysis.

11.16. Speculate why glycogen phosphorylase might be so strongly up-regulated under heat stress conditions.

11.17. Discuss the impact of the half-lives of mRNAs and enzymes on the dynamics of the heat stress response.

11.18. Carry out simulations that combine alterations in gene expression. Assume that changes in gene expression as shown in **Table 11.5** can be directly translated into corresponding changes in enzyme activities. Explore what happens if only one or two enzyme activities are changed at a time. Summarize your findings in a report.

11.19. Combine changes in gene expression with heat-induced changes in enzyme activities, as observed by Neves and François [35]. Assume at first a constant supply of external glucose. In a second set of simulations, consider glucose depletion.

11.20. Several studies in the literature [2, 31, 32] have analyzed the expression of various genes coding for enzymes of the glycolytic pathway and the trehalose cycle using methods described in the text, as well as other techniques. Summarize the highlights of these investigations in a report.

11.21. Temperature-induced changes in enzymes occur very rapidly, while the expression of genes and the synthesis of new proteins take some while. Discuss the implications of these timescale issues for models of the heat stress response.

11.22. Set up a model to test the hypothesis that trehalose may be present in free or bound form. Use the model (11.5). First set β and *Bolus* = 0 to model the absence of both input and degradation. Study what happens if $k_a = 0.01$ or $k_a = 5$. Set *Bolus* = 5 for 3 time units and study the effect. Set *Bolus* = 5 for 5 time units and activate the trehalase reaction by setting $\beta = 0.15$ and $h = 0.1$.

11.23. Use the model (11.5) under heat stress conditions and compare the results with those under optimal temperature. In a similar study, start with heat conditions and then give a second glucose bolus under either heat or cold conditions.

TABLE 11.5: FOLD INCREASES IN ENZYME ACTIVITIES USED FOR MODELING THE HEAT STRESS RESPONSE

Gene of enzyme or transporter	Variable	Catalytic or transport step	Heat-induced fold increase in enzyme activity
HXT	X_8	Glucose uptake	8
HXK1/2, GLK	X_9	Hexokinase/glucokinase	8
PFK1/2	X_{10}	Phosphofructokinase	1
ZWF1	X_{11}	G6P dehydrogenase	6
PGM1/2	X_{12}	Phosphoglucomutase	16
UPG1	X_{13}	UDPG pyrophosphorylase	16
GSY1/2	X_{14}	Glycogen synthase	16
GPH	X_{15}	Glycogen phosphorylase	50
GLC3	X_{16}	Glycogen use	16
TPS1	X_{17}	α, α-T6P synthase	12
TPS2	X_{18}	α, α-T6P phosphatase	18
NTH	X_{19}	Trehalase	6

11.24. Read some of the literature on heat shock proteins and design a strategy for incorporating them into one of the models discussed here or into a different type of heat stress response model.

11.25. Explore the role of sphingolipids in heat stress and design a strategy for incorporating them into one of the models discussed here or into a different type of heat stress response model.

11.26. Draft a model for the network in Figure 11.1 and develop a strategy for implementing it. Without actually designing this model, make a to-do list of steps and information needed.

REFERENCES

[1] Hohmann S & Mager WH (eds). Yeast Stress Responses. Springer, 2003.

[2] Voit EO. Biochemical and genomic regulation of the trehalose cycle in yeast: review of observations and canonical model analysis. *J. Theor. Biol.* 223 (2003) 55–78.

[3] Paul MJ, Primavesi LF, Jhurreea D & Zhang Y. Trehalose metabolism and signaling. *Annu. Rev. Plant Biol.* 59 (2008) 417–441.

[4] Shima J & Takagi H. Stress-tolerance of baker's-yeast (*Saccharomyces cerevisiae*) cells: stress-protective molecules and genes involved in stress tolerance. *Biotechnol. Appl. Biochem.* 53 (2009) 155–164.

[5] Hottiger T, Schmutz P & Wiemken A. Heat-induced accumulation and futile cycling of trehalose in *Saccharomyces cerevisiae. J. Bacteriol.* 169 (1987) 5518–5522.

[6] Hannun YA & Obeid LM. Principles of bioactive lipid signaling: lessons from sphingolipids. *Nat. Rev. Mol. Cell Biol.* 9 (2008) 139–150.

[7] Jenkins GM, Richards A, Wahl T, et al. Involvement of yeast sphingolipids in the heat stress response of *Saccharomyces cerevisiae. J. Biol. Chem.* 272 (1997) 32566–32572.

[8] Cowart LA, Shotwell M, Worley ML, et al. Revealing a signaling role of phytosphingosine-1-phosphate in yeast. *Mol. Syst. Biol.* 6 (2010) 349.

[9] Gasch AP, Spellman PT, Kao CM, et al. Genomic expression programs in the response of yeast cells to environmental changes. *Mol. Biol. Cell* 11 (2000) 4241–4257.

[10] Vilaprlnyo E, Alves R & Sorrlbas A. Minimization of biosynthetic costs in adaptive gene expression responses of yeast to environmental changes. *PLoS Comput. Biol.* 6 (2010) e1000674.

[11] Palotai R, Szalay MS & Csermely P. Chaperones as integrators of cellular networks: changes of cellular integrity in stress and diseases. *IUBMB Life* 60 (2008) 10–18.

[12] Hahn JS, Neef DW & Thiele DJ. A stress regulatory network for co-ordinated activation of proteasome expression mediated by yeast heat shock transcription factor. *Mol. Microbiol.* 60 (2006) 240–251.

[13] Dickson RC, Nagiec EE, Skrzypek M, et al. Sphingolipids are potential heat stress signals in *Saccharomyces. J. Biol. Chem.* 272 (1997) 30196–30200.

[14] Görner W, Schüller C & Ruis H. Being at the right place at the right time: the role of nuclear transport in dynamic transcriptional regulation in yeast. *Biol. Chem.* 380 (1999) 147–150.

[15] Lee P, Cho BR, Joo HS & Hahn JS. Yeast Yak1 kinase, a bridge between PKA and stress-responsive transcription factors, Hsf1 and Msn2/Msn4. *Mol. Microbiol.* 70 (2008) 882–895.

[16] KEGG (Kyoto Encyclopedia of Genes and Genomes) Pathway Database. http://www.genome.jp/kegg/pathway.html.

[17] BioCyc Database Collection. http://biocyc.org/.

[18] Galazzo JL & Bailey JE. Fermentation pathway kinetics and metabolic flux control in suspended and immobilized *Saccharomyces cerevisiae. Enzyme Microb. Technol.* 12 (1990) 162–172.

[19] François J & Parrou JL. Reserve carbohydrates metabolism in the yeast *Saccharomyces cerevisiae. FEMS Microbiol. Rev.* 25 (2001) 125–145.

[20] Nwaka S & Holzer H. Molecular biology of trehalose and the trehalases in the yeast *Saccharomyces cerevisiae*. *Prog. Nucleic Acid Res. Mol. Biol.* 58 (1998) 197–237.
[21] Paiva CL & Panek AD. Biotechnological applications of the disaccharide trehalose. *Biotechnol. Annu. Rev.* 2 (1996) 293–314.
[22] Lillford PJ & Holt CB. *In vitro* uses of biological cryoprotectants. *Philos. Trans. R. Soc. Lond. B Biol. Sci.* 357 (2002) 945–951.
[23] Hayashibara. The sugar of life. *Time*, November 22 (2010) Global 5; see also http://www.hayashibara.co.jp/.
[24] Savageau MA. Biochemical Systems Analysis: A Study of Function and Design in Molecular Biology. Addison-Wesley, 1976.
[25] Torres NV & Voit EO. Pathway Analysis and Optimization in Metabolic Engineering. Cambridge University Press, 2002.
[26] Voit EO. Computational Analysis of Biochemical Systems: A Practical Guide for Biochemists and Molecular Biologists. Cambridge University Press, 2000.
[27] BRENDA: The Comprehensive Enzyme Information System. http://www.brenda-enzymes.org/.
[28] Curto R, Sorribas A & Cascante M. Comparative characterization of the fermentation pathway of *Saccharomyces cerevisiae* using biochemical systems theory and metabolic control analysis. Model definition and nomenclature. *Math. Biosci.* 130 (1995) 25–50.
[29] Schlosser PM, Riedy G & Bailey JE. Ethanol production in baker's yeast: application of experimental perturbation techniques for model development and resultant changes in flux control analysis. *Biotechnol. Prog.* 10 (1994) 141–154.
[30] Polisetty PK, Gatzke EP & Voit EO. Yield optimization of regulated metabolic systems using deterministic branch-and-reduce methods. *Biotechnol. Bioeng.* 99 (2008) 1154–1169.
[31] Vilaprinyo E, Alves R & Sorribas A. Use of physiological constraints to identify quantitative design principles for gene expression in yeast adaptation to heat shock. *BMC Bioinformatics* 7 (2006) 184.
[32] Voit EO & Radivoyevitch T. Biochemical systems analysis of genome-wide expression data. *Bioinformatics* 16 (2000) 1023–1037.
[33] Eisen MB, Spellman PT, Brown PO & Botstein D. Cluster analysis and display of genome-wide expression patterns. *Proc. Natl Acad. Sci. USA* 95 (1998) 14863–14868.
[34] Fonseca L, Sánchez C, Santos H & Voit EO. Complex coordination of multi-scale cellular responses to environmental stress. *Mol. BioSyst.* 7 (2011) 731–741.
[35] Neves MJ & François J. On the mechanism by which a heat shock induces trehalose accumulation in *Saccharomyces cerevisiae*. *Biochem. J.* 288 (1992) 859–864.
[36] Griffin TJ, Gygi SP, Ideker T, et al. Complementary profiling of gene expression at the transcriptome and proteome levels in *Saccharomyces cerevisiae*. *Mol. Cell. Proteomics* 1 (2002) 323–333.
[37] Stanford_Gene_Database. http://genome-www.stanford.edu/yeast_stress/data/rawdata/complete_dataset.txt.
[38] Sur IP, Lobo Z & Maitra PK. Analysis of *PFK3*—a gene involved in particulate phosphofructokinase synthesis reveals additional functions of *TPS2* in *Saccharomyces cerevisiae*. *Yeast* 10 (1994) 199–209.
[39] Thevelein JM & Hohmann S. Trehalose synthase: guard to the gate of glycolysis in yeast? *Trends Biochem. Sci.* 20 (1995) 3–10.
[40] Lee CWB, Waugh JS & Griffin RG. Solid-state NMR study of trehalose/1,2-dipalmitoyl-*sn*-phosphatidylcholine interactions. *Biochemistry* 25 (1986) 3739–3742.
[41] Xie G & Timasheff SN. The thermodynamic mechanism of protein stabilization by trehalose. *Biophys. Chem.* 64 (1997) 25–34.

FURTHER READING

Eisen MB, Spellman PT, Brown PO & Botstein D. Cluster analysis and display of genome-wide expression patterns. *Proc. Natl Acad. Sci. USA* 95 (1998) 14863–14868.
Fonseca L, Sánchez C, Santos H & Voit EO. Complex coordination of multi-scale cellular responses to environmental stress. *Mol. BioSyst.* 7 (2011) 731–741.
Gasch AP, Spellman PT, Kao CM, et al. Genomic expression programs in the response of yeast cells to environmental changes. *Mol. Biol. Cell* 11 (2000) 4241–4257.
Görner W, Schüller C & Ruis H. Being at the right place at the right time: the role of nuclear transport in dynamic transcriptional regulation in yeast. *Biol. Chem.* 380 (1999) 147–150.
Hannun YA & Obeid LM. Principles of bioactive lipid signaling: lessons from sphingolipids. *Nat. Rev. Mol. Cell Biol.* 9 (2008) 139–150.
Hohmann S & Mager WH (eds). Yeast Stress Responses. Springer, 2003.
Vilaprinyo E, Alves R & Sorribas A. Use of physiological constraints to identify quantitative design principles for gene expression in yeast adaptation to heat shock. *BMC Bioinformatics* 7 (2006) 184.
Vilaprinyo E, Alves R & Sorribas A. Minimization of biosynthetic costs in adaptive gene expression responses of yeast to environmental changes. *PLoS Comput. Biol.* 6 (2010) e1000674.
Voit EO & Radivoyevitch T. Biochemical systems analysis of genome-wide expression data. *Bioinformatics* 16 (2000) 1023–1037.
Voit EO. Biochemical and genomic regulation of the trehalose cycle in yeast: review of observations and canonical model analysis. *J. Theor. Biol.* 223 (2003) 55–78.

Physiological Modeling: The Heart as an Example

12

When you have read this chapter, you should be able to:

- Understand the basics of the anatomy and physiology of the heart
- Discuss the electrochemical processes during contraction and relaxation
- Formulate black-box models of heartbeats
- Analyze the role of perturbations and pacemakers in oscillators
- Explain the biology and mathematical formulation of action potentials
- Understand the principles and challenges of multiscale modeling

The heart is an incredible machine. A little bigger than a fist and weighing just about half a pound, it pumps blood without ceasing, beating roughly 100,000 times every day, between two and three billion times in a normal lifetime. It keeps on working without us thinking about it, but beware if it skips even a few beats in a row! Many societies have given the heart exalted roles, associating it with life, love, and stress. Just think: what would Valentine's Day be without that heart shape? The fact that it is really quite different from the heart's true anatomy (**Figure 12.1**) is easily forgotten on February 14th.

Virtually unlimited sources of general and specific information are readily available about all aspects of the heart, from its cells and tissues to its anatomy, from its normal electrical and mechanical functioning to the plethora of possible problems and pathologies. The easiest access to matters of the heart is probably the large body of textbooks, as well as an enormous variety of websites (for example, [1–5]), which we will not specifically cite in the following.

The role of the heart is to move blood through the body and to supply the cells with most of what they need. The blood carries oxygen and nutrients, and transports unwanted chemicals away. The heart accomplishes this feat by a regular pumping action that moves blood through a network of about 60,000 miles of arteries, veins, and capillaries. Every minute, it pumps about 5 liters of blood, for a total of over 7000 liters per day. If the heart stops, we are in trouble. Coronary heart disease is responsible for about half of all deaths in Western countries, and about 700,000 people die from heart attacks every year in the United States alone.

This chapter discusses several aspects of the function of the heart with mathematical models in the form of vignettes. Some of the models are very simple, others are more complicated, and some of the simulation models that have been developed for academic and commercial purposes are extremely complex, and we will only mention them. To some degree, the usefulness of these models correlates with their complexity, and the most sophisticated simulation models have become accurate enough to allow realistic studies of the normal heart and some of its diseases.

Nonetheless, the simpler models are of interest as well, because they give us insights into the heart's macroscopic oscillatory patterns and its basic electrical and chemical properties, without overwhelming us with too much detail.

The strategy of the chapter is as follows. It begins with a general background section that is rather superficial and by no means does justice to the amazing features of the heart; a more detailed, very accessible introduction can be found in [1]. Next, we introduce some simple black-box models that describe and analyze heartbeat-like oscillations. Looking a little deeper into the physiology of the heart, we find that its function is directly associated with oscillations in the dynamics of calcium. Thus, the next step in our modeling efforts is to describe calcium oscillations with a minimalistic ordinary differential equation model and to explore how far such a model can lead us in understanding heart function. Indeed, we will see that quite simplistic models are sufficient to describe stable calcium oscillations. However, these models are of course very limited and, for instance, do not account for the **electrochemical** processes that we know to exist in the heart. We therefore dig deeper, trying to understand how chemistry and the laws of **electrophysiology** govern the contractions in individual heart cells, and how the activities of all heart cells are coordinated in a fashion that results in proper heart pumping. We stop short of the most sophisticated simulation models of the heart, because they are naturally very complicated. Nonetheless, we do briefly discuss their underlying principles. Thus, we will start at the macroscopic level, dive into metabolism, electrophysiology, chemistry, and genetics, and ultimately return to the effects of molecular changes on the functioning of the whole heart.

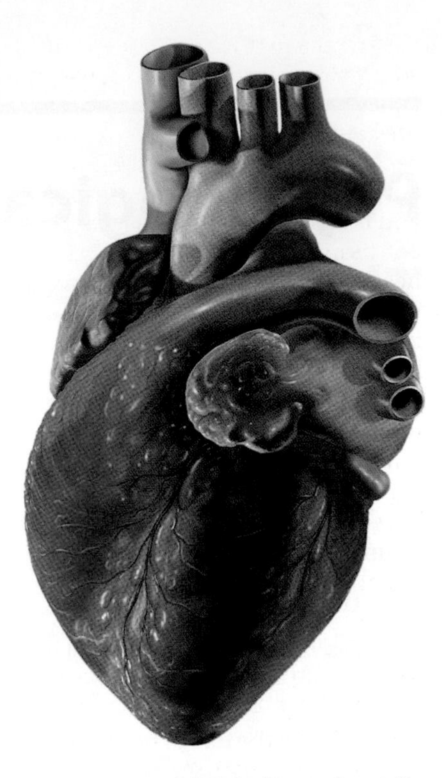

Figure 12.1 Model of the human heart. The heart is a complicated muscle that encloses four chambers. With each beat, the heart dynamically changes its shape and thereby pumps blood to the lungs and throughout the body. (Courtesy of Patrick J. Lynch under the Creative Commons Attribution 2.5 Generic license.)

This gradual procedure of starting at a high level with a rather coarse model and then studying selected aspects with a finer resolution and with an increasing level of sophistication mimics the way we learn many things when growing up. Early on, we distinguish static items from things that move, we learn to differentiate between living and engineered machines that move, we become able to distinguish cars from trucks, and cheap cars from expensive cars, and in some cases we learn to discern the year a car was made even if the differences from one year to the next are very subtle. In modeling, a similar manner of learning has been proposed in biomechanics, where one might begin with a macroscopic, high-level assessment of locomotion, and subsequently embed increasingly finer-grained models into these first coarse models [6]. In this biomechanical context, the high-level models are called templates. They help the modeler organize anchors, which are more complex models representing anatomical and physiological features in a detailed fashion and can be incorporated into the template models in a natural manner.

Numerous models of the heart have been developed over the past century that could serve as templates or anchors. In fact, many investigators have devoted their careers to heart modeling. Some giants in the field include James Bassingthwaighte, Peter Hunter, Denis Noble, Charles Peskin, and Raimond Winslow. Moreover, the electrochemical processes in heart muscle cells show a number of resemblances with nerve cells, and descriptions of the dynamics of beating heart cells have been inspired directly by experimental and computational research on nerve cells, which began more than half a century ago with the pioneering work of Alan Lloyd Hodgkin and Andrew Huxley and was extended and further analyzed by leaders in the field of neuroscience, including Richard FitzHugh, Jin-Ichi Nagumo, and John Rinzel (see, for example, [7]). Indeed, 50 years ago, Noble [8, 9] showed with the first computer model of its kind that action potentials in the heart can be explained with equations similar to those proposed by Hodgkin and Huxley.

HIERARCHY OF SCALES AND MODELING APPROACHES

It is quite obvious that the function of the heart spans many scales in size, organization, and process speed. Beginning with the most evident scale of a whole organ, the heart is a peristaltic pump that moves blood. This pump is composed of tissues, which need to work in a tightly coordinated fashion. The tissues consist of several types of cells, which have their specific functions within the concerted effort of making the heart beat. Within and between the cells, there is a lot of controlled movement of molecules and especially **ions**, which are directly associated with the electrical activity of the heart. Inside the cells, we find metabolites and proteins, some of which are found in many cells, while others are specific. Finally, there are genes, some of which are only expressed in heart cells and can cause problems if they are mutated.

Ideally, complementary models of all these aspects would seamlessly span the entire spectrum from electrical and chemical features to intracellular mechanical and metabolic events, from cells to cell aggregates and tissues, and from the different tissues to a complete picture of the normal physiology and pathology of the heart. However, such a complete multiscale story has not been written, and for the demonstration in this chapter, we will select some aspects, while unfortunately having to minimize—or even omit—others. For instance, we will not discuss in any length very important issues of tissue mechanics and blood flow, three-dimensional models of the anatomy and physiology of the heart, four-dimensional changes in the heart over short and long time horizons, or mechanisms of adaptation and remodeling after infarction. Good reviews of pertinent research questions, modeling needs, and accomplishments at different hierarchical levels of organization have been presented within the framework of the Physiome, an international effort to understand normal and abnormal functioning of the heart through computational modeling and to translate the insights gained from modeling into advances in clinical applications (see, for example, [10–15]). Like other large-scale efforts, such as the Virtual Liver Network [16], the Lung Physiome [17], and the Blue Brain Project [18], the target of model development for these complex, multiscale systems is not necessarily a single all-encompassing supermodel that would account for every detail, but rather a set of complementary models of different dimensionality and sophistication that explain certain aspects. The reason for not targeting a single model is that the relevant time and size scales in the heart and other organs are so vastly different that it seems impossible, at least with current methods, to merge them in a meaningful fashion [10]. Nonetheless, the various models should be able to talk to each other, which the Physiome Project facilitates by using cell- and tissue-specific mark-up languages (XMLs), such as CellML, SBML, and FieldML [13, 19, 20].

12.1 Basics of Heart Anatomy

Before we discuss different modeling approaches, we need at least a coarse description of our subject. The human heart is a muscle that contains four chambers, namely, the left and right **atria** and the left and right **ventricles** (**Figure 12.2**).

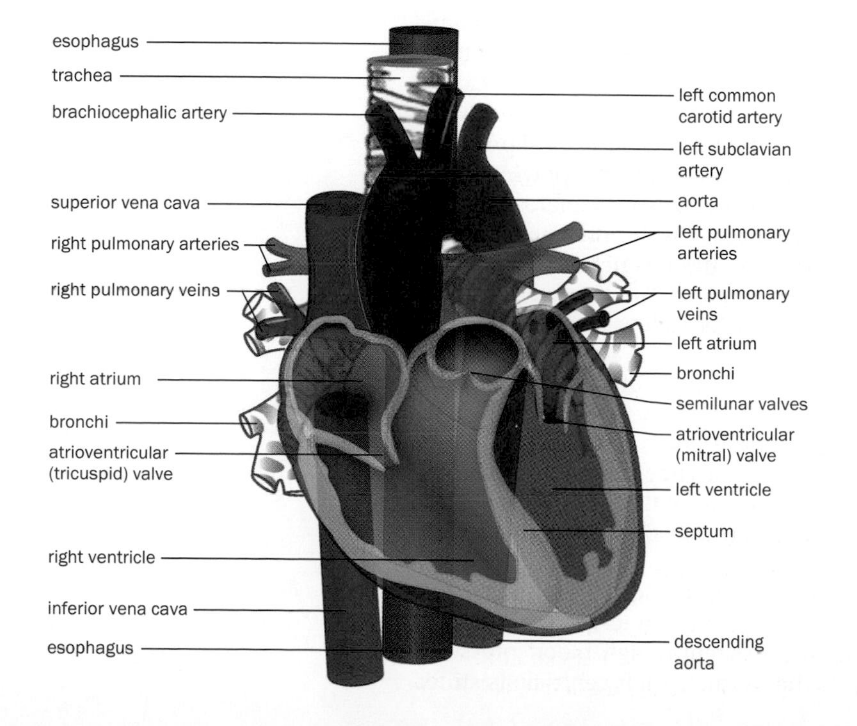

Figure 12.2 Diagram of the parts of the human heart. In healthy humans, the left chambers (atrium and ventricle) are completely separated from the right chambers. During the cardiac cycle, blood runs through the four chambers in a well-coordinated fashion. (Courtesy of ZooFari under Creative Commons Attribution-Share Alike 3.0 Unported license.)

The left and right sides of a human heart become strictly separated during embryonic development, so there is normally no direct blood flow between them. With each heartbeat, blood flows through the chambers in a well-controlled, cyclic manner that consists of two phases called **diastole** and **systole**, which are Greek terms for dilation and contraction, respectively. Because the process is cyclic, it does not really matter where we start. So, let us begin with the phase of early diastole. At this point, the heart is relaxed and begins to receive blood in two states. The first is deoxygenated, that is, relatively low in oxygen content, because it returns from all parts throughout the body, where oxygen had diffused from the bloodstream into the various tissues. The second state of blood is oxygenated. It comes from the lungs, where hemoglobin molecules in the red blood cells have been loaded with oxygen.

The right atrium exclusively receives the deoxygenated blood through two branches of the vena cava, which collect blood from the various tissues of the body. Meanwhile, the left atrium exclusively receives oxygenated blood through the pulmonary vein that comes from the lungs. The atria fill quickly, and because the atrioventricular (AV) valves are open (see Figure 12.2), so do the ventricles. These are in a relaxed state where their wall tissue is thin and the plane of the AV valves shifts back up toward the base of the heart. A following contraction by the atria pushes additional blood into the ventricles. During the second phase, the ventricles begin to contract and their walls thicken. The AV valves close and the semilunar outflow valves open. Deoxygenated blood is sent through the pulmonary artery on its way to the lungs, while oxygenated blood enters the aorta, the largest artery in the body, from where it is distributed throughout the body, first through increasingly smaller arteries, and finally through tiny capillaries that are found in all parts of the body and release oxygen to adjacent cells. The capillaries turn into small and then larger veins, which ultimately return the deoxygenated blood to the right atrium. The total volume of the heart does not change significantly during the heartbeat.

12.2 Modeling Targets at the Organ Level

At the macroscopic level of an organ, the heart acts like a peristaltic pump—or perhaps it is better to say that the heart is a combination of two pumps that drive different portions of the circulation of blood. Unlike most engineered pumps, the heart is highly sensitive to its mechanical surroundings and constantly adjusts influx and efflux in response to the current requirements of the body and various stresses. These adjustments occur through autonomic controls that do not require conscious input. In fact, very few people are capable of conscientiously affecting their heartbeat, although mental stresses such as worry or fear can easily alter it.

The adjustment mechanism of the heart to different pumping demands is characterized by the **Frank–Starling law** [21]. This law is derived from a **nonlinear partial differential equation (PDE)**, developed in the nineteenth century and called the Young–Laplace equation, which generically relates pressure differences in fluids to the shape of the wall retaining the fluid. Broadly, it states that increased pressure requires increased wall thickness to maintain a stable wall tension. Applying these relationships to the heart, the Frank–Starling law formalizes how increasing the venous blood input to the ventricle stretches the ventricular wall, enhances contractility, and elevates the diastolic pressure and volume of the ventricle, which in turn leads to an increased stroke volume [22]. Thus, the heart uses an adaptive response mechanism that adjusts each ventricular output to its inflow and ensures the balance between the outputs of the two ventricles over time.

The balancing of venous input and **cardiac** output occurs without external regulation. The heart accomplishes this control task by adjusting the force of contraction proportionally to changes in the lengths of its muscle fibers. This adjustment is related to a decreased spacing of **filaments** within the **myocytes**, the formation of cross-bridges between filaments during each contraction of the heart muscle, and an increased sensitivity of the filaments to calcium (see later in this chapter and [23]).

The physical relationship between pressure and wall thickness expressed in the Frank–Starling law has been used to explain the gradual thickening of arteries and of the left-ventricular wall in response to persistently high blood pressure. The unfortunate consequence of this trend is that a thicker left ventricle is stiffer

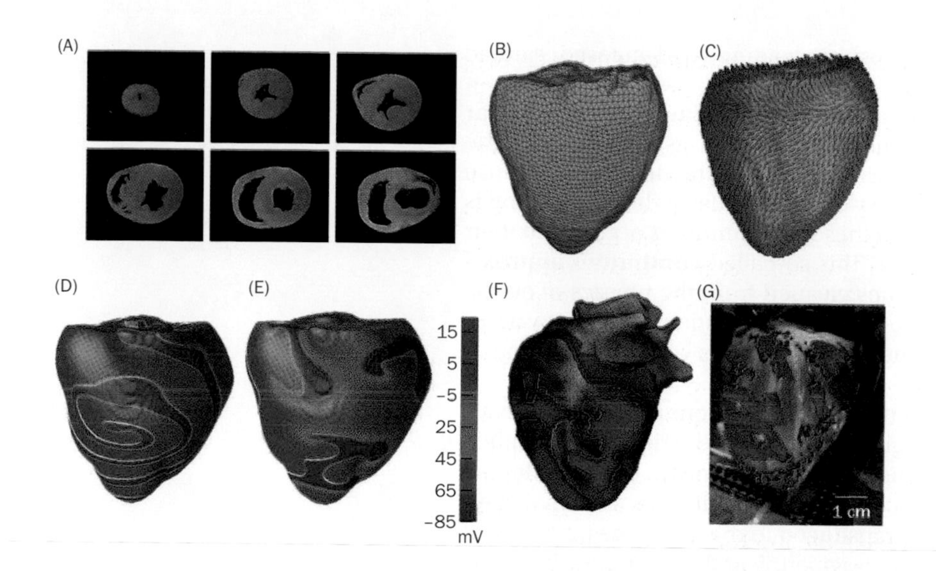

Figure 12.3 Images of ventricular tachycardia and fibrillation, which are potentially life-threatening arrhythmias that originate in one of the ventricles of the heart. (A) Transverse sections from a sophisticated MRI scan of a rabbit heart, proceeding from apex to base. (B) High-resolution reconstruction of the rabbit ventricular structure. (C) Orientation of fibers on the rabbit ventricular structure. (D) Simulation of ventricular tachycardia, in the form of a single spiral wave, in the rabbit ventricular structure. (E) Simulation of ventricular fibrillation, in the form of multiple spiral waves, in the rabbit ventricular structure. (F) Simulation of ventricular fibrillation in a reconstructed canine heart. (G) Experimental ventricular fibrillation in a canine wedge preparation recorded through optical mapping. Images of arrhythmias in motion can be found at the interactive website http://thevirtualheart.org. (Courtesy of Elizabeth M. Cherry, Alfonso Bueno-Orovio and Flavio H. Fenton, Cornell University. With permission from Elsevier.)

than normal, so that filling it with blood requires elevated pressure. Over time, this altered pressure can lead to a decline in performance during diastole and eventually to diastolic heart failure.

Beyond the important, yet somewhat generic, insights gained from the Frank–Starling law, modern methods of **fluid mechanics**, engineering, computing, and modeling permit detailed and very sophisticated analyses of the spatial organization and three-dimensional architecture of the heart and their impact on proper functioning. These analyses are crucial for an understanding of healthy and perturbed blood flow in the atria and ventricles, as well as in the **coronaries**—the arteries and veins that provide the heart muscle with oxygen and nutrients and remove breakdown products. These vessels are rather narrow and prone to blockage, which can lead to atherosclerosis and heart attacks. Modeling blood flow as well as the mechanical deformation of the heart during the cardiac cycle requires systems of PDEs and **finite element approaches**, which subdivide a three-dimensional body into very small cubes and quantify changes within and between these cubes over time. Finite element models faithfully reflect the anatomy of the heart, including the orientation of fibers within the layers of the heart wall. They are fundamental to predicting whole-organ function based on features of the heart tissue that vary from person to person, for instance when the wall thickens owing to disease [15, 24]. Researchers at the École Polytechnique Fédérale de Lausanne (EPFL) have developed a very sophisticated three-dimensional supercomputer model of coronary blood flow that is able to predict heart attacks [25].

In addition to mechanical features and issues of blood flow, comprehensive organ-level models need to address the electrical activation of heart contractions, which initiate at the sino-atrial node and spread to the Purkinje fibers and the entire **myocardium** (heart muscle), as discussed in detail later. Fascinating pictures and movies of electrical activity waves can be found in [13, 26] and on many websites; an example is shown in **Figure 12.3**. Finally, all cells in the heart need energy, which requires efficient transport of metabolites and oxygen between the coronaries and the myocardium, and this needs to be taken into account in realistic organ models.

12.3 Modeling Targets at the Tissue Level

The pumping action of the heart is accomplished with contractile tissues that contain specialized cells, called cardiac cells, myocardial cells, or **cardiomyocytes**. These cells are in many ways similar to skeletal muscle cells, but, in contrast to the latter, the coordinated pumping action of the heart does not require our conscious input. Also, the heart muscle is very resistant to fatigue, because the cardiomyocytes contain many mitochondria, which are responsible for respiration and energy supply, and a large number of myoglobins, which are proteins that are able to store oxygen. Furthermore, the heart muscle is richly equipped with coronary blood vessels, which provide oxygen and nutrients to its cells. If the heart does not receive sufficient amounts of oxygen, an infarct results, leading to decreased cardiac function, tissue damage, and possibly death.

Models of the heart at the tissue level account for structural details and for properties of electrical **conductance**. Prominent structural aspects include the mechanics of the deformation of the heart tissue during contractions [11]. Because the heart contains roughly 10^{10} myocytes, a key concept of tissue-level modeling is the approximation of the discrete aggregate of these large numbers of cells and their interactions with a collective representation. This so-called **continuum approximation** permits the application of approaches gleaned from the physics of elastic materials, which deals with forces, stresses, strains, deformations, and distortions, and helps explain shape changes during the cardiac cycle as well as the spread of electrical signals.

The electrical signals affecting the heart tissue and its billions of cardiomyocytes are coordinated by a primary pacemaker region in the wall of the right atrium, called the **sinoatrial (SA) node**. We will discuss this node several times in the following, so it suffices to state here that the SA node cells beat on their own, and that their oscillation in electric membrane charge is transduced to the **excitable** myocytes of the atria and ventricles. This traveling signal, originating at the SA node, ultimately makes the heart contract in a well-controlled manner. The SA node consists of a group of modified cardiac myocytes that spontaneously contract at the order of 100 times per minute.

The electrical signals generated in the SA node cause the atria to contract. They also spread along dedicated conduction channels in the walls of the atria and quickly reach the **atrioventricular (AV) node**, which constitutes the electrical connection between the atria and ventricles. Signals from the SA node reach the AV node with a delay of about 0.12 s. This delay may sound trivial, but is very important because it ensures the correct timing between the ejection of blood from the atria and the contraction of the ventricles. The AV node contains a control system that slows down conduction if the node is stimulated too frequently. This mechanism, together with a **refractory period** where further triggering is not possible (see later), prevents contractions of the ventricles that are too rapid and could be hazardous. Signals from SA and AV nodes subsequently run through the conductive **bundle of His**, named after the Swiss cardiologist Wilhelm His, and to the **Purkinje fibers**, which branch out throughout the walls of the ventricles. The bundle and Purkinje fibers are specialized to conduct electrical signals rapidly to the myocardium. AV node and Purkinje cells can fire autonomously, but because their rate is slow, the electrical signals are normally dominated by the SA node.

The beating rate of the SA node is constantly modified by impulses from the **autonomic nervous system (ANS)**, which generally acts as a control system that maintains homeostasis in the body, so that the ultimate heart rate under resting conditions is only about 70 beats per minute, but may go up to 200 beats during strenuous exercise. ANS activities occur without conscious control, sensation or thought. The heart rate can also be affected by noradrenaline (also known as norepinephrine), the most common neurotransmitter of this system, and the hormone adrenaline (also known as epinephrine). We all know that sudden excitement or stress can lead to a burst in adrenaline (and noradrenaline) and sudden increases in heartbeat that get us ready for "fight or flight." Alas, phenomena like the ANS-controlled pacemaker activity in the SA node are usually more complex than they seem at first. As an analogy, consider the act of focusing on near or distant objects, which requires the combined voluntary actions of the extra-ocular muscles to move the eyes and of smooth muscles, controlled by the ANS, to change the shape of the lens. Clearly, the focusing task combines voluntary and involuntary actions. Similarly, the heartbeat is not entirely controlled by the ANS, and, for instance, mobilization of the cardiovascular system is driven in advance of exercise to ready the musculature for motor activity.

As in the case of whole-organ studies, the spatial and mechanical properties of the heart tissue are of great importance at the tissue level, and computational (finite element) analyses are crucial for understanding the design principles underlying a successful electrical propagation from the pacemaker cells to the surrounding, excitable heart muscle [27]. For instance, it is important that the electrical coupling between pacemaker cells and excitable cells in the ventricles suppresses electrical interactions at inappropriate times. This suppression depends on a variety of factors, such as the intercellular coupling among the heart's pacemaker cells and differences in their membrane properties, as well as the communication among the surrounding cells, which depends on the number and spatial orientation of gap junctions connecting them. Thus, analyzing the spatial organization of the heart at the tissue level is important for a deep understanding of the correct propagation of electrical signals throughout the heart, the fine-tuned timing of contractions of the chambers, the pathological responses of the heart close to areas of infarction, and even for correct interpretations of ECG readings (see later).

12.4 Modeling Targets at the Cell Level

The cells of the heart have intriguing specific features, but of course they also contain generic components that would have to be included in a comprehensive model of the heart. Not surprisingly, heart cells express different genes than other cells, their metabolism requires more energy, they contain proteins some of which are uncommon in other cells, and some of their signaling systems are germane to the function of the heart. Like skeletal muscle cells, most cells of the heart muscle are contractile, and the mechanics of the responsible filaments constitutes an interesting target for modeling. The electrical activity of these cells is tightly associated with the flux of ions, so that ion transport and electrochemical models demand special attention (see later). Finally, specifically adapted genomic, proteomic, and metabolic models are required for understanding the particular features of the heart. Some interesting modeling work in this context is presented in [28, 29].

Of crucial importance is that the heart contains a number of different cell types, two of which are of particular relevance here, namely, cells of the SA node and excitable cardiomyocytes. The cells in the SA node are modified cardiomyocytes and constitute only about 1% of all heart cells, but they are of immense importance. SA node cells exhibit a number of physiological differences in comparison with ventricular cells. For instance, while they do contain contractile filaments, they do not contract. The most important distinguishing feature is the ability of SA node cells to beat (**depolarize**) spontaneously (discussed in detail later), which has led to SA node cells being termed **pacemaker cells**. This electrical depolarization, which is caused by ion fluxes across the cell membrane, makes the membrane more positive and results in an **action potential**, which ultimately spreads throughout the heart muscle, as we will discuss in more detail later. The spontaneous depolarization in SA node cells occurs at a rate of about 100 times per minute, which, however, is constantly modified by the ANS. Thus, SA and AV node cells are clearly different from most other cardiomyocytes, but there are also significant differences within the latter group, for instance, with respect to the characteristics of their **resting potentials** and the shapes of their action potentials [30].

Interestingly, the transition between cells in the SA node and the peripheral atrial cells is smooth, with one cell type gradually changing into the other within a region surrounding the SA node. While this spatial transition might seem to be a subtle detail, it has turned out to be very important. Three-dimensional tissue reconstructions have shown how cell types and their communication channels are distributed in these transition regions and how this spatial heterogeneity is crucial for proper signal propagation. In fact, even within the SA node, differences can be found between central and peripheral regions in terms of both morphology and action potentials [27].

The spreading of the electrical impulse from cell to cell occurs through gap junctions, which are specialized sets of half-channels connecting the cytoplasm of two neighboring cells and permitting the free flow of ions and some other molecules (see Chapter 7). Gap junctions are of utmost importance to the heart, because they allow the efficient passing of ion-based electrical signals from one cell to another and thereby enable the coordinated depolarization and contraction of the heart muscle. In addition to the spreading of electrical charge through gap junctions, the electrical signal is transduced through the bundle of His and Purkinje fibers, which consist of specialized cell types that conduct five to ten times as fast as signal transduction through gap junctions.

The receivers of the electrical impulses from the SA and AV nodes and the Purkinje fibers are the regular, excitable cardiomyocytes. These elongated cells slightly differ in structure and electrical properties between the atria and the ventricles, but their important common property is that they are contractile. This ability to contract is due to large numbers of protein filaments, or **myofibrils**, which are a few micrometers long and similar to those found in skeletal muscle cells. The contraction itself is a complex process, which is ultimately accomplished by the sliding motion of two myofibrils, **actin** and **myosin**. In simplified terms, the myosin filaments contain numerous hinged heads, which form cross-bridges with the thinner actin filaments (**Figure 12.4**). By changing the angle of the hinges, the actin and myosin filaments slide against each other in a ratcheting manner, for which the energy is provided by hydrolysis of adenosine triphosphate (ATP). As a result of the sliding action, the entire cell contracts and becomes shorter and thicker. A review of some of the pertinent modeling activities is given in [31].

In reality, the myofibrils are components of a complicated multiprotein complex that includes the important proteins titin and troponin. The details of the sliding motion are quite involved, and it suffices here to state that they are coupled to oscillations in the cytosolic calcium concentration (see later and, for example, [1, 32]). When the calcium concentration increases, the cell contracts. When the concentration decreases, the filaments slide in opposite directions, and the cell relaxes. This tight correspondence is called **excitation–contraction (EC) coupling** [14]. Because the filament movement requires a lot of energy, cardiomyocytes contain many mitochondria.

Nobel Laureate Andrew Huxley proposed the first sliding filament model. While this model is not as prominent as his model of nerve action potentials, it has served as a basis for an entirely different class of models than the electrical models of nerve and heart cells.

Importantly, the contraction process in cardiomyocytes is initiated by electrical impulses and the flow of ions, as we will discuss later in greater detail. This electrical excitation in a cardiomyocyte is followed by a refractory period during which the cell normally does not respond to further electrical excitations. After this period, the cells are ready to fire again.

The membranes of cardiomyocytes change in their electrical charge once during every heartbeat. As a consequence, an electrical charge spreads throughout the heart and to the surrounding tissues, and ultimately causes very subtle electrical changes on the skin. These changes can be measured over a period of time with electrodes attached to the skin and are interpreted with methods of electrocardiography. Because the electrical impulses move as waves throughout the heart and are modified by the tissues of the body, the shape of the resulting

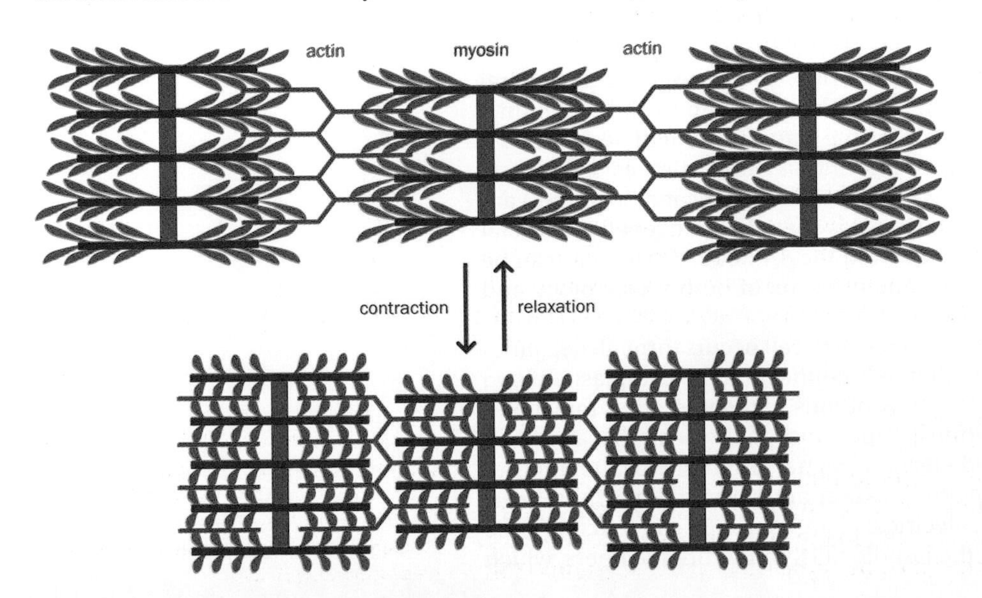

Figure 12.4 Simplified representation of the sliding action of myofilaments, which causes a cardiomyocyte to contract. The myosin filaments (brown) contain hinged heads (tan) that form cross-bridges with the thinner actin filaments (green). A change in the angle of the hinges leads to sliding and contraction. In reality, this process is much more complicated.

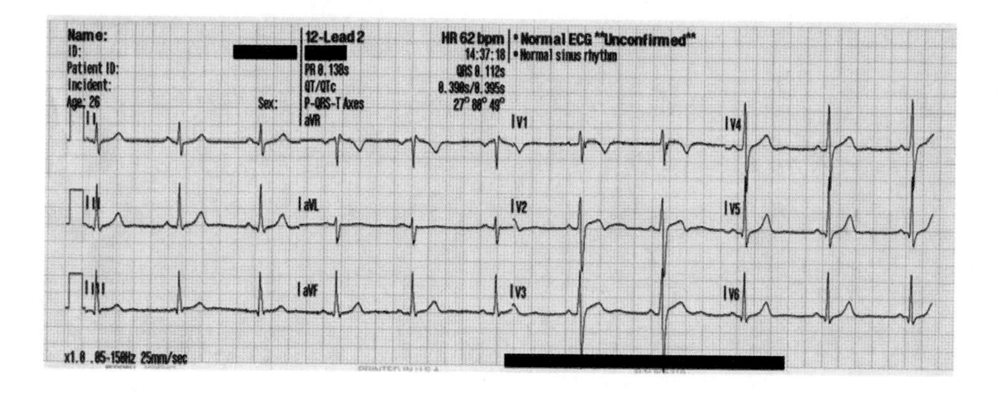

Figure 12.5 A 12-lead ECG of a healthy young man. Electrical currents generated during the cardiac cycle spread throughout the body and can be measured by an array of electrodes placed on the skin. See the text for further explanation. (Courtesy of Wikimedia Commons, File:12leadECG.jpg.)

electrocardiogram (ECG; or **EKG** from the German analog) depends on the three-dimensional architecture of the heart and the exact locations of the electrodes on the body. In order to assess heart function as comprehensively as feasible, it is therefore common practice to place several sets of electrodes at defined locations on the chest and limbs. The ECG displays the voltages between pairs of electrodes, and their spatial as well as temporal interpretation gives clues regarding the overall health status of the heart, as well as abnormal rhythms that may be due to imbalances in electrolytes or to damage to the heart tissue in specific locations; some details will be discussed later. The ECG can also identify, and to some degree locate, tissue damage in the case of a myocardial infarction (MI—a heart attack). Since the ECG only measures electrical activity, it does not truly measure the heart's ability to pump sufficient quantities of blood. The amounts of blood flowing through the atria and ventricles can be measured with an ultrasound-based echocardiogram or with methods of nuclear medicine. **Figure 12.5** shows a 12-lead ECG of a healthy young man.

Thus, we started with organ-level models, pointed to modeling needs at the tissue level, discussed features of heart cells, and have now returned to their effects as they are manifest at the whole-body level.

SIMPLE MODELS OF OSCILLATIONS

The information we have discussed so far is already sufficient for some very simple mathematical heart models. Let us start by focusing on the most obvious—the oscillatory pattern. This oscillation can be observed in many forms, for instance in the ECG, but also in the increasing and decreasing volume of blood in a ventricle, cyclic changes in electrical voltage, or rhythmic changes in the deformation of the ventricular wall. All these derivative oscillations ultimately originate from the autonomous depolarization and repolarization of the cells in the SA node, which result in a relatively regular oscillation in voltage (**Figure 12.6**). Here the term "autonomous" means that the cells exhibit sustained oscillations without needing repeated triggering events.

An intriguing aspect of these oscillations is that small perturbations are easily tolerated. In other words, if something temporarily disturbs or modulates the rhythm or the amplitude of the oscillation, the healthy heartbeat soon returns to the original rhythm. Of course, this tolerance is absolutely necessary for the heart to operate in a healthy way. Just imagine going to a horror movie, where a gruesome zombie attack makes your heart race. For a moment, you may enjoy the excitement, but you would certainly like your heart to go back to normal after a while and without conscious prodding.

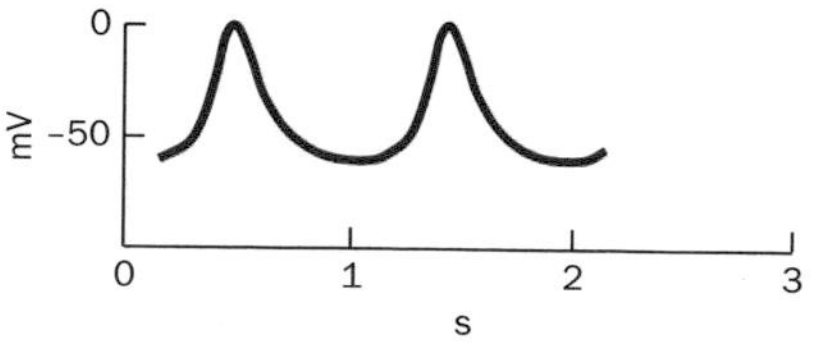

Figure 12.6 Oscillatory change in the voltage of an SA node cell. Different cell types and cells in different locations of the heart exhibit distinctly different shapes (see Figure 12.12), and these are again very different from the various patterns in an ECG (see Figure 12.5) [1, 29].

12.5 Black-Box Models of Oscillations

As we discussed in Chapter 4, tolerance to perturbations is a hallmark feature of stable limit-cycle oscillations. We mentioned different formats, including the van der Pol oscillation [33, 34], which was proposed many decades ago as a model for heartbeats, and return here to the very flexible limit cycles based on S-systems [35].

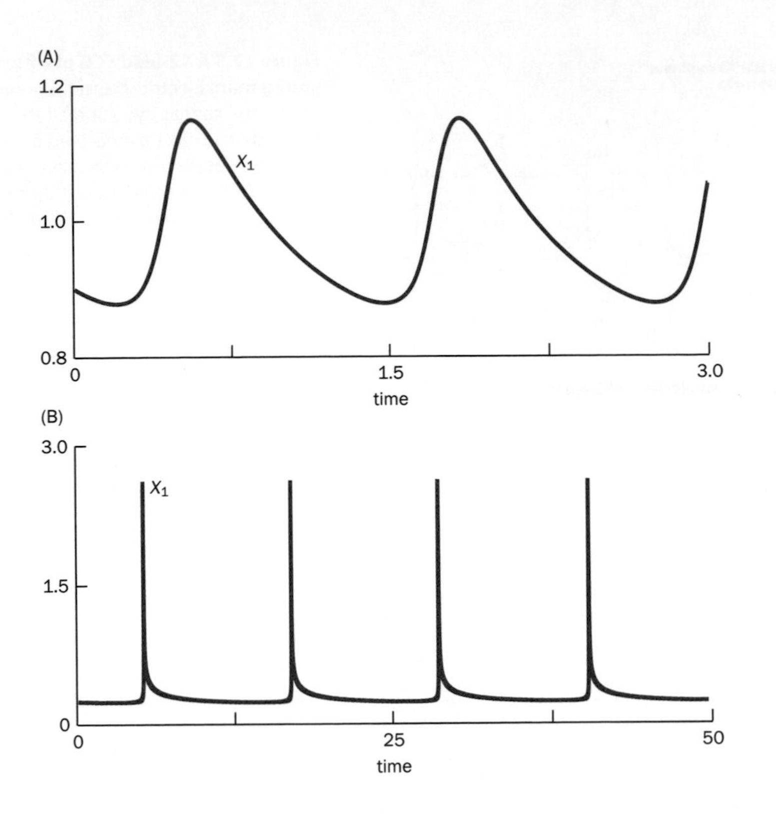

Figure 12.7 Stable S-system oscillations. (A) Variable X_1 of System A in (12.1), which resembles the oscillations in the SA node (see Figure 12.6). (B) Variable X_1 of System B in (12.1), which is similar to a simplified ECG trace (see Figure 12.5).

Specifically, we consider the two examples in **Figure 12.7**, which correspond to the model equations

$$
\begin{aligned}
&\text{System A:} \quad && \dot{X}_1 = 2.5(X_2^{-6} - X_1^{3}X_2^{-3.2}), && X_1 = 0.9, \\
& && \dot{X}_2 = 1.001 \cdot 2.5(1 - X_1^{-5}X_2^{-3}), && X_2 = 1.2; \\
&\text{System B:} \quad && \dot{X}_1 = 1.005(X_2^{-8} - X_1^{7}X_2^{5}), && X_1 = 0.25, \\
& && \dot{X}_2 = X_1^{6}X_2^{5} - X_1^{-1}X_2^{-4}, && X_2 = 2.5.
\end{aligned} \tag{12.1}
$$

Systems A and B lead to oscillations that are vaguely reminiscent of the dynamics in the SA node (see Figure 12.6) and of ECG traces (see Figure 12.5), and we can easily confirm that both oscillations are indeed stable, namely, by briefly perturbing them and watching them return to their original oscillations.

If stable oscillations are repeatedly "poked" with the "wrong" frequency, bad things can happen. Mathematically speaking, a stable limit cycle that is exposed to an oscillating stimulus pattern may become chaotic: although everything in the system is entirely deterministic, the oscillatory behavior is no longer predictable, unless one actually solves the equations. The analogy with heart physiology is evident: repeated spurious electrical signals can affect the system, and the result may be an unpredictable beating pattern. Specific examples are blockages in the AV node or the fast conducting bundles, which may lead to repeated additional electrical signals that can interfere with the usual electrical patterns in the heart and cause arrhythmia and even death. At the same time, one must be careful with the interpretation of irregularities: while chaotic oscillations may be a sign of disease, a fetal heartbeat that is too regular is often a sign of trouble [36–38]. Indeed, loss of variability in oscillations has been associated with different pathologies, frailty, and aging (see Chapter 15 and [39, 40]).

As a numerical example, suppose the stable oscillatory System B in (12.1) is subjected to a small regular oscillation. We model this idea by adding $0.01(s + 1)$ to both equations, where s is the value of a sine function, which we discussed in the form of ordinary differential equations (ODEs) (see (4.25) and (4.72) in Chapter 4), so that $s + 1$ oscillates between 0 and 2. In addition, we introduce an extra parameter a, which permits us to change the frequency of the sine oscillation. Thus, our heartbeat together with spurious electrical signals is given by the following four-variable model:

$$
\begin{aligned}
\dot{X}_1 &= 1.005(X_2^{-8} - X_1^7 X_2^5) + 0.01(s+1), & X_1 &= 0.25, \\
\dot{X}_2 &= X_1^6 X_2^5 - X_1^{-1} X_2^{-4} + 0.01(s+1), & X_2 &= 2.5, \\
\dot{s} &= ac, & s(0) &= 0, \\
\dot{c} &= -as, & c(0) &= 1.
\end{aligned}
\tag{12.2}
$$

If $a = 1$, the frequency of the sine system is the same as before. Larger values mean faster oscillations, smaller values slower oscillations. If $a = 0$, the sine system does not oscillate at all (the reader should confirm this with mathematical arguments and with simulations!), which permits a good comparison between systems with and without spurious stimuli. The system is in this case almost the same as in (12.1), except that both equations have an extra term +0.01, which is apparently negligible. The oscillations of this system (without spurious oscillations) are shown in **Figure 12.8A** for a long time horizon. In Figure 12.8B–F, the frequency parameter is varied; the sine oscillation itself is visualized for clarity with a ten-fold magnification of its amplitude. If $a = 10$, the sine oscillation is so fast that it does not change the appearance of the heart oscillation (this is not shown, but is very similar to Figure 12.8A). In Figure 12.8B, with $a = 1$, the heartbeat is not quite even. Figure 12.8C, with $a = 0.5$, spells trouble, with quite arrhythmic oscillations, which seem to recover after about 700 time units. However, the heart rate is much increased. Further lowering the parameter a seems to remedy the situation, but if we look

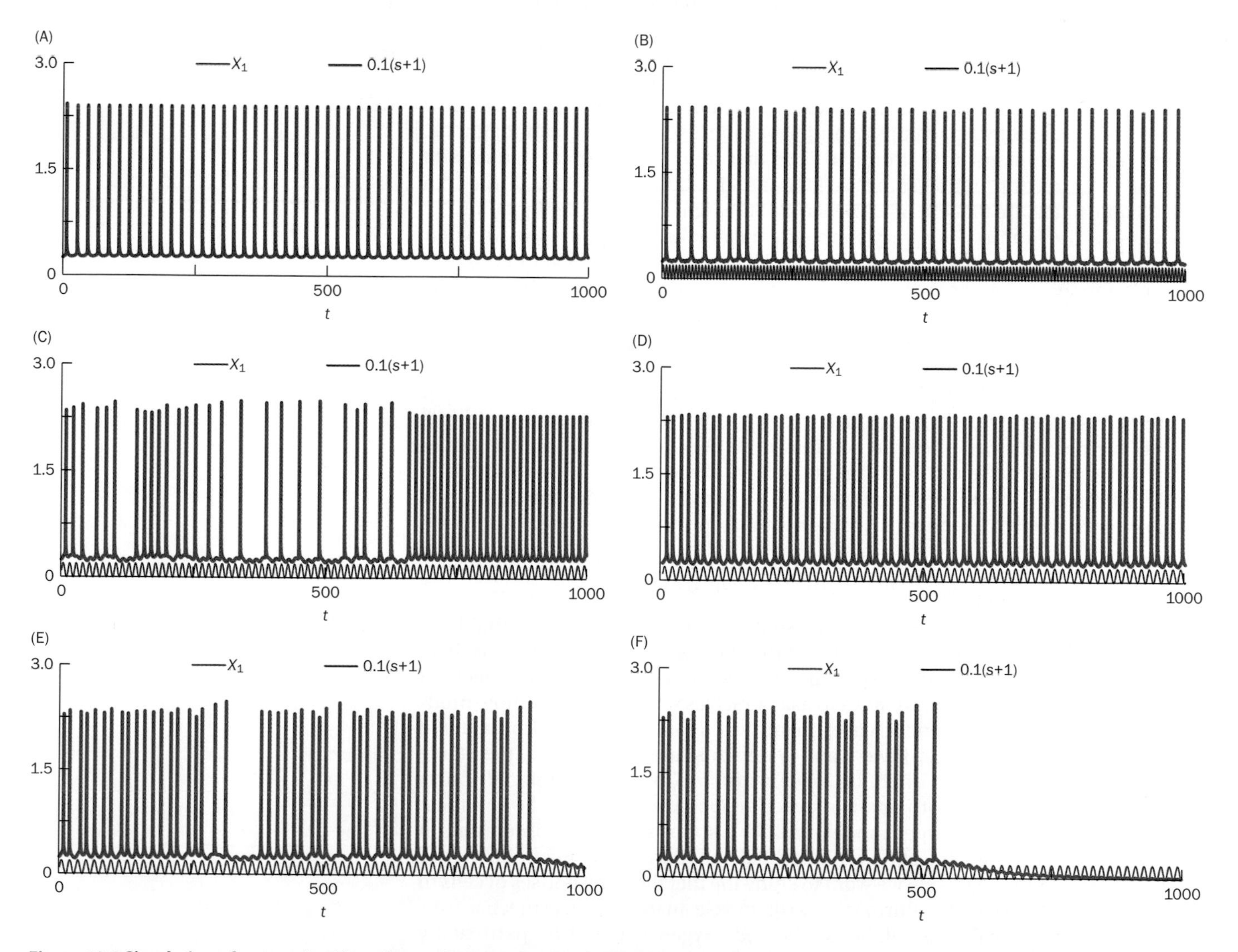

Figure 12.8 Simulation of a perturbed heartbeat, ultimately leading to fibrillation and death. If a stable oscillator is regularly "poked," for instance with a sine function, the results may be negligible or detrimental, depending on how the frequencies of the two oscillators relate to each other. The frequency of the sine oscillator is determined by a. See the text for details.

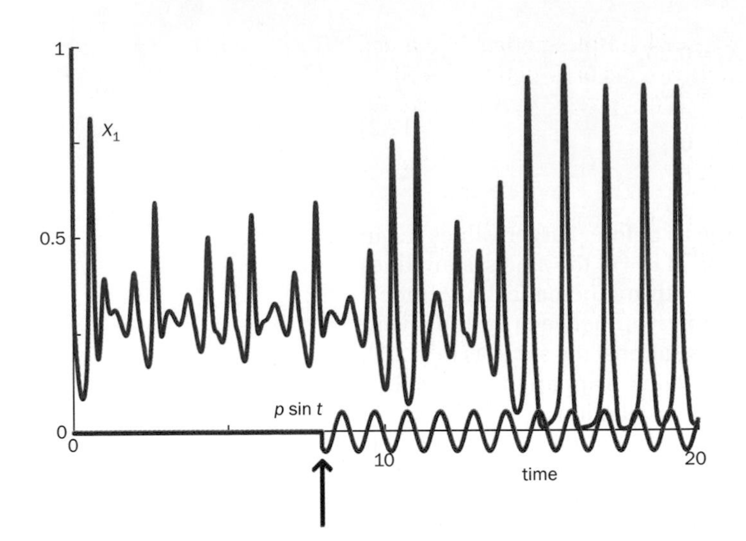

Figure 12.9 Effectiveness of a simple pacemaker. An irregular heartbeat, here modeled as a chaotic Lotka–Volterra (LV) oscillator, becomes regular through the effect of an appropriate sine function. Only the first variable of the LV system is shown, along with a scaled sine function, which oscillates between −0.05 and 0.05, starting at $t = 8$. For better visibility, the sine function is not shifted in this graph, although it is shifted by 1 in the model.

closely, the heart goes into double beats. For $a = 0.40$, the heart starts skipping beats, and even very slight further decreases in a cause the heartbeat to stop. The smaller the value of a, the faster this happens (see Figure 12.8D–F).

We can see that even simple models are able to capture the essence of the heartbeat and point to some very serious problems. Because these models are so simple, they are well suited for rigorous mathematical analysis. Indeed, many researchers have studied limit cycles in this and other contexts and explored their vulnerability with respect to input oscillations of an unfavorable frequency (see, for example, [41, 42]).

One may also use simple models to explore whether a highly irregular heartbeat can be made regular by a pacemaker, and how robust such a correction might be. Pacemakers are implanted electronic devices that send out electrical signals with the goal of correcting troublesome oscillations. If adjusted correctly, a modern pacemaker can ensure that the heart beats with the right speed at the right time.

To assess the function of a very simplistic pacemaker, we model the irregular heartbeat with the chaotic Lotka–Volterra (LV) oscillator from Chapter 10. To resemble the timing of a heartbeat, we multiply all equations by 50, which results in a rather erratic heartbeat with about 60–70 beats per minute. We model the overly simplistic pacemaker with a sine oscillator, represented as a pair of ODEs ($\dot{s} = q$ $\dot{c} = -qs$; see above and Chapter 4) and add to each equation in the LV-system a term of the form $p(s + 1)$. An example result, with $q = 6$, is shown in **Figure 12.9**. Initially we set $p = 0$, which corresponds to the unaffected heartbeat. At $t = 8$, we reset $p = 0.05$. We can see that it takes several beats, but the pacemaker indeed tames the chaos. In reality, pacemakers are much more sophisticated and respond to different demands of the body that require the heart beat faster or more slowly. Exercise 12.5 explores this topic in more detail.

12.6 Summary of Black-Box Oscillation Models

Rather simple oscillation models can represent some of the phenomena that have been observed in well-functioning and diseased hearts, such as arrhythmias, beat skipping, the function of pacemakers, and the potentially detrimental effects of spurious electrical signals. So, we have to ask ourselves whether these simple models are sufficient. As is so often the case, the answer depends on the purpose of the modeling effort. For instance, if we want to demonstrate arrhythmias, the stable black-box oscillation models might be most instructive, because nothing distracts from the simplicity of their mathematical structure. Or, if we plan to use the heartbeat as an input module for a larger model, and we are not specifically interested in mechanistic details, a small ordinary differential equation model might be a good choice. As an example, suppose we want to study the metabolic responses of cells to oxygen supply, which changes during the cardiac cycle. In this case, the mechanisms that fundamentally lead to oscillations in blood oxygen may not be particularly relevant. In fact, since larger models always require more effort in terms of parameter estimation and diagnostics, investing more time on the oscillator may not be worth the effort. Also, as mentioned earlier, a lot can be learned mathematically about small limit cycle oscillators, and many books have used simple models like those above for analyzing different types of oscillations, questions of structural stability, and bifurcation points in nerves, hearts, and other oscillating systems, which would not be possible in larger models [7, 41, 42].

While small models have strong advantages, they obviously have limitations as well. For instance, if we want to dissect a disease pattern into possible molecular, physiological, or electrochemical causes, the small models simply cannot provide adequate answers. It is not even possible to ask the right questions, because these models do not contain the relevant components, such as a membrane, concentrations of ions, or gated channels, which are very important for proper functioning, as we will see later in this chapter. In order to explain the processes that are ultimately responsible for a regular heartbeat, and for the purposes of studying and treating heart disease, we need to dig deeper, down to the real molecular and physical drivers, formulate them quantitatively, and then use mathematical models and computation to integrate results from molecular and electrochemical levels to gain insight into macroscopic function or failure. We begin with an exploration of calcium dynamics, which is crucial for proper function, develop a model for this particular aspect, and then work toward more comprehensive models that include chemical and electrical processes. To venture toward such more sophisticated models, we need to understand more about the biology of the heart: what makes the heart tick?

12.7 From a Black Box to Meaningful Models

We have seen that it is surprisingly easy to construct oscillations and even limit-cycle oscillations as black-box models. The question is now whether we can construct a simple yet biologically meaningful model that exhibits sustained and robust oscillations. One approach, which we will pursue here, is to study the movement of ions during the cardiac cycle.

The chain of contraction and relaxation events in heart cells, including SA node cells, is closely associated with the periodic movement of calcium ions between three locations, namely the extracellular fluid, the cytosol, and an intracellular compartment, called the **sarcoplasmic reticulum (SR)**. The cycle begins when the membrane of a cell depolarizes, which is associated with an influx of sodium and a relatively small amount of calcium from the extracellular space into the cytosol. The calcium binds to proteins in the membrane of the SR, where it triggers a mechanism called **calcium-induced calcium release (CICR)**, which results in the flow of large amounts of calcium from the SR into the cytosol [43, 44]. In regular cardiomyocytes, this calcium binds to troponin and leads to the sliding action of the myofibrils actin and myosin, as discussed before, and this sliding action causes contraction of the myocyte. Calcium now disengages from the myofibrils. Most of it is pumped back into the SR, and some leaves the cell through specialized calcium exit carriers that are located on the cell's membrane and simultaneously export calcium and import sodium. The cell relaxes and is ready for a new contraction.

In regular cardiomyocytes, the initial depolarization is triggered by electrical signals coming from the SA node or from neighboring cells, as discussed earlier. In contrast, the membrane in an SA node cell depolarizes spontaneously, starting from a minimum charge of about −60 mV (see Figure 12.6). While the initial trigger is different, the release of calcium from the SR also happens in the SA node [44]. Thus, both SA node cells and ventricular cardiomyocytes depend on the movement of calcium among three compartments.

The question then becomes whether a mathematical model of the three-compartment calcium dynamics can in principle lead to sustained, stable oscillations, as observed in cardiomyocytes. Because contractions in heart and skeletal muscle, as well as calcium bursting patterns in organs such as the pancreas, are of great clinical importance, it is not surprising that there is plenty of literature on models of calcium oscillations. As a sort of proof of principle, we only look at one minimalistic model that captures the calcium dynamics nicely and quite easily, and serves our purposes of providing a biologically motivated oscillator. This model was proposed by Somogyi and Stucki [45, 46], and many more complex models followed later; a review is given in [47].

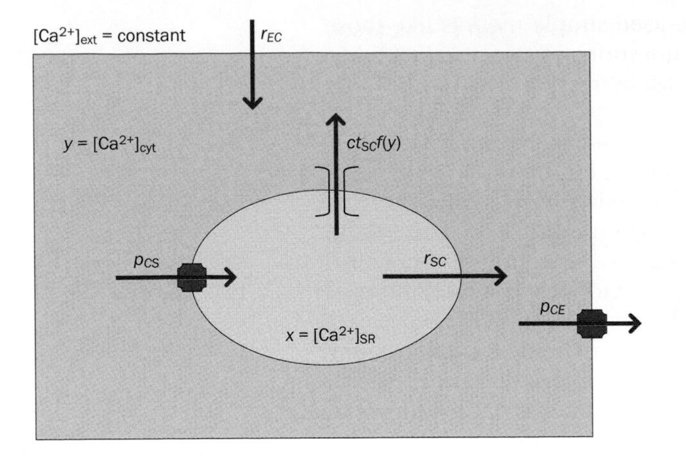

Figure 12.10 Simplified diagram of calcium flow associated with a myocyte. The illustration shows the processes and symbols used in the Somogyi–Stucki model, which leads to stable calcium oscillations. The red area represents the cytosol of the cell, while the pale area represents the sarcoplasmic reticulum. The notation is explained in the text.

The Somogyi–Stucki model describes a dynamical system that consists of three physical spaces and calcium movements among them (**Figure 12.10**). The spaces are the cytosol, the SR inside the cell, and the extracellular space. The model assumes that the extracellular space contains such a high concentration of calcium, $[Ca^{2+}]_{ext}$, in comparison with the cytosol that it is considered an independent, constant variable. The calcium concentrations in the SR and the cytosol are denoted by $x = [Ca^{2+}]_{SR}$ and $y = [Ca^{2+}]_{cyt}$, respectively, pump rates by p, simple transport rates by r, and channel transport by ct, with subscripts indicating source and target. According to Somogyi and Stucki, the following assumptions are made in the construction of the model:

1. The extracellular space serves as an inexhaustible source for an influx of $Ca^{2+} = [Ca^{2+}]_{ext}$ at a constant rate r_{EC}.
2. Ca^{2+} is pumped from the cytosol into the extracellular space at a constant rate p_{CE}.
3. Ca^{2+} is pumped from the cytosol into the SR at a constant rate p_{CS}.
4. Ca^{2+} leaks from the SR into the cytosol proportionally to the concentration difference between the SR and the cytosol and at a rate r_{SC}.
5. Ca^{2+} is actively pumped from the SR into the cytosol. The process depends on the concentration difference between the SR and the cytosol, and its rate is a sigmoidal function of the concentration in the SR.

The equations for this small dynamic system are easily set up according to the guidelines discussed in Chapters 2 and 4. Namely, because we are interested in time-dependent processes, we use differential equations, and if we ignore spatial details, for reasons of simplicity, these equations are ODEs. Thus, we formulate an ODE for the two dependent variables, x and y, and we include on the right-hand sides all relevant variables, functions, and rates. To implement this strategy, we make the simplifying assumption that all processes can be adequately represented as linear functions. This assumption is typical for transport processes and for processes whose turnover rates are proportional to their substrate concentrations. The only exception is $f(y)$, which describes the channel transport. For this function, Somogyi and Stucki use a Hill function with Hill parameter h, which saturates at the maximal rate $(ct)_{SC}$. Thus, the nonlinear channel transport process is modeled as

$$f(y) = (ct)_{SC} \frac{y^h}{K^h + y^h}. \tag{12.3}$$

Taken together, x is affected by transport into and out of the SR, and y is affected by transport into and out of the cytosol, and the mathematical formulation is

$$\begin{aligned} \dot{x} &= p_{CS}y - r_{SC}(x-y) - (ct)_{SC}\frac{y^h}{K^h+y^h}(x-y), \\ \dot{y} &= r_{EC} + r_{SC}(x-y) + (ct)_{SC}\frac{y^h}{K^h+y^h}(x-y) - p_{CE}y - p_{CS}y. \end{aligned} \tag{12.4}$$

As in so many cases, it is not easy to pull suitable parameter values out of a hat (see Chapter 5). Since we are not really interested in parameter estimation here, we simply take values similar to those used by Somogyi and Stucki [45, 46], namely,

$$p_{CS} = 4.5, \quad r_{SC} = 0.05, \quad (ct)_{SC} = 3,$$
$$h = 4, \quad K = 0.1,$$
$$r_{EC} = 0.075, \quad p_{CE} = 1.5.$$

With these values, the system becomes

$$\dot{x} = 4.5y - 0.05(x-y) - 3\frac{y^4}{0.1^4 + y^4}(x-y),$$
$$\dot{y} = 0.075 + 0.05(x-y) + 3\frac{y^4}{0.1^4 + y^4}(x-y) - 6y. \qquad (12.5)$$

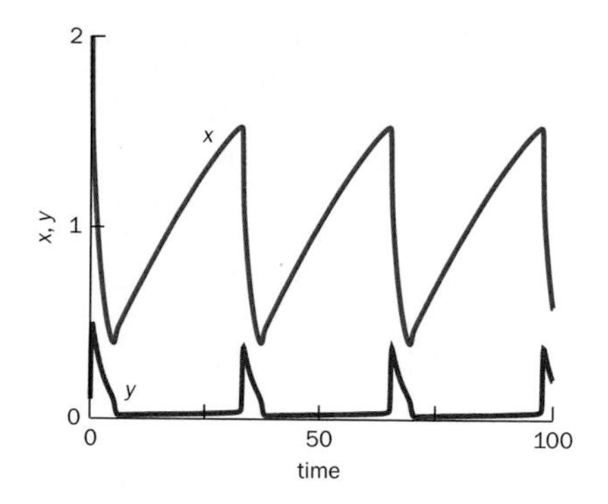

Figure 12.11 Oscillations generated by the Somogyi–Stucki system (12.5). Even without external cues, the system oscillates in a stable manner. The variables x and y represent calcium concentrations in the SR and the cytosol, respectively.

The initial values turn out to be rather unimportant, and we set $x_0 = 2$, $y_0 = 0.1$. The result of a simulation with these conditions is shown in **Figure 12.11**. We see that this rather simple model of the calcium dynamics in a cell is indeed able to oscillate. In fact, perturbations demonstrate that it oscillates all by itself in a stable manner. Our proof of principle is complete.

ELECTROCHEMISTRY IN CARDIOMYOCYTES

The oscillations in the heart are ultimately due to ion movements and electrical charges, but we have not really discussed how these two are related. The key to understanding these relationships lies in transport processes at the cell membrane and the membrane of the SR. A superb description of these cross-membrane processes can be found in [48]; much of the following is excerpted from this source.

The transport processes across the membranes are tied to two distinctly different, yet intimately related, gradients, namely, a concentration gradient and an electrical (charge) gradient. While the cell contains many types of molecules, the concentrations of interest here are those of ions. To refresh our memory, ions are atoms or molecules carrying electrical charges that result from a difference between their numbers of electrons and protons. In the case of an atom, the dense nucleus contains protons, which are positively charged, and neutrons, which are electrically neutral. The nucleus is surrounded by a cloud of negatively charged electrons. In a negatively charged ion, also called an anion, the cloud contains more electrons than there are protons in the nucleus. The most important examples here include ionized atoms such as chloride ion and some larger molecules such as negatively charged proteins. Positively charged ions, called cations, have one or several more protons in the nucleus than electrons in the shells of their electron clouds. The most important examples for membrane activities are sodium, potassium, and calcium ions.

In the case of a cell that is surrounded by a watery medium and separated from the medium by a porous membrane, ions can be transported in and out through the membrane either by means of specific transport mechanisms or, to some degree, through leakage. As we know from many real-life situations, differences within gradients tend to equalize if left alone. Expressed in the terminology of physics, the total energy of the system approaches the minimum that is achievable in the given situation. Hot coffee in a cool room gets cold, and eventually reaches the same temperature as its surroundings. Sugar poured into a glass of water dissolves through random molecular movements, and its concentration eventually becomes homogeneous. This natural tendency to equalize concentration differences also applies to concentrations inside and outside a cell. However, in the process of equalizing the concentrations of ions, it is possible that the charges become unbalanced. Thus, the total energy of the system is composed of two components: the chemical energy associated with molecular movement within the concentration gradient and the electrical energy associated with charges. Nature tends to minimize the sum of these two components, and, given enough time, the sum becomes zero. We will discuss this aspect later more rigorously.

The gatekeeper of ion movements between the cytosol and the extracellular space is the cell membrane. It consists of a lipid bilayer that creates a substantial barrier for polar molecules, and ions in particular, because of strong attractive electrostatic forces between ions and water molecules. This impediment to free ion flow is crucially important, because it allows an excitable cell, such as a cardiomyocyte, to create and retain concentrations and electric charges in its cytosol that differ drastically from the conditions in the surrounding extracellular fluid. The cell manages the creation of concentration gradients across membranes, and thus the storage of potential energy and the possibility of electrical signal transduction, through transmembrane **channel proteins** and **carrier proteins** that are often specialized to transport particular ions or water-soluble organic molecules into or out of the cell.

Channel proteins form hydrophilic, water-filled pores that cross the lipid bilayer and allow select ions to travel through the membrane. This travel is very fast, with up to 100 million ions being able to pass through a channel within 1 s [48]. With such speed, channel transport can be 100,000 times faster than transport facilitated by carrier proteins, a property that is of great import in electric signaling. The most characteristic feature of channels is that transport must always occur down the concentration gradient. In the case of ions, this passive transport results in a change in membrane potential, which in turn may influence further ion transport. Because most cells exhibit a negative charge toward the inside of the membrane, the entry of positively charged ions is favored. A fascinating feature of channel proteins is their fine-tuned specificity, which is a consequence of their molecular structure. A pertinent example is the potassium leak channel, which conducts K^+ ions 10,000 times better than it does Na^+. This difference is intriguing, because a sodium atom is slightly smaller than a potassium atom. Nature solved this problem with a potassium leak channel protein that creates a specially charged ion selectivity filter, which almost exclusively prevents sodium from traversing the membrane [48].

In contrast to simple aqueous pores, carrier proteins and ion channels are controllable by the cell. Specifically, they may be opened and closed through electrical or mechanical means, as well as through the binding of ligands. Most important among the channels are **voltage-gated** and receptor-gated channels that are respectively controlled by voltage and by binding of molecules such as the neurotransmitter acetylcholine.

Carrier proteins, also called **pumps** or permeases, are transporters that bind a specific ion and, while undergoing a transformational change, actively facilitate its transport through the membrane. Carrier proteins are specific with respect to the molecules they transport and require energy, which may be supplied, for instance, in the form of ATP. Their dynamics resembles that of enzyme-catalyzed reactions and shows a concentration dependence resembling that of a Michaelis–Menten process (see Chapters 4 and 8), with a characteristic binding constant K_M and saturation at a maximal rate V_{max} for high ion concentrations. Some carrier proteins also permit passive transport in either direction, as long as it occurs down the concentration gradient, but their main role is active transport against the electrochemical gradient.

Carrier proteins come in several variations. In the first case, uphill transport (against the concentration gradient) is coupled to downhill transport of another molecule. If both the ion and the other molecule are transported in the same direction, the transport protein is called a symporter. If they are transported in opposite directions, we talk about an antiporter. For example, moving an ion down the gradient releases free energy that can be used to move a larger molecule, such as a sugar, up the gradient. The second type of a carrier protein is a uniporter. In this case, only one ion is transported and the process is coupled to the hydrolysis of ATP. It is also possible that an antiporter is combined with ATP hydrolysis. As a further possibility, which, however, is not of relevance here, bacteria may use light as an energy source for ion transport against the gradient.

The most important transport process in our context is the **Na^+–K^+ pump**, which hydrolyzes ATP. Because Na^+ is pumped out of the cell, against a steep electrochemical gradient, while K^+ is pumped in, the transporter is an antiporter, and, since it uses ATP, it is sometimes called a Na^+–K^+ ATPase. The pump can also run in reverse, where it generates ATP from ADP. The Na^+–K^+ pump functions in such a manner that during each cycle three Na^+ ions are pumped out, while only two K^+ ions are pumped in. As a consequence, the pump changes the charge balance and is therefore called electrogenic, that is, it alters the membrane potential. However, the electrogenic change is small in comparison with other processes and contributes little to the electrical state of the cell's membrane.

A second crucial antiporter protein for the transport of ions during the cardiac cycle is the sodium–calcium exchanger (**Na^+–Ca^{2+} exchanger**), which removes calcium from the cell, in particular after an action potential. It allows sodium to flow across the plasma membrane, down its electrochemical gradient, and thereby transports calcium ions in the opposite direction. The removal of calcium is very fast, with several thousand calcium ions crossing the membrane per second. This speed is necessary for effective signal transduction, but it is also expensive: the removal of each calcium ion in this process requires the import of three sodium ions. The transport through the Na^+–Ca^{2+} exchanger is electrogenic, and strong depolarization of the membrane can actually reverse the direction of ion exchange, which is possible under the influence of some toxins and drugs. Alterations in the activity of the Na^+–Ca^{2+} exchanger can have serious consequences and lead to ventricular arrhythmias [49].

An additional pump that removes calcium is the plasma membrane Ca^{2+} ATPase (also known as the PMCA), which has a higher affinity for calcium than the Na^+–Ca^{2+} exchanger, but a much lower transport capacity. Because of its higher affinity, the Ca^{2+} ATPase is effective when the calcium concentration is low, and thus it complements the Na^+–Ca^{2+} exchanger. The Ca^{2+} ATPase is well understood in terms of its structure and function [48].

Normally, the Na^+–K^+ exchanger sets up a sodium gradient that drives the Na^+–Ca^{2+} exchanger to extrude calcium. The foxglove plant (*Digitalis purpurea*) produces the toxic substance digitalis, which is used as a medication to control arrhythmias of the heart. Digitalis functions by poisoning the Na^+–K^+ exchanger and thereby decreasing the Na^+ gradient, which in turns reduces Ca^{2+} extrusion. The accumulated intracellular Ca^{2+} promotes contraction [50].

The channels are the mediators for transfer of ions into and out of cardiomyocytes, and because ions are electrically charged, their transport can lead to changes in the electrical state of the cell and, specifically, its membrane. How is it possible that these transport processes lead to autonomous oscillations?

12.8 Biophysical Description of Electrochemical Processes at the Membrane of Cardiomyocytes

Ion transport and changes in the cell's electrical state constitute one of the rather rare cases in biological systems analysis where processes are so close to processes in physics that the corresponding laws of physics can be used directly for explanations, or at least as an inspiration. Here we borrow heavily from electrical theory. As discussed before, nature tends to minimize energy, which in the case of excitable cells consists of two components, namely, the chemical energy associated with molecular movement within the concentration gradient and the electrical energy associated with charges. Given enough time without outside influences, the sum of the two sources of energy approaches zero.

Formulating this fundamental phenomenon in the terminology of physics leads to the **Nernst equation**, which quantifies electrochemical gradients. One begins with the free-energy change per mole ion, ΔG_{conc}, which is associated with the concentration gradient across the membrane. Nernst found that this quantity can be computed as $-RT\ln(C_o/C_i)$, where R is the gas constant, T is the absolute temperature in Kelvin, and C_o and C_i are the concentrations of ions outside and inside the cell, respectively. Moving an ion across the membrane into the cell causes a charge-related free-energy change of size ΔG_{charge}. This change depends on the charge z of the ion and the voltage V between the inside and outside of the cell. Specifically, the change in energy ΔG_{charge} is given as zFV per mole ion, where F is Faraday's constant.

A resting cell approaches a state where the total energy is minimal. This state is therefore characterized by

$$-RT\ln\left(\frac{C_o}{C_i}\right)+zFV=0. \tag{12.6}$$

It is typically more useful to turn this equation around in order to compute the equilibrium potential or voltage at rest as

$$V=\frac{RT}{zF}\ln\left(\frac{C_o}{C_i}\right). \tag{12.7}$$

Furthermore, the logarithm with base 10 ($\log_{10}$) is often preferred over the natural logarithm (ln), and if we are considering a human cell, the temperature is more or less constant at 37°C, which corresponds to 310.15 K. Thus, with $R = 1.987$ cal K^{-1} mol^{-1}, $F = 2.3 \times 10^4$ cal V^{-1} mol^{-1} and ln 10 = 2.3026, the equilibrium potential for a singly charged ion is

$$V\approx 61.7\log_{10}\left(\frac{C_o}{C_i}\right)\ [\text{mV}]. \tag{12.8}$$

For the example of a typical cardiomyocyte, the intra- and extracellular potassium concentrations, $[K]_i$ and $[K]_o$, are about 140–150 and 4–5 mM, respectively, so that the equilibrium potential for potassium, V_K, is roughly between –90 and –100 mV.

The same type of computation holds for other ions, and the laws of physics allow us to compute the total potential by an appropriate summation of concentrations, if we account for the fact that different ions have different permeabilities through the membrane. Specifically, the membrane potential V_M is the equilibrium potential at which the voltage gradients and the concentration gradients of all ions are in balance, so that there is no flux of ions in either direction. This membrane potential is formulated as the **Goldman–Hodgkin–Katz equation**, which for a combined potential including K^+, Na^+, and Cl^- reads

$$V_M=\frac{RT}{zF}\ln\left(\frac{P_K[K^+]_o+P_{Na}[Na^+]_o+P_{Cl}[Cl^-]_i}{P_K[K^+]_i+P_{Na}[Na^+]_i+P_{Cl}[Cl^-]_o}\right). \tag{12.9}$$

Note that the subscripts *i* and *o* for chloride are exchanged, since chloride is an anion, and that the different quantities *P* quantify the permeability for each ion and depend on the properties of the membrane [7].

Although they are very important, ions like hydrogen and magnesium are often ignored in this computation, because their concentrations are orders of magnitude smaller than those of K^+, Na^+, and Cl^- [48]. The membranes of most excitable cells are much more permeable to K^+ than to Na^+ and Cl^-, which results in a resting membrane potential that is closer to the equilibrium potential for K^+ than to that for Na^+ and Cl^- [51].

12.9 Resting Potentials and Action Potentials

When an excitable cardiomyocyte is relaxed, its membrane is negatively charged at about –90 mV. This state is called the resting potential and describes the electrical potential across the cell membrane, when the outside is considered 0 mV. Electrical stimulation of sufficient strength, or a change in ion balance, changes this state by causing ion channels to open. When positively charged Na^+ and Ca^{2+} ions enter the cell, the membrane begins to depolarize, that is, its charge becomes more positive. Intriguingly, only relatively few Na^+ ions are needed to trigger depolarization. After a brief period of time, K^+ ions begin to leave the cell, the cell repolarizes (its membrane potential becomes more negative), and the cell returns toward its negative charge. Na^+/Ca^{2+} and Na^+/K^+ exchangers reestablish the resting potential.

One cycle of up-and-down changes in voltage constitutes an action potential. While some aspects of action potentials are the same for excitable cardiomyocytes and for cells of the SA node, there are also significant differences. We discuss the two cell types separately.

An intriguing feature of SA node cells is that they do not have a true resting potential (**Figure 12.12A**). Instead, they spontaneously depolarize and therefore can generate repeated action potentials, a phenomenon called **automaticity**. In comparison with other cardiomyocytes, which cycle through five phases during each action potential (see below), SA node cells have only three phases (4, 0, and 3). Phase 4 is characterized by the most negative membrane potential, about –60 mV. This potential is less negative than in ventricular cardiomyocytes and sufficiently high for the activation of slow inward-pointing, mixed sodium–potassium currents, called **funny currents**. These funny currents initiate the depolarization of the membrane and thereby control the spontaneous activity of the SA node [52]. Once the potential reaches about –50 mV, transient T-type calcium channels open and Ca^{2+} rushes into the cell, thereby depolarizing the cell even further. At –40 mV, additional L-type calcium channels open and increase the Ca^{2+} influx further. At the same time, potassium ceases to move out of the cell. During the following phase 0, more Ca^{2+} enters the cell through L-channels, while the T-channels begin to close. Overall, the depolarization takes longer than in other cardiac cells. In phase 3, the SA node cells **repolarize**. Potassium channels open, leading to an increased outward flow of K^+. L-channels close, and the depolarizing inward Ca^{2+} currents subside. The action potentials in atrioventricular (AV) node cells are similar to those in the SA node.

Of note is that the controlling funny currents can also be activated by cyclic nucleotides such as cyclic adenosine monophosphate (cAMP) and by **G-protein-coupled receptor** signaling (see Chapter 9). This metabolic signaling option provides a mechanism for the autonomic nervous system to stimulate the beating of the heart [53].

In contrast to SA and AV node cells, ventricular and atrial cardiomyocytes, as well as Purkinje cells, have a true resting potential, which is associated with diastole and characterized by a selective permeability to different ions. Specifically, the membrane is permeable to K^+ ions but essentially impermeable to other ions. As a consequence, the resting potential of about –90 mV is close to the equilibrium potential with respect to K^+. Cells in this resting state generate an action potential only when externally stimulated. Under normal conditions, this stimulus ultimately originates at the SA node and spreads throughout the heart muscle, as we discussed before. Once triggered by adjacent cells, the action potential in these cells develops very quickly, and its overall appearance is distinctly different from that of SA node cells (Figure 12.12B). Starting again with phase 4, the cells have a resting membrane potential of about –90 mV. When the cell receives an action potential from a neighboring cell, it begins to depolarize (phase 0). As soon as the charge exceeds a threshold of about –70 mV, the process is followed by further, rapid depolarization that is mediated by an increase in fast inward Na^+ currents. During this phase, the cell begins to contract. At the same time, the K^+ channels close. Subsequently, the fast Na^+ channels are inactivated, and outward K^+ channels begin to open, leading to a brief hyperpolarization, before the cell begins to repolarize (phase 1). At this time, contraction is still in progress. As we discussed before, the sodium influx changes the membrane such that calcium begins to flow in as well, thereby triggering an amplification through the calcium-induced calcium release mechanism and resulting in the flow of large amounts of calcium from the sarcoplasmic reticulum into the cytosol [43], which we also mentioned before and which we will discuss in detail later. Owing to the influx of calcium, the initial repolarization is delayed, which can be seen in a plateau of the action potential (phase 2). The contraction ends, and the cell begins to relax. The full repolarization occurs during phase 3, where the potassium efflux is high, the Ca^{2+} channels are inactivated, and the membrane potential becomes more negative. To restore the resting potential, Na^+–K^+ and Na^+–Ca^{2+} exchangers, as well as the plasma membrane Ca^{2+} ATPase, become active. To retain a long-term balance, the amount of Ca^{2+} leaving the cell equals the amount of trigger Ca^{2+} that entered at the start of the cycle. Between phase 0 and most of phase 3, the so-called effective refractory period, the cell cannot be excited again. This mechanism normally prevents spurious signals from spreading throughout the heart

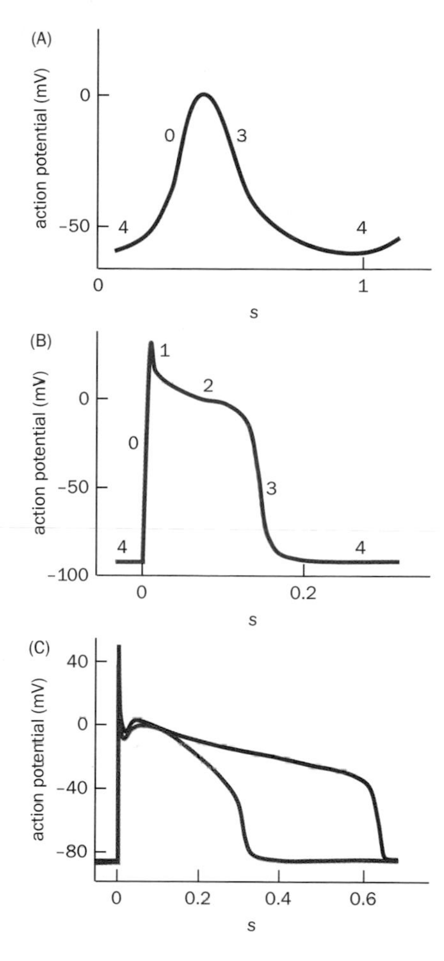

Figure 12.12 Action potentials in heart cells [1, 29, 54]. (A) Action potential of a spontaneously depolarizing SA node cell. (B) Action potential of a typical ventricular cardiomyocyte. Phases 1 and 2 of the latter are missing in the action potential in the SA node. (C) Canine cardiac action potential (green) in a healthy animal and in an animal with congestive heart failure (red).

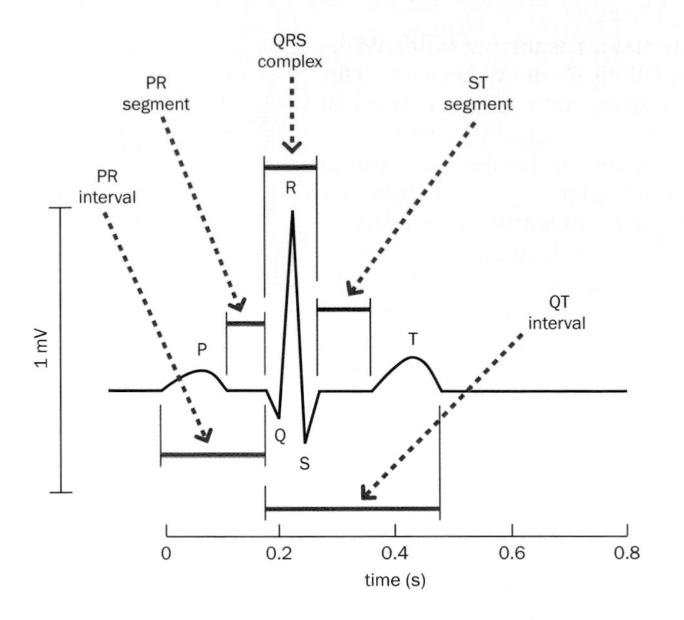

Figure 12.13 Diagram of an ECG trace, showing depolarization and repolarization of the atria and ventricles in the heart. Standardization of the recording speed allows measurements of the heart rate from the intervals between different waves. See the text for further explanation.

muscle and ensures that there is sufficient time to fill the heart with blood and eject it in a controlled manner.

It should be noted that the actual shapes of action potentials vary among different regions of the heart, which is important for the correct propagation of electric activity and normal cardiac function throughout the heart muscle [29]. Some heart diseases are associated with a change in the pattern of a normal action potential (Figure 12.12C).

The different phases of the action potential can be directly related to features of an ECG (**Figure 12.13**). The PR interval between the onset of the P wave and Q has a length of 120–200 ms and measures the time between the onset of atrial and ventricular depolarization. Within this interval, the P wave reflects the depolarization spreading from the SA node throughout the atria, which lasts for 80–100 ms. During the brief isoelectric period afterwards, the impulse travels to the AV node. The QRS complex is the phase of rapid ventricular depolarization, which lasts for 60–100 ms. The ST segment represents the isoelectric period during which the entire ventricle is depolarized. It corresponds more or less to the plateau phase in the ventricular action potential. The T wave represents ventricular repolarization. This process takes longer than the depolarization during the P wave. The QT interval corresponds roughly to the average ventricular action potential and ranges between 200 and 400 ms, but is shorter at high heart rates. Even seemingly minute abnormalities in any of these constituents of the ECG trace are often signs of various pathologies.

12.10 Models of Action Potentials

The first scientists to study action potentials in excitable cells quantitatively were Alan Hodgkin and Andrew Huxley. They did their legendary work on axons of squid neurons in the early 1950s [55] and were awarded the Nobel Prize for their efforts in 1963. A key to their success was the recognition of a direct analogy between the functioning of a nerve cell and a simple electrical circuit. A few years later, Noble [8, 9, 56] and others [57, 58] demonstrated that the Hodgkin–Huxley model for nerve cells could be adapted to represent cardiac action potentials, which have a much longer repolarization phase than neurons.

Hodgkin and Huxley exploited the analogy with electrical circuits to apply well-established laws from the theory of electricity to action potentials. They interpreted the lipid bilayer of the membrane as a capacitor, ion channels as conductors, electrochemical gradients driving the flow of ions as batteries, and ion pumps as current sources. To refresh our memory, let us revisit the basic terms of electricity.

- The most fundamental, subatomic property that describes relative amounts of electrons and protons in atoms and molecules is the electrical charge q. Of particular importance here are ions, which have either an excess or a shortage of electrons compared with neutral atoms. The electrical charge of a macroscopic entity is equal to the sum of the charges of all its particles. Thus, in excitable cells, the electrical charge is the net positive or negative charge carried by the ions in the system. A charge creates an electrical field in the surrounding space, which exerts a force on other electrically charged objects.

- An electrical conductor is a material that contains movable electrical charges. In electrical circuits, conductors typically consist of metals such as copper, while the conductor in our case consists of watery solutions containing ions and the channels that allow ions to cross the lipid bilayer.
- Electric current is the actual movement or flow of electric charge. It is typically denoted by I and, because it describes the change in charge at a given point, it is given as

$$I = \dot{q}. \tag{12.10}$$

- The electrical potential is the potential energy per unit of charge in an electric field. The difference between the electrical potentials at two points is called voltage (denoted by V) or voltage drop and quantifies the ability to move electrical charge though a resistance. Sometimes this potential difference is called the electromotive force, or EMF.
- Capacitance, denoted by C, is a measure of the amount of stored electrical charge in relation to an electric potential. It is related to charge and voltage as

$$C = \frac{q}{V}. \tag{12.11}$$

- Electrical conductance and resistance are properties of materials and are inverse to each other. Resistance R describes how much the flow of charged particles, such as ions, is impeded, while conductance g measures how easy it is for charge to flow through some material. It is the inverse of resistance:

$$g = \frac{1}{R}. \tag{12.12}$$

Famous physicists such as Maxwell, Faraday, Ohm, Coulomb, and Kirchhoff discovered laws relating the various quantities in electrical systems. Of particular relevance here are three.

- Ohm's law states that the current through a conductor is directly proportional to the voltage across the two points, and inversely proportional to the resistance between them. Thus, for time-dependent V and I, we have

$$V = IR = \frac{I}{g}. \tag{12.13}$$

- Kirchhoff's two most important laws address the conservation of charge and energy at any point in an electrical circuit (with the exception of a capacitor). Thus, if there are k currents flowing toward or away from a point, their sum is zero. In mathematical terms,

$$\sum_k I_k = 0. \tag{12.14}$$

- Similarly, the sum of all k voltages around any closed electrical circuit is zero, if the positive and negative signs are assigned correctly. Thus,

$$\sum_k V_k = 0. \tag{12.15}$$

Because an electric current describes the flow of electric charge, the total current equals the sum of all currents in the system. In other words, the currents in a system with k resistances are additive in the sense that

$$I=\sum_k I_k=\sum_k\left(\frac{V}{R_k}\right)=V\sum_k g_k. \tag{12.16}$$

These relationships look rather static, but they also apply in a system whose electrical state changes over time. Of particular importance here, dynamic changes in voltages and currents across the membrane of a cardiac cell can lead to the generation of a time-dependent action potential. To describe this action potential mathematically, we acknowledge explicitly that voltage and currents are time-dependent by defining $V(t)$ as the difference between internal and external voltages: $V(t) = V_i(t) - V_e(t)$. $V(t)$ is a time-dependent variable, which we may differentiate with respect to time. Using (12.10), we obtain

$$\dot{V}=\frac{1}{C}\dot{q}=\frac{I}{C}. \tag{12.17}$$

According to the laws of electricity, the sum of the capacitive current I and the ionic current I_{ion} at the membrane must be zero at every time point t. Substituting this conservation law into (12.17) and rearranging yields

$$C\dot{V}+I_{\text{ion}}=0. \tag{12.18}$$

This is our fundamental equation. However, in order to account for the various ion fluxes, which sometimes run in opposite directions, it is now necessary to split the voltage and ionic current into partial voltages and currents that are associated with Na^+, K^+, and a pool L, which accounts for all other ions, such as calcium, chloride, and magnesium. The symbol L was chosen to suggest leakage. Thus, we define V_{Na}, V_{K}, and V_{L}, representing contributions to the resting potential, along with the corresponding currents $I_{\text{Na}} = g_{\text{Na}}(V - V_{\text{Na}})$, $I_{\text{K}} = g_{\text{K}}(V - V_{\text{K}})$, and $I_{\text{L}} = g_L(V - V_{\text{L}})$. Recalling the relationship (12.13) between current, conductance, and voltage, we can rewrite (12.18) as

$$C\dot{V}=-g_{\text{eff}}(V-V_{\text{eq}}). \tag{12.19}$$

Here, the effective conductance $g_{\text{eff}} = g_{\text{Na}} + g_{\text{K}} + g_{\text{L}}$ is composed of the conductances related to Na^+, K^+, and the collective pool L for other ions. The resting (or equilibrium) potential is composed of individual contributions of conductances and voltages as

$$V_{\text{eq}}=\frac{g_{\text{Na}}V_{\text{Na}}+g_{\text{K}}V_{\text{K}}+g_{\text{L}}V_{\text{L}}}{g_{\text{eff}}}. \tag{12.20}$$

These terms are substituted into (12.19), and one may furthermore account for the membrane current, which is 0 at equilibrium or is equal to I_{app} applied to the system in typical experiments. It is customary in this field not to express relationships in absolute potentials or voltages but instead to consider a relative potential v that describes the deviation from the resting potential:

$$v=V-V_{\text{eq}}. \tag{12.21}$$

With this convention, we obtain an ODE describing dynamic changes in voltage as

$$C\dot{v}=-g_{\text{Na}}(v-v_{\text{Na}})-g_{\text{K}}(v-v_{\text{K}})-g_{\text{L}}(v-v_{\text{L}})+I_{\text{app}}. \tag{12.22}$$

Note that the capacitance C and the different conductances g are properties of the material and are therefore constant over time in most electrical circuits. In the case of the heart, the capacitance is related to the thickness of the membrane, which is not entirely constant in a cardiac cell, but C is usually considered constant, which is probably a reasonable approximation over the short period of time of an action potential. Importantly, the conductances correspond to channels in membranes, and some of these are ion-gated or voltage-gated and are therefore functions of the electrochemical milieu of the cell. In other words, both the voltage and the conductances are functions of time and far from constant.

Hodgkin and Huxley formulated the dependences of conductances on voltage essentially as black-box models that were inspired by experimental data, but do not have a direct interpretation in terms of biological mechanisms. In a stroke of genius, they defined

$$\begin{aligned} g_{\mathrm{K}} &= \bar{g}_{\mathrm{K}} n^4, \\ g_{\mathrm{Na}} &= \bar{g}_{\mathrm{Na}} m^3 h, \end{aligned} \tag{12.23}$$

and considered g_{L} as constant, because it was seen as negligible in comparison with g_{Na} and g_{K}. The terms $\bar{g}_{\mathrm{K}}$ and $\bar{g}_{\mathrm{Na}}$ were defined as constant conductivity parameters, while the quantities n, m, and h, which may be interpreted as characteristics of the potassium and sodium gates, were given as the solutions of differential equations that also depend on voltage:

$$\begin{aligned} \dot{n} &= \alpha_n(v)(1-n) - \beta_n(v)n, \\ \dot{m} &= \alpha_m(v)(1-m) - \beta_m(v)m, \\ \dot{h} &= \alpha_h(v)(1-h) - \beta_h(v)h. \end{aligned} \tag{12.24}$$

The terms n^4 and m^3h are proportional to the fractions of open ion channels or the probability that these channels are open. If these quantities are 1 or 0, then all corresponding channels are open or closed, respectively. The power of n was chosen to reflect four subunits of the potassium channel, which at the time was a postulate. The dependences on voltage in the Hodgkin–Huxley model are typically formulated as

$$\begin{aligned} \alpha_n(v) &= 0.01(10-v)(e^{(10-v)/10}-1)^{-1}, & \beta_n(v) &= 0.125e^{-v/80}, \\ \alpha_m(v) &= 0.1(25-v)(e^{(25-v)/10}-1)^{-1}, & \beta_m(v) &= 4.0e^{-v/18}, \\ \alpha_h(v) &= 0.07e^{-v/20}, & \beta_h(v) &= (e^{(30-v)/10}+1)^{-1}, \end{aligned} \tag{12.25}$$

with initial values specified as $v(0) = 10.5$, $n(0) = 0.33$, $m(0) = 0.05$, $h(0) = 0.6$, and $C = 1$, $v_{\mathrm{Na}} = 115$, $v_{\mathrm{K}} = -12$, $v_{\mathrm{L}} = 10.6$, $\bar{g}_{\mathrm{Na}} = 120$, $\bar{g}_{\mathrm{K}} = 36$, and $g_{\mathrm{L}} = 3$ (see, for example, [7, 59]). The apparently arbitrary form of the equations for n, m, and h comes from arguments describing how these quantities decay over time in a first-order fashion.

With these definitions and settings, the dynamics at the membrane is described as a system of four ODEs. The first represents the action potential itself, (12.22), while the remaining three are equations describing time-dependent changes in conductances and are defined as in (12.24) and the associated functional dependences (12.25). Putting the four equations into a numerical solver and specifying the initial values, we quickly obtain the fruit of our labor (**Figure 12.14**): an action potential in a nerve cell that shows rapid depolarization and repolarization. The numerical solution also shows the dynamics of the auxiliary variables n, m, and h, even though they do not really have a direct meaning, except insofar as they represent features of the different ion channels (**Figure 12.15**).

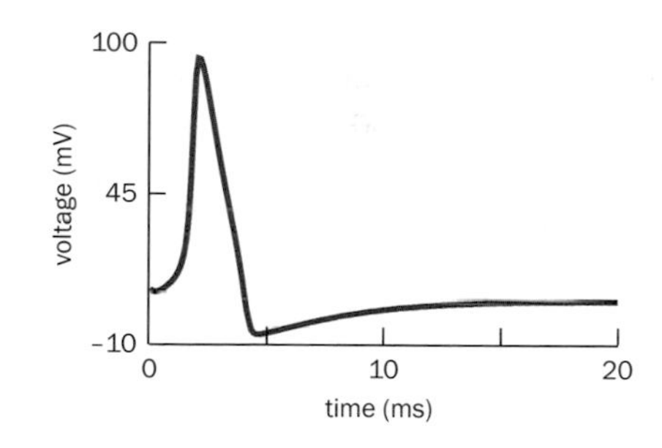

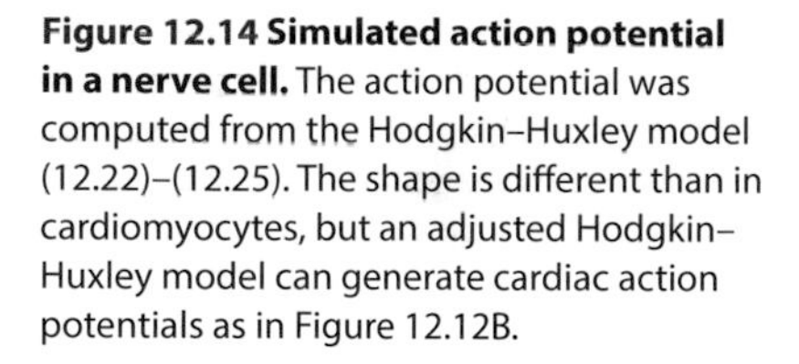

Figure 12.14 Simulated action potential in a nerve cell. The action potential was computed from the Hodgkin–Huxley model (12.22)–(12.25). The shape is different than in cardiomyocytes, but an adjusted Hodgkin–Huxley model can generate cardiac action potentials as in Figure 12.12B.

The appearance of this action potential reflects that of nerve cells. It is much faster than in cardiomyocytes, and the pronounced plateau during repolarization in phase 2 (see Figure 12.12B) is entirely missing. Furthermore, the nerve model exhibits a slight undershoot before returning to the resting potential. Nonetheless, variations in Hodgkin and Huxley's equations, such as adaptations in the representation of the potassium current, were shown to lead to longer action potentials and repolarization plateaus as we find them in cardiomyocytes [8, 9, 56–58].

Before we continue, let us summarize the key points of this section. The electrochemical changes during contraction and relaxation in a nerve cell—and by extension in a cardiac cell—can be modeled in analogy to an electrical circuit, and Hodgkin and Huxley showed that the physical laws describing such a circuit can be applied to this physiological system. The most significant difference between a standard electrical circuit and a cell is that some of the channels in the cell membrane are gated, that

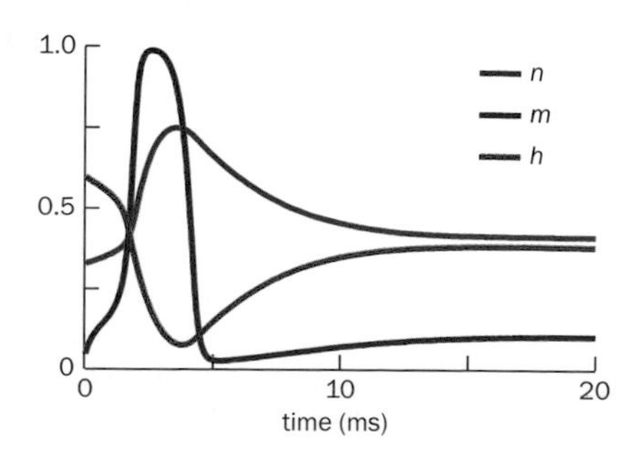

Figure 12.15 Temporal responses of gating variables in the Hodgkin–Huxley model (12.24). Close inspection shows that the variables do not return to their starting values after the action potential.

is, controlled by voltage or ions, whereas conductance in a typical electrical circuit is constant. As a consequence, there is a logical cycle of influences where ion concentrations affect voltage and voltage affects the movement of ions and thereby the chemical and electrical balances in the system. Hodgkin and Huxley modeled the voltage dependence of the channels based on mathematical assumptions regarding the functional dependences and by fitting experimental data. The amazing outcome is that their model works very well under a number of conditions.

12.11 Repeated Heartbeats

We now have an action potential model at our disposal, (12.22)–(12.24), but it fires only once. If we simply run the simulation longer, nothing happens. The system has approached a steady state and stays there. How does it lead to pacemaker activity and a heartbeat? The vast majority of the heart cells receive electrical signals from the SA node, through the AV node and the Purkinje fibers, which are mediated through gap junctions. Thus, their triggers are changes in the action potential of their membrane. The pacemaker cells, in turn, autonomously fire in a process of depolarization that involves influxes and effluxes of ions, and we have seen before that depolarization begins with funny currents, but really proceeds when Ca^{2+} rushes into the cytosol.

Let us recall that n, m, and h in the Hodgkin–Huxley model describe voltage-dependent gates and that their conductance is affected by changes in voltage and thus by changes in ions such as calcium. If we look more closely at the dynamics of n in the action potential model, we see that its final value is about 0.415, while its initial value is about 0.33 (see Figure 12.14). What happens if we artificially reset the value of n to the initial value after the action potential has fired? This is a simple simulation experiment and, *voilà*, the action potential fires again. In **Figure 12.16**, we repeatedly reset n at times $t = 20$, 40, and 60. We can see that the artificially induced potentials are a little lower than the first potential, but that the shape of the subsequent beats is identical. Would we receive the same peaks in all cases if we not only reset n, but also m and/or h?

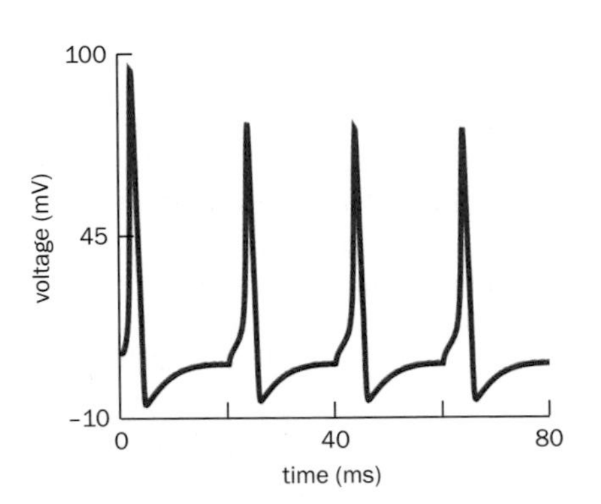

Figure 12.16 Creating repeated heartbeats. Artificial resetting the value of n in (12.24) to the initial values at times 20, 40, and 60 results in a beating pattern of repeated action potentials.

Because a four-variable system is difficult to study mathematically, Fitzhugh and Nagumo developed a simpler variant of the Hodgkin–Huxley model, in which they separated the variables into fast and slow variables [60, 61]. The result had only two variables, and it turns out that their model is a variant of the van der Pol oscillator that we discussed at the beginning of this chapter and in Chapter 4! We have come full circle and in the process given some biological justification to van der Pol's expectation [34] that his oscillator might describe heart rhythms and maybe even serve as a pacemaker. Many researchers have performed very elegant analyses of the mathematical structure of the Fitzhugh–Nagumo equations and their bifurcation structure; the interested reader is referred to [7, 59].

To generate the repeated action potential, we cheated a little bit by manually resetting n at times 20, 40, and 60. Obviously, the real SA node does not need our input for every beat. So we should ask whether it is possible to replace this artificial repeated triggering with a more natural pacemaker. One candidate could be our calcium oscillation model, because we can easily argue that the calcium oscillations are associated with depolarization and repolarization. The strategy is therefore simple: we set up a simulation program that contains the Hodgkin–Huxley model (12.22)–(12.25) and also the calcium oscillation model (12.5) and then couple the two systems. For instance, let us simulate a situation where the calcium concentration in the cytosol, y, affects the conductance of h. The simplest, but not the only, way of achieving the coupling is simply to add y to the equation for h. Revisiting the output of the calcium model, we see that the value of y is mostly low, close to 0, but spikes regularly up to about 0.4. These low values make sense, because the calcium level in the cytosol is about 10,000 times lower than calcium levels in the extracellular space. The simulation indeed shows the desired effect: without artificial intervention, the action potential spikes and depolarizes in a repeated manner (**Figure 12.17**), as we see it in an ECG. Of course, our success is no proof that the stitched-together model is an appropriate model of a beating heart, but it demonstrates how complex problems might be approached in a modular fashion, at least as a proof of principle and a means for investigating the internal logic of a system.

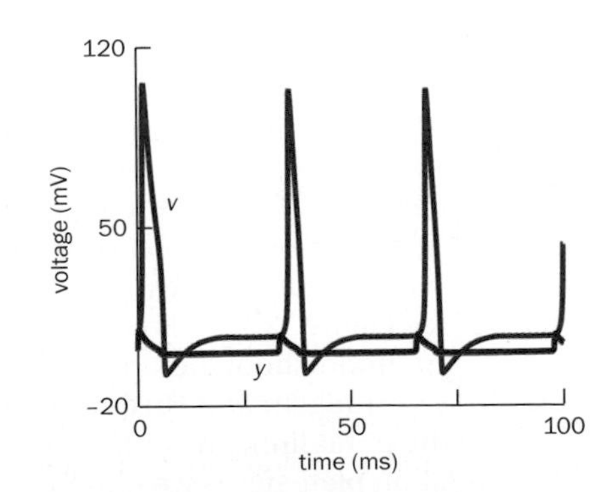

Figure 12.17 Creating self-sustained heartbeats. The combined Hodgkin–Huxley/Somogyi–Stucki model generates repeated action potentials without external cues. Note that y is shown with twenty-fold magnification.

ISSUES OF A FAILING HEART

Although the heart is incredibly resistant to fatigue and faithfully keeps on pumping, age and disease eventually take a toll. Uncounted things in the complex systems that work within the heart with unparalleled precision for most of our lives can go wrong, and the result is often heart failure, which presently affects about 5 million individuals in the United States, of whom about 20% die within 1 year of diagnosis. Indeed, heart failure is the leading cause of death in the Western world.

It is very clearly beyond the scope of this chapter to elaborate on all, or even the most prevalent, physiological changes in the heart that might result in heart failure. Instead, as a glimpse into multiscale modeling, the following sections discuss just one thread that connects the intracellular level with failure at the organ level. In line with our emphasis throughout the chapter, we focus on action potentials and the dynamics of calcium.

The first class of heart problems is related to abnormalities in the pacemaker system, which can lead to different arrhythmias. For instance, in sick sinus syndrome, which is often related to senescence of the SA node, the heart rate is too slow, which can lead to dizziness, sluggishness, and shortness of breath. The slow heartbeat (bradycardia; *bradys* is Greek for slow) may alternate with fast atrial arrhythmia (tachyarrhythmia; *tachys* is Greek for fast), leading to brady–tachy syndrome. A delay in conduction between the SA node and the AV node can lead to a first-degree heart block, which often does not lead to noticeable symptoms, because the AV node can serve as a sort of back-up pacemaker. By contrast, a block below the AV node, namely in the bundle of His and the Purkinje system, may be fatal if untreated. An implanted pacemaker may alleviate such a block and prevent cardiac arrest.

A fast heartbeat sometimes originates in one of the ventricles, leading to a life-threatening arrhythmia called ventricular tachycardia. A very common cause of this condition is scarring of the heart muscle due to an earlier heart attack. The scar does not conduct electrical signals, so that the signal propagates on both sides around the damaged tissue. Earlier, we mentioned that three-dimensional modeling with finite element methods has become sophisticated enough to mimic this disease on the computer, and the result of such an analysis was shown in Figure 12.3.

A seemingly minute, but indeed significant, change in the proper functioning of the heart during aging is a slowing down of the calcium fluxes between the sarcoplasmic reticulum (SR) and the cytosol [62]. If the calcium dynamics in individual myocytes becomes less efficient, the consequences are felt in the contractions of the heart: it does not fill with blood and empty as quickly as it used to. The slower calcium dynamics often goes hand in hand with an age-related stiffening of the arteries, and together they can lead to shortness of breath, especially during strenuous activities. These systemic connections can be explained by the mechanical features of the heart: perfusion is reduced, and this reduction leads to an accumulation of fluid beneath the skin, called edema, and to the entire spectrum of problems associated with congestive heart failure.

Another change in cardiomyocyte activity that is associated with slower calcium dynamics is a lengthening of the action potential during repolarization (see Figure 12.12C). The longer action potential keeps the ion gates in the cell membrane open for a slightly longer period of time, thereby allowing more calcium to enter and exit the cytosol between beats and giving the weakened SR more time to release or take up calcium. The downside is a higher risk of unbalanced calcium flow and a slower response to increased demands on heart activity. Age also affects the contractile elements in the myocytes, and in particular the heavy chains of myosin, which are responsible for forming cross-bridges with actin (see Figure 12.4). Some of these changes can be detected in the expression of genes, which is different in young and old individuals. Finally, as with nearly all cells, cardiomyocytes are susceptible to unstable oxygen molecules, called free radicals, that create oxidative stress within the cells and may impair energy metabolism and other functionalities by damaging proteins, DNA, membranes, and other complex structures in the cell. In cardiomyocytes, free radicals are doubly problematic, because of the high energy demands of these cells and because free radicals can damage the calcium pumps in the membrane of the SR, which connects this aging mechanism again to reduced calcium dynamics. During aging, many cardiomyocytes die, and even in healthy septuagenarians, the number of cardiac cells may be decreased by 30%.

To the detriment of many elderly people, some of the effects of aging on the heart form a vicious cycle that gradually diminishes efficiency. When the calcium dynamics slows down and the arteries stiffen, the heart has to work harder to compensate. As in other muscles, the harder work leads to enlargement of the cardiomyocytes, which in turn leads to a thickening of the heart walls, as described by the Frank–Starling law. This normal thickening is exacerbated by diseases such as coronary heart disease and high blood pressure. Together, the metabolic and anatomic changes make the heart more susceptible to cardiovascular conditions such as left ventricular hypertrophy and atrial fibrillation. The latter is a common cardiac arrhythmia that involves the two atria: the heart muscle quivers instead of contracting in a coordinated fashion. Strong fibrillations may lead to heart palpitations and congestive heart failure.

12.12 Modeling Heart Function and Failure Based on Molecular Events

The models that we have developed in this chapter are snapshots of heart function at different levels. Considering these snapshots collectively will give an impression of how difficult it is to connect heart function and failure to specific molecular events through comprehensive, true multiscale models, as they are envisioned in the Physiome [13] and Virtual Physiological Rat [63] Projects. Nonetheless, the following will discuss in broad strokes how our models can inform models at higher levels in order to connect intracellular calcium-associated events with macroscopic heart problems at the organ level. This line of reasoning with models was proposed by Raimond Winslow and his collaborators over a span of many years. Most of the technical details will have to be skipped here since they involve a lot of mathematics, but the interested reader may start diving into this field with reviews such as [64–66].

We begin our journey at the molecular scale, and specifically at the cell membrane of the cardiomyocyte, the sarcolemma. The sarcolemma contains invaginations into the cytosol that effectively increase its surface area. The invaginations, called T tubules, contain calcium channels that functionally connect the sarcolemma and the membrane of the tubular network of the SR (**Figure 12.18**). This functional connection occurs in spatially restricted microdomains of about

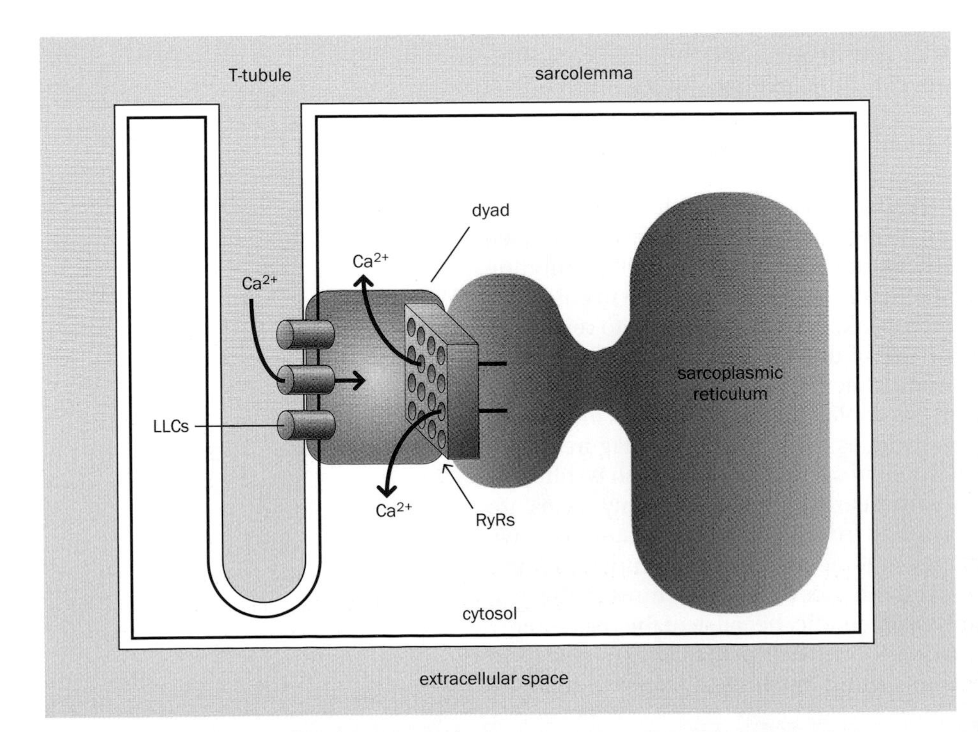

Figure 12.18 Simplified illustration of a dyad and its role in the cardiac cycle [67]. During the action potential, calcium from the extracellular space enters the dyad through L-type Ca^{2+} channels (LCCs) in the T tubules. The influx activates arrays of ryanodine receptors (RyRs), which are located on the opposite side of the dyad on the sarcoplasmic reticulum (SR). The opening of the RyRs leads to a calcium flux from the SR into the dyad, from where it diffuses into the cytosol, increasing its calcium concentration greatly. The cytosolic calcium binds to myofilaments, which contract and shorten the cell. (Data from Tanskanen AJ, Greenstein JL, Chen A, et al. *Biophys. J.* 92 [2007] 3379–3396.)

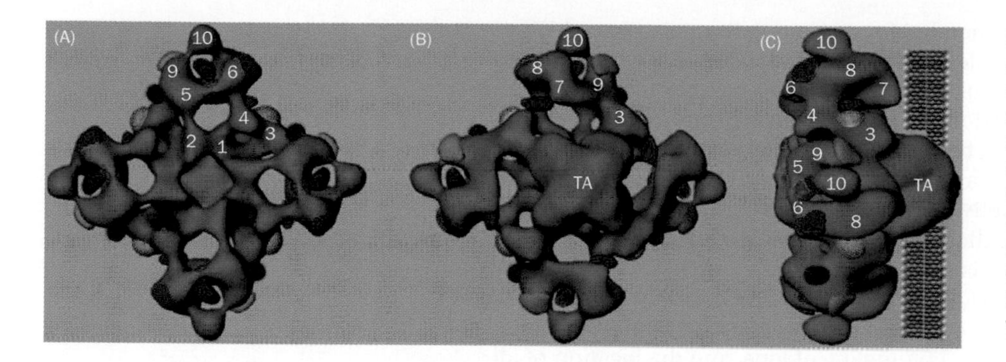

Figure 12.19 Three-dimensional structure of a ryanodine receptor. (A) View from the cytoplasm to the lumen of the sarcoplasmic reticulum (SR). (B) View in the opposite direction. (C) Side view. The numbers identify different globular domains and TA denotes the transmembrane assembly. (From Zissimopoulos S & Lai FA. Ryanodine receptor structure, function and pathophysiology. In Calcium: A Matter of Life or Death [J Krebs & M Michalak, eds], pp 287–342. Elsevier, 2007. With permission from Elsevier.)

10 nm width, called **dyads**. Each cardiomyocyte contains roughly 12,000 such dyads. From the lumen of the T tubules, the dyads can receive calcium through specific **L-type Ca^{2+}** channels (LCCs) that peek through the sarcolemma. They can also receive calcium from the SR through ryanodine receptor (RyR) proteins (**Figure 12.19**). The role of this structural set-up is to facilitate the calcium-induced calcium release (CICR) that we discussed before. In response to membrane depolarization, the LCCs open and thereby produce a small flux of Ca^{2+} ions from the T tubule into the dyad. The Ca^{2+} ions cause the RyRs on the opposite site of the dyad to open, and the result is a strong calcium flow from the SR into the cytoplasm that ultimately leads to cardiomyocyte contraction. This LCC–RyR excitation–contraction mechanism is very fast and occurs in a range of micro- and milliseconds [14]. With a length in the nanometer range and a volume of only about 10^{-19} liter, the dyads may be very small, but they are "at the heart of the heart." To allow relaxation of the filament complex, intracellular calcium is returned to the SR by SERCA (SR Ca^{2+} ATPase) pumps and also pumped out of the cell through sodium–calcium exchangers or plasma membrane Ca^{2+} ATPase. The reduction in calcium concentration causes the bonds between the myosin and actin filaments to loosen.

Experiments have shown that not every RyR is associated with an LCC, but that there is a stochastic distribution of one LCC among about five RyRs (**Figure 12.20**). A careful computational analysis [64] showed that the particular shape of the RyR protein restricts the movement of Ca^{2+} ions and funnels them to their target binding sites. In contrast to a random process such as Brownian motion, this guiding action enormously increases the efficiency with which an external excitation by calcium ultimately leads to myocyte contraction.

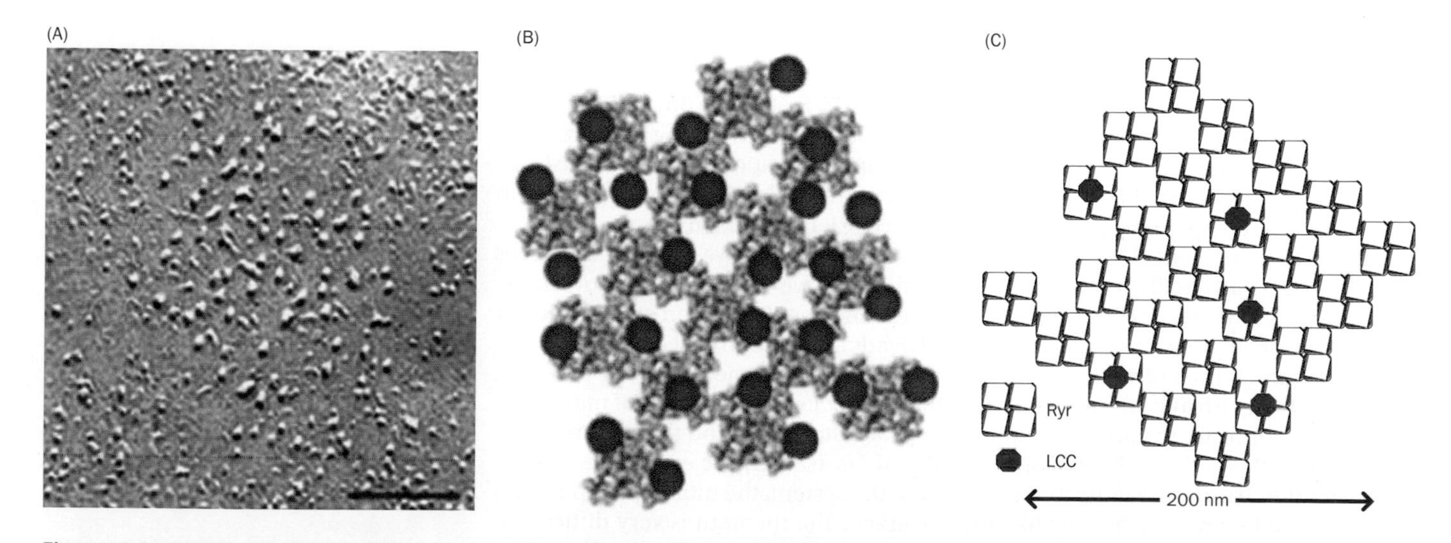

Figure 12.20 Spatial distribution of RyRs on the surface of the sarcoplasmic reticulum and of fewer LCCs on the adjacent membrane of the T-tubule, which is part of the sarcolemma. (A) Electron micrograph of Ca^{2+} release units in myocytes of a chicken heart. (B) Organization of cardiac RyR units into ordered arrays. (C) Schematic representation of RyR and LCC units on opposing sides of a dyad. ((B) from Yin C-C, D'Cruz LG & Lai FA. *Trends Cell Biol* 18 [2008] 149–156. With permission from Elsevier. (C) adapted from Tanskanen AJ, Greenstein JL, Chen A, et al. *Biophys. J.* 92 [2007] 3379–3396 and from Winslow RL, Tanskanen AJ, Chen M & Greenstein JL. *Ann. NY Acad. Sci.* 1080 [2006] 362–375. With permission from Elsevier. With permission from John Wiley & Sons.)

The number of calcium molecules needed for mediating the signal from LCC to RyR is small. Even at the peak of the calcium flux, only about 10–100 free Ca^{2+} ions are found in each dyad. This low number has two consequences for any model representation: the process is stochastic and the noise level is high—one may find an average number of 50 ions in a dyad, but the number in any given dyad could also be 20 or 120. A situation leading to such a random occupancy is typically represented as a stochastic process [68, 69]. We briefly discussed stochastic processes in Chapters 2 and 4, comparing and contrasting them with deterministic dynamic processes. The key feature of stochastic processes is that random numbers affect their progression. The simplest stochastic processes are probably flipping a coin and rolling a die. Typical models for stochastic processes are Poisson processes and Markov models.

The defining feature of the dyad is the channeling of ions, and the location of all Ca^{2+} ions in the vicinity of dyads is therefore of great importance. As we discussed in Chapters 2 and 4, the consideration of spatial aspects in a dynamical process typically calls for a representation in the form of a partial differential equation (PDE). Thus, to determine whether a Ca^{2+} ion will enter a dyad, a detailed model should quantify the probability $P(x, t)$ that a Ca^{2+} ion is close enough to and properly aligned with a dyad position x at time t. From a mathematical point of view, the combination of free motion and stochastic binding events would suggest as the default model a representation of the movement of individual calcium molecules in the form of a stochastic PDE, which is very complicated. Indeed, such detailed focus on each and every dyad, or maybe even every Ca^{2+} ion, would be extremely involved, because of the roughly 12,000 dyads per cell and the billions of Ca^{2+} ions flowing between the extracellular space, the cytosol, and the SR. As a consequence, simulations would take weeks to complete, even on a high-performance computer, and computation time would become prohibitive even for modeling one single cardiomyocyte, let alone the entire heart.

It may be infeasible to model processes with such resolution and detail, but discussion of the intricate molecular details and biologically observed mechanisms is important because it indicates how a stochastic model design would proceed in principle and also because it allows us to assess where one might make valid simplifying assumptions without compromising realism. To the rescue in our specific case comes the very fact that makes a direct simulation so complicated, namely, the large number of dyads and the fact that they are more or less randomly distributed along the T tubules. A reasonable first simplification in such a situation is to ignore spatial considerations, with the argument that one may study a local area of the cell membrane, a single T tubule, or even a single dyad and subsequently scale-up the results statistically to the thousands of dyads in each cell. The important advantage is that this simplification step circumvents the need for PDEs to describe the detailed mechanisms in favor of a model consisting of ordinary differential equations (ODEs) that use average numbers of ions in dyads instead of a specific number in each dyad. Before we make this step, we must convince ourselves that we do not lose too much resolution. A strict mathematical justification would require setting up a detailed model and comparing it with the simplifying approximation. Indeed, Winslow's group performed a careful analysis demonstrating that a simplified model based on a single dyad retained important features of the CICR mechanism.

We cannot present the details of this analysis here, but we can ask ourselves which aspects are really important to us. In the end, we need to know how many calcium ions enter the collection of dyads. Does the large population of dyads buffer the stochastic fluctuations among individual dyads and make their collective behavior predictable? The affirmative answer comes from a fundamental law of statistics, known as the **central limit theorem**. This theorem states that the sum of many random events of similar character follows a normal distribution and that the standard deviation of this distribution becomes smaller if more and more events are considered. In other words, the more items are part of the system, the more the population of items will behave predictably like their average. The theorem is very difficult to prove, but Exercise 12.17 suggests some easy computational experiments that might convince you that it is correct.

Unfortunately, even the simplified model of one dyad is still too complicated as a module in a multiscale model of a whole cardiomyocyte. A second opportunity for simplification arises from a technique called **separation of timescales**. This generic technique is useful if a system involves different types of processes, some of which are very fast and others comparatively slow. The argument is that on the timescale of the fast processes, the slow processes are essentially constant and their dynamics can simply be ignored. On the timescale of the slow processes, it is often reasonable to expect that much faster processes assume their steady-state level essentially instantaneously and are therefore almost always constant. The result of this separation is a set of two or more simpler models, and one only analyzes the one whose timescale is of particular interest in a given situation. In the case of calcium transport associated with the dyads, the processes at LCCs and RyRs are comparable in speed, but much faster than the release of large quantities of calcium from the SR. A possible strategy is therefore to focus only on the LCC and RyR as a coupled system that can be represented as a low-dimensional system of ODEs [66, 70].

Experimental work has shown that opening and closing of each LCC depend on the local calcium concentration. Specifically, the data suggest that Ca^{2+} binding induces the channel to switch from an active mode (mode normal) to an inactive mode (mode Ca) where transitions to the open state are very rare. A first model describing the opening and closing dynamics assumed that four subunits could independently close each LCC [71, 72]. Thus, five active states ($C_0, \ldots, C_4$) were defined, with the subscript representing the number of permissive subunits. Similarly, five inactive states were defined. As a result, any given LCC could be in one of 12 states, namely, 10 closed states (5 active and 5 inactive) and 2 open states (active or inactive). In line with experimental results, the transitions among active states and among inactive states were made voltage-dependent, while transitions between closed and open states were made voltage-independent. The transitions from active to inactive states were formulated to depend on the Ca^{2+} concentration, with an increase in calcium causing the LCC to shift to a gating mode permitting only infrequent opening.

Rigorous analysis of this model justified a simplification from five to two pairs of active and inactive states. It also showed that the transition probability from an inactive closed state to an open state was negligibly small. Thus, the result was a simplified LCC model consisting of five states: two closed active states C_1 and C_2, one open state O, which is accessible from C_2, and two closed (Ca^{2+} inactivated) states I_1 and I_2 accessible from C_1 and C_2, respectively (**Figure 12.21A**).

The rates of transitions between the two active states and between the two inactivated states are fast (the rates denoted by f in Figure 12.21A). By contrast, inactivation by Ca^{2+} and activation are slower (the rates denoted by s). This difference in rates suggested a separation of timescales and thus a possible further simplification to only three states (Figure 12.21B), using the argument that the fast transitions become less relevant on the timescale of the slow transitions.

It is straightforward to set up mass-action equations for the two systems. Implementing both in the same program allows us to judge the loss in resolution due to

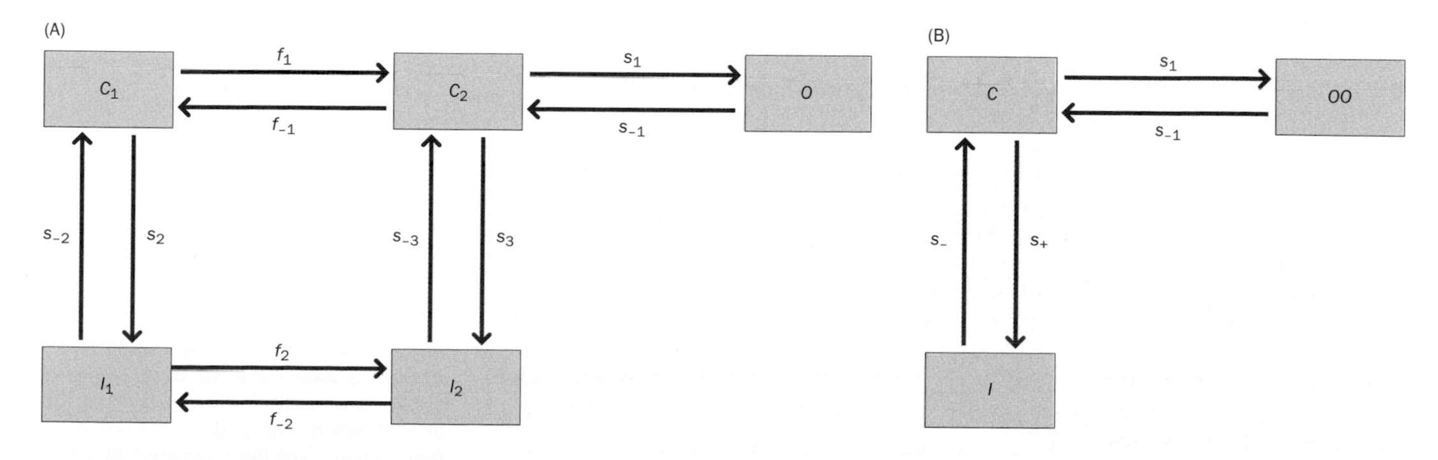

Figure 12.21 Original and simplified models of calcium dynamics in dyads. Computational validation experiments are needed to assess the degree to which the simpler three-state model in (B) reflects the responses of the more complex five-state model in (A). (Adapted from Hinch R, Greenstein JL & Winslow RL. *Prog. Biophys. Mol. Biol.* 90 [2006] 136–150. With permission from Elsevier.)

model reduction. The two models, inspired by Hinch and co-workers [66, 70], have the following forms:

Five-state model

$$
\begin{aligned}
\dot{C}_1 &= 50C_2 + 0.5I_1 - 40C_1 - C_1, & C_1 &= 0.60,\\
\dot{C}_2 &= 40C_1 + I_2 - 50C_2 - 2C_2 + 0.5O - 2C_2, & C_2 &= 0.49,\\
\dot{I}_1 &= 40I_2 + C_1 - 60I_1 - 0.5I_1, & I_1 &= 0.79,\\
\dot{I}_2 &= 60I_1 + 2C_2 - 40I_2 - I_2, & I_2 &= 1.18,\\
\dot{O} &= 2C_2 - 0.5O + perturb, & O &= 1.95;
\end{aligned}
\tag{12.26}
$$

Three-state model

$$
\begin{aligned}
\dot{C} &= 1.1I - 2C + 1.1OO - 2C, & C &= 1.09,\\
\dot{I} &= 2C - 1.1I, & I &= 1.97,\\
\dot{OO} &= 2C - 1.1OO + perturb, & OO &= 1.95.
\end{aligned}
\tag{12.27}
$$

The two systems have parameters that lead to similar steady states for $C_1 + C_2$ and C, $I_1 + I_2$ and I, as well as O and OO. The simulation begins close to these states. At time $t = 2$, we perturb the open state (O, OO) by setting the parameter $perturb = 1$. At $t = 4$, we send the signal $perturb = -1$, and at $t = 6$, we return to normal by setting $perturb = 0$. The result is shown in **Figure 12.22A**. We see that the two systems respond quite similarly.

For the second part of the simulation, we multiply all fast (f) parameters by 0.05, which makes them similar in magnitude to the slow (s) parameters, or by 0.005, which makes them even slower. Figure 12.22B and C demonstrate that the systems return to similar, although not exactly the same, steady states as before, but that the transient dynamics diverge a little more than before (Figure 12.22A). This one set of simulations suggests that the three-state model is doing quite well. In reality, one would execute many more and different types of perturbations with actual parameter values and compare the transients of the five- and three-state systems. If the differences in results can be deemed negligible for all relevant scenarios, the three-state model is a sufficient representation of the larger, five-state system.

The more realistic model of Hinch and colleagues [66] is substantially more complicated than this illustration, because the voltage and calcium dependences of the transitions are explicitly taken into account, parameters are measurable quantities, obtainable from specific experiments, and the channel current is represented with the Goldman–Hodgkin–Katz equation (12.9) that we discussed before. Even in this more complex implementation, validation studies have attested that the reduction to the three-state model was of acceptable accuracy.

Both experiments and mechanistic models indicate that RyR inactivation is the primary mechanism of termination of calcium release from the SR. This inactivation process has characteristics similar to the LCC activation process, and model construction begins again with five variables for different states of the RyR, which can be reduced to three states, with similar arguments of timescale separation as for the LCC model.

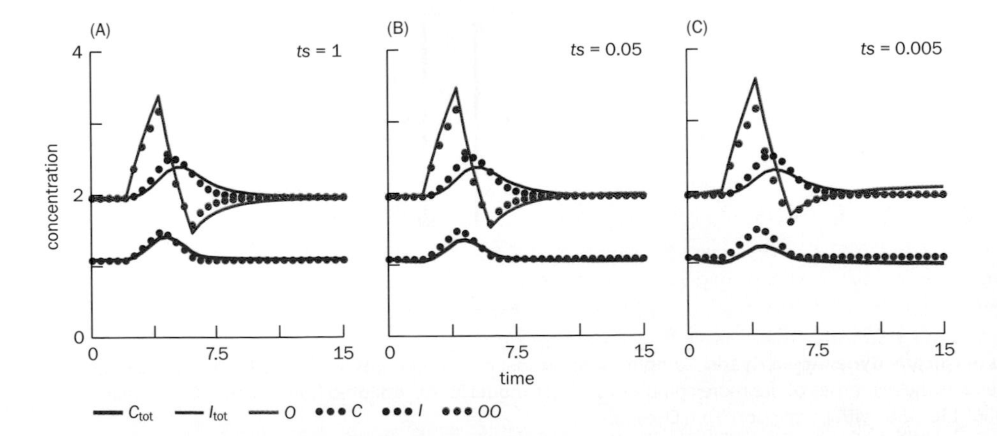

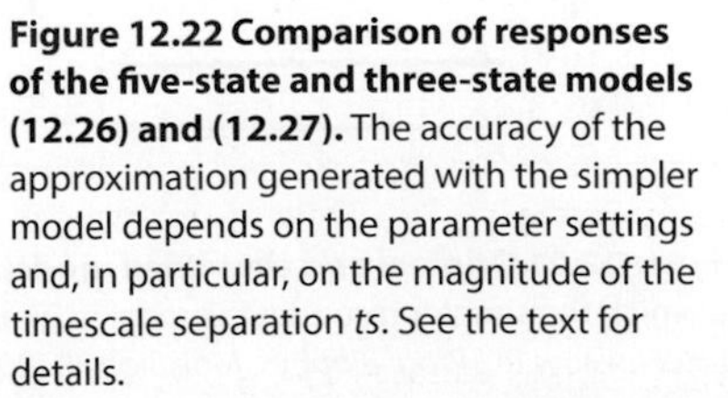

Figure 12.22 Comparison of responses of the five-state and three-state models (12.26) and (12.27). The accuracy of the approximation generated with the simpler model depends on the parameter settings and, in particular, on the magnitude of the timescale separation *ts*. See the text for details.

The next step toward a more comprehensive model is the merging of the LCC and RyR models. A minimal representation for each such calcium release unit (CaRU) consists of one LCC, a closely apposed RyR, and the dyadic space. One LCC opening may activate four to six RyRs (see Figure 12.20). The Ca^{2+} flux from the dyadic space to the cytoplasm is assumed to be governed by simple diffusion. Based on this assumption, it is possible to establish an approximate relationship between the concentration of calcium in the dyadic space and that in the cytoplasm. This relationship, which also depends on the currents through the LCC and RyR, is very important because it greatly reduces the number of differential equations in the model: assuming that the relationship is valid, it is no longer necessary to compute the calcium concentration for each dyad. The CaRU model can have nine different states, resulting from combinations of the three LCC states and the three RyR states, but can again be simplified to three states under the assumption that a separation of timescales is legitimate.

Hinch and collaborators showed that all the different pieces (component models) can be merged into a whole-cell Ca^{2+} regulation model that describes the electrical, chemical, and transport processes in a cardiomyocyte very well [66]. This model contains the CaRU system and the different pumps and ion channels that we discussed before, as well as mechanisms of energy balance. Needless to say, this model is rather complicated. Because of the strategy of simplifying details, followed by validating each simplification, the model is not 100% realistic, but it seems sufficiently accurate, as well as amenable to solution and analysis in a reasonable amount of computer time.

Indeed, this model is accurate enough to allow the representation of some heart diseases. Of particular importance is sudden cardiac death, which may have a wide variety of root causes. Here, the models permit one chain of causes and effects, leading from changes in LLCs and RyRs to a prolongation of the action potential (see Figure 12.12C), reduced Ca^{2+} flux, increased Ca^{2+} relaxation time, and subsequent arrhythmia (cf. [73]). Interestingly, some of the molecular events can be seen in the expression of ion channels and of genes that code for transporters and Ca^{2+} handling proteins [29]. Such changes in expression can be represented as corresponding changes in model parameters (see Chapter 11), and the model does indeed lead to a prolonged action potential.

The models thus connect genetic, molecular, physiological, and clinical phenomena in a stepwise, causative fashion. The same models can furthermore be extended into a whole-heart model, which demonstrates how molecular aberrations at the cellular level may lead to macroscopic changes, such as arrhythmias, ventricular tachycardia, and fibrillation. For such a whole-heart model, one must account for the shape of the heart and the complex orientation of fibers within the heart muscle, as we discussed early in the chapter. At this point, it is no longer possible to avoid or simplify the PDEs or finite element representations that permit a detailed description of how electric impulses move along the surface of the heart (see Figure 12.3). Nevertheless, because of the simplified, yet validated, submodels, in particular for the LCC and RyR, it is actually feasible to construct detailed whole-heart models that capture the entire chain of causes and effects. In the chain of events discussed above, changes in channels, transporters, and genes affecting calcium dynamics within individual cardiomyocytes lead to alterations in the electrical conductance in the heart tissue, which may lead to arrhythmia and myocardial infarction.

OUTLOOK FOR PHYSIOLOGICAL MULTISCALE MODELING

This chapter has served two purposes. First, of course, it has shed light on the complexity of the heart and on diverse modeling approaches. It is clear that no single modeling strategy is superior to all the others, and that model structure and complexity are directly related to the questions to be asked. For a full understanding, the vastly differing aspects of the organ must all be understood and integrated. These include the electrochemical processes that we have focused on in this chapter, but also the structural, mechanical, physical, and morphological phenomena that are crucial for the conductance of electrical signals and the beating and pumping action of the heart. They should also allow investigations of the energy needs and of changes in the heart due to exercise, diet, aging, and disease.

A tool for such a comprehensive task might ultimately be a template-and-anchor model [6] that uses the high-level blood pumping function of the heart as the template and embeds into this template electrophysiological, biochemical, and biophysical submodels of much finer granularity as anchors. More likely will be a set of complementary models that focus on one or two levels and inform models at other levels, as has been proposed in the Physiome and Virtual Physiological Rat Projects [10, 13, 63].

As a second purpose of this chapter, the discussion has shown how it is possible to concatenate different models so that one may causally connect morphological, physiological, and clinical changes in an organ to altered aspects in molecular structures and events, and to variations in the expression state of the genome [64–66]. Such a chain of models requires that large-scale systems be divided into functional modules, which are analyzed by focusing on events at their specific spatial, temporal, and organizational scales. Within the human body, such systems of systems are abundant. The liver consists of more or less homogeneous lobes, which themselves consist of hepatocytes that produce over 1000 important enzymes and are essential for the production of proteins and other compounds needed for digestion and detoxification. Initially ignoring the requirements and means of metabolite transport, much of the function of the liver can be studied on the level of individual hepatocytes. However, moving to higher levels of functionality, one must address the overarching functions of the tissues and the whole organ, such as the regenerative potential of the liver with all its spatial and functional implications. The Virtual Liver Network approaches this multiscale topic with different types of computational models [16, 74].

Like the liver, other organs consist primarily of large numbers of similar modules, which suggests analogous modeling approaches [17, 63, 75]: the kidney is composed of nephrons, and the lung consists of bronchi, bronchioles, and alveoli. While this conceptual and computational dissection of large systems into smaller modules is of enormous help for modeling purposes, caution is necessary, because the dissection sometimes eliminates crucial interactions between components. A prime example is the brain, which consists of billions of neurons, whose individual function we understand quite well, but also of highly regulated brain circuits and other physiological systems, which are often ill characterized. We have a good grasp of the signal transduction processes occurring at synapses, where neurons communicate with each other, but the sheer number of neurons, their biochemical and physiological dynamics, and the complex distribution of synapses throughout the brain lead to emerging properties such as adaptation, learning, and memory that we are still unable to grasp (see Chapter 15). Beyond organs, we need to be recognizant of the fact that problems in one system easily spread to other systems. For instance, a full understanding of cancer without consideration of the immune system is without doubt incomplete. One thing is clear: micro- and macro-physiology will continue to present us with grand challenges, but also with a glimpse of where we might want to target our research efforts in the years to come.

EXERCISES

12.1. As a nerdy Valentine's gift, produce two- and three-dimensional hearts, using one of the many available heart curves [76–78], such as

(1) $(x^2+y^2-1)^3-x^2y^3=0$;

(2) $x=\sin t\cos t\ln|t|,\ \ y=\sqrt{|t|}\cos t,\ \ -1\le t\le 1$;

(3) $(x^2+\frac{9}{4}y^2+z^2-1)^3-x^2z^3$
$-\frac{9}{80}y^2z^3=0,\quad -3\le x,y,z\le 3.$

Customize your hearts by changing or adding parameters of the models.

12.2. (a) Explore the role of the parameter k in the van der Pol oscillator in Chapter 4; see (4.73)–(4.76). Report your findings.

(b) What happens if both equations in (4.76) are multiplied by the same positive number? What happens if the multiplier is negative? What happens if the two equations are multiplied by different numbers?

(c) Study the effects of larger perturbations on v. Study the effects of perturbations on w; interpret what such perturbations mean.

(d) Study simultaneous perturbations in both v and w.

(e) Study the effects of regularly iterated perturbations (given to the system at regular intervals by an outside pacemaker).

12.3. (a) Initiate the stable oscillator System B in (12.1) with different initial values. Begin with $X_1 = 0.05$, $X_2 = 2.5$, and $X_1 = 0.75$, $X_2 = 2.5$. Try other combinations. Report and interpret your findings.

(b) Modulate the oscillator by slightly changing some of the exponents or by multiplying the entire systems or some terms by constant (positive or negative) factors. Report your findings.

(c) Multiply both equations in the oscillator System A in (12.1) by the same X_1 or X_2, raised to some positive or negative power, such as X_2^{-4}. Visualize the results with time courses and phase-plane plots in which X_2 is plotted against X_1. Describe in a report how the shape of the oscillation changes.

12.4. (a) In the series of examples of spurious sine function impulses, we changed the frequency. As a consequence, the oscillations became chaotic, and eventually the entire heartbeat stopped. Explore what happens if the frequency is retained (for instance, $a = 10$, $a = 1$, $a = 0.5$, or $a = 0.0001$) but the amplitude of the oscillation is altered.

(b) Explain why it is not clinically sufficient that a pacemaker generates a regular impulse pattern of, say, 70 beats per minute. What else is needed?

12.5. Explore further the degree to which a highly irregular heartbeat can be made regular by a simplistic pacemaker (see the discussion of Figure 12.9). Multiply all equations of the chaotic Lotka–Volterra (LV) oscillator in Chapter 10 by 50, which should result in a rather irregular heartbeat with about 60–70 beats per minute. Formulate a sine oscillator (pacemaker) as a pair of ODEs. ($\dot{s} = qc$, $\dot{c} = -qs$). Add to each equation in the LV system a term of the form $p(s + 1)$. Confirm with $p = 0$ the chaotic arrhythmia. Perform three series of experiments, with the goal of regularizing the arrhythmia. In the first series, fix $q = 6$, increase p from 0 in small steps, and study the resulting heartbeat. In the second series, set p at some small value and vary q. In the third series, vary both. Finally, select a combination of p and q that seems to work well, and then alter the demand of the body for more or fewer heartbeats by multiplying all LV equations with the same value from the interval [0.5, 1.5].

12.6. Describe what the setting of $[Ca^{2+}]_{ext}$ as an independent variable in Section 12.7 entails. What is the importance of its numerical value? What assumptions have to be made and justified to assign the status of an independent rather than a dependent variable? What would have to be changed if $[Ca^{2+}]_{ext}$ were considered a dependent variable? Explain biologically and mathematically.

12.7. Adding the two equations in (12.4) together yields the much simpler equation $\dot{x} + \dot{y} = r_{EC} - p_{CE}y$. Explain the biological and mathematical meaning of this equation. What would it mean (biologically and mathematically) if $r_{EC} = p_{CE} = 0$, $r_{EC} - p_{CE} = 0$, and $r_{EC} - p_{CE} \neq 0$?

12.8. Confirm numerically that the oscillation in the system (12.5) is stable. Is (12.5) a closed system? Explain.

12.9. Mine the literature or the Internet for three-dimensional molecular structures of channel proteins in cardiomyocytes.

12.10. The Na^+ concentration is usually much higher outside the cell (about 150 mM) than inside the cell (between 5 and 20 mM). Furthermore, while there is a lot of calcium in cells, much of it is bound, and the free calcium concentrations are much lower than those of Na^+, with values of $[Ca^{2+}]_i = 10^{-4}$ mM and $[Ca^{2+}]_o$ between 1 and 3 mM. Compute the equilibrium potentials for Na^+ and Ca^{2+} and discuss their contributions to the membrane potential.

12.11. The form of the equations for n, m, and h in the Hodgkin–Huxley model derives from the following argument. Let x be one of the three quantities; then we can write its differential equation in simplified form as

$$\dot{x} = \alpha_x(1 - x) - \beta_x x.$$

Show that this equation has the solution

$$x(t) = x_\infty + (x_0 - x_\infty)e^{-t/\tau_x},$$

where $x(0) = x_0$, $x_\infty = \alpha_x/(\alpha_x + \beta_x)$, and $\tau_x = 1/\alpha_x + \beta_x)$. Interpret x_∞.

12.12. Study the effects of initial conditions (for v, n, m, and h) on the shape of the action potential. Use as settings for other parameters $v_{Na} = 120$ mV, $v_K = -12$ mV, $v_L = 10.6$ mV, $C = 1$, and $I_{app} = 1$.

12.13. Compare the results of the Hodgkin–Huxley model in the text with two other representations for the gating functions n, m, and h. Use in both cases the parameters $v(0) = 12$, $n = 0.33$, $m = 0.05$, $h = 0.6$, $C = 1$, $v_{Na} = 115$, $v_K = -12$, and $v_L = 10.6$. Study the voltage and gating functions and discuss repeated beats.

(a) Explore the following functions (adapted from [79]) with $\bar{g}_{Na} = 120$, $\bar{g}_K = 3.6$, and $g_L = 0.3$:

$$\alpha_n(v) = \frac{0.01(v+50)}{1 - e^{-(v+50)/10}},$$

$$\beta_n(v) = 0.125e^{-(v+60)/80},$$

$$\alpha_m(v) = \frac{0.1(v+35)}{1 - e^{-(v+35)/10}},$$

$$\beta_m(v) = 4.0e^{-(v+60)/18},$$

$$\alpha_h(v) = 0.07e^{-(v+60)/20},$$

$$\beta_h(v) = \frac{1}{1 + e^{-(v+30)/10}}.$$

(b) Use the following functions with $\bar{g}_{Na} = 120$, $\bar{g}_K = 3.6$, and $g_L = 0.3$:

[38] Ferrario M, Signorini MG, Magenes G & Cerutti S. Comparison of entropy-based regularity estimators: Application to the fetal heart rate signal for the identification of fetal distress. *IEEE Trans. Biomed. Eng.* 33 (2006) 119–125.
[39] Lipsitz LA. Physiological complexity, aging, and the path to frailty. *Sci. Aging Knowledge Environ.* 2004 (2004) pe16.
[40] Lipsitz LA & Goldberger AL. Loss of "complexity" and aging. Potential applications of fractals and chaos theory to senescence. *JAMA* 267 (1992) 1806–1809.
[41] Glass L & Mackey MC. From Clocks to Chaos: The Rhythms of Life. Princeton University Press, 1988.
[42] Thompson JMT & Stewart HB. Nonlinear Dynamics and Chaos, 2nd ed. Wiley, 2002.
[43] Endo M. Calcium release from the sarcoplasmic reticulum. *Physiol. Rev.* 57 (1977) 71–108.
[44] Vinogradova TM, Brochet DX, Sirenko S, et al. Sarcoplasmic reticulum Ca^{2+} pumping kinetics regulates timing of local Ca^{2+} releases and spontaneous beating rate of rabbit sinoatrial node pacemaker cells. *Circ. Res.* 107 (2010) 767–775.
[45] Somogyi R & Stucki J. Hormone-induced calcium oscillations in liver cells can be explained by a simple one pool model. *J. Biol. Chem.* 266 (1991) 11068–11077.
[46] Stucki J & Somogyi R. A dialogue on Ca^{2+} oscillations: an attempt to understand the essentials of mechanisms leading to hormone-induced intracellular Ca^{2+} oscillations in various kinds of cell on a theoretical level. *Biochim. Biophys. Acta* 1183 (1994) 453–472.
[47] Bertram R, Sherman A & Satin LS. Electrical bursting, calcium oscillations, and synchronization of pancreatic islets. *Adv. Exp. Med. Biol.* 654 (2010) 261–279.
[48] Alberts B, Johnson A, Lewis J, et al. Molecular Biology of the Cell, 6th ed. Garland Science, 2015.
[49] Sipido KR, Bito V, Antoons G, et al. Na/Ca exchange and cardiac ventricular arrhythmias. *Ann. NY Acad. Sci.* 1099 (2007) 339–348.
[50] Ojetti V, Migneco A, Bononi F, et al. Calcium channel blockers, beta-blockers and digitalis poisoning: management in the emergency room. *Eur. Rev. Med. Pharmacol. Sci.* 9 (2005) 241–246.
[51] Sherwood L. Human Physiology: From Cells to Systems, 9th ed. Cengage Learning, 2016.
[52] DiFrancesco D. The role of the funny current in pacemaker activity. *Circ. Res.* 106 (2010) 434–446.
[53] Lakatta EG, Maltsev VA & Vinogradova TM. A coupled system of intracellular Ca^{2+} clocks and surface membrane voltage clocks controls the timekeeping mechanism of the heart's pacemaker. *Circ. Res.* 106 (2010) 659–673.
[54] Winslow RL, Rice J, Jafri S, et al. Mechanisms of altered excitation–contraction coupling in canine tachycardia-induced heart failure, II: Model studies. *Circ. Res.* 84 (1999) 571–586.
[55] Hodgkin A & Huxley A. A quantitative description of membrane current and its application to conduction and excitation in nerve. *J. Physiol.* 117 (1952) 500–544.
[56] Noble D. From the Hodgkin–Huxley axon to the virtual heart. *J. Physiol.* 580 (2007) 15–22.
[57] Brady AJ & Woodbury JW. The sodium–potassium hypothesis as the basis of electrical activity in frog ventricle. *J. Physiol.* 154 (1960) 385–407.
[58] Fitzhugh R. Thresholds and plateaus in the Hodgkin–Huxley nerve equations. *J. Gen. Physiol.* 43 (1960) 867–896.
[59] Izhikevich EM. Dynamical Systems in Neuroscience: The Geometry of Excitability and Bursting. MIT Press, 2007.
[60] Fitzhugh R. Impulses and physiological states in theoretical models of nerve membrane. *Biophys. J.* 1 (1961) 445–466.
[61] Nagumo JA & Yoshizawa S. An active pulse transmission line simulating nerve axon. *Proc. Inst. Radio Eng.* 50 (1962) 2061–2070.
[62] Mubagwa K, Kaplan P, Shivalkar B, et al. Calcium uptake by the sarcoplasmic reticulum, high energy content and histological changes in ischemic cardiomyopathy. *Cardiovasc. Res.* 37 (1998) 515–523.
[63] The Virtual Physiological Rat Project. http://virtualrat.org/.
[64] Winslow RL, Tanskanen A, Chen M & Greenstein JL. Multiscale modeling of calcium signaling in the cardiac dyad. *Ann. NY Acad. Sci.* 1080 (2006) 362–375.
[65] Cortassa S, Aon MA, Marbán E, et al. An integrated model of cardiac mitochondrial energy metabolism and calcium dynamics. *Biophys. J.* 84 (2003) 2734–2755.
[66] Hinch R, Greenstein JL & Winslow RL. Multi-scale models of local control of calcium induced calcium release. *Prog. Biophys. Mol. Biol.* 90 (2006) 136–150.
[67] Tanskanen AJ, Greenstein JL, Chen A, et al. Protein geometry and placement in the cardiac dyad influence macroscopic properties of calcium-induced calcium release. *Biophys. J.* 92 (2007) 3379–3396.
[68] Xing J, Wang H & Oster G. From continuum Fokker–Planck models to discrete kinetic models. *Biophys. J.* 89 (2005) 1551–1563.
[69] Smith GD. Modeling the stochastic gating of ion channels. In Computational Cell Biology (CP Fall, ES Marland, JM Wagner & JJ Tyson, eds), pp 285–319. Springer, 2002.
[70] Hinch R, Greenstein JL, Tanskanen AJ, et al. A simplified local control model of calcium-induced calcium release in cardiac ventricular myocytes. *Biophys. J.* 87 (2004) 3723–3736.
[71] Imredy JP & Yue DT. Mechanism of Ca^{2+}-sensitive inactivation of L-type Ca^{2+} channels. *Neuron* 12 (1994) 1301–1318.
[72] Jafri MS, Rice JJ & Winslow RL. Cardiac Ca^{2+} dynamics: the roles of ryanodine receptor adaptation and sarcoplasmic reticulum load. *Biophys. J.* 74 (1998) 1149–1168.
[73] Phillips RM, Narayan P, Gomez AM, et al. Sarcoplasmic reticulum in heart failure: central player or bystander? *Cardiovasc. Res.* 37 (1998) 346–351.
[74] Hoehme S, Brulport M, Bauer A, et al. Prediction and validation of cell alignment along microvessels as order principle to restore tissue architecture in liver regeneration. *Proc. Natl Acad. Sci. USA* 107 (2010) 10371–10376.
[75] Physiome Project: Urinary system and Kidney; http://physiomeproject.org/research/urinary-kidney
[76] Beutel E. Algebraische Kurven. Göschen, 1909–11.
[77] Heart Curve. Mathematische Basteleien. http://www.mathematische-basteleien.de/heart.htm.
[78] Taubin G. An accurate algorithm for rasterizing algebraic curves. In Proceedings of Second ACM/IEEE Symposium on Solid Modeling and Applications, Montreal, 1993, pp 221–230. ACM, 1993.
[79] Peterson J. A simple sodium–potassium gate model. Chapter 2: The basic Hodgkin–Huxley model. cecas.clemson.edu/~petersj/Courses/M390/Project/Project.pdf.

FURTHER READING

Alberts B, Johnson A, Lewis J, et al. Molecular Biology of the Cell, 6th ed. Garland Science, 2015.

Bassingthwaighte J, Hunter P & Noble D. The Cardiac Physiome: perspectives for the future. *Exp. Physiol.* 94 (2009) 597–605.

Crampin EJ, Halstead M, Hunter P, et al. Computational physiology and the Physiome Project. *Exp. Physiol.* 89 (2004) 1–26.

Keener J & Sneyd J. Mathematical Physiology. II: Systems Physiology, 2nd ed. Springer, 2009.

Klabunde RE. Cardiovascular Physiology Concepts, 2nd ed. Lippincott Williams & Wilkins, 2011.

Nerbonne JM & Kass RS. Molecular physiology of cardiac repolarization. *Physiol. Rev.* 85 (2005) 1205–1253.

Noble D. From the Hodgkin–Huxley axon to the virtual heart. *J. Physiol.* 580 (2007) 15–22.

Winslow RL, Tanskanen A, Chen M & Greenstein JL. Multiscale modeling of calcium signaling in the cardiac dyad. *Ann. NY Acad. Sci.* 1080 (2006) 362–375.

Systems Biology in Medicine and Drug Development

13

When you have read this chapter, you should be able to:

- Identify molecular causes of human variation
- Understand disease as a dynamic deviation in parameter space
- Describe the concepts of transforming an average disease model into a personalized disease model
- Understand the drug development process
- Construct and analyze antibody–target binding models
- Construct and analyze pharmacokinetic models
- Analyze differential equation models of disease models

ARE YOU UNIQUE?

13.1 Biological Variability and Disease

If you and I and everybody else were exactly the same, life would be rather boring, but medicine and drug development would be much easier. In a simplistic analogy, if one knows how to exchange a windshield wiper motor in one 2005 Ford Mustang, one knows how to do it in another 2005 Ford Mustang. Not so with complex diseases. Two people may present with exactly the same symptoms, but react in a distinctly different manner to the same dose of the same drug. As a consequence, and in contrast to car repair, it is not sufficient to test a drug regimen on only one person—instead it is necessary to perform large, complex, and expensive clinical trials that often require many years of investigation. Then again, in spite of the clear differences, we humans cannot be all that different from each other. After all, medicine has become quite successful, and most prescription drugs actually work as advertised.

So, how different are we really? Let us start with our genes. As we discussed in Chapter 6, each of us has roughly three billion nucleotide pairs. For most of them, there are no differences between two people and even among the entire human population. If indeed there is a genetic difference, say in base 5,000,000, between any two people on Earth, then this base is the location of a **SNP** (pronounced 'snip' and meaning **single nucleotide polymorphism**, which roughly translates into "many different forms in a given base of the DNA"). According to the best current estimates, SNPs occur within the entire human population about every 300 bases, which implies that you and I differ genetically by far, far less than 1% [1]. This does not sound like much, but SNPs account for about 90% of all genetic variation, with other changes occurring in deletions, multiplications of DNA, translocations of DNA from one chromosome to another, and other rather rare alterations. Just considering SNPs, is there a sufficient repertoire of differences to make 6 or 7 billion people unique? A quick and very rough computation leaves no doubt that we will not run out of unique people any time soon. If every 300th nucleotide is changeable, and if we simplistically assume that the probability of a single nucleotide (A, T, C, or G) is the same at each SNP, then there are 10 million locations for SNPs, with a choice of four nucleotides at each location. Try to compute $4^{10,000,000}$. It is a truly enormous number that dwarfs the number of all humans who have ever lived. It even dwarfs by far the number of all atoms in the observable universe, which astronomers believe to be about 10^{80}, give or take a few zeros. Adding to this genetic interpersonal diversity, SNPs are not the only determinants of our individuality. While genes are generally considered the blueprint of an organism, we know that even organisms with exactly the same genetic material are not identical. First, there are identical twins, who may look indistinguishable, but may have rather different traits and personalities. Also, we have learned from the cloning of animals such as Dolly the Sheep and CopyCat the Cat that the "offspring" differs from the mother, for instance, in the coloration pattern.

If genes are not telling the whole story, what else is there? One relatively new field of biology focuses on **epigenetics**, which studies differences in the activation and deactivation patterns of genes among otherwise similar cells or organisms (see Chapter 6). These differences are often the consequence of modifications at the gene level, which, however, are not encoded in the DNA sequence. Instead, the DNA is methylated or otherwise modulated with residues at certain locations, which tends to affect the rate of gene transcription. Another possibility is a change in histone proteins, around which the DNA is wrapped, and which may again affect the rate of gene activation or deactivation. Other variations in an organism are, of course, related to the environment, including exposure to harmful chemicals, diet, and all our lifestyle choices.

All in all, we are different, and no two people are the same. This simple fact makes medicine and the development and application of prescription drugs challenging. Fortunately, the situation is not entirely hopeless. First, not every point mutation or SNP has a major effect. Much of our DNA does not code for proteins, and mutations in coding or noncoding sections may not always be critical. As a result of a mutation, an amino acid may be different in the protein for which the gene codes, but the incorrect amino acid may be located outside the active site of the protein, where no great harm is done. It is also possible that the protein is slightly impaired but that the many control mechanisms in the body offer effective compensation. Maybe most importantly, if we concentrate on one particular disease, very many features and processes elsewhere in the body become irrelevant. Because these facts come to our rescue, we are usually successful treating a child's ear infection, irrespective of her size, weight, race, or hair color. Nonetheless, many diseases are so severe, and prescription drugs may have such worrisome side effects, that a new focus in the medical field is on personalized "precision" medicine that seeks to develop treatments customized to each individual patient.

13.2 Modeling Variability and Disease

The situation of "heterogeneity in medicine" can be readily illustrated with an example from metabolism, where, in a typical case, an inter-personal difference will be manifest in the activity of some enzyme or modulator. It is easy to set up small dynamic models to get a feel for what such a difference may involve (see Chapters 2 and 4). As an example, let us study a generic branched pathway with some regulatory signals (**Figure 13.1**). The mathematical form is not really important for this type of illustration, and we translate the diagram into a mixed **Michaelis–Menten/power-law** model:

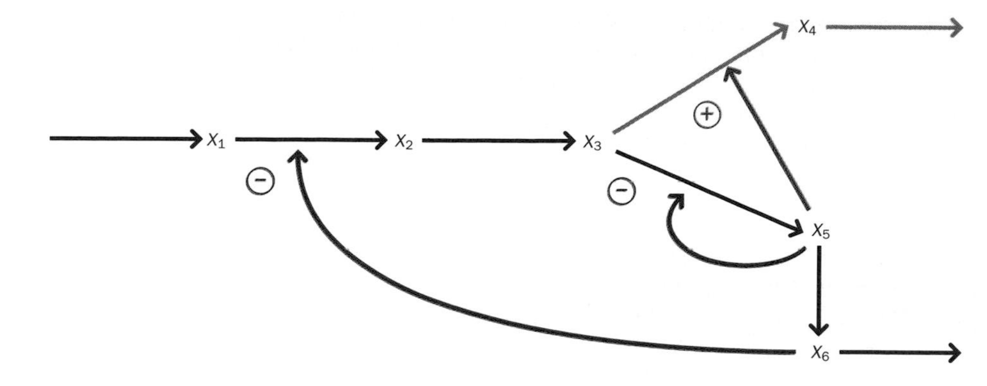

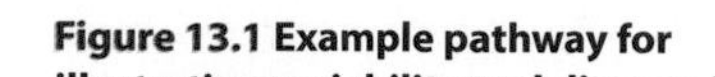

Figure 13.1 Example pathway for illustrating variability and disease. The generic pathway consists of five metabolites, two feedback signals (red), and one activation signal (green). The fluxes to and from X_4 (gray) are very small.

$$
\begin{aligned}
\dot{X}_1 &= Input - \frac{V_{\max 1} X_1}{K_{M1}(1 + X_6/K_{I6}) + X_1}, \\
\dot{X}_2 &= \frac{V_{\max 1} X_1}{K_{M1}(1 + X_6/K_{I6}) + X_1} - \frac{V_{\max 2} X_2}{K_{M2} + X_2}, \\
\dot{X}_3 &= \frac{V_{\max 2} X_2}{K_{M2} + X_2} - \beta_{31} X_3^{h_{331}} X_5^{h_{351}} - \beta_{32} X_3^{h_{332}} X_5^{h_{352}}, \\
\dot{X}_4 &= \beta_{31} X_3^{h_{331}} X_5^{h_{351}} - \beta_4 X_4^{h_{44}}, \\
\dot{X}_5 &= \beta_{32} X_3^{h_{332}} X_5^{h_{352}} - \frac{V_{\max 5} X_5}{K_{M5} + X_5}, \\
\dot{X}_6 &= \frac{V_{\max 5} X_5}{K_{M5} + X_5} - \beta_6 X_6^{h_{66}}.
\end{aligned}
\qquad (13.1)
$$

Parameter values for the model were chosen almost arbitrarily and are given in **Table 13.1**. We begin our analysis by computing the steady state, which, with three significant digits, is $(X_1, \ldots, X_6) = (2.57, 4.00, 3.23, 3.70, 3.30, 3.13)$. For a typical simulation, we initiate the system at this state and double the input during the time period [2, 3]. A plot of the results is shown in **Figure 13.2**. As is common for this type of pathway structure, the variables rise and fall in order of their position and then return to the original steady state. Notably, X_4 responds very slowly, because the flux going through this pool is very small in comparison with the others.

It is difficult to predict intuitively how such a system responds to perturbations in parameter values, but it is easy to check computationally. For instance, we may perform a series of simulations in which we raise one parameter at a time by 20% and study the effect on the steady state. Interestingly, we find that most perturbations are very well buffered. Usually, only one or two of the steady-state values deviate by more than 10% from their original values, and deviations of between 20% and 30% only occur in the variables directly affected by the alteration. As a specific example, if $V_{\max 5}$ is raised by 20%, then X_3 and X_5 decrease by about 21%, but all other variables are essentially unchanged. Exercises 13.1–13.4 explore this example further. In reality, a deviation of 20% often goes unnoticed and does not lead to disease or even symptoms.

While quite simplistic, this cursory analysis implies that regulated systems have a strong buffering capacity. Indeed, a Japanese group made the same observation, when they exposed *Escherichia coli* cells to various genetic and environmental perturbations [4]. While the gene activities and enzymes close to the perturbation were often strongly affected, the remaining metabolic profile was by and large unaffected, owing to a compensatory rerouting of fluxes in the cells. In this light, the reader

TABLE 13.1 PARAMETER VALUES FOR THE PATHWAY IN FIGURE 13.1 AND (13.1)

Input	$V_{\max 1}$	K_{M1}	K_{I6}	$V_{\max 2}$	K_{M2}	β_{31}	h_{331}	h_{351}
2	4	1	2	5	6	0.005	0.8	0.2
β_{32}	h_{332}	h_{352}	β_4	h_{44}	$V_{\max 5}$	K_{M5}	β_6	h_{66}
2	0.4	−0.4	0.005	0.9	8	10	1	0.6

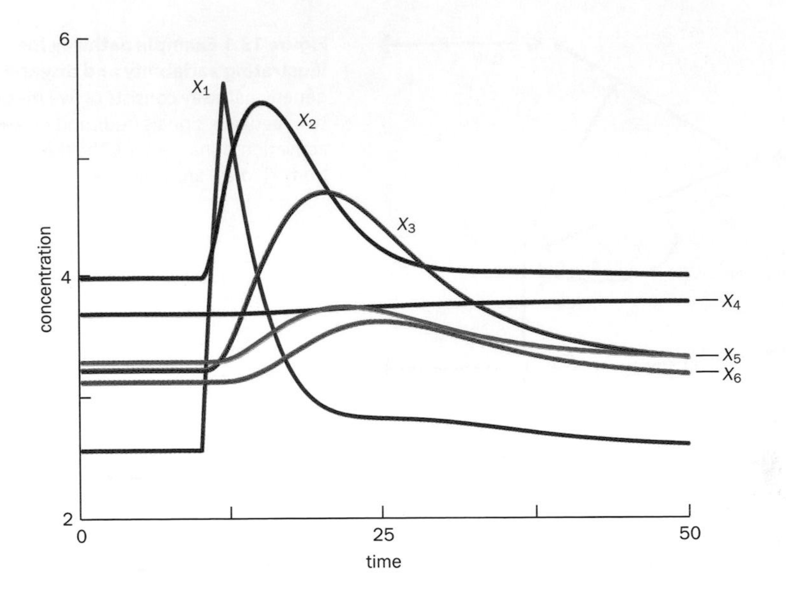

Figure 13.2 Typical scenario simulation of the pathway in Figure 13.1. The simulation used the equations in (13.1) and the parameter values in Table 13.1. From time $t = 10$ to time $t = 12$, the input to the system is doubled from 2 to 4.

should revisit the discussions regarding robustness of **small-world networks** (see Chapter 3).

The analysis of perturbations and their ultimate effects provides several insights. First, most steady-state values are not much affected by changes in any of the parameters. Second, an enzyme alteration at the beginning or center of the pathway may affect immediately associated metabolites, but the end products are often only mildly influenced. A direct implication is that, in many disease-related cases, changes in the intermediate metabolites may not be of relevance. Third, the dynamic response to a change in input or enzyme activity may consist of a transient metabolic deviation, which, however, may not be of long-term pertinence with respect to disease. What is not evident from the example is that larger and more complex systems are usually less prone to local perturbations, such as an altered enzyme activity, than smaller, less regulated systems.

PERSONALIZED MEDICINE AND PREDICTIVE HEALTH

Even if it seems that biomedical systems can compensate for altered components and processes quite well, we know of course that there are many diseases that cause long-term pain and suffering. In a relatively small number of cases, these diseases are caused by a single variation at the genome level. Perhaps the best example is sickle cell anemia, which is caused by a point mutation in the hemoglobin gene *HBB*. While there are many variants, the most common is very well understood: it consists of the substitution of the hydrophilic glutamic acid with the hydrophobic amino acid valine at the sixth amino acid position of the HBB polypeptide chain. So, we know exactly what is wrong, but we cannot do much about it.

In contrast to single-gene diseases like sickle cell anemia and cystic fibrosis, most disease pathologies result from a combination of genetic predisposition (as manifest in several gene alterations) and epigenetics, as well as environmental and possibly other factors, which nowadays is collectively called the exposome [5]. These combinations make it difficult to answer questions such as: How can it be that some nonsmokers develop lung cancer while some long-term chain-smokers do not? Why do people respond differently to the same treatment? These types of questions are considered to be at the core of two new and related approaches to medicine: **personalized** or **precision medicine** and **predictive health.**

The key goal of personalized medicine is the prevention and treatment of disease in a fashion that is custom-tailored to the individual. For instance, instead of determining the dose of a prescription drug simply based on body weight, and using average information obtained from large populations to assess the efficacy of the drug per kilogram of body weight, the dose is computed based on the personal characteristics of the patient. These include obvious information such as body weight, sex, and age, but also body mass index, gene markers, relevant enzyme profiles, metabolic rate, and a host of other **biomarkers** that could be of relevance to the disease and its treatment. Ideally, such a personalized, precise treatment in the future will maximize efficacy and minimize undesired side effects.

Predictive health is a conceptual continuation of personalized medicine. Its key goal is to maintain health rather than to treat disease. Thus, predictive health aims to predict disease in an individual before it manifests itself. The main targets are chronic diseases such as diabetes, cancer, and neurodegeneration. These diseases often begin with a long, silent phase, where a person does not even perceive symptoms. However, this early phase constitutes a measurable deviation from the norm, and although the person is apparently healthy, her or his biochemistry, physiology, gene expression, or protein and metabolite profiles already contain early warning signs indicating that the person is on an undesirable trajectory toward a chronic disease. If we had a reliable and comprehensive catalog of such warning signs, and if healthy people received early screening and treatment, while they were still apparently healthy, then chronic diseases could potentially be prevented or managed from their early stages on. There is plenty of indication that such a treatment option would be much cheaper than treatments of chronic disease that are typical in today's world, and efforts are underway to promote predictive health [6].

One intriguing challenge for any type of personalized medicine and predictive health is that it is not at all easy to define what exactly health or disease means. We have seen above that there can be a great deal of variability among healthy individuals, but we also know that the transition step from health to disease does not have to be large. A straightforward distinction between health and disease may be a different set of important parameters, such as blood pressure, cholesterol, or some genetic predisposition (see above and [7]). An extension of this view is that such changes collectively may disturb the normal state of **homeostasis** and lead to **allostasis**, which in the language of physiological modeling corresponds to a different, and in this case undesirable, steady state [8]. A more recent view is that disease may be the trade-off with the physiological robustness of the human body or the loss of complexity in the dynamics of a physiological subsystem (see [9, 10] and Chapter 15).

13.3 Data Needs and Biomarkers

Two components are required to realize ideas of personalized medicine and predictive health: first, lots of data; second, computational methods that are capable of integrating and analyzing these data. The data must cover many physiological aspects of individuals when they are healthy, and follow them throughout their healthy life and possibly into disease. The data from a single person will not shed sufficient light on complex disease development processes, but a growing collection of many personal trajectories, some staying within the healthy range and some diverting toward disease, will reveal an increasingly comprehensive picture of how disease develops and may suggest which early and very early signs (biomarkers) should be considered in regular screening examinations of healthy individuals. In more general terms, these data and their analysis will permit new and sharper definitions of what it means to be healthy, premorbid (with early signs of disease but without too severe symptoms), or diseased [3].

The tracing back of a disease to disease predictors, late biomarkers, and ultimately early biomarkers (**Figure 13.3**) may seem to lie far in the future. However, in some instances, it has already become standard. A good example is an elevated blood sugar level, which in itself does not constitute a disease but is often the precursor of type 2 diabetes. Another example is amniocentesis, a means of diagnosing a fetus within the womb by testing a small amount of amniotic fluid for genetic abnormalities and infections. Even though a baby might appear healthy in an ultrasound examination, the amniocentesis might foreshadow diseases later in life, based on genetic markers.

Biomarkers can also be indicators of whether a particular drug treatment might be effective, before the drug is actually administered [11]. In the new field of **pharmacogenomics**, analyses of expressed genes, especially those from the class encoding cytochrome P450 enzymes, are used to predict the sensitivity of individuals to specific drugs. A beautiful example is a prognostic genome analysis in children with acute lymphoblastic leukemia (ALL). It had been known that some children with

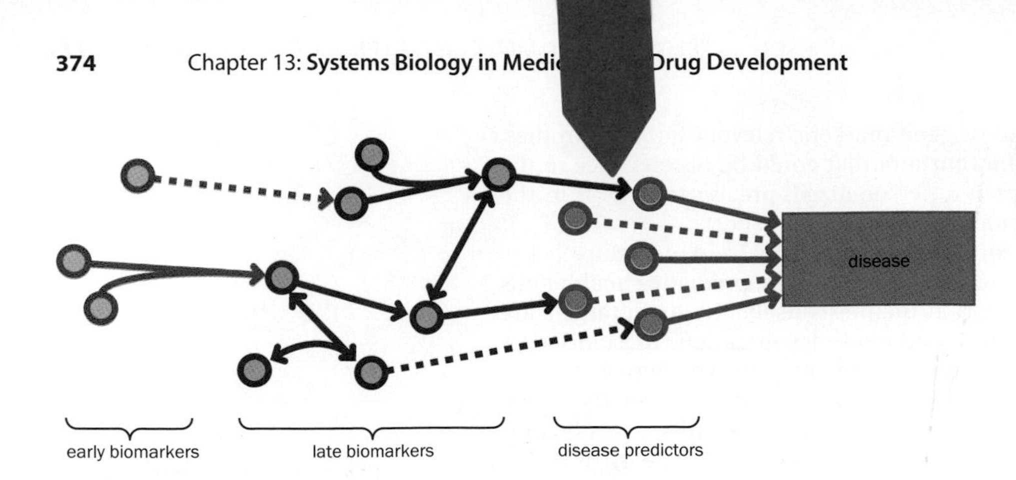

Figure 13.3 Hierarchical biomarker network. Disease is usually preceded by disease predictors. These are most likely preceded by late and early biomarkers.

ALL responded very well to standard drug treatments, while others developed severe side effects that greatly compromised the efficacy and applicability of the drugs. Scientists at St. Jude Children's Research Hospital in Memphis, Tennessee, discovered that a microarray test was able to classify young ALL patients before treatment into responders and nonresponders [12]. In this particular case, 45 differently expressed genes were found to be associated with success or failure of all four drug treatments tested, and 139 genes were identified in children responsive to one drug, but not the others.

As in the example of ALL, the first step toward rational predictions of disease and treatment outcome will be a statistical association analysis, where certain patterns in the data are predominantly observed in cases of disease or drug resistance, while other patterns are more often associated with health and a successful drug treatment. As an extension of standard association analyses of statistics, it is possible to train an **artificial neural network (ANN)** or some other machine learning technique to consider many markers and to classify individuals as healthy or at-risk. Because ANNs can be large, they have the potential of identifying at-risk patients earlier and more accurately than standard association methods of statistics. An example is the computer-assisted ProstAsure test for prostate cancer [13].

The identification of associations between biomarkers and disease is invaluable, but not the ultimate goal of predictive health. Building upon association data, the next step is the construction of causal models, which not only associate certain biomarkers with disease, but also demonstrate that a particular disease is actually caused by a variation in these biomarkers. Once this causality has been established, the biomarker is no longer just a diagnostic or prognostic symptom, but becomes a true drug target. The following sections indicate how modeling and systems biology of the future might contribute to our understanding of personalized health, disease, and treatment.

13.4 Personalizing Mathematical Models

Let us begin by painting a picture, in very broad strokes, of how today's medicine works. **Figure 13.4** shows a simplified diagram. The left side depicts, in reddish colors, the typical paths of current knowledge discovery. The two main sources of input are epidemiological studies, which study large populations in order to associate diseases with risk factors, and modern biology, which characterizes biochemical and physiological mechanisms that might lead to disease if they are altered and quantifies these mechanisms with pertinent parameters. The results of these two approaches ultimately enter clinical trials, which lead to prescriptions of treatments that are, by the nature of the process, based on population averages. Process descriptions and parameters have also been the basis for the design of disease models that, owing to the nature of the underlying data, again address average individuals. These models can be diagnosed and analyzed, as we discussed in Chapter 4, and may lead to better characterizations of a disease, as well as to a better understanding of averaged treatment options.

Systems biology enters the field with two components. Experimental systems biology offers new and very powerful means of identifying physiological mechanisms and their parameters. It will also be the foundation of information gathering in personalized precision medicine and predictive health, as discussed before.

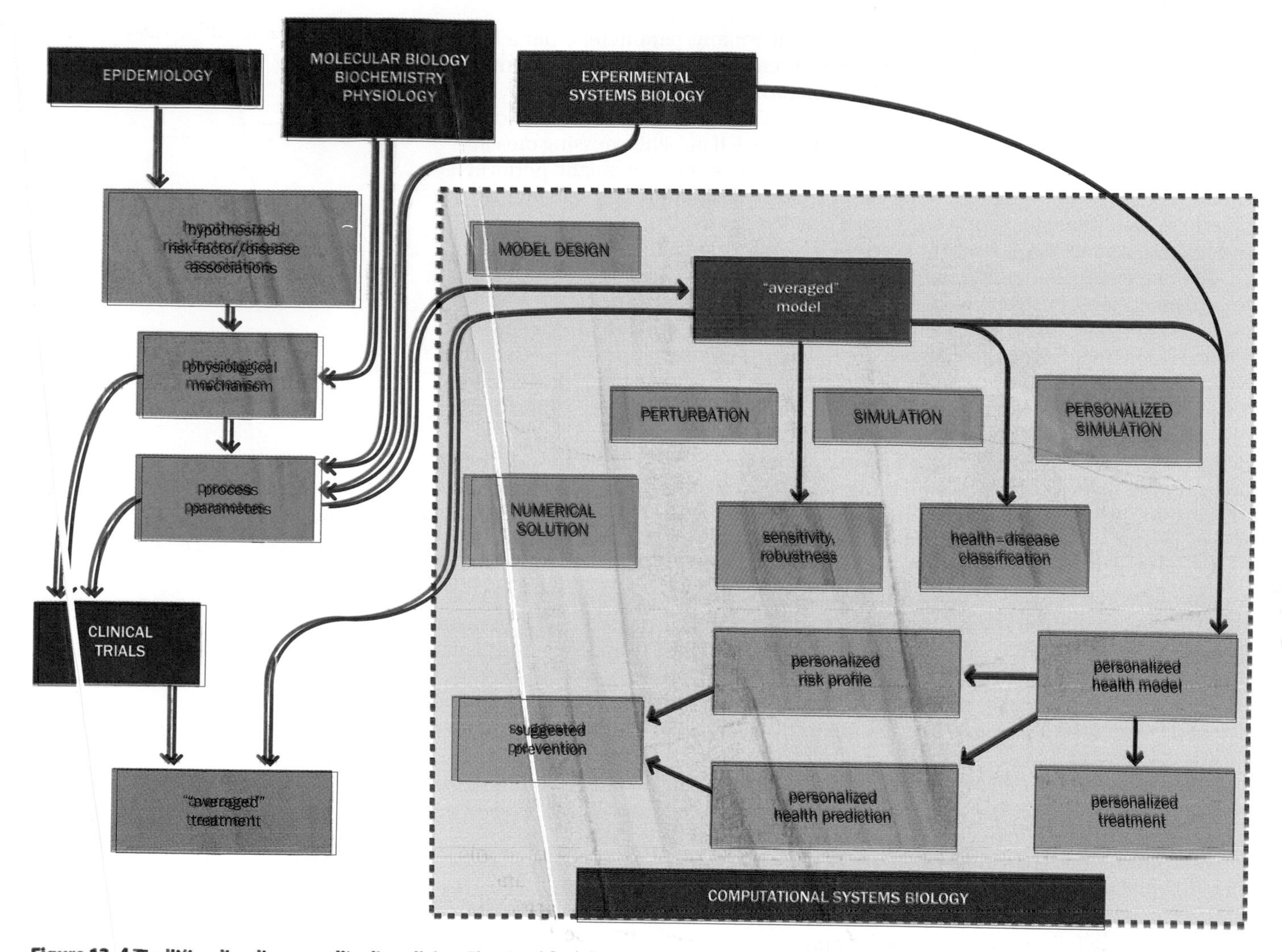

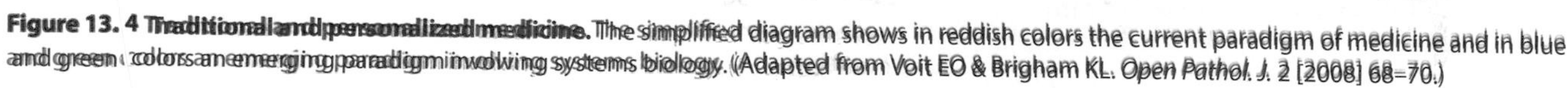
Figure 13. 4 Traditional and personalized medicine. The simplified diagram shows in reddish colors the current paradigm of medicine and in blue and green colors an emerging paradigm involving systems biology. (Adapted from Voit EO & Brigham KL. *Open Pathol. J.* 2 [2008] 68–70.)

Computational systems biology will use averaged disease models, personalize them, based on individualized data, yield personal risk profiles, and suggest possible prevention strategies.

The computational aspect of personalized medicine and predictive health requires the personalization of models [7, 14]. Conceptually, this step is fairly simple, and companies such as Entelos [15] have achieved successes in simulating virtual patients. In generic terms, one begins with a disease model that had been constructed and parameterized with the best available data and information currently available. This model reflects population averages, rather than one single individual. Suppose now that many biomarkers had been measured for a specific person and that they do not wildly fluctuate over time. A comprehensive model should directly or indirectly permit these biomarkers to be entered into the model. For instance, an altered enzyme activity might be translated into the K_M value of some enzymatic step. A deactivated gene could, at least in a first approximation, be translated into the decreased activity of a protein, which would in turn correspond to a decreased transport rate, altered catalytic conversion, a muffled signal transduction process, or some other effect on the model. Because the biomarkers were measured in a specific person, their values are most likely different from the average over an entire population. Thus, the original average model is customized by changing each average biomarker value to the measured, personal biomarker value, and every person can be seen as a unique point in a huge, dynamically changing parameter space.

In the simple example at the beginning of the chapter, we modified parameters by increasing them by 20%. In some sense, this exercise corresponded to an abstract personalization. Here, we simultaneously change many parameters, namely, all those for which personalized biomarker measurements are available. Presumably, this personal biomarker profile is not complete, which requires us to retain average values for the remaining parameters, in the hope that the individual is more or less normal in aspects that were not specifically measured. If possible, missing measurements can later be substituted with personal values, or one might perform a simulation study, for instance, using **Monte Carlo simulation** (see Chapter 2), to determine how influential the average parameters really are. If the system is not sensitive to these parameters, it might not even be necessary to obtain personalized values.

The personalization of a model opens new avenues for various prevention and treatment options. First, one will probably simulate the model from the current "initial" state forward. Such a simulation shows whether the individual converges to some healthy steady state or whether the individual biomarker profile leads to the slow, but unrelenting accumulation of molecules such as cholesterol (see Exercises 13.2 and 13.3). In addition to the steady state, the model also indicates the speed of change and shows whether any transients during the scenario simulation could be dangerous. This baseline simulation is expanded into scenario simulations, where one attempts to intercept undesired trajectories, such as the accumulation of harmful substances. It is expected that a reliable model will identify successful strategies. It may also help screen for interventions that at first seem a good idea, but in fact do not really work, for one unexpected reason or another.

The collection of many personalized health and disease trajectories will begin to shape a deeper understanding of causes, symptoms, and progressions of disease processes. It will show that some abnormal biomarkers alone are not dangerous, but that they may forecast disease if they are combined with other altered biomarkers. Ultimately, a comprehensive collection of trajectories will help us determine effective early biomarkers that precede late biomarkers and precursors of disease, as shown in Figure 13.3.

Looking at the complexity of metabolic and physiological systems, one might come to the conclusion that personalized medicine is a dream without much of a chance of realization. Three arguments counter this pessimistic view.

First, physicians have actually been pursuing the same strategy since the beginnings of medicine: they gained their knowledge from population averages and learned to customize it toward a patient. Maybe the biomarkers were initially simpler, consisting of blood pressure, temperature, and past personal trends such as weight loss. However, in more recent times, physicians have been using many more and specific biomarkers, and the only true difference from a computerized system of personalized medicine is that doctors perform the integration of data in their heads, aided by their knowledge and experience. Looking into the future, this integration will become increasingly more difficult, because the number of measurable biomarkers will continue to climb rapidly, eventually overwhelming even the most astute human mind. It seems that the only remedy against this overload will be computer-aided data integration. One should add that computer predictions will eventually become so complex that the physician is no longer able to explain how a suggestion from a computer was obtained.

The second argument is the success of computational expert systems in select areas of medicine, such as the identification and severity of poisonings from symptoms [16] and emergency medicine [17]. While these systems are based on methods of artificial intelligence, which are quite different from systems modeling with differential equations, their successes suggest that computer-assisted diagnostics and treatment are not out of the question in medicine. In the much simpler case of car mechanics, computerized diagnostics has become an accepted standard.

Third, while most of our computational models of molecular systems are presently too coarse and insufficiently comprehensive to facilitate personalized medicine, specific numerical, model-based predictions are already possible in other medical fields. For instance, before cosmetic surgeries are performed, they are often simulated first in a virtual environment that permits personalization. Similarly, a group at Emory University and the Georgia Institute of Technology have developed an image-based software system for blood flow reconstruction surgery on the heart that is custom-tailored to a specific patient [18]. The computational platform allows surgeons to perform, in a simulation, alternative surgical scenarios,

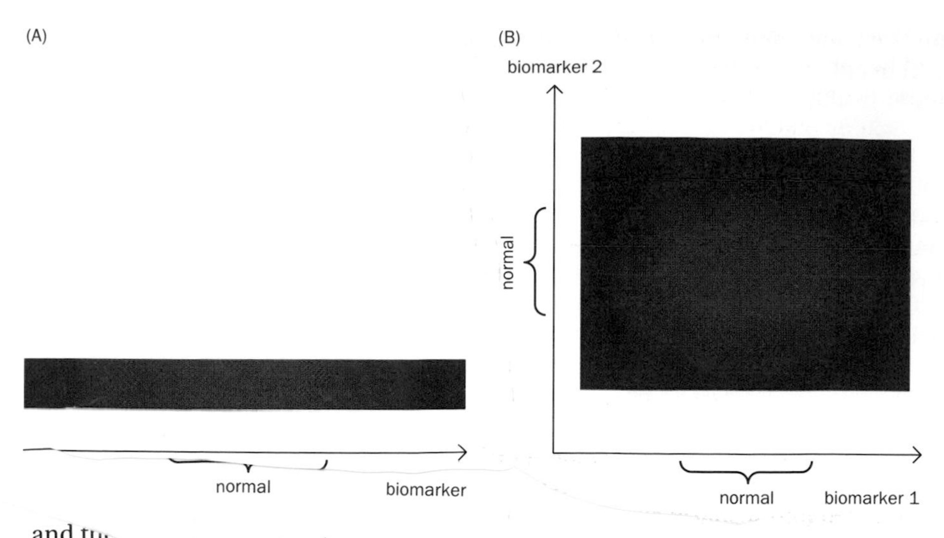

Figure 13.5 Health is related to biomarkers within their normal ranges. (A) For a single biomarker, a normal range is relatively easy to define. (B) For several biomarkers, the healthy domain may be different from a simple combination of normal ranges.

and the system computes the postoperative hemodynamic flow properties of each resulting reconstruction.

It is clear that the complexity of physiological systems in health and disease will render it very challenging to design and implement precise models in specific individuals. Nonetheless, in most cases, it is not necessary to target the entire human physiology at once. Instead, it will be feasible to move gradually from what we have already accomplished in medicine to increasingly detailed systems of disease processes.

The starting point is a consideration of what health and disease really mean. Precise definitions are difficult, but it is sufficient here to begin with the pragmatic definition of a normal range of a biomarker. We all know that a blood pressure of 220 over 120 is probably not good in the long run, and that the body core temperature has to remain within a rather tight range for us to function normally. Generically, one may define a normal range for a biomarker as shown in **Figure 13.5A**, which acknowledges that the boundary between normal and abnormal is fuzzy. The situation becomes more complicated when we consider two biomarkers simultaneously (Figure 13.5B). In many cases, combinations of extreme values of both biomarkers are no longer healthy, so that the green domain is not a rectangle but an ellipse or some other shape with truncated corners (**Figure 13.6A**). The consequence is that the normal domain of a biomarker is no longer absolute. In Figure 13.6B, the orange and blue "patients" have exactly the same value for biomarker 1, but blue is healthy, while orange is not. The purple patient is (borderline) healthy for biomarkers 1 and 2, but the combination is no longer in the healthy domain. The turquoise patient, by contrast, shows an unhealthy reading for biomarker 1, but is still considered healthy. Expressed differently, the outcome of a single biomarker test can lead to **false-positive** or **false-negative** outcomes. In the first case, the test falsely suggests disease, whereas in the second case, the test falsely

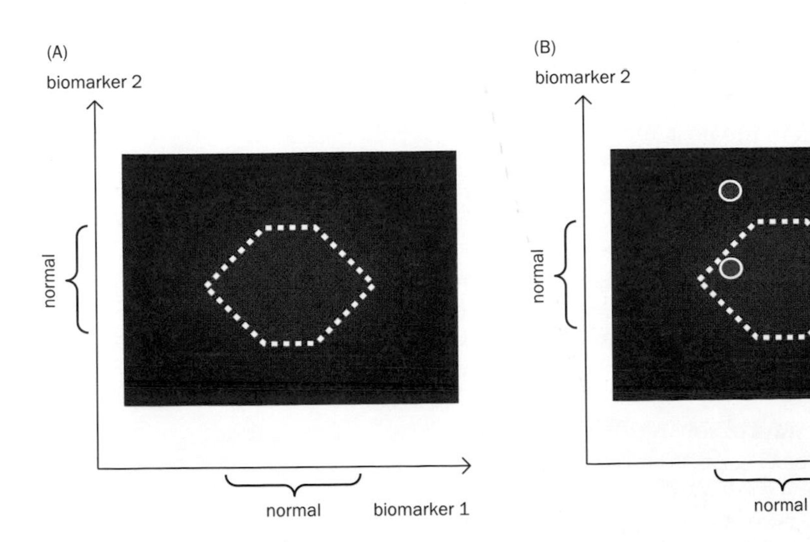

Figure 13.6 Definition of health and disease. For simplicity, the healthy domain in (A) is defined by the dotted hexagon. (B) shows cases where consideration of a single biomarker leads to "false positives" or "false negatives." The issue is much more pronounced when many more biomarkers are involved.

indicates health. Although simplified, this discussion indicates that individual biomarkers are seldom absolute predictors of health or disease and that their values must be seen within the context of a larger health system where biomarkers are interdependent on one another.

Today's physicians know the normal ranges for many biomarkers, as well as some of the more important combinations of a few biomarkers. With this knowledge, it is possible to develop models that take account of information from many sources and integrate it in a meaningful manner. Over time, such models will improve in quality and reliability, and eventually will be able to take hundreds of biomarkers and classify whether their combinations are healthy, premorbid, or indicative of disease. Furthermore, model-based integration of repeated measurements of biomarker combinations over the lifetime of an individual will show whether this individual is on a healthy trajectory or whether he or she is moving dangerously close to the realm of disease. Analyzing and interpreting thousands of such trajectories together will create a rational basis for predictive health [3].

THE DRUG DEVELOPMENT PROCESS

Between the dawn and dusk of the twentieth century, the average human life expectancy in the United States went up from about 45 to 77 years [19, 20]. Much of this enormous extension of life can be credited to vast improvements in medicine and the availability of drugs and biomedical devices, along with better hygiene, plenty of food, and a deeper understanding of physiology, molecular biology, and microbiology. While the maximal lifespan for humans is unknown, current estimates lie between 120 and 130 years, and as more and more individuals are becoming centenarians, the emphasis of extending life is shifting toward good health at an advanced age. Realizing this emphasis will only be possible with continued developments and improvements in medicine and drug development.

Unfortunately, the rate at which new drugs are introduced has been slowing down considerably in recent years. It is not the case that the pharmaceutical industry has run out of drug targets. In fact, high-throughput biology has been suggesting new drug targets galore. Rather, the main reason seems to be that "Big Pharma" is leery of the length and cost of the drug development process, which may take 10–20 years and require an investment of between one and two billion dollars. Patents protect new drugs for only about 15–20 years, depending on the country, before cheaper generic drugs become competitors. In areas such as HIV/AIDS, the investment in a drug must be recouped much faster, because new formulations and cocktails may begin to challenge a drug after just a few years. Adding to this situation is the enormous attrition rate of **new chemical entities (NCEs)**, that is, of active ingredients, during the drug development process [21, 22]. Out of 100 NCEs, only one to five make it into preclinical trials, and only a few percent of these survivors will ultimately be approved. Obviously, these numbers depend on many aspects of the disease and the NCE, but indicate that the *a priori* probability of success of an NCE is low.

Reasons for the low success rate are plentiful and include the fact that the pharmaceutical industry has begun to attack very complex diseases, often even with more resources for research and development than in the past, and that the standards set forth by the Food and Drug Administration (FDA) in the United States and similar health agencies elsewhere are becoming more and more stringent. The reasons for dismissal of an NCE are manifold as well, but two of them stand out. First, 30–60% of the dismissed drugs are abandoned owing to issues of efficacy and bioavailability. In other words, the human body does not accept the drug in sufficient quantities to permit a therapeutic effect. Another 30% have to be eliminated because of toxicity and serious side effects, while 7% are not commercially viable [19]. Thus, the landscape of drug discovery, development, and marketing is treacherous and requires significant resources.

So, what exactly are the problems, and are there ways of solving them? Certainly, nobody intends to cut corners or to simplify the testing procedures. Instead, the challenge to be solved may be posed as (1) increasing the overall success rate of the drug development pipeline and (2) accelerating the elimination of an NCE from the pipeline. The former is rather obvious. The latter is important because the later

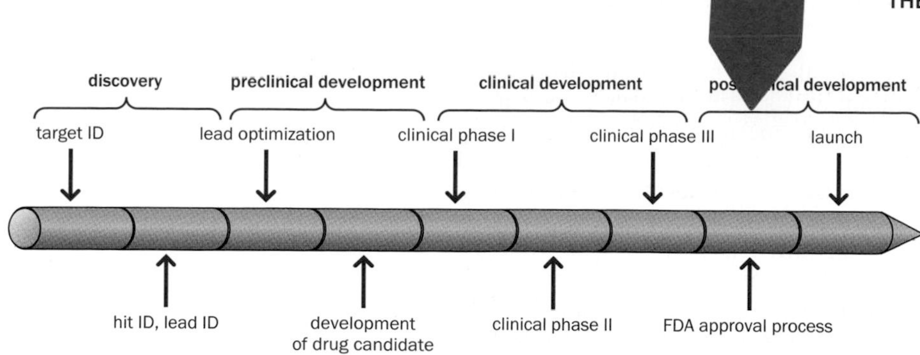

Figure 13.7 The drug development pipeline. The pipeline consists of distinct phases, which any drug has to survive before it can be used for treatment. The process is very costly and time-consuming. Without compromising safety, costs and time can only be saved if drug candidates with undesirable properties are screened out early.

phases of the pipeline, and in particular the phases of clinical trials, may cost hundreds of millions of dollars. Thus, if a problematic new NCE can be eliminated early, substantial amounts of resources are saved. The drug pipeline is depicted in **Figure 13.7**. It consists of four major phases that are readily subdivided into further segments. The outcome of testing at each segment of each phase may suggest abandoning further consideration of the NCE.

The first phase is concerned with early-stage research and discovery, target identification, the identification of small-molecule **hits**, and validation. At the beginning, a specific biological target is identified as being associated with a particular disease. The target could be a receptor, a signaling molecule, an enzyme, or some other protein that plays a crucial role in the disease. The hit one is hoping for is a molecule that interacts with the target in such a fashion that the disease process is interrupted. In the case of a receptor, a hit molecule may be an artificial ligand that attaches to the receptor and blocks it from accepting the disease-associated ligand. In the case of an enzyme, the hit might be an inhibitor. It is not surprising that many possible hits must be investigated. Often, these hits come from extensive combinatorial molecular libraries that a pharmaceutical company has assembled. In most cases, only one or a few of the hit molecules will continue in the drug development pipeline as **leads** for further testing. Hit and lead identifications are nowadays executed in a high-throughput fashion that utilizes many of the methods of modern molecular biology and experimental systems biology [23]. In addition to the initial identification, the target, hit, and lead must be independently validated. This step provides proof of concept that the target may be worth pursuing with the identified leads. For instance, if the target is a receptor, then binding affinities, length of receptor occupancy, and accessibility of the receptor to the lead compound must be investigated.

Validated hits with favorable characteristics that show potential for an improved medication become leads in the next phase of preclinical drug development. The first component of this phase is lead optimization. The lead is subjected to further tests of medicinal chemistry and, in particular, to an optimization process that investigates small molecular variations, which may affect receptor binding or the strength of inhibition with respect to a target enzyme. An important guideline is the exploration of **quantitative structure–activity relationships (QSARs)**, which attempt to assign specific features to chemical structures that might be used as side groups of the lead molecule [24, 25]. In the case of proteins, a similar approach is the creation of peptoids, which are artificial peptide-like compounds that can be designed to work as drugs with desirable properties [26]. Such drugs could be intriguing because of the preeminent role of specific peptide stretches in protein–protein interactions, molecular recognition, and signaling.

The second component of lead optimization is the assessment of dose, efficacy, and acute toxicology in animals. Compounds that are found to be effective in animals and do not lead to acute toxicity are moved forward in the drug development pipeline as drug candidates. The further investigation of drug candidates constitutes the second step of preclinical drug development. Here, toxicity over longer time horizons is assessed, along with **pharmacokinetic** tests that reveal how long the compound resides within some tissue and the bloodstream. The half-lives in different organs may differ drastically, because, for instance, metabolism in fat is typically slow, while it is the role of the liver to detoxify the body as quickly as possible. In the context of pharmacokinetics, one sometimes speaks of investigating ADME: absorption, distribution, metabolism, and excretion of a drug. Drug candidates surviving the muster of preclinical assessment are moved forward to clinical testing [27].

The clinical phase may easily consume half of all costs of the entire drug development process and half of the time. It is commonly subdivided into three subsequent types of clinical trials. The first, phase I, is primarily concerned with safety issues. The compound, which will already have been found to be safe in animal studies, is administered to a few dozen healthy human volunteers over a time span of 1–2 years. If toxicity or other undesired side effects are detected, the testing is stopped and the drug is not permitted to move forward in its current form. Throughout the clinical phase, studies also investigate different formulations and dosing regimens. In phase II, both efficacy and safety are tested on a relatively small cohort of 100–300 patients. This phase again lasts for 1–2 years. Finally, in phase III, the drug candidate is used in a randomized clinical trial involving a few thousand patients. The foci of this 2- to 3-year-long study are again efficacy and safety. Because volunteers, patients, and hospitals are involved in clinical trials, this phase of the drug development pipeline is very costly. Roughly 20% of drug candidates survive the three phases of clinical trials.

Once all preclinical and clinical testing has yielded satisfactory results, the drug candidate must be approved by the FDA in the United States and/or corresponding health agencies in other countries. Drug applications to the FDA require vast amounts of documentation about the positive and negative results of all former tests and trials, and approval typically takes 1–2 years. The final, post-approval phase of the drug pipeline contains the launch. It consists primarily of the sharing of information with physicians, hospitals, and trade magazines, as well as direct advertising and marketing. This phase can cost $100 million or more [28].

THE ROLE OF SYSTEMS BIOLOGY IN DRUG DEVELOPMENT

Results of deliberations by think tanks have suggested that academia and the pharmaceutical industry should form partnerships, in which academicians would primarily focus on drug discovery and possibly some of the preclinical studies, while the pharmaceutical industry would take the lead in the expensive clinical trials. However the roles of all potential contributors to the process are split up, it is more important here to ask how systems biology might be able to help improve drug discovery and development. Several roles of systems biology are already emerging, and the industry has been embracing them to various degrees for quite a while. Indeed, the potential of systems biology in drug discovery has been recognized for some time (see, for example, [29–31]).

Looking at the drug development pipeline shown in Figure 13.7 from a systems point of view, two distinct types of improvements would obviously contribute to desirable solutions. First, it would be desirable to increase the number of promising hits. And second, it would be highly desirable to find out as early as possible whether particular hits or leads would later be doomed. This activity would not reduce overall attrition, but it would screen out hits and leads before their testing became very expensive, which happens particularly in the clinical development phase.

Regarding increased numbers of hits, there is no doubt that methods of experimental systems biology and of high-throughput data acquisition and analysis are very promising tools for the separate or combined identification of targets, hits, and leads, as well as for global toxicity studies. All major pharmaceutical industries are already heavily engaged in parallelized "omics," robotic screening, and bioinformatics, and some smaller companies have sprung up that concentrate on genomic analyses that are made available to larger companies; examples include Selventa [32] and GNS Healthcare [33]. Proteomic studies are also being pursued, especially within the realm of signaling systems, where sophisticated methods are beginning to permit crisp assessments of the phosphorylation states of specific signaling proteins [34, 35].

In addition to searching for NCEs, the pharmaceutical industry has begun to invest heavily in the exploration of **biologics** [36, 37]. In contrast to synthetically produced, small-molecule NCEs, these biologics are derived from naturally occurring biological materials. They often have a favorable safety profile, because they are less likely to elicit an immune response, and they break down into natural molecules such as amino acids. In fact, it is sometimes even therapeutically beneficial to supply an unchanged biomolecule to a patient who shows a corresponding deficiency. For example, insulin is given to patients with type 1 diabetes and factor VIII is prescribed for hemophilia A. In other cases, natural compounds are used as templates that are chemically modified, in order, for instance, to improve the binding to a receptor. Because it is possible to attach a wide variety of residues to the natural base compounds, the number of potentially efficacious biologics is enormous. As a consequence, the biologics sector of the pharmaceutical industry has in recent years been growing at an annual rate of 20% and now is a multi-billion-dollar market. Monoclonal antibodies constitute the best-known category of biologics. Other examples include engineered proteases, for instance, from the blood clotting cascade, and aptamers [38]. Aptamers are artificial oligonucleotides that do not code for a peptide or interact with DNA but are designed to bind and thereby deactivate a target, such as the active domain of a disease-linked protein. Modern methods of experimental systems biology make it relatively easy and cheap to produce aptamers in vast variations. Overall, the role of experimental systems biology in drug discovery seems quite evident, and we will not discuss it further.

13.5 Computational Target and Lead Identification

Different branches of computational systems biology have also begun to be helpful for target and lead identification (see, for example, [39]). For instance, methods of molecular modeling and combinatorial chemistry permit the computational assessment of molecules and active groups and are becoming increasingly reliable predictors of the efficacy and toxicity of permutated drugs and their side groups, as well as of biologics. These predictions are, in some sense, based on a new generation of sophisticated QSARs that have been in use for many years. Systems biology is also used increasingly for receptor binding and dosing studies with newly created molecules. Not to be overlooked, computational methods of intelligent data and **text mining** are rapidly moving to the forefront. With thousands of biomedical journals being published on a regular basis, automated alerts and sophisticated searches have become a necessity. In addition to simple key words or phrases, modern search engines are rapidly becoming better in semantic searches that may even include images.

Many aspects of systems biology look promising for reducing the cost of drug discovery through effective early screening and hit elimination. The key to moving the elimination of NCEs and biologics closer to the front of the drug pipeline is a solid understanding of the mechanisms of action of a hit or lead with respect to a particular target. Because disease processes are complex, such enhanced understanding greatly benefits from computational models. Beautiful examples of very comprehensive disease models, for instance for diabetes and rheumatoid arthritis, have been developed by Entelos [15] and utilized by larger pharmaceutical companies. Entelos and other companies of this type construct these models with methods discussed in earlier chapters of this book, by mining the literature exhaustively, curating the results very carefully, and converting all pertinent information into processes and customized parameters for simulations of virtual patients. These examples are real-world extensions of the sandbox analysis that we used in the beginning of the chapter (see (13.1)) for illustrative purposes.

While comprehensive models show clear potential, even simple models can be of great value. For instance, it is not always necessary to model a signal transduction pathway in great detail, and it may suffice instead to construct a black-box model that is used as a surrogate for a comprehensive systems model [36]. In the following, we discuss three examples. The first pertains to receptor dynamics, which plays a major role for very many drugs. The second example discusses compartment modeling and pharmacokinetics, which have already been mentioned in Chapter 4. Finally, we will show how dynamic pathway models may help weed out NCEs and contribute to an understanding of safety and toxicity.

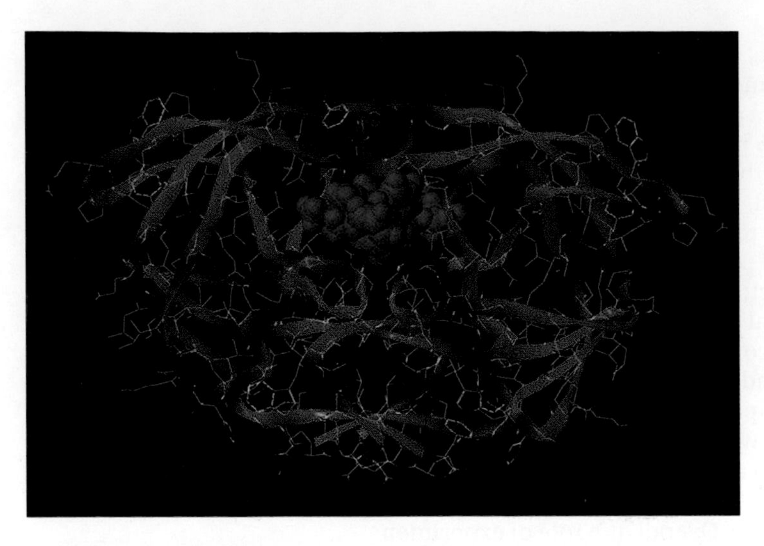

Figure 13.8 Many drugs work by binding to proteins. Here, the FDA-approved drug indinavir docks into the cavity of an HIV protease in a lock-and-key mechanism (Protein Data Bank: PDB 1ODW and 1HSG). (Courtesy of Juan Cui and Ying Xu, University of Georgia.)

13.6 Receptor Dynamics

Essentially all prescription drugs need to be taken up by cells, and very many of them are associated with membrane-bound receptors (**Figure 13.8**). It is therefore of great interest to study binding characteristics, and even relatively small mathematical models can yield valuable insights. As an example, let us consider two scenarios involving therapeutic antibodies. In the first, very much simplified, case, adapted from a study by the pharmaceutical industry [37], a monoclonal antibody M is injected intravenously and in the bloodstream binds to a target T, which could be a signaling peptide or a hormone. On encountering one another, M and T form a complex C, which prevents T from exerting a negative effect on the system (**Figure 13.9**). The goal of the treatment is to reduce T to a low level, for example, 20%. Ignoring all spatial aspects, it is easy to set up a diagram and a dynamic model of the system (**Figure 13.10**; see also Exercises 13.15–13.17). It shows that the target is synthesized and also cleared by the body. The antibody is also cleared, although with a different rate. The complex in this simple example is assumed to be relatively stable and, owing to its size, not cleared by the body within the time horizon of interest [36]. The default assumption for setting up a model of the system is mass action kinetics, which leads to the following representation:

$$\begin{aligned} T' &= k_T^+ - k_T^- T - k_{\text{on}} TM + k_{\text{off}} C, \\ M' &= k_{\text{off}} C - k_{\text{on}} TM - k_M^- M, \\ C' &= k_{\text{on}} TM - k_{\text{off}} C. \end{aligned} \tag{13.2}$$

Note that the injection of antibody is not modeled explicitly. Instead, at time t_0, the value of M is set equal to a value corresponding to the antibody

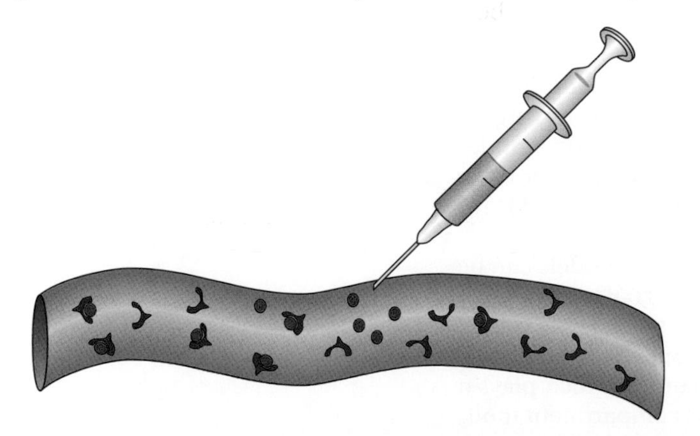

Figure 13.9 Treatment of a disease with biologics. Monoclonal antibodies (green circles) are injected into the bloodstream, where they bind reversibly to their target molecules.

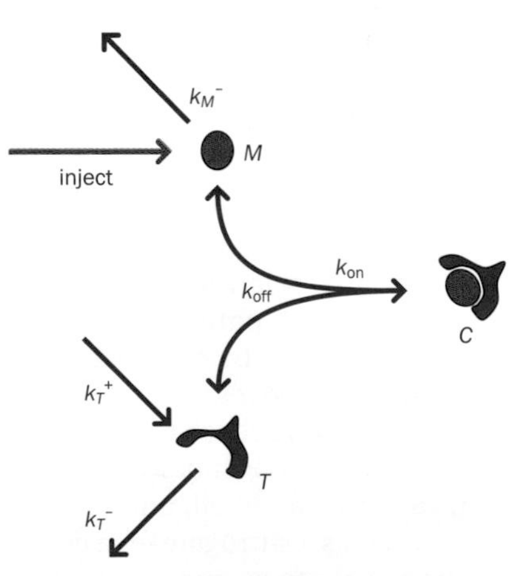

Figure 13.10 Diagram of a monoclonal antibody M binding to a target T. The binding yields a complex C that renders the target dysfunctional. Rate constants are defined in (13.2) and the text.

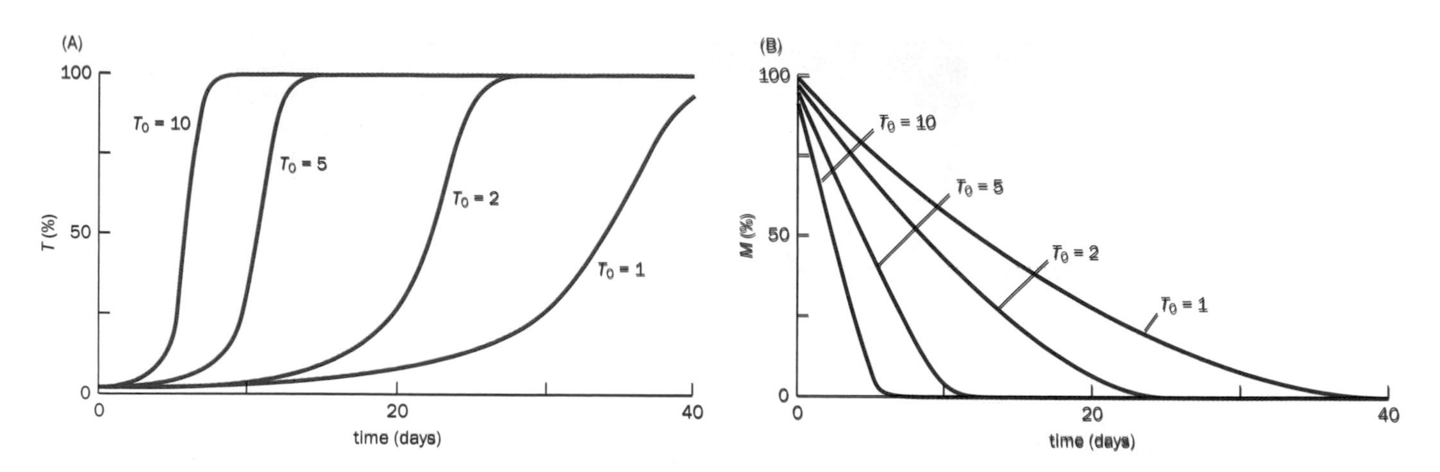

Figure 13.11 Simulation results with the model (13.2) of antibody binding. For a fixed initial monoclonal antibody concentration M_0, both (A) the rebounding dynamics of the free target T as a percentage of the initial target concentration T_0 and (B) the free antibody concentration M as a percentage of M_0 depend strongly on T_0. Note that time is given in days, rather than hours.

concentration in the bloodstream. We simulate the system with initial values $(T(t_0), M(t_0), C(t_0)) = (T_0, M_0, C_0) = $ (different values, 100, 0.001) and parameter values $k_T^+ = T_0 k_T^-$, $k_T^- = \ln(2)/10$, $k_M^- = \ln(2)/(21 \cdot 24)$, $k_{\text{on}} = 0.036$, and $k_{\text{off}} = 0.00072$, given for a timescale of hours, as suggested in [37, 40]. Let us check the rationale for these values. The synthesis rate of T, k_T^+, is set such that the system without any antibodies or complexes is in a steady state. This assumption is often (although not always) true. Setting all terms containing M or C to zero and equating the right-hand side of the first equation in (13.2) to zero results in the given definition of the synthesis rate. The degradation rate of T, k_T^-, corresponds to a half-life of 10 hours. The clearance rate of M corresponds to a half-life of 3 weeks. The rates of complex formation and dissociation are typical for these types of complexes.

It is now easy to run simulations over a typical period of say 40 days, with different values for $T(t_0)$. Some results are shown in **Figure 13.11**. They show that the concentration of the target drops immediately and then recovers after some while. The higher the initial target concentration, the faster T recovers. The figure also shows the disappearance of free antibody, which is due to binding to T and also to clearance.

The second example is similar in spirit, but it is now assumed that the system consists of a receptor R, a ligand molecule L that naturally binds to the receptor, thereby contributing to the disease process, and a therapeutic antibody A. An important question for drug design is whether the antibody should be developed to bind to the receptor or to the ligand. In either case, the association of ligand and receptor would be impeded, so does it really matter? A sketch of the situation (**Figure 13.12**) accounts for both scenarios; in reality, binding occurs only to the ligand or only to the receptor. The structural diagram, again accounting for both

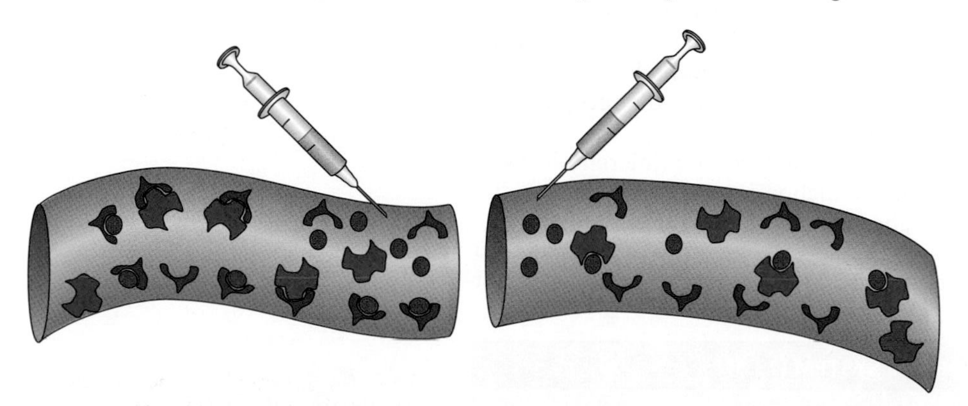

Figure 13.12 Alternative options for antibody treatment. The monoclonal antibody (green circles), which is injected into the bloodstream, binds reversibly either to the receptor (left) or to the ligand (right) associated with the target disease.

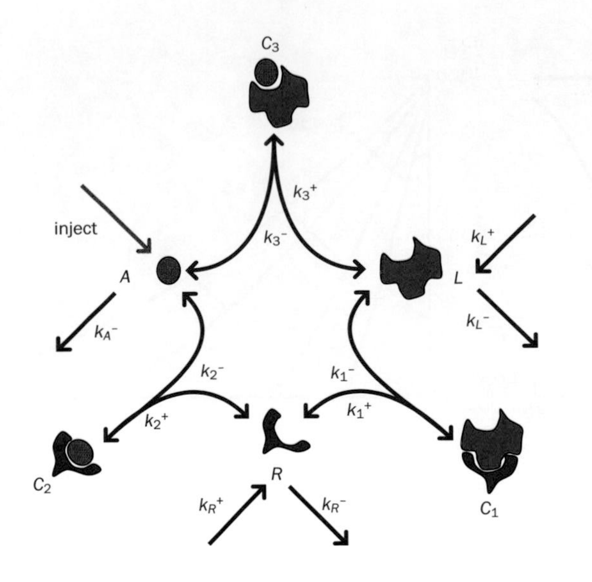

Figure 13.13 Diagram of a monoclonal antibody *A* binding to a target receptor *R* or to a target ligand *L*. Three types of complexes arise: between receptor and ligand (C_1), and either between receptor and antibody (C_2) or between ligand and antibody (C_3). Rate constants are defined in (13.3) and the text.

situations, is given in **Figure 13.13**, and the equations describing the diagram are as follows:

$$
\begin{aligned}
R' &= k_R^+ - k_R^- R - k_1^+ RL + k_1^- C_1 - k_2^+ RA + k_2^- C_2, \\
L' &= k_L^+ - k_L^- L - k_1^+ RL + k_1^- C_1 - k_3^+ LA + k_3^- C_3, \\
A' &= -k_A^- A - k_2^+ RA + k_2^- C_2 - k_3^+ LA + k_3^- C_3, \\
C_1' &= k_1^+ RL - k_1^- C_1, \\
C_2' &= k_2^+ RA - k_2^- C_2, \\
C_3' &= k_1^+ LA - k_3^- C_3.
\end{aligned}
\qquad (13.3)
$$

For the case of receptor binding by the antibody, the blue boxes are zero, whereas the pink boxes are zero for ligand binding. Parameter values are given in **Table 13.2**; they are inspired by actual measurements [37, 40]. The initial values R_0, L_0, and A_0 are at first set to 4, 100, and 100, respectively, but are varied in different simulations.

As an illustration, we explore the system for the case where the antibody binds to the ligand. Thus, $R_0 = 4$, $A_0 = 100$, and L_0 is varied for different scenarios. Representative results (**Figure 13.14**) suggest the following. For low numbers of ligands, the antibody reduces their concentration effectively, and the ligand concentration remains below 30% of the initial value for about 2 days. By contrast, if the initial ligand concentration is at the order of 10, the antibodies reduce it initially only to about 40%, and within 2 days the concentration has reached 60%. For even higher concentrations, for example, $L_0 = 100$, the ligand concentration is almost back to 100% within 2 days. In addition, there is a strong accumulation of ligand–antibody complex, which might cause undesirable side effects of the treatment. These results suggest that designing an antibody to a ligand can be effective if the ligand concentration is relatively low. If the concentration is high, the antibody is no longer efficacious. Exercise 13.18 explores whether an antibody against the receptor is to be preferred in this case.

In addition to changing R_0 or L_0, we can also easily explore different doses of antibody. For instance, one might surmise that a higher dose might counteract the decreasing efficacy of the treatment for higher ligand concentrations. To test this hypothesis, we consider the case $L_0 = 10$, and quadruple the dose of antibodies to 400. In comparison with the lower dose, the ligand concentration is better

TABLE 13.2: PARAMETER VALUES FOR THE SYSTEM (13.3)

k_R^+	k_R^-	k_L^+	k_L^-	k_A^+	k_1^+	k_1^-	k_2^+	k_2^-	k_3^+	k_3^-	C_i
$R_0 k_R^-$	0.0693	$L_0 k_L^-$	0.1155	0.0014	4	1	0.036	0.036	0.0018	0.036	0.001

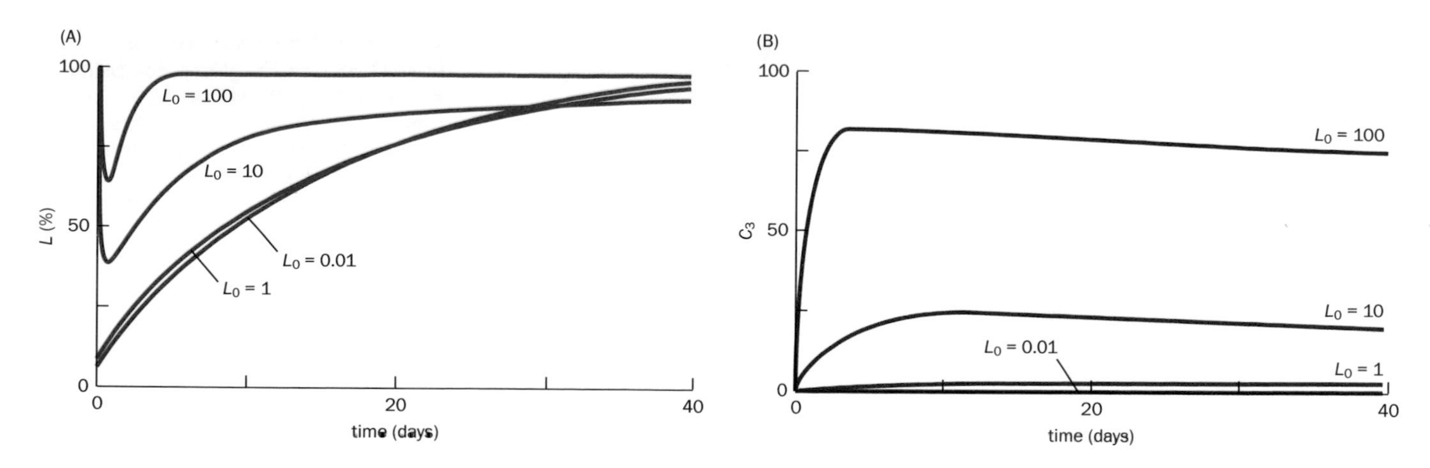

Figure 13.14 Simulation results with a model of different types of antibody binding (13.3). For fixed initial monoclonal antibody and receptor concentrations A_0 and R_0, both (A) the dynamics of free-ligand rebounding (L, as a percentage of the initial target concentration L_0) and (B) the concentration of the complex, C_3, depend critically on L_0.

controlled, but the concentration of the complex is again high (**Figure 13.15**). An alternative to a higher dose is repeated injection of antibodies. Exercise 13.19 asks you to explore variations on this theme.

13.7 Pharmacokinetic Modeling

Computational systems biology has been playing a significant role in drug development for several decades by supporting pharmacokinetic and pharmacodynamic assessments of the levels of therapeutic agents in different tissues and organs of the body (see, for example, [41]). Two classes of approaches use either relatively simple **compartment** models [42] or a particularly powerful expansion of these toward **physiologically based pharmacokinetic (PBPK) models** [2, 43].

The base concepts for pharmacokinetic analyses and simulations are relatively simple, at least conceptually. The body is represented as a set of homogeneous compartments, and a drug that was administered orally or through an injection migrates among these compartments. Some compartments may excrete the drug (lungs through exhalation, kidneys through urine, the gastrointestinal (GI) tract through feces), and the liver, in particular, may break it down through enzymatic activity. The key question is how long the drug resides in these different compartments and in what concentrations, because this information is important for developing an effective dosing regimen.

To obtain an impression of the principles of this type of approach, let us consider a simple **compartment model** that consists of just two compartments: the bloodstream (B) and the liver (L) (**Figure 13.16**). In this simplified system, B receives the drug through injection and later also from other organs, which, however, are not modeled here explicitly; the input rate is k_{OB}. B loses some of the metabolite to other organs (with rate k_{BO}) and exchanges the metabolite with the liver, with rates k_{BL} and

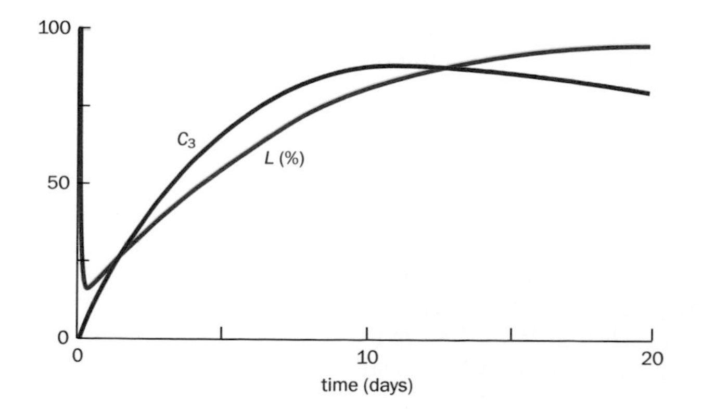

Figure 13.15 Simulation result with the model (13.3) for increased antibody loads. If the initial antibody concentration A_0 is increased to 400, the free ligand rebounds more slowly (compare with the curve $L_0 = 10$ in Figure 13.14), but the concentration of the complex, C_3, is much higher.

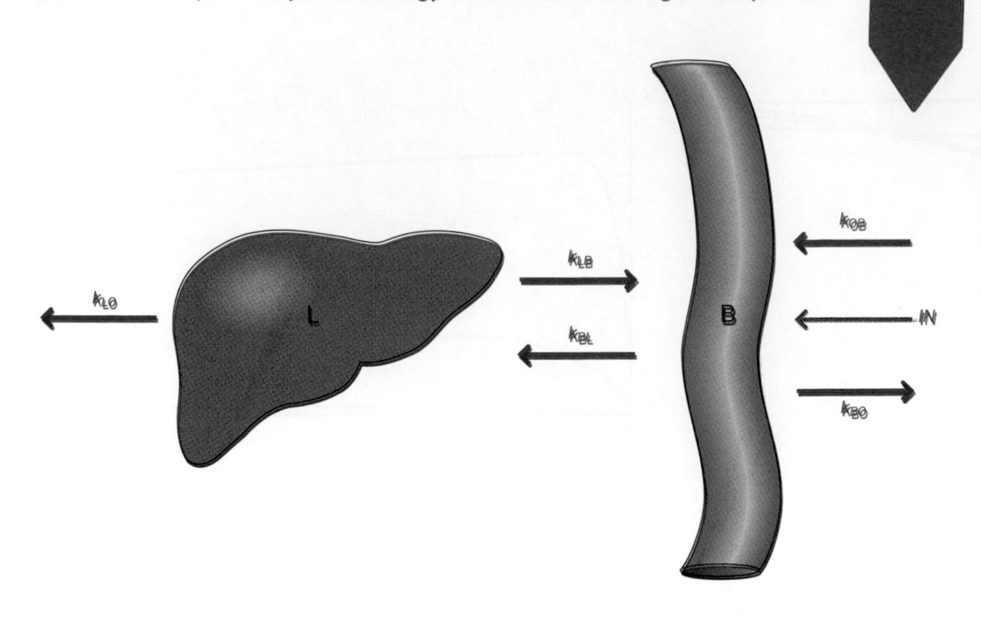

Figure 13.16 A two-compartment model for the exchange of a metabolite between blood (B) and liver (L). Several processes govern this exchange. They can be formulated as a mathematical model such as (13.4) in a straightforward manner (see Chapter 2). The parameters are explained in the text.

k_{LB}, respectively. The liver delivers some of the metabolite to the intestines or gall-bladder with rate k_{LO}.

Calling the concentration of the metabolite in the liver L and that in the blood B, a linear model with possible additional input IN reads

$$\begin{aligned} \dot{B} &= k_{OB} + k_{LB}L - (k_{BO} + k_{BL})B + IN, \\ \dot{L} &= k_{BL}B - (k_{LO} + k_{LB})L. \end{aligned} \tag{13.4}$$

Mathematically, k_{OB} and IN could be merged, but we keep them separate to distinguish normal physiology, where B is continually augmented with material from the rest of the body, and with additional input from an intravenous injection.

Let us suppose for a moment that no metabolite is transported from the bloodstream into the liver ($k_{BL} = 0$). The equation for L then has the form $\dot{L} = -kL$, with rate $k = k_{LO} + k_{LB}$, which immediately signals that L is losing material in an exponential fashion (see Chapters 2 and 4). As a second situation, imagine no influx through k_{OB} but a bolus injection IN at some time point. In other words, after the brief injection of IN, there is no further input and the only active processes are the exchanges between the two compartments and efflux from both. After a short initial phase of equilibration between L and B, the dynamics now consists merely of material shuttling between or leaving the pools, and these losses occur in an exponential fashion, which corresponds to straight lines in a logarithmic plot (**Figure 13.17A**). This solution has the explicit form

$$\begin{aligned} B(t) &= c_{11}e^{\lambda_1 t} + c_{12}e^{\lambda_2 t}, \\ L(t) &= c_{21}e^{\lambda_1 t} + c_{22}e^{\lambda_2 t}, \end{aligned} \tag{13.5}$$

where λ_1 and λ_2 are the eigenvalues of the system matrix **A** and $c_{11}, \ldots, c_{22}$ are constants that depend on the numerical specifications of the model (see Chapter 4). Because the eigenvalues might be real or imaginary numbers, the functions in (13.5) could in principle decrease toward zero or oscillate.

The most interesting situation occurs if none of the rate constants are zero and the system is in a steady state, but then receives an injection IN. For instance, if $k_{OB} = 1.2$, $k_{LB} = 0.5$, $k_{BO} = 0.9$, $k_{BL} = 0.8$, $k_{LO} = 0.3$, and $IN = 0$, the system happens to have the steady state $B = L = 1$. Suppose that we inject 5 units of IN within the time interval $3 \leq t \leq 4$. Since B is the primary recipient, it increases, but not in a linear fashion, because some of the new material immediately starts moving into L, which increases subsequently. After some while, IN has been cleared from the system, which returns to its steady state (Figure 13.17B). Many scenarios can easily be tested in this fashion (see Exercises 13.20–13.23).

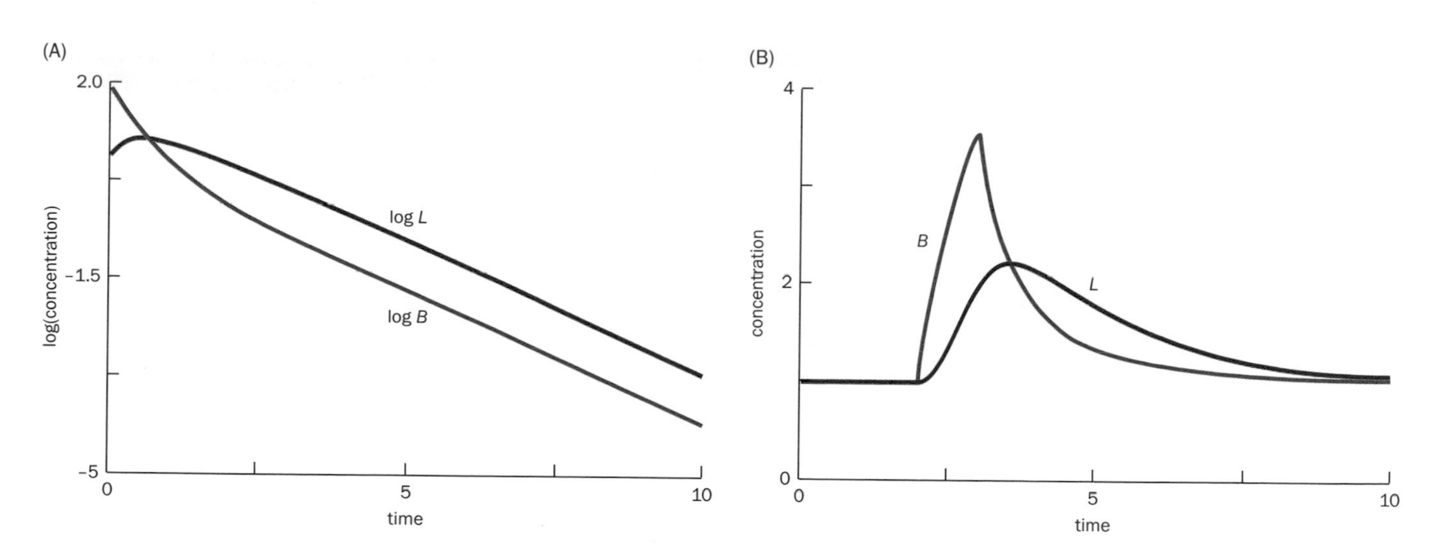

Figure 13.17 Responses of the two-compartment model (13.4) for the exchange of a metabolite between blood (B, with concentration *B*) and liver (L, with concentration *L*). (A) After a one-time input, the material equilibrates and afterwards is lost in an exponential fashion (note the logarithmic vertical axis). (B) The system operates at a steady state with constant influx, effluxes, and exchanges between B and L. Between times 3 and 4, B receives an injection with additional material, which moves between compartments and is ultimately cleared from the system.

If we realistically assume that the liver enzymatically degrades the metabolite, the equation of L is augmented with additional degradation terms, maybe in the form of **Michaelis–Menten, Hill,** or **power-law functions**, which make the equation nonlinear (see Exercise 13.24). The nonlinearities make the model more realistic but preclude some strictly mathematical analyses; however, they are not a hindrance for all kinds of simulations. As so often, the true challenge of this expansion lies in an accurate determination of suitable functions and parameter values.

More sophisticated and realistic than simple compartment models are PBPK models that consist of multiple compartments representing organs or tissues, which are functionally connected through the circulatory system (**Figure 13.18**). The compartments are quantitatively represented with measurable characteristics such as volume and perfusion rate. A drug enters the body by mouth, inhalation, or injection and is transported to and among the various tissues via the bloodstream. Organs such as the liver metabolize some of the drug. The rest cycles through the body and is eventually excreted through feces and urine or, in some cases, by exhalation or sweating, which causes the concentration eventually to decrease to insignificantly low levels. Typical questions are: How long will it take before a drug reaches a target tissue? How long will it remain in this tissue in therapeutically relevant doses? What is the best route and dosing regimen for drug administration? These questions are not easy to answer intuitively, because some tissues are very strongly perfused (for example, the brain and liver), while others are not (for example, fat). Also, many prescription drugs are lipophilic, so that they have a much higher affinity to fat than, for instance, muscle. Furthermore, the drug usually cycles through the system several times, and its residence time in the various organs depends critically on its lipid content and other physicochemical properties. PBPK models can be used not only to study the dynamics of the original drug, but also the accumulation of toxic substances, including breakdown products of the drug.

Probably the most important feature of PBPK models is that they can be utilized as a mechanistic tool for scaling drug scenarios from animal experiments with mice, rats, or dogs to humans. The same models are set up for the different species, but with parameter values that are adjusted, for instance, to the volumes of organs, tested for one or two species, and then used to make predictions about the drug dynamics in humans. A typical case study is presented in **Box 13.1**. It is clear that the success rate of the clinical drug development phase rises substantially if one can rely on PBPK models that were set up for animals as reliable predictors of efficacy, safety, and toxicity in humans.

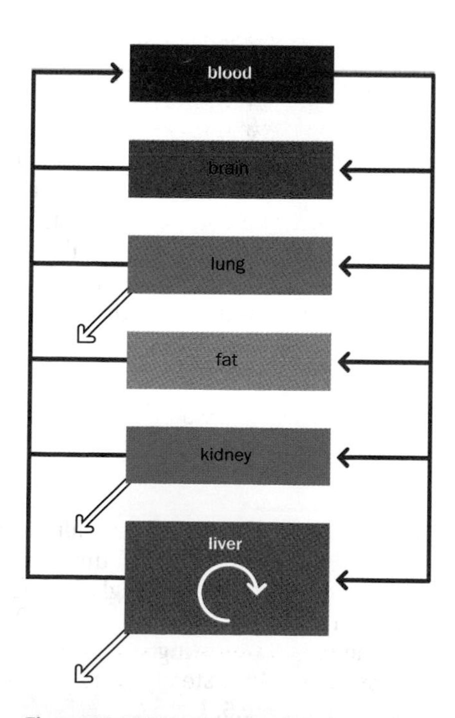

Figure 13.18 Diagram of a generic physiologically based pharmacokinetic (PBPK) model. Blood circulates among organs and tissues, carrying a drug of interest. The liver metabolizes some of the drug, and the lung, kidney, and liver "lose" drug by exhalation or excretion.

BOX 13.1: PHARMACOKINETICS OF METHOTREXATE

Methotrexate (MTX) is a powerful prescription drug that inhibits folate metabolism, which is crucial for the production and repair of DNA and therefore the growth of cell populations. MTX is used to treat various cancers and autoimmune diseases and induces medical abortions. It is on the World Health Organization's List of Essential Medicines, which contains the most important medications needed in a basic health system [44]. Being a very powerful medication, it is not surprising that MTX can have serious, even life-threatening, side effects in the liver, kidneys, lungs, and numerous other tissues. Any treatment with MTX therefore requires a well-balanced dosing regimen.

Bischoff and colleagues [45, 46] studied the pharmacokinetics of MTX with a PBPK model that was originally implemented for mice, but then also tested in rats and dogs. Model predictions were subsequently compared with plasma concentrations in humans, where they reflected the actual observations quite well. Bischoff's papers are quite old, but illustrate the principles of pharmacokinetic modeling very well. The following analysis is directly adapted from this work [46].

The specific model diagram is shown in **Figure 1**. It is fundamentally the same as Figure 13.16, but the authors found it necessary to account for multistep processes of bile secretion ($B_1, \ldots, B_3$) and transport through the gut lumen ($G_1, \ldots, G_4$).

The model is set up as explained in the text. In fact, the equations for plasma, muscle and kidney are identical to (13.6)–(13.8). The equation for liver is very similar, but additionally contains terms for input from the gastrointestinal (GI) tract and also for biliary secretion, which is modeled as the Michaelis–Menten process B_0. Thus, it has the form

$$V_L\dot{L} = (q_L - q_G)\left(P - \frac{L}{r_L}\right) + q_G\left(\frac{G}{r_G} - \frac{L}{r_L}\right) - B_0, \quad (1)$$

where

$$B_0 = \frac{k_L\left(\frac{L}{r_L}\right)}{K_L + \frac{L}{r_L}}. \quad (2)$$

The movement through the bile duct is modeled with the equations

$$\dot{B}_i = \tau^{-1}(B_{i-1} - B_i) \quad \text{for} \quad i = 1,2,3. \quad (3)$$

Here, τ is the holding time in each section of the bile duct, and B_0 is the Michaelis–Menten term in (2). The dynamics in the gut tissue is represented as

$$V_G\dot{G} = q_G\left(P - \frac{G}{r_G}\right) + \frac{1}{4}\sum_{i=1}^{4}\left(\frac{k_G G_i}{K_G + G_i} + bG_i\right). \quad (4)$$

The gut tissue receives material from the four compartments $G_1, \ldots, G_4$ of the gut lumen (see Figure 1). These transport processes are represented with a Michaelis–Menten process that accounts for saturation as well as a nonsaturable effect. The concentrations of MTX in these compartments are modeled individually and for the entire gut lumen as

$$\begin{aligned} \frac{V_{GL}}{4}\dot{G}_1 &= B_3 - k_f V_{GL} G_1 - \frac{1}{4}\left(\frac{k_G G_1}{K_G + G_1} + bG_1\right), \\ \frac{V_{GL}}{4}\dot{G}_i &= k_f V_{GL}(G_{i-1} - G_i) - \frac{1}{4}\left(\frac{k_G G_i}{K_G + G_i} + bG_i\right), \quad i = 2,3,4, \\ (\dot{G}L) &= \frac{1}{4}\sum_{i=1}^{4}\dot{G}_i. \end{aligned} \quad (5)$$

By entering the equations into a solver like PLAS or MatLab and using the parameter values in **Table 1**, it is now possible to test the model against actual measurements. As a first example, **Figure 2** shows the results of a simulation where 3 mg MTX per gram of body weight were injected intravenously into a mouse. The injection may be modeled with an input (*IN*) to the equation for plasma or, more simply, by setting the initial value of *P* as 66 (3 mg times body weight divided by the volume of plasma) and all other initial values close to zero (for example, 0.001). Note that results of PBPK analyses are traditionally displayed with a logarithmic scale on the *y*-axis. Similar to the simple two-compartment model (Figure 13.17), one observes that all concentration profiles, after the initial phase, become more or less linear, which corresponds to exponential decay in all tissues. Exercises 13.25 and 13.26

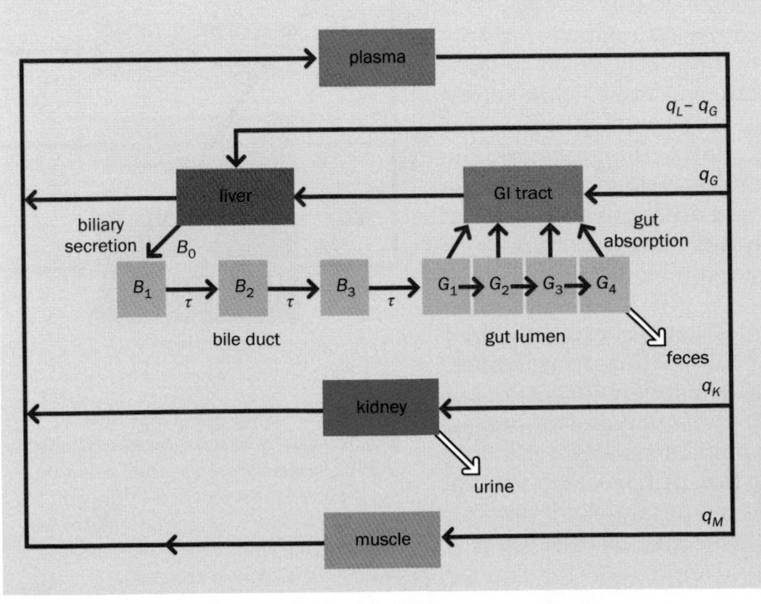

Figure 1 Diagram for the movement of methotrexate among different compartments within the body. Note that secretion from the liver is subdivided into three compartments of the bile duct and that the gut lumen is subdivided into four sections. (Adapted from Bischoff KB, Dedrick RL, Zaharko DS & Longstreth JA. Methotrexate pharmacokinetics. *J. Pharmacol. Sci.* 60 [1971] 1128–1133.)

TABLE 1: MODEL PARAMETERS FOR A METHOTREXATE CASE STUDY FOR THREE SPECIES*

Parameter	Model component	Mouse	Rat	Human
Body weight (g)		22	200	70,000
Volume (mL)				
Plasma	V_P	1	9.0	3000
Muscle	V_M	10.0	100	35,000
Kidney	V_K	0.34	1.9	280
Liver	V_L	1.3	8.3	1350
Gut tissue	V_G	1.5	11.0	2100
Gut lumen	V_{GL}	1.5	11.0	2100
Plasma flow rate (mL/min)				
Muscle	q_M	0.5	3.0	420
Kidney	q_K	0.8	5.0	700
Liver	q_L	1.1	6.5	800
Gut tissue	q_G	0.9	5.3	700
Partition coefficient				
Muscle	r_M	0.18	0.11	0.15
Kidney	r_K	3.0	3.3	3.0
Liver	r_L	7.0	3.3	3.0
Gut tissue	r_G	1.2	1.0	1.0
Kidney clearance (mL/min)	k_K	0.2	1.1	190
Bile secretion parameters				
Clearance (mL/min)	k_L/K_L	0.5	3.0	200
Residence time (min)	τ	2.0	2.0	10
Gut lumen parameters				
Transit time (min)		100	100	1000
Reciprocal transit time (min^{-1})	k_f	0.01	0.01	0.001
Maximal velocity	k_G	0.2	20	1900
Michaelis constant	K_G	6.0	200	200
Rate constant, absorption (mL/min)	b	0.001	0.0	0.0

*Adapted from Bischoff KB, Dedrick RL, Zaharko DS & Longstreth JA. Methotrexate pharmacokinetics. *J. Pharmacol. Sci.* 60 (1971) 1128–1133.

address a simulation of 300 mg MTX per gram of body weight and a variant modeling strategy where the injection is represented as a bolus.

As a second example, let us consider an injection of 6 mg MTX per gram of body weight into a rat. The results are shown in **Figure 3**. The overall appearance is similar to that of the mouse experiment, even though the concentration values are of course different. Exercise 13.27 analyzes the MTX distribution among tissues in humans.

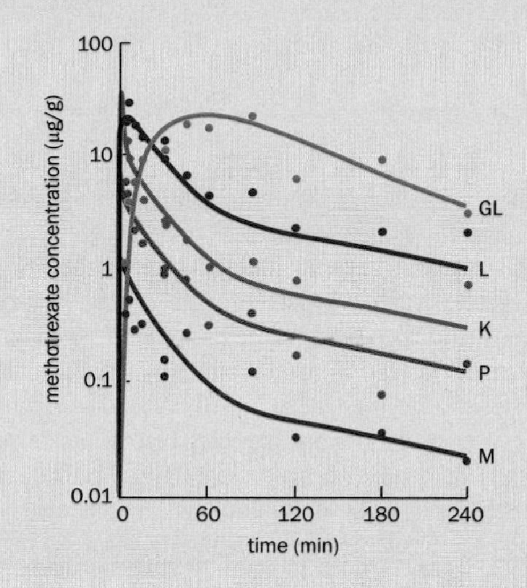

Figure 2 Methotrexate distribution among mouse tissues over time. Comparison of model predictions (lines) and experimental data (dots) for intravenous injection of 3 mg methotrexate per gram of body weight in a mouse. GL, gut lumen; K, kidney; L, liver; M, muscle; P, plasma. (Data from Bischoff KB, Dedrick RL, Zaharko DS & Longstreth JA. Methotrexate pharmacokinetics. *J. Pharmacol. Scl.* 60 [1971] 1128–1133.)

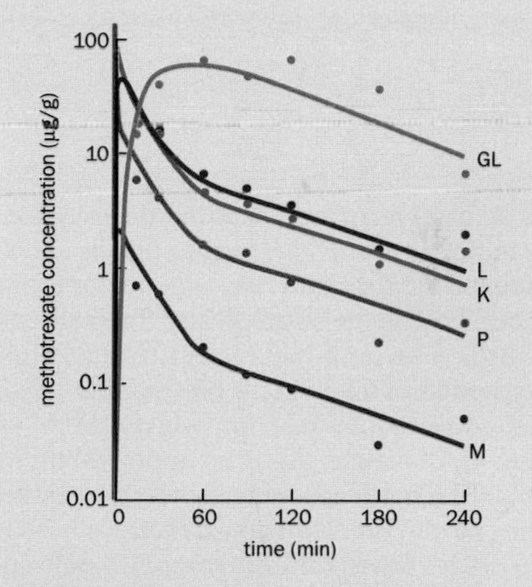

Figure 3 Methotrexate distribution among rat tissues over time. Comparison of model predictions (lines) and experimental data (dots) for intravenous injection of 6 mg methotrexate per gram of body weight in a rat. GL, gut lumen; K = kidney; L, liver; M, muscle; P, plasma. (Data from Bischoff KB, Dedrick RL, Zaharko DS & Longstreth JA. Methotrexate pharmacokinetics. *J. Pharmacol. Sci.* 60 [1971] 1128–1133.)

One somewhat tricky technical issue with PBPK models is that the typical measurement unit in biochemistry, physiology, and pharmaceutical science is a concentration, but the various organs and tissues have very different volumes, so the same amount of a drug moving, say, from the bloodstream to the lung tissue all of a sudden corresponds to a different concentration. PBPK models deal with this issue by multiplying concentrations by volumes, thereby computing the drug dynamics in absolute amounts. Outside this feature, most PBPK models use standard mass-action kinetics and Michaelis–Menten functions, but they could of course also be formulated with other rate functions, such as power laws.

As an important example, the dynamics of a drug in the blood plasma is typically written as

$$V_P\dot{P} = IN + \frac{q_L}{r_L}L + \frac{q_K}{r_K}K + \frac{q_M}{r_M}M - (q_L + q_K + q_M)P. \tag{13.6}$$

The left-hand side shows the time derivative of the concentration P of the drug in plasma, multiplied by the volume V_P of plasma, and is thus the time derivative of the actual amount of drug. On the right-hand side, we have amounts of drug entering (positive signs) and leaving (negative signs) the bloodstream. Beyond the input (IN), typically an intravenous injection, we see terms with two new types of parameters: flow rates between compartments, traditionally denoted by q and a subscript identifying the compartment, and so-called partition coefficients, which are denoted by subscripted r parameters and represent the fact that only portions of a tissue actively take up or release the drug. The quantities denoted by capital letters are concentrations of the drug in liver (L), kidney (K) and muscle (M). Depending on the application, other compartments, such as brain, lung, and fatty tissue, could also be considered.

The equations for other organs are constructed similarly, but are usually simpler. For instance, the drug concentration (or amount) in muscle is formulated as

$$V_M\dot{M} = q_M\left(P - \frac{M}{r_M}\right). \tag{13.7}$$

Similarly, the typical equation for the kidneys is written as

$$V_K\dot{K} = q_K\left(P - \frac{K}{r_K}\right) - k_K\frac{K}{r_K}. \tag{13.8}$$

It contains an extra term representing the excretion of the drug through urine. The case study in Box 13.1 provides further details and a few case-specific twists.

In principle, PBPK models may account for arbitrarily many compartments, and there is no real limit to their complexity. Typically, all transport steps are modeled as first-order processes, and metabolism within organs is not considered, with the possible exception of the liver, so that standard PBPK models consist of systems of linear differential equations, for which there are many methods of analysis (see Chapter 4). Of course, this is an approximation.

All in all, PBPK models have been very successful in the pharmaceutical industry because they permit predictions of dynamically changing drug distributions among organs and tissues based on animal experiments and on individual human-specific parameters like organ volumes, partition coefficients, and flow rates that can be estimated for humans without exposing them to the new drug.

13.8 Pathway Screening with Dynamic Models

Most dynamic pathway models do not account explicitly for different spatial locations such as organs. Instead, they attempt to capture quantitatively how a drug alters the movement of metabolites through a diseased pathway system and hopefully leads to a more desirable steady-state metabolic profile. The specific goal is typically an increase or decrease in a specific target metabolite. In addition to representing the primary pathway metabolites, dynamic models permit analyses of side effects within the pathway, such as the accumulation of toxic substances or the depletion of vital metabolites. Sufficiently accurate pathway models are thus very versatile and can play a role in efficacy, safety, toxicity, and dosing assessments [47]. Furthermore, similar to pharmacokinetic models, they have the potential to aid extrapolations from animal to human test results. Sophisticated models in the future will be capable of capturing large portions of human physiology and metabolism in sufficient detail. They will become reliable enough to screen out seemingly promising but ultimately ineffective leads early on in the drug development pipeline and thereby save valuable resources. Pathway models for drug discovery are reviewed in [48].

As an illustration of the role of pathway models in target screening, consider a very simplified model of purine metabolism (**Figure 13.19**). This biosynthetic pathway is responsible for generating sufficient amounts of purines, which are components of DNA and RNA and can also be converted into a number of other crucial metabolites. Here, we assume that the cells are not growing, so that hardly any new DNA is needed, and that the RNA pool is in equilibrium, where as much is produced as degraded.

For our simplistic illustration, we focus on a disorder called hyperuricemia, which is an elevation of the uric acid concentration in the blood and can be the result of various perturbations within the purine pathway. A common disease associated with hyperuricemia is gout. The model corresponding to the pathway in Figure 13.19 was inspired by earlier modeling efforts [14, 49–51], and, although it is much simpler than the original model, it allows us to illustrate some salient points. The mathematical representation in power-law form is as follows:

$$
\begin{aligned}
\dot{P} &\equiv v_{in} - v_{PG} - v_{PI} - v_{PH},\\
\dot{I} &\equiv v_{PI} + v_{PHI} + v_{GI} - v_{IG} - v_{IH},\\
\dot{G} &\equiv v_{PG} + v_{IG} - v_{GI} - v_{GX},\\
\dot{H} &\equiv v_{PH} + v_{IH} - v_{PHI} - v_{HX} - v_{Hout},\\
\dot{X} &\equiv v_{GX} + v_{HX} - v_{XU},\\
\dot{U} &\equiv v_{XU} - v_{Uout},
\end{aligned}
\tag{13.9}
$$

where the fluxes, in self-explanatory notation, take the form

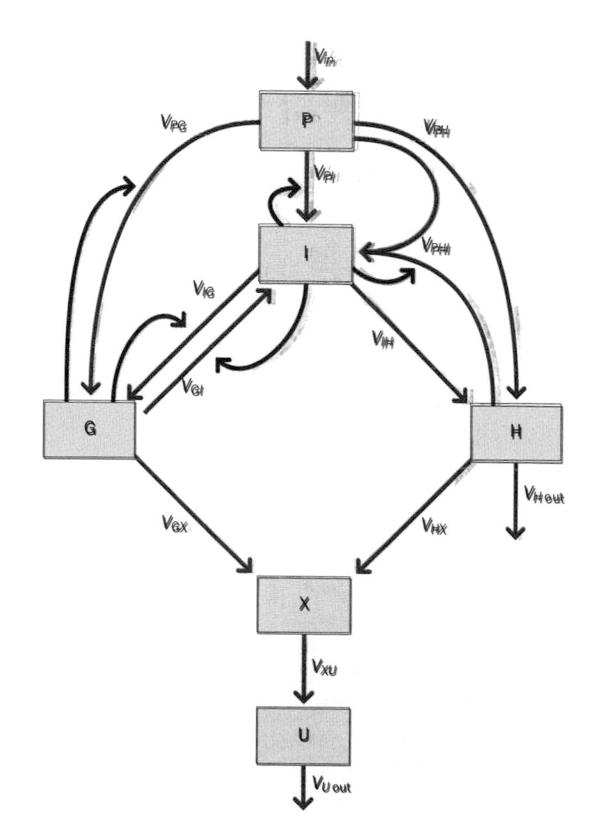

Figure 13.19 Simplified representation of purine metabolism. The diagram, and the corresponding model (13.6), (13.7), consist of metabolites (boxes), enzymatic conversions and transport steps (blue arrows), and inhibitory signals (red arrows). P, phosphoribosyl pyrophosphate; I, inosine monophosphate; G, guanylates; H, hypoxanthine and inosine; X, xanthine; U, uric acid.

$$\begin{aligned}
&v_{in} = 5, && \\
&v_{PG} = 320P^{1.2}G^{-1.2}, && v_{GX} = 0.01G^{0.5}, \\
&v_{PI} = 0.5P^{2}I^{-0.6}, && v_{IH} = 0.1I^{0.8}, \\
&v_{PH} = 0.3P^{0.5}, && v_{HX} = 0.75H^{0.6}, \\
&v_{PHI} = 1.2PI^{-0.5}H^{0.5}, && v_{H\,out} = 0.004H, \\
&v_{IG} = 2I^{0.2}G^{-0.2}, && v_{XU} = 1.4X^{0.5}, \\
&v_{GI} = 12G^{0.7}I^{-1.2}, && v_{U\,out} = 0.031U.
\end{aligned} \tag{13.10}$$

With these functions and numerical settings, the steady state is approximately (P, I, G, H, X, U) = (5, 100, 400, 10, 5, 100) (cf. [51]).

We discuss two perturbations within this pathway that lead to hyperuricemia. The first is hyperactivity of the enzyme phosphoribosylpyrophosphate synthase (PRPPS), which catalyzes the input step v_{in}. The second perturbation is due to a severe deficiency in the enzyme hypoxanthine–guanine phosphosribosyltransferase (HGPRT), which in the simplified model affects v_{PG} and v_{PHI}. In both cases, the concentration of uric acid (U) is significantly increased, among other problems. It is easy to simulate these conditions with the model in (13.9) and (13.10).

To represent PRPPS hyperactivity, we simply multiply v_{in} by an appropriate factor, such as 2. In response, the steady-state profile changes to about (P, I, G, H, X, U) = (13, 260, 1327, 23, 14, 170). All metabolites are elevated, including U, which is not surprising, because more material is flowing into the system. For the simulation of HGPRT deficiency, we multiply v_{PG} and v_{PHI} by a number between 0 and 1, where 0 would correspond to total absence of enzyme activity and 1 is the normal case. If the enzyme has only two-thirds of its normal activity, the situation is actually not so bad: the steady-state profile in this case is about (P, I, G, H, X, U) = (6, 103, 370, 11, 6, 109), which is different from the normal profile, but not by all that much. The situation deteriorates critically if the activity drops much further, leading to a disease that is known as Lesch–Nyhan syndrome and is accompanied by severely compromised brain function. If we reduce the activity in the model to 1%, the metabolite profile becomes (P, I, G, H, X, U) = (11, 97, 168, 21, 12, 157), where the guanylate levels are low and U is high.

The pertinent question here is whether it is possible to interfere effectively with any of the enzymes in order to compensate the changes in the metabolite profile that are caused by a disorder like PRPPS hyperactivity. The most straightforward strategy would be to inhibit PRPPS, but that is currently not possible. One active ingredient on the market is allopurinol, which is structurally similar to hypoxanthine and inhibits the enzyme xanthine oxidase, which oxidizes both hypoxanthine to xanthine and xanthine into uric acid. Multiplying v_{HX} and v_{XU} by 0.33, reflecting 33% remaining enzyme activity, does indeed control the uric acid level. The resulting profile is (P, I, G, H, X, U) = (10, 259, 1124, 66, 54, 109). However, as is also observed in clinical studies, hypoxanthine and xanthine accumulate, along with the other metabolites in the system.

It is rather evident that an effective remedy should remove disease-associated excess material from the system. Uricase is an enzyme that catalyzes the oxidation of uric acid to 5-hydroxyisourate and has been formulated as a drug for hyperuricemia. To test its efficacy, we multiply $v_{U\,out}$ by a factor representing the efficacy of the drug. If the activity of $v_{U\,out}$ is increased by 50%, the uric acid concentration indeed comes within 10% of the normal value, and the metabolite profile is (P, I, G, H, X, U) = (13, 252, 1215, 22, 13, 110). Alas, as in the previous scenario, P, I, and G accumulate quite significantly.

The enzyme information system BRENDA [52] indicates that the key initial enzyme of purine biosynthesis, amidophosphoribosyltransferase (EC 2.4.2.14), has a variety of inhibitors. Would it be useful to design a drug based on one of these? We can easily explore a hypothetical inhibitor affecting the corresponding process v_{PI}. For example, reducing the activity to 5% leads to the metabolite profile (P, I, G, H, X, U) = (22, 235, 1475, 11, 7, 118). As in the previous case, the uric acid is more or less under control. Also as before, P, I, and G are greatly increased.

In contrast to the allopurinol treatment, hypoxanthine and xanthine are now close to normal.

Are there other options? The diagram of the pathway contains two exit routes: one through U and one through H. Increasing $v_{H\,\text{out}}$ by 100 does indeed control U, along with X and H, but P, G, and I still accumulate. Could it potentially be effective to reroute excess P so that intermediates do not accumulate? Indeed, if we increase v_{PH} by a factor of 8 in addition to increasing $v_{H\,\text{out}}$ by 100, essentially all excess material is funneled out of the system and the resulting profile is almost normal, namely, $(P, I, G, H, X, U) = (5, 110, 426, 11, 6, 111)$. The model obviously cannot tell whether such a combined strategy can be developed as a useful pharmaceutical option or if it leads to complications elsewhere in the body. However, it demonstrates that modeling can be a powerful tool for screening hypotheses (see Exercises 13.28 and 13.29).

Dynamic models in pharmaceutical research are not limited to metabolism and transport between organs, the two systems chosen here because of their relative simplicity. Chemical, hormonal, electrical, and mechanical signaling, as well as aspects of higher-level systems physiology, will play increasingly important roles as diseases of the brain and of the cardiovascular, endocrine, and immune systems become sufficiently well understood for mechanical investigations at the molecular level. An example of such efforts is the Physiome Project [53], which has as its central theme the creation of integrated computational models of complete organ systems and, eventually, organisms.

13.9 Emerging Roles of Systems Biology in Drug Development

The examples shown in this chapter indicate that the role of computational systems biology in drug development has already begun to shift from "potential" to "actively contributing." In many cases, the contributions are still modest, but they do suggest the growing contribution of systems biology to this field. In addition to molecular and QSAR modeling, systems biology should be expected to play an increasingly important role in exploring the following [41]:

- the feasibility of an envisioned disease-disrupting drug treatment
- the choice of a target (ligand, receptor, complex; upstream or downstream)
- the efficacy of a therapeutic molecule
- the balance between safety and efficacy
- the toxicity of NCEs and biologics
- the consequences of properties of a therapeutic molecule (half-life, affinity, attainable concentration, etc.)
- the best therapeutic modality (antibodies, peptide mimics, engineered enzymes, RNA interference (RNAi), etc.)
- the route of administration
- the optimal dosing regimen
- the need to account for diffusion and transport (within tumors, through membranes, across the blood–brain barrier, etc.)
- the effects of clearance
- the extrapolation of test results between animals and from animals to humans

Considering the fact that the young field of systems biology is already involved in many areas of drug discovery, it is fair to expect that systems biology will move into the mainstream of pharmaceutical research and development in the foreseeable future. At the same time, considerations of health will shift more from treating to preventing disease, and the application of prescription drugs will become increasingly personalized. Systems biology will provide powerful tools to assist with these shifts.

EXERCISES

13.1. Using the example in (13.1), perform the same type of analysis by decreasing one enzyme activity at a time. Are the results symmetric in a sense that they mirror the results of increasing activities?

13.2. In the example in (13.1), the flux through X_4 is very small. Suppose that this flux represents clearance of an undesired metabolite by the kidneys, and that a patient with kidney disease can eliminate X_4 at only 10% capacity or not at all. Implement these two scenarios in the model and report the results.

13.3. Using the example in (13.1), study the effects of a decreased efflux from X_6. Compare the results with those of decreasing the clearance of X_4.

13.4. Using the example in (13.1), determine the most influential combination of two parameters with respect to X_6. Develop a strategy for testing the effects of three or four simultaneous perturbations (between 5% and 50%) in enzyme activities in the model. Look in the modeling chapters for alternatives to exhaustive analyses, which become infeasible owing to combinatorial explosion.

13.5. Define health and disease without expressing one in terms of the other.

13.6. In a model of a metabolic health and disease system, are biomarkers to be represented by parameters, independent or dependent variables, two of the three, or all three? Explain your answer and provide examples where applicable.

13.7. Search the literature for examples of biomarkers that are not genomic or related to enzymes. Discuss how they would enter mathematical models of some health and disease system under investigation.

13.8. Discuss whether false-positive or false-negative interpretations of biomarker readings, such as those corresponding to the pink and green points in Figure 13.6, become more of a problem or less of a problem in a larger system of biomarkers.

13.9. Search the literature for reports on drug treatments whose efficacy can be predicted from genome analyses.

13.10. Search the literature for nongenomic biomarkers that are predictive of disease or the success of some specific medication.

13.11. Discuss the role of cytochrome P450 enzymes as possible biomarkers for disease, disease resistance, or the efficacy of drugs. Formulate your findings as a report.

13.12. Using the simple model (13.1) at the beginning of the chapter, define health and disease states and analyze parameter settings within the disease domain. Is it possible to simulate situations where the analysis of a single biomarker suggests a healthy state, whereas the consideration of several biomarkers suggests disease? Either show examples or prove that such a situation is not possible.

13.13. Again using the simple model (13.1), study the long-term transient effects of a decreased clearance of X_4. For instance, if a concentration of X_4 above 12 or 40 (roughly 3 or 10 times the normal level, respectively) is deleterious, how long does it take before the patient should be considered sick?

13.14. Search the literature for data on the success or attrition of NCEs and biologics in different disease areas (inflammation, cardiovascular, oncology, etc.). Identify at which stages of the drug development process the failure rate is the highest and the lowest.

13.15. Expand the simple antibody–target model (13.2) to include clearance of the complex C. Perform simulations and compare the results with corresponding results of the model without clearance of C. Explore different half-lives for the clearance process.

13.16. Expand the simple antibody–target model (13.2) to allow for multiple doses of antibody injection. Perform simulations with different dosing regimens. Determine an optimal dosing schedule that keeps the target under control, while minimizing the total dose and adhering to a patient-friendly schedule.

13.17. Chabot and Gomes [36] state that the half-life of a collection of different clearance processes can be computed as

$$\frac{1}{t_{1/2;\,\text{overall}}} = \frac{1}{t_{1/2;\,\text{proteolysis}}} + \frac{1}{t_{1/2;\,\text{kidney}}} + \frac{1}{t_{1/2;\,\text{endocytosis}}} + \frac{1}{t_{1/2;\,\text{other}}}.$$

Prove this statement mathematically or design an ordinary differential equation model to test it computationally. Compare the collective half-life with the half-lives of the individual clearance processes.

13.18. Use the receptor–ligand–antibody model (13.3) to study the dynamics of ligands if the antibody binds to the receptor. Compare your results with those in the text.

13.19. Use the receptor–ligand–antibody model (13.3) to allow for multiple doses of antibody injection. Perform simulations with different dosing regimens.

13.20. Simulate constant input *IN* in the compartment model (13.4).

13.21. Simulate three equal daily doses of a drug that enters the compartment system (13.4) as input *IN*. Can you set the doses such that the drug concentration in blood is constant? Or that it at least stays within some predefined range?

13.22. What happens if the efflux parameters k_{BO} and k_{LO} in the compartment model (13.4) are zero? Make a prediction before you simulate.

13.23. Develop a compartment model like (13.4) that contains fatty tissue in addition to blood and liver. Set the parameters such that they reflect very low perfusion in fat. Perform representative simulations.

13.24. Develop a compartment model like (13.4) that accounts for degradation of the toxic substance in the liver. Compare the results for degradation processes in the form of a Michaelis–Menten rate law and a power-law function.

13.25. Simulate an injection of 300 mg MTX per gram of body weight and compare the results with the data in **Table 13.3**.

13.26. Using the information from Box 13.1, explore the differences between modeling the drug injection either as a change in the initial value of P or as an input bolus IN, which is set at 66 for one time unit, 33 for two time units, or 22 for three time units.

13.27. Using the information from Box 13.1, simulate the dynamic MTX distribution among tissues in humans, given an injection of 1 mg per kg of body weight. Compare the results with the measurements in **Table 13.4**. Note that measurements for other tissues are not available.

13.28. Explore other possible treatments of hyperuricemia with the simplified purine model (13.9).

13.29. Using the simplified purine model (13.9), compare the efficacy of combined treatments (at least at two target sites) with inhibition or activation of individual steps. Is it possible that drugs are synergistic in a sense that their combination is stronger than the sum of the two individual treatments?

TABLE 13.3: MEASURED CONCENTRATIONS (μg/g) OF MTX IN RESPONSE TO AN INJECTION OF 300 mg MTX PER GRAM BODY WEIGHT IN MOUSE*

Time after injection (min)	20	50	70	80	150	180	200
Plasma	100	25.6	18.2	10	2.2	1.5	4.3
Gut lumen	1292	1668	1978	2154	2346	426	256

*From Bischoff KB, Dedrick RL, Zaharko DS & Longstreth JA. Methotrexate pharmacokinetics. *J. Pharmacol. Sci.* 60 (1971) 1128–1133.

TABLE 13.4: MEASURED CONCENTRATIONS (μg/g) OF MTX IN PLASMA FOLLOWING AN INJECTION OF 1 mg MTX PER GRAM BODY WEIGHT IN TWO HUMANS*,†

Time after injection (min)	15	45	60	120	150	240	300	360
Experiment 1	1.10	0.821	0.693	0.453	—	0.284	—	0.211
Experiment 2	1.91	1.31	1.06	—	0.560	—	0.23	—

*From Bischoff KB, Dedrick RL, Zaharko DS & Longstreth JA. Methotrexate pharmacokinetics. *J. Pharmacol. Sci.* 60 (1971) 1128–1133.
†The data were obtained in two experiments.

REFERENCES

[1] Human Genome Project Information. SNP Fact Sheet. https://ghr.nlm.nih.gov/primer/genomicresearch/snp.

[2] Reddy MB, Yang RS, Clewell HJ III & Andersen ME (eds). Physiologically Based Pharmacokinetic Modeling: Science and Applications. Wiley, 2005.

[3] Voit EO. A systems-theoretical framework for health and disease. Abstract modeling of inflammation and preconditioning. *Math. Biosci.* 217 (2008) 11–18.

[4] Ishii N, Nakahigashi K, Baba T, et al. Multiple high-throughput analyses monitor the response of *E. coli* to perturbations. *Science* 316 (2007) 593–597.

[5] Miller GW. The Exposome: A Primer. Academic Press, 2014.

[6] Emory/Georgia Tech Predictive Health Institute. http://predictivehealth.emory.edu/.

[7] Voit EO & Brigham KL. The role of systems biology in predictive health and personalized medicine. *Open Pathol. J.* 2 (2008) 68–70.

[8] Sterling P. Principles of allostasis: optimal design, predictive regulation, pathophysiology and rational therapeutics. In Allostasis, Homeostasis, and the Costs of Adaptation (J Schulkin, ed.), pp 17–64. Cambridge University Press, 2004.

[9] Kitano H, Oda K, Kimura T, et al. Metabolic syndrome and robustness tradeoffs. *Diabetes* 53 (Suppl 3) (2004) S6–S15.

[10] Lipsitz LA. Physiological complexity, aging, and the path to frailty. *Sci. Aging Knowledge Environ.* 2004 (2004) pe16.

[11] van't Veer LJ & Bernards R. Enabling personalized cancer medicine through analysis of gene-expression patterns. *Nature* 452 (2008) 564–570.

[12] Foubister V. Genes predict childhood leukemia outcome. *Drug Discov. Today* 10 (2005) 812.

[13] Babaian RJ, Fritsche HA, Zhang Z, et al. Evaluation of ProstAsure index in the detection of prostate cancer: a preliminary report. *Urology* 51 (1998) 132–136.

[14] Voit EO. Computational Analysis of Biochemical Systems: A Practical Guide for Biochemists and Molecular Biologists. Cambridge University Press, 2000.

[15] Entelos, Inc. http://www.Entelos.com.

[16] Darmoni SJ, Massari P, Droy J-M, et al. SETH: an expert system for the management on acute drug poisoning in adults. *Comput. Methods Programs Biomed.* 43 (1994) 171–176.

[17] Luciani D, Cavuto S, Antiga L, et al. Bayes pulmonary embolism assisted diagnosis: a new expert system for clinical use. *Emerg. Med. J.* 24 (2007) 157–164.
[18] Sundareswaran KS, de Zelicourt D, Sharma S, et al. Correction of pulmonary arteriovenous malformation using image-based surgical planning. *JACC Cardiovasc. Imaging* 2 (2009) 1024–1030.
[19] Kola I & Landis J. Can the pharmaceutical industry reduce attrition rates? *Nat. Rev. Drug Discov.* 3 (2004) 711–715.
[20] Central Intelligence Agency. CIA World Factbook: Life Expectancy at Birth. https://www.cia.gov/library/publications/the-world-factbook/fields/2102.html#us.
[21] Research and development costs for new drugs by therapeutic category. A study of the US pharmaceutical industry. DiMasi JA, Hansen RW, Grabowski HG, Lasagna L. Pharmacoeconomics (1995) 152–169. PMID: 10155302.
[22] An analysis of the attrition of drug candidates from four major pharmaceutical companies. Waring MJ, Arrowsmith J, Leach AR, et al. *Nat. Rev. Drug Discov*. 14 (2015) 475–486. PMID: 26091267.
[23] Goodnow RAJ. Hit and lead identification: integrated technology-based approaches. *Drug Discov. Today: Technologies* 3 (2006) 367–375.
[24] Andrade CH, Pasqualoto KF, Ferreira EI & Hopfinger AJ. 4D-QSAR: perspectives in drug design. *Molecules* 15 (2010) 3281–3294.
[25] Funatsu K, Miyao T & Arakawa M. Systematic generation of chemical structures for rational drug design based on QSAR models. *Curr. Comput. Aided Drug Des.* 7 (2011) 1–9.
[26] Zuckermann RN & Kodadek T. Peptoids as potential therapeutics. *Curr. Opin. Mol. Ther.* 11 (2009) 299–307.
[27] Wishart DS. Improving early drug discovery through ADME modelling: an overview. *Drugs R D* 8 (2007) 349–362.
[28] DiMasi JA, Hansen RW & Grabowski HG. The price of innovation: new estimates of drug development costs. *J. Health Econ.* 835 (2003) 1–35.
[29] Kitano H. Computational systems biology. *Nature* 420 (2002) 206–210.
[30] Hood L. Systems biology: integrating technology, biology, and computation. *Mech. Ageing Dev.* 124 (2003) 9–16.
[31] Voit EO. Metabolic modeling: a tool of drug discovery in the post-genomic era. *Drug Discov. Today* 7 (2002) 621–628.
[32] Selventa. http://www.selventa.com/.
[33] GNS Healthcare. https://www.gnshealthcare.com/.
[34] Huang PH, Mukasa A, Bonavia R, et al. Quantitative analysis of EGFRvIII cellular signaling networks reveals a combinatorial therapeutic strategy for glioblastoma. *Proc. Natl Acad. Sci. USA* 104 (2007) 12867–12872.
[35] Sachs K, Perez O, Pe'er S, et al. Causal protein-signaling networks derived from multiparameter single-cell data. *Science* 308 (2005) 523–529.
[36] Chabot JR & Gomes B. Modeling efficacy and safety of engineered biologics. In Drug Efficacy, Safety, and Biologics Discovery: Emerging Technologies and Tools (S Ekins & JJ Xu, eds), pp 301–326. Wiley, 2009.
[37] Mayawala K & Gomes B. Prediction of therapeutic index of antibody-based therapeutics: mathematical modeling approaches. In Predictive Toxicology in Drug Safety (JJ Xu & L Urban, eds), pp 330–343. Cambridge University Press, 2010.
[38] Nimjee SM, Rusconi CP & Sullenger BA. Aptamers: an emerging class of therapeutics. *Annu. Rev. Med.* 56 (2005) 555–583.
[39] Arakaki AK, Mezencev R, Bowen N, et al. Identification of metabolites with anticancer properties by computational metabolomics. *Mol. Cancer* 7 (2008) 57.
[40] Gomes B. Industrial systems biology. Presentation at Foundations of Systems Biology in Engineering, FOSBE'09, Denver, CO, August 2009.
[41] Bonate PL. Pharmacokinetic–Pharmacodynamic Modeling and Simulation, 2nd ed. Springer, 2011.
[42] Jacquez JA. Compartmental Analysis in Biology and Medicine, 3rd ed. BioMedware, 1996.
[43] Lipscomb JC, Haddad S, Poet T, Krishnan K. Physiologically-based pharmacokinetic (PBPK) models in toxicity testing and risk assessment. In New Technologies for Toxicity Testing (M Balls, RD Combes & N Bhogal N, eds), pp 76–95. Landes Bioscience and Springer Science + Business Media, 2012.
[44] World Health Organization (WHO). WHO Model List of Essential Medicines. http://www.who.int/medicines/publications/essentialmedicines/en/.
[45] Bischoff KB, Dedrick RL & Zaharko DS. Preliminary model for methotrexate pharmacokinetics. *J. Pharmacol. Sci.* 59 (1970) 149–154.
[46] Bischoff KB, Dedrick RL, Zaharko DS & Longstreth JA. Methotrexate pharmacokinetics. *J. Pharmacol. Sci.* 60 (1971) 1128–1133.
[47] Kell DB. Systems biology, metabolic modelling and metabolomics in drug discovery and development. *Drug Discov. Today* 11 (2006) 1085–1092.
[48] Yuryev A (ed.). Pathway Analysis for Drug Discovery: Computational Infrastructure and Applications. Wiley, 2008.
[49] Curto R, Voit EO & Cascante M. Analysis of abnormalities in purine metabolism leading to gout and to neurological dysfunctions in man. *Biochem. J.* 329 (1998) 477–487.
[50] Curto R, Voit EO, Sorribas A & Cascante M. Validation and steady-state analysis of a power-law model of purine metabolism in man. *Biochem. J.* 324 (1997) 761–775.
[51] Curto R, Voit EO, Sorribas A & Cascante M. Mathematical models of purine metabolism in man. *Math. Biosci.* 151 (1998) 1–49.
[52] BRENDA: The Comprehensive Enzyme Information System. http://www.brenda-enzymes.org/.
[53] IUPS Physiome Project. http://www.physiome.org.nz/.

FURTHER READING

Auffray C, Chen Z & Hood L. Systems medicine: the future of medical genomics and healthcare. *Genome Med.* 1 (2009) 2.
Bassingthwaighte J, Hunter P & Noble D. The Cardiac Physiome: perspectives for the future. *Exp. Physiol.* 94 (2009) 597–605.
Bonate PL. Pharmacokinetic–Pharmacodynamic Modeling and Simulation, 2nd ed. Springer, 2011.
Gonzalez-Angulo AM, Hennessy BT & Mills GB. Future of personalized medicine in oncology: a systems biology approach. *J. Clin. Oncol.* 28 (2010) 2777–2783.
McDonald JF. Integrated cancer systems biology: current progress and future promise. *Future Oncol.* 7 (2011) 599–601.
Kell DB. Systems biology, metabolic modelling and metabolomics in drug discovery and development. *Drug Discov. Today* 11 (2006) 1085–1092.
Kitano H. Computational systems biology. *Nature* 420 (2002) 206–210.
Kitano H, Oda K, Kimura T, et al. Metabolic syndrome and robustness tradeoffs. *Diabetes* 53 (Suppl 3) (2004) S6–S15.
Lipsitz LA. Physiological complexity, aging, and the path to frailty. *Sci. Aging Knowledge Environ.* 2004 (2004) pe16.
Reddy MB, Yang RS, Clewell HJ III & Andersen ME (eds). Physiologically Based Pharmacokinetic Modeling: Science and Applications. Wiley, 2005.

Voit EO. A systems-theoretical framework for health and disease. Abstract modeling of inflammation and preconditioning. *Math. Biosci.* 217 (2008) 11–18.
Voit EO & Brigham KL. The role of systems biology in predictive health and personalized medicine. *Open Pathol. J.* 2 (2008) 68–70.
Voit EO. The Inner Workings of Life. Vignettes in Systems Biology. Cambridge University Press, 2016.
West GB & Bergman A. Toward a systems biology framework for understanding aging and health span. *J. Gerontol. A Biol. Sci. Med. Sci.* 64 (2009) 205–208.
Weston AD & Hood L. Systems biology, proteomics, and the future of health care: toward predictive, preventative, and personalized medicine. *J. Proteome Res.* 3 (2004) 179–196.

Design of Biological Systems

14

When you have read this chapter, you should be able to:

- Recognize the importance of studying design principles
- Identify concepts and basic tools of design analysis
- Discuss different types of motifs in biological networks and systems
- Set up static and dynamic models for analyzing design principles
- Describe methods for manipulating biological systems toward a goal
- Understand the concepts and basic tools of synthetic biology
- Characterize the role of systems and synthetic biology in metabolic engineering

This chapter intentionally carries a potentially ambiguous title, because it discusses two topics that fit under the same heading and are closely related to each other, but are also genuinely distinct. The first aspect addresses questions of the natural **designs** of biological systems, as we observe them: Are there structural patterns or **motifs** that we find time and again in biological networks? If so, what are their roles and advantages? Are there **design principles** that are prevalent in dynamic biological systems? Why are some genes regulated by an inducer and some by a repressor? Does it really matter? What is the advantage of a feedback signal inhibiting the first step of a linear pathway rather than the second or third step?

The second aspect covered in this chapter addresses how we can alter the existing design of a biological system in a goal-oriented fashion or even create new biological systems from scratch. This artificial design serves two purposes. First, it adheres to Richard Feynman's famous quote "What I cannot create, I do not understand." In other words, by creating first small and then larger biological modules that function as predicted, we can demonstrate that we comprehend the principles that underlie the organization and function of natural systems. Or, if the artificial system behaves differently than expected, it indicates that there are aspects we do not understand correctly. Second, the targeted alteration of existing biological systems or the *de novo* creation of novel systems offers the potential of using biology for new purposes. This second aspect of design is nowadays called **synthetic biology** and can be seen as applied systems biology or a true engineering approach to biology. Presently, the most prevalent application is metabolic engineering, which addresses the microbial production of valuable compounds such as human insulin or the large-scale microbial production of bulk materials such as ethanol or citric acid. A different type of application is the artificial creation of microbial systems as very sensitive sensors for environmental toxins or as drivers of environmental remediation for soils containing metals or radionuclides.

Of course, the natural and artificial sides of biological design studies have quite a degree of overlap. First, they both deal with the structure, function, and control of biological systems, with one being a manifestation of evolution and the other the result of clever molecular biology and engineering. Second, both directly or indirectly involve optimization. Natural designs are often assumed to be optimized by the forces of selection, although this assumption cannot really be proven. It seems indeed to be the case that natural designs are typically (always?) superior to comparable alternatives, but who is to say that evolution has stopped at the beginning of the third millennium of our era? Should one really doubt that organisms will evolve further and that they will have out-competed today's species in a few million years? You could even ask yourself: am I optimal?

Optimization is certainly center-stage in the design of synthetic biological systems. In terms of *de novo* designs, we may initially be happy if we succeed with the creation of a biological module that functions at all, but it is our human nature immediately to start improving and trying to optimize such a creation. Even more pronounced is the role of optimization in the manipulation of existing systems, for instance, in the cheap production of biofuel. Is it possible to increase the yield in yeast-based alcohol fermentation? What are the optimal strain, medium, and mode of gene expression? Can we make the production cheaper by using organisms such as *Escherichia coli* instead of yeast? Is it possible to use autotrophic organisms like plants or algae that are able to convert freely available sunlight and carbon dioxide into sugars, and to manipulate them into converting the sugars into ethanol, butanol, or harvestable fatty acids? How could we maximize yield?

No matter whether the investigation addresses natural or artificial systems, the exploration and exploitation of biological design principles requires a mix of biology, computation, and engineering, and thus of systems biology. So, let us embark on the study of biological design, be it natural or created artificially, exploring optimality or at least superiority with respect to reasonable alternatives.

NATURAL DESIGN OF BIOLOGICAL SYSTEMS

If every two biological systems had entirely different structures and modes of operation, the study of biology would have a lot of similarity with casual stargazing, where we may find patterns remotely resembling people, animals, or objects, but where there is no rhyme or reason for why such patterns happen to be there. As humans have done for thousands of years, we simply describe and name them. We could study biological systems one at a time and classify them by some features, but we would not gain much insight into the rationale behind these features and their specific roles. Thus, when we talk about investigating the design of a biological system, we are really trying to address design patterns, design principles, or motifs that show up time and again with relatively minor variations. Instead of comparing biology with stargazing, we should therefore compare biological designs with technological and engineered designs. A good watchmaker can explain the role of every spring or cogwheel in a particular location of a mechanical watch, even if he or she has never seen this particular watch before. The shapes of cars are far from coincidental, and style and uniqueness are extremely constrained by functionality, aerodynamics, desirable features, comfort, and price. After all, why do almost all passenger cars have four wheels, while most large trucks have more? Why are brake lights red? ...ms biology must eventually come to a point where we understand in detail why ...genes are turned on and off in particular sequences, why cells and organ... ...red in the way they are, and what is behind the details of the function ...gans, tissues, cells, and intracellular components in particular

...uctural Patterns

...ciples is an interesting line of research that began ... taken off only recently (see, for example, [5–8]). ...d to discover designs that are used over and over again in biology, and second, we have to figure out why these designs are apparently preferred over other imaginable designs.

In some cases, design patterns are easy to detect, especially if they are associated with macroscopic features. Birds have wings, fish have fins, spiders have eight legs, and most mammals have hair. In other cases, such discoveries are much more difficult. It took quite a bit of molecular biology to figure out the almost ubiquitous structure of lipid bilayers or the genetic code that seems to be universal among essentially all earthly organisms. In yet other cases, the design patterns are so intricate that they easily escape casual observation. As a specific example, we have discovered with sophisticated computational analyses that genomes are not arbitrarily arranged in gene sequences, but have much more structure. Bacterial genes are arranged in **operons**, regulons, and über-operons [9], and it appears that operons involved in multiple pathways tend to be kept in nearby genome locations [10] (**Figure 14.1**). Species show their own codon bias, and bacterial genomes consist of almost unique nucleotide compositions and combinations that resemble molecular fingerprints or barcodes [11] (see Chapter 6). These types of patterns were only discovered with the help of very large datasets and custom-tailored computer algorithms.

It has been known for some while that the expression of a gene is correlated with the expression of other genes. But are there patterns? Microarray studies, taken at several time points after some stimulus, suggest that groups of genes may be co-expressed, while the expression of other genes is inversely correlated or correlated with some time delay. In many cases, the clusters of co-expressed genes can be explained with the coordinated function of the proteins for which they code, but in other cases they remain a puzzle (see, for example, [12]).

Like gene sequences, metabolic and signaling pathways show commonalities among different cells and organisms, some of which have been known for a long time. Feedback inhibition by the end product is usually exerted on the first step of a linear pathway, and one can show—with mathematical rigor—that this is indeed a design that is superior to other modes of inhibition, such as inhibition of the last step, or of several steps, in a reaction sequence (cf. [1, 13]). Signaling is often accomplished with cascades of proteins that are phosphorylated and dephosphorylated at three levels (Chapter 9). Why three? We have recently learned to look for generic patterns in the connectivity patterns of natural and artificial networks (see Chapter 3). For instance, most metabolites are associated with only a few reactions, and relatively few metabolites are associated with many different production and degradation steps [14]. This same pattern of only a few highly connected hubs within a large network shows up time and again, in biology, the physical world, and society. Just study the flight routes of a major airline!

Interestingly, but not really surprisingly, many networks we find in biology are very robust. In a beautiful demonstration of this feature, Isalan and collaborators

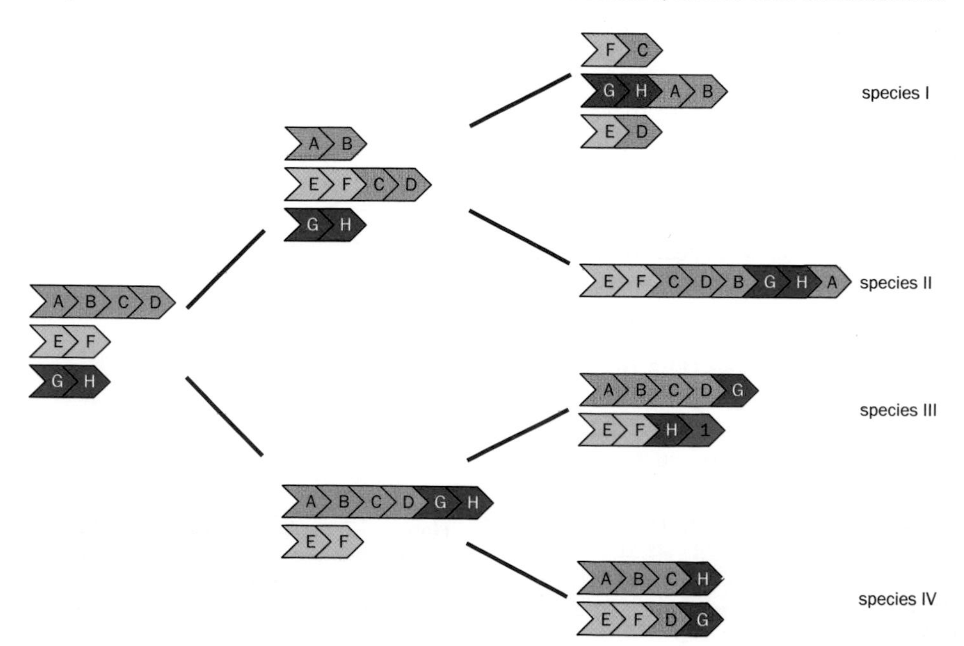

Figure 14.1 Conceptual representation of operons and über-operons. Operons are contiguous, co-transcribed, and co-regulated sets of genes. While they are usually poorly conserved during evolution, the rearrangement of genes often maintains individual genes in specific functional and regulatory contexts, called über-operons. The ancestral genome (left) is assumed to contain three clusters of similarly regulated and functionally related genes, A–H, in three separate genomic locations. During evolution, the clusters are rearranged into new clusters. Further genome rearrangements may occur, but some clusters and individual genes are retained in the same neighborhoods. (From Lathe WC III, Snel B & Bork P. *Trends Biochem. Sci.* 25 [2000] 474–479. With permission from Elsevier.)

randomly rewired the transcriptional network of *E. coli* and found that only about 5% of these scrambled networks showed growth defects [15]. A similar change in the connectivity of gene transcription networks seems to have occurred naturally in the evolution of yeast, and it could be that such flexibility is a general feature of gene transcription systems. It is intriguing to ponder whether it might be possible to use rewiring as a precisely targeted tool in synthetic biology [16].

Another important aspect of biological systems is that they employ similar components time and again in a modular and hierarchical manner. This use of the same motifs found within different networks and at different organizational levels renders complex biological networks simpler than they might otherwise have been, and indeed offers hope that we might eventually understand critical aspects of the complexity of biological systems [17].

The identification and rigorous characterization of design principles requires methods of formalizing complex systems and their organization. This task may be accomplished in different ways. One school of thought has been focusing on network graphs and their properties, while another has been looking for design principles in fully regulated dynamic systems and the corresponding differential equation models. Interestingly, both approaches ultimately employ a similar strategy, namely comparisons between the observed system structures and reasonable alternatives. In the case of graphs, the properties of an observed network are compared with the corresponding properties of a randomly connected network. For instance, one might look for the number of nodes that have more links than one would statistically expect in a random network (see Chapter 3). In the case of dynamic systems, an observed design is compared with a very similar dynamic system structure that differs in just one particular feature of interest, such as a regulatory signal.

14.2 Network Motifs

The first step in an analysis of static network motifs is the abstraction of the network as a graph. As Chapter 3 discussed in detail, a typical analysis of such a graph reveals global network features, such as the small-world property, which is related to the clustering coefficient, the degree distribution, and the average shortest pathway length between any two nodes [18]. Beyond connectivity patterns, the search for design principles in graph models focuses on network motifs, which are specific, small subgraphs that are encountered more often than probability theory would predict in a randomly connected graph [5, 19].

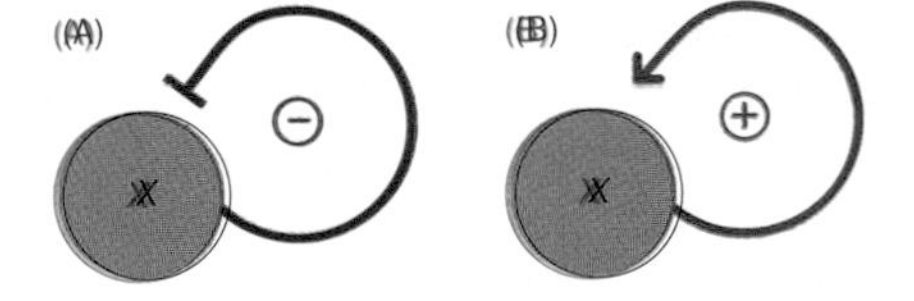

Figure 14.2 Network motifs can be very simple. Shown here are negative (A) and positive (B) auto-regulation motifs. The regulation arrows may encompass several steps.

The simplest network motif, containing just one node, is **auto-regulation**, which may be negative or positive (**Figure 14.2**). It should be noted that this motif allows for more than a single node: the arrow feeding back may consist of a chain of processes. For instance, negative auto-regulation is frequently found in transcription factors that repress their own transcription. While the transcription factor is the only node explicitly considered, the system in reality also contains at least one gene, mRNA, ribosomes, and other molecular components. Negative auto-regulatory loops are very stable and reduce fluctuations in inputs. However, if the feedback is delayed, the system may also oscillate and eventually lose stability. Positive auto-regulation permits a cell to assume multiple internal states, and, in particular, toggling between two stable states. We discussed one example in Chapter 9. Other examples have been reported for the genetic switch between lytic and lysogenic responses in the bacteriophage λ as well as in signaling cascades and cell cycle phenomena (see, for example, [20, 21]). We discuss a toggle switch in more detail in the case study in Section 14.9.

A slightly more complicated, quite prevalent network motif is the **bi-fan**, which consists of two source nodes sending signals to two target nodes (**Figure 14.3**) [19]. This design is interesting, because it allows temporal regulation [22]. In the case of signal transduction, it can also sort, filter, de-noise, and synchronize signals. The interactions between the two signals arriving at a target may be of different types. In the case of an AND gate, the signal is only transduced to T_1 (or T_2) if both sources, S_1 and S_2, fire, while in the case of an OR gate, a single source, S_1 or S_2, is sufficient. An OR gate confers reliability onto the system, because one source is sufficient, while the AND gate permits the detection of signal coincidence and can be employed to prevent a target from being turned on spuriously when only one signal is present.

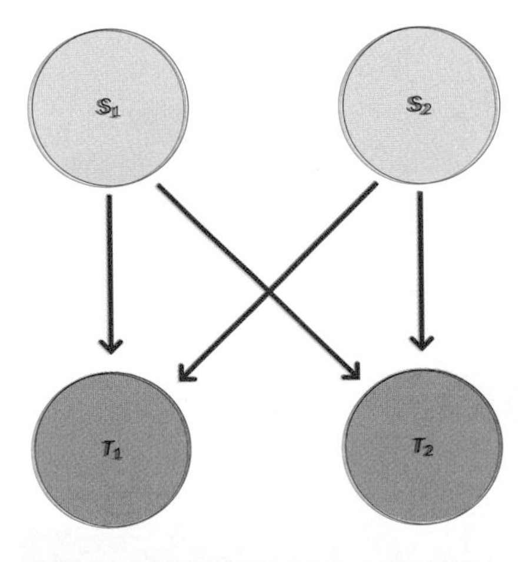

Figure 14.3 The bi-fan is a very prevalent motif. In this type of small network, two source nodes (S_1 and S_2) send signals to two target nodes (T_1 and T_2).

It is not surprising that small motifs can be combined, both in nature and in theory, to form larger regulatory structures. These larger structures can be implemented in parallel or by stacking smaller motifs. They can also be complemented with additional components and links, thereby creating a potentially huge repertoire of possible responses. A well-known example of moderate complexity is the cross-talk between signaling cascades, which can create sophisticated response patterns such as in-band filters, where the system responds only to a signal within a strictly bounded range, or an EXCLUSIVE OR function, where the system responds only if exactly one of two sources is active, but not both [23].

While they are very important and instructive, small motifs often do not reveal characteristic topological features of larger networks [24]. At the same time, searching for all possible motifs of even moderate size within a large network is not a trivial matter; in fact, it is currently not computationally feasible to identify all motifs of 10 or more nodes within realistically sized networks of a few thousand nodes and links. A strategy for addressing the situation is to reduce the scope of the search. For instance, Ma'ayan and collaborators successfully searched large networks exclusively for cycles consisting of between 3 and 15 directional, nondirectional, or bidirectional links. Particularly important among their findings was that biological and engineered networks contain only small numbers of **feedback** loops of the types shown in **Figure 14.4**. Among them, all-pass-through feedback loops that are coherent in a sense that all arrows point in the same direction pose the risk of destabilizing the system, while anti-coherent cycles that contain sinks and sources are stable. Lipshtat and collaborators showed that feedback loops in signal transduction systems can exhibit bistable behavior, oscillations, and delays [22]. Consistent with the argument of instability in coherent feedback loops, they found that hubs in natural networks are mostly either source or sink nodes, but not pass-through nodes. They also observed that networks with fewer nodes involved in feedback loops are more stable than corresponding networks with more nodes involved in feedback loops. They concluded that the prudent placement of feedback loops might be a general design principle of complex biological networks.

Similar to some of the feedback loops are **feedforward** loop motifs that contain a direct and an indirect path (**Figure 14.5**). Each arrow may be activating or inhibitory, which for the case of three nodes leads to eight combinations with different response patterns [5]. Four of these are coherent, in the sense that the indirect path overall sends the same activating or inhibitory signal as the direct path, while the other four combinations are incoherent, because they contain opposite signals. As an example, the incoherent motif in Figure 14.5C indicates inhibition along the direct path, but double inhibition, and thus activation, along the indirect path. Both coherent and incoherent patterns have been observed in natural systems, and the motif with all positive signals (Figure 14.5A) is most prevalent among them.

An interesting example of a feedforward loop is a mechanism for sensing stress signals in *E. coli* and many other bacteria [17]. When the bacterium senses the stress, the feedforward loop triggers the production of flagellum proteins (Chapter 7), which continues for about an hour, even if the signal disappears. This time period is

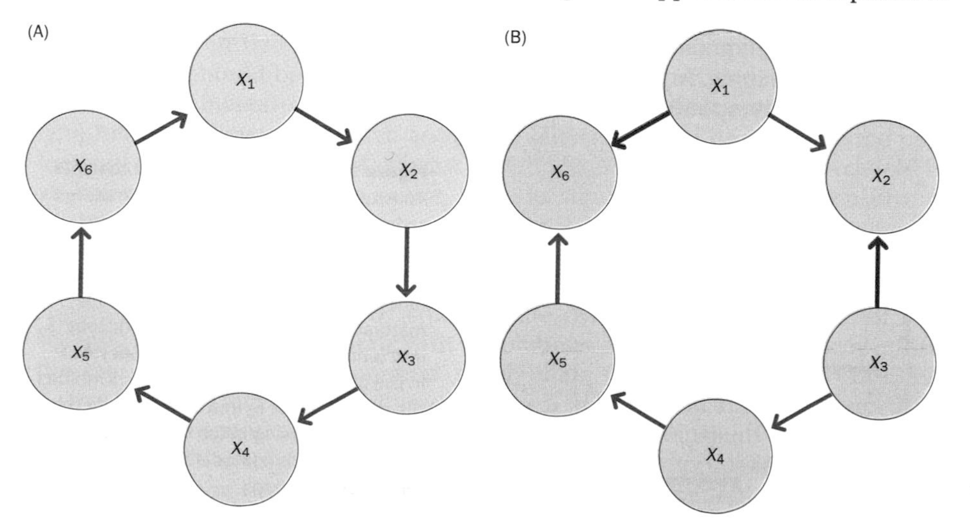

Figure 14.4 Subtle differences in network structure can have significant consequences on the stability of a network. The coherent all-pass-through feedback loop in (A) poses the risk of destabilization, while the anti-coherent network in (B) is stable, with two source nodes (X_1 and X_3), two sink nodes (X_2 and X_6), and two pass-through nodes (X_4 and X_5).

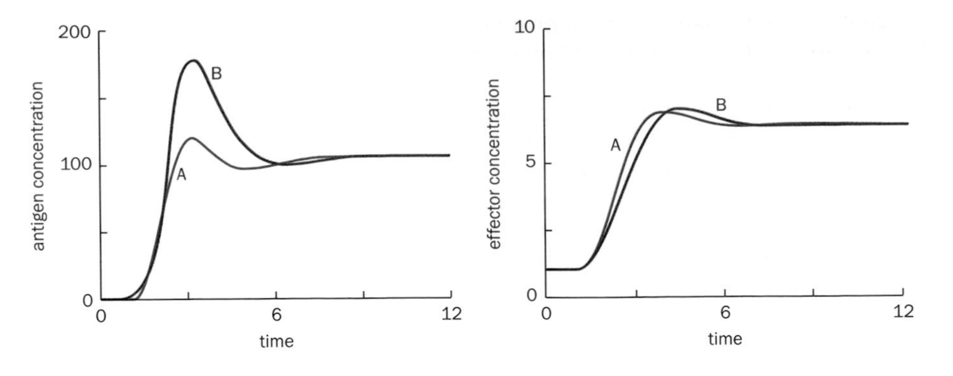

Figure 14.7 Comparison of time courses of two designs of immune cascades. Dynamic responses of the observed system design (Model A) and a hypothetical, alternative, design (Model B), which lacks inhibition of effector lymphocyte maturation, to a fourfold source antigen challenge at time $t = 1$.

overshoots and the effector concentration responds more quickly. For other source antigen levels, the numerical features are different, but the superiority of the observed design (Model A) persists. Thus, lower peak levels are inherent functional characteristics of suppression. Algebraic analysis furthermore indicates that Model A is less sensitive to potentially harmful fluctuations in the environment [25]. Overall, the observed design has preferable features over the *a priori* also reasonable design without suppression.

Since its inception, MMCC has been applied to a variety of systems, primarily in the realm of metabolism and signaling, and variations have been proposed to study entire design classes at once (for example, [13]), characterize design spaces [26, 27], or establish superiority of a design in a statistical sense [28]. A particularly intriguing example is Demand Theory [2, 29, 30], which rationalizes and predicts the mode of gene regulation in bacteria, based on the microbes' natural environments, and thus reveals strong natural design principles (see Chapter 6). Another recent example is the two-component sensing system in bacteria, which we have already discussed in Chapter 9. A careful analysis of this system [31, 32] led to a deeper understanding of its natural design and provided clues for how to manipulate a system with methods of synthetic biology. For instance, Skerker and collaborators [33] explored the large amount of available sequence data available for *E. coli*'s envZ–OmpR two-component sensing system and identified individual amino acid residues that always varied together as a pair. They speculated that these pairs were critical for sensor specificity. Subsequently, they substituted the specificity-determining residues of one sensor with those of a different sensor, which indeed allowed the sensor's response to be triggered by the "wrong" signal.

14.4 Operating Principles

As a variation on the theme of design principles, one may also ask questions regarding **operating principles** for systems with given designs. An example is the controlled comparative analysis of different regulation patterns in a branched pathway (**Figure 14.8**), where the goal might be to increase the yield v_{40} by altering the ratios of fluxes flowing into different branches, v_{10}/v_{12} and/or v_{23}/v_{24} [34]. It turns out

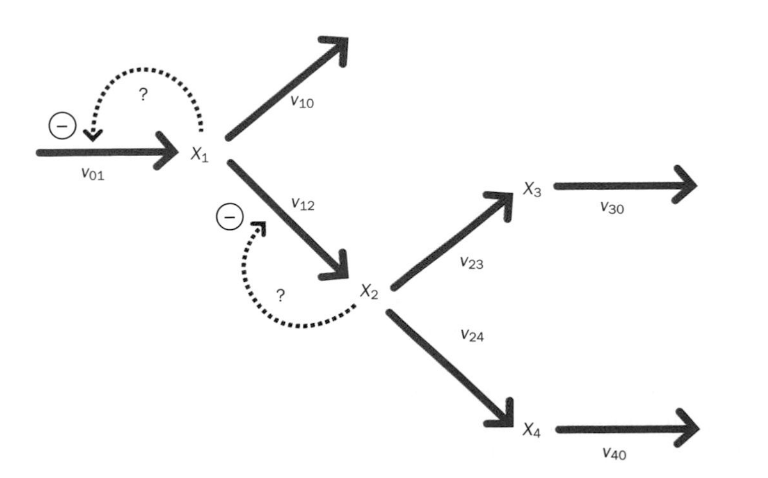

Figure 14.8 Optimal operating strategies depend on both structural and regulatory pathway features. Depending on the existence of no, one, or two feedback inhibition signals (indicated by question marks), the optimal strategy for increasing the flux v_{40} by manipulation of the flux ratios v_{10}/v_{12} and v_{23}/v_{24} is different.

that a variety of manipulations exhibit equal yields if the pathway is not regulated, but that some manipulations are significantly more effective than others if some of the intermediates exert feedback inhibition.

The discovery and analysis of design and operating principles is an ongoing effort. A different aspect than we have discussed so far is the search for rules related to spatial phenomena. For instance, if a protein is to be localized in the cell nucleus, it often contains a high proportion of hydrophobic surface regions and a charge that is complementary to that of the nuclear pore complexes, which control transport between the nucleus and cytoplasm [35]. It may also be that some intracellular gradients and heterogeneous concentration patterns are associated with the spatial design issue. A special issue of *Mathematical Biosciences* [8] contains a collection of papers representing the state of the art.

GOAL-ORIENTED MANIPULATIONS AND SYNTHETIC DESIGN OF BIOLOGICAL SYSTEMS

14.5 Metabolic Engineering

Between the discovery and analysis of natural design principles on the one hand and the creation of novel biological systems from component parts (that is, synthetic biology) on the other lies a large field of activities that attempt to manipulate, and subsequently optimize, existing systems toward a desired goal. The most well-developed line of research in this context is the **optimization** of metabolic pathway systems in microbes, which falls under the umbrella of **metabolic engineering**. There is no clear dividing line between manipulating existing systems and changing them to such an extent that they can be considered new, and so both metabolic engineering and synthetic biology flow into each other and can be seen as modern endeavors in a long-term effort to alter and control biological systems that reaches back about as far as recorded history and the first humans dabbling in agriculture. After all, the breeding of plants and animals is a slow form of genetic and metabolic engineering and thus of pushing biological systems toward a desired goal.

While excruciatingly slow by modern standards, the success of breeding efforts over hundreds of years has been stunning, and one can only be amazed on comparing modern corn or carrots with the original wild types from which they were derived. The selection of the most desirable plant and animal species proceeded throughout human history at a modest pace until the second half of the twentieth century, when it started speeding up tremendously. This huge jump in the pace of development and success occurred when we became knowledgeable enough to explain the intuitive concepts of Mendelian genetics on the basis of a theoretical foundation, and when we began to decipher the structure and function of genes, and later the control of their expression.

Once the role of genes became clearer, effective strategies to generate higher yield emerged. For instance, by exploiting nature's own method of improving design, the method of directed or adaptive evolution was invented to increase the efficient production of proteins (see, for example, [36]). In its natural form, evolution is slow, but methods of bioinformatics overcame this problem, because it became feasible with efficient algorithms to compare pertinent gene sequences from many different organisms or strains that code for the same target protein. Such a comparison shows which amino acids vary a lot and which are essentially unaltered among different candidate sequences. Considering that most mutations are deleterious, the immutable amino acids are more likely to have an important role, whereas other amino acids can probably be changed without much consequence. By focusing on the important locations, it is possible to engineer bottleneck enzymes in target pathways and to improve productivity significantly. A beautiful example for this strategy is the alteration of an isoprenoid pathway in *E. coli*, where productivity increased roughly 1000-fold within a few years [37]. Another instance is the optimization of a biosynthetic pathway by multiplex automated genome engineering, which allowed the creation of over four billion combinatorial genomic variants per day and the isolation of variants with over fivefold increased yields within just three days [38].

EMERGING APPLICATIONS

15.1 From Neurons to Brains

Both experimental biology and biomathematics have long been fascinated with nerves and brains [2, 3]. It was over half a century ago that Hodgkin and Huxley proposed their pioneering modeling work on squid neurons, for which they received the Nobel prize (see [4] and Chapter 12). Yet, in spite of a lot of research, we still do not know how our memory works or what determines whether a person will be afflicted with Alzheimer's disease later in life.

By necessity, modeling approaches in neurobiology have mostly addressed either small systems in detail or large systems in a much coarser fashion. Uncounted variations on models for individual action potentials have been proposed, and careful mathematical analyses have been able to explain the basis of their functioning [5–7]. Using action potentials as input, biochemical models have been developed that describe the dynamics of **neurotransmitters** in individual neurons, as well as the transduction of signals across the synapse between two neurons (see, for example, [8–12]). At a higher level of organization, models of neuronal signal transduction have focused on the function and interactions of different brain modules, but by doing so they have had to ignore most details regarding the cellular and intracellular events that are the basis for neuronal signaling [13].

Naturally, complex and detailed models are expected to permit more realistic analyses than minimal models. Yet, very simplified models have their own appeal [14, 15]. The extreme in simplification is a biologically inspired construct, called an **artificial neural network (ANN)**, in which neurons are coded exclusively by their activation state as on or off, 1 or 0, and this state is transduced throughout the network in a discrete fashion (see Chapters 4, 9, and 13). At each step, the 1's and 0's reaching a neuron are multiplied by numerical weights, and the sum of the weighted inputs determines whether this neuron will be on or off for the next step of the discrete process. This drastic simplification of a realistic network of biological neurons has the advantage that large arrays of thousands of artificial neurons can be studied very effectively. Intriguingly, it has been shown that, over time, ANNs can be trained to recognize patterns by changing the weights with which each neuron processes its inputs [16, 17]. As a result, ANNs are capable of representing an unlimited spectrum of responses, and extensive research has led to applications of ANNs to all kinds of classification and optimization tasks that no longer have anything to do with neuroscience. For instance, ANNs can be trained to find and read zip codes on envelopes and distinguish a sloppily written 7 from a 1.

At the other end of the spectrum between simplicity and complexity, the culmination of simulating realistic interaction networks of neurons is presently the Blue Brain Project [18], for which a large group of scientists is collecting detailed micro-anatomical, genetic, and electrical data of one specific human brain section, the neocortical column, and integrating them into three-dimensional simulation models that account for very large populations of over 200 types of neurons (**Figures 15.1** and **15.2**). The sheer size of the project requires enormous supercomputing power, and the results are fascinating. A major issue with this approach is that the simulation results have become so complex that their analysis is a significant challenge in itself.

Figure 15.1 Image of a model reconstruction of the rat neocortical column. (Courtesy of the Blue Brain Project, EPFL.)

The different approaches to understanding the brain highlight distinct aspects of neurons and neural networks and ignore others. The grand challenge for neurosystems biology is now the creation of conceptual and computational multiscale models that integrate electrical, biochemical, genetic, anatomical, developmental, and environmental aspects associated with the brain in a manner that offers deep insights into fundamental functions such as learning, memory, and adaptation, as well as the development and progression of neurodegenerative diseases and addiction [19]. The first step will be to bridge the gap from single-neuron models to detailed, realistically sized neural networks. This step is by no means trivial, since not only are neural networks in the brain very complex in structure, but they can dynamically change in activity, for instance, between wakefulness and sleep. Switches between these states are often rapid, and the rules guiding them are largely

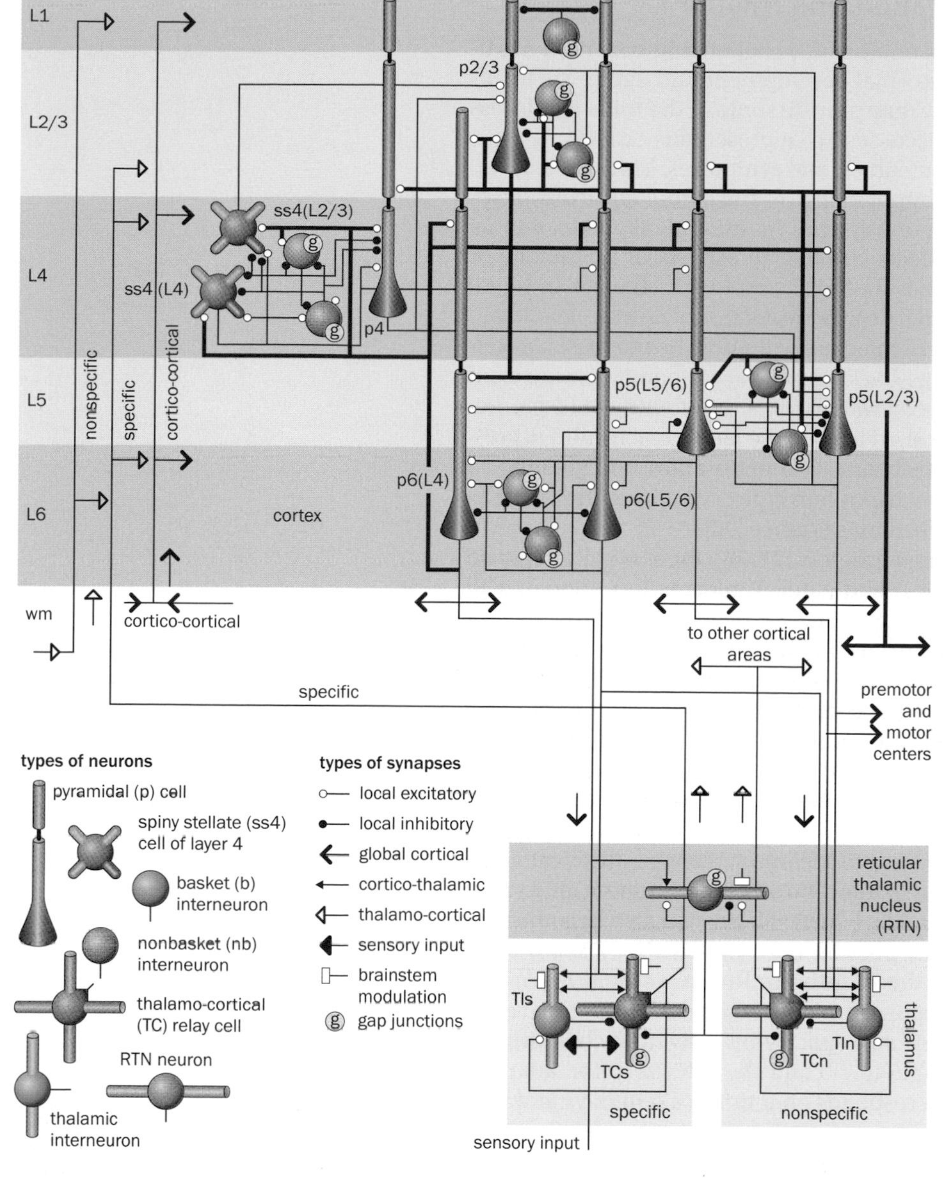

Figure 15.2 Toward a wiring diagram of the brain. Simplified Blue Brain representation of the cortical laminar structure in the human brain. (From de Garis H, Shuo C, Goertzel B & Ruiting L. *Neurocomputing* 74 [2010] 3–29. With permission from Elsevier.)

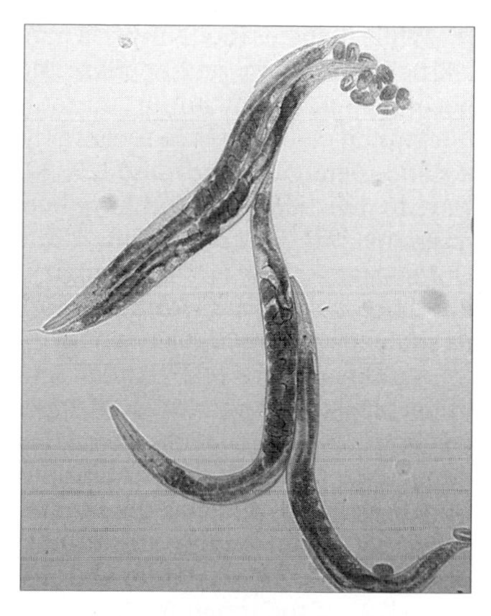

Figure 15.3 Accounting for all neurons. *Caenorhabditis elegans* is a simple worm with exactly 959 cells, 302 of which are neurons. (Courtesy of kdfj under the Creative Commons Attribution-Share Alike 3.0 Unported license.)

unknown [20]. Initial attempts in this direction are models that study bursting patterns in networks of maybe a dozen to 100 neurons [21]. A complementary approach is the creation of computer languages that are custom-tailored to analyses of how anatomical features, synapses, ion channels, and electrochemical processes govern the functionality of the brain [10].

A good starting point for studying realistic neural networks might be a relatively simple model organism such as the fascinating worm *Caenorhabditis elegans*, which consists of exactly 959 somatic cells, of which exactly 302 are neurons (**Figure 15.3**). Researchers in Japan have even created a database of synaptic connectivities in this organism [22]. Another candidate is the California sea slug *Aplysia californica*, which has become a model for neurobiological studies, because it has only about 20,000 neurons, but exhibits interesting responses to stimulation, such as the ejection of red ink (**Figure 15.4**). Eventually, models of these and more complex neural networks will be extended and become reliable enough to permit targeted interrogations and analyses of brains in higher organisms. They will lead to deeper understanding and might help us explain neurodegenerative disease, addiction, epilepsy, and other menacing diseases, as well as normal, physiological phenomena such as intelligence and conscience (see, for example, [23, 24]).

Figure 15.4 The sea slug *Aplysia californica* has become a model organism for studying learning and memory. Here, the slug releases a cloud of ink to confuse an attacker. (Courtesy of Dibberi under the Creative Commons Attribution-Share Alike 3.0 Unported license.)

15.2 Complex Diseases, Inflammation, and Trauma

The brain is certainly fascinating, but it is not the only challenging system in medicine. Many complex diseases, and even normal development and aging, are insufficiently understood, and systems biology may provide some of the tools to address them in a holistic manner. This section discusses two representative cases.

Diabetes has relatively clear causes, outcomes, and symptoms, but is a complex, systemic disease that may be exacerbated by physiological and pathological responses of other organs and reactions by the immune system. An example is reduced renal clearance, which directly or indirectly affects virtually all aspects of physiology or pathology, beginning with a reduced removal of toxins and with changes in blood pressure, volume, and flow, which may affect cardiovascular functioning, leading to easy bruising and bleeding with a chance of infection, vomiting, headaches, and seizures. Mathematical models of diabetes go back a long way [25] and, coming from humble beginnings, they have reached a level where they are beginning to be predictive. Indeed, simple diabetes models have shed light on the kinetics of insulin *in vivo*, demonstrated the importance of pancreatic beta-cell compensation, and identified a particular composite model parameter, the disposition index, as representative of the ability of the pancreas to compensate for insulin resistance [26].

While diabetes modeling has made great progress [27, 28], there is still a lot to be learned, and as we gather more biological and clinical information, models will become increasingly important tools of integration. The beginnings of this type of integration can be seen in recent physiologically based simulation models of diabetes and metabolic syndrome [29, 30] that capture the function of regulatory pathways in these diseases and have been used to connect diabetes to the risk of heart problems [31]. Given that type 2 diabetes afflicts roughly 20 million people in the United States alone and is a primary cause of heart and kidney disease, as well as of blindness, the rewards from developing a deeper understanding of the disease and its sequelae are beyond any doubt.

Another disease-related and very intriguing modeling challenge is just now in the initial phase of being addressed. It pertains to infection and inflammation, and to the ensuing immune responses, which are usually beneficial, but can also become very dangerous. Mammals, including humans, possess two systems that protect them from infecting pathogens: the **innate** and the **adaptive immune systems**. According to current understanding, the former recognizes unique molecular features of pathogens with specific receptors and responds very quickly by activating countermeasures [32]. If the innate response is unable to contain the infection, the adaptive immune system is called up. This system responds on a timescale of days and ultimately equips the organism with a memory of earlier pathogen infections and immune responses. Although the adaptive system operates at a much slower speed, it interacts directly with the innate system: T cells of the adaptive response and macrophages of the innate immune system are part of a positive-feedforward loop, in which they activate each other through signaling proteins called **cytokines**. The innate immune cells are needed to trigger responses in the adaptive system, while the activated adaptive immune cells effect the amplification of anti-pathogenic responses in the innate cells [33]. Positive-feedback loops are potentially dangerous, because they run the risk of spinning out of control. In the body, this risk is internally controlled by regulatory T cells that normally suppress the innate response in order to preclude excessive responses.

Thus, the innate and adaptive immune systems constitute an integrated defense system, with each component depending on the other for amplification and regulation. This fine-tuned system works very well, as long as all components are operational. However, if one system component is compromised, the system can derail. For instance, if the $CD4^+$ T-cell count is very low, the regulation of the innate immune system is lacking, and it is possible that inflammatory cytokines will dramatically accumulate into a cytokine storm that can be strong enough to kill the host through tissue damage and septic shock (**Figure 15.5**). As an example, a cytokine storm in the lung can become severe enough to block the airways through the accumulation of fluids and immune cells. Under normal conditions, cytokines move through the circulatory system in picomolar concentrations. However, these levels can increase 1000-fold during infection or trauma, especially in otherwise-healthy young

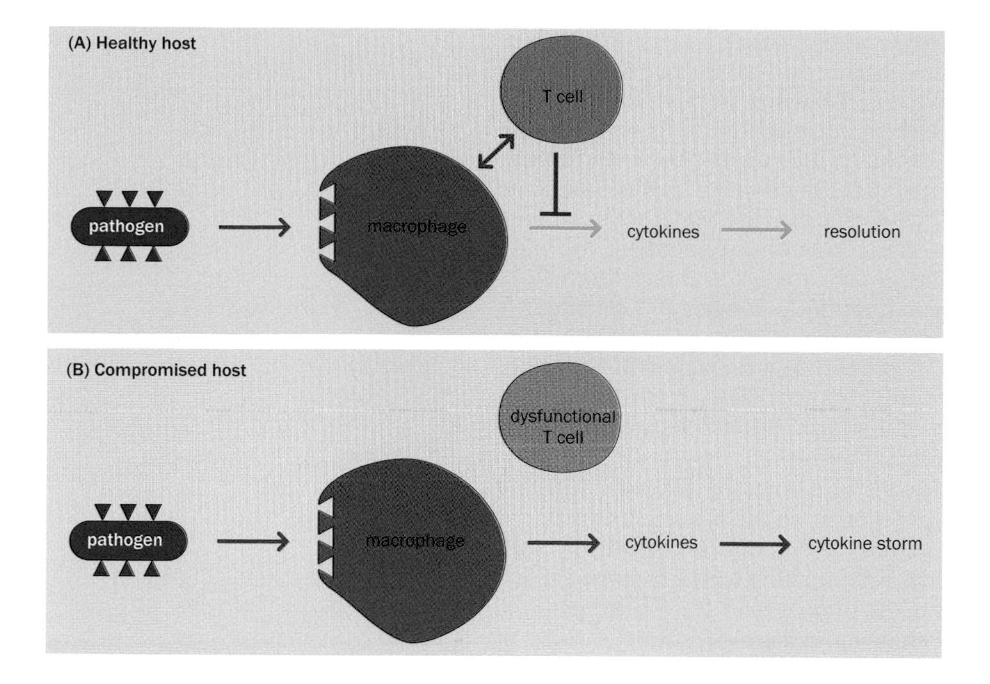

Figure 15.5 A brewing storm? (A) In healthy hosts, T cells of the adaptive immune system interact with macrophages of the innate immune system and probably suppress the production of cytokines. (B) In compromised hosts with insufficient functional T cells, the production of cytokines is uncontrolled and can result in a lethal cytokine storm. (Adapted from Palm NW & Medzhitov R. *Nat. Med.* 13 [2007] 1142–1144. With permission from Macmillan Publishers Limited, part of Springer Nature.)

individuals. There is speculation that the high number of deaths among young adults during the 1918 influenza pandemic and during the 2005 bird flu outbreak might have been due to cytokine storms.

Of course, as so often in biology, the diagrams of inflammation in Figure 15.5 are woefully over-simplified. In reality, the body produces numerous families of cytokines, including interleukins, chemokines, and interferons. These families are diverse and form complex interaction networks, and we are even beginning to recognize that the early distinction between "pro-inflammatory" and "anti-inflammatory" cytokines no longer makes real sense in many instances, because a pro-inflammatory cytokine may trigger the production of an anti-inflammatory cytokine, so that the pro-inflammatory cytokine indirectly becomes anti-inflammatory. Cytokine storms have been estimated to involve more than 150 different cytokines, coagulation factors, and radical oxygen species, and the ultimate response to an alteration of one cytokine species within such a complex network depends on much more than the type of cytokine; in particular, it seems that the preconditioning history of the network is of extreme importance [34]. Furthermore, cells of the innate immune system initiate both innate and adaptive immune responses [33] through the release of cytokines, which makes intuitive predictions of responses very difficult. To make matters even more complicated, there is plenty of crosstalk among different cell types and their signaling activities. These large numbers of different players and interactions clearly suggest that methods of systems biology will be needed to elucidate immune responses and cytokine storms [34].

Infection is often the first trigger for calling up the immune system, which remedies the situation. Unfortunately, the response of the immune system itself can cause damage and inflammation in associated tissues. In fact, it is possible that most of the infecting pathogens are killed but a vicious cycle continues, in which tissue damage causes more inflammation, which in turn causes more damage. Again, this positive feedback is driven by cytokines and immune cells [35].

Intriguingly, the point has been made that inflammation is a unifying factor tying together diseases that at first appear to be very different, such as atherosclerosis, rheumatoid arthritis, asthma, cancer, AIDS, metabolic syndrome, Alzheimer's disease, chronically non-healing wounds, and sepsis [34, 35]. At the same time, inflammation is associated with natural aging processes as well as the beneficial effects of exercise and rehabilitation. When seen in such broad contexts, inflammation per se is not a detrimental process, but the reflection of a complex communication network, with which the organism signals potentially harmful perturbations. Under healthy conditions, the system works remarkably well, but if it becomes too unbalanced for an extended period of time, disease is the consequence.

While the inflammatory processes in their entirety may appear to be overwhelmingly complex, they constitute a homeostatic control system whose basic structure is ubiquitous and actually quite simple. Focusing on this simplicity permits mathematical analyses that can yield interesting insights [34]. Specifically, the inflammatory response may be characterized as a balance of processes (**Figure 15.6**). The simplified representation accommodates several entry points into the acute inflammatory response and furthermore includes a damage signal that can feed back in a detrimental manner. Under normal conditions, a pathogen evokes a strong inflammatory response and simultaneously a weaker anti-inflammatory response. The inflammation triggers an attack on the pathogen and secondarily activates the anti-inflammatory machinery. At the same time, pro-inflammatory agents, such as tissue necrosis factor (TNF), can cause tissue damage or dysfunction. Anti-inflammatory agents, such as transforming growth factor β (TGF-β1), suppress inflammation and stimulate healing of damaged tissues. Tissue damage, if not controlled, can lead to further inflammation.

Given a simplified base structure as in Figure 15.6, it is not too difficult to imagine that the outcome of a pathogen attack is determined by the balance between the different processes in the system. For instance, if the stimulation of the anti-inflammatory machinery is sufficiently strong, inflammation will be suppressed and tissues have a chance to heal: in Figure 15.6, the processes in green overcome the red processes. By contrast, if the processes shown in red are too powerful, one can see that not only is tissue damaged, but a vicious cycle continues between inflammation and further tissue damage, even after the pathogen has been defeated, as we discussed earlier. The diagram also shows that inflammation can be triggered by tissue damage, which may be due to trauma or disease.

Recent years have seen enormous progress in this field, and we are beginning to understand enough details of the inflammatory network to think about initiating formal systems biological investigations. These investigations should ideally include preclinical studies, clinical trials, in-hospital care, and the long-term use of treatment [34]. The challenges faced when attempting to model inflammation are manifold. Not only are the normal inflammatory processes and their regulation complicated, but components of the inflammatory response system also interact with non-inflammatory physiological systems, which makes it difficult to deduce the architecture and control of the system from biological and clinical observations. Furthermore, individual inflammatory responses depend critically on the preconditioning state of the organism, that is, on the current health status as well as the health history of the affected organism [34].

Distinct approaches have been employed to model inflammation (for reviews, see [34, 36]). They have included mechanistically based ordinary and partial differential equation models and simulations, and different stochastic and hybrid formulations, as well as rule- and agent-based models, which will be discussed in a later section of this chapter. A few examples of inflammation analyses are presented in the next few paragraphs.

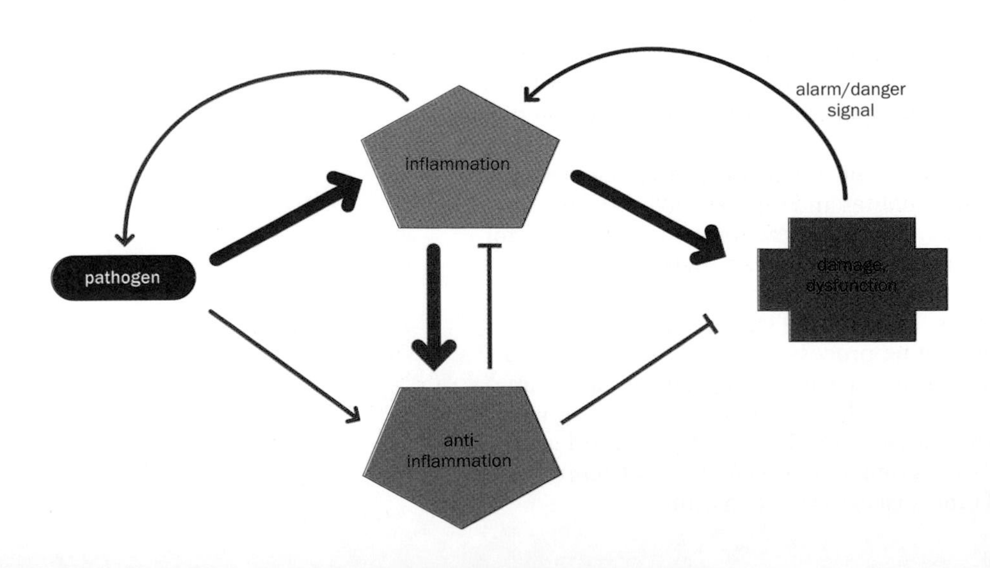

Figure 15.6 Inflammation as an imbalance. This very simplified diagram demonstrates the fine balance between counteracting processes leading either to inflammation or to recovery from an infection (red and green, respectively). (Adapted from Vodovotz Y, Constantine G, Rubin J, et al. *Math. Biosci.* 217 [2009] 1–10. With permission from Elsevier.)

A mechanistic mathematical model was constructed to elucidate processes of bacterial infection, cytokine dynamics, the acute inflammatory response, global tissue dysfunction, and a therapeutic intervention [37]. The model was subsequently used to simulate a clinical trial designed to demonstrate the benefits of a therapy with different doses of a neutralizing antibody directed against TNF in patients with sepsis. The simulation predicted cases with favorable and unfavorable outcomes and also revealed why an actual clinical trial had failed and how the trial should have been designed to succeed.

An equation-based model was developed to investigate whether different shock states following an infection could be explained by a single inflammatory response system [38]. Supported by different sets of experimental mouse data, the model demonstrated different trajectories of the inflammatory process to either resolution or shock, which depended on the individual cellular and molecular state of each mouse. This modeling effort demonstrated three aspects. First, it proved that the complex biology of inflammation can indeed be captured by mechanistic models. Second, it made clear that the shock response to different physiological challenges depends critically on the preconditioning state of the organism. And finally, the model presented trends in the progression of the inflammation process that were personalized and predictive with respect to survival.

Another example of the usefulness of differential equations in the analysis of inflammation is a model for the receptor-mediated inflammatory response in humans exposed to toxins produced by bacteria inside the body [14, 39]. This model is interesting because it integrates genomic information and extracellular signaling processes into the bacterial infection process. By changing its parameter settings, the single model allows for the characterization of different outcomes, including healthy resolution responses, persistent infection responses, aseptic inflammation, and desensitization.

An interesting example of agent-based inflammation modeling is a hierarchical concatenation of interacting submodels that collectively integrate basic scientific and clinical information on acute inflammation into a modular, multiscale architecture [40]. The individual submodels address pertinent cellular response mechanisms to infection and inflammation, the permeability of gut epithelia, which are the ultimate barrier against microbial pathogens, the inflammatory response of the gut to ischemia, and responses in the lung. The model successfully matched clinical observations on the pathogenesis of multiple organ failure and the underlying crosstalk between gut and lung.

Cancer has been a target of modeling for a long time, and some of the early approaches still have relevance. The initial concepts and models considered the process of **carcinogenesis** (cancer formation) as a multistage or multihit process (**Figure 15.7**). This process is thought to begin with normal cells which, as a result of some initiation event, become premalignant. The premalignant cells can remain in this stage of promotion for a long time. A transformation event converts one of the premalignant cells into a malignant cell, which replicates during a progression phase into a tumor. Radiation, exposure to toxins, and a number of other processes can function as possible initiation and transformation events [41]. In this view of carcinogenesis, a single cell moving through all phases is seen as sufficient to lead to a tumor. Early mathematical models assessed the probabilities or rates of transitions within this multistage process, and also allowed for cell replication and death in the various phases [42, 43].

A more modern view of carcinogenesis still contains the same phases, but recognizes that tumor formation is much more complicated. First, it has become clear that cancer cell populations are heterogeneous and contain cells with different degrees of carcinogenic potential [44]. Thus, killing most of the cancer is not necessarily the best

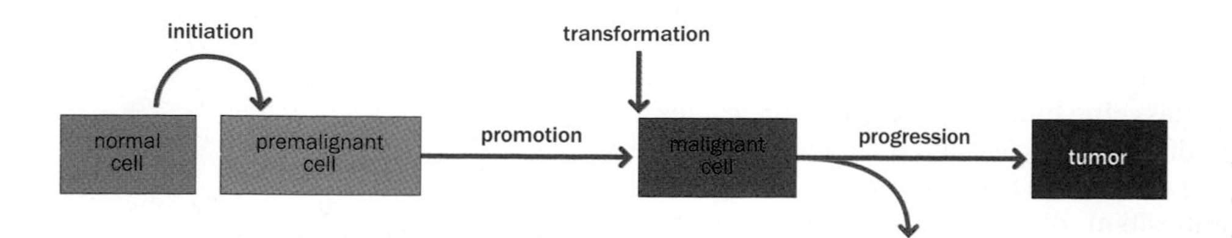

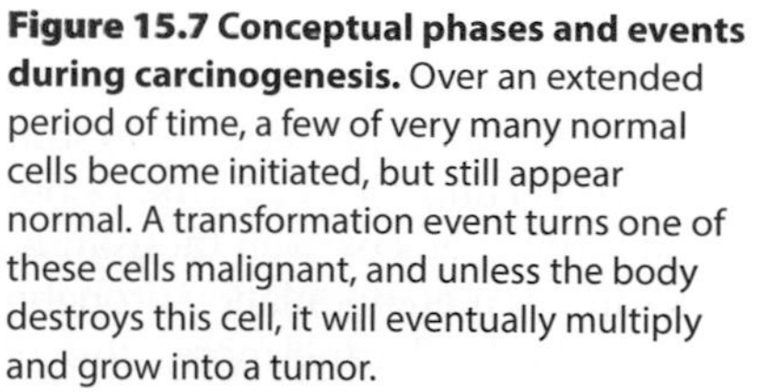

Figure 15.7 Conceptual phases and events during carcinogenesis. Over an extended period of time, a few of very many normal cells become initiated, but still appear normal. A transformation event turns one of these cells malignant, and unless the body destroys this cell, it will eventually multiply and grow into a tumor.

strategy if the most virulent cells survive the treatment and subsequently form an even more aggressive tumor. Instead, a better strategy may call for controlling the tumor rather than attempting to remove it [45]. Second, several interesting studies (see, for example, [46]) have demonstrated that essentially all middle-aged and older adults harbor microscopic tumors, most of which stay dormant for a long time, if not for the person's entire life. As one impressive example, less than 1% of the adult population is ever diagnosed with thyroid cancer, but 99.9% of all individuals autopsied for other reasons were found to have thyroid cancer lesions. Third, recent studies show that small tumors can have distinctly different fates (the orange arrow in Figure 15.7). Some grow into full-size tumors and maybe even metastasize, but others will grow and shrink in an oscillatory pattern, some will regress on their own, and some may stay dormant at a very small size that per se is not harmful [44, 47]. What determines these diverging trajectories is not entirely understood, but it seems that a change in the balance between stimulators and inhibitors may flip a switch during the progression phase, upon which the tumor starts to grow and to recruit blood vessels. This recruitment process, called angiogenesis, apparently constitutes at least one significant bottleneck in tumor formation [48].

Future models of carcinogenesis will probably retain the backbone of the **multistage cancer model**, but will incorporate it into systems models that take account of what we learn at the genomic level about small-RNA regulation (see Chapter 6), of immune responses to the emergence of cancer cells, of the recruitment of progenitor stem cells, of the recruitment of blood vessels, and of the interactions between cancer cells and healthy tissues. It is clear that these models must either be of an integrating multiscale nature or address some aspect that can subsequently be embedded in a larger systems model. The models will be useful for forecasting the dynamics of carcinogenesis and the efficacy of different treatment options. They may also reveal overarching principles that differentiate tumor development from normal physiology.

In more generic terms of modeling human development, health, and disease, some research teams have begun to consider disease, aging, and frailty as the consequence of a loss of complexity, arguing that some highly periodic processes become diseased if they are too regular and do not exhibit the characteristics of fractals or even chaos [49, 50]. An example is a very regular fetal heartbeat, which very often is a cause for concern [51–53].

15.3 Organisms and their Interactions with the Environment

It is complicated enough to design models for systems within individual cells and organisms, but the degree of complexity jumps again when it is our task to investigate organisms in their environments. Any comprehensive study of this type immediately requires multiple scales in time, space, and organization; deterministic processes mingle with stochastic effects; and data and other pieces of information are so diverse and heterogeneous that it often seems overwhelmingly difficult to merge and integrate them in our minds. It seems again that system approaches will eventually be our best bet. This section discusses two examples in which systems thinking and computational modeling show very high potential.

Metapopulations consist of several populations, each of a different species, that live in the same environment and interact with each other. Although we tend to think of populations in large expanses of the environment, metapopulations are similarly intriguing in very small ecological niches, such as localized parcels of soil, the human oral cavity, or even the gut of a termite. Whatever the context, microbial metapopulations come in enormous sizes. It has been estimated that 5×10^{30} prokaryotic cells inhabit the Earth [54]. A single gram of soil can contain enough microbial DNA to stretch almost 1000 miles [55]. A lake may contain 20,000 different species of bacteria [56]. Humans harbor 10 times as many bacteria as their own cells. Very many of these live in the gut, where their density can reach 10^{12} cells per milliliter. This number is enormous, but the bacteria are so small in size that the gut **microbiome** makes up only between 1% and 2% of a human's body mass. The human gut microbiome can easily contain hundreds or even thousands of different species, and the totality of different genes in this metapopulation probably exceeds our own genome size by several orders of magnitude. Furthermore, the composition of the gut microbiome changes dynamically and drastically in response to age, diet, diseases, and of course the ingestion of antibiotics [57]. Because the gut microbiome is directly or indirectly associated with a number of diseases, the US National Institutes of Health in 2007 launched an NIH Roadmap effort to use genomic technologies to explore the role of microbes in human health and disease [58]. Similar projects have been launched elsewhere [59, 60].

Similarly important, ubiquitous metapopulations can be found in **biofilms**, which cover moist surfaces from teeth to the insides of sewage pipes (**Figure 15.8**). Like the gut microbiome, biofilms consists of hundreds or thousands of different species, and waste products from one species are used as nutrients by others. Intriguingly, the different species in metapopulations, and the individual organisms within these species, compete for the same physical space and the same resources, yet they depend on each other and often demonstrate a strict division of labor. Thus, in contrast to the typical competition in shared environments, different microbes assume distinctly different and complementary roles, to a point where a species cannot survive without the others [57].

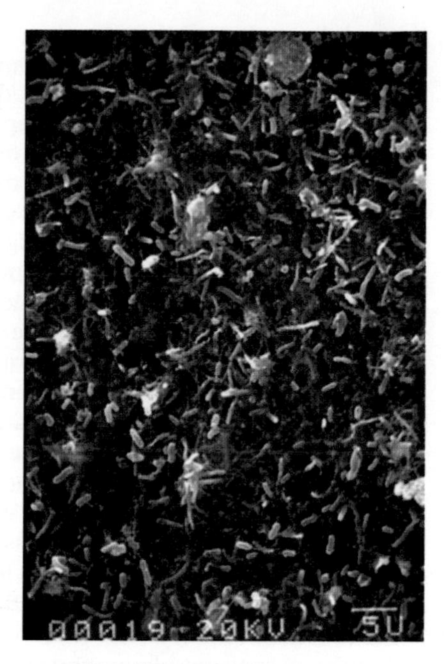

Figure 15.8 Scanning electron microscope image of a microbial biofilm. This metapopulation developed on the inside of untreated water tubing within some dental equipment. Such water lines often contain bacteria, as well as protozoa and fungi. (From Walker JT & Marsh PD. *J. Dentistry* 35 [2007] 721–730. With permission from Elsevier.)

Understanding microbial metapopulations faces significant challenges, the most restrictive of which might be the fact that the vast majority of species in these populations are unknown. They cannot even be cultured in the laboratory and are therefore difficult to characterize. Even if a few representative species can be grown in the laboratory, their artificial environment is so deprived in comparison with their natural surroundings that the extrinsic validity of subsequent results often seems doubtful.

An interesting strategy for overcoming this problem was the introduction of **metagenomics** [61]. Here, samples of entire microbial ecosystems are taken directly from their natural environment and, without any attempt to culture them, analyzed as if they came from a single species. While this procedure is now technically possible, the challenge shifts to extracting reliable information from the mixture of millions of new genomic sequences that in some unknown fashion belong to thousands of species. Needless to say, metagenomics stretches current methods of bioinformatics to the limit. As a consequence, a full assembly of genomes is presently only possible for mixtures of a very few species [51].

In addition to metagenomic studies, techniques are now becoming available for community proteomics, which has the potential of yielding unprecedented insights into the composition and function of microbial metapopulations. These techniques have already been applied to a variety of systems, including sludge, biofilm, soil, and gut microbiomes [62], but, as for many other proteomics techniques, the experimental and computational analysis is challenging.

Various application areas have witnessed a time trend from molecular biology to bioinformatics and subsequently to systems biology. The same trend holds for microbial metapopulations [63, 64], where it is clear that systemic analyses will be very challenging, fascinating, and absolutely necessary.

At the other end of the size spectrum, the oceans are without doubt among the most challenging multiscale systems that await more systematic analyses than have been possible in the past. They provide most of the oxygen we breathe and provide the ultimate storage for water and a lot of food. As with many subspecialties of biology, many aspects of the dynamics of and in oceans have been addressed with reductionist methods and models of specific details, yet it has so far not been possible to integrate the huge range of spatial, temporal, and organizational scales into reliable computational models. Nonetheless, we can see the rudimentary beginnings of systemic analyses in many places.

A first step beyond observational and wet sciences is the rise of ecoinformatics, which uses methods of bioinformatics to manage, analyze, and interpret the enormous quantities of data that have relevance for ecological systems and, in particular, the oceans [65]. The challenges are great, not merely because of the sheer volume of information, but also owing to the highly heterogeneous nature of this information as well as a reluctance to adopt worldwide data standards and experimental and analytical protocols. Such challenges are currently being addressed with the creation of data warehouses, the collection and structuring of metadata, laboratory information management systems (LIMS), data pipelines, customized mark-up languages, specialized ontologies, and semantic web tools, including computer-based reasoning systems. However, efficient storage and access to data alone will not be sufficient.

Beyond direct data analyses and automated inferences, computational means of information integration, simulation, and interpretation are needed. In other words, ecoinformatics must be complemented with dynamical systems ecology.

Recognizing this need, the US Department of Energy sponsors a program, Systems Biology for Energy and Environment, that identifies the Earth's living systems as a potential source of capabilities that can be put to use to meet national challenges and has the ultimate scientific goal of understanding the principles underlying the structural and functional design of living systems [66]. The program focuses on energy, environmental remediation, and carbon cycling and sequestration. Within the latter focus area, the role of the oceans clearly plays a crucial role, because marine cyanobacteria (blue–green algae) are responsible for half of the global carbon dioxide fixation. There is little doubt that the well-being of these organisms depends directly on nutrients, pH, temperature, salinity, contamination, and the surrounding multispecies communities, but the exact parameters of optimal, suboptimal, and barely tolerable growth conditions are as yet unknown.

An example of how molecular and systems biology have begun to reach out to investigations of global ocean processes is a study on marine cyanobacteria, which are among the most abundant organisms on Earth (**Figure 15.9**). These small creatures are able to convert solar energy into biomass and thereby account for 20–30% of the Earth's photosynthetic productivity. Furthermore, they recycle about 40% of the entire carbon that cycles between the oceans, the atmosphere, the Earth's crust, soils, microorganisms, plants, and animals. Researchers at Harvard's Department of Systems Biology found that cyanobacteria of the genus *Synechococcus* fix carbon and convert it into sugar within spatially well-placed internal structures, called carboxysomes [67]. A single protein is responsible for the optimal spatial organization of these carboxysomes, and mutations in the corresponding gene lead to a photosynthetic productivity that is decreased to 50% of that of the wild-type production. Thus, a systemic, causal connection has been elucidated between the action of a specific gene, the protein it codes for, the functionality of a microorganism, and a truly global phenomenon. This same causal connection could become the starting point for the design of genetically modified algae or bacteria for the production of hydrogen, using methods of synthetic biology.

A different starting point for systemic studies of oceans is the careful investigation of coexisting marine populations using methods described in Chapter 10 (**Figure 15.10**). More and more detailed data for such studies are being collected worldwide for different marine systems. For instance, Burkepile and Hay [68] measured the effects of single-species and mixed-species herbivorous fish populations on the colonization and progression of algal communities and their impact on coral growth. In carefully controlled experiments with otherwise comparable systems, they demonstrated convincingly that species diversity is critical for the health of coral systems and the resilience of the oceans following disturbances. Dam and colleagues [56] used Lotka–Volterra models to study dynamic interaction patterns among lake metapopulations consisting of almost 20,000 species.

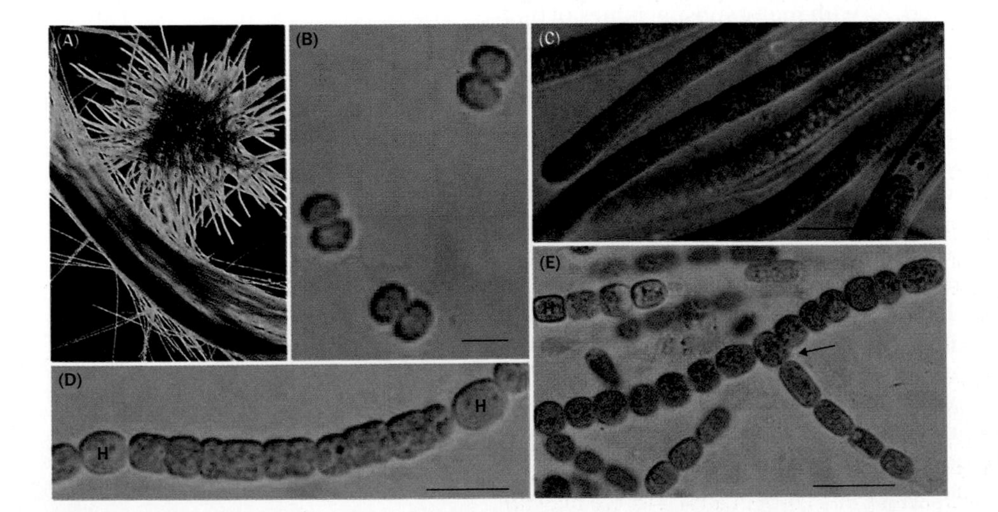

Figure 15.9 Different strains of cyanobacteria (blue–green algae) may have very diverse shapes. (A) Bloom-forming, filamentous, and colony-forming *Trichodesmium thiebautii* (scale bar 100 μm). (B) Unicellular *Synechococcus PCC 6301* (scale bar 1 μm). (C) Filamentous *Symploca PCC 8002* (scale bar 10 μm). (D) Filamentous, nonbranching, and heterocystous (H) *Nostoc PCC 7107* (scale bar 10 μm). (E) Filamentous, heterocystous (H) and branching (arrow) *Fischerella PCC 7521* (scale bar 15 μm). (From Cox PA, Banack SA, Murch SJ, et al. *Proc. Natl Acad. Sci. USA* 102 [2005] 5074–8. With permission from National Academy of Sciences, USA.)

Figure 15.10 Corals and other invertebrates compete for space. Data characterizing interacting populations are currently being collected and create a starting point for systems-biological model analyses. (From Burkepile DE & Hay ME. Coral reefs. In Encyclopedia of Ecology [SE Jorgensen & B Fath, eds], pp 784–796. Elsevier, 2008. With permission from Elsevier.)

MODELING NEEDS

To predict future modeling needs, it might be prudent to revisit the present state of the art, consider how we got here, and then try to extrapolate trends into the coming years. There is no doubt that computational modeling has come a long way. Thirty or forty years ago, computing was done on mainframe computers that required manual feeding of punch cards with instruction code. There was no Internet, no parallel computing, no user-friendly software. Much has happened since those dark days. Computational power and speed have doubled every 18 months since the 1950s, access to computing has increased millions of times, and today's higher-level languages such as MATLAB®, Mathematica®, and **SBML** have facilitated the development and implementation of computational ideas among users most of whom would certainly have been intimidated by the need to master computer languages like Fortran, let alone assembly language, just a few decades ago.

While progress has been amazingly fast, it is clear that we still have a long way to go. But the bottleneck is not just computational speed and convenience. In fact, outside certain areas, such as the prediction of protein structure and function and very large-scale simulation and optimization studies, it is not necessarily the technical aspects of computing that stand in the way of the next significant advance. New concepts and radically new modeling techniques might be needed. Rainwater runs down a pile of dirt in a very efficient fashion, yet capturing the process in a realistic mathematical model requires enormous effort. Why is that? Spiders with rather small brains are able to construct complex webs, and animals respond to situations never encountered before with appropriate and effective actions, but we are hardly capable of representing such activities adequately in systems of differential equations. To realize model-aided personalized medicine, it is not sufficient to characterize the direct effects of a disease or treatment, but also their secondary and tertiary effects, along with interactions between different treatments and the multitudinous personal characteristics of an individual.

These challenges are indeed great, and it might well be possible that mathematical modeling and systems biology are in need of a significant paradigm shift, comparable to the one that physics experienced in the early 1900s when quantum theory opened the door to concepts unimaginable just a few decades before. But before we throw our hands in the air and resign ourselves to waiting for a "new math" to appear, before we discard modeling as simplistic and powerless and instead suggest doing another set of laboratory experiments, let us just see what the alternatives are. It is undisputed that realistic systems consist of hundreds or thousands of important components and that these make modeling very complicated. But the same hundreds or thousands of important components make every other approach even more complicated. Yes, we may not be able to keep track of hundreds of dynamically changing systems components in a computer-aided model analysis right now, but we will *never* be able to manage them in our unaided minds. Statistical association models may be sufficient for identifying averages, but they will not suffice to address specific cases, as required for instance in personalized medicine. Beyond these considerations of magnitude, many experiments simply cannot be executed, because they are technically impossible, practically infeasible, or unethical. We cannot test a new drug on every human, and it is probably not a good idea to introduce a new strain of algae in the oceans on a trial basis.

Thus, while there is a clear mismatch between the grand challenges to be solved and the tools we have presently at our disposal, the magnitude of the tasks before us should not be a reason to abandon systems biology but rather one to increase our efforts—because we have no choice. Besides, it will be fascinating and fun to witness the field tackle these issues. Step by step, sometimes evolutionary and sometimes revolutionary, sometimes elegantly and sometimes with brute force, we will expand and strengthen our arsenal of modeling methods, address small problems, then larger problems, first connect two biological levels and later on more. As a warm-up, we will study hundreds of isolated processes, then systems, and ultimately realistic and relevant interacting systems of systems. The journey is open-ended, but it invites us to take the next steps. And while we move along, new methods will allow us to increase speed and efficiency, and we will learn which types of techniques tend to work and which not. Ultimately, some smart student, uninhibited by tradition, may indeed design an entirely new mathematical approach. It is moot to speculate what this might look like, but it is possible to assemble a partial wish list of topics that need to be addressed. Some of these appear to be high-priority candidates.

Interestingly, our current models are often either too simple or not simple enough. If our goal is to explore and manipulate a specific system, drastic simplifications may hurt us by making predictions unreliable. For instance, if the task is to increase product yield in a population of microorganisms, accounting for as many details as possible is more likely to lead to success than simplifications of the model. Similarly, it seems that the consideration of a larger number of biomarkers will be advantageous in personalized medicine. At the same time, as models become larger and more complex, they become more difficult to analyze, understand, and predict, while simpler models often offer insights that would be overwhelmed and overshadowed by too much detail. Thus, starting from the level of complexity of today's models, some models will have to increase and some decrease in scope and complexity, depending on the purposes of the modeling effort. And so the circle closes: the introduction to modeling in Chapter 2 made it clear that the goals and purposes of a model dictate its structure and features, and this statement still holds. The following sections will discuss first extensions into greater complexity and then simplifications that might lead us one day to one or more theories of biology.

15.4 Multiscale Modeling

The hallmark of systems biology is a global, integrated approach to biological challenges. Yet, if we look at our current modeling activities, we must admit that they often seem like reductionism in a different form. Many models focus on one set of components, whether these be genes, neurons, or different species within the same environment, and on one or maybe two organizational levels within the multiscale hierarchy of biology. We are generating models of gene expression and of physiology, but it is rare to find models spanning the two levels, as well as all significant aspects in between. For instance, imagine the task of predicting the effect on climate of a new strain of marine algae with increased photosynthetic activity. One could presumably estimate the total volume of additional carbon dioxide fixation, but how would this change influence other components of the oceanic ecosystem, cloud formation, and light absorption, and feed back on the new algae and their competitors? We are far from making such predictions with any degree of reliability.

In addition to the challenge of designing comprehensive multiscale models, we need effective methods of analyzing them and of exploring their conceptual boundaries. If we were to truly list each and every explicit and implicit assumption made in a complex model, we would see how limited the scope of each model really is. Pointing out these limitations is not a critique, but a means of taking an inventory and of showing how much further we have to go.

In the context of multiscale analyses, we will need effective means of integrating coarse- and fine-grained information. Models of phenomena such as complex diseases will have to be modular and expandable. No models in the near future will capture realistic systems in their entirety, which means that we must develop techniques that functionally and validly connect the well-defined core of a modeled system with an ill-defined outside. We need to characterize local effects of a model on global responses in its environment and capture how these global effects feed back to the modeled system. We must also find ways to connect different timescales in an effective manner. We must invent efficient techniques for merging deterministic and stochastic features that change dynamically and in spatially heterogeneous environments.

15.5 A Data-Modeling Pipeline

In the past, there was a clear distinction between what biologists did and what modelers did. While division of labor has been one of the greatest inventions, there is a real risk of losing information and insight when systems biology is strictly separated into experiments and model analyses. As a case in point, the biological literature contains thousands of diagrams showing how one component activates or inhibits another component or process in a system. As soon as a modeler picks up the diagram and tries to convert it into a model, the biologist's intuition and much undocumented tangential information are lost, which is a real pity. An initial step toward circumventing this problem would be a more systematic warehousing of data together with contextual information, as we discussed in the context of ecoinformatics. However, it is difficult to see how data management alone can solve the problem, because intuition is hard to code. Thus, a data-modeling pipeline is needed that loses as little information as possible. Two solutions come to mind. First, the topic could be discussed in a team of biologists and modelers, which, however, would require that almost every biologist have access to a modeler. Second, the modeling community could create tools that would allow the biologist to enter the information embedded in the diagram, as well as other tangential and intuitive information, into a pipeline that leads toward a computational model. Two rudimentary approaches in this direction are **agent-based modeling (ABM)** and **concept-map modeling**.

ABM approaches started to become popular not long ago [40, 69]. They originated in the business world, but are also suitable for the simulation and analysis of any autonomous interacting agents, which could be individuals, machines, parts of cells, molecules, or a variety of other entities. ABM is a very general stochastic modeling technique, where the behavior of discrete, identifiable agents is governed by features and rules. The features may be almost any quantifiable characteristics, and the rules may be simple reactive decisions or the results of complex adaptive artificial intelligence [70]. The agents may be of different types, operate on different length- and timescales, and follow different sets of rules. All agents operate in a common virtual environment, where they interact with each other according to rules and protocols, and the state of each agent is updated in discrete time steps. In biology, typical agent-based models try to emulate population behaviors in two- or three-dimensional space, such as schooling, flocking, or swarming in fish, birds, and ants, where individuals are autonomous, yet influenced by their neighbors to such a degree that the collective result has been termed swarm intelligence [71]. The simplest models of agent-based behavior are probably the cellular automata that we discussed in Chapter 9. Models of this type, as well as corresponding partial differential equation models, have been used for many decades to describe the formation of patterns, for instance, in developmental biology (see, for example, [72–74]). Free software supporting ABM includes NetLogo [75].

We have already mentioned that ABM has found interesting applications in the area of inflammation. Another typical application of ABM is tumor growth. An example is a model in which cells are characterized by their gene expression and phenotype, which includes the presence of growth factor receptors [76]. The agents (cells) are placed on a two- or three-dimensional grid and allowed to proliferate, migrate according to chemotactic signals, turn quiescent, or undergo apoptosis. Simulations with the model lead to experimentally testable hypotheses. For comprehensive references on agent-based and cellular automata models in immunology and cancer research, see [40, 77]. A specific example is illustrated in **Figure 15.11**.

While agent-based models are very appealing, especially for complex phenomena occurring in spatially heterogeneous environments, they have significant drawbacks compared with process-based models in the form of differential or difference equations. First, the computational demands are often high, because many rules have to be called up very frequently [76, 77]. A possible solution of this problem is parallel implementation of the system code. Second, the estimation of parameter values is often complicated [76]. Finally, it is very difficult to analyze ABMs mathematically. While it is feasible to execute very large numbers of simulations, general questions about the structure of the simulated system are hard to answer. For instance, outside comprehensive trial and error, it is difficult to determine whether a system is able to exhibit stable limit-cycle oscillations.

In concept-map modeling, the components and processes within a diagram are formulated as symbolic equations [78]. To some degree, this process can be automated if one uses **canonical models** in default representations (see Chapter 4). In

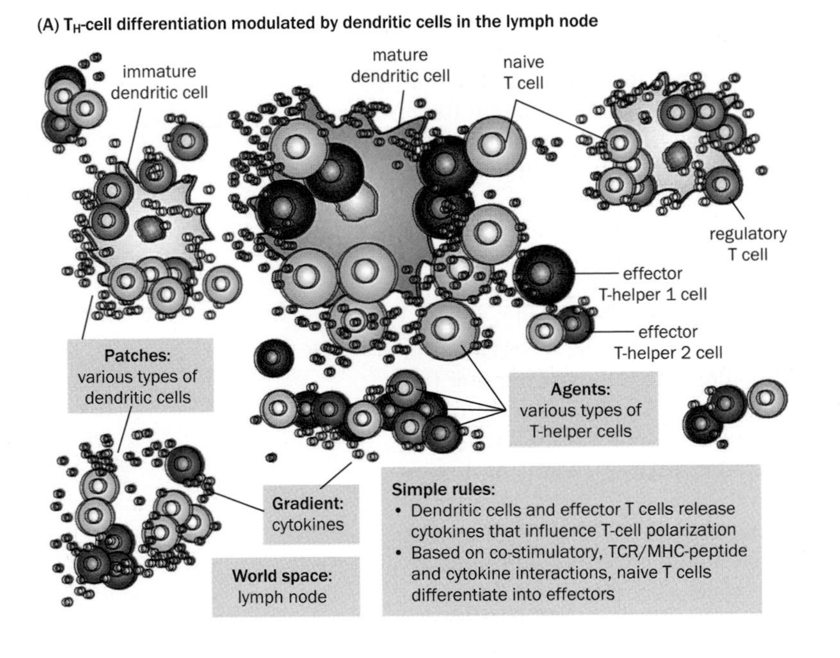

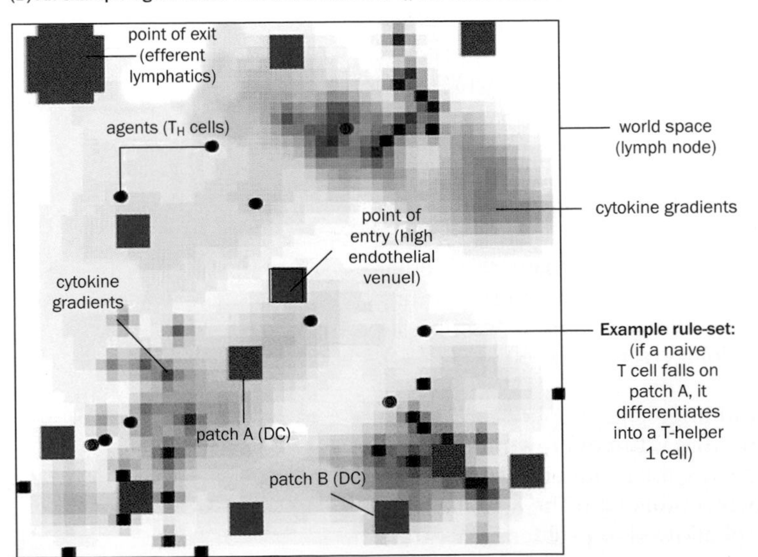

Figure 15.11 Definition of an agent-based model (ABM) for the differentiation of T-helper cells in a lymph node. Differentiation is the result of interactions between naive T cells and dendritic cells. The process also involves cytokines. (A) The process in ABM terminology. (B) Result of an ABM simulation, using NetLogo [63]. (From Chavali AK, Gianchandani EP, Tung KS, et al. *Trends Immunol.* 29 [2008] 589–599. With permission from Elsevier.)

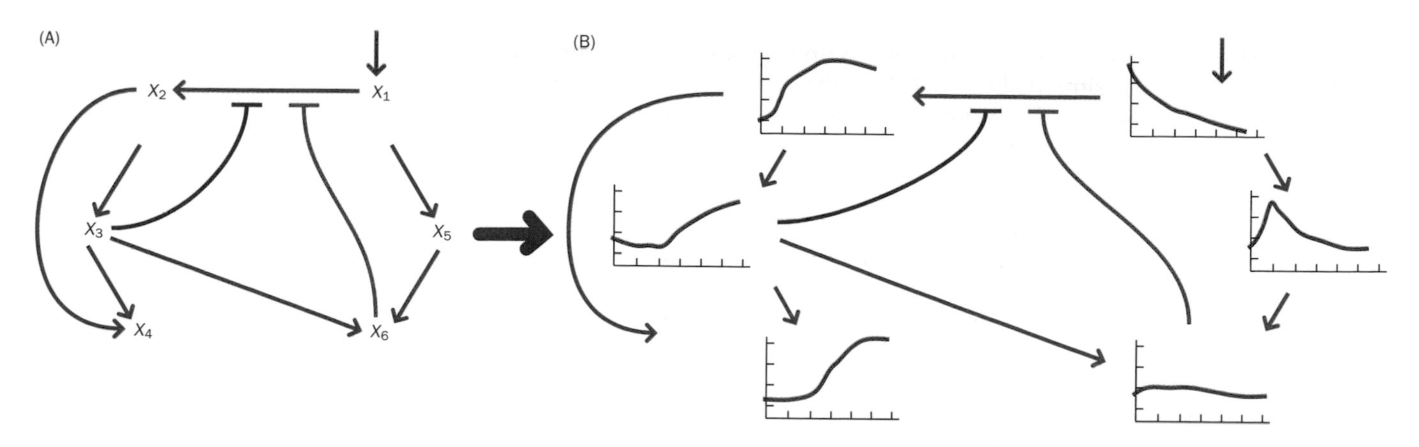

Figure 15.12 Concept-map modeling. (A) A typical diagram of a biological system is augmented by coarse, measured or expected dynamic responses to a typical input (green arrow). (B) The result is a system of time courses (green lines) that can be formulated as symbolic equations. Its parameter values can be estimated using inverse methods (see Chapter 5). Blue arrows indicate flow of material. Red and orange blunted lines indicate known and hypothesized inhibition signals, respectively.

the second step, the biologist sketches the known or expected response of each system component to a perturbation, such as an increase in input or the knock-down of a gene (**Figure 15.12**). For instance, the biologist might expect some component to increase slowly, then speed up, and finally reach a saturation level. This information might have been measured or might reflect the biologist's experience and intuition. The measured or alleged responses are converted into numerical time series that show how much and how fast each component is expected to change and what the transient response might look like. Most experienced biologists will have good intuition regarding such responses in systems that they have been investigating for many years. The alleged time courses are now converted into parameter values of the symbolic model that had been formulated from the components and the connectivity of the original diagram. The method for this conversion is inverse parameter estimation, as discussed in Chapter 5. Additional information from other sources can be added to the model as well. Ideally, the result is a numerical model that captures the biologist's intuition of the features of all system components. Under opportune conditions, it is even possible to avoid parameters and execute nonparametric studies directly based on data [79].

Whether parametric or nonparametric, it is now easy to perform model simulations. These might try to assess the likelihood of a hypothesized mechanism (such as the orange inhibitory signal in Figure 15.12) or simulate tested and yet untested scenarios. The results are compared with the biologist's expectations or with new data. Confirmatory simulation results are of course welcome, but counterintuitive results are at least as valuable, because they point to an incomplete understanding of some detail. A thorough analysis of ill-fitting simulations provides concrete hints of where the problems are most likely to lie. These problems might be structural and due to missing or ill-placed processes or regulatory signals. They may also be numerical in the sense that rate constants or other parameters are assigned very different values than what they should have. The analysis may also suggest the replacement of the original canonical model formulation with more complicated functions. Importantly, the process generates a computational structure that mimics the biological phenomenon and is incomparably easier to interrogate than the actual biological system. This interrogation will often lead to new hypotheses that are to be tested in the laboratory. An example is [8].

TOWARD A THEORY OF BIOLOGY . . . OR SEVERAL THEORIES?

According to the tenets of mathematical logic, a theory consists of a formal language that permits the formulation of rules of inference and a set of sentences or theorems that are assumed or proved to be true. The assumed sentences are often called axioms, which form a set of common beliefs; an example is the axiom that every integer has a successor. If these axioms are not accepted as true, then further inferences are usually moot, because one cannot decide whether they are true or false. Valid application of the rules of inference to axioms and proven theorems leads to new theorems that can be proved to be true, as long as one accepts the axioms.

Faced with such rigor, does it even make sense to strive for a theory of biology? What could serve as the language, what would be the rules of inference, what kinds of axioms could one formulate? Are there absolute truths in biology? Would the effort even pay off in the end if we could actually reach our goal? Indeed, it might seem quite esoteric to pose a biological theory as a long-term goal, when so many practical and pressing questions await answers. But, as the Prussian philosopher and psychologist Kurt Lewin once said, "there is nothing so practical as a good theory" ([80], p 288). To convince ourselves of the veracity of such a bold statement, just look at mathematics and physics. The laws of mathematics and physics have helped us understand the world, and the theories of physics are the foundations of engineering, with all its enormous implications and applications. If we had similarly strong laws in biology, or at least in certain subspecialties, we would no longer have to question the logic of observed phenomena and could save plenty of time and effort if the laws would allow us to make reliable inferences.

What might such biological axioms or laws look like? Some of them we actually have already. A famous candidate for a biological axiom is the quote *omnis cellula e cellula* (every cell comes from a cell) of Rudolf Virchow, a German cell physiologist of the nineteenth century. Or consider the genetic code that assigns amino acids to certain codons, and which is essentially ubiquitous among all organisms on Earth (see Chapter 7 for this code and a few exceptions). Without this "codon law," we would have to analyze every new gene sequence! Every time we discovered a new organism, we would not only have to sequence its genome, but also have to determine how this specific organism translated its gene sequences into peptides and proteins. We know about the Central Dogma, and even though details of this dogma have been challenged, for instance in the context of RNA viruses, it is a very powerful tool that we have learned to take for granted within its domain of applicability: if we knock out a gene, we do not expect the corresponding protein to be present. Mendelian genetics is a good first approximation to the processes of inheritance, and it is simply necessary to define the boundaries of this "theory" a little bit more stringently than the Austrian monk Gregor Mendel did. Many biologists consider evolution a mature theory. Within a more limited scope, Michael Savageau's demand theory (see [81] and Chapter 14) has the character of a theory. Hiroaki Kitano suggested a theory linking disease to the robustness of the physiological system [29, 82]. A small group of scientists have begun to create a theory of chemical reaction networks [83–85]. There are certainly laws governing the development of an embryo, and we rely on these laws almost as much as on physical laws. There are even laws at higher levels, such as the famous phenomenon of behavioral conditioning that was studied by the Russian physiologist and psychologist Ivan Petrovich Pavlov. Much effort from different angles is currently being devoted to discovering fundamental laws that span several biological scales (see, for example, Chapter 14).

Up to now, the search for general principles has required us to boil down complex, realistic systems and their models to essential features such as network motifs. This reduction of complexity presently is an art that has eluded strict general guidelines, which suggests that systematic methods of valid model reduction should be high on our list of priorities. In a parallel line of thinking, efforts are underway to create mathematical concepts and techniques of topological data analysis that facilitate the detection and analysis of highly nonlinear and often hidden geometric structures in massive datasets. Specific tasks to be addressed with these methods include how one might infer such high-dimensional structures in data from low-dimensional representations and how discrete data points might be assembled into global structures [86].

In the near future, any budding theories of biological subdisciplines might be modest in scope [81, 83, 87], but they will eventually span larger domains and later lead to a comprehensive understanding of processes such as carcinogenesis and the dynamics of metapopulations. We may also discover that the number of hard, unbreakable laws in biology is limited, but that it is quite possible to identify **fuzzy** or **probabilistic** laws. As we discussed in Chapter 14, a substantial number of researchers have begun to focus on the formulation of motifs, design principles, and operating principles, under which many organisms operate. Many of these will be encountered much more frequently than expected from a random design, but not in all cases. Thus, some of these principles may hold almost always, whereas other patterns may simply stand out more often than not, and only in certain contexts. These limits in scope will become parts of theories and theorems, as is common in mathematics, where certain statements are only true for certain sets of objects.

A strong theory of subareas of biology, for instance in gene regulation, would have tremendous implications. In analogy with the applications of physical theory in engineering, a biological theory of gene regulation would enormously streamline synthetic biology by permitting reliable predictions of which manipulations would actually lead to the desired goals. In many cases, we are already making these predictions, and they are often actually correct (Chapter 14). With a theory and with knowledge of its boundaries, the degree of certainty would come close to 100%.

Beyond all speculations regarding the future, a few trends are clear. First, we live in an unprecedented and incredibly exciting time for biology, modeling, and systems biology. Second, we have barely scratched the surface, and a book like this may cause a good chuckle 40 or 50 years from now. Hopefully by that time, our generation of systems biologists will have earned and deserved respect for how much we accomplished with the little we had. Above all, there is a huge and fascinating world out there that invites us to explore it.

REFERENCES

[1] Meadows DL & Meadows DH (eds). Toward Global Equilibrium: Collected Papers. Wright-Allen Press, 1973.

[2] Izhikevich EM. Dynamical Systems in Neuroscience: The Geometry of Excitability and Bursting. MIT Press, 2007.

[3] Keener J & Sneyd J. Mathematical Physiology. I: Cellular Physiology, 2nd ed. Springer, 2009.

[4] Hodgkin A & Huxley A. A quantitative description of membrane current and its application to conduction and excitation in nerve. *J. Physiol.* 117 (1952) 500–544.

[5] Fitzhugh R. Impulses and physiological states in theoretical models of nerve membrane. *Biophys. J.* 1 (1961) 445–466.

[6] Herz AV, Gollisch T, Machens CK & Jaeger D. Modeling single-neuron dynamics and computations: a balance of detail and abstraction. *Science* 314 (2006) 80–85.

[7] Nagumo JA & Yoshizawa S. An active pulse transmission line simulating nerve axon. *Proc. IRE* 50 (1962) 2061–2070.

[8] Qi Z, Miller GW & Voit EO. Computational systems analysis of dopamine metabolism. *PLoS One* 3 (2008) e2444.

[9] Qi Z, Miller GW & Voit EO. The internal state of medium spiny neurons varies in response to different input signals. *BMC Syst. Biol.* 4 (2010) 26.

[10] Gleeson P, Croo S, Cannon RC, et al. NeuroML: a language for describing data driven models of neurons and networks with a high degree of biological detail. *PLoS Comput. Biol.* 6 (2010) e1000815.

[11] Kikuchi S, Fujimoto K, Kitagawa N, et al. Kinetic simulation of signal transduction system in hippocampal long-term potentiation with dynamic modeling of protein phosphatase 2A. *Neural Netw.* 16 (2003) 1389–1398.

[12] Nakano T, Doi T, Yoshimoto J & Doya K. A kinetic model of dopamine- and calcium-dependent striatal synaptic plasticity. *PLoS Comput. Biol.* 6 (2010) e1000670.

[13] Tretter F. Mental illness, synapses and the brain—behavioral disorders by a system of molecules with a system of neurons? *Pharmacopsychiatry* 43 (2010) S9–S20.

[14] Voit EO. Mesoscopic modeling as a starting point for computational analyses of cystic fibrosis as a systemic disease. *Biochim. Biophys. Acta* 1844 (2013) 258–270.

[15] Qi Z, Yu G, Tretter F, et al. A heuristic model for working memory deficit in schizophrenia. *Biochem. Biophys. Acta* 1860 (2016) 2696–2705.

[16] McCulloch W & Pitts W. A logical calculus of the ideas immanent in nervous activity. *Bull. Math. Biophys.* 7 (1943) 115–133.

[17] Hornik K, Stinchcombe M & White H. Multilayer feedforward networks are universal approximators. *Neural Netw.* 2 (1989) 359–366.

[18] Blue Brain Project: http://bluebrain.epfl.ch/.

[19] Dhawale A & Bhalla US. The network and the synapse: 100 years after Cajal. *HFSP J.* 2 (2008) 12–16.

[20] Destexhe A & Contreras D. Neuronal computations with stochastic network states. *Science* 314 (2006) 85–90.

[21] Rubin JE. Emergent bursting in small networks of model conditional pacemakers in the pre-Botzinger complex. *Adv. Exp. Med. Biol.* 605 (2008) 119–124.

[22] Oshio K, Iwasaki Y, Morita S, et al. Database of Synaptic Connectivity of *C. elegans* for Computation. http://ims.dse.ibaraki.ac.jp/ccep/.

[23] Tretter F, Gebicke-Haerter PJ, Albus M, et al. Systems biology and addiction. *Pharmacopsychiatry* 42 (Suppl. 1) (2009) S11–S31.

[24] Human Connectome Project. http://www.humanconnectomeproject.org/.

[25] Bergman RN & Refai ME. Dynamic control of hepatic glucose metabolism: studies by experiment and computer simulation. *Ann. Biomed. Eng.* 3 (1975) 411–432.

[26] Bergman RN. Minimal model: perspective from 2005. *Horm. Res.* 64 (Suppl 3) (2005) 8–15.

[27] Ali SF & Padhi R. Optimal blood glucose regulation of diabetic patients using single network adaptive critics. *Optim. Control Appl. Methods* 32 (2011) 196–214.

[28] De Gaetano A, Hardy T, Beck B, et al. Mathematical models of diabetes progression. *Am. J. Physiol. Endocrinol. Metab.* 295 (2008) E1462–E1479.

[29] Kitano H, Oda K, Kimura T, et al. Metabolic syndrome and robustness tradeoffs. *Diabetes* 53 (Suppl 3) (2004) S6–S15.

[30] Tam J, Fukumura D & Jain RK. A mathematical model of murine metabolic regulation by leptin: energy balance and defense of a stable body weight. *Cell Metab.* 9 (2005) 52–63.

[31] Eddy D, Kahn R, Peskin B & Schiebinger R. The relationship between insulin resistance and related metabolic variables to coronary artery disease: a mathematical analysis. *Diabetes Care* 32 (2009) 361–366.

[32] Palm NW & Medzhitov R. Not so fast: adaptive suppression of innate immunity. *Nat. Med.* 13 (2007) 1142–1144.

[33] Zhao J, Yang X, Auh SL, et al. Do adaptive immune cells suppress or activate innate immunity? *Trends Immunol.* 30 (2008) 8–12.

[34] Vodovotz Y, Constantine G, Rubin J, et al. Mechanistic simulations of inflammation: current state and future prospects. *Math. Biosci.* 217 (2009) 1–10.

[35] Vodovotz Y. Translational systems biology of inflammation and healing. *Wound Repair Regen.* 18 (2010) 3–7.

[36] Vodovotz Y, Constantine G, Faeder J, et al. Translational systems approaches to the biology of inflammation and healing. *Immunopharmacol. Immunotoxicol.* 32 (2010) 181–195.

[37] Clermont G, Bartels J, Kumar R, et al. *In silico* design of clinical trials: a method coming of age. *Crit. Care Med.* 32 (2004) 2061–2070.

[38] Chow CC, Clermont G, Kumar R, et al. The acute inflammatory response in diverse shock states. *Shock* 24 (2005) 74–84.

[39] Foteinou PT, Calvano SE, Lowry SF & Androulakis IP. Modeling endotoxin-induced systemic inflammation using an indirect response approach. *Math. Biosci.* 217 (2009) 27–42.

[40] An G. Introduction of an agent-based multi-scale modular architecture for dynamic knowledge representation of acute inflammation. *Theor. Biol. Med. Model.* 5 (2008) 11.

[41] Feinendegen L, Hahnfeldt P, Schadt EE, et al. Systems biology and its potential role in radiobiology. *Radiat. Environ. Biophys.* 47 (2008) 5–23.

[42] Moolgavkar SH & Knudson AGJ. Mutation and cancer: a model for human carcinogenesis. *J. Natl Cancer Inst.* 66 (1981) 1037–1052.

[43] Moolgavkar SH & Luebeck EG. Multistage carcinogenesis and the incidence of human cancer. *Genes Chromosomes Cancer* (2003) 302–306.

[44] Enderling H, Hlatky L & Hahnfeldt P. Migration rules: tumours are conglomerates of self-metastases. *Br. J. Cancer* 100 (2009) 1917–1925.

[45] Folkman J & Kalluri R. Concept cancer without disease. *Nature* 427 (2004) 787.

[46] Black WC & Welch HG. Advances in diagnostic imaging and overestimations of disease prevalence and the benefits of therapy. *N. Engl. J. Med.* 328 (1993) 1237–1243.

[47] Zahl PH, Maehlen J & Welch HG. The natural history of invasive breast cancers detected by screening mammography. *Arch Intern. Med.* 168 (2008) 2311–2316.

[48] Abdollahi A, Schwager C, Kleeff J, et al. Transcriptional network governing the angiogenic switch in human pancreatic cancer. *Proc. Natl Acad. Sci. USA* 104 (2007) 12890–12895.

[49] Lipsitz LA & Goldberger AL. Loss of "complexity" and aging. Potential applications of fractals and chaos theory to senescence. *JAMA* 267 (1992) 1806–1809.

[50] Lipsitz LA. Physiological complexity, aging, and the path to frailty. *Sci. Aging Knowledge Environ.* 2004 (2004) pe16.

[51] Beard RW, Filshie GM, Knight CA & Roberts GM. The significance of the changes in the continuous fetal heart rate in the first stage of labour. *Br. J. Obstet. Gynaecol.* 78 (1971) 865–881.

[52] Williams KP & Galerneau F. Intrapartum fetal heart rate patterns in the prediction of neonatal acidemia. *Am. J. Obstet. Gynecol.* 188 (2003) 820–823.

[53] Ferrario M, Signorini MG, Magenes G & Cerutti S. Comparison of entropy-based regularity estimators: Application to the fetal heart rate signal for the identification of fetal distress. *IEEE Trans. Biomed. Eng.* 33 (2006) 119–125.

[54] Wooley JC, Godzik A & Friedberg I. A primer on metagenomics. *PLoS Comput. Biol.* 6 (2010) e1000667.

[55] Trevors JT. One gram of soil: a microbial biochemical gene library. *Antonie Van Leeuwenhoek* 97 (2010) 99–106.

[56] Dam P, Fonseca, LL, Konstantinidis, KT & Voit EO. Dynamic models of the complex microbial metapopulation of Lake Mendota. *NPJ Syst. Biol. Appl.* 2 (2016) 16007.

[57] Turroni F, Ribbera A, Foroni E, et al. Human gut microbiota and bifidobacteria: from composition to functionality. *Antonie Van Leeuwenhoek* 94 (2008) 35–50.

[58] NIH launches Human Microbiome Project. US National Institutes of Health Public Release, 19 December 2007. www.eurekalert.org/pub_releases/2007-12/nhgr-nlh121907.php.

[59] Canadian Microbiome Initiative. http://www.cihr-irsc.gc.ca/e/39939.html.

[60] International Human Microbiome Consortium. http://www.human-microbiome.org/.

[61] Rondon MR, August PR, Bettermann AD, et al. Cloning the soil metagenome: a strategy for accessing the genetic and functional diversity of uncultured microorganisms. *Appl. Environ. Microbiol.* 66 (2000) 2541–2547.

[62] Wilmes P & Bond PL. Microbial community proteomics: elucidating the catalysts and metabolic mechanisms that drive the Earth's biogeochemical cycles. *Curr. Opin. Microbiol.* 12 (2009) 310–317.

[63] Medina M & Sachs JL. Symbiont genomics, our new tangled bank. *Genomics* 95 (2010) 129–137.

[64] Vieites JM, Guazzaroni ME, Beloqui A, et al. Metagenomics approaches in systems microbiology. *FEMS Microbiol. Rev.* 33 (2009) 236–255.

[65] Jones MB, Schildhauer MP, Reichman OJ & Bowers S. The new bioinformatics: integrating ecological data from the gene to the biosphere. *Annu. Rev. Ecol. Evol. Syst.* 37 (2006) 519–544.

[66] Genomic Science Program: Systems Biology for Energy and Environment. http://genomicscience.energy.gov/.

[67] Savage DF, Afonso B, Chen A & Silver PA. Spatially ordered dynamics of the bacterial carbon fixation machinery. *Science* 327 (2010) 1258–1261.

[68] Burkepile DE & Hay ME. Impact of herbivore identity on algal succession and coral growth on a Caribbean reef. *PLoS One* 5 (2010) e8963.

[69] Bonabeau E. Agent-based modeling: methods and techniques for simulating human systems. *Proc. Natl Acad. Sci. USA* 14 (2002) 7280–7287.

[70] Macal CM & North M. Tutorial on agent-based modeling and simulation. Part 2: How to model with agents. In Proceedings of the Winter Simulation Conference, December 2006 (LF Perrone, FP Wieland, J Liu, et al. eds), pp 73–83. IEEE, Monterey, CA, 2006.

[71] Robinson EJH, Ratnieks FLW & Holcombe M. An agent-based model to investigate the roles of attractive and repellent pheromones in ant decision making during foraging. *J. Theor. Biol.* 255 (2008) 250–258.
[72] Ermentrout GB & Edelstein-Keshet L. Cellular automata approaches to biological modeling. *J. Theor. Biol.* 160 (1993) 97–133.
[73] Wolpert L. Pattern formation in biological development. *Sci. Am.* 239(4) (1978) 154–164.
[74] Gierer A & Meinhardt H. A theory of biological pattern formation. *Kybernetik* 12 (1972) 30–39.
[75] Wilensky U. NetLogo. http://ccl.northwestern.edu/netlogo/.
[76] Zhang L, Wang Z, Sagotsky JA & Deisboeck TS. Multiscale agent-based cancer modeling. *J. Math. Biol.* 58 (2009) 545–559.
[77] Chavali AK, Gianchandani EP, Tung KS, et al. Characterizing emergent properties of immunological systems with multi-cellular rule-based computational modeling. *Trends Immunol.* 29 (2008) 589–599.
[78] Goel G, Chou IC & Voit EO. Biological systems modeling and analysis: a biomolecular technique of the twenty-first century. *J. Biomol. Tech.* 17 (2006) 252–269.
[79] Faraji M & Voit EO. Nonparametric dynamic modeling. *Math. Biosci.* (2016) http://dx.doi.org/10.1016/j.mbs.2016.08.004.
[80] Lewin K. Resolving Social Conflicts and Field Theory in Social Science. American Psychological Association, 1997.
[81] Savageau MA. Demand theory of gene regulation. I. Quantitative development of the theory. *Genetics* 149 (1998) 1665–1676.
[82] Kitano H. Grand challenges in systems physiology. *Front. Physiol.* 1 (2010) 3.
[83] Horn FJM & Jackson R. General mass action kinetics. *Archive Rat. Mech. Anal.* 47 (1972) 81–116.
[84] Feinberg M. Complex balancing in general kinetic systems. *Arch. Rat. Mech. Anal.* 49 (1972) 187–194.
[85] Arceo CP, Jose EC, Marin-Sanguino A, Mendoza ER. Chemical reaction network approaches to biochemical systems theory. *Math. Biosci.* 269 (2015) 135–152.
[86] Ghrist R. Barcodes: the persistent topology of data. *Bull. Am. Math. Soc.* 45 (2008) 61–75.
[87] Wolkenhauer O, Mesarovic M & Wellstead P. A plea for more theory in molecular biology. In Systems Biology—Applications and Perspectives (P Bringmann, EC Butcher, G Parry, B Weiss *eds*), Springer, 2007.

FURTHER READING

Bassingthwaighte J, Hunter P & Noble D. The Cardiac Physiome: perspectives for the future. *Exp. Physiol.* 94 (2009) 597–605.
Calvert J & Fujimura JH. Calculating life? Duelling discourses in interdisciplinary systems biology. *Stud. Hist. Philos. Biol. Biomed. Sci.* 42 (2011) 155–163.
del Sol A, Balling R, Hood L & Galas D. Diseases as network perturbations. *Curr. Opin. Biotechnol.* 21 (2010) 566–571.
Gonzalez-Angulo AM, Hennessy BT & Mills GB. Future of personalized medicine in oncology: a systems biology approach. *J. Clin. Oncol.* 28 (2010) 2777–2783.
Hester RL, Iliescu R, Summers R & Coleman TG. Systems biology and integrative physiological modelling. *J. Physiol.* 589 (2011) 1053–1060.
Ho RL & Lieu CA. Systems biology: an evolving approach in drug discovery and development. *Drugs R D* 9 (2008) 203–216.
Hood, L. Systems biology and P4 medicine: Past, present, and future. *Rambam Maimonides Med. J.* 4 (2013) e0012.
Keasling JD. Manufacturing molecules through metabolic engineering. *Science* 330 (2010) 1355–1358.
Kinross JM, Darzi AW & Nicholson JK. Gut microbiome–host interactions in health and disease. *Genome Med.* 3 (2011) 14.
Kitano H. Grand challenges in systems physiology. *Front. Physiol.* 1 (2010) 3.
Kitano H, Oda K, Kimura T, et al. Metabolic syndrome and robustness tradeoffs. *Diabetes* 53 (Suppl 3) (2004) S6–S15.
Kreeger PK & Lauffenburger DA. Cancer systems biology: a network modeling perspective. *Carcinogenesis* 31 (2010) 2–8.
Kriete A, Lechner M, Clearfield D & Bohmann D. Computational systems biology of aging. *Wiley Interdiscip. Rev. Syst. Biol. Med.* 3 (2010) 414–428.
Likic VA, McConville MJ, Lithgow T & Bacic A. Systems biology: the next frontier for bioinformatics. *Adv. Bioinformatics* 2010 (2010) 268925.
Lucas M, Laplaze L & Bennett MJ. Plant systems biology: network matters. *Plant Cell Environ.* 34 (2011) 535–553.
McDonald JF. Integrated cancer systems biology: current progress and future promise. *Future Oncol.* 7 (2011) 599–601.
Noble D. The aims of systems biology: between molecules and organisms. *Pharmacopsychiatry* 44 (Suppl. 1) (2011) S9–S14.
Palsson B. Metabolic systems biology. *FEBS Lett.* 583 (2009) 3900–3904.
Sperling SR. Systems biology approaches to heart development and congenital heart disease. *Cardiovasc. Res.* 91 (2011) 269–278.
Tretter F. Mental illness, synapses and the brain—behavioral disorders by a system of molecules with a system of neurons? *Pharmacopsychiatry* 43 (2010) S9–S20.
Vodovotz Y. Translational systems biology of inflammation and healing. *Wound Repair Regen.* 18 (2010) 3–7.
Werner HMJ, Mills, GB, Ram PT. Cancer systems biology: a peak into the future of patient care? *Nat. Rev. Clin. Oncol.* 11 (2014) 167–176.
West GB & Bergman A. Toward a systems biology framework for understanding aging and health span. *J. Gerontol. A Biol. Sci. Med. Sci.* 64 (2009) 205–208.
Weston AD & Hood L. Systems biology, proteomics, and the future of health care: toward predictive, preventative, and personalized medicine. *J. Proteome Res.* 3 (2004) 179–196.
Wolkenhauer O, Mesarovic M & Wellstead P. A plea for more theory in molecular biology. *Ernst Schering Res Found Workshop* (2007) 117–137.

Glossary

The chapter numbers refer to the first substantial discussion of the topic.

Actin
A protein that is one unit of an actin filament in eukaryotic cells. Thin actin filaments and thicker **myosin** filaments are key components of the contractile machinery in skeletal and heart muscle cells. (Chapter 12)

Action potential
A dynamic pattern in voltage exhibited by the membranes of **excitable** cells, including neurons and heart cells. Typical action potentials consist of a fast **depolarization** and a slower repolarization phase. Excitable cells exhibit a **resting potential** between two action potentials. Action potentials are directly associated with neuronal signal transduction, muscle contraction, and heartbeats. (Chapter 12)

***Ad hoc* model**
In contrast to a **canonical model**, a mathematical model that is constructed for a specific purpose and without adhering to a general **modeling** framework. (Chapter 4)

Adaptive
1. In immunology, a system of specialized cells, primarily B and T lymphocytes, that learn to recognize foreign antigens like pathogens, eliminate them or prevent their growth, and remember them in cases of re-infections. The system is activated by the **innate immune system**.
2. In systems theory, a natural **system** or computational **model** that changes its characteristics in response to outside stimuli. (Chapter 9)

Affinity
A measure of the strength of an interaction between atoms or molecules or for the tendency of an atom or molecule to enter into a chemical reaction with unlike atoms or molecules. (Chapter 5)

Agent-based modeling (ABM)
A computational modeling method in which each component (agent) acts according to specific sets of rules. ABM is particularly useful for spatial modeling in heterogeneous environments. (Chapter 15)

Algorithm
A series of computational instructions designed for solving a specific task; usually implemented as computer code or software. (Chapter 5)

Allostasis
An abnormal state of an organism that differs from normal **homeostasis** and is associated with inferior functioning or disease. (Chapter 13)

Allosteric (regulation)
Regulation of enzyme activity through binding of a ligand to the enzyme outside its active site. (Chapters 7 and 8)

Amino acid
One of 20 molecular building blocks used to generate **peptides** and proteins according to the **genetic code**. (Chapter 7)

Antagonism/Antagonistic
Literally, "working against each other." **Nonlinear** interaction between two **system** components that lessens both components' individual effects. Antonym to **synergism/synergistic**. (Chapter 11)

Antibody
See **Immunoglobulin**. (Chapter 7)

Apoptosis
A natural or induced cellular mechanism resulting in the self-destruction of a cell. Also called "programmed cell death." (Chapter 11)

Approximation
The replacement of a mathematical function with a simpler function. Usually the two have the same value, slope, and possibly higher derivatives at one chosen point, the **operating point**. (Chapter 2)

Arrhythmia
An irregular heartbeat or contraction pattern. (Chapter 12)

Artificial neural network (ANN)
A **machine learning algorithm**, inspired by natural neural networks, that can be trained to classify data and model complex relationships between inputs and outputs. For instance, upon successful training, an ANN may be able to distinguish healthy and cancer cells, based on images or **biomarker** profiles. (Chapter 13)

Atrioventricular (AV) node
A relatively small group of cells with **pacemaker** potential in the wall of the right **atrium** of the heart. Normally, the AV node receives electrical signals from the **sino-atrial node** and directs them through the **bundle of His** and the **Purkinje fibers** throughout the **ventricles**. (Chapter 12)

Atrium (pl. Atria)
One of the two upper, smaller chambers of the heart. See also **Ventricle**. (Chapter 12)

Attractor
In the field of dynamical systems, a point or set of points to which **trajectories** converge. The most typical attractors are stable steady-state points and stable **limit cycles**. In certain **chaotic** systems one talks of 'strange attractors.' The antonym is a **repeller**. (Chapter 4)

Automaticity
Term for the ability of **sino-atrial (SA) node** cells in the heart to **depolarize** spontaneously, thereby serving as **pacemaker cells**. (Chapter 12)

Autonomic nervous system
Part of the brain that, largely without conscious input, controls the function of most inner organs. (Chapter 12)

Auto-regulation
Regulation of an **adaptive** biological **system** by itself. An example is blood pressure regulation in the brain. (Chapter 14)

Base pair (bp)
A matching couple of two **nucleotides** on complementary DNA or RNA strands, such as cytosine–guanine (DNA and RNA) and adenine–thymine (DNA) or adenine–uracil (RNA), which are connected through hydrogen bonds. (Chapter 6)

Basin of Attraction
A subset of the state space of a model with the following property: If a simulation starts at any point within this subset, it will converge to the same attractor, which may be a steady-state point, limit cycle, or chaotic attractor. (Chapter 9)

Bayesian (network) inference
A statistical method for deducing from data the most probable connectivity (although not **causality**) between **nodes** in a **network** or

graph that does not contain cycles. See also **Directed acyclic graph (DAG)**. (Chapter 3)

Behavior (of a system)
A collective term for changes in the **state** of a system, often over time and in response to a **perturbation** or **stimulus**. (Chapter 2)

Bi-fan
A **network motif** where two source **nodes** send **signals** to each of two target nodes. (Chapter 14)

Bifurcation (in a dynamical system)
A critical value of a **parameter** in a **system** (usually of **differential equations**), where the behavior of the system changes **qualitatively**, for instance, from a **stable steady state** to an unstable state surrounded by a stable **limit cycle oscillation**. (Chapter 4)

Biochemical systems theory (BST)
A **canonical** modeling framework that represents biological **systems** with **ordinary differential equations**, in which all processes are described as products of **power-law functions**. See also **Generalized mass action (GMA) system** and **S-system**. (Chapter 4)

Biofilm
A microbial **metapopulation** consisting of many species and collectively forming a thin layer. (Chapter 15)

Bioinformatics
A field of study at the intersection of biology and computer science that organizes, analyzes, and interprets datasets and provides computational support for **-omics** studies. (Chapter 3)

Biologics
Pharmaceutically active molecules, such as **antibodies**, that are derived from naturally occurring biological materials. (Chapter 13)

Biomarker
A biological or clinical measurement that is associated with, or can be used to predict, a specific event (such as a disease). Some biomarkers are causative, while others are merely symptomatic. (Chapter 13)

Bistability
The property of a system to possess two **stable steady states** (and at least one unstable steady state). (Chapter 9)

Boolean model of a system
Typically a **discrete** model, in which the **state** of a **multivariate** system at time $t + 1$ is determined by a logical function of the states of the system components at time t. (Chapter 3)

Bootstrap
A statistical method for determining desired features of a dataset of size n. It is based on constructing and analyzing many samples of size n, whose data are randomly chosen one-by-one from the dataset and replaced before the next datum is chosen. See also **Jackknife**. (Chapter 5)

Branch-and-bound method
One of several sophisticated **search algorithms** that find **global optima** by systematically and iteratively subdividing the **parameter** space and discarding regions that cannot contain the solution. (Chapter 5)

Bundle of His
A collection of heart cells specialized for the rapid conduction of electrical signals, functionally connecting the **atrioventricular node** and the **Purkinje fibers**. (Chapter 12)

Calcium-induced calcium release (CICR)
An amplification mechanism in **cardiomyocytes**, in which a small amount of calcium input to the cell leads to a large efflux of calcium from the **sarcoplasmic reticulum** into the cytosol, where it leads to contraction. (Chapter 12)

Canonical model
A mathematical modeling format that follows strict construction rules. Examples are **linear** models, **Lotka–Volterra** models, and models within **biochemical systems theory**. (Chapter 4)

Carcinogenesis
The process of cancer development. (Chapter 15)

Cardiac
Related to the heart. (Chapter 12)

Cardiomyocyte
A contractible heart muscle cell. See also **Excitable cell** and **Pacemaker cell**. (Chapter 12)

Carrier protein
A protein transporting specific **ions** and other small molecules between the inside and outside of a cell. The required energy is often supplied in the form of ATP. Also called a pump or permease. (Chapter 12)

Carrying capacity
The number of individuals an environment can sustainably accommodate. See also **crowding** and **logistic growth**. (Chapter 10)

Causality analysis
A statistical assessment of the likelihood that one component (or condition) of a **system** directly controls or causes changes in another component (or condition). (Chapter 3)

cDNA
Complementary DNA; generated from mRNA by the action of the **enzymes** reverse transcriptase and DNA polymerase. (Chapter 6)

Cell cycle
The sequence of intracellular events that ultimately lead to the division and duplication of a cell. (Chapter 3)

Central Dogma (of Molecular Biology)
The general (and somewhat simplistic) concept underlying the unidirectional information transfer from DNA to mRNA (**transcription**) and then to protein (**translation**). (Chapter 6)

Central limit theorem
A fundamental law of statistics stating that the sum of many random variables typically approaches a normal (Gaussian) distribution (bell curve). (Chapter 12)

Channel protein
A protein spanning a cell membrane and forming a hydrophilic pore that permits very fast travel of specific **ions** across the membrane. (Chapter 12)

Chaos/chaotic
Without order or predictability. In the terminology of **systems** theory, **deterministic** chaos refers to a phenomenon of some **nonlinear dynamic systems** that are completely deterministic and numerically characterized, yet exhibit erratic, unpredictable **behaviors**. Two **simulations** of these systems with different, but arbitrarily close, **initial values** eventually show no more similarities in their responses than two simulations with very different initial values. (Chapter 4)

Chaperone
A protein that assists in the correct folding of proteins and the assembly of macromolecular structures. Chaperones also protect the three-dimensional **conformation** of proteins in conditions of environmental stress. Heat-shock proteins are chaperones that protect proteins against misfolding due to high temperatures. (Chapter 7)

Chemical master equation
A detailed **ordinary differential equation** description of a chemical **reaction** that accounts for the **stochastic** nature of every molecular event associated with this reaction. (Chapter 8)

***Cis*-regulatory DNA**
A region of DNA that regulates the expression of a gene located on the same DNA molecule, for instance, by coding for a **transcription factor**, or serving as operator. (Chapter 6)

Clique
A part of an undirected **network** or **graph** where all **nodes** are neighbors, that is, directly connected to each other. (Chapter 3)

Closed system
In contrast to an **open system**, a mathematical **model** without input or loss of material. (Chapter 2)

Clustering (of genes)
The sorting of genes by some feature, such as a similar sequence or co-expression, usually by means of a **machine learning algorithm**. (Chapter 11)

Clustering coefficient
A quantitative measure for the connectivity of a **network**, either throughout the entire network or within the neighborhood of a given **node**. (Chapter 3)

Codon
A triplet of **nucleotides** that, when translated, represent an **amino acid**. (Chapter 6)

Coefficient
A (typically constant) multiplicative factor (number) in some term of an equation or a matrix. See also **Linear**. (Chapter 4)

Compartment
A defined space that is separated from its surroundings, for instance, by a membrane. See also **Compartment model**. (Chapter 4)

Compartment model
Usually a **linear ordinary differential equation model** of the **dynamics** of material moving between **compartments**, such as the organs and the bloodstream in an organism. See also **Pharmacokinetic model**. (Chapter 4)

Complexity
An ill-defined conglomerate of features of a **system** that includes its number of components, processes, and regulatory **signals**, as well as the degree of **nonlinearity** and unpredictability of the system. (Chapter 4)

Concept-map modeling
A strategy for converting biological diagrams and coarse input-response information into **dynamic models**. (Chapter 15)

Conductance (electrical)
A measure of how easily electricity can move through some material. (Chapter 12)

Conformation
The three-dimensional arrangement of atoms in a large molecule like a protein. The conformation can change with the molecule's chemical and physical environment. See also **Tertiary structure**. (Chapter 7)

Constraint (optimization)
Any type of threshold that a component, subject to **optimization**, is not permitted to cross; a condition that variables must satisfy in an optimization problem. (Chapter 3)

Continuous
Antonym to **discrete**. The feature of a function or set not to have holes or interruptions; in systems biology usually with respect to time, and sometimes with respect to space. (Chapter 4)

Continuum approximation
An approximation of the mechanical features of an array of many similar objects with the mechanics of a continuous material; used, for instance, in modeling heart tissue. (Chapter 12)

Control coefficient
A quantitative measure of **steady-state sensitivity** in **metabolic control analysis**. (Chapter 3)

Controlled comparison
A mathematical technique for revealing **design principles**, based on comparing the **designs** of two similar **systems** in a minimally biased fashion. (Chapter 14)

Cooperativity
The phenomenon that **enzymes** or **receptors** may have multiple binding sites, which may increase (positive cooperativity) or decrease (negative cooperativity) their **affinity. Hill models** are often used to represent this phenomenon. See also **Allosteric** (regulation). (Chapter 7)

Coronaries
Arteries and veins supplying the heart muscle with blood circulation. (Chapter 12)

Correlative model
A **model** that relates one quantity to another, without explaining what causes the relationship or what the relationship means. See also **Explanatory model**. (Chapter 2)

Crosstalk
Information exchange between two processes or pathways; typically in **signal transduction**. (Chapter 9)

Crowding
A collective term for impeding interactions within a growing population, such as the competition for resources that lead to a slowing down of population growth, often until the **carrying capacity** is reached. (Chapter 10)

Curated (database)
Referring to the fact that information (in the database) was checked by an expert. (Chapter 8)

Cytokine
A signaling protein involved in responses to infections and inflammation. (Chapters 7 and 15)

Damped oscillation
An oscillation that subsides over time. See also **Limit cycle** and **Overshoot**. (Chapter 4)

Data mining
The process of extracting patterns from large datasets and forging them into useful information, by combining methods from **machine learning**, artificial intelligence, statistics, and database management. See also **Text mining**. (Chapter 8)

Degree distribution
Statistical distribution of the number of **edges** per **node** in a **graph** or **network**. (Chapter 3)

Denature
To change or destroy the structure of a macromolecule, typically a protein or nucleic acid, by applying extreme physical or chemical stress. Denatured proteins are often removed by the **proteasome**. (Chapter 11)

Dependent variable
A **system** variable that is potentially affected by the **dynamics** of the system. See also **Independent variable**. (Chapter 2)

Depolarization
A rapid change, from negative to positive voltage, in the membrane potential of neurons, muscle cells, and heart cells during an **action potential**; it is followed by **repolarization**. (Chapter 12)

Design
1. Of a natural **system**. Similar to a network **motif**, the structural and regulatory composition of a **dynamic** system as it is observed. The observed design is sometimes analyzed in comparison with hypothetical alternatives, in order to reveal **design** and **operating principles**.
2. Of an artificial biological system. The core subject of **synthetic biology**.
3. Of an experiment. A set of statistical techniques assuring that results from the experiment will likely be statistically significant. (Chapter 14)

Design principle
A structural or **dynamic** feature of a **system** that, due to its functional superiority, is found more often than alternative designs. See also **Motif** and **Operating principle**. (Chapter 14)

Deterministic (model)
A model that does not allow or account for **random** events. See also **Stochastic** model and deterministic **Chaos**. (Chapter 2)

Diastole
Greek term for dilation and used to describe the relaxed state of the heart. See also **Systole**. (Chapter 12)

Difference equation
A **discrete** function in which the **state** of a **system** is defined as a function of its state at one or more previous time points. See also **Iteration** and **Boolean model**. (Chapter 4)

Differential equation
An equation or system of equations containing functions and their derivatives. Most **dynamical systems** are represented as differential equations. See also **Ordinary differential equation** and **Partial differential equation**. (Chapter 4)

Differentiation
1. In developmental biology, the natural process of converting a cell with many possible roles or fates, such as a fertilized egg or stem cell, into a cell with one or a few specific functions. (Chapter 11)
2. In mathematics, the process of computing a derivative (slope) of a function. (Chapter 4)

Directed acyclic graph (DAG)
A **graph** in which all **edges** have a unique direction, and which does not allow **feedback** or the return to any **node**, once this node had been left. (Chapter 3)

Discrete
Antonym to **continuous**. The feature of a function or set to exhibit interruptions. For instance, a discrete timescale is defined for distinct time points t_1, t_2, t_3, ..., but not in between, like the display of a digital watch. (Chapter 2)

DNA chip
Similar to a **microarray**, a tool for comparing gene expression in different cell types or under different conditions. In contrast to a microarray, a DNA chip contains artificially constructed, short nucleotide probes. (Chapter 6)

Dyad
A small space between the cell membrane and the **sarcoplasmic reticulum** in **cardiomyocytes**, where **calcium-induced calcium release** is controlled. (Chapter 12)

Dynamical system or model
A **system**, or **model** of a system, whose **state** and **behavior** changes over time. (Chapter 2)

Dynamics (of a model)
The collection of changes in **model** features over time. (Chapter 2)

Edge (in a graph)
The directed or undirected connection between two **nodes** in a **graph** or between a node and the **environment** of the **graph**. (Chapter 3)

Eigenvalue
One of a group of complex numbers that characterize various aspects of a linear or linearized **system**, such as its **stability** at a **steady state** and the different speeds with which different variables evolve over time. (Chapter 4)

Elasticity
A quantitative measure of the **sensitivity** of an **enzyme** or **reaction** to changes in the local environment, for instance, induced by changes in substrate; used in **metabolic control analysis**. Equivalent to the concept of a **kinetic order** in **biochemical systems theory**. (Chapter 3)

Electrocardiogram (ECG, EKG)
A reading from several pairs of electrodes, attached to the torso and limbs, of the electrical activity of the heart. (Chapter 12)

Electrochemical
Related to electrical events associated with **ions**. (Chapter 12)

Electrophysiology
Study of the normal functioning of electrical events in the body. (Chapter 12)

Elementary mode analysis (EMA)
A method for determining independent reaction paths within a **stoichiometric network**. (Chapter 14)

Emergent property
A property of a **system** that cannot be assigned to the features of a single component, but only becomes possible from **nonlinear** interactions between the system components. See also **Superposition**. (Chapter 1)

Environment (of a model, graph or system)
Components and processes outside the explicit definition of a **model**, **graph** or **system**. (Chapter 2)

Enzyme
A protein that catalyzes (facilitates) a biochemical **reaction**. (Chapter 7)

Epidemic
An infectious disease that sickens large percentages of a population. An extremely widespread epidemic is called a pandemic. (Chapter 10)

Epigenetics
The relatively new field of studying heritable features that are not due to changes in the **nucleotide** sequence of an organism's DNA. (Chapter 6)

Error (in modeling and statistics)
See **Residual**. (Chapter 5)

Evolutionary algorithm
One of several **machine learning** methods that, inspired by natural evolution, search for optimal and near-optimal solutions. See also **Genetic algorithm** and **Optimization**. (Chapter 5)

Excitable (cell)
A neuron or muscle cell that responds to **electrochemical** stimulation. See also **Action potential**. (Chapter 12)

Excitation–contraction (EC) coupling
A fundamental process in muscle physiology, where an electrical stimulus (usually an **action potential**) is converted into an increase in calcium concentration within the cytosol of a muscle cell, which causes a sliding motion between **actin** and **myosin filaments** and leads to contraction of the cell. (Chapter 12)

Exon
A segment of a gene sequence that is transcribed into mRNA and subsequently translated into protein. See also **Intron**. (Chapter 6)

Explanatory model
A conceptual or mathematical **model** that explains a phenomenon (for example, development) based on events at a lower level of organization (for example, gene expression). (Chapter 2)

Exponential growth
A model of growth, or **growth law**, that can be described by an exponential function with a positive **growth rate**. The corresponding process with a negative rate is called exponential decay. (Chapter 10)

Expressed sequence tag (EST)
A gene fragment of about 300 nucleotides that has been sequenced from **cDNA**. (Chapter 6)

Extrapolation
The prediction of **model** responses outside the range for which actual data are available. (Chapter 2)

False negative
The result of a test indicating a negative outcome even though the outcome is in actuality positive. For instance, a test may suggest the absence of disease even though the person has the disease. See also **False positive**. (Chapter 13)

False positive
The result of a test indicating a positive outcome even though the outcome is in actuality negative. For instance, a test may suggest cancer even though the person does not have cancer. See also **False negative**. (Chapter 13)

Feature selection
A technique in statistics and **machine learning** for reducing the number of **variables** to a subset of most influential variables. (Chapter 5)

Feedback
A **signal** affecting a preceding reaction step or process in a generic pathway. (Chapter 2)

Feedforward
A **signal** affecting a downstream reaction step or process within a generic pathway. (Chapter 14)

Filament (muscle cell)
A microfiber composed of proteins like **actin** and **myosin**, which endows muscle cells with contractility. (Chapter 12)

Finite element approach
A set of methods for assessing the dynamics of fluids, materials, or solid bodies, based on subdividing these into very small units (elements) and studying the behavior of and between these elements. Often used to analyze **partial differential equations** as well as changes within bodies that have complicated shapes. (Chapter 12)

Fitness
1. In genetics, a quantitative measure of superiority of a gene, its gene product, or the organism possessing the gene.
2. In **Optimization**, a quantitative measure of superiority of one solution over another. (Chapter 5)

Fitting (of a model)
The automatic or manual adjustment of **parameter** values so that a **model** matches observation data as closely as possible. See also **parameter estimation**, **optimization**, **residual**, and **validation**. (Chapter 5)

Fluid mechanics
The study of the movement of particles within liquids, often using **partial differential equations** and **finite element approaches**. (Chapter 12)

Flux
The amount of material (number of molecules) flowing through a pool (**variable**), **pathway**, or **system** during a defined time period. (Chapter 3)

Flux balance analysis (FBA)
A modeling approach for (large) metabolic **networks**, which at a **steady state** balances metabolic **fluxes** entering and leaving **nodes**, according to the **stoichiometry** of each reaction, and which assumes that the cell or organism optimizes some feature, such as its growth rate. See also **Metabolic flux analysis**. (Chapter 3)

Frank–Starling law
A relationship between blood volume, pressures, forces, and wall tensions in the heart. (Chapter 12)

Funny current
Relatively slow, mixed sodium–potassium current, entering a cell of the **sino-atrial node** and triggering spontaneous **depolarization**. (Chapter 12)

Fuzzy (logic)
A set of logical rules that are applied in a **probabilistic** (rather than a **deterministic**) fashion. (Chapter 15)

Gain
The **sensitivity** of a system response with respect to a small change in an **independent variable**. (Chapter 4)

Gap junction
A channel that connects the cytosols of two neighboring cells. The channel consists of two protein complexes that form half-channels spanning the cell membranes. The two half-channels meet in the extracellular space between the neighboring cells. See also **Transmembrane protein**. (Chapter 7)

Gate (logic)
A logical conjunction between two signals. In the AND gate, the signal is only propagated if both incoming signals are active. (Chapter 14)

Gene annotation
The assignment of a protein or function to a gene sequence. See also **Gene ontology**. (Chapter 6)

Gene ontology (GO)
A widely accepted, controlled vocabulary for **gene annotation**. (Chapter 6)

Generalized mass action (GMA) system
A modeling format within **biochemical systems theory**, based on **ordinary differential equations**, in which every process is represented with a product of **power-law functions**. See also **S-system**. (Chapter 4)

Genetic algorithm
A **machine learning** method that, inspired by the process of genetic inheritance, recombination, and mutation, searches for optimal and near-optimal solutions. See also **Evolutionary algorithm** and **Optimization**. (Chapter 5)

Genetic code
The assignment of each natural amino acid to one or more **codons** of **nucleotides** in DNA or RNA. Because of the near-universality of the genetic code, amino acid sequences in peptides and proteins can reliably be predicted from their coding genes. (Chapter 6)

Genetic engineering
The targeted manipulation of genes toward a goal of (industrial) interest. See also **Metabolic engineering** and **Synthetic biology**. (Chapter 14)

Genome
The totality of DNA in an organism. In addition to genes coding for proteins, much of the DNA is used for other purposes, including information regarding regulatory RNA. See also **Proteome** and **-ome**. (Chapter 6)

GFP
Green fluorescent protein. A protein whose DNA sequence can be inserted within the protein-coding region of a target gene, permitting the localization of gene expression through a fluorescence assay. Fluorescent proteins with many other colors are available now. (Chapter 6)

Global optimum
The best possible point (with respect to some defined criterion or metric). Finding the global optimum for a **nonlinear system** is a difficult **optimization** problem of computer science. See also **Local optimum** and **Branch-and-bound method**. (Chapter 5)

Goldman–Hodgkin–Katz equation
An equation describing the equilibrium potential of a cell membrane as a function of internal and external **ion** concentrations and permeability constants. (Chapter 12)

G-protein
One of several specific proteins on the inside of the cell membrane that interact with **G-protein-coupled receptors** in numerous **signal transduction** processes. (Chapter 9)

G-protein-coupled receptor (GPCR)
A class of membrane-associated **receptors** that are involved in uncounted **signal transduction** and disease processes. (Chapter 9)

Graph
A visual or mathematical representation of a **network** consisting of **nodes** (**vertices**) and **edges** connecting some of the nodes with each other. (Chapter 3)

Graph theory
A branch of mathematics concerned with **graphs**. Graph theory has found rich applications in biological **network** analysis. (Chapter 3)

Growth law
A typically simple function describing changes in the size of a cell, organism or population over time. The best-known examples are **exponential** and **logistic growth**, but many others exist. (Chapter 10)

Growth rate
Parameter in a **growth law** characterizing the speed of growth. (Chapter 10)

Half-life
The time it takes for the amount of some material to degrade from 100% to 50%, under the assumption of exponential decay. (Chapter 4)

Heat shock protein
One of a class of specific proteins that are activated under heat stress and protect other proteins from unfolding or **denaturing**. (Chapter 11)

Hill function
A type of **kinetic rate law** that is well suited for **modeling allosteric** processes and **cooperativity** involving n binding sites of an **enzyme**. The number n is called the Hill coefficient; $n > 1$ means positive cooperativity, $n < 1$ negative cooperativity. For $n = 1$, the Hill function becomes a **Michaelis–Menten rate law**. Often used to describe sigmoidal behavior. (Chapter 8)

Hit (drug development)
A natural or artificially created molecule, investigated during drug discovery and development, that interacts with a drug target and interrupts a disease process. (Chapter 13)

Hodgkin–Huxley model
The first mathematical **model** describing electrical activity, transmembrane currents, and action potentials in neurons and other excitable cells. (Chapter 12)

Homeostasis
Maintenance of the normal (steady) **state** of an organism or physiological **system**, which is achieved through the action of often complex internal control systems. See also **Allostasis**. (Chapter 13)

Homogeneous
Possessing the same features throughout. (Chapter 10)

Hub
A particularly well connected, and usually important, **node** within a **graph, network**, or **system**. (Chapter 3)

Hybridization
A core technique in numerous experimental methods for the identification of identical or similar DNA or RNA **nucleotide** sequences; based on the fact that nucleotide strands specifically bind to their complementary sequence through **base pairs**, such as cytosine and guanine. (Chapter 6)

Hysteresis
A property of **nonlinear systems** with two or more **stable steady states** where the **transient** toward one of these states depends on the recent history of the **dynamics** of the system. (Chapter 9)

Identifiability
A property of a **system** and a dataset indicating whether the system structure or **parameters** can be uniquely determined from the dataset. See also **Sloppiness**. (Chapter 5)

Immunoglobulin
Also called an **antibody**. A protein produced by white blood cells that is either attached to a B cell of the immune system or circulates through the blood and other bodily fluids. Immunoglobulins detect and neutralize invaders such as bacteria, viruses, and foreign proteins. (Chapter 7)

Immunoprecipitation
A technique for concentrating, isolating, and purifying proteins from a cell extract or protein mixture by means of specific **antibodies**. The protein–antibody complex is extracted from the solution, the antibodies are removed, and the protein can be analyzed. (Chapter 7)

Independent variable
A system variable that is not affected by the **dynamics** of the **system** itself, but potentially by the environment or the experimenter. In many cases, an independent variable is constant during a given experiment. See also **Dependent variable**. (Chapter 2)

Initial value
A numerical start value that needs to be defined for every **dependent variable** in a system of **differential equations**, before the system can be solved numerically. (Chapter 4)

Innate (immune system)
A generic defense system in a variety of multicellular organisms that protects the host from infection by pathogens. In vertebrates, cells of the innate immune system, such as macrophages and neutrophils, rush to the site of infection and destroy invaders. The innate immune system activates the **adaptive immune system**, which learns to recognize and remember pathogens. (Chapter 15)

Intron
A segment of a gene sequence that is transcribed into RNA, but later spliced out from the mature RNA. In protein coding genes, introns are not translated into protein. See also **Exon**. (Chapter 6)

Ion
An atom or molecule that carries an electrical charge due to a difference between its numbers of electrons and protons. (Chapter 12)

Irreversible (reaction)
Feature of a (biochemical) **reaction** not permitting a back-reaction. Absolutely irreversible reactions are rare, but the reverse direction is often negligible because of unfavorable energetics. See also **Reversible reaction**. (Chapter 8)

Isozyme
One of several different **enzymes** that catalyze the same biochemical **reaction**(s), sometimes with different efficiency and sensitivity to inhibitors or modulators. (Chapter 11)

Iteration
A recurring step or sequence of steps in an **algorithm** or **simulation**, usually with slightly changed settings. (Chapter 2)

Jackknife
Similar to the **bootstrap** method, the jackknife is a statistical method for determining desired features of a dataset. It is based on the repeated estimation of the desired feature when one or more data points are randomly left out. (Chapter 5)

Jacobian
A **matrix**, consisting of partial derivatives, that **approximately** represents the **dynamics** of a **nonlinear system** close to some **operating point**. (Chapter 4)

Kinetic order
1. In elemental chemical kinetics, the number of molecules of the same species undergoing a **reaction**.
2. In **biochemical systems theory**, the exponent of a **variable** in a **power-law function**, quantifying the effect of the variable on the process represented by the function; equivalent to the concept of **elasticity** in **metabolic control analysis**. (Chapter 4)

Kinetic rate law
A function describing how the **dynamics** of a biochemical **reaction** is affected by substrates, enzyme, and modulators. See also **Michaelis–Menten kinetics** and **Hill function**. (Chapter 8)

Kinetics (in biochemistry)
Pertaining to the **dynamics** of a biochemical **reaction**. (Chapter 8)

Latin square or hypercube sampling
A statistical design and sampling method that circumvents large numbers of analyses that would be necessary if every combination of numbers in a complete grid were used. Instead, only one number per row and per column of the grid is selected. In higher dimensions, the grid is replaced with a cube or hypercube, where the latter is the analog to a cube in more than three dimensions. (Chapter 5)

Lead (drug development)
A potential drug molecule, such as a **hit** with improved potency, that interrupts a disease process and has desirable properties related to efficacy, **pharmacokinetics**, and side effects. (Chapter 13)

Least-squares method
A statistical **regression** technique that optimally matches a **model** to a given dataset. Optimality is defined based on the minimum of the sum of squared differences between each data point and the corresponding model value. See also **Residual**. (Chapter 5)

Limit cycle
An **attractor** or **repeller** in the form of an **oscillation** in a **dynamical system** whose amplitude does not change over time. In the case of a **stable** (unstable) limit cycle, nearby **oscillations** eventually converge to (diverge away from) the attractor (repeller). (Chapter 4)

Linear
The mathematical property of a function or **system** to be describable with sums of variables that are multiplied with constant **coefficients** (numbers). If f is a linear function of x, and x_1 and x_2 are any values of x, then the **superposition principle** $f(x_1 + x_2) = f(x_1) + f(x_2)$ holds. The antonym is a **nonlinear** function or system, where the above relationships do not hold. (Chapter 2)

Linearization
The **linear approximation** of a system at a chosen **operating point**. The linearization is characterized by the **Jacobian** matrix. (Chapter 4)

Local minimum (optimum)
A **state** of a **system** that is better (with respect to some defined criterion or metric) than all states close by. However, farther away, the system may have even better states; see also **Global optimum**. Local optima often prevent computer **algorithms** from finding the global optimum. (Chapter 5)

Logistic growth
A sigmoidal model of growth, or **growth law**, that is initially exponential but ultimately approaches an upper value, called the **carrying capacity**. The deviation from **exponential growth** is often attributed to **crowding**. (Chapter 10)

Lotka–Volterra model
A **canonical model**, based on **ordinary differential equations** and used primarily in ecological **systems** analysis, where all processes are represented as products of two variables and a rate **coefficient**. Each equation j may also contain just the jth variable with a coefficient. (Chapter 4)

Machine learning
A computational branch of the field of artificial intelligence, which trains **algorithms** to classify data according to given criteria and to reveal patterns within complex datasets. See also **Parameter estimation**, **Data mining**, and **Text mining**. (Chapter 5)

MAPK
Mitogen-activated protein kinase. One of a class of **enzymes** at the core of many **signaling cascades** in eukaryotes. Typically, phosphorylation events of different MAPKs occur at three hierarchical levels, thereby transducing a signal, for instance, from the cell membrane to the nucleus. (Chapter 9)

Markov model
Usually a **discrete**, **probabilistic** mathematical **model**, in which the **state** of a **system** is determined by the previous state of the system and a (fixed) set of transition probabilities for moving the system from any one state to any other state. (Chapter 4)

Mass spectrometry (MS)
An experimental technique used to characterize mixtures of molecules based on their mass and an applied electrical charge. (Chapter 7)

Mass-action (kinetics)
A simple quantitative description of the **dynamics** of chemical **reactions**, consisting of the product of a **rate constant** and all substrates involved in the reaction. If two substrate molecules of the same species are needed for the reaction, the substrate enters the mass-action function with a power of 2. See also **Kinetics** and **Generalized mass action system**. (Chapter 8)

Matrix (in linear algebra; pl. Matrices)
An array of numbers that typically correspond to the **coefficients** in a set of **linear** equations; see also **Vector**. Many aspects within the fields of linear algebra, **multivariate** calculus, **systems** analysis, and statistics are based on vectors and matrices. (Chapter 4)

Metabolic burden
The cost to the organism for maintaining certain levels of mRNAs, proteins, and metabolic products, especially if they are expendable. (Chapter 14)

Metabolic control analysis (MCA)
A framework of theorems and methods, based on **sensitivity analysis**, for the assessment of control in metabolic, genetic, and signaling **systems** at **steady state**. (Chapter 3)

Metabolic engineering
A set of techniques from molecular biology, genetics, and engineering for manipulating a culture of microorganisms toward a desired goal, such as the production of a valuable **amino acid**. (Chapter 14)

Metabolic flux analysis
A collection of experimental (labeling) and computational (**flux balance analysis**-like) techniques for the assessment of flux distributions in metabolic **networks**. (Chapter 8)

Metabolic pathway
Any prominent sequence of (typically **enzyme**-catalyzed) reaction steps, involving the synthesis or degradation of organic molecules. Metabolic pathways are often used to organize the complexity of metabolic **networks**. (Chapter 8)

Metabolic profile
A comprehensive set of metabolite concentrations characterizing a **metabolic pathway** or organism at one time point or a series of time points. (Chapter 8)

Metabolomics
The simultaneous study of many or all metabolites in a system. See also **-ome** and **-omics**. (Chapter 8)

Metagenomics
Methods of genomics applied to a **metapopulation**. A collective term for simultaneous **genome** analyses of many (coexisting) species in environments such as soil, the oceans, sewage pipes, or the human gut. (Chapter 15)

Metapopulation
A collection of coexisting populations of different species. Commonly used in the context of microbial species. (Chapter 15)

Michaelis–Menten mechanism (rate law)
A **conceptual** and mathematical **model** (**rate law**) of an **enzyme**-catalyzed **reaction**, in which substrate and enzyme reversibly first form an intermediate complex, before the product is generated and the enzyme recycled. A suitable **quasi-steady-state assumption** converts the describing system of **mass-action differential equations** into a simple function that quantifies the rate of product formation in terms of the substrate concentration. Its two parameters are V_{max}, the maximal rate of product formation, and the Michaelis constant K_M, which corresponds to the substrate concentration for which this rate is $V_{max}/2$. A mathematical generalization is the **Hill function**. (Chapter 8)

Microarray (technology)
A set of methods for comparing gene expression in different cell types or under different conditions. Each microarray consists of a solid surface that holds very small probes of known DNA (or sometimes other molecules). See also **DNA chip** and **RNA-Seq**. (Chapter 6)

Microbiome
A microbial **metapopulation**, usually living within the human body; for instance, in the oral cavity, vagina, or gut. (Chapter 15)

MicroRNA
A short **noncoding** RNA of between 20 and 30 nucleotides that serves a number of regulatory roles, including the silencing of genes. See also **Small RNA**. (Chapter 6)

Model (in biology)

1. In biological experimentation, the use of a particular species as a representative for a larger domain of organisms, such as a "mouse model."
2. Conceptual: a formalized idea of how components of a **system** might be related to and interact with each other.
3. Mathematical: a set of functions or (**differential**) **equations** that in some manner represent a biological phenomenon.
4. Computational: the use of **algorithms** supporting the **modeling** of a biological phenomenon. (Chapter 2)

Modeling (in biology)
The conversion of a biological phenomenon into a mathematical or computational construct and the subsequent analysis of this construct. (Chapter 2)

Module
A more or less confined or autonomous (sub-)**model** that is part of a larger model. (Chapter 2)

Molecular dynamics
A computationally intensive **simulation** technique for exploring the location and motion of each atom in a large molecule such as a protein. (Chapter 7)

Monte Carlo simulation
A statistical technique for exploring possible, likely, and worst-case **behaviors** of **complex systems** that is based on thousands of **simulations** with different randomized combinations of model settings. (Chapter 2)

Motif
A recurring structural feature of a **network** or **system** that is found more often than one would expect in a corresponding, **random**ly composed system. See also **Design principle**. (Chapter 14)

Multiscale modeling
The construction of **models** that simultaneously capture essential **system** features and **behaviors** at different temporal and organizational scales, such as gene expression, **physiology**, and aging. (Chapter 12)

Multistage cancer model
One of several mathematical **models** formalizing the assumption that a normal cell experiences several **discrete** changes before becoming a cancer cell. (Chapter 15)

Multivariate
Simultaneously depending on several **variables**. (Chapter 2)

Myocardium
Heart muscle. (Chapter 12)

Myocyte
Muscle cell. In the heart, it is called a cardiac myocyte or **cardiomyocyte**. See also **Myofibril**. (Chapter 12)

Myofibril
The basic contractile unit of a **myocyte**, consisting of **actin**, **myosin**, and other proteins. (Chapter 12)

Myosin
A motor protein that forms filaments in eukaryotic cells. Together with **actin** filaments, myosin filaments are key components of the contractile machinery in skeletal and heart muscle cells. (Chapter 12)

Na^+–Ca^{2+} exchanger
A transport protein for ions, which exchanges calcium inside an **excitable cell** against sodium; in particular, after an **action potential**. (Chapter 12)

Na^+–K^+ pump
A transport protein that pumps Na^+ out of an **excitable cell**, while K^+ is pumped in. The pump can also run in reverse. (Chapter 12)

Nernst equation
An equation governing properties of **electrochemical** gradients. It describes free energy in terms of differences in **ion** concentrations across a membrane. (Chapter 12)

Network
A collection of **nodes** (**vertices**) and connecting **edges**. (Chapter 3)

Network motif
See **Motif**. (Chapter 14)

Neurotransmitter
A chemical such as dopamine or serotonin that is responsible for the transmission of specific **signals** between neurons across a synapse. (Chapter 15)

New chemical entity (NCE)
A potential drug molecule, intensely investigated in the early phases of drug development, that had not been approved by the FDA in any prior application. (Chapter 13)

NMR spectroscopy
Abbreviation for nuclear magnetic resonance spectroscopy. A technique that exploits the fact that atomic nuclei with an odd number of protons and neutrons spin on their own axes. This spin is assessed with strong magnets and can be interpreted to yield the structure of molecules, such as proteins. The same method can be used for noninvasive *in vivo* analyses of labeled metabolites. (Chapter 7)

Node
A **variable** or pool within a **network** or **graph**, which is connected by **edges** with other nodes in a graph, or with the **environment** of the graph. Also called a **vertex**. (Chapter 3)

Noise
Summary expression for unexplained, apparently **random** errors in data. See also **Residual**. (Chapter 5)

Noncoding DNA/RNA
DNA or RNA that is ultimately not translated into protein. (Chapter 6)

Nonlinear
Antonym of **linear**. The property of a curved appearance of a function or a system. If f is a nonlinear function of x, and if x_1 and x_2 are values of x, then $f(x_1 + x_2)$ is generally not equal to $f(x_1) + f(x_2)$. See also **Superposition** and **Synergism**. (Chapter 2)

Nonlinear system
Any mathematical **model** that exhibits **nonlinear** features (possibly in addition to **linear** features). (Chapter 4)

Northern blot
A technique for measuring gene expression based on produced mRNAs. See also **RT-PCR** and **Western blot**. (Chapter 6)

Nucleoside
A building block of DNA or RNA, consisting of a base and a sugar (deoxyribose or ribose, respectively). If a phosphate group is attached as well, a nucleoside becomes a **nucleotide**. (Chapter 6)

Nucleotide
A building block of DNA or RNA, consisting of a base, a sugar (deoxyribose or ribose, respectively), and a phosphate group. DNA bases are adenine (A), cytosine (C), guanine (G), and thymine (T). In RNA, thymine is replaced by uracil (U). (Chapter 6)

Nullcline
A set of points in a **phase-plane plot** containing all situations where a particular **variable** has a slope ("cline") of zero ("null") and therefore does not change. Nullclines divide the space of **system** responses into sections of qualitatively similar **behaviors**. (Chapter 10)

Objective function
A function that is maximized (or minimized) in an **optimization** task. In **parameter estimation**, the objective function is often the sum of squared errors between a **model** and the modeled data. See also **Residual**. (Chapter 5)

-ome, -omics
Suffix added to words in order to indicate comprehensiveness or totality. Thus, the **proteome** consists of all proteins in a cell or organism. One of the oldest -omes is the rhizome, the totality of all roots of a plant. -ome is similar to the suffix -oma, which is used to describe

tumors. Some new word concoctions using -ome and -omics are quite silly. See also **Genome**. (Chapter 6)

Ontology
A controlled vocabulary with generally accepted definitions of terms within a limited domain of knowledge. A well-known example is **gene ontology**. See also **Gene annotation**. (Chapter 6)

Open system
In contrast to a **closed system**, a mathematical **model** with input and/or output. (Chapter 2)

Operating point
The point at which a function and an **approximation** have the same value, slope, and possibly higher derivatives. (Chapter 4)

Operating principle
The dynamic analog to a **design principle** or **motif**: a natural strategy for solving a task that is superior to alternative strategies. (Chapter 14)

Operon
The clustered arrangement of (bacterial) genes under the control of the same promoter. Operons often lead to co-expression of functionally related proteins. (Chapter 6)

Optimization
The mathematical or computational process of making a **system** behave as closely as possible to a given objective, for example, by minimizing the error between some output of the model and the desired output. See also **Objective function**, **Parameter estimation**, and **Residual**. (Chapters 5 and 14)

Orbit
A closed **trajectory**. If a dynamical system starts at some point of an orbit, it will reenter this point infinitely often again. An example is a **limit cycle**. (Chapter 4)

Ordinary differential equation (ODE)
A type of **differential equation**, or system of differential equations, in which all derivatives are taken with respect to the same **variable**, which is usually time. See also **Partial differential equation**. (Chapter 4)

Oscillation
The behavior of a function (or **dynamical system**), where the value of the function (or of a system variable) alternates between phases with positive and negative slopes. See also **Damped oscillation**, **Overshoot**, and **Limit cycle**. (Chapter 4)

Overfitting
The fact that a **model** with (too) many **parameters** permits a good representation of a **training set** of data, but not of other datasets. (Chapter 5)

Overshoot
The **dynamics** of a function or **system** that temporarily exceeds its ultimate final value. (Chapter 4)

Pacemaker cell
Cells, primarily in the **sino-atrial node** of the heart, that **depolarize** spontaneously and therefore set the pace of the heartbeat. The signal from the sino-atrial node eventually reaches all **cardiomyocytes**. See also **Atrioventricular node**, **Bundle of His**, and **Purkinje fiber**. (Chapter 12)

Parameter (of a model)
A numerical quantity in a mathematical **model** that is constant throughout a given computational experiment or **simulation**. Its value may be altered in a different experiment. See also **Parameter estimation**. (Chapter 2)

Parameter estimation
A diverse set of methods and techniques for identifying numerical values of model **parameters** that render the **model** representation of a given dataset acceptable. (Chapter 5)

Partial differential equation (PDE)
An equation (or system of equations) that contains functions and derivatives with respect to two or more variables, which, for instance, may represent time and one or more space coordinates. See also **Differential equation** and **Ordinary differential equation**. (Chapter 4)

Path length
The number of consecutive **edges** connecting one **node** with another node within a **graph**. Of interest is often the longest minimal path length in a graph. (Chapter 3)

Peptide
A chain of up to about 30 **amino acids** that is deemed too short to be called a protein, but otherwise has the same properties. (Chapter 7)

Peptide bond
The chemical (covalent) bond between **amino acids** in a **peptide** or protein, in which the carboxyl group of one amino acid reacts with the amino group of the neighboring amino acid. (Chapter 7)

Personalized (precision) medicine
An emerging approach to health and disease that uses **genomic**, **proteomic**, **metabolic**, and physiological **biomarkers** for the custom-tailored prevention and treatment of an individual's disease. (Chapter 13)

Perturbation
A usually rapid shift in conditions applied either to a biological **system** or to a mathematical **model**. (Chapter 2)

Petri net
A specific formalism for modeling **networks** and **systems**, consisting of two distinct types of **variables**: pools and reactions. (Chapter 3)

Pharmacogenomics
The study of interactions between gene expression and drugs. (Chapter 13)

Pharmacokinetics/Pharmacokinetic model
The **dynamics** of a drug within an organ, body, or cell culture; a mathematical representation of this dynamics. A notable special case is a **physiologically based pharmacokinetic (PBPK)** model, which is a hybrid between a **compartment model** and a set of **kinetic** models within the **compartments**. (Chapter 13)

Phase diagram
A rough sketch of domains in which a **system** exhibits different **qualitative behaviors**. For instance, the system may exhibit **oscillations** in one domain but not in another. See also **Phase-plane plot**. (Chapter 10)

Phase-plane analysis/plot
A graphical representation of a **dynamical system**, typically with two **dependent variables**, where one variable is plotted against the other, rather than plotting both variables as functions of time. The plot is divided by **nullclines** into domains of qualitatively similar **behaviors**. See also **phase diagram**. (Chapter 10)

Phylogeny, Phylogenetics
Pertaining to studies of the relatedness of species throughout evolution. (Chapter 6)

Physiologically based pharmacokinetic (PBPK) model
A **compartment model** with biologically meaningful **compartments**, such as organs, tissues, or the bloodstream. Often used to assess the fate of drugs within a higher organism. The dynamics of drugs within compartments may be represented with kinetic models. (Chapter 13)

Polysaccharide
A linear or branched polymer consisting of many sugar (carbohydrate) molecules of the same or different type. (Chapter 11)

Post-translational modification (PTM)
Any chemical change to a **peptide** or protein after it has been translated from mRNA and that is not encoded in the RNA sequence. (Chapter 7)

Power function/Power-law function
A mathematical function consisting of the product of a (non-negative) **rate constant** and one or more positive-valued **variables** that

are each raised to some real-valued exponent, which in **biochemical systems theory** is called a **kinetic order**. See also **Canonical model**. (Chapter 4)

Power-law approximation
Linear approximation of a function within a logarithmic coordinate system, which always results in a (possibly **multivariate**) **power-law function**. (Chapter 4)

Power-law distribution
Specific **probability distribution** that follows a **power-law function** with negative exponent. An example of recent prominence is the distribution of **nodes** with certain numbers of **edges** in **networks** exhibiting the **small-world property**. (Chapter 3)

Prediction (of a model)
A feature of a **system**, such as a numerical or approximate value of a system **variable**, that has not been measured experimentally but is computed with the model. (Chapter 2)

Predictive health
A novel paradigm of health care whose goal is maintenance of health rather than treatment of disease. Predictive health heavily relies on personalized assessments of **biomarkers** that begin when the individual is still apparently healthy and follows him or her into old age and, possibly, chronic disease. (Chapter 13)

Probabilistic
Affected by **random** (**stochastic**) influences. (Chapter 2)

Probability distribution
A succinct, collective characterization of all values that a **random variable** may assume, along with their likelihoods. (Chapter 2)

Profile (of metabolites or proteins)
A collection of features characterizing many or all metabolites (or proteins) of a biological **system** at one time point or at a series of time points. See also **Time-series data**. (Chapter 8)

Proteasome
A cellular organelle consisting of a large protein complex and responsible for the disassembly of proteins that had been tagged by **ubiquitin**. The cell reuses the **peptides** and **amino acids** generated in the process. (Chapter 7)

Proteome
A collective term for all proteins in a cell, organism, or otherwise defined biological sample. See also **Genome** and **-ome**. (Chapter 7)

Pump (in a cell membrane)
A **carrier protein**, also called a permease, that transports specific **ions** across a cell membrane. (Chapter 12)

Purkinje fibers
Modified heart muscle cells that rapidly conduct electrical impulses from the **sino-atrial** and **atrioventricular nodes** to the excitable **cardiomyocytes** in the **ventricles**. See also **Bundle of His**. (Chapter 12)

Qualitative analysis
A type of mathematical analysis that depends only minimally on specific numerical settings, such as **parameter** values, and therefore tends to yield rather general results. (Chapter 10)

Qualitative behavior (of a model)
A **model** response that is not characterized by specific numerical features, but rather as a member of a distinct class of features, such as **damped oscillations** or **limit cycles**. (Chapter 2)

Quantitative structure–activity relationship (QSAR)
A specific property of a molecule, such as drug activity, that is experientially or computationally inferred from its chemical composition. (Chapter 13)

Quasi-steady-state assumption (QSSA)
The assumption that an intermediate complex between a substrate and an **enzyme** is essentially constant during a **reaction**. The assumption permits the explicit formulation of **Michaelis–Menten** and **Hill rate laws** as functions, rather than systems of **differential equations**. (Chapter 8)

Quorum sensing (QS)
A mechanism with which microbes collectively respond to their sensing that many other microbes in their environment send out chemical signals. (Chapter 9)

Random
Containing an aspect that is unpredictable, possibly because of limited information. (Chapter 2)

Random variable
Variable in a **stochastic** context, whose specific value is determined by a **probability distribution**. (Chapter 2)

Rate constant
A positive or zero-valued quantity, usually in chemical and biochemical **kinetics**, representing the turnover rate of a **reaction**. See also **Kinetic order**, **Elasticity**, and **Mass-action kinetics**. (Chapter 4)

Reaction (in biochemistry)
The conversion of a molecule into a different molecule, a process that is often catalyzed by an **enzyme**. See also **Reversible**, **Irreversible**, **Kinetic**, and **Rate constant**. (Chapter 8)

Receptor
A protein on the surface of (or spanning) a cell membrane that senses external signals which, if strong enough, trigger a **signal transduction** process. Prominent examples are **G-protein-coupled receptors** and **Toll-like receptors**. See also **Transmembrane protein**. (Chapter 7)

Receptor tyrosine kinase (RTK)
A member of an important class of **transmembrane proteins** that transfer a phosphate group from a donor to a substrate and by doing so transduce a **signal** from the outside of a cell to the inside. (Chapter 7)

Recursion
In mathematics, a process characterized by a temporal sequence of **states** of a **system** that are defined in terms of the states of the system at one or more earlier time points. (Chapter 10)

Redox
Contraction of the terms reduction and oxidation. Used as an adjective referring to the **state**, or change in state, of a chemical **reaction** system with respect to losses or gains of electrons. For example, the physiological redox state is changed under oxidative **stress**. (Chapter 11)

Reductionism
The philosophy that the functioning of a **system** can be understood by studying its parts and the parts of parts, until the ultimate building blocks are revealed. **Systems biology** maintains that reductionism must be complemented with a reconstruction phase that integrates the building blocks into a functioning entity. (Chapter 1)

Refractory period
A time period, typically toward the end of an **action potential**, during which an **excitable cell** cannot be excited again. (Chapter 12)

Regression
A statistical technique for matching a **model** to experimental data in such a way that the **residuals** are minimized. (Chapter 5)

Repeller
In the field of dynamical systems, a set of points from which **trajectories** diverge. The most typical repellers are unstable steady-state points and unstable **limit cycles.** The antonym is **attractor**. (Chapter 4)

Repolarization
As part of an **action potential**, the return in voltage across the membrane of a neuron, muscle cell, or **cardiomyocyte** from a **depolarized** state to the relaxed state, which in most cells corresponds to the **resting potential**. (Chapter 12)

Repressilator
A system of three genes (A, B, and C), where A represses B, B represses C, and C represses A. (Chapter 14)

Residual (error)
The remaining difference (error) between a **model** and data, after the model has been **fitted**. See also **Regression** and **Least-squares method**, and **Parameter estimation** (Chapter 4)

Resting potential
The electrical potential across the cell membrane of a relaxed, **excitable cell**, such as a neuron or **cardiomyocyte**. In neurons, the resting potential is about –70 mV, whereas it is about –90 mV in cardiomyocytes. See also **Action potential**. (Chapter 12)

Retrovirus
An RNA virus that needs to be reverse-transcribed by the host into complementary DNA, before it can be integrated into the host **genome**. See also **transcription**. (Chapter 11)

Reverse engineering
A diverse set of methods for inferring structures, processes, or mechanisms from data. An example is **Bayesian network inference**. (Chapter 3)

Reversible (reaction)
A (biochemical) **reaction** that may occur in forward and backward direction, usually with a different rate. See also **Irreversible reaction**. (Chapter 8)

Ribosome
A complex, consisting of specific RNAs and proteins, that facilitates the translation of mRNA into protein. (Chapter 7)

Riboswitch
An RNA-based regulatory element of gene expression in bacteria. (Chapter 14)

Ribozyme
A ribonucleic acid (RNA) that catalyzes certain chemical reactions. The word is composed from "ribonucleic acid" and **enzyme**. (Chapter 6)

RNA-Seq
A relatively recent high-throughput method for measuring gene expression, based on **cDNA** sequencing. See also **Microarray**. (Chapter 6)

Robust
An often vaguely defined term for the tolerance of a **system** toward **perturbations** or changes in **parameters**. See also **Sensitivity** and **Gain**. (Chapter 4)

RT-PCR
Abbreviation for:
1. Reverse transcription followed by a quantitative polymerase chain reaction. A method for quantifying mRNA and thus gene expression. See also **Northern blot**.
2. Real-time polymerase chain reaction. A fluorescence-based method for monitoring in real time how much DNA had been polymerized since the beginning of the experiment. Also called RT-qPCR, where q means quantitative. (Chapter 6)

Sarcoplasmic reticulum (SR)
A compartment in **cardiomyocytes** that can contain large quantities of calcium. Controlled calcium movement between the sarcoplasmic reticulum and the cytosol is ultimately responsible for cell contraction. (Chapter 12)

SBML (Systems Biology Markup Language)
A standardized computer language for **modeling** and sharing biological **systems** and data. (Chapters 1 and 15)

Scale-free
Property of a **graph** or **network** to have similar features, independent of the number of **nodes**. (Chapter 3)

Scatter plot
Graphical representation of two (or three) features of some entity as coordinates of points in a coordinate system, which often looks like a cloud of dots. An example consists of math and physics scores of a class of high-school students. For each student, the math score could be the *x*-coordinate and the physics score the *y*-coordinate of a unique point that represents this student. (Chapter 5)

Search algorithm
A set of computing instructions for finding an optimal solution, consisting of **iterations** of searching and comparing with the so-far-best solutions. See also **Optimization** and **Parameter estimation** (Chapter 4)

Second messenger.
A small molecule, such as nitric oxide, calcium, cAMP, inositol trisphosphate, or the sphingolipid ceramide, that is involved in **signal transduction**. (Chapter 9)

Secondary structure (RNA)
The folding and partial base matching of a single-stranded RNA with itself, creating accurately matched regions, interrupted by loops and possibly ending in unmatched ends. (Chapter 6)

Sensitivity analysis
A branch of general **systems** analysis that quantifies how much a system feature changes if a **parameter** value is slightly altered. (Chapter 4)

Separation of timescales
A modeling technique for **systems** containing processes that run at distinctly different speeds. By assuming that very slow processes are essentially constant and that very fast processes are essentially at **steady state**, one restricts a **model** to the one or two most relevant timescales. (Chapter 12)

Signal
Any molecular, chemical or physical event that can be sensed and transmitted by a cell or organism, usually through a **receptor**. (Chapter 9)

Signal transduction (signaling)
A collective term for the management and integration of **signals** received by the **receptors** of a cell. In many cases, **signaling cascades** relay an external signal to the **genome**, where specific genes are up- or down-regulated in response. (Chapter 9)

Signaling cascade
Usually a hierarchically arranged set of proteins, which, by becoming phosphorylated or dephosphorylated, transduce a chemical or physical **signal** from the cell surface to the **genome**. (Chapter 9)

Silencing (of a gene)
The ability of certain small RNAs to suppress gene expression. (Chapter 6)

Simulated annealing
A **parameter estimation** technique inspired by a method of heating and cooling metals in order to reduce impurities and defects. (Chapter 5)

Simulation
The process of using a mathematical **model** with different **parameter** and **variable** settings, with the goal of assessing its possible range of responses, including best, worst, and most likely scenarios, or of explaining scenarios observed in, or predicted for, a biological **system**. (Chapter 2)

Single nucleotide polymorphism (SNP)
A point mutation in a specific **nucleotide** that distinguishes individuals. Because multiple **codons** can be translated into the same amino acid, the change is often neutral. (Chapter 13)

Sino-atrial (SA) node
A relatively small group of **pacemaker cells** located between the **atria** and **ventricles** of the heart, which **depolarize** spontaneously and thereby set the speed of a heartbeat. (Chapter 12)

SIR model
An established prototype **model** for assessing the spread of communicable diseases. *S*, *I*, and *R* stand for susceptible, infective, and (from the disease process) removed individuals. Thousands of variants have been proposed. (Chapter 2)

Sloppiness
The result of a **parameter estimation** process showing that many combinations of parameter values for a given **model** can **fit** the same datasets equally (or nearly equally) well. See also **Identifiability**. (Chapter 5)

Small RNA
Relatively short RNA sequences of 50–200 nucleotides that form specific loop structures and can bind to proteins or mRNAs, thereby affecting their function. See also **MicroRNA**. (Chapter 6)

Small-world (property of networks)
A special type of connectivity within networks, where a few **hubs** are heavily connected and the remaining **nodes** are more loosely connected. Thus, most nodes are not neighbors of one another, but most nodes can be reached from any other node by a small number of steps. The number of nodes with a given degree of connectivity follows a **power-law distribution** with negative exponent. (Chapter 3)

Smoothing
A mathematical technique for reducing **noise** in experimental data and for bringing the data closer to a **continuous** trend line. (Chapter 5)

Sphingolipid
A member of a specific class of lipids that are crucial constituents of membranes and also possess signaling function, for instance, during **differentiation**, aging, **apoptosis**, and a number of diseases, including cancer. (Chapter 11)

S-system
A **modeling** format within **biochemical systems theory**, in which every **ordinary differential equation** consists of exactly two products of **power-law functions** (one of which may be zero): one collectively representing all augmenting processes and one representing all degrading processes. See also **Biochemical systems theory** and **Generalized mass action system**. (Chapter 4)

Stability
Local stability refers to a **steady state** or **limit cycle** and indicates that a **system** will return to the steady state or limit cycle after small **perturbations**. Structural stability refers to the **robustness** of the system **behavior** to slight, persistent changes in **parameter values**. See also **Bifurcation**. (Chapter 4)

State (of a system or model)
A comprehensive collection of numerical features characterizing a **system** or **model** at a given point in time. (Chapter 2)

Static (model)
A **model** that describes a **network** or **system** without accounting for changes over time. (Chapter 3)

Statistical Mechanics
A body of concepts and theorems describing the mechanical and energetic features of large assemblies of molecules or other particles through appropriate averaging. Statistical mechanics relates the macroscopic bulk behavior of materials to the microscopic features of molecules. (Chapter 8)

Steady state
The **state** of a **dynamical system** where no **dependent variable** changes in magnitude, although material may be flowing through the system. At a steady state, the time derivatives in all **differential equations** describing a system are equal to zero, and thus the equations become algebraic equations. See also **Homeostasis** and **Stability**. (Chapter 2)

Stimulus
An external **perturbation** to a **system** or **model**. Stimuli are often used to study possible **behaviors** of the system. (Chapter 2)

Stochastic
Containing **probabilistic** or **random** aspects; often considered the antonym of **deterministic**. (Chapter 2)

Stochastic simulation
A **simulation** with a **model** that is subject to **random** events.

Stochastic simulation algorithm
A specific **algorithm** for **modeling kinetic reaction** processes that involve small numbers of molecules and therefore do not permit the use of methods of **statistical mechanics** and **kinetics**. (Chapter 8)

Stoichiometry
The connectivity of a metabolic network, augmented with an account of the relative quantities of reactants and products in each **reaction**. (Chapter 3)

Strange Attractor
The domain of a **chaotic** system to which **trajectories** converge and that they do not leave once they are in the attractor. See also **Attractor**. (Chapter 4)

Stress
A situation that moves a cell or organism away from its normal **physiological** state or range. See also **Allostasis**. (Chapter 11)

Stress response element (STRE)
A short sequence of (maybe five) **nucleotides** in the promoter region of genes involved in responses to external **stresses**, such as heat, osmotic pressure, strong acid, or the depletion of required chemicals in a cell's environment. (Chapter 11)

Subpopulation
A subset of a population characterized by some common feature, such as age, gender, race, or a particular disease status. (Chapter 10)

Superposition (principle)
A feature characteristic of **linear**, but not **nonlinear**, **systems**: the response of a system to two simultaneous stimuli is equal to the sum of the responses to the same two stimuli administered separately. If the superposition principle holds, a system can be analyzed by performing many **stimulus**–response experiments and adding the observed stimulus–response relationships. See also **Synergism**. (Chapter 4)

Synergism/synergistic
Literally, "working together." The phenomenon that the combined action of two **system** components (or people) has a stronger effect than the sum of their individual actions. Synergism is a feature of **nonlinear** systems and violates the **superposition principle**. Antonym of **antagonism/antagonistic**. (Chapter 1)

Synthetic biology
A diverse set of experimental and computational manipulation techniques with the goal of creating new (usually microbial) biological **systems**. (Chapter 14)

System
An organized collection of dynamically interacting components.

Systems biology
A branch of biology that uses experimental and computational techniques to assess dynamic biological phenomena within larger contexts. (Chapter 1)

Systole
Greek term for contraction and used to describe the contracted state of the heart. See also **Diastole**. (Chapter 12)

Taylor approximation
A general type of **approximation** for functions of one or more **variables** around a specific **operating point**; resulting in a (possibly **multivariate**) **linear**, polynomial or **power-law representation**. (Chapter 4)

Tertiary structure (of a protein)
The three-dimensional shape of a protein molecule. See also **Conformation**. (Chapter 7)

Text mining
Use of a **machine learning algorithm** to extract information from large amounts of text. See also **Data mining**. (Chapter 13)

Time-series data
Experimental or artificial data, consisting of values of **variables** at (many) subsequent time points. (Chapter 5)

Toggle switch (genetics)
A phenomenon or **model** in which a gene is either fully expressed or silent, but does not allow intermediates. (Chapter 14)

Toll-like receptor (TLR)
A member of a class of **signaling** proteins that recognize a variety of foreign macromolecules, especially those on the cell walls of bacteria, and trigger the initiation of an immune response. See also **Receptor** and **Transmembrane protein**. (Chapter 7)

Training set
A portion of a dataset used by a **machine learning algorithm** to distinguish different classes of outputs and to establish classification rules. See also **Validation**. (Chapter 5)

Trajectory
The collective change in the **state** of a **system** over time; usually between a **stimulus** and some later state, such as a steady state. (Chapter 10)

Transcription
The process of creating a matching messenger RNA (mRNA) from a DNA sequence. (Chapter 6)

Transcription factor
A protein affecting the expression of a specific gene by binding to the corresponding DNA. (Chapter 6)

Transcriptome
The totality of RNAs transcribed from the **genome** of an organism. See also **-ome**. (Chapter 6)

Transfer RNA (tRNA)
One of several specific RNAs that facilitate the attachment of the correct amino acid to a growing **peptide** or protein, while it is being translated from mRNA. This **translation** process occurs within a ribosome. (Chapter 6)

Transient
A collective term for behaviors of a **system** between the time of a stimulus and the time the system reaches a **steady state**, a **stable oscillation**, or some other **attractor** or point of interest. (Chapter 4)

Translation
The process of generating a protein or **peptide** whose **amino acid** sequence corresponds to the **genetic code** of an mRNA. The process occurs within a ribosome and uses **transfer RNA**. See also **Post-translational modifications**. (Chapter 6)

Transmembrane protein
A member of a large class of proteins that span the membrane of a cell and thereby allow communication between the outside and inside of the cell. Transmembrane proteins are involved in **signal transduction** and a large number of diseases. See also **Gap junction** and **G-protein-coupled receptor.** (Chapter 7)

Transposon
A stretch of DNA that can move from one location in a **genome** to another. (Chapter 6)

Trial (stochastic process)
One of many evaluations of a **stochastic model**. (Chapter 2)

Two-component (signaling) (TCS) system
While not exclusively limited to this definition, usually a specific type of **signaling** mechanism, with which microorganisms sense their environment. See also **Receptor** and **Signal transduction**. (Chapter 9)

Two-dimensional gel electrophoresis
An experimental method for separating proteins by their size and electrical charge. (Chapter 7)

Ubiquitin
A small protein that, when attached to other proteins, tags them for disassembly in the **proteasome** and later recycling of the resulting **peptides** and **amino acids**. The tagging process is called ubiquitination. (Chapter 7)

Uncertainty
Degree to which an experimental result is not exactly known, owing to experimental inaccuracies or other challenges; see also **Variability**. Advanced measurement techniques might reduce uncertainty, but not variability. (Chapter 5)

Validation
The process of testing (and confirming) the appropriateness of a **model** with data that had not been used for the construction of the model. See also **Training set**, **Identifiability**, and **Sloppiness**. (Chapter 2)

Variability
Degree to which an experimental result is affected by differences among individual cells, organisms, or other items under investigation; see also **Uncertainty**. Advanced measurement techniques might reduce uncertainty, but not variability. (Chapter 2)

Variable
A symbol representing a component of a **system**, which may or may not change over time. See also **Dependent** and **Independent variable**. (Chapter 2)

Vector
1. In mathematics, a convenient arrangement of variables or numbers in a one-dimensional array. Often used in conjunction with a **matrix**. Many aspects within the fields of linear algebra, **multivariate** calculus, and statistics are based on vectors and matrices. (Chapter 4)
2. In genetic engineering, an artificially constructed virus or plasmid used to introduce foreign DNA into a bacterium or host cell. (Chapter 14)

Ventricle
One of the two larger chambers of the heart. See also **Atrium**. (Chapter 12)

Vertex (pl. Vertices)
Synonym of **node**. (Chapter 3)

Voltage-gated
Property of a membrane channel to open and close in response to the membrane potential. (Chapter 12)

Western blot
A technique for measuring gene expression based on produced protein. See also **Northern blot**. (Chapter 6)

XML
Abbreviation for eXtensible Mark-up Language. Computer language with a defined vocabulary that permits the structuring of data and facilitates translation into different languages and software packages. See also **Ontology** and **SBML**. (Chapters 1 and 15)

X-ray crystallography
A technique for determining the structure of a protein. The protein must first be crystallized. An X-ray shot through the crystal scatters in a characteristic fashion that allows the deduction of the molecular structure of the protein. (Chapter 7)

Index

Note: Page numbers followed by F refer to figures, those followed by T refer to tables, and those followed by B refer to information boxes

A

B

C

F

G

H

I

J

K

L

M

N

O

P

Q

R

S

T

U

V

W

X

Y

Z